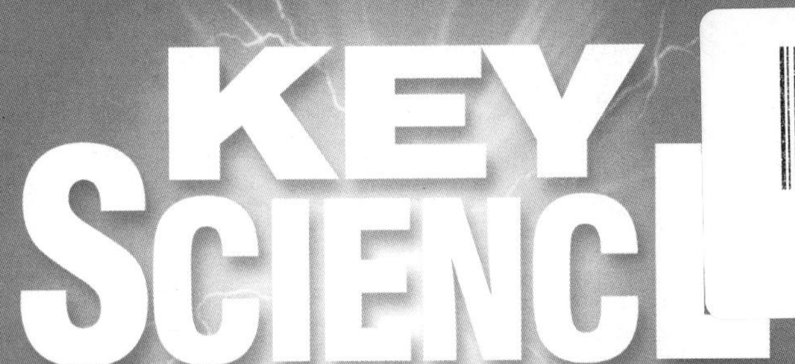

Physics

JIM BREITHAUPT BSc MSc

Sixth Form Centre Manager
Wigan and Leigh College

First published in 1994 by:
Stanley Thornes (Publishers) Ltd
Second edition 1997

This edition published in 2001 by:
Nelson Thornes Ltd
Delta Place
27 Bath Road
CHELTENHAM
GL53 7TH
United Kingdom

01 02 03 04 05 / 10 9 8 7 6 5 4 3 2 1

A catalogue record of this book is available from the British Library

ISBN 0 7487 6243 4

Illustrations by Barking Dog Art, Peters and Zebransky, Harry Venning and Cauldron Design Studio
Page make-up by Tech Set Ltd, Gateshead, Tyne & Wear

Printed in Spain by Graficas Estella

Related titles
Key Science: Biology (0 7487 6241 8)
Key Science: Chemistry (0 7487 6242 6)
Key Science: Physics Extension File (0 7487 6255 8)

Acknowledgements

Dr Jerry Wellington and Mr Jon Scaife have provided some excellent material on information technology, and I thank them for their contributions to the text.

I would like to thank the following organisations and people who have supplied photographs.
Acorn Computers: 23.5K; Allsport: 18.4A (Vandystadt); BBC Weather Centre: 5.1A; British Coal Open Cast: 6.2A; British Coal Board: 6.2B; British Gas plc: 4.4A; British Library: 1.5E; Bodleian Library, Oxford: 1.5D; Biophoto Associates: 11.5Gb; B Sky B: 15.3Ha,b; Dunlop Slazenger: 16.1I; Ecoscene: 6.3H (John Farmer); Electricity Association: 4.4B; Ford Motor Company Limited: 17.7C; Heather Angel: 11.4E; Fisher Scientific UK: 21.3D; G H Zeal Limited: 5.1D; Hulton Deutsch Collection: 17.5D, 22.2Ia; Image Bank: x top (Jay Freis), bottom (Gary Bistram); ICI Chemicals and Polymers: 3.4D; James Davis Photography: 2.1D; Leslie Garland Picture Library: 5.2C; JET Information Office: 6.2G; J Sainsbury plc: 5.4C; Last Resort Picture Library: p261; Mark Boulton: 11.3A, 14.4A, p193; 15.2B, 16.1A, 17.6A, 17.6C, 17.6D, p278; 21.2G, 21.4D, 23.1A, 23.5Ec, 23.8Ca,b; Mary Evans Picture Library: 1.5C, 5.3B, 5.3C, 5.3D, p97; 19.2A; Mountain Camera: 3.4G, 5.8C (John Cleare); Martyn Chillmaid: 3.5D, 5.1G, 7.3Aa, 17.8A, 17.8B, 17.8C, 19.1F, 20.3Fa, 20.4A, 21.2A, 21.2F, 21.4J, p339; 22.3F 23.5G, 23.5Ic, 23.5Ja; Magnum Photos: 9.5B (Shunkichi Kikuchi); MIRA: 18.2A; National Medical Slide Bank: 4.1D; NASA: p2, 1.2A, 1.2Bc, 1.3C 1.3D, 1.6B, 2.1C (Jeff Hester and Paul Scowen/ Arizona State University), 16.1E, 18.8A, 18.8G, 20.2I; Natural History Photographic Agency: 11.1E; Naval Research Laboratory, Washington D.C.: 1.A; Novosti Press Agency (APN): p118; Phillips Research Laboratories: p143, 22.4B, 22.4D; Professor Thomas D. Rossing: 15.4A; Robert Harding: 12.4F (David Hughes), 17.6B; Science Photography: viii (Alfred Pasieka), 1.2Ba,b,d (John Sanford), 1.4A (Frank Zullo), 2.1Ab,c, (Kim Gordon), 2.2D, 13.2E, (NASA), 2.4D (Dr M B Hursthouse), 5.1H (Chris Priest and M Clarke), 5.7A, 6.1A (Malcolm Fielding), 6.2F (Mere Words), 6.3B, 6.3F, 23.8Ga (Martin Bond), 6.3D (Professor David Hall), 7.2C (Dr Mitsuo Ohtsuki), 9.1B, 10.10A 12.2H (Hank Morgan), 6.3E (Simon Fraser), 9.1G, (Lawrence Berkeley Laboratory), 9.3C (Phillipe Plailly), 9.3F (U.S. Department of Energy), 10.4Aa, 15.2E (Alexander Tsiaras), 9.8 (Novosti), 10.1C (Carl Anderson), 10.1Db (Tim Beddow) 10.2A (SLAC), 12.4C (Jonathon Watts), 14.2A (Dr Beer-Gabel/ CNRI), 14.5A (Andrew Syred), 15.2C (Dr Ray Clark and Mervyn Goff), 15.4C (Department of Physics, Imperial College), 17.2D (Labat/Lanceau, Jerrican), p290 (D Nunjk), 22.2D (A Bartel), 22.2Ga (J-L Charmet), 23.5Jb (Dr Jeremy Burgess), 23.7C (Astrid and Hans Frieder Michler), 1.2Db, 12.2E, 15.2F, 19.2B, p340; Shell Photographic Library: 6.2D; Sue Boulton: 23.8Cc; Sygma: 13.3B (Les Stone); 20.1G (Houston Post); Still Pictures: 17.1A; Topham Picturepoint: 14.3A, 16.3E; Transport Research Laboratory: p212; Tony Stone Images: 4.2B right (Bruce Ayers), 15.1A; United Kingdom Atomic Energy Authority: 6.2E, 9.1D, 9.3D, 9.4Ab,c; United States Information Service: 12.2G; Woodmansterne Limited: 1.5B (Jeremy Marks), 12.2D; Westland Group plc: 19.3H; Wood Mackenzie: 4.4A (UKCS Maps); Zefa Pictures: 11.2Eb, 13.2D, 18.5H, 20.2F

Every effort has been made to contact copyright holders to clear permission for reproduction of copyright material. The publishers should like to apologise if any such material has not been fully acknowledged and shall endeavour to rectify the situation at the earliest opportunity.

I thank the following examining groups for permission to reproduce questions from examination papers: University of London Examinations and Assessment Council,Midland Examining Group, Northern Examinations and Assessment Board, Northern Ireland Council for the Curriculum Examination and Assessment , Southern Examining Group, Welsh Joint Education Committee.

The examining groups bear no responsibility for the answers to questions taken from their papers contained in this publication.

The production of a science text book from manuscripts involves considerable effort, energy and expertise from many people, and I wish to acknowledge the work of all members of the publishing team.

Finally I thank my family for the tolerance which they have shown and the encouragement which they have given during the preparation of this book.

Jim Breithaupt

Contents

Theme E ▸ Forces

Theme F ▸ Electricity

Preface

Key Science: Physics is a comprehensive and up-to-date textbook designed to meet the requirements of the six examining groups for all their GCSE science syllabuses. The textbook can be used for the physics component of the Single or Double Award and for all GCSE Science: Physics syllabuses.

Topics are differentiated into core material for Single or Double Science and extension material for Science: Physics (blue/grey margin). The examining groups have chosen different extension topics in addition to the Programme of Study for Key Stage 4; therefore teachers and students need to be familiar with the specific requirements of the syllabus. *Key Science: Physics* contains the extension topics for all boards.

Key Science: Physics is organised in 6 Themes covering 23 major topics. Each topic is sub-divided into numbered sections covering all aspects of the topic. Each section includes special features, usually located in the margin, such as:

- **First Thoughts,** setting the scene and reminding students of the background knowledge to the section,
- **It's a Fact,** sharpening interest and enhancing text,
- **Key Scientist,** providing a historical perspective on how scientific ideas have developed,
- **A key** , in the margin indicates an opportunity to exercise one or more of the key skills: application of number (AoN), communication (Comm) and information technology (IT). These opportunities are listed on p xxiv of the Teachers' Guide,
- **A star** , in the margin suggests the topic can be used for the development of scientific ideas and the weighing of evidence. These suggestions are listed on p xxvi of the Teachers' Guide,
- **A computer mouse** , in the margin indicates an appropriate opportunity to use computer software or hardware. These opportunities are listed at www.keyscience.co.uk.
- **Commentary and Summaries,** within each section of a topic and at the end of each section, helping students review their work and highlighting the key points to learn and understand,
- **Checkpoints,** testing knowledge and understanding through sets of short questions at regular intervals at the end of each section,
- **Theme Questions,** based on examination questions.

Answers to all numerical questions and Checkpoints and a comprehensive **Index** are at the end of the book.

Support for the text book

Key Science Physics is supported by an *Extension File* which contains a *Teacher's Guide*, a bank of photocopiable material and key diagrams from the text in colour on CD-ROM. There are over 60 photocopiable Activities for Sc1 and Assignments for Sc4 and GCSE physics topics. There are 60 Exam File questions for homework and revision, with mark schemes provided.

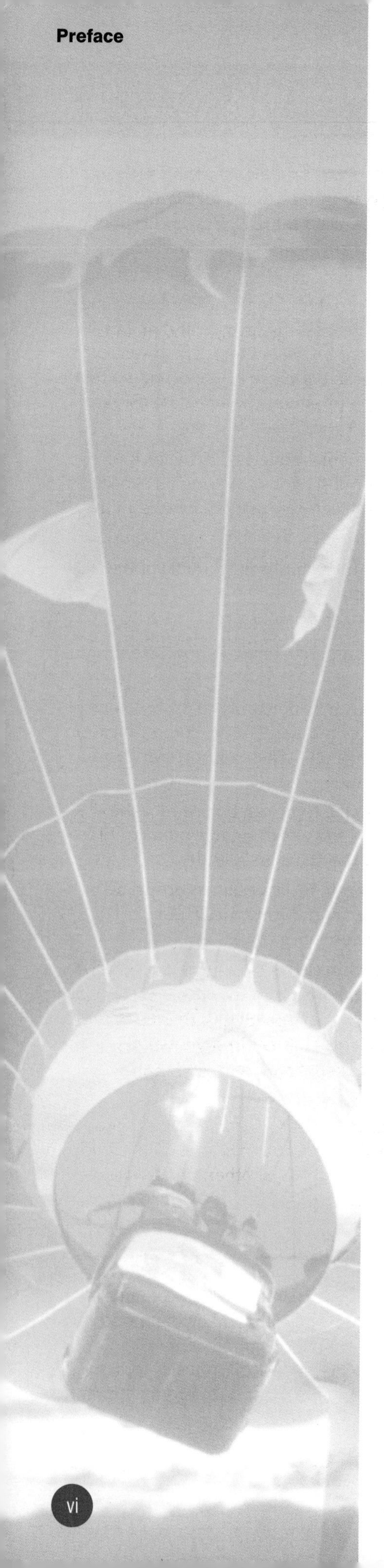

The *Teacher's Guide* offers advice on how students can use the activities provided in *Key Science Physics* and in the *Extension File* to gain evidence for their key skills portfolio. It also gives advice on topics that can be used for developing the theme of ideas and evidence in science. Guidance on the use of computers in science is provided. Detailed notes at www.keyscience.co.uk include references to the internet, appropriate software and hardware for computer-assisted learning.

Student's Note

I hope that this book will provide all the information you need and enough practice to enable you to do well in the tests and examinations in your GCSE course. However, examinations are not the most important part of your course. I hope also that you will have a real interest in science, which will continue after your course has finished. Finally, I hope that you enjoy using the book. If you do, all the effort will have been worthwhile.

Jim Breithaupt

I do not know what I may appear to the world; but to myself I seem to have been only like a boy, playing on the seashore, and diverting myself, in now and then finding a smoother pebble or a prettier shell than ordinary, while the great ocean of truth lay all undiscovered before me.

(Isaac Newton, 1642–1727)

No amount of experimentation can ever prove me right; a single experiment may at any time prove me wrong.

(Albert Einstein, 1879–1955)

The test of science is its ability to predict. Had you never visited the earth, could you predict the thunderstorms, the volcanoes, the ocean waves, the auroras, and the colorful sunset? . . .

(Richard Feynman 1918–1988)

Adventures in science

The Mirror of Galadriel

In the book *Lord of the Rings*, amazing things happen to Gandalf and the hobbits. They risk their lives on a perilous journey to save their world. They journey through secret tunnels and unknown lands and they witness astonishing events. In the early parts of their travels, they are invited to look into a magic mirror – a silver bowl filled with water. It shows 'things that were, and things that are, the things that yet may be.' But the viewer can't tell which it is showing!

Imagine you have travelled in time from two centuries ago to the present day. What would you make of television – a silver bowl showing events from the past and the present? Television is a product of the scientific age in which we live.

Science has many more amazing things to reveal. Imagine food 'grown' in a factory or round-the-world flight in a few hours or 'intelligent' computers that need no programming. These and many more projects are the subject of intense scientific research now. Fact is stranger than fiction.

Can you believe what you see?

Look at the picture in Figure 1A. *What can you see?* Some people see an old lady; others see a young woman. Different people looking at the same picture see different images.

Witnesses of an event often report totally different versions of the event. Invent a harmless incident and try it out in front of unsuspecting witnesses without warning them. Then ask them to say what happened. Each person will probably have a different story to tell.

In science, there are countless events to observe. The story or 'account' of an event in science should not differ from one observer to another. Otherwise, who could you believe?

Figure A What can you see?

How science works

Science involves finding out how and why things happen. Natural curiosity sometimes provides the starting point for a scientific investigation. Another investigation might set out to solve a certain problem and therefore have a definite aim. Even then, unexpected results may emerge to stimulate our natural curiosity.

The history of science has many examples where an important discovery or invention came about unexpectedly. The electric motor was invented when an electric generator was wired up wrongly at the Vienna Exhibition of 1873. X-rays, radioactivity and penicillin are further examples of unexpected scientific discoveries.

Discoveries in science are unpredictable. No-one can predict which research projects will lead to exciting new developments because no-one knows what new areas of science lie waiting to be discovered beyond the frontiers of knowledge.

Physics as a key branch of science

Physics is a fundamental branch of science because its principles form the foundation of other branches of science. For example, conservation of energy is a key principle of physics that forms a starting point in a wide range of scientific studies from atomic bonding to weather forecasting.

What are the distinctive features of physics? Although the subject may be described as the study of matter and energy, understanding natural phenomena is a central aim of physics. In the early years of the twentieth century, physicists discovered that the atom consists of a positively charged nucleus surrounded by electrons. Further experiments showed that each nucleus is composed of neutrons and protons. Now physicists probing the nucleus have found that neutrons and protons are composed from even smaller particles.

Figure B ▲ Electron microscope picture of a human cell

Such 'pure' research leads to a better understanding of natural phenomena and can often lead to important applications in other branches of science. The use of physics in other branches of science is perhaps one of the reasons why physics is at the centre of the Scientific Age we live in. The discovery of the electron in the late nineteenth century led to the development of the electron microscope which has revolutionised cell biology.

New technologies have been created as a result of discoveries in physics. The transistor was invented over 30 years ago. Now thousands of transistors can be made on a single chip called an integrated circuit. Use of integrated circuits has brought about a revolution in the way information is transmitted and used. Even day-to-day activities such as shopping or visiting the bank have altered as a result of the chip revolution.

A scientific race

On 18 March 1987, more than two thousand scientists crammed into a conference room in New York to listen to progress reports on research into superconductivity. The conference had to be relayed to hundreds of scientists in adjacent rooms.

Superconducting materials have no electrical resistance. Electric cables and electric motors made from superconductors would be far more efficient than those made from ordinary conductors. Superconducting computers and magnets would be much more powerful. However, before 1987, most scientists thought that superconductors would work only at very low temperatures. Materials that are superconducting at room temperature would be a technological breakthrough.

The first superconductors were discovered in 1911 when certain metals were cooled to below −270°C. In 1986, two European scientists, Georg Bednorz and Alex Müller, reported they had made a superconductor at 30°C higher (−240°C). Scientists round the world turned their attention to superconductors. Teams of scientists in different countries raced each other to find superconductors which would work at higher temperatures. By March 1987, the temperature limit had been raised to −179°C. By then the political leaders of the major industrial nations realised how important superconductors could become. The race continues because the benefits would be immense. The importance of Bednorz and Müller's work starting the race was recognised when they received the 1987 Nobel Prize for physics. Since then, the temperature limit has been raised to −130°C but the challenge of finding room temperature superconductors remains.

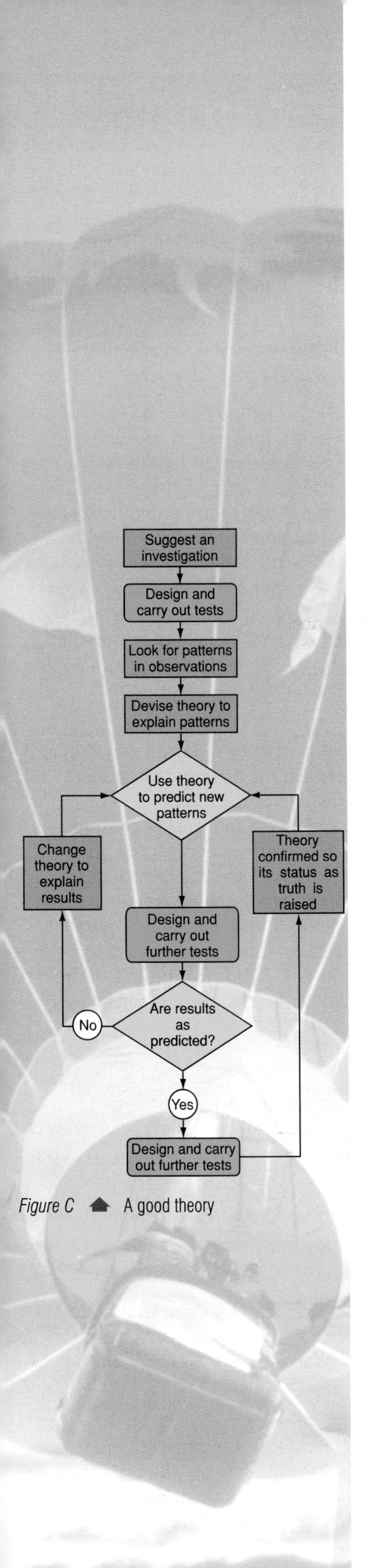

Figure C ▲ A good theory

Models and theories

In 1987, scientists in laboratories throughout the world worked round the clock to try to make new superconductors. Other scientists were busy with computers trying to work out why the new materials were superconducting. They made models of how the atoms of the new materials might be fixed together. They tested and changed the models using their computers until they found a model that explained what was observed in the laboratories. Understanding how and why new superconductors work is important since it can give vital clues to better superconductors. However, the fact that room temperature superconductors remain undiscovered shows the need in science for patience and persistence as well as imagination and creativity.

A good theory or model is one which gives predictions that turn out to be correct. The predictions are tested in the laboratory. If they do not pass the test, the theory or model must be altered or even thrown out. The more tests a model or theory passes, the more 'faith' scientists have in it. However, complete faith is never possible. Someone, somewhere might one day make a discovery which can't be explained by the theory.

Beyond the frontiers

Who knows where a new discovery can lead? When Michael Faraday discovered how to generate electricity, he was asked 'What use is electricity, Mr Faraday?' He is reported to have replied 'What use is a new baby?' No one knows what might become of a new invention or a new discovery. More than 150 years later, life without electricity for most of us is unthinkable. Michael Faraday would have been truly astonished. The microchip is another discovery that has revolutionised our lives in recent years. The 2000 Nobel Prize for physics was awarded jointly to physicists Jack Kilby for his invention of the microchip 25 years earlier and Zhores Alferov and Herbert Kroemer for their work on developing the materials now used in CD players, laser printers and fibre-optic communications. The electronic devices that have revolutionised our lives in recent years all followed from the invention of the microchip. None of us know for sure what changes will take place in the next 25 years due to the microchip revolution. Science will undoubtedly have many surprises in store. New and unexpected discoveries are likely to be made. Discoveries may unfold as a result of current research. We cannot predict what discoveries lie ahead but we know from past discoveries that science can improve our lives profoundly.

Careers in science

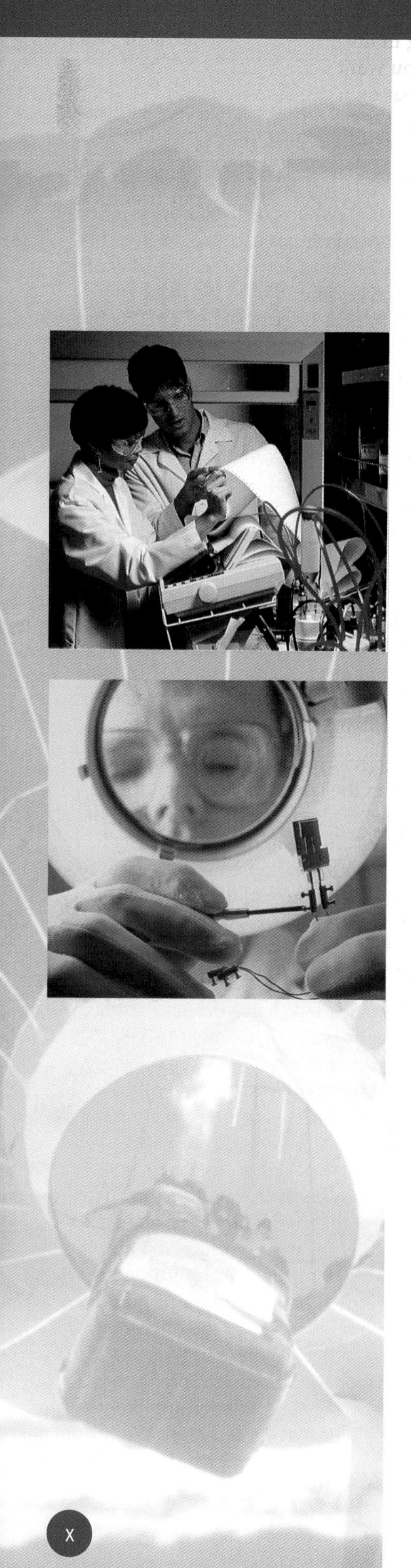

Whatever career you choose to follow, you will almost certainly need to use your GCSE science skills and knowledge. Science is a feature of everyday life, at work or at home. Industry, hospitals, transport and agriculture are examples of major sectors of the economy that depend heavily on science. At home, almost all the things you do make use of science in some way.

Your GCSE science course is designed to give you a good scientific background so that you can, if you wish, carry on with science studies after GCSE.

To work in science requires specific personal qualities in addition to academic qualifications. Scientists are very creative and imaginative people and the work of an individual scientist can bring huge benefits to everyone. For example, Alexander Fleming's discovery of penicillin has saved countless lives. But do not be misled into thinking that the life of a scientist is one of continual discoveries. Scientists have to be very patient and methodical to discover anything; they have to be good at working together and at communicating their ideas to each other and to other people. The qualifications needed to become a scientist are outlined on p. xii; the qualities needed to become a scientist are just the same as you need in your GCSE course – enthusiasm, hard work, imagination, awareness and concern.

What jobs are done by scientists? In industry, scientists design, develop and test new products. For example, scientists in the glass industry are developing amazingly clear glass for use as **optical fibres** in communication links. In medicine, scientists are continually finding applications for scientific discoveries; for example, medical scientists have developed a high-power ultrasonic transmitter for destroying kidney stones, thus avoiding a surgical operation. These are just two examples of the work of scientists. You will find scientists at work in research laboratories, industrial laboratories, forensic laboratories, hospitals, schools, on field trips, expeditions, radio, TV and lots of other places. Scientists have to be very versatile as science is a very wide and varied field.

The skills and knowledge you gain through studying science will enable you to gain the benefits of new technologies. Ask your parents what aspects of life have **not** changed since they were children – they may be stuck for an answer! New technologies force the pace of change and if you do not learn to use them, you will not share their benefits. Studying science encourages you to develop an open mind and to seek new approaches. That is why a wide range of careers involve further studies in science.

For many careers further studies in science are essential, for example, medicine, dentistry, pharmacy, engineering and computing. For other careers such as law, business studies, administration, the armed forces and retail management, studying science after GCSE will be helpful. Thus by continuing your science studies after GCSE, you keep many career options open, which is important when you are considering your choice of career.

The next steps after GCSE

Read this section carefully, bearing in mind that your working life will probably be about forty years. If you want your life ahead to be interesting; if you want to make your own decisions about what you do; if you want to make the most of your talents; then you should continue your studies. After GCSE, you can continue in full-time education at school or at college, or you can train in a job through part-time study. If you take a job without training, you will soon find that your friends who stayed on have much better prospects.

Most students aiming for a career in science or technology continue full-time study for two years, taking either GCE AS and A-levels or an AVCE course in Science. Successful completion of a suitable combination of these courses can then lead to a degree course at a university or a college of higher education or, alternatively, straight into employment.

The A-level route to higher education requires successful completion of a two-year full-time course, usually consisting of 4 or 5 AS-level subjects studied in the first year followed by at least 3 of these subjects studied to full A-level standard in the second year. Students taking A- or AS-level Physics usually also take A- or AS-level Mathematics or a Mathematics support course. A-levels in Physics and Mathematics and one other subject lead to a wide range of degree courses in science, engineering, medicine and management studies.

The AVCE (Advanced Vocational Certificate of Education) science route is also a two-year full-time course at advanced level, leading to a qualification equivalent to double award A-level science. All students study some biology, chemistry and physics and can then specialise. The course is assessed continuously through tests and laboratory work.

Key Skills are a feature of all post-16 courses just as at GCSE. Employers, college admissions tutors and university admissions tutors expect all students to develop their key skills in Application of Number, Communications and ICT after GCSE. Credit for university admissions is gained as a result of successful completion of key skills. If you are following a course of 4 or 5 AS subjects, you will probably be able to incorporate key skills in your AS studies where appropriate with little extra effort. Physics is an ideal subject for gaining key skill credits in this way because application of number, the use of ICT and communications are activities in any physics course. That's why physicists and physics students are highly valued people.

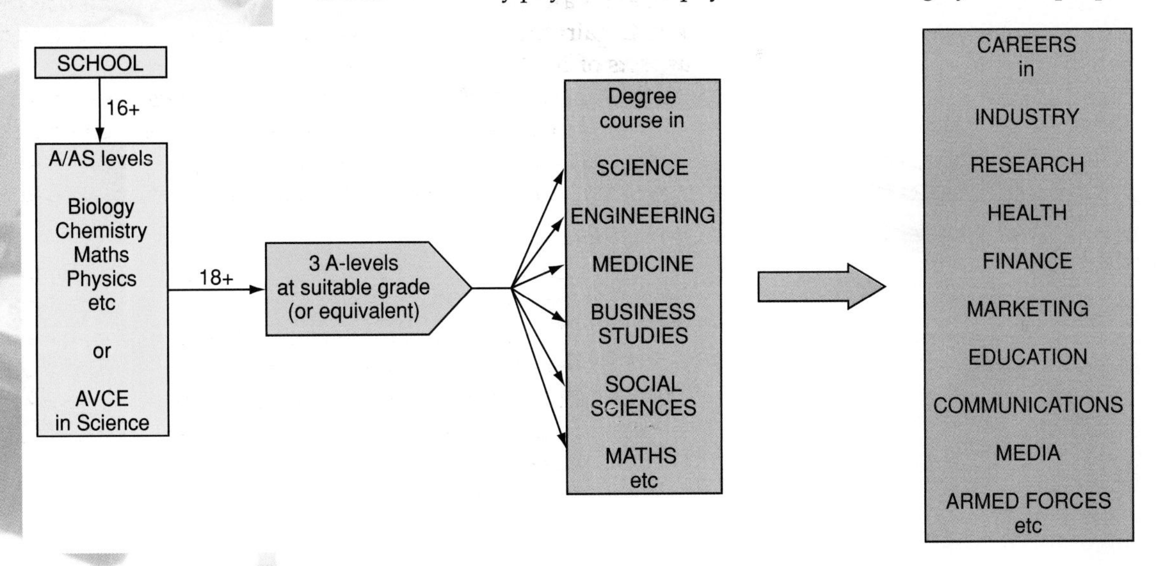

Figure D ▲ Career routes from GCSE

About the Earth

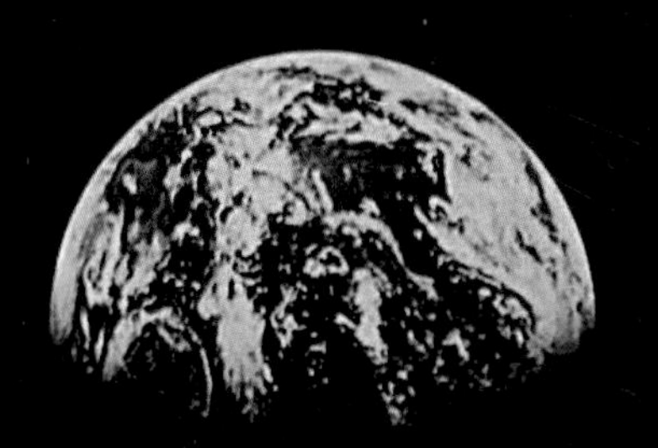

Why has life evolved on a minor planet called Earth that orbits a middle-aged star called the Sun?

This star is part of one of many galaxies that make up the universe. Earth and the other planets in orbit round the Sun formed from molten rock billions of years ago. As Earth cooled, its outer layers solidified. Continents and oceans developed and a huge variety of life forms evolved. One particular life form is capable of making a scientific study of its environment!

Topic 1 The Earth in space

1.1 ▶ The Sun

FIRST THOUGHTS

The Sun and the Moon are the two most prominent objects in the sky. Why does the Moon's appearance change? What causes an eclipse? Read on to find out why.

We live near the edge of a fiery ball of gas we call the Sun. If the Sun died out, we would soon perish. All our energy comes, directly or indirectly, from the Sun. All forms of life need energy. So life on Earth depends on the Sun.

The Sun is a star. Stars are huge balls of hot gas that emit light. The Sun appears much bigger and brighter than any other star because we are so close to it.

Figure 1.1A ▲ Solar activity

If the Sun is represented by a football, the Earth on the same scale would be a grain of rice 30 m away.

How far away is the Sun? Light from the Sun takes about eight minutes to reach Earth compared with about 4.2 years from the next nearest star, Proxima Centauri. The distance to the Sun's nearest neighbour is truly astronomical.

How big is the Sun? Its diameter is 110 times greater than Earth's. The distance from Earth to the surface of the Sun is about 100 solar diameters. Earth is one of the nine planets in orbit round the Sun. We can see the other planets because they reflect sunlight.

What keeps Earth in orbit round the Sun? The Sun's **gravity** stretches out into space and keeps the planets moving along their orbits. If the Sun's gravity was suddenly removed, Earth would shoot off into the cold depths of space.

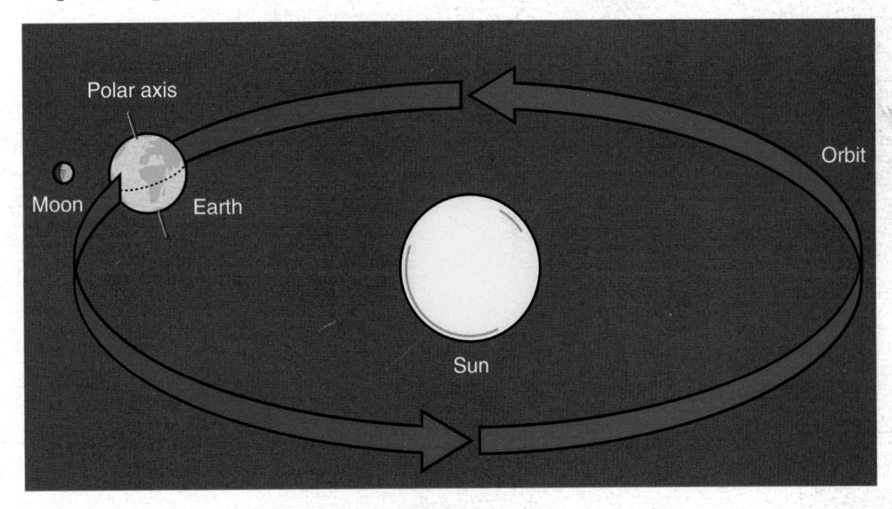

Figure 1.1B ▲ The Earth's orbit

SUMMARY

Stars are huge balls of hot gas that emit light. Earth is a small planet in orbit round a star that we call the Sun. Planets do not emit light but they reflect sunlight.

1.2 The Moon

Phases of the Moon

Our nearest neighbour in space, the Moon, orbits the Earth as Earth orbits the Sun. Even without a telescope, you can see that the Moon's surface has bright and dark areas and that it is covered with craters. We can see the Moon because it reflects sunlight. It orbits Earth once every $27\frac{1}{4}$ days, always keeping the same face towards Earth.

To Earth-bound observers, the Moon passes through a cycle of phases each month. This is because the amount of its sunlit face visible from Earth changes as it moves round Earth.

See www.keyscience.co.uk for more about the Earth and the Moon.

Figure 1.2A The Moon

It's a fact!

Ocean tides are caused by the Moon's gravity. Even though the Moon is so far from Earth, its pull is enough to raise the ocean by a few metres at opposite sides of Earth. Earth spins once each day so the two tidal bulges on opposite sides of the Earth give two tides per day. The tidal bulge nearest the Moon is due to the Moon's pull on the ocean. On the opposite side of the Earth, the tidal bulge is because the ocean is thrown outwards slightly by the rotation of the Earth–Moon system.

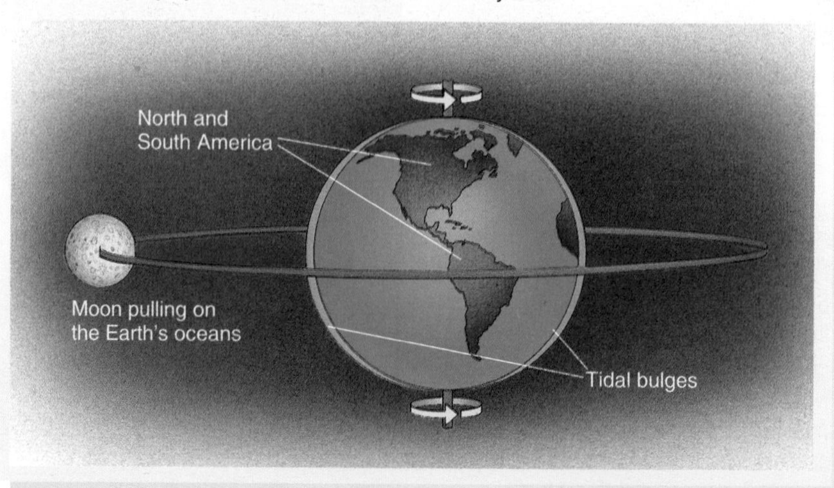

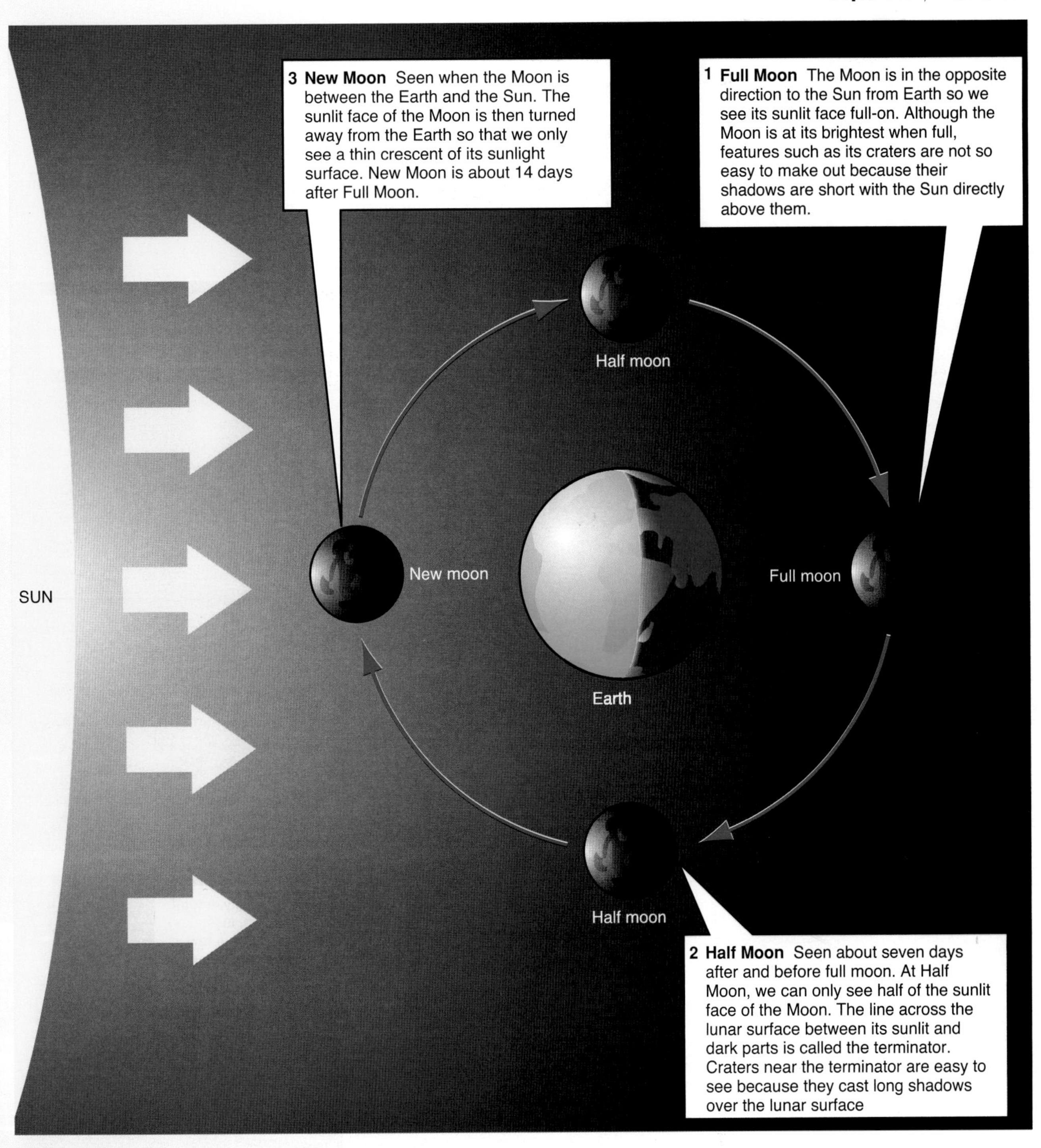

3 New Moon Seen when the Moon is between the Earth and the Sun. The sunlit face of the Moon is then turned away from the Earth so that we only see a thin crescent of its sunlight surface. New Moon is about 14 days after Full Moon.
1 Full Moon The Moon is in the opposite direction to the Sun from Earth so we see its sunlit face full-on. Although the Moon is at its brightest when full, features such as its craters are not so easy to make out because their shadows are short with the Sun directly above them.
Half moon
SUN
New moon
Earth
Full moon
Half moon
2 Half Moon Seen about seven days after and before full moon. At Half Moon, we can only see half of the sunlit face of the Moon. The line across the lunar surface between its sunlit and dark parts is called the terminator. Craters near the terminator are easy to see because they cast long shadows over the lunar surface

(a) 4-day crescent Moon

(b) 10-day Moon

(c) Full Moon

(d) 21-day Moon

Figure 1.2B ▲ Phases of the Moon

Eclipses of the Sun and the Moon are predictable events. In ancient times, they were thought to foretell catastrophes and disasters! Eclipses happen when either Earth or the Moon passes into the other's shadow.

Eclipses

Lunar eclipses

Lunar eclipses occur when the Moon passes through Earth's shadow. Figure 1.2C shows the idea. The shadow of Earth reaches far out into space, well beyond the Moon. Lunar eclipses take several hours and can be seen from any point on the night-time half of Earth.

Figure 1.2C A lunar eclipse

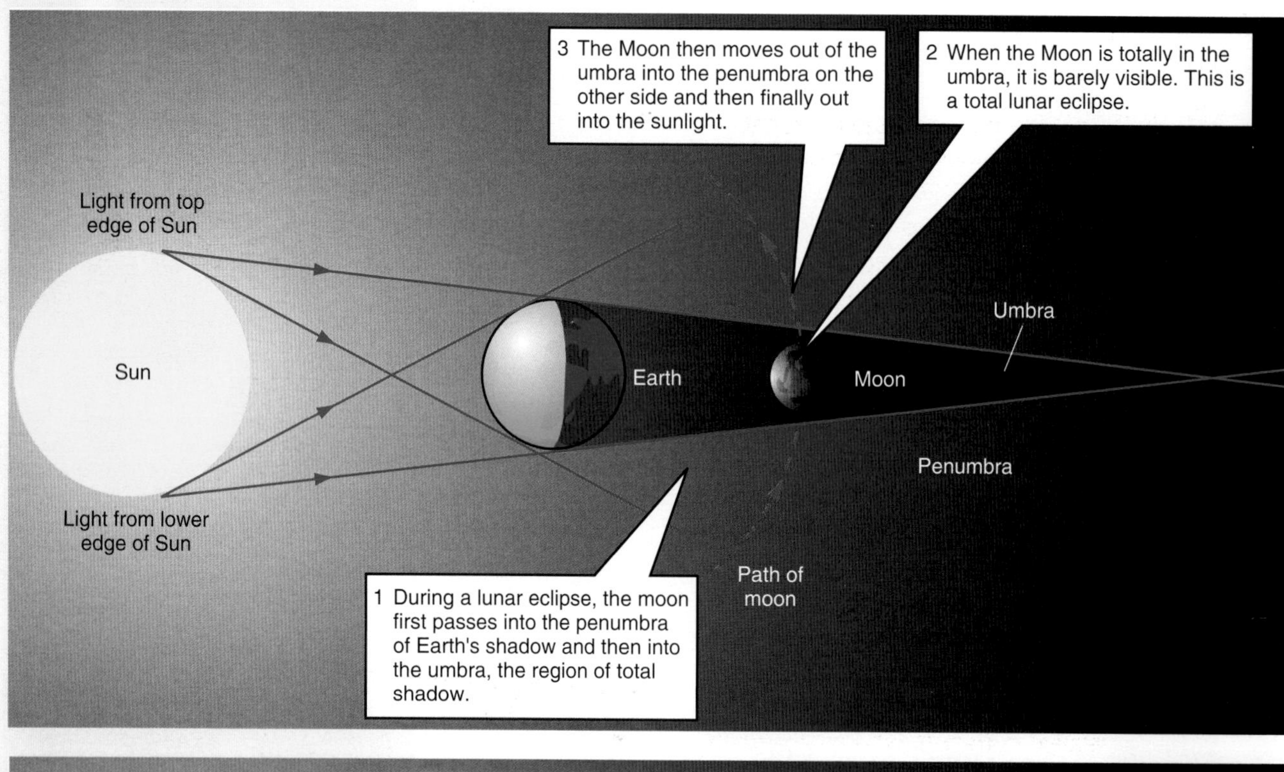
3 The Moon then moves out of the umbra into the penumbra on the other side and then finally out into the sunlight.
2 When the Moon is totally in the umbra, it is barely visible. This is a total lunar eclipse.
Light from top edge of Sun
Sun
Earth
Moon
Umbra
Penumbra
Light from lower edge of Sun
Path of moon
1 During a lunar eclipse, the moon first passes into the penumbra of Earth's shadow and then into the umbra, the region of total shadow.

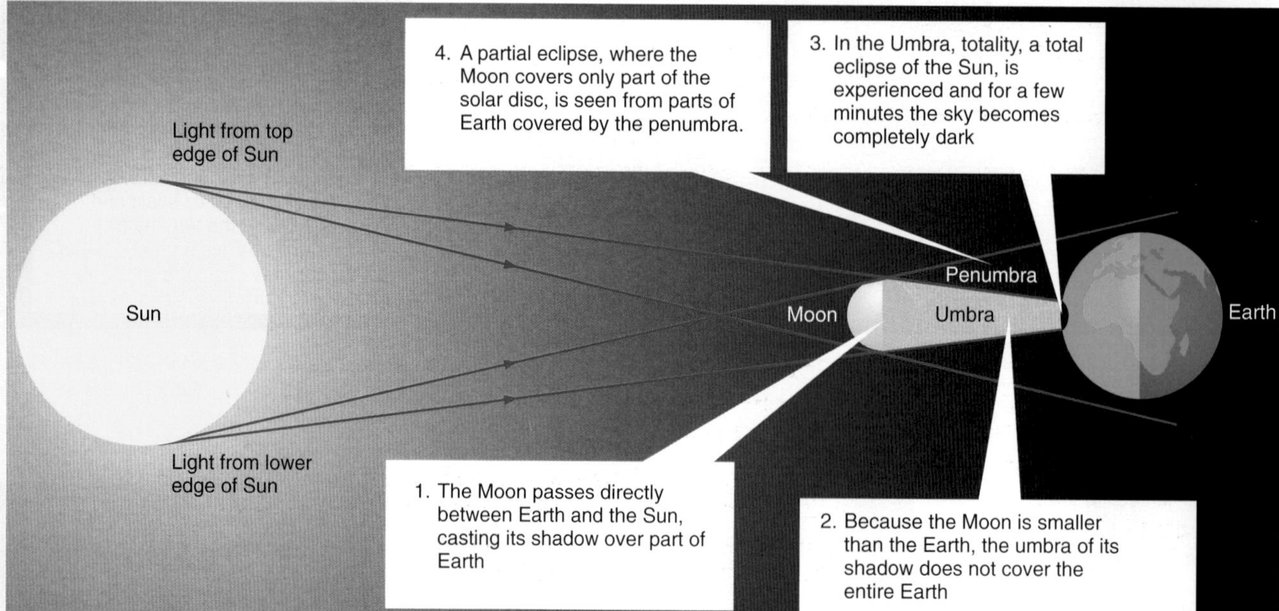
4. A partial eclipse, where the Moon covers only part of the solar disc, is seen from parts of Earth covered by the penumbra.
3. In the Umbra, totality, a total eclipse of the Sun, is experienced and for a few minutes the sky becomes completely dark
Light from top edge of Sun
Sun
Penumbra
Moon
Umbra
Earth
Light from lower edge of Sun
1. The Moon passes directly between Earth and the Sun, casting its shadow over part of Earth
2. Because the Moon is smaller than the Earth, the umbra of its shadow does not cover the entire Earth

Figure 1.2D (a) A solar eclipse

Solar eclipses

Solar eclipses take place when our view of the Sun is blocked by the Moon. Figure 1.2D shows how this happens.

SUMMARY

Phases and eclipses are caused by the relative positions of the Sun, Earth and Moon. Eclipses happen when the Sun, Earth and Moon are exactly in line with one another.

Sometimes the umbra does not reach Earth and then an **annular eclipse** is seen from Earth. This is where the solar disc forms a ring or annulus round the Moon. Total solar eclipses occur just as often as lunar eclipses but because totality covers only a small area of Earth, they are not seen as often.

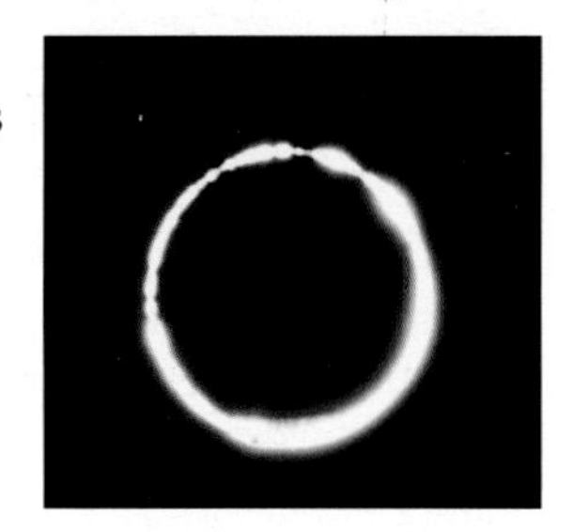

Figure 1.2D (b) ▲
Annular eclipse

CHECKPOINT

▶ 1 In the picture opposite, the crescent moon is drawn incorrectly. Explain why the drawing is incorrect.

▶ 2 (a) Sketch the phases of the Moon.
(b) Show the relative positions of the Moon, Earth and Sun at Half Moon.

▶ 3 The Moon goes 360° round Earth once every 27.25 days. How far does it go round in
(a) 1 day, (b) 1 week?

▶ 4 Look for the Moon at the same time of night over successive nights. Does the Moon go round Earth from east to west or the other way?

▶ 5 (a) Make a sketch to show the relative positions of the Moon, Earth and Sun during a solar eclipse.
(b) The Moon's orbit is tilted slightly relative to Earth's orbit as shown in the figure opposite. Use this to explain why solar eclipses do not occur every New Moon when the Moon is between Earth and the Sun.

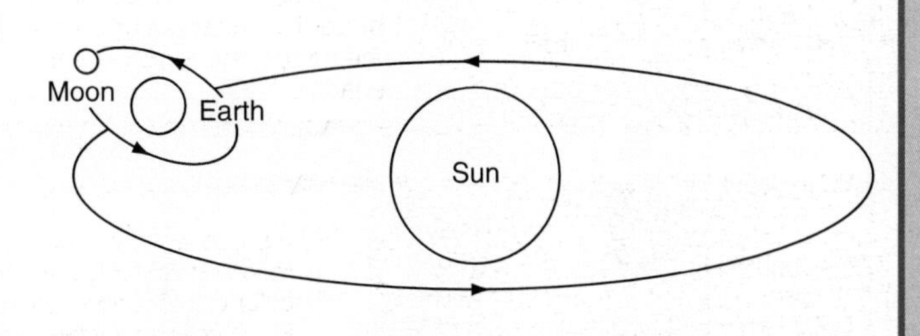

1.3 ▶ The planets

FIRST THOUGHTS

How can you find a planet in the night sky? In this section, you can find out where to look and what to look for.

The Sun and the nine planets and their moons make up most of the solar system. It may surprise you to learn that other planets have moons. Jupiter has at least 14 moons!

Astronomers of long-ago mapped the stars and used their imagination to link groups of stars together. These groups are called **constellations**. We see them now much as they appeared thousands of years ago. One of the best known constellations is Orion, which was thought to resemble a warrior.

Seen from Earth, the other planets change their positions among the constellations from night to night. This is because they orbit the Sun and so the direction to each planet, from Earth, changes continuously. The word planet is the Greek word for wanderer. The astronomers of Ancient Greece named the planets after their gods who lived in the heavens.

How do the other planets compare with Earth? Figure 1.3A shows how they compare in terms of appearance, size and distance from the Sun. Here are one or two tips to help planet hunters!

Mercury is hard to find in the sky because it is so near the Sun. It can only be seen for an hour or so before sunrise and after sunset.

Venus is sometimes so bright in the night sky that it casts shadows. It moves further from the Sun than Mercury so we sometimes see it for several hours before sunrise and after sunset. Its brilliance after sunset is the reason why it is sometimes called the evening star.

Mars is closest to Earth when it is seen in the opposite direction to the Sun, as in Figure 1.3B. This is called **opposition** and it happens about once every two years. The progress of Mars through the constellations appears to reverse as Earth catches up with it near opposition.

0 3.2 6.0 8.3 12.6 43 79 159 249 324

Time for light to travel from Sun (in minutes)

SUN

Comparison of the planets	Mercury	Venus	Earth	Mars	Jupiter	Saturn	Uranus	Neptune	Pluto
Diameter of planet (in Earth diameters)	0.39	0.97	1	0.53	11.2	9.5	3.7	3.5	0.4
Distance from Sun (in Sun-Earth distances)	0.39	0.72	1	1.52	5.20	9.53	19.2	30	39
Length of year (in Earth years)	0.24	0.61	1	1.88	11.9	29.5	84	165	250
Length of day (in Earth days)	59	243	1	1.03	0.41	0.44	0.67	Not known	Not known
Atmosphere	None	Mostly carbon dioxide	Mostly nitrogen and oxygen	Mostly carbon dioxide	Mostly hydrogen	Mostly hydrogen	Not known	Not known	Not known
Appearance	Cratered	Cloud covered	Blue / green some cloud	Red with white caps at poles	Red/orange with bright and dark zones	Yellow/orange with bright and dark zones and rings	Greenish disc with bright and dark zones	Pale blue disc	Too far to see
Composition	Rock	Rock	Rock	Rock	Fluid	Fluid	Fluid	Fluid	Probably rock

Earth's diameter = 12 735 km. All measurements in terms of the Earth Sun's diameter = 1.39 million kilometres = 1.39×10^6 km

1 Astronomical Unit (AU) = Sun-Earth separation = 150 million kilometres = 150×10^6 km

Figure 1.3A ▲ The solar system

It's a fact!

You can find some of the planets for yourself in the night sky. Newspapers often feature a regular Guide to the Night Sky telling readers when and where to look for the planets.

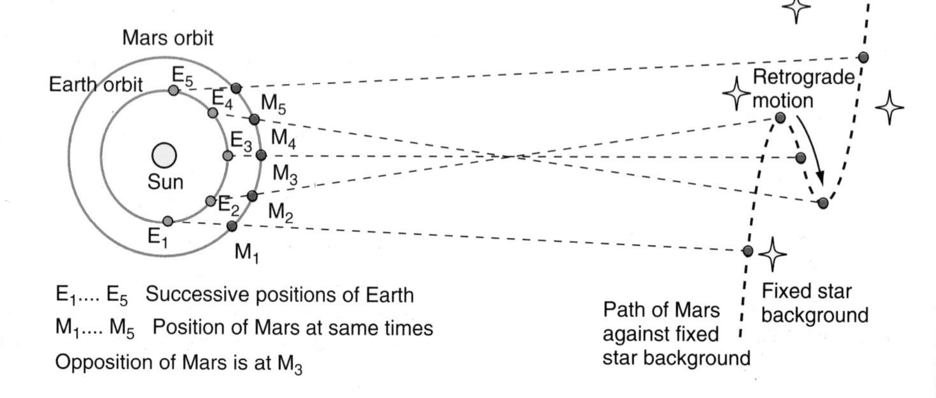

Figure 1.3B ▶ The retrograde motion of Mars

Figure 1.3C ▲ Jupiter

Figure 1.3D ▲ (a) The rings of Saturn

Jupiter was first observed through a telescope by Galileo who discovered four moons in orbit round Jupiter. Another surprise from Jupiter is the presence of the Great Red Spot on its surface. This is thought to be a whirlwind. It was first observed by Galileo and is still active.

Saturn is an amazing sight seen through a telescope as it is circled by huge, thin rings which stretch out far above its equator. The rings are made of lots of micro moons, each orbiting Saturn while their own gravity keeps them in a thin disc. Saturn's axis is tilted at 29° to its orbit. This means that its rings are tilted and we see them at their best once every 14.5 years. Saturn has a number of moons too; the largest, Titan, is bigger than our Moon.

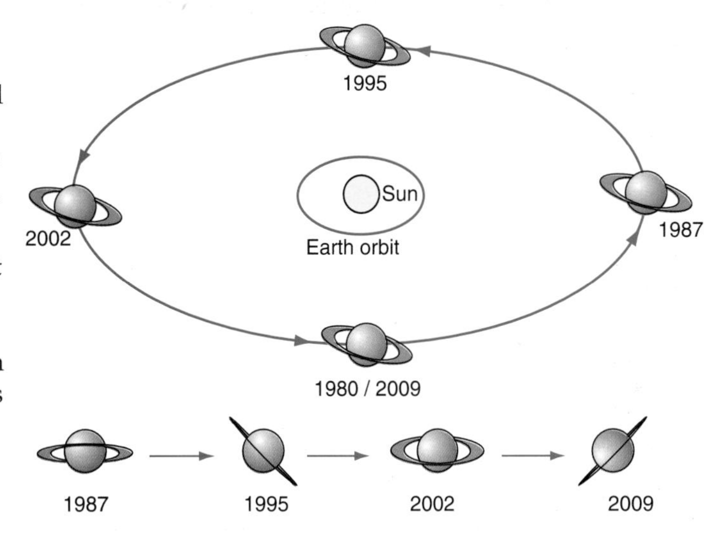

(b) The orbit of Saturn

EXTENSION FILE ASSIGNMENT

SUMMARY

The four inner planets, Mercury, Venus, Earth and Mars, are solid and much smaller than Jupiter, Saturn, Uranus and Neptune. These outer planets are giant balls of fluid. The outermost planet, Pluto, is thought to be small and solid.

The three planets beyond Saturn are too dim to see without the aid of a telescope.

Uranus was discovered in the eighteenth century. In 1977, astronomers discovered that Uranus has a ring system. The rings are too faint to be seen directly but they were first detected when Uranus passed in front of a star. Just before and after Uranus blocked the star out, the star blinked several times as each ring covered the star.

Neptune was discovered in 1846, as a result of detailed calculations on the orbit of Uranus. Astronomers found that Uranus did not move quite as predicted and they reasoned that an outer planet was responsible. They worked out where this outer planet ought to be and, sure enough, there it was!

Pluto, the outermost planet, was discovered from its effect on Neptune's orbit, although this was not until 1930. Its orbit is not circular so at times it is closer to the Sun than Neptune is.

Are there any more planets out there? Astronomers have known since 1801 about the minor planets, called asteroids, between Mars and Jupiter. The largest of these, Ceres, is about 1000 km in diameter although the others are much smaller. Some astronomers think there is a tenth planet beyond Pluto.

The **force of gravity** between a planet and the Sun keeps the planet in its orbit. There is a pull of gravity between any two objects. The force of the Sun's gravity on a planet continually changes the direction of motion of the planet, making it go round the Sun (see p.275 for more about circular motion). The Sun's gravitational pull decreases with increasing distance, becoming zero at infinity. The further a planet is from the Sun, the smaller its speed must be – otherwise, the Sun's gravity would be unable to keep it in its orbit. This is why the time taken for planets to orbit the Sun increases with distance from the Sun.

CHECKPOINT

1. Use the table in Figure 1.3A to answer the questions below.
   (a) List the names of planets which are not solid.
   (b) (i) Why is Mercury much hotter than the Earth?
   (ii) Why is Mars much colder than the Earth?
   (c) Which planet is closest to the Earth?
   (d) Which two planets are closest to each other?
2. Copy and complete the table below to compare the following planets.

	Mercury	*Venus*	*Earth*	*Mars*
Diameter				
Atmosphere				
Appearance				
Days per year				

3. Work out the least distance and greatest distance, in terms of the distance from Sun to Earth, between
   (a) Earth and Mars,
   (b) Earth and Venus.
4. (a) What is the least possible distance between Jupiter and Earth?
   (b) Why does the brightness of Saturn change as it moves along its orbit?
5. Which planet do you consider to be most like the Earth? State and explain your reasons for this choice.

1.4 Comets

FIRST THOUGHTS

In primitive civilisations, comets and meteors were regarded as signs of events to come. After working through this section, you will see that comets and meteors are entirely natural.

The appearance of a comet in the sky causes great excitement. In 1986, space probes were sent up from Earth to meet Halley's Comet and scientists learned a great deal about comets as a result. Before this, comets were thought to be huge chunks of frozen rock, but the space probes revealed a pitch black surface with jets of gas bursting through.

Comets orbit the Sun in non-circular orbits called ellipses. Their orbits usually stretch far beyond Pluto, taking many years to complete. As a comet approaches the inner part of the solar system, the Sun heats it up and turns some of its solid matter into a long tail of gases. The tail always points away from the Sun, driven away by the force of solar radiation. Comets far from the Sun are dark and impossible to see. But as a comet moves nearer the Sun, it gets hot and a glowing tail of gas forms which makes the comet visible.

Figure 1.4A Comet Hale-Bopp returned to the inner solar system in 1997 after a journey of several thousand years

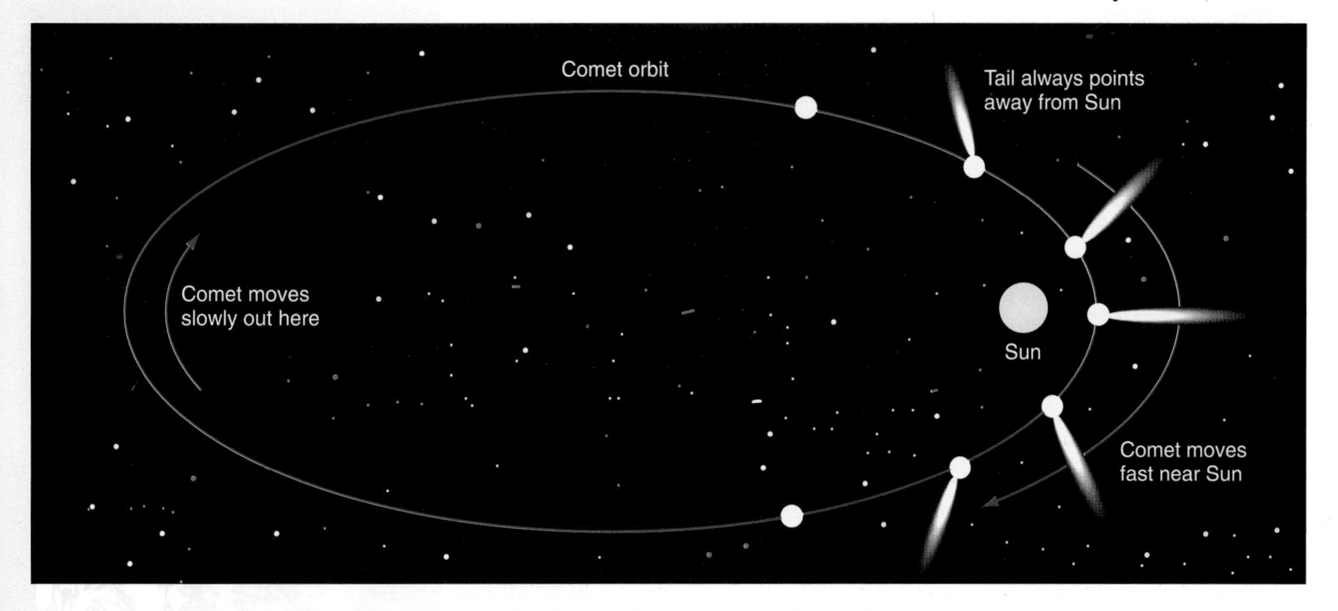

Figure 1.4B ▲ Comet paths

Halley's Comet can be seen from Earth every 76 years as it passes through the inner solar system. In 1910, its return was visible in daylight but in 1986, its return was less easy to see because it was in the same direction as the Sun when at its brightest. In 1682, Edmund Halley studied a bright comet and was the first person to realise that it had been seen several times before at intervals of about 76 years. He correctly predicted it would return in 1758 and so the comet was named after him.

Meteors

Meteors, sometimes called shooting stars, can be breathtaking to see. They look like tiny balls of fire shooting across the sky. This happens when a particle of matter from space enters our atmosphere at high speed. Air resistance makes the particle so hot that it burns up. Sometimes, if the meteor is large enough, it does not burn up completely and falls to Earth as a meteorite. The largest known meteorite, thought to have fallen in prehistoric times, has a mass of more than 60 000 kg!

SUMMARY

Comets orbit the Sun in elliptical orbits, taking many years to complete each orbit. Meteors are particles of matter from space that burn up because they enter Earth's atmosphere at high speeds.

CHECKPOINT

1. Why are comets visible only when they are near the Sun?
2. When is Halley's Comet next expected to return?
3. Use two drawing pins and some thread to draw an ellipse as shown in the figure opposite. This gives you an accurate sketch of a comet orbit with the Sun at one of the pinpoints. Mark the Sun on your sketch and show a comet with its tail near the Sun.
4. (a) How does Earth's atmosphere help to protect us from meteorites?
   (b) The moon and Mercury are both heavily cratered. Why?
5. A bright comet is shown on the Bayeux Tapestry which depicts the Norman Conquest of England in 1066. Could this comet have been Halley's Comet? Explain your answer.

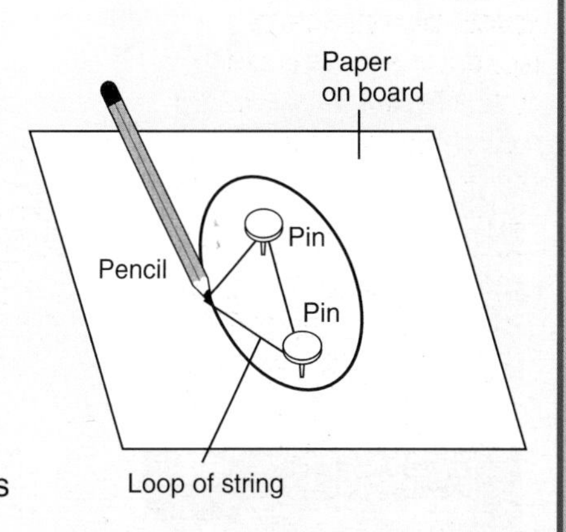

1.5 Star watch

FIRST THOUGHTS

The stars and the Sun have been used for timekeeping and navigation for centuries. Why do we have leap years? How can we use the stars to navigate? Astronomy has always been a very practical subject.

Earth spins round as it orbits the Sun. Its axis points in a fixed direction in space so that the **Pole Star** is always directly above Earth's North Pole. Anyone who is north of the equator can find due north by looking for the Pole Star.

In the northern hemisphere, if you watch the stars for a few hours, you will see that they move round the Pole Star. The stars appear to move round the sky once every 24 hours because Earth is spinning at a steady rate of one revolution every 24 hours. In Ancient Greece, astronomers imagined the stars were attached to an invisible spinning globe, the **celestial sphere**. Earth's equator projected onto this sphere is called the **celestial equator**.

The Earth spins at a steady rate of 15° per hour.

See www.keyscience.co.uk for more about stars.

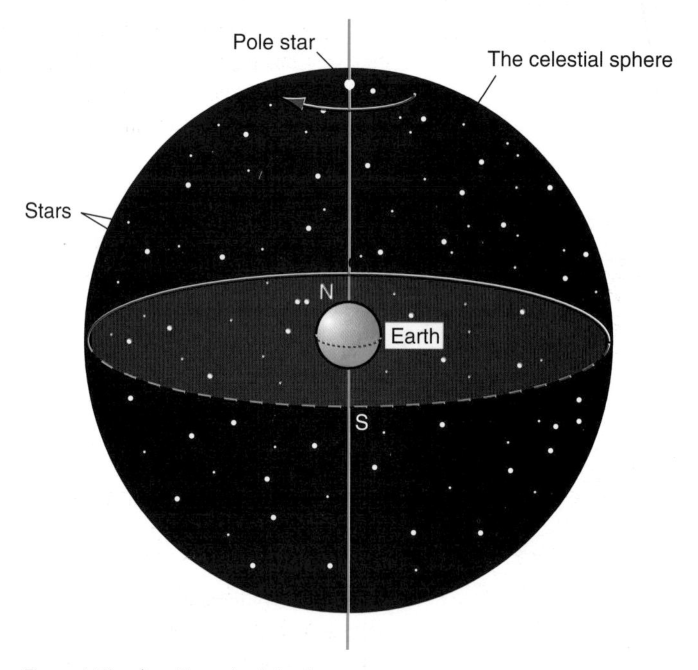

Figure 1.5A ▲ The celestial sphere

Science, religion and the solar system

KEY SCIENTISTS

■ Ancient Greece

Astronomy is the oldest branch of science. In ancient civilisations, it was studied for practical reasons such as keeping a calendar. In Ancient Greece in about 500 BC Pythagoras devised a model to explain the motions of the stars across the sky. In this model, Earth is at the centre of an invisible spinning sphere on which the stars are fixed. The sphere carries the stars across the sky as it turns.

The Pythagorean model is a geocentric model which means that Earth is at the centre. This was very agreeable to the prophets and priests of Ancient Greece because their gods lived in the heavens above Earth.

An alternative model was put forward by Aristarchus in about 300 BC. He placed the Sun at the centre and relegated Earth into orbit around the Sun as no more than a spinning planet. *Why was this imaginative model rejected in favour of Pythagoras' model?*

Figure 1.5B ▲ Astronomical clock

Figure 1.5C ▲ Nicolaus Copernicus

NICOLAI COPERNICI

net, in quo terram cum orbe lunari tanquam epicyclo contineri diximus. Quinto loco Venus nono mense reducitur. Sextum denique locum Mercurius tenet, octuaginta dierum spacio circu currens. In medio uero omnium residet Sol. Quis enim in hoc

pulcherrimo templo lampadem hanc in alio uel meliori loco po neret, quàm unde totum simul possit illuminare? Siquidem non inepte quidam lucernam mundi, alij mentem, alij rectorem uo cant. Trimegistus uisibilem Deum, Sophoclis Electra intuentē omnia. Ita profecto tanquam in solio re gali Sol residens circum agentem gubernat Astrorum familiam. Tellus quoque minime fraudatur lunari ministerio, sed ut Aristoteles de animalibus ait, maximā Luna cū terra cognationē habet. Concipit interea à Sole terra, & impregnatur annuo partu. Inuenimus igitur sub hac

Figure 1.5D ▲ A page from *De Revolutionibus* showing Copernicus' model of the solar system

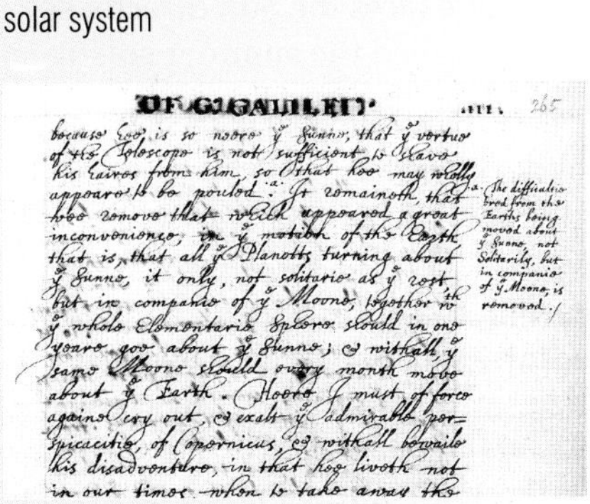

Figure 1.5E ▲ A page from Galileo's notebook describing the discovery of Jupiter's moons

- If Earth spins, objects on its surface should be thrown off.
- If Earth moves round the Sun, the stars should change position in the constellations according to Earth's positions.
- In the Pythagorean model Earth is the most important body in the universe. Aristarchus' model would remove Earth from its special place.

In fact, the first two effects are very small but they can be detected using modern measuring equipment. An object weighs about 0.3% less at the equator than at the poles because of Earth's rotation. Also, nearby stars do change position very slightly in the constellations as Earth moves round the Sun. For example, Proxima Centauri, the nearest star, shifts its position in the constellation of the Centaur by 0.0004° as Earth moves from one side of the Sun to the other.

Aristarchus didn't have our modern methods for making such small measurements so he had no experimental support for his model. Keeping the geocentric model suited the rulers of Ancient Greece because it helped to maintain their privileged position over their subjects.

A more sophisticated geocentric model was developed by Ptolemy in 120 AD. He imagined that the Sun circled Earth and the other planets circled the Sun. This explained the retrograde motion of Jupiter and Mars and it remained the accepted model of the Universe for almost 2000 years.

The modern view

Ptolemy's model dominated astronomy until the sixteenth century when a Polish monk called Nicolaus Copernicus decided to improve it. He wanted to simplify Ptolemy's model and his research led him to rediscover the ideas of Aristarchus. He published his ideas in a book, De Revolutionibus. His work is often credited with starting modern science!

In fact, Copernicus did not publish his book until 1543, the last year of his life. He must have realised that his ideas would challenge the authorities. This was a time of bitter conflict in Europe as Protestant reformers challenged the authority of the Church. Those who challenged the Church's teaching publicly were harshly treated. The Inquisition was set up by the Church to curb dissent. In Spain, more than 2000 people were burned to death by the Inquisition.

Faced with such hostility, those who supported the new ideas were in great personal danger. In 1591, Giordano Bruno, a Dominican friar, was imprisoned for his theory that the universe was infinite, made up of stars with their own planets. He was tried and burned at the stake in 1600.

About this time, an Italian professor of mathematics, Galileo Galilei, began to take a scientific interest in the Copernican model. In 1609, he used a new invention called the telescope to study the sky. He discovered mountains and craters on the Moon and sunspots on the face of the Sun. His most important discovery was the moons of Jupiter. The four moons in orbit round Jupiter form a system like the planets in orbit round the Sun. Galileo made many important discoveries in other branches of science and mathematics. Although he was the most important scientist of his generation, the Church instructed him in 1616 to stop teaching the Copernican model.

For several years, he concentrated on other work but the appearance of three comets in 1618–1619 re-awakened his interest. In 1632, he published his case for the Copernican model in a book *Dialogue on the Two Chief Systems of the World*. His book provoked immediate hostility and in 1633 he was condemned by the Inquisition and was sentenced to life imprisonment in his country house in Italy where he died in 1642.

Galileo's discoveries in astronomy had shown the power of observation in understanding the natural world. He showed how to apply mathematics to understand motion. Isaac Newton, born in Protestant England in 1642, used Galileo's ideas to develop the mathematical principles of physics. Newton's theory of gravity left no doubt that Copernicus was right: Earth does move round the Sun.

Keeping time

One day is the time taken for Earth to turn on its axis once. Once every 24 hours, we see the Sun rise in the east and set in the west as Earth spins round.

Before the invention of accurate mechanical clocks in the 18th century, the position of the Sun was used to tell the time of day.

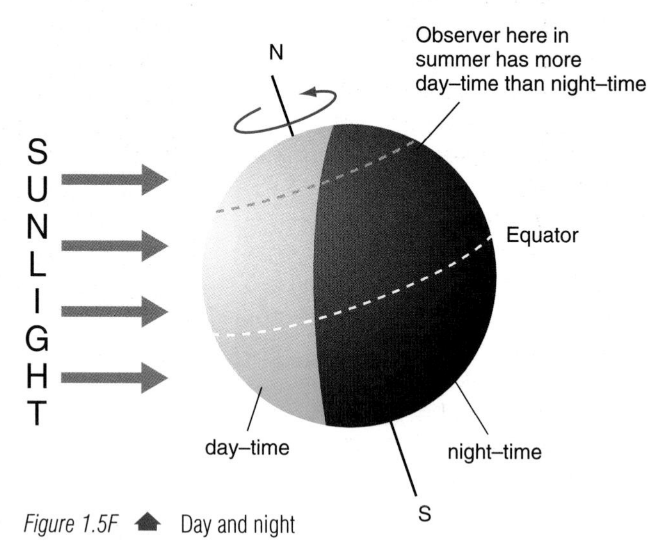

Figure 1.5F Day and night

EXTENSION FILE
ASSIGNMENT

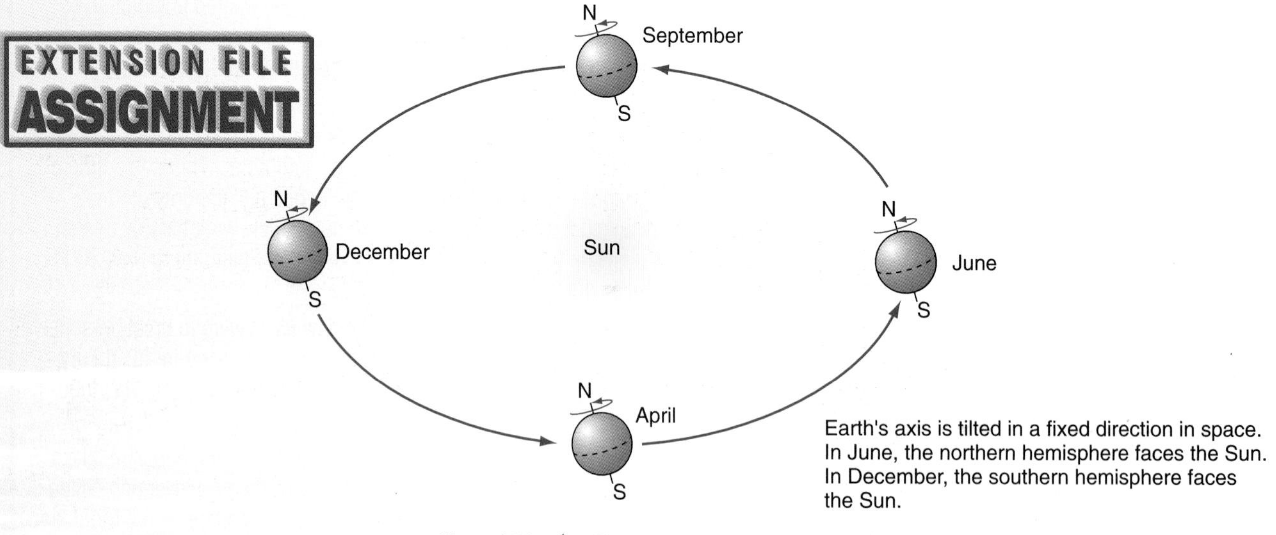

Figure 1.5G The seasons

One year is the time taken for Earth to orbit the Sun once. The year is divided up into seasons because Earth's axis is tilted in a fixed direction in space. On 21 June, the northern hemisphere faces the Sun directly so 21 June is the northern mid-summer. This is called the **summer solstice**.

Why do we get more daylight in the summer than in winter? This is because our hemisphere is tilted towards the Sun in summer. So each summer's day, we spend more time on the daylight side of Earth than on the night side.

Why do we have leap years? The reason is that the Earth turns $365\frac{1}{4}$ times each year as it orbits the Sun. By including an extra day every fourth year, we can ensure that the year changes on average at a fixed point on the Earth's orbit. *What do you think would happen if we didn't bother with the extra day?* Nothing too dramatic at first but mid-summer and mid-winter would happen 1 day later every 4 years. *What date would mid-summer become 100 years from now?*

Over a few weeks, watch from home how the position of the setting sun changes on the horizon. If you're an early riser, watch the changing sunrise too! In Britain, at mid-summer, the Sun rises north of due east and climbs high in the sky before setting north of due west. At mid-winter, the Sun rises south of due east and sets south of due west.

Our calendar was established by Julius Caesar but even with leap years every fourth year it proved inaccurate by the Middle Ages. This is because there are not quite 365.25 days in a year. By the Middle Ages, the seasons had changed by ten days. So in 1582, Pope Gregory XIII issued a papal bull making 5 October into 15 October. He also decreed that the first year of a century would not be a leap year unless it was divisible by 400. Protestant England ignored the new calendar until 1752, catching up by making 2 September into 14 September. However, the late conversion was made up for by changing the start of the year from 25 March to 1 January!

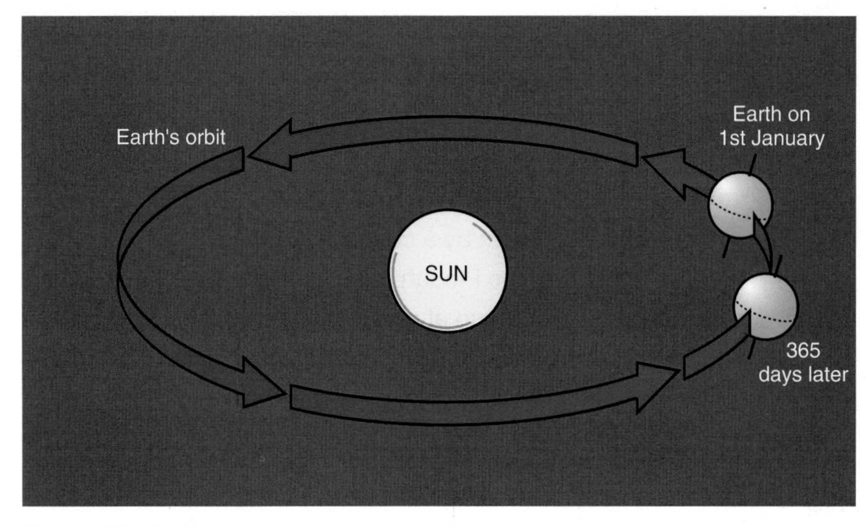

Figure 1.5H ▲ Leap year

The best place on Earth

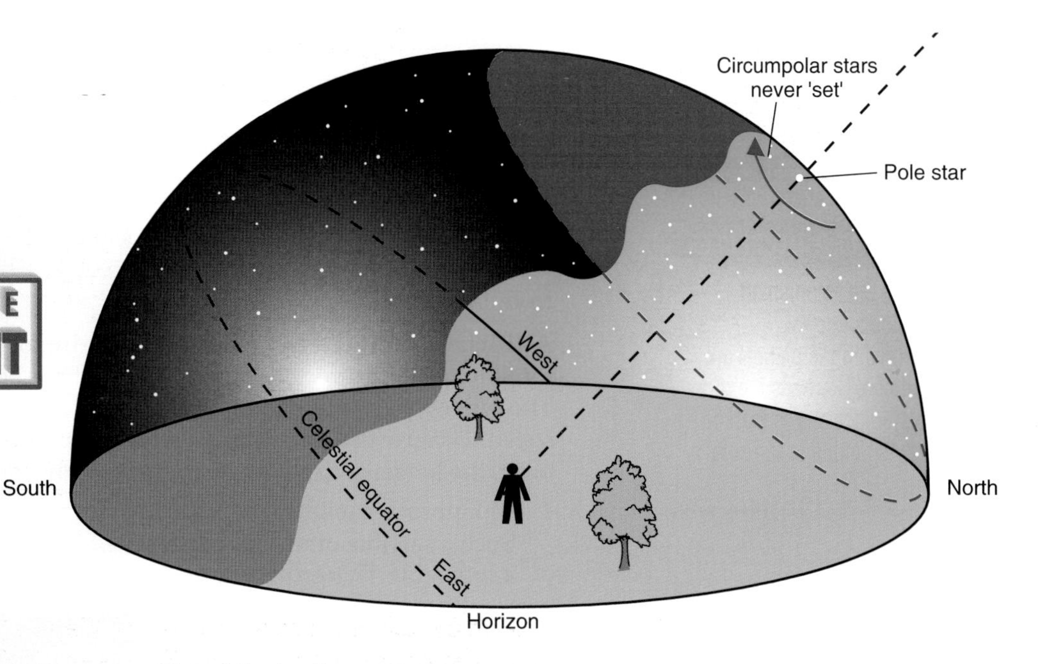

Figure 1.5I ▲ Circumpolar stars

EXTENSION FILE
ASSIGNMENT

Where is the best place to be on Earth to see all the stars? Anyone on the equator sees every star at some time. That's why observatories with huge telescopes are based near the equator.

Circumpolar stars are stars that never set. They are always above the horizon. The latitude of the observer determines which stars are circumpolar. From the UK, latitude 55° north of the equator, any star more than 35° north of the equator is circumpolar and so never sets. However, stars more than 35° south of the equator can never be seen from the UK. Stars between these latitudes rise and set each day. Whether you can see them at night depends on the time of night, the time of year and the latitude of the observer.

SUMMARY

Earth turns on its axis once per day and orbits the Sun once per year. The stars move across the sky due to Earth's rotation.

CHECKPOINT

1 (a) Why is the Pole Star always in the same direction?
(b) Why is it not possible to see the Pole Star in the day time?
(c) The Plough is part of the constellation Ursa Major which is near the Pole Star.
(i) What is meant by a constellation?
(ii) Why does the Plough change its position during the night whereas the Pole Star does not?

2 (a) Explain why the Sun at mid-day in winter is much lower in the sky than at mid-day in Summer.
(b) Why is it winter in the Southern hemisphere when it is summer in the Northern hemisphere?
(c) Mid-summer is on June 21st. If we did not have leap years in the calendar, what would happen to the date of mid-summer after a century?

3 (a) Earth moves 360° around the Sun in 1 year. How far does it move in (i) 1 day, (ii) 1 week?
(b) Why does the Sun rise earlier in London than in New York?
(c) The Sun does not set on mid-summer's day inside the Arctic Circle. Why?

4 Orion is a prominent constellation in winter in northern Europe. It cannot be seen in summer at night in northern Europe.
(a) Sketch a circle to represent the Earth's orbit around the Sun. On your sketch, label two positions on the orbit for the Earth at mid-summer and at mid-winter. Draw an arrow to indicate the direction towards Orion.
(b) Use your diagram to explain why Orion cannot be seen in summer from northern Europe.

1.6 Journey into space

FIRST THOUGHTS

Some of the astonishing discoveries made by spaceflights to the planets are outlined here. Perhaps the most important discovery is that life on Earth is unique in the solar system.

Gravity keeps objects on the Earth, including the molecules of the Earth's atmosphere. Energy is needed to lift an object above the atmosphere from the surface.

Artificial satellites are carried into space by rockets or space shuttles. A satellite must be directed into orbit once it has been carried into space. The pull of gravity due to the Earth then keeps it in orbit at a steady speed. Because gravity weakens with increasing distance, the higher the orbit, the longer the satellite takes to orbit the Earth.

1 Monitoring satellites are put into a low orbit passing over the poles. Such a satellite orbits the Earth several times each day because it is in a low orbit. This enables it to scan the entire surface each day.

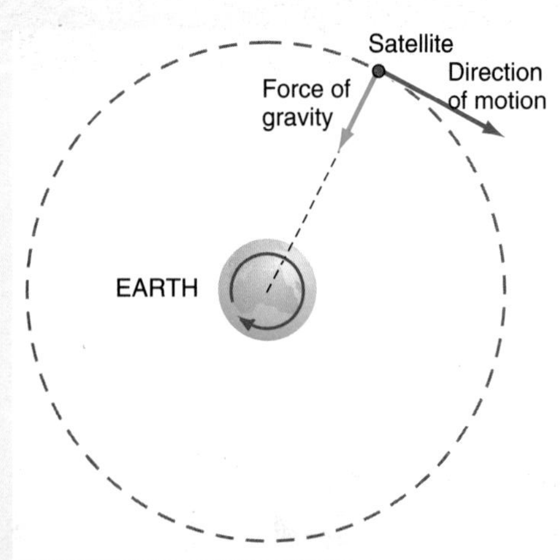

Figure 1.6A A satellite in orbit

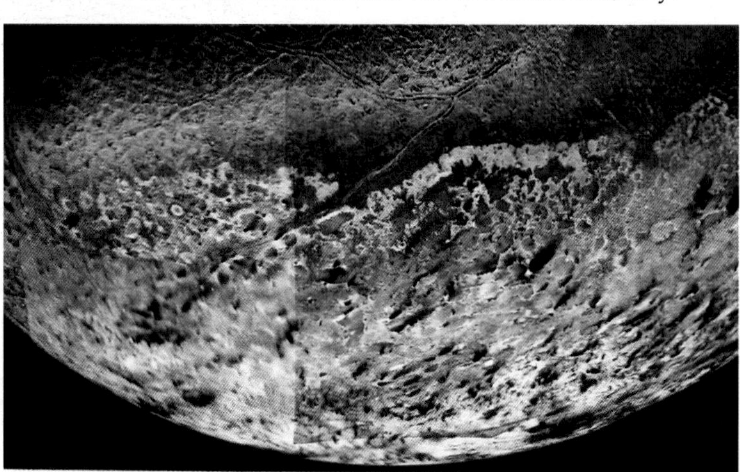

Figure 1.6B Triton, the largest moon of the planet Nepture

Science at work

There are hundreds of satellites in space. Communications satellites take 24 hours to orbit Earth so they are always at the same point over the Earth's surface. Using these satellites, TV pictures can be sent round the world.

See www.keyscience.co.uk for more about astronomy and space.

2 Communication satellites are in much higher orbits above the equator. A communication satellite orbits the Earth exactly once every 24 hours. It therefore stays permanently above the same point on the equator. This is important since satellite dishes pointing towards it to receive signals do not need to be moved once aligned with the satellite.

Space probes have been sent to most of the planets. The information they have sent back has increased our knowledge of the planets enormously and will be useful when space colonies are eventually established. Space missions to the Moon and space probes to Mars found no signs of life but meteorite evidence described on p 167 indicates life may have once existed on Mars. Another possibility is beneath the ice surface of Europa, one of Jupiter's moons, which may be hiding a sub-surface ocean containing life.

A trip round the Solar System

Space missions to the planets take years to plan and carry out. The flight paths of the Voyager and Pioneer probes were carefully worked out in advance so that each probe reached its target in the shortest possible time.

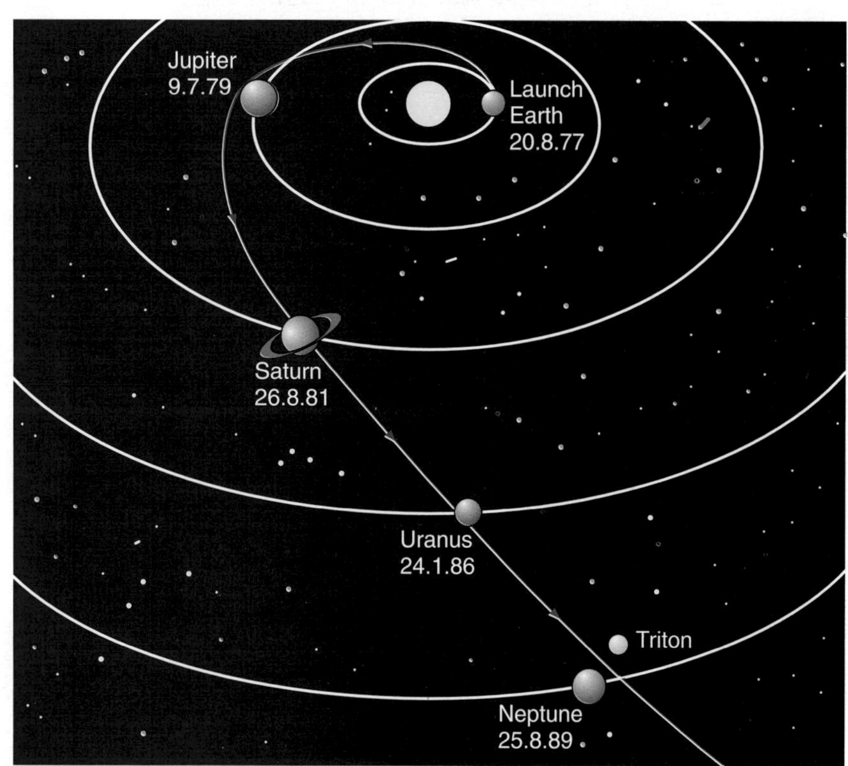

Figure 1.6C ▶ The flight path of the Voyager 2 probe. As Voyager 2 approached each planet, the gravitational pull of the planet swung it round onto a flight path for the next planet. Voyager 2 sent back amazing pictures of these planets, their moons and the ring systems of Saturn and Uranus. Astonishing discoveries were made from its pictures, including volcanoes on Jupiter's moon, Io, rings round Neptune, and most incredible of all, Triton. This is the largest of Neptune's moons, a freezing, volcanic and colourful world.

SUMMARY

Earth is the only planet that supports life. Mars is likely to be the first planet on which space colonies are established. Space missions need to be planned very carefully and space probes take many months to reach their targets.

CHECKPOINT

1. Find out the year in which the following space events happened.
   (a) The first artificial satellite was launched.
   (b) The first person set foot on the Moon.
2. Why is Mars the most Earth-like of the planets?
3. Why would life on Venus be very harsh for visitors from Earth?
4. Make a sketch of a possible flight path from Earth to Mars, showing their orbits to scale.
5. Why would it be impossible to land on any of the giant planets?

Topic 2 Beyond the solar system

2.1 Stars and galaxies

FIRST THOUGHTS

After working through this section, you will see that the Sun is just one of the countless stars in one of countless galaxies that make up the universe.

The Sun is one of millions of stars that form the Milky Way galaxy. The universe is made up of many galaxies, each one containing millions of stars. Light takes more than 10 000 million years to reach us from the most distant galaxies.

Radio telescopes like that at Jodrell Bank, England have been used to show that our galaxy is disc-shaped with spiral arms. Light takes about 100 000 years to cross the Milky Way galaxy and the Sun is about 30 000 light years from the galactic centre. On a clear night, the Milky Way can be seen stretching across the sky.

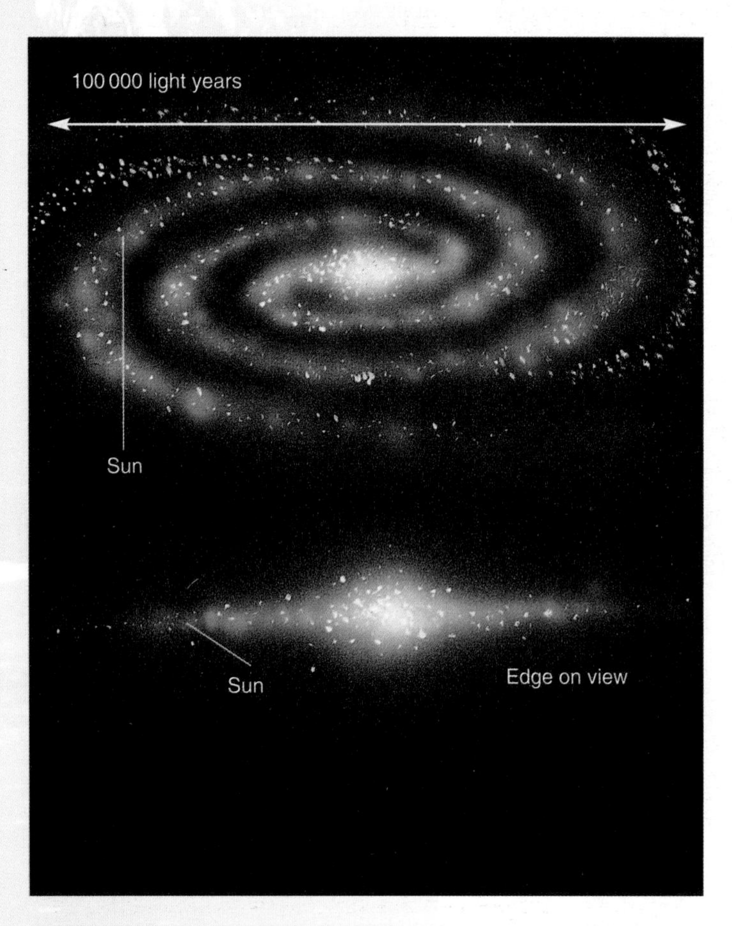

Figure 2.1A (*a*) The Milky Way

(*b*) The Andromeda galaxy

(*c*) The Sombrero Hat galaxy

See www.keyscience.co.uk for space images.

What differences can you see when you look from one star to another?
Are the stars all the same? Even though they appear only as points of light astronomers can work out their size and temperature, how far away they are and how much energy they emit each second. All this information and much more is obtained by studying their light in detail and by using the laws of physics. For example, we know that if a glowing object is made hotter and hotter, its colour changes from red to orange to blue. So a blue star must be much hotter than a red star.

The birth and death of a star

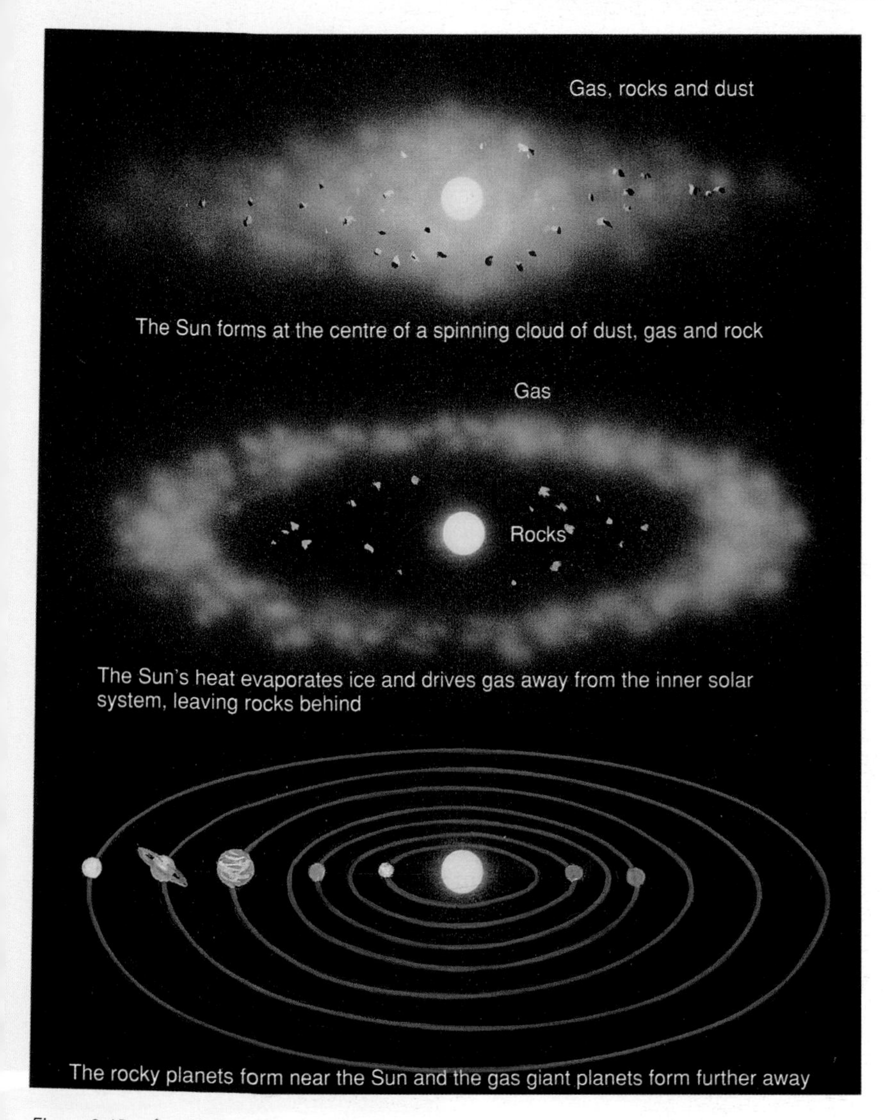
Gas, rocks and dust
The Sun forms at the centre of a spinning cloud of dust, gas and rock
Gas
Rocks
The Sun's heat evaporates ice and drives gas away from the inner solar system, leaving rocks behind
The rocky planets form near the Sun and the gas giant planets form further away

Figure 2.1B ▲ How the planets formed

Stars are thought to form out of clouds of gas and dust. The particles gather together under their own gravity which makes them more and more concentrated. As the 'protostar' becomes more and more dense it gets hotter and hotter. The temperature becomes so great that the atoms **fuse** together and release more energy. So it gets hotter and brighter. A star is born!

The planets formed from the same cloud of gas and dust as the Sun. Most of the cloud formed the Sun but relatively small amounts condensed outside the Sun to form the planets. The Sun's gravity prevents the planets from escaping into space.

Stars like the Sun will radiate energy for billions of years before they run out of atoms to fuse together. The Sun is about half way through its lifecycle. When the Sun does start to run short of fuel, it will swell up into a huge red ball, known as a **red giant**, stretching beyond Mercury. The star Betelgeuse in the constellation of Orion is a red giant. After the red giant stage, a star collapses in on itself and becomes a hot, dense white star, known as a **white dwarf**, until it gradually fades away. Big stars explode near the end of their lifecycle. Such an event is known as a **supernova** and can outshine an entire galaxy while it lasts.

Figure 2.1C ▲ The Crab Nebula

The Crab Nebula is thought to be the remnants of a supernova explosion that was first observed in about the eleventh century. In 1987, a star in the southern hemisphere exploded and became the biggest supernova to be seen for four centuries. Astronomers pinpointed this supernova to Sandaluk II, a star which existed in the Andromeda galaxy, more than 10 million light years from Earth.

What remains after a supernova occurs? Astronomers think that the explosion compresses the central core into an extremely dense object called a neutron star (see 8.2 for neutrons). If the neutron star is massive enough, it would be a **black hole**. This is formed if its own gravity is so strong that nothing can escape from it, not even light.

Figure 2.1D ⬆ A uranium mine

Stars, birthplace of the chemical elements

The energy emitted by a star is produced in its core. Matter in the core is extremely dense and very hot because it is compressed by the weight of the outer layers. The temperature in the core is so great that nuclear fusion occurs. In this process, the nuclei of the atoms collide with such force that they fuse to form heavier nuclei. Energy is released as radiation in this process, maintaining the core temperature and keeping the fusion process going. The outward flow of radiation prevents further compression of the core so the star is stable.

Stars like the Sun release energy as a result of hydrogen atoms fusing to form helium atoms. When the Sun runs out of hydrogen in its core, it will become a red giant, releasing energy as a result of fusing helium atoms into heavier atoms. However, atoms heavier than iron are not formed by this process because such atoms would split up inside the core (see 9.7 for more about nuclear fusion).

Atoms of the chemical elements heavier than iron are formed when a massive star ends its life in a supernova explosion. The enormous force of a supernova explosion makes light atoms fuse into very heavy atoms and scatters matter far into space at high speed. Eventually, new stars will be formed from this matter due to its own gravity. Each star may be accompanied by planets formed at the same time. These planets would be composed of all the known elements because the matter from which they formed contained all known types of atoms.

The heaviest known natural element is uranium. It is an unstable element with a half-life of 4500 million years. The presence of uranium in the Earth is evidence that the Solar System must have formed from the remnants of a supernova.

EXTENSION FILE
ASSIGNMENT

The search for extra-terrestial intelligence

Astronomers know that molecules of carbon-based chemicals are present in space. Life on Earth probably developed from such chemicals created from gases in the atmosphere by lightning. Meteorites from Mars may have carried primitive life forms to Earth. See p 167. Although there is no other evidence for life on other planets in our own Solar System, life could have developed on planets in distant solar systems. If so, intelligent species may exist and they could be transmitting radio waves. Radio astronomers have searched without success for such radio transmissions for several decades.

The search for extra-terrestial intelligence, known as SETI, continues as astronomers develop more sensitive receivers. Radio astronomers have mostly concentrated on the range of frequencies from about 1000 to 10000 MHz as there is little natural radiation in this waveband from stars and galaxies and little absorption due to the atmosphere occurs in this wave band. In addition, infra-red telescopes in orbit above the Earth are being used to spot earthlike planets in distant solar systems. Based on the number of stars in the Milky Way galaxy and the lifespan of an intelligent civilisation on an earth-like planet, astronomers reckon there might just be a few intelligent species in our own galaxy.

SUMMARY

The universe is made up of many galaxies. Each galaxy contains millions of stars. The Sun is one of many stars in the Milky Way galaxy. Earth is a planet in orbit round the Sun. Life on Earth has existed for a tiny fraction of the age of the universe. Does life exist elsewhere?

CHECKPOINT

1 List the following objects in order of increasing size:

The Solar System The Sun The Andromeda galaxy Earth

2 Study the galaxies in Figure 2.1A.
(a) Why is the Milky Way referred to as a spiral galaxy?
(b) Do you think the other two galaxies are also spiral galaxies? Explain your answers.
(c) The picture of the Milky Way is not a photograph. Why is it not possible to take a photograph of the Milky Way like this?

3 (a) List the following objects in order of increasing age:

The Sun The Milky Way Earth The Crab Nebula

(b) The main stages in the lifecycle of the Sun are listed below in incorrect order. Rewrite these stages in the correct order.

red giant present white dwarf protostar

4 Here are four categories of astronomical objects:

A stars **B** planets **C** galaxies **D** satellites

State to which category each object listed below belongs:
(a) the Moon
(b) Andromeda
(c) Pluto
(d) the red giant, Betelgeuse

5 (a) What is a supernova?
(b) Outline the reason why astronomers think that the Solar System formed from the debris of a supernova.

2.2 The Big Bang

FIRST THOUGHTS

How did the galaxies form? Did they originate at the same time in the past? Read on to find out how the universe started.

Scientists think that the Universe is expanding. The most distant galaxies are known to be over 10 billion light years away and rushing away from Earth almost as fast as light. Looking at them is like looking back in time because the light from them takes billions of years to reach Earth.

All the distant galaxies are moving away from each other as the Universe expands. Billions of years ago, they must have been very close together. In fact, astronomers think that the Universe originated in a massive explosion, **the Big Bang**. The rate of expansion is thought to be gradually getting slower. Maybe the Universe will stop expanding and collapse in a few billion years time!

Evidence for the Big Bang

- ***Red shift***

Observation The spectrum of light from a distant galaxy is shifted towards the red part of the spectrum. Figure 11.5H shows the wavelength of light waves for each colour of the spectrum. The waves of light from a distant galaxy are longer than light from a nearby galaxy or star. See Fig 2.2A on the next page.

The wavelength of light increases from blue to red across the spectrum. A red shift is a shift towards longer waves and is caused by the source of light moving away.

Cause The light waves from any star or galaxy moving away from Earth are longer because the light source is moving away from us. The faster the light source moves away, the longer are the waves we receive from it.

See www.keyscience.co.uk for images of newly-discovered galaxies.

Laboratory source of light

Dark lines due to absorption of light in the source

Pattern of absorption lines shifted to red end of spectrum

Light from a receding galaxy

Figure 2.2A ▲ Red shift

Conclusion The further a galaxy is from Earth, the greater its red shift is. This was discovered by the astronomer Edwin Hubble about 70 years ago. He made measurements on many galaxies. His discovery is known as **Hubble's law**:

1 the distant galaxies are all receding from Earth,

2 the speed of recession of a galaxy is proportional to its distance from Earth.

The further a galaxy is from us, the faster it is receding.

Interpretation Hubble's law can be explained by the **theory of the expanding Universe** in which every point is moving away from every other point at a steady rate. The further apart two points are, the faster they are moving away from each other. A simple and effective demonstration of this idea is to mark some dots on a balloon and then observe the balloon as it is inflated; any two dots move apart at a rate which is greater the further apart they are.

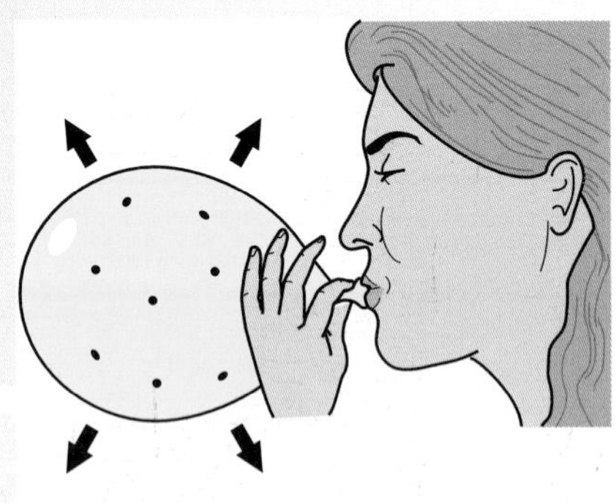

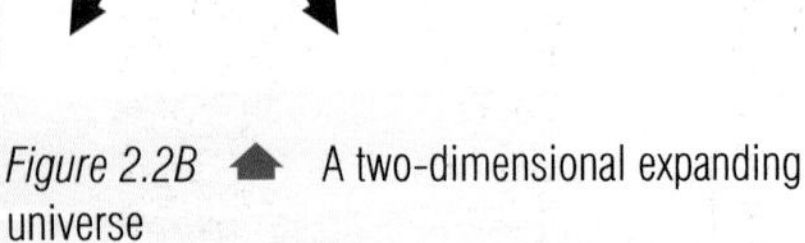

Figure 2.2B ▲ A two-dimensional expanding universe

• *Microwave background*

Controversy Hubble's law is a scientific fact. After it was discovered, two theories were put forward to explain it. Some astronomers could not accept the expanding Universe theory and put forward the 'steady state' theory based on the idea that the Universe now is the same as in the past and in the future; some parts were considered to be expanding and other parts contracting. The discovery of microwave background radiation was a major reason for acceptance of the expanding Universe theory.

Discovery Further experimental evidence in support of the Big Bang was discovered in 1965 when microwave radiation was detected in all directions in space. This is explained as background radiation, released in the Big Bang, which became longer and longer in wavelength as the Universe expanded. In 1989 the Cosmic Background Explorer (COBE) was launched into orbit round the Earth to map out this background radiation; its results suggested slight unevenness referred to as 'ripples' in line with predictions from the Big Bang. Improved measurements made using more advanced equipment in 1999 from a high altitude balloon confirmed this unevenness and indicate that the Universe is likely to expand forever..

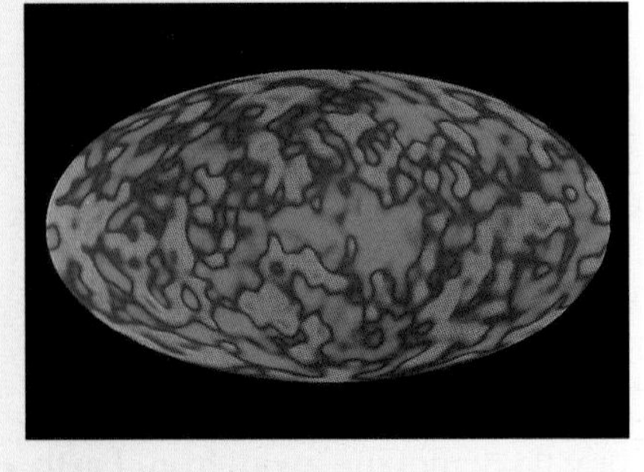

Figure 2.2C ▲ Image from COBE, the Cosmic Background Explorer

SUMMARY

The Universe is thought to be still expanding. It is thought to have originated over 10 billion years ago in the Big Bang. This theory is supported by evidence from Hubble's law, the existence of background microwave radiation and the explanation of the Universe's hydrogen to helium ratio. The Universe might continue to expand forever or it might reverse, pulled back by its own gravity if its total mass is sufficiently large. The most recent evidence suggests expansion forever and no Big Crunch!

- ***The light elements***

Hydrogen and helium are thought to have been created in the Big Bang. Clouds of these two elements in space then formed galaxies and eventually stars formed. Measurements of the proportion of hydrogen to helium in very old stars show that 10 hydrogen atoms were formed for every helium atom. This agrees with calculations on the formation of these two elements in the Big Bang. Hydrogen and helium inside stars fuse to form heavier elements. Very heavy elements such as uranium are formed in supernovae.

Past, present and future

The Universe is thought to have originated in the Big Bang. Hubble's measurements on red shifts have been repeated and used to estimate how long ago the Big Bang occurred. This gives about 10–15 billion years for the age of the Universe. This estimate is uncertain because astronomers are not sure what the total mass of the Universe is. The more mass there is, the slower the expansion because of the inward pull of gravity.

The future is just as uncertain. A high-mass Universe would stop expanding because of its own gravity and then contract, ending in the so-called **Big Crunch**. A low-mass Universe would continue to expand for ever. Because astronomers are not sure at present what the total mass of the Universe is, the future of the Universe is uncertain, although measurement of the ripples in the background microwave radiation indicate expansion forever, the so-called **Big Yawn!**

CHECKPOINT

1 (a) State the colours of the spectrum.
(b) Which colour of light has: (i) the longest wavelength, (ii) the shortest wavelength?
(c) The light from a distant galaxy is red-shifted. What does this statement mean?

2 The diagram opposite shows two galaxies X and Y of the same diameter seen from Earth.
(a) Galaxy X is further away than galaxy Y. How is it possible to make this deduction from the picture?
(b) Which galaxy produces the larger red shift?

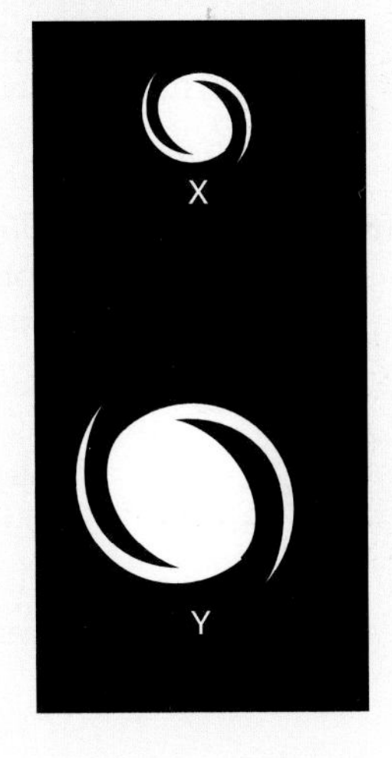

3 In a demonstration of the expansion of a one-dimensional universe, 11 students stand side by side along a straight line, spaced 1 m apart, as shown in the diagram. The students are told to increase the spacing by 50 cm every 10 seconds.

What is the average speed of separation of two students who are:
(a) next to each other?
(b) at opposite ends of the line?

4 (a) What is Hubble's law?
(b) What was deduced about the Universe from Hubble's law?
(c) What additional evidence supports this deduction?

5 (a) What is a light year?
(b) The furthermost galaxies are thought to be over 12 000 million light years away, moving away almost at the speed of light. Use this information to estimate the age of the Universe.

Theme Questions

TOPIC 1

1 (a) Copy and complete the following sentences, using the words in italics below. At midday, the Sun reaches its ____ point in the sky. At this point, its direction is due ____ . Each day, the Sun ____ in the east and ____ in the west. The daily motion of the Sun across the sky is caused by the Earth ____ at a steady rate about an axis through its ____ .

sets south poles turning rises highest

(b) In the figure below, which position (A, B, C or D) shows the correct position of the Earth at (i) mid-winter in the northern hemisphere, (ii) mid-spring in the northern hemisphere?

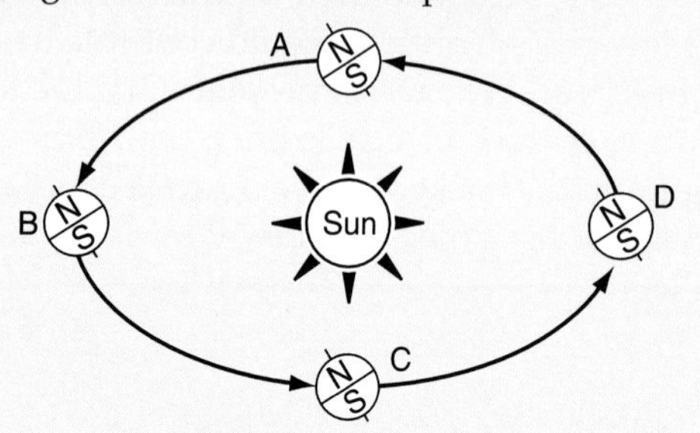

(c) Huge flocks of birds can often be seen in Britain in early winter, migrating from arctic countries. Why? What happens to these birds after each winter?

2 (a) Copy the figure below and use your sketch to show the position of the Moon when it appears as (i) a full moon, (ii) a new moon.

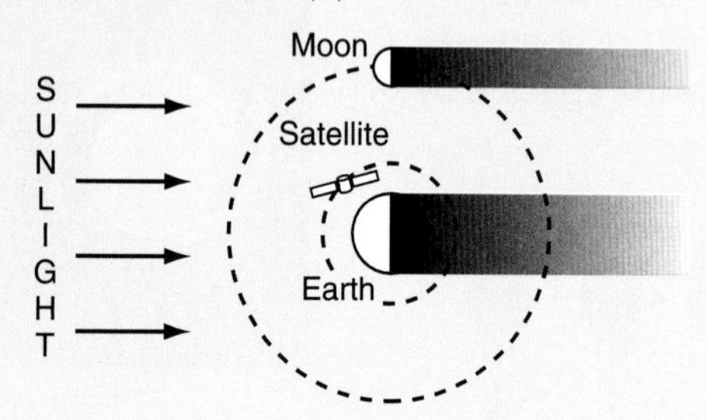

(b) A lunar eclipse takes place when the Sun, the Earth and the Moon lie in a straight line. Why do we not see a lunar eclipse every full moon?

(c) A satellite orbits the Earth once every two hours as shown above. Solar panels on board the satellite convert solar energy into electricity for the satellite's circuits.

(i) What happens to the electricity supply from these panels when the satellite passes into the Earth's shadow?

(ii) The panels are used to charge batteries on board the satellite. Why are the batteries needed?

3 The universe is about 12 000 million years old. The solar system was formed about 5000 million years ago. The earliest traces of humans are 200 000 years old. Suppose the age of the universe is scaled down to 24 hours. On this scale:

(a) how long ago was the solar system formed?

(b) how long ago did human life begin?

4 (a) Imagine you are planning to send a space probe to Venus.

(i) Why is it necessary to use a huge rocket to launch the probe into space?

(ii) The figure below shows the orbits of Earth and Venus. Copy the diagram and show a possible flight path from Earth to Venus.

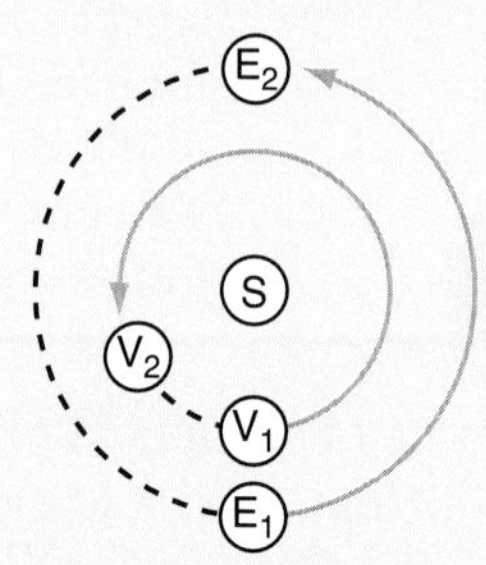

E_1 and V_1 show Earth and Venus at least distance apart, E_2 and V_2 show them six months later

(b) The atmosphere of Venus is much hotter than the Earth's atmosphere. Why?

5 (a) The Earth is an insignificant planet in orbit around a middle-aged star called the Sun. Five centuries ago, the Earth was believed to be at the centre of the universe. Outline one piece of evidence that the Earth goes round the Sun, not the Sun round the Earth.

(b) Galileo Galilei did not agree with the teaching of the Church that the Earth was at the centre of the universe. Galileo was forced to admit his 'error' in public. With your friends, write and act a short play about Galileo's show trial.

6 Table 1 gives data for some planets around the star Epsilon Zeta.

Table 1

Planet name	Alpha	Beta	Gamma	Delta
Orbit size (million km)	40	80	160	320
Surface temperature (K)	500	600	300	200
Surface gravity (N/kg)	2	12	9	4

(a) (i) Name the force which keeps the planets in orbit around the star.

(ii) Which planet orbits the star in the smallest time? Explain your choice.

(b) (i) Explain why you would expect the planet closest to the star to be the hottest.
(ii) Suggest why the average surface temperature on Beta is greater than on Alpha.

(c) (i) Suggest which of the planets is most likely to be able to support human life. Give **two** reasons for your choice.
(ii) State two other conditions which are necessary for this planet to support human life.

(MEG, Salters)

TOPIC 2

7 Read this passage carefully and then answer the following questions.

The Sun is our nearest star. In the Sun, energy is generated by a thermo-nuclear reaction in which hydrogen is converted into helium at a core temperature of about 2×10^7 °C. The life of our Sun is estimated to be 10^{10} years and it is likely that the Sun will eventually become a white dwarf. Stars with considerably more mass than our Sun have core temperatures which are more than 2×10^7 °C and may eventually become black holes.

The distance from the Earth to Proxima Centauri, our next nearest star, is 4.2 light years and the distance to the nearest spiral galaxy, Andromeda, is nearly 2.5 million light years.

Astronomers noticed that light from distant stars showed a *red shift* and that the size of the red shift was greatest for stars furthest away from the Earth. These observations give rise to the theory that distant galaxies are moving away from the Earth and the further away the galaxies are, the faster they are moving.

The speed at which a galaxy is moving relative to the Earth and its distance from the Earth can be related by the following expression known as Hubble's law.

$$\frac{\text{Speed of galaxy away from Earth}}{\text{Distance of galaxy from the Earth}} = \text{Hubble constant}$$

(a) How does the core temperature of a large star differ from that of our Sun?

(b) (i) Describe the main stages in the life of a star similar to our Sun, starting with the present situation and ending with the formation of a white dwarf.
(ii) What difference would there be if the star had a mass of 20 times that of our Sun?

(c) (i) What do you understand by the term *red shift* (line 15)?
(ii) How is this red shift interpreted by astronomers?
(iii) Using a value of 20 km/s per million light years for the Hubble constant, calculate the approximate speed at which the Andromeda galaxy is moving away from the Earth.

(London)

8 The *red shifts* of some galaxies are shown below. They are given in percentages, where 100% would correspond to movement at the speed of light.

constellation where galaxy is observed	*distance* (millions of light years)	*red shift*	*speed* (km per second)	*speed/distance* (km per second / millions of light years)
Ursa Major	900	+5.3%	15 000	17
Corona Borealis	1200	+7.3%		
Bootes	2400	+14%		
Hydra	3600	+23%		

(a) (i) What is meant by *red shift*?
(ii) What causes it to occur?

(b) (i) The red shifts shown are positive. What would negative values indicate?

(c) The graph shows the relation between red shift and the speed of the galaxy (relative to Earth).

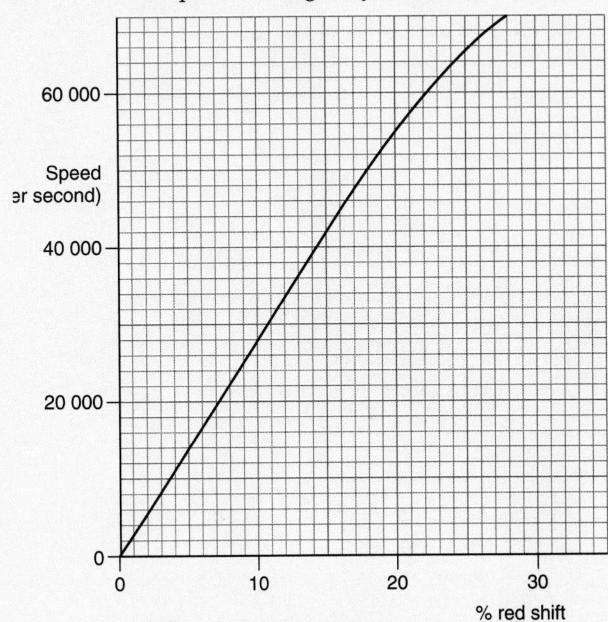

Use the graph to estimate the speeds of the four galaxies shown in the table. Copy the table and put your results in the fourth column. The first one has been done for you.

(d) The astronomer Edwin Hubble suggested that the speed a galaxy moves (relative to Earth) is proportional to its distance from the Earth.
(i) Use results in the table to find, for each galaxy, the value of *speed/distance*. Keep to the units shown in the table. Put these values in the fifth column of the table.
(ii) What do you notice about the values of *speed/distance* in the fifth column?
(iii) How does this confirm Hubble's theory?

(e) Galaxies would be travelling at nearly the speed of light at the 'edge' of the Universe. The speed of light is 300 000 km/s. Use your answers from (d) to estimate the size of the Universe in millions of light years. Show your working clearly.

(MEG, Nuffield)

Theme B

Matter and Energy

Earth and all the other planets in the universe are made of matter. How many different kinds of matter are there? They are countless: far too many for one book to mention, but in Theme B we start to answer the question. What does matter do: how does it change?

Energy enables changes to take place. The study of energy in its different forms and its effect on matter is a major part of physics.

The ideas about matter and energy in this theme are used in other branches of science, such as in the study of chemical reactions and energy in the study of biological processes, such as photosynthesis and respiration.

Topic 3 Matter

3.1 The states of matter

FIRST THOUGHTS

What is a state of matter? How do states of matter differ? These are questions for you to explore in this topic.

Figure 3.1A ▲ Ice skaters

Everything you see around you is made of matter. The skaters, the skates, the ice and all the other things in Figure 3.1A are different kinds of matter.

The skaters are overcoming friction as they glide across the frozen pond. *How do they manage this*? They change one kind of matter, ice, into another kind of matter, water. A thin layer of water forms between a skate and the ice, and this reduces friction and enables the skater to glide across the pond. The water beneath the skate refreezes as the skater moves on. *How are skaters able to melt ice? Why does it refreeze behind them?* Read on to find out.

The different kinds of matter are **solid** and **liquid** and **gaseous** matter. These are called the **states of matter**. Table 3.1 summarises the differences between the three chief states of matter. The symbols (s) for solid, (l) for liquid and (g) for gas are called **state symbols**. In chemistry you will use another state symbol, (aq), which means 'in aqueous (water) solution'.

Table 3.1 ▼ States of matter

State	Description
Solid (s)	Has a fixed volume and a definite shape. The shape is usually difficult to change.
Liquid (l)	Has a fixed volume. Flows easily; changes its shape to fit the shape of its container.
Gas (g)	Has neither a fixed volume nor a fixed shape; changes it volume and shape to fit the size and shape of its container. Flows easily; liquids and gases are called fluids. Gases are much less dense than solids and liquids.

SUMMARY

The three chief states of matter are:
- solid (s): fixed volume and shape,
- liquid (l): fixed volume; shape changes,
- gas (g): neither volume nor shape is fixed.

Liquids and gases are fluids.

3.2 Density

FIRST THOUGHTS

Which is heavier – a tonne of lead or a tonne of feathers? Don't confuse mass and density.

Table 3.1 tells you that gases are much less dense than solids and liquids. What does dense mean? What is **density**? The two lengths of car bumper shown in Figure 3.2A have the same volume. You can see that they do not have the same mass. The steel bumper is heavier than the plastic bumper. This is because steel is a more **dense** material than the plastic; steel has a higher **density** than the plastic has.

$$\text{Density} = \frac{\text{Mass}}{\text{Volume}}$$

The unit of density is kg/m^3 or g/cm^3. The density values of some common substances are shown in Table 3.2.

Tips

Never leave out the unit when you give a numerical answer. Note the following:

mass; $1\ kg = 1000\ g$

length; $1\ m = 100\ cm = 1000\ mm$

area; $1\ m^2 = 10^4\ cm^2 = 10^6\ mm^2$

volume; $1\ m^3 = 10^6\ cm^3$

The litre is still in common use; 1 litre = $1000\ cm^3$; 1 millilitre (ml) = $1\ cm^3$. Prove for yourself that $1\ g/cm^3$ for the density of water is the same as $1000\ kg/m^3$.

Table 3.2 Density

Substance	Density (g/cm^3)
Air	1.2×10^{-3}
Aluminium	2.7
Copper	8.92
Ethanol	0.789
Gold	19.3
Hydrogen	8.33×10^{-5}
Iron	7.86
Lead	11.3
Methane	6.67×10^{-4}
Oxygen	1.33×10^{-3}
Silver	10.5
Water	1.00

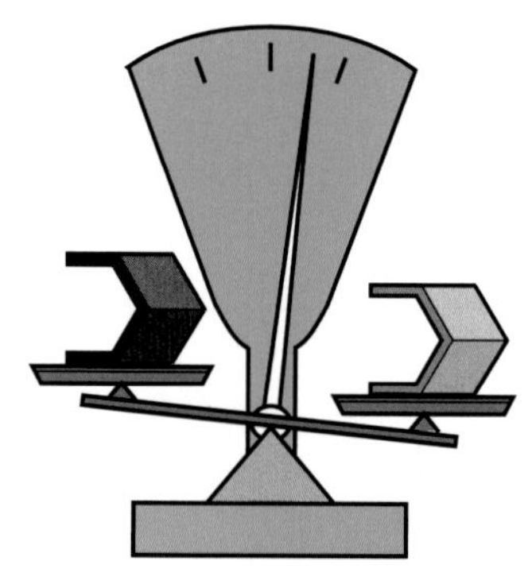

Figure 3.2A Two objects with the same volume

You can see that the gases are much less dense than any of the solid or liquid substances.

Measuring the density of a liquid

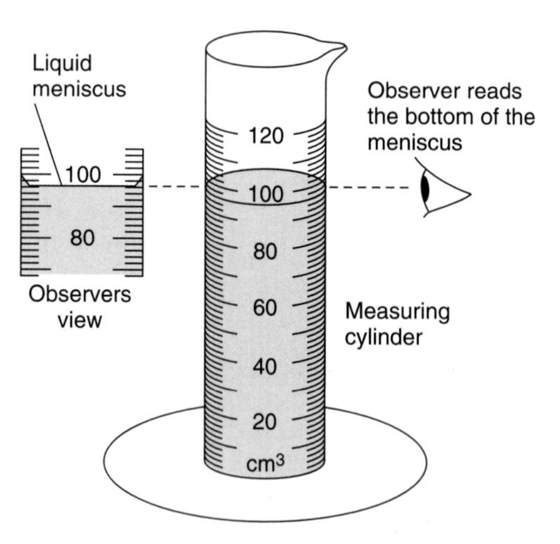

Figure 3.2B Using a measuring cylinder

- Use a measuring cylinder as in Figure 3.2B to measure the volume of the liquid.
- Use a top pan balance to measure the mass of an empty beaker sufficiently large to contain the liquid in the measuring cylinder. Pour the liquid into the empty beaker and use the balance to measure the total mass of the beaker and the liquid. The mass of liquid in the beaker can then be worked out by subtracting the mass of the empty beaker from the mass of the beaker and the liquid.

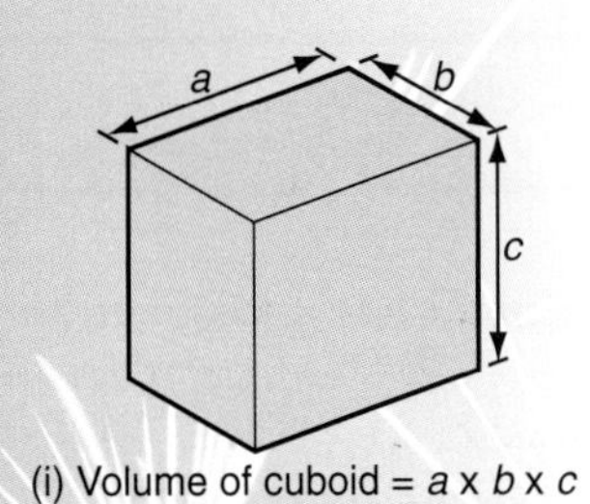
a
b
c

(i) Volume of cuboid = $a \times b \times c$

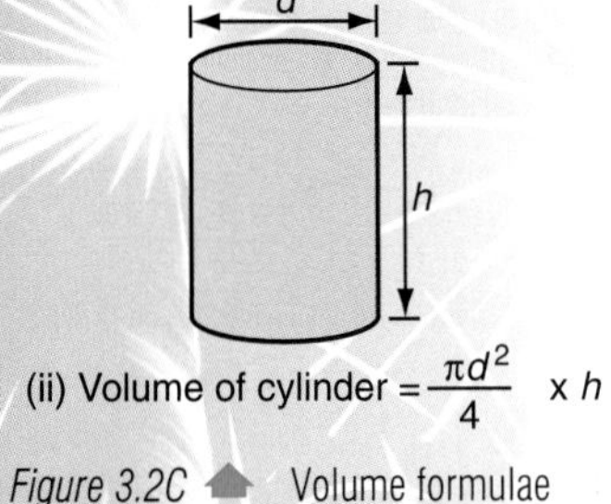
d
h

(ii) Volume of cylinder = $\frac{\pi d^2}{4} \times h$

Figure 3.2C Volume formulae

SUMMARY

$$\text{Density} = \frac{\text{Mass}}{\text{Volume}}$$

$$\text{Mass} = \text{Volume} \times \text{Density}$$

$$\text{Volume} = \frac{\text{Mass}}{\text{Density}}$$

The density triangle:

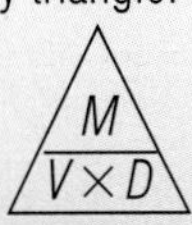
M
V × D

To find the quantity you want, cover up that letter. The other letters in the triangle show you the formula.

- Work out the density of the liquid from the equation

$$\text{Density} = \frac{\text{Mass}}{\text{Volume}}$$

Measuring the density of a solid object

- Use a top pan balance to measure the mass of the object.
- For a regular solid such as a cuboid or a cylinder, work out its volume from its dimensions which can be measured using a millimetre scale. The formulae for a cuboid and a cylinder are given in Figure 3.2C.
- For an irregular solid, measure its volume using a measuring cylinder and a displacement can, as in Figure 3.2D. Water is the most suitable liquid to use provided the solid does not dissolve in it.
- Work out the density of the liquid from the equation

$$\text{Density} = \frac{\text{Mass}}{\text{Volume}}$$

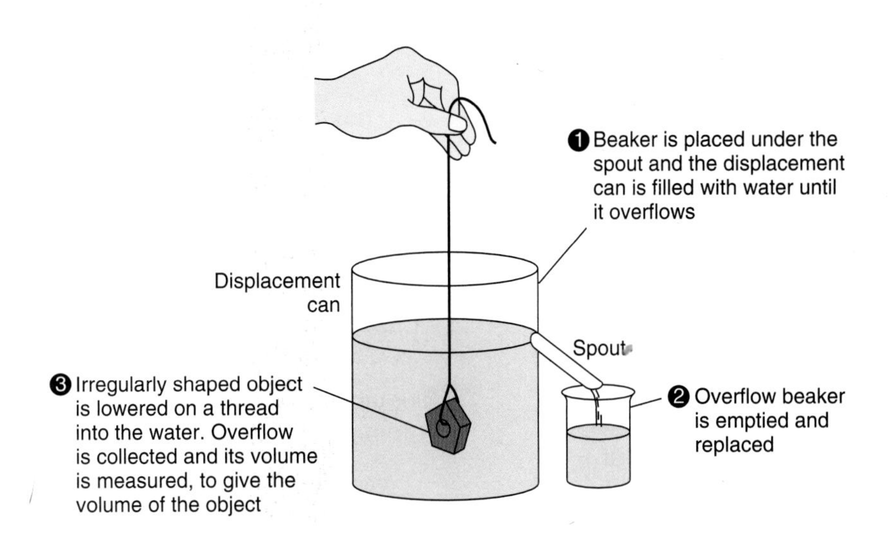
1 Beaker is placed under the spout and the displacement can is filled with water until it overflows
Displacement can
Spout
3 Irregularly shaped object is lowered on a thread into the water. Overflow is collected and its volume is measured, to give the volume of the object
2 Overflow beaker is emptied and replaced

Figure 3.2D Measuring the volume of an irregularly shaped object

CHECKPOINT

1. A worker in an aluminium plant taps off 300 cm^3 of the molten metal. It weighs 810 g. What is the density of aluminium?
2. An object has a volume of 2500 cm^3 and a density of 3.00 g/cm^3. What is its mass?
3. Mercury is a liquid metal with a density of 13.6 g/cm^3. What is the mass of 200 cm^3 of mercury?
4. 50.0 cm^3 of metal A weigh 43.0 g
   52.0 cm^3 of metal B weigh 225 g
   Calculate the density of each metal. Say whether they will float or sink in water.
5. (a) A rectangular concrete paving slab is 0.8 m long, 0.6 m wide and 0.05 m thick. Calculate its volume in (i) m^3, (ii) cm^3.
   (b) The mass of the concrete slab is 60 kg. Calculate the density of the concrete in kg/m^3.
6. A block of gold is 0.10 m in length, 0.08 m in width and 0.05 m in thickness.
   (a) Calculate the total surface area and the volume of this gold block.
   (b) Gold has a density of 18 g/cm^3. Calculate the mass of this block of gold.

3.3 Change of state

FIRST THOUGHTS

Melting and boiling take place without change of temperature. Evaporation occurs over a range of temperatures.

See www.keyscience.co.uk for more about states of matter.

Matter can change from one state into another. Some changes of state are summarised in Figure 3.3A.

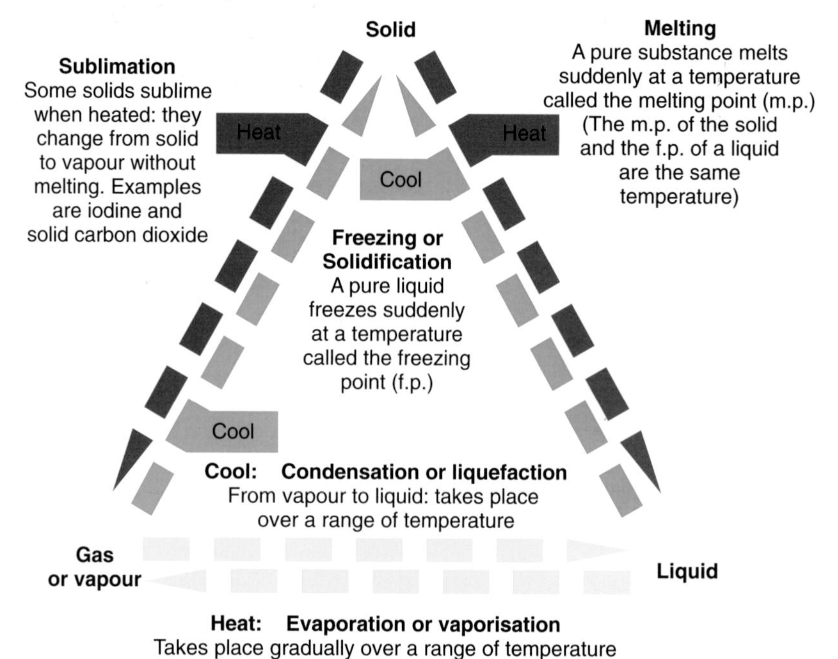

Figure 3.3A Changes of state

Sometimes gases are described as vapours. A liquid evaporates to form a vapour. A gas is called a vapour when it is cool enough to be liquefied.

$$\text{Vapour} \xrightarrow{\text{Either cool or compress without cooling}} \text{Liquid}$$

The hotter a liquid is, the faster it evaporates. At a certain temperature, it becomes hot enough for vapour to form in the body of the liquid and not just at the surface. Bubbles of vapour appear inside the liquid. When this happens, the liquid is boiling, and the temperature is the boiling point of the liquid.

SUMMARY

Matter can change from one state into another.

The changes of state are:

- melting
- freezing or solidification
- evaporation or vaporisation
- condensation or liquefaction
- sublimation

CHECKPOINT

1 Copy the diagram opposite. Fill in the names of the changes of state. (Some of them have two names.)

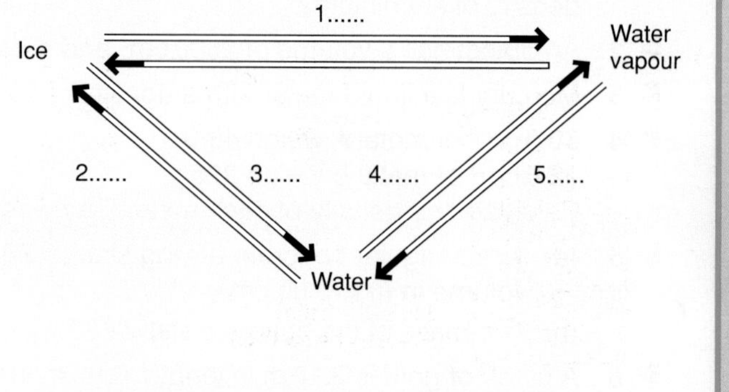

2 Give the scientific name for each of these changes.
   (a) A puddle of water gradually disappears.
   (b) A mist appears on your glasses.
   (c) A mothball gradually disappears from your wardrobe.
   (d) The change that happens when margarine is heated.

3.4 Finding melting points and boiling points

FIRST THOUGHTS

You can identify substances by finding their melting points and boiling points. Accuracy is essential, as you will see in this section.

Finding the melting point of a solid

Figure 3.4A shows an apparatus which can be used to find the melting point of a solid. To get an accurate result:

- first note the temperature at which the solid melts,
- allow the liquid which has been formed to cool and note the temperature at which it freezes.

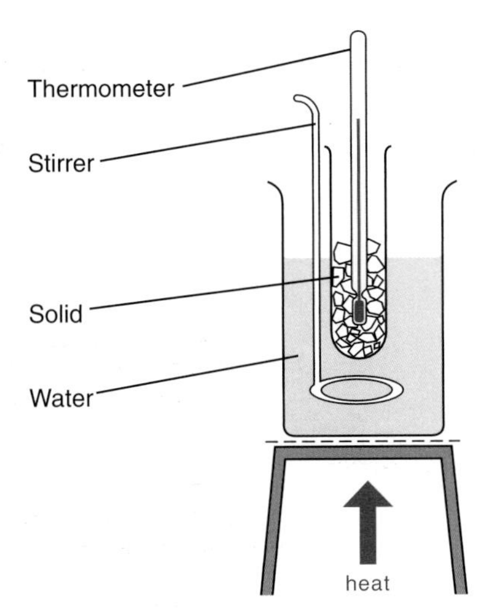

Figure 3.4A Finding the melting point of a solid (for solids which melt above 100 °C, a liquid other than water must be used)

The apparatus shown in Figure 3.4A will work between 20 °C and 100 °C. For solids with melting points above 100 °C, a liquid with a higher boiling point than water must be used. For liquids which freeze below room temperature, a liquid with a lower freezing point than water must be used. A mixture of ice and salt can be used down to −18 °C (see Figure 3.4B).

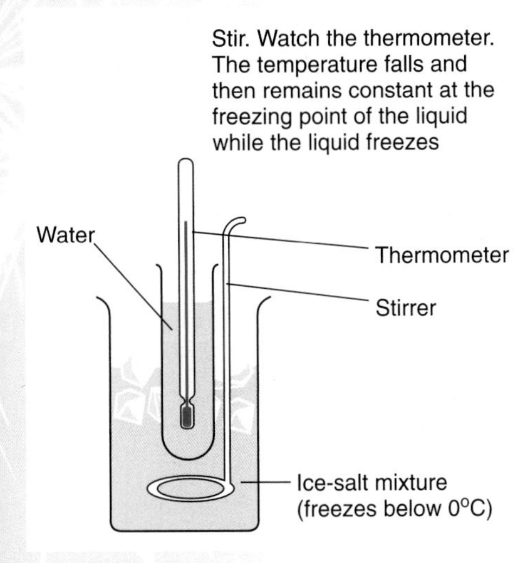

Figure 3.4B Finding the freezing point of water

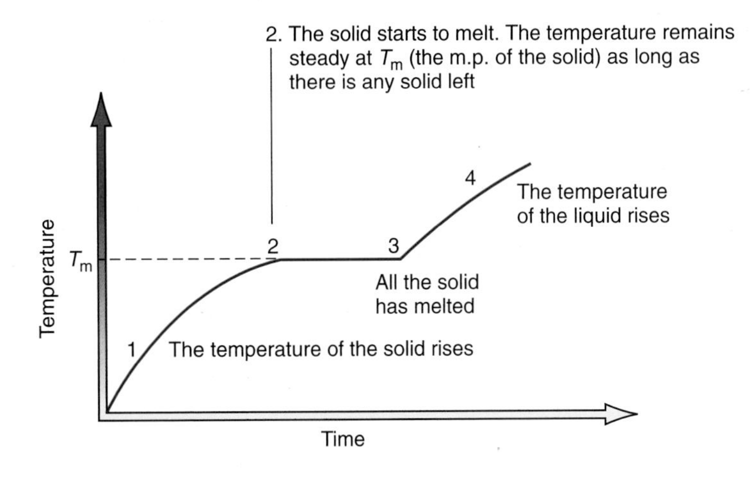

Figure 3.4C A graph of temperature against time when a pure solid melts

If the solid is a pure substance, a graph of temperature against time as the solid melts will look like Figure 3.4C.

SUMMARY

The apparatus shown in Figure 3.4A or 3.4B can be used to find the melting point of a solid and the freezing point of a liquid.

- The temperature of a pure solid stays constant while it is melting.
- The temperature of a pure liquid stays constant while it is freezing.

If you have a pure solid and you do not know what it is, you can use its melting point to find out. Scientists have drawn up lists of pure substances with their melting points. You find the melting point of the unknown solid and compare it with the listed melting points. *Which of the solids could be X?*

Solid	Melting point (°C)
Unknown solid X	116
Benzamide	132
Butanamide	116
Ethanamide	82

Figure 3.4D A truck spreading salt on an icy road in winter

The melting point can be used to identify an unknown pure solid. The presence of an impurity lowers the melting point.

If a solid is not pure, the melting point will be low, and the impure solid will melt gradually over a range of temperatures. Look at Figure 3.4D and explain why the ice on the road melts.

Finding the boiling point of a liquid

Figure shows an apparatus which can be used to find the boiling point of a non-flammable liquid. For a flammable liquid, a distillation apparatus, see *KS: Chemistry*, must be used.

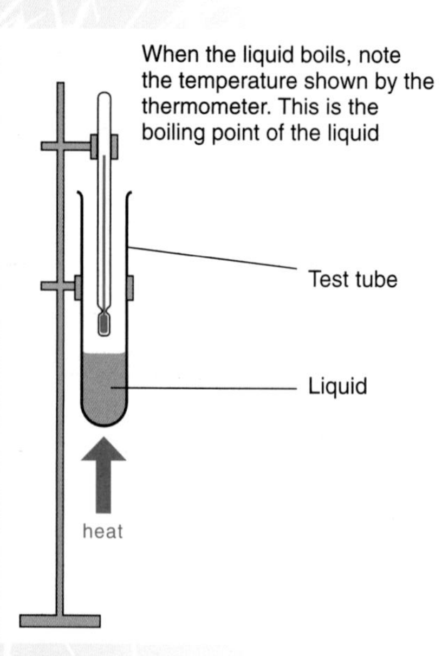
When the liquid boils, note the temperature shown by the thermometer. This is the boiling point of the liquid
Test tube
Liquid
heat

Figure 3.4E Finding the boiling point of a non- flammable liquid

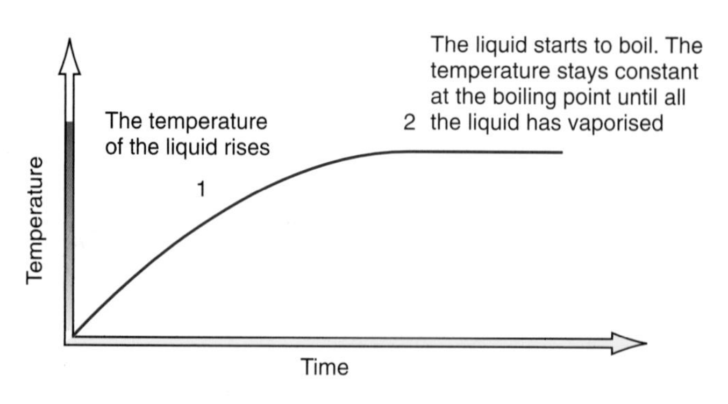
The liquid starts to boil. The temperature stays constant at the boiling point until all the liquid has vaporised
2
The temperature of the liquid rises
1
Temperature
Time

Figure 3.4F A graph of temperature against time when a pure liquid is heated

The apparatus in Figure 3.4E can be used to find the boiling point of a liquid. The temperature of a pure liquid stays constant while it is boiling. Dissolving a solid in a liquid raises the boiling point.

While a pure liquid is boiling, the temperature remains steady at its boiling point. All the heat going into the liquid is used to vaporise the liquid and not to raise its temperature. Figure 3.4F shows a graph of temperature against time when a pure liquid is heated.

A mixture of liquids, such as crude oil, boils over a range of temperature. If a solid is dissolved in a pure liquid, it raises the boiling point.

Figure 3.4G ▲ Why will he have difficulty in brewing a strong cup of tea?

The boiling point of a liquid depends on the surrounding pressure. If the surrounding pressure falls, the boiling point falls. The boiling point of water on a high mountain is lower than 100 °C. An increase in the surrounding pressure raises the boiling point (see Figure 3.4H).

The boiling point of a liquid is raised in a pressure cooker if the pressure is raised.

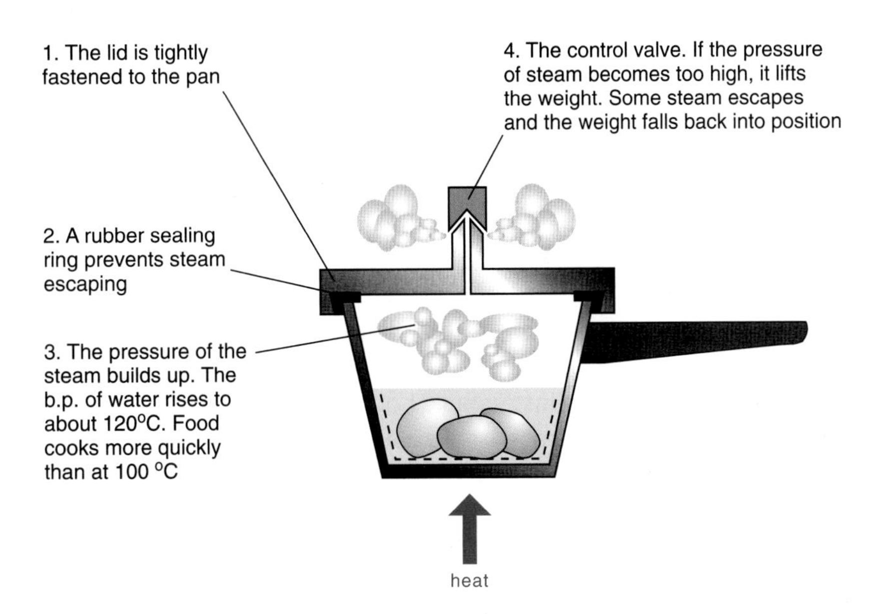

1. The lid is tightly fastened to the pan
4. The control valve. If the pressure of steam becomes too high, it lifts the weight. Some steam escapes and the weight falls back into position
2. A rubber sealing ring prevents steam escaping
3. The pressure of the steam builds up. The b.p. of water rises to about 120°C. Food cooks more quickly than at 100 °C
heat

Figure 3.4H ▲ How a pressure cooker works

SUMMARY

The boiling point of a liquid is stated at standard pressure.

- At lower pressures, the boiling point is lower.
- At higher pressure, the boiling point is higher.

Boiling points are stated at standard pressure (atmospheric pressure at sea level). Boiling points can be used to identify pure liquids. You take the boiling point of the liquid you want to identify; then you look through a list of boiling points of known liquids and find one which matches. Which of the liquids could be the unknown substance X?

No two substances have both the same boiling point and the same melting point.

Liquid	Boiling point (°C)
Substance X	111
Benzene	80
Methylbenzene	111
Naphthalene	218

CHECKPOINT

1 A pupil heated a beaker full of ice (and a little cold water) with a Bunsen burner. She recorded the temperature of the ice at intervals until the contents of the beaker had turned into boiling water. The table shows the results which the pupil recorded.

Time (minutes)	0	2	4	6	8	10	12	14	16
Temperature (°C)	0	0	0	26	51	76	100	100	100

(a) On graph paper, plot the temperature (on the vertical axis) against time (on the horizontal axis).

(b) On your graph, mark the m.p. of ice and the b.p of water.

(c) What happens to the temperature while the ice is melting?

(d) What happens to the temperature while the water is boiling?

(e) The Bunsen burner gives out heat at a steady rate. Explain what happens to the heat energy (i) when the beaker contains a mixture of ice and water at 0 °C, (ii) when the beaker contains water at 100 °C and (iii) when the beaker contains water at 50 °C.

2 Bacteria are killed by a temperature of 120 °C. One way of sterilising medical instruments is to heat them in an autoclave (a sort of pressure cooker: see the figure below). The table shows the effect of pressure on the boiling point of water.

Boiling point of water (°C)	*Pressure (kPa)*
80	47
90	68
100	101
110	140
120	195
130	273

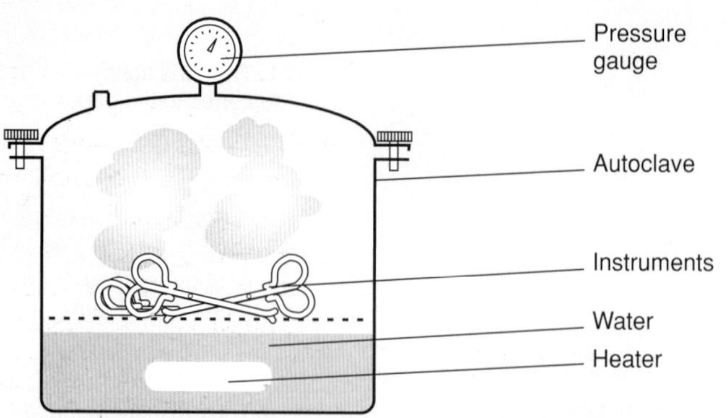

(a) Why are the instruments not simply boiled in a covered pan?

(b) What pressure must the autoclave reach to sterilise the instruments?

(c) What is the value of standard pressure in kPa (kilopascals)?

3.5 Properties and uses of materials

FIRST THOUGHTS

Properties of materials
- Hardness
- Toughness
- Strength
- Flexibility
- Elasticity
- Solubility
- Density (Topic 3.2)
- Melting point (Topic 3.4)
- Boiling point (Topic 3.4)
- Conduction of heat (Topics 5.5 and 5.6)
- Conduction of electricity (Topic 20)

Types of matter from which things are made are called materials. Different materials are used for different jobs. The reason is that their different **properties** (characteristics) make them useful for different purposes. You will see a list of properties in the margin. Some of them have been mentioned earlier in this topic. Conduction of heat and electricity will be covered in later topics. Now let us look at the rest.

Hardness

It is difficult to change the shape of a hard material. A hard material will dent or scratch a softer material. A hard material will withstand impact without changing. Table 3.3 shows the relative hardness of some materials on a 1–10 scale.

Table 3.3 ▼ Relative hardness of materials

Material	Relative hardness	Uses
Diamond	10.0	Jewellery, cutting tools
Silicon carbide	9.7	Abrasives
Tungsten carbide	8.5	Drills
Steel	7–5	Machinery, vehicles, buildings
Sand	7.0	Abrasives, e.g. sandpaper
Glass	5.5	Cut glass can be made by cutting glass with harder materials.
Nickel	5.5	Used in coins; hard-wearing
Concrete	5–4	Building material
Wood	3–1	Construction furniture
Tin	1.5	Plating steel food cans

Toughness and brittleness

A tough material dents without breaking.

A brittle material breaks without denting.

Construction workers on building sites wear 'hard hats' to protect themselves from falling objects. A hard hat is designed to absorb the energy of an impact. The hat material is tough, that is, it is difficult to break, although it may be dented by the impact. In comparison, a brick is difficult to dent and will shatter if dropped onto a concrete floor. The brick is **brittle**. Glass is another brittle material. These materials cannot absorb the energy of a large force without cracking. If a still larger force is applied the cracks get bigger and the materials shatter.

Figure 3.5A ▲ Hardness

Composite materials

To make brittle materials tougher you have to try to stop them cracking. Mixing a brittle material with a material made of fibres, e.g. glass fibre or

paper, will often do this. The fibres are able to absorb the energy of a force and the brittle material does not crack. Plaster is a brittle material. Plasterboard is much tougher. It is made by coating a sheet of plaster with paper fibres. It is a **composite material**. Glass-reinforced plastic (GRP) is a mixture of glass fibre and a plastic resin.

Paper
Made of fibres

Plaster
Brittle

Figure 3.5B ▲ Plasterboard

Figure 3.5C ▲ A GRP canoe

Figure 3.5D ▲ Corrugated plastic roof

Concrete has great compressive strength, but it can crack if it is stretched. For construction purposes, **reinforced concrete** is used. Running through this composite material are steel rods which act like the glass fibres in GRP.

The shape of a piece of material alters its strength. Corrugated cardboard is used for packaging. Corrugated iron sheet and corrugated plastic sheet are used for roofing.

SUMMARY

Materials may be:

- hard – resistant to impact, difficult to scratch,
- tough – difficult to break, will 'give' before breaking,
- brittle – will break without 'giving',
- composite – made of more than one substance.

Strength

A strong material is difficult to break by applying force. The force may be a stretching force (e.g. a pull on a rope), or a squeeze (e.g. a vice tightening round a piece of wood), or a blow (e.g. a hammer blow on a lump of stone). A material which is hard to break by stretching has good **tensile strength**; a material which is hard to break by crushing has good **compressive strength**. The tensile strength of a material depends on its cross-sectional area.

Flexibility

While a material is pulled it is being stretched: it is under **tension**. While a material is squashed it is being compressed: it is under **compression**. When a material is bent, one side of the material is being stretched while the opposite side is being compressed. A material which is easy to bend without breaking has both tensile strength and compressive strength. It is **flexible**.

Figure 3.5E ▲ It's flexible

An elastic material regains its shape after all the forces applied to it have been removed.

Elasticity

You can change the shape of a material by applying enough force. When you stop applying the force, some materials retain their new shapes; these are **plastic** materials. Other materials return to their old shape when you stop applying the force; these are **elastic materials**.

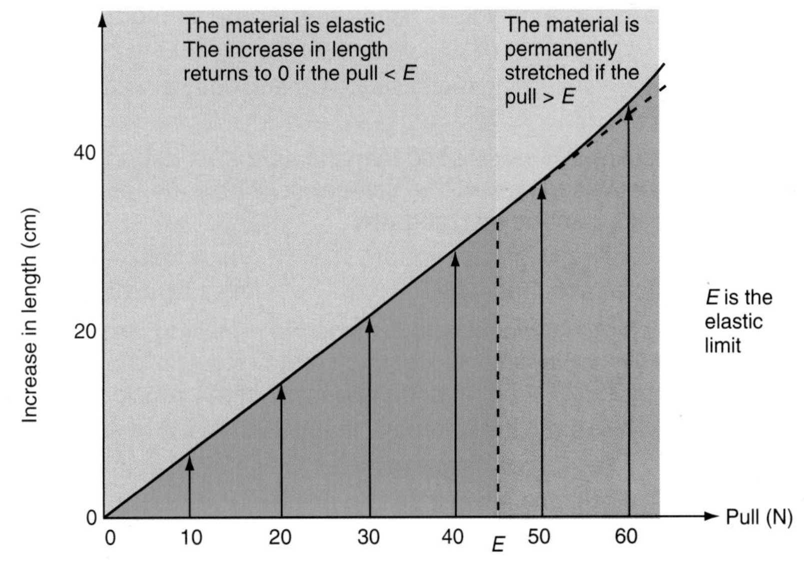

Figure 3.5F ▲ Increasing length

When you pull an elastic material, it stretches – increases in length. At first, when you double the pull, you double the increase in length. As the pull increases, however, you reach a point where the material no longer returns to its original shape. This pull is the **elastic limit** of the material. Increasing the pull still more eventually makes the material break (see Figure 3.5F).

The force needed to extend a steel spring is proportional to its change in length. The change in length is referred to as the **extension**. The link between force and extension for a steel spring is known as **Hooke's law**. It is usually written as an equation,

$$\text{Force} = \text{constant} \times \text{extension}$$

A graph of these measurements for a steel spring is a straight line. See 16.1 for more about springs.

SUMMARY

Materials may have tensile strength (resistance to stretching). compressive strength (resistance to pressure), flexibility (both tensile and compressive strength) or elasticity (the ability to return to their original shape after being stretched).

CHECKPOINT

1 (a) Name two materials than can be used to drill through steel.
(b) Why can a steel blade slice through tin?
(c) Name a material that is used to make wood smooth.
(d) Why are diamond-tipped saws used to slice through concrete?
(e) Name two materials which can be used to cut glass.
(f) Why is nickel a better coinage material than tin?

2 (a) Name two tough materials. Say what they are used for.
(b) Name two brittle materials. Say what can be done to make them tougher.

3 Use the information in the table below to answer this question.

Cross-sectional area (mm^2)	1	2	4	6	8	10
Breaking force (N)	0.5	1	2	3	4	5

(a) On a piece of graph paper, plot the breaking force (on the vertical axis) against the cross-sectional area (on the horizontal axis).
(b) From the shape of the graph, say how the breaking force alters when the cross-sectional area (i) doubles, (ii) increases by a factor of 10.
(c) Write a sentence saying how the tensile strength of the material depends on the cross-sectional area.

4 Name a material to fit each of the following descriptions: hard, soft, strong, flexible, tough, brittle, composite.

5 Do the following materials need good tensile strength or good compressive strength? a tent rope, a tow bar, a stone wall, an anchor chain, a concrete paving stone, a building brick, a rope ladder.

6 Classify the following materials as either plastic or elastic: plasticine, pottery clay, a rubber band, a balloon, 'potty putty'.

7 The choice of material used to make a product depends on the conditions under which the product is used and on the materials available. For each product listed below, state what type of material it is made from and why.
(a) a car tyre
(b) cables to support a lift
(c) a tent
(d) disposable beakers

8 (a) A paint manufacturer is considering replacing metal paint tins with moulded plastic containers of the same dimensions. Each empty steel tin has a mass of 0.320 kg. The density of the metal is 7800 kg/m^3 and the density of the plastic is 1400 kg/m^3.
(i) Work out the volume of the metal in each metal tin.
(ii) Work out the mass of a plastic container of the same volume as the metal tin.
(iii) State one advantage and one disadvantage of using plastic containers.
(b) If the density of the paint is 2500 kg/m^3, work out the mass of paint that has a volume of 0.001 m^3 (1 litre). What would be the total mass of each type of container filled with 0.001 m^3 of paint?

9 In an experiment to test the stretching of a copper wire, the wire was hung vertically from a fixed clamp and was then stretched by hanging load weights from its lower end. The table below shows how its length increased as the load increased.

Load (N)	0	10	20	30	40	50	55	60	65	70
Extension (mm)	0	1.5	3.0	4.5	6.0	7.7	8.8	10.1	13.5	Snapped

(a) Plot a graph of load against extension.
(b) Work out the load needed to extend the wire by exactly 1.0 mm from its initial length.
(c) How much force would be needed to break a cable made of 20 strands of this wire?

Topic 4 Energy

4.1 What is energy?

FIRST THOUGHTS

What is energy? Studying the energy changes described in this section will help you to understand.

A polar trek

Imagine you are a polar explorer about to set off for a few days on an expedition from base camp. *What sort of preparations would you make for your expedition?* You would obviously need suitable weatherproof clothing, a supply of food and a suitable means of transport. The clothing must keep you warm and dry; food is essential so you can keep working and transport saves you getting tired trudging across ice and snow. If your clothing was not very effective or your food ran out or your transport broke down, your survival would be at stake. Clearly, polar explorers must make very careful preparations for their expeditions! They must plan to make the most of their resources to keep warm and to keep working.

Energy is what makes thing work (i.e. makes things move or makes changes happen).

Energy is needed to warm things up.

Figure 4.1A A polar expedition

Forms of energy

Objects can have energy in different forms. For example, winding up the mainspring of a clockwork toy car stores energy in the spring. When the spring unwinds, it drives the wheels and the car moves. The wound-up spring is said to contain **potential energy**. Any moving object is said to have **kinetic energy**. When the car is moving, potential energy in the spring is being changed into kinetic energy of the car. In other words, energy is changing from one form into another when the car is moving.

Potential energy is energy stored due to position. A weight raised above the ground has potential energy due to its position. A spring in tension has potential energy because its position differs from when there is no tension in it.

Figure 4.1B Gaining potential energy

Another example of energy changing from one form into another is when you use a torchbulb. The torch battery pushes electric current through the torchbulb when the torch is switched on. As a result, the torchbulb emits light energy. The battery converts **chemical energy into electrical energy** and the torchbulb converts the electrical energy into **light energy**.

It's a fact!

Space invaders

Hornets that raid beehives are cooked to death by the bees. Scientists have discovered that an invading hornet is quickly engulfed by hundreds of bees that generate heat through shivering their wings. The temperature in the ball of bees rises and the hornet dies. The bees survive because they are able to withstand higher temperatures than the hornet.

Can you think of an example where light energy is changed into the energy of chemical bonds? This is what happens when a plant makes food. Light is needed for a plant to make sugars from carbon dioxide and water. This process is called **photosynthesis**. The light energy is converted into the chemical energy of the sugar molecules. The energy in the food made by the plants is passed on to animals that eat the plants.

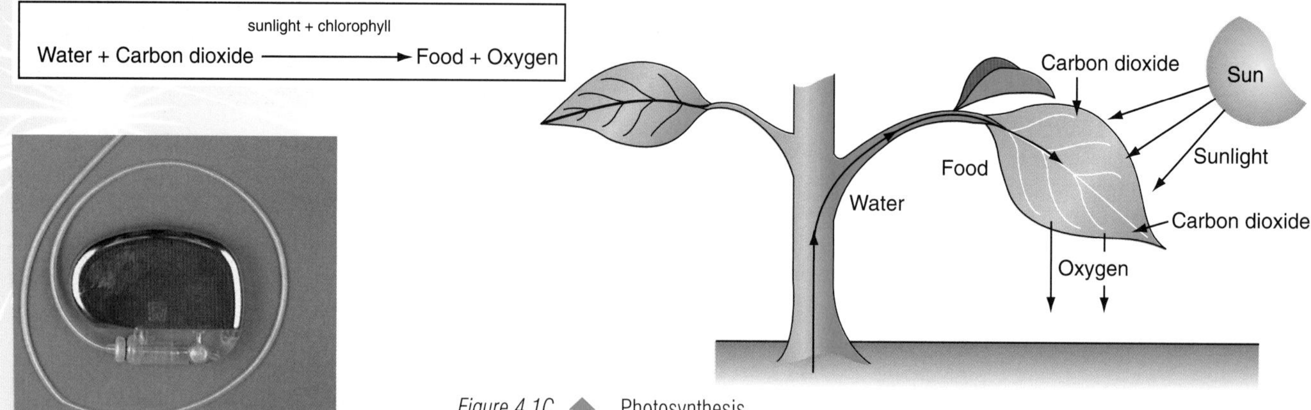

Figure 4.1C ▲ Photosynthesis

Figure 4.1D ▲ A heart pacemaker

We obtain our food from plants or from animals fed on plants. We all need food to keep going, so the next time you charge round the sports field, remember you got your energy from sunlight. Not all the energy from the food is used to keep you going. Your muscles produce **heat when they are at work and if you are a very noisy sportsperson, some is used to create sound energy**.

Nuclear energy is used to keep heart pacemakers going. Ordinary batteries are unsuitable, even though they can be made small enough. This is because they don't last long enough and the user would need to be 'opened up' every time the battery needed changing. Using a small radioactive source gives enough energy to operate a pacemaker for years. The pacemaker converts nuclear energy into electrical energy in the form of pulses to keep the heart beating steadily.

SUMMARY

Energy is needed to make things work or to warm things up. Objects can possess energy in different forms. Energy can be changed from any form into any other form.

CHECKPOINT

1. Which of the following need energy? Explain your answers.
   - (a) Running up a flight of steps.
   - (b) Holding a book out in front of you.
   - (c) Balancing on a tightrope.
   - (d) Floating in the sea.
2. Each of the following devices is designed to convert energy from one form into another form. For each device, write down what form of energy it uses and what this is changed into.
   - (a) A hair dryer
   - (b) A door bell
   - (c) A jet engine
   - (d) A radio
   - (e) A candle
   - (f) An electric kettle
   - (g) A bicycle
   - (h) A windmill
3. Write down the name of a device that is designed to change:
   - (a) electrical energy into light energy.
   - (b) electrical energy into potential energy,
   - (c) electrical energy into heat.
4. The power for a clockwork radio is obtained from a small dynamo powered by a clockwork spring.
   - (a) What form of energy is stored in a clockwork spring after it has been wound up?
   - (b) What form of energy does a dynamo produce?
   - (c) Why is a clockwork radio popular in under-developed countries?

4.2 Conservation of energy

FIRST THOUGHTS

Why do we believe that energy can neither be created nor destroyed? Read on to find out why.

Strike a match and you will cause chemical energy to be released as the match burns. The substance on the match head ignites and reacts with oxygen from the air to release chemical energy. *What happens to the chemical energy?* The burning match produces heat and light which are different forms of energy. This is an example where one form of energy is changed into two or more different forms of energy.

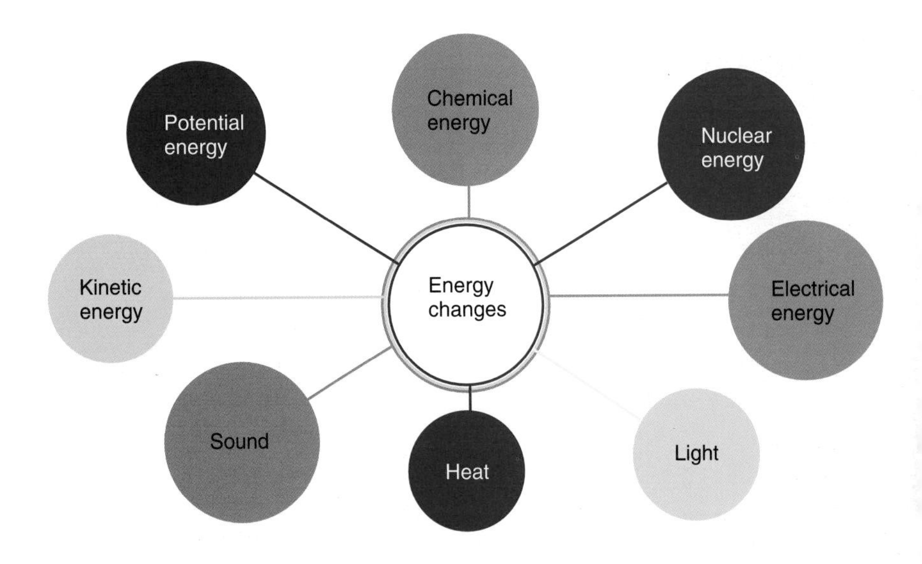

Figure 4.2A ▲ Energy links

Most energy changes involve energy changing from one form into two or more different forms. Devices designed to change energy into a particular form often produce other forms of energy as well. For example, an electric torch is designed to change chemical energy (in a battery) into light energy. However the torch bulb releases heat which warms the bulb holder up. In fact, powerful lamps like stage lights get really hot when in use.

Figure 4.2B ▲ Some energy changes

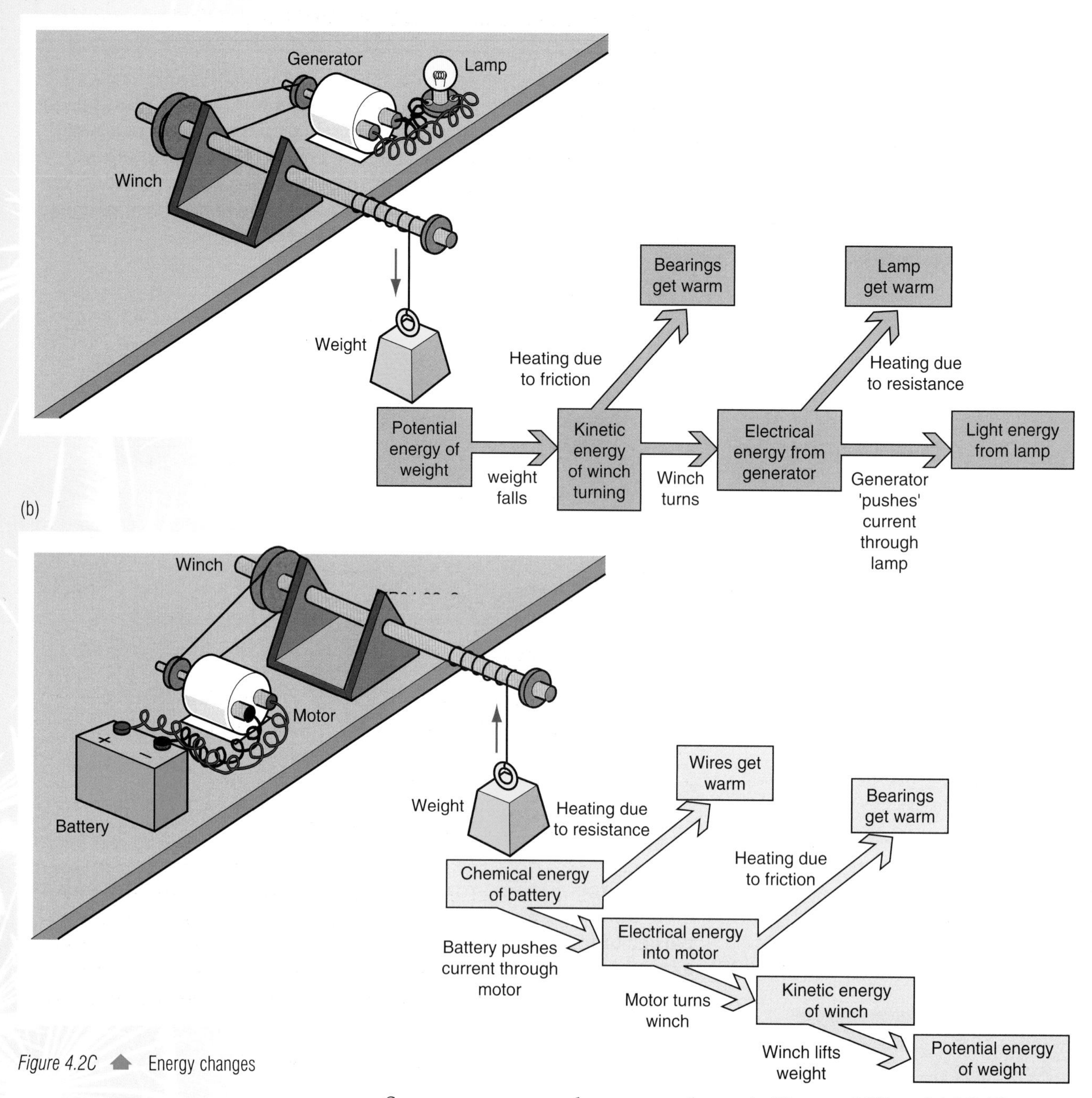

Figure 4.2C ▲ Energy changes

It's a fact!

An enterprising inventor at a nineteenth century exhibition in America almost convinced visitors that his 'motor' could produce energy from nowhere. For days on end, the motor continued to turn with no apparent energy supply – until someone pedalling away was discovered in the room below the motor!

Some more energy changes are shown in Figures 4.2B and 4.2C. The energy flow diagrams shown are examples of **Sankey diagrams**. Each system shown involves changing energy from one form into different forms. *What do these changes have in common with each other?*

1. **Energy must be present in one form or another to start with.**

In other words, energy can't be produced form nowhere. Fuels such as coal and oil contain chemical energy produced millions of years ago by the growing plants absorbing sunlight.

2. **Energy changing from one form into another involves either heating or working.**

In other words, energy changes occur when something is forced to happen (i.e. making something work) or when there is a temperature difference. Heat is energy 'on the move' due to temperature difference. Look again at Figure 4.2C and check that all the arrows for energy changes show 'working' or 'heating'.

When energy changes from one form into other forms, how much energy is there at the end of the change compared with the start? Can you think of a situation where the energy changes back to the form it started in?

Squash players use very bouncy balls which rebound very effectively. Try releasing such a ball from a fixed height above a flat surface and see how high it rebounds. You ought to find out that a really bouncy ball almost regains its initial height. Figure 4.2D shows the energy changes after an ideal ball is released. The potential energy of the ball is recovered fully after the bounce. This example shows quite clearly that no energy is lost.

The Principle of Conservation of Energy is a very important principle (VIP) in science. Make sure you know what it means.

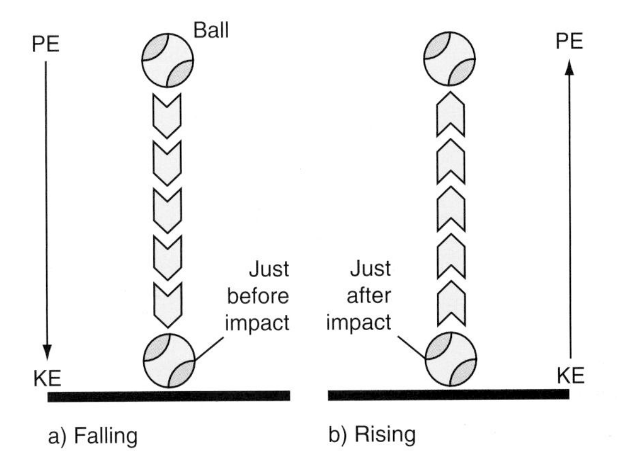

Figure 4.2D A bouncy ball

EXTENSION FILE ACTIVITY

How would dropping a cricket ball or a hockey ball onto a flat surface compare with dropping a squash ball? This time there would be little bounce and the ball would not return to its initial height. Only part of the initial potential energy is recovered after the bounce. *What happens to the potential energy that is lost?* The impact of the ball on the floor heats the ball and floor slightly and sound energy is also produced.

When energy changes from one form into other forms, the total amount of energy after the change is equal to the total amount of energy at the start. **Energy is always conserved**. This means that when it is changed from any one form into any other, the total amount of energy after the change is always equal to the amount at the start. Scientists have tested the conservation of energy principle many times. But their experiments have always led to the same conclusion:

Energy is always conserved.

SUMMARY

- Energy can change from any one form into other forms.
- In any energy change, energy is conserved even though it may be changed into many different forms.

CHECKPOINT

1. Which of the following energy changes are due to something being forced to move and which are due to heating?
   (a) Changing chemical energy into potential energy when a book is lifted.
   (b) Using electrical energy to warm water in a kettle.
   (c) Using chemical energy to warm your hands by rubbing them together.
   (d) Drying damp clothes by hanging them outside on a washing line.
2. Each of the following situations involves energy changing from one form into several other forms. Describe the energy changes in each situation.
   (a) Clapping your hands together loudly.
   (b) Using an electric motor to raise a weight.
   (c) Throwing a ball into the air.
   (d) Using a gas cooker to boil an egg.

4.3 Measuring energy

FIRST THOUGHTS

Energy and power as measurable quantities are introduced here. They are used extensively in later topics so read on carefully.

When energy changes into different forms, how much energy goes into each form? To answer this question, we need to measure how much energy is needed to heat things or to force them to move.

Imagine you have the job of carrying lots of bricks up a ladder. A certain amount of energy is needed to lift one brick up by one metre. To lift two bricks by one metre or one brick by two metres requires twice as much energy. To lift ten bricks up by five metres requires 50 times as much energy as lifting one brick up by one metre. We could measure energy in units of the amount needed to lift one brick by one metre. But the brick would need to be a 'standard' brick with a known weight.

The **joule** (J) is the scientific unit of energy. One joule is the energy needed to lift an object of weight one newton by one metre. So if each brick in the above example has a weight of 20 N, the energy needed to lift one brick by 1 m would be 20 J. *How much energy would be needed to lift 100 bricks by 5 m?*

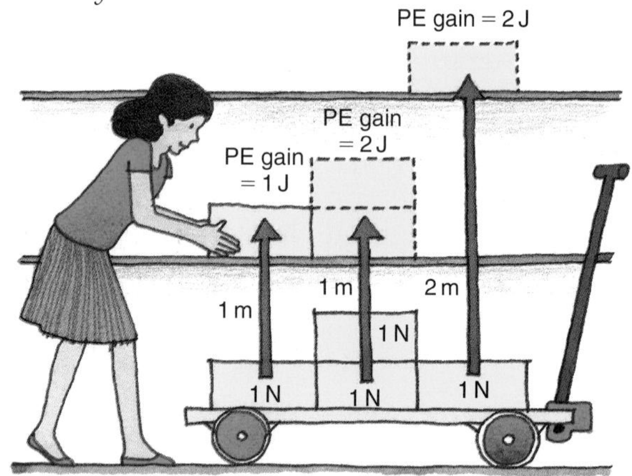

Figure 4.3A Using joules

Energy is always measured in joules, no matter what form the energy is in. Energy in any given form can be used to lift a known weight by a measured height. In this way, energy in any given form can be measured in joules. For example, you could measure the energy stored in a catapult by measuring how high the catapult can throw a known weight.

It's a fact!

James Joule was a nineteenth century scientist who investigated lots of energy changes. On honeymoon in Switzerland he and his wife visited waterfalls to find out if the water at the bottom was warmer than at the top. He reasoned that the loss of potential energy due to the fall ought to produce heat which would warm the water. In fact, using an accurate thermometer he discovered the water at the bottom was cooler than at the top. He realised that this was because of the cooling effect of the water spray.

An energetic challenge

In games lessons, Simon and Kevin are always trying to outdo each other. Simon has challenged Kevin to a rope-climbing competition to see who can shin up the rope the fastest. Their friends time them and use a metre rule to measure how high they climb.

Here are the results:

	Kevin	*Simon*
Time taken (s)	6.2	5.5
Height gain (m)	3.5	3.5

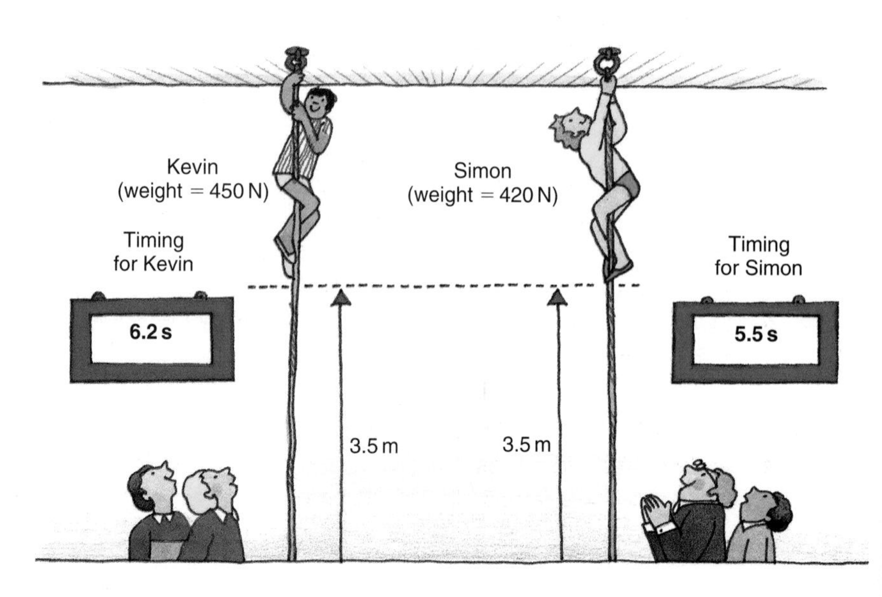

Figure 4.3B
Who is more powerful?

The meaning of work

Work is done when an object is moved by a force. The amount of work done is defined as the force × the distance moved in the direction of the force. Energy is the capacity to do work. When a force F moves an object through a distance d in the direction of the force, the work done by the force is $F \times d$.

Work done (or energy transferred) = force × distance moved (in the direction of the force)

Simon claims victory but Kevin insists that they must take account of body weight. Off they go to the science laboratory where there is a set of bathroom scales marked in newtons.

Kevin's weight = 450 N

Simon's weight = 420 N

Who *gains more potential energy climbing the rope*?

Kevin gained 450 J for every metre gain of height (= 450 N × 1 m). So he gained 1575 J in total (= 450 N × 3.5 m) to climb the rope.

How much potential energy does Simon gain in total?

Who is the more energetic of the two?

Simon claims his muscles are more powerful since he climbed faster.

How much potential energy did Simon gain per second? How much did Kevin gain per second? Is Simon's claim correct?

Power is the rate at which energy is used.

The unit of power is the watt **(W) which is equal to one joule per second (J/s).**

Machines designed to do jobs are rated in terms of how much power they use. For example, a 3000 W electric winch uses 3000 J of electrical energy each second when it is switched on. Heaters and lamps are also rated in watts. A 1.5 kW heater uses 1500 J of electrical energy each second. *How long would it take a 100 W lamp to use 1500 J of electrical energy?*

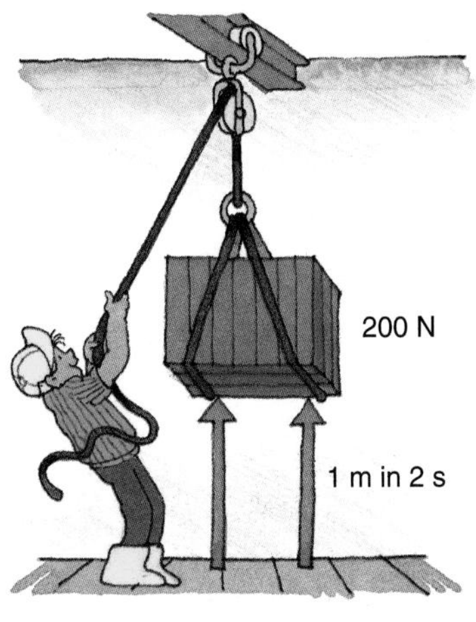

Figure 4.3C ▲ A 100 watt worker

Energy used (in joules) = Power supplied (in watts) × Time taken (in seconds)

SUMMARY

Energy is always measured in joules. Power is the rate at which energy is used. The unit of power is the watt which equals 1 joule per second. The energy used by a device can be worked out if its power and time of use are known.

CHECKPOINT

1. How much energy is used in each of the following situations?
   (a) A 60 W hairdryer is used for 60 seconds.
   (b) A 150 W lamp is used for 2 hours.
   (c) A 3 kW electric kettle is used for 5 minutes.
   (d) Two 60 W lamps are used together for 30 minutes.
2. In an investigation to estimate the power of a gas cooker ring, a student finds that 1 litre of water in a pan heated by gas reached boiling point in 250 seconds. The student then times how long a 2.0 kW electric kettle takes to heat 1 litre of water to boiling point from the same initial temperature. She finds the kettle takes 200 seconds to do this.
   (a) Do you think the gas cooker ring is more powerful than the electric kettle?
   (b) How much energy was supplied to the kettle to heat the water?
   (c) What was the power of the gas cooker ring?
3. A student of weight 500 N is timed as she steps on and off a box of height 0.15 m. She does 100 steps in exactly 5 minutes. Calculate:
   (a) her gain in potential energy each step,
   (b) the average power output of her legs.
4. Your heart never stops pumping blood round your body throughout your life. The power of a typical human heart is about 1 W. How much energy is used by a typical heart in one day?

4.4 Energy on the move

FIRST THOUGHTS

Electricity is a very convenient means of moving energy from where it is generated to where it is needed. How does it compare with other means of transporting energy?

How is energy supplied to your home? Most homes are connected to mains electricity cables that supply electricity produced in power stations. Natural gas is supplied to homes through a network of underground pipes called the gas mains. The gas is pumped through the pipes to your home from a gas terminal. This is where the gas is fed into the system from tankers or from undersea pipe lines joined to gas rigs.

Electricity and gas are the main sources of power to most homes because they are convenient.

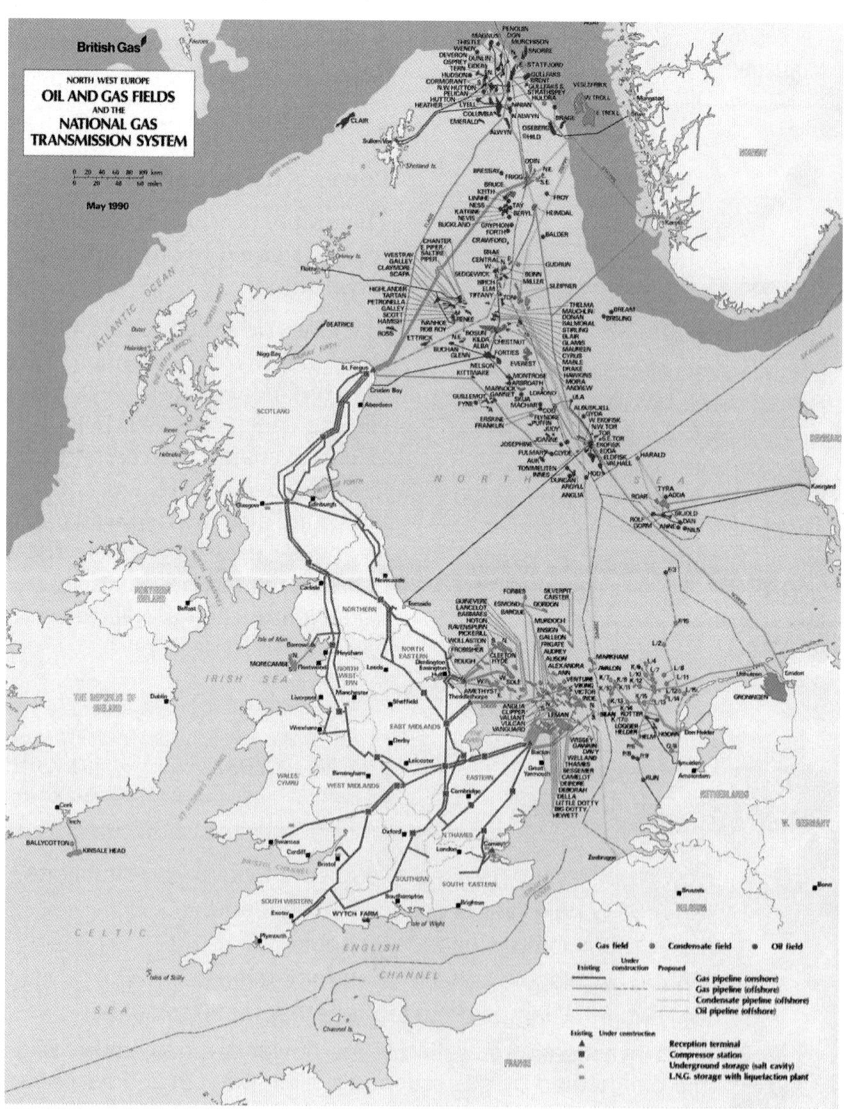

Figure 4.4A ▲ Supplying natural gas

Pressure is needed to force gas through a pipe. The pressure in a gas pipe is applied at a pumping station. Valves are needed to make sure the gas does not flow back towards the terminal. The pressure inside a mains gas pipe must not be too high, otherwise gas would leak from the pipe.

When natural gas is burned, carbon dioxide and carbon monoxide gases are produced and released into the atmosphere. This also happens when coal or oil is burned. Electricity is 'clean' in this respect since electrical appliances do not give **greenhouse gases** such as carbon dioxide. However, power stations that burn fossil fuels (i.e. coal, oil or gas) release these gases to produce electricity.

It's a fact!

Gas supplied to homes and offices is used for heating and cooking. More than fifty years ago, gas was also used for lighting. Gas lamps in streets were lit by a lamplighter who was employed to turn the lamps on and off every day.

Figure 4.4B A domestic electricity meter

Measuring electrical energy

Every home using mains electricity is connected via an electricity meter. The meter registers the electrical energy used in **kilowatt hours** (kW h).

One kilowatt hour is the amount of electrical energy used by a 1 kW electric heater in 1 hour. For example, a 3 kW electric heater used for 5 hours would use 15 kilowatt hours of electrical energy. This would increase the meter reading by 15 'units'. Prove for yourself that 1 kilowatt hour is equal to 3.6 MJ.

For low voltage circuits, a **joulemeter** may be used to measure electrical energy. Figure 4.4C shows a joulemeter connected into a circuit to measure the electrical energy used by a torchbulb. The joulemeter must be read before the switch is closed and then again when the switch is reopened. The difference between the two readings gives the number of joules supplied.

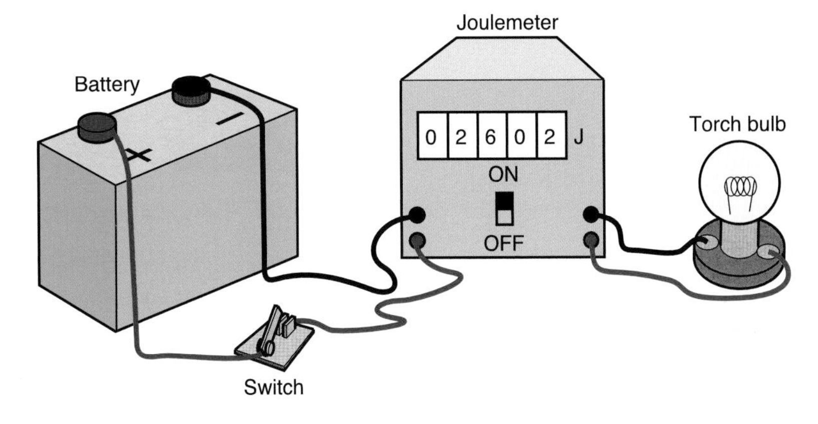

Figure 4.4C Using a joulemeter

SUMMARY

Electricity and gas are used to supply energy to most homes. A complete circuit is needed to supply electricity. The blood system is like a complete circuit, used to supply energy to the muscles.

Human energy

Our bodies need food to obtain energy. We need energy to enable our muscles to do work. The blood system carries glucose from the digestive system to the muscles. Here glucose reacts to supply energy to the muscles. The heart forces blood through the arteries and veins of the blood system.

CHECKPOINT

1. Jasmin and Imram are discussing the relative merits of gas and electricity for home heating. Jasmin thinks that electricity is best because it is 'clean'. Imram isn't sure about this argument. What do you think?
2. (a) A gas leak can cause an explosion. Natural gas is odourless so another gas is added to make it smell. Why?
   (b) Why is it important not to light a match or switch a light on if you suspect a gas leak at home?
3. Alan is using a joulemeter to measure the energy used by an electric heater. The joulemeter reads 2602 J initially. He closes the battery switch for exactly 5 minutes, then opens the switch. The meter now reads 17 587 J.
   (a) How much energy was used?
   (b) Work out the power of the heater in watts.
4. The electricity mains and the gas mains both move energy from one place to another.
   (a) What carries the energy in each case?
   (b) What happens to the energy carriers in each case after releasing the energy?

Topic 5 Heat

5.1 Temperature

FIRST THOUGHTS

What do the butcher, the baker and the candlestick maker have in common? Read on to find out.

Figure 5.1A The weather forecast

Don't confuse heat with temperature. Heat is energy transfer due to temperature difference. Temperature is a measure of hotness.

Our everyday lives are affected by temperature in lots of ways. The clothes you choose to wear each day, the food you eat, how well you feel; all these depend in some way on temperature. When you read the weather forecast and learn that tomorrow's outdoor temperature is expected to be as high as 30 °C, you can look forward to a hot and sunny day!

Temperature scales

Temperature is a measure of hotness. **Fixed points** are used to define a scale of temperature. These points are 'degrees of hotness' that can be set up precisely. They are usually melting points or boiling points of pure substances.

The Celsius scale

The Celsius scale, denoted by °C, is defined by two fixed points, which are;

1 **Ice Point** at 0 °C; this is the temperature at which pure ice melts.

2 **Steam point** at 100 °C; this is the temperature at which pure water boils at standard atmospheric pressure.

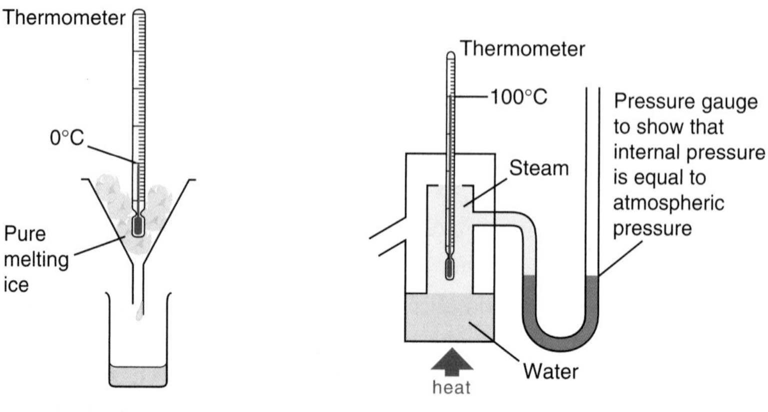

Figure 5.1B Ice point

Figure 5.1C Steam point

It's a fact!

Body temperature for a healthy person is about 37 °C. Someone in a fever might get as hot as 40°C but if their temperature gets as high as 45 °C, there's not much hope! The human body has a remarkable control system, keeping each of us at about 37 °C, no matter where we are.

Thermometers

What is it that the butcher, the baker and the candlestick maker have in common? The answer is that they all need to measure temperature. Thermometers are used to measure temperature.

The butcher has a cold room to keep meat in. A thermometer is used to make sure the room stays cold. If the room became warm, the meat in it would go off and would have to be destroyed.

The baker has an oven to bake bread. If the oven gets too hot, the bread is burnt. A thermometer probe inside the oven is used to give a temperature reading on a meter outside.

The candlestick maker melts wax and pours it into moulds to make the candles. He uses a thermometer to make sure the molten wax is not too hot.

Thermometers are made in all shapes and sizes but they all must be calibrated (i.e. marked with a scale). This is usually done in °C, marking the ice point reading as 0° and the steam point reading as 100°. The scale between these two marks is then divided into 100 degrees.

Liquid-in-glass thermometers

These are based on the principle that liquids expand much more than solids when heated. Mercury and coloured ethanol (alcohol) are the usual thermometer liquids. When the thermometer bulb is warmed up, the liquid in it expands but the glass bulb expands very little. As a result, the liquid is forced out of the bulb and along the capillary tube in the stem. The temperature is measured from the position of the end of the liquid thread in the stem.

Clinical thermometers

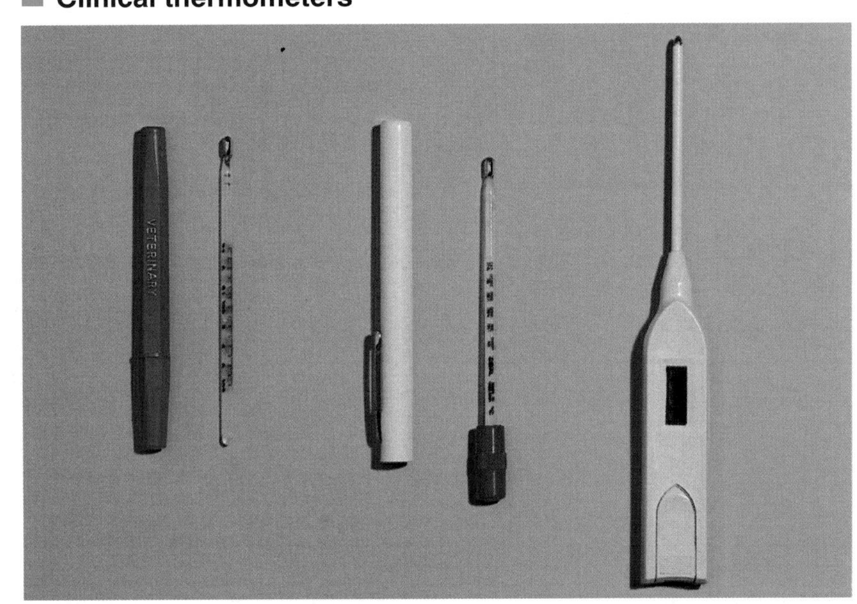

Figure 5.1D ▲ A range of clinical thermometers

Figure 5.1E ▲ A clinical thermometer

Clinical thermometers are designed to measure body temperature accurately. They are liquid-in-glass thermometers with a scale from about 35 °C to 45 °C in graduations of 0.1 °C. The capillary tube is constricted near the bulb, as shown in Figure 5.1E. This is to prevent the liquid returning to the bulb after the thermometer is removed from the patient. In this way, the user doesn't have to take the reading while the thermometer is actually in the patient. After the thermometer has been removed from the patient and read, it is given a quick flick to make the liquid return to the bulb.

All thermometers should read 0 °C at ice point and 100 °C at steam point.

Thermocouple thermometers

These are electrical thermometers that use the fact that when two different metals are placed in contact, a voltage develops between them. This voltage varies with temperature. An iron wire and a copper wire can be used to make a thermocouple thermometer. One of the junctions is kept in melting ice at 0 °C while the other one is used as the temperature probe. The voltmeter can be calibrated directly in °C.

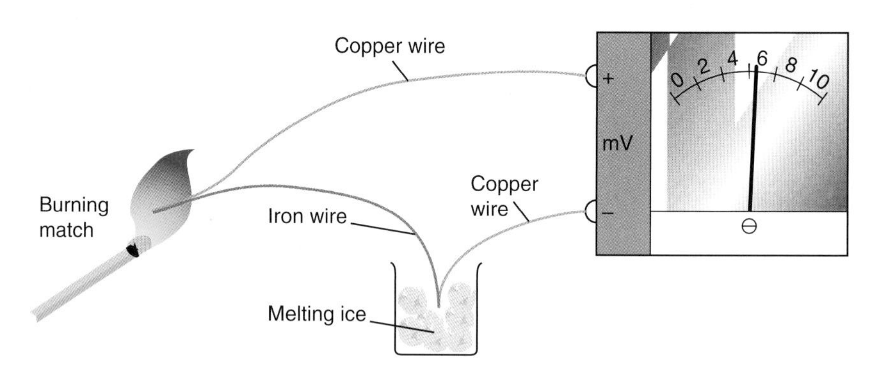

Figure 5.1F A thermocouple thermometer

Figure 5.1G Thermistors

Thermistor thermometers

Thermistor thermometers use electrical resistors called thermistors. The resistance of a thermistor changes with temperature. This is why it can be used to measure temperature. The thermistor is usually plugged into an electronic circuit designed to give a read-out in °C on a digital display. The circuit usually has two variable resistors to set the ice point read-out at 0 °C and the steam point read-out at 100 °C.

Temperature strips

Temperature strips change colour when warmed. Each strip is designed so that its colour alters from blue to red as it is warmed.

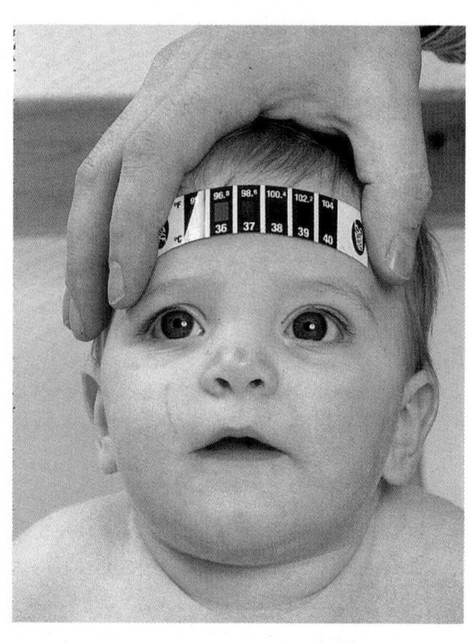

Figure 5.1H A forehead temperature strip

SUMMARY

The Celsius scale is defined in terms of ice point (0 °C) and steam point (100 °C). All thermometers must be calibrated. Different types of thermometers are used for different purposes.

CHECKPOINT

1. Say what type of thermometer you would choose;
   (a) to measure the body temperature of an adult,
   (b) to measure the body temperature of a baby,
   (c) to measure the temperature in a greenhouse,
   (d) to monitor the water temperature at the outflow of a solar heating panel,
   (e) to monitor the soil temperature in a plant experiment.
2. The following measurements were made when a thermocouple thermometer was being compared with a mercury-in-glass thermometer;

Voltmeter reading (mV)	0	1.5	3.0	4.5	6.0	7.5
M-in-G reading (°C)	0	20	40	60	80	100

   (a) Plot a graph of the voltmeter reading (on the vertical axis) against the mercury-in-glass thermometer reading.
   (b) Use your graph to work out the voltmeter reading for 50°C on the mercury-in-glass thermometer.
3. Weather stations often use specially-designed thermometers to record the highest and the lowest temperature each day. The figure shows the construction of one of each type.
   (a) Why is mercury used in the maximum thermometer and alcohol used in the minimum thermometer?
   (b) What was the highest temperature reached by the maximum thermometer?
   (c) What was the lowest temperature reached by the minimum thermometer?
   (d) How could you reset each thermometer?

Iron 'index'
Mercury
Iron 'index'
Alcohol

5.2 Thermal expansion

FIRST THOUGHTS

Thermal expansion can be a nuisance. It can also be put to good use. This section will tell you how.

Materials expand, if allowed, when heated. Gases are capable of expanding much more than solids or liquids. Figure 5.2A shows how the expansion of gas may be demonstrated; warming the test tube by hand is enough to make the air in it expand and push the thread of water up the tube. Hot-air balloonists use this principle to go up in the air. A burner under the balloon causes the balloon to fill with hot air which then lifts the balloon.

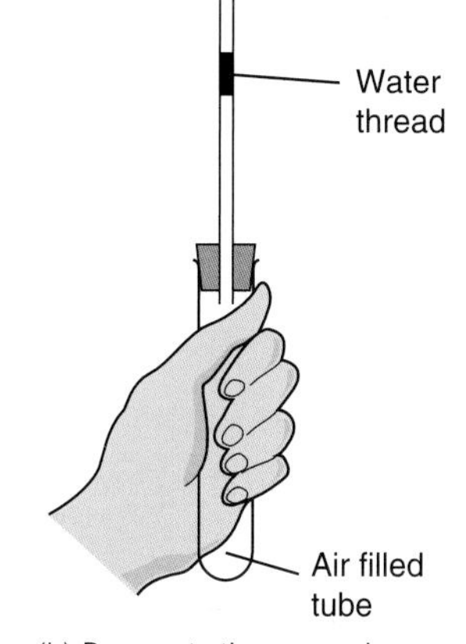

(b) Demonstrating expansion

Liquid-in-glass thermometers make use of the thermal expansion of mercury or alcohol. Water is a very unusual liquid because it contracts when heated from 0 to 4°C; above 4°C, it expands like most other liquids. Water pipes can split if they freeze up in winter. This is due to the water trying to expand as it cools from 4°C to freezing point. If the ends of the pipe are already frozen up, tremendous pressure can build up in the middle as the water freezes and this may split the pipe.

(a) Filling a balloon with hot air

Figure 5.2A ▲ Expansion of a gas

It's a fact!

In freezing weather, ponds freeze from the top. This is because water at the surface gets less dense as it gets colder so it stays at the surface. Beneath the surface, the water is warmer, keeping the fish and the pond life alive.

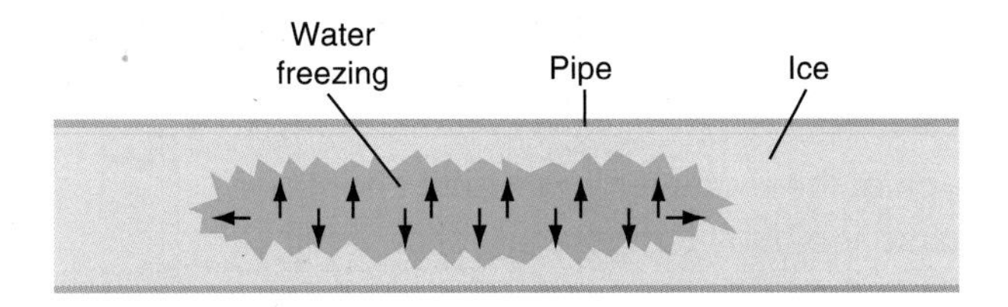

Figure 5.2B Water freezing

Expansion gaps are necessary in buildings to allow for thermal expansion. Outdoor temperatures can change by as much as 50 °C from winter to summer. Building materials expand due to rise of temperature. What do you think would happen if there were no expansion gaps? Motorway bridges are made in concrete sections with expansion gaps between the sections. The expansion gaps are filled with soft material like rubber to stop chunks of rock falling in. Railway tracks too are designed to allow for thermal expansion. Gaps between rails are essential or else the tracks would buckle in hot weather.

Expansion can be very useful. For example, train wheels are fitted with steel tyres by heating the tyre. This makes it expand slightly so it can then be fitted on to the wheel. Then, as the tyre cools, it contracts to make a very tight fit on the wheel.

Figure 5.2C Expansion gaps in a motorway bridge

The linear expansivity of a solid is the expansion of a 1 m length of the solid when the temperature rises by 1 °C.

The value of linear expansivity for a solid is used to work out the expansion of a bar of the solid when its temperature rises. To do this the following equation is used:

$$\underset{\text{(in m)}}{\text{expansion}} = \underset{\text{(in m)}}{\text{initial length}} \times \underset{\text{(per °C)}}{\text{linear expansivity}} \times \underset{\text{(in °C)}}{\text{temperature rise}}$$

Here are the values of linear expansivity for some different solids.

Solid	Steel	Brass	Aluminium	Invar
Linear expansivity (per °C)	1.1×10^{-5}	1.9×10^{-5}	2.6×10^{-5}	0.1×10^{-5}

Invar is an alloy of nickel and steel used to make accurate measuring scales. Why do you think Invar is preferred to steel or aluminium for such a purpose?

The expansion of a solid depends on:
- its length
- its temperature rise
- the type of material.

The bimetallic strip

Different materials expand by different amounts. A brass bar would give more expansion than an iron bar of the same length for the same temperature rise. Brass expands more than iron.

The bimetallic strip consists of a strip of brass bonded to a strip of steel. When the temperature rises, the brass expands more than steel so the bimetallic strip bends. This is used in **thermostats** designed to control heaters or operate fire alarms.

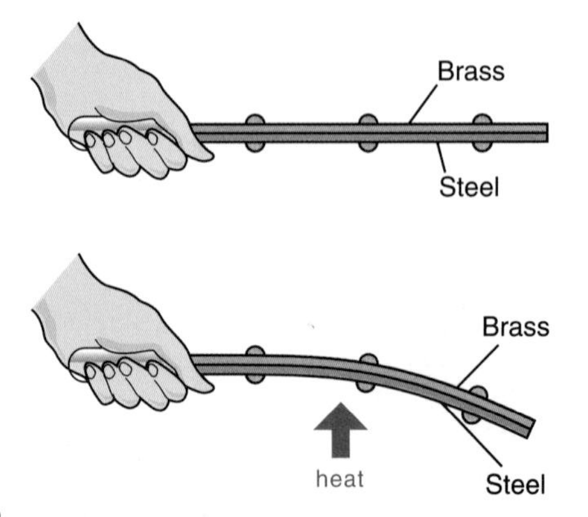

Figure 5.2D The bimetallic strip (a) Cold (b) Hot

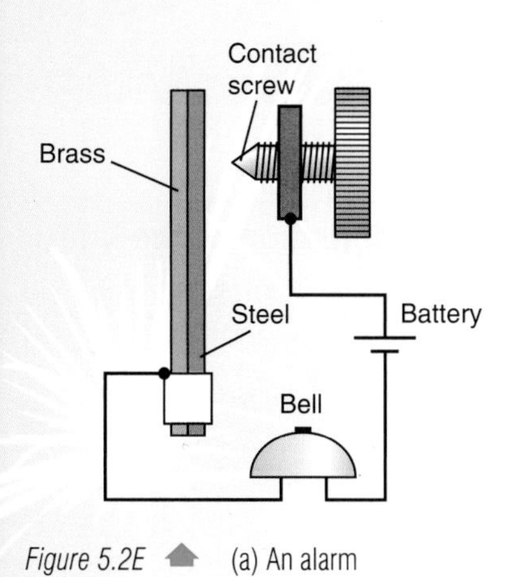

Figure 5.2E ▲ (a) An alarm

- In the alarm circuit shown in Figure 5.2E(a), the strip bends towards the contact screw when the temperature increases. The strip bends until it touches the contact screw which sets the alarm off.
- In a heater control, the strip bends away from the contact when it gets too hot. So the heater is switched off when the temperature reaches a certain point.
- **A radiator thermostat** makes use of the expansion of oil when it becomes warm.

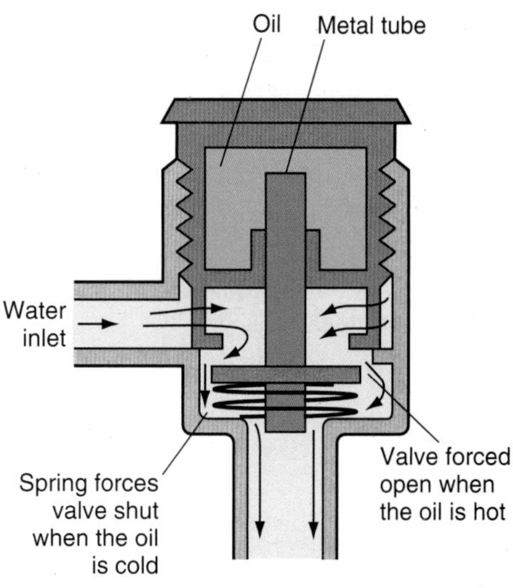

Figure 5.2E ▲ (b) A radiator thermostat

Figure 5.2E (b) shows how it works. In a car radiator, the valve stays closed until the engine has warmed up the thermostat. The metal tube is forced out of the oil chamber as the oil becomes hotter. The movement of the tube opens the valve, allowing the hot water to flow to the radiator.

Expansion of gases

Figure 5.2F shows how to measure the volume of a gas at different temperatures. As the water bath is warmed, the gas trapped in the glass tube becomes hotter and expands, pushing the liquid thread up the tube. The length of the gas column increases. This is a measure of the gas volume.

Suppose the gas volume is measured at ice point and at steam point. Typical measurements are plotted on the graph in Figure 5.2F. The straight line drawn between the two points cuts the temperature axis at –273 °C no matter which gas or how much gas is in the tube. This is the **absolute zero of temperature**, the lowest possible temperature.

The absolute scale of temperature is based on absolute zero. The scale is used by scientists because it is a measure of the average kinetic energy of a molecule. The unit of temperature on this scale, the kelvin (denoted by K) is named after Lord Kelvin who first developed the idea of the absolute temperature scale. The scale is defined so that temperature values in °C can be converted to kelvins by adding 273.

> Temperature in kelvins = Temperature in °C + 273

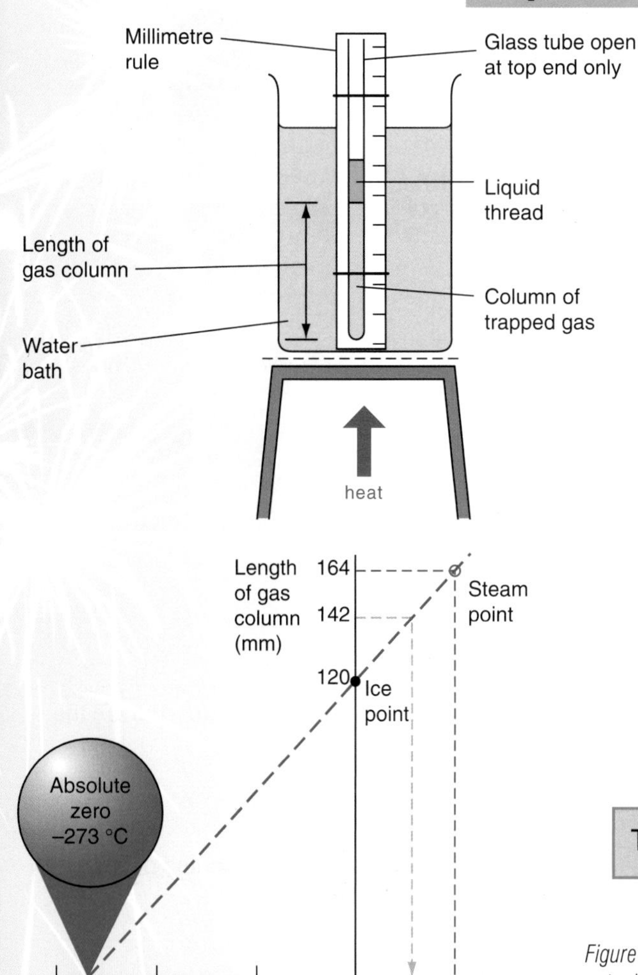

Figure 5.2F ◀ Measuring the expansion of a gas (a) Using a water bath (b) Typical results

The gas laws

Charles' law

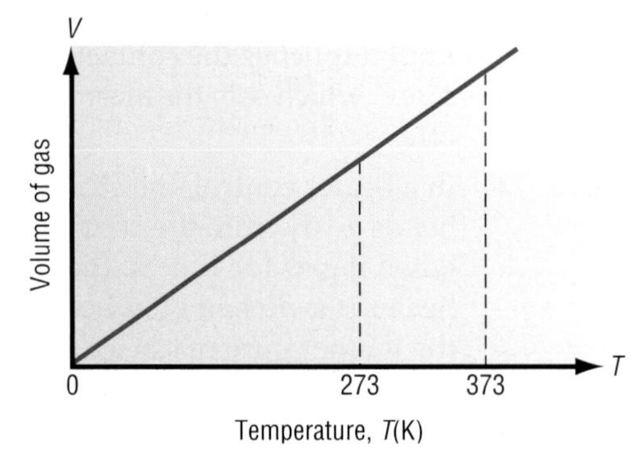

Figure 5.2G Charles' law

Equation of the line

$V = \text{constant} \times T$

or

$\frac{V}{T} = \text{constant}$

Absolute zero is the lowest possible temperature.

The tube of gas in Figure 5.2F may be used as a thermometer. By measuring the length of the gas column, the temperature can be read off the graph. For example, if the length is midway between ice point and steam point, the temperature must be 323 K (= 50 °C).

In more general terms, the straight line drawn through absolute zero may be represented by an equation.

Volume = constant × Absolute temperature
or $V = \text{constant} \times T$
(provided the pressure and mass of gas are constant)

This equation, known as **Charles' law**, can be used to work out the absolute temperature for any given volume if the constant in the equation is known.

Worked example The volume of a fixed mass of gas at constant pressure in a tube at ice point (273 K) is 80 mm^3. Work out the absolute temperature when the volume is 87 mm^3.

Solution

Since $V = \text{constant} \times T$

then $\frac{V}{T} = \text{constant}$

for any values of V and T provided the mass of gas and the pressure are the same.

Remember to work in kelvins.

so $\frac{V}{T} = \frac{V_0}{T_0}$

where $V_0 = 80$ mm^3, $T_0 = 273$ K, $V = 87$ mm^3 and T is to be worked out

hence $\frac{87}{T} = \frac{80}{273}$

therefore $T = \frac{87 \times 273}{80} = 297$ K

Boyle's law

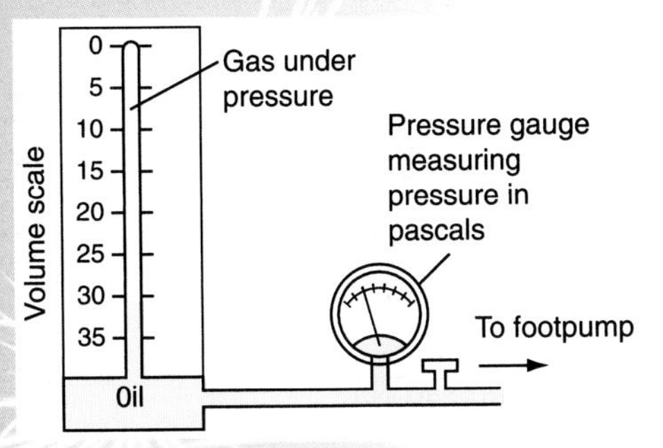

Figure 5.2H ▲ Testing Boyle's law

Pressure (Pa)	100	120	150	180
Volume (cm³)	36	30	24	20
Pressure×Volume (Pa cm³)	3600	3600	3600	3600

The volume of a gas depends on its pressure as well as its temperature. Figure 5.2H shows how to measure the volume of gas at different pressures while the temperature is maintained constant. The measurements show that the product (pressure × volume) is constant. This is known as Boyle's law, named after Robert Boyle who first discovered the law in 1662.

> Pressure × Volume = constant
> or pV = constant
> (provided the temperature and mass of gas are constant)

The pressure law

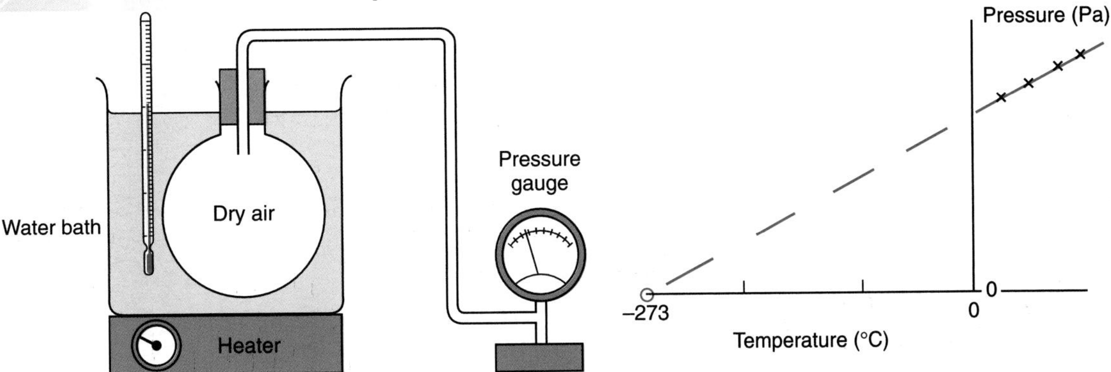

Figure 5.2I ▲ Measuring pressure against temperature

The pressure of a fixed volume of gas increases as the temperature increases. Figure 5.2I shows how to measure gas pressure at different temperatures. The results show that the gas pressure is proportional to the absolute temperature of the gas. This may be written as an equation.

> Pressure = constant × Absolute temperature
> or $p = \text{constant} \times T$
> (provided the volume and mass remain constant)

The combined gas law

The three separate gas laws may be written in a single equation which can be used for any change of pressure, volume or temperature.

> $$\frac{\text{Pressure} \times \text{Volume}}{\text{Absolute temperature}} = \text{constant}$$
> or $$\frac{pV}{T} = \text{constant}$$
> (provided the mass of gas is constant)

You can derive Boyle's Law, Charles' Law and the pressure law from $\frac{pV}{T}$ = constant.

SUMMARY

Gases, solids and most liquids expand, if allowed, when heated. The expansion of a solid is greater the higher the temperature rise. Different solids expand by different amounts. Bimetallic strips are used in thermostats to switch appliances on or off. The combined gas law may be used to work out changes of pressure, volume and temperature of a gas.

Worked example In a chemical reaction, 5.00 cm^3 of gas are collected in a test tube at 110 kPa pressure and a temperature of 290 K. Work out the volume of this gas at a pressure of 100 kPa and a temperature of 273 K.

Solution The combined gas law may be written as

$$\frac{p_1V_1}{T_1} = \frac{p_2V_2}{T_2}$$

where $p_1 = 110\,\text{kPa}$ and $p_2 = 100\,\text{kPa}$

$V_1 = 5.00\,\text{cm}^3$ $V_2 =$ to be worked out

$T_1 = 290\,\text{K}$ $T_2 = 273\,\text{K}$

Substituting these values into the equation above gives

$$\frac{110 \times 5.00}{290} = \frac{100 \times V_2}{273}$$

Rearranging $$V_2 = \frac{110 \times 5.00 \times 273}{290 \times 100} = 5.18\,\text{cm}^3$$

CHECKPOINT

1 Explain;
(a) why expansion gaps are necessary between concrete sections in buildings,
(b) why these gaps are usually filled with soft material.

2 This figure shows a gas thermostat. If the oven gets too hot, the valve closes a little to lessen the flow of gas. Explain why this happens.

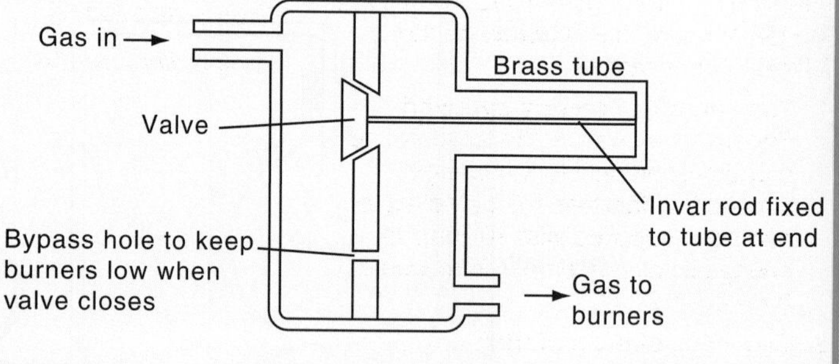

3 In an experiment to measure thermal expansion of iron, a 50 cm iron bar was heated from 20°C to 320°C and its length increased by 1.5 mm.

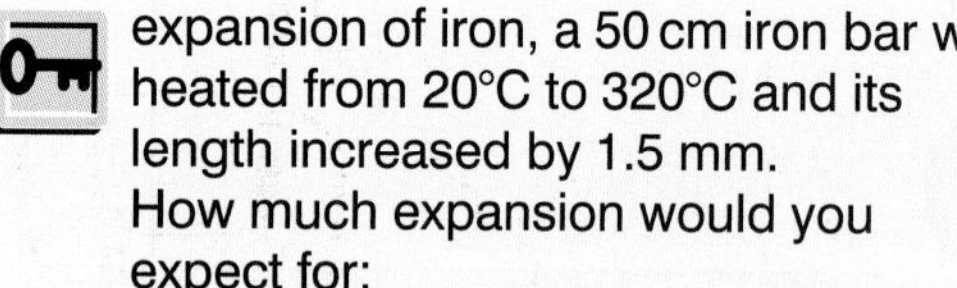

How much expansion would you expect for:
(a) a 50 m rail heated from 20°C to 320°C
(b) a 50 m rail heated from 20°C to 40°C?
(c) Work out the linear expansivity of iron.

4 In a chemistry experiment, 25 cm^3 of a gas is collected in a syringe at a pressure of 120 kPa.
(a) What is this volume in m^3?
(b) What would the volume of this amount of gas be if its pressure was changed to 100 kPa without change in temperature?

5 A cycle pump contains 100 cm^3 of air at atmospheric pressure of 100 kPa. The outlet of the pump is then blocked.
(a) The air pressure in the pump is then raised to 125 kPa by reducing the volume of air in the pump without change in temperature. Calculate the volume of air in the pump when its pressure is 125 kPa.
(b) Calculate the pressure if the volume is reduced further to 75 cm^3.

6 Work out the unknown quantity in each of the following changes involving a fixed mass of gas.

	Initial conditions			*Final conditions*		
	Pressure (kPa)	*Volume* (m^3)	*Temperature* (K)	*Pressure* (kPa)	*Volume* (m^3)	*Temperature* (K)
(a)	100	3.0	300	100	?	400
(b)	100	3.0	300	150	2.0	?
(c)	100	3.0	300	?	1.5	400
(d)	100	3.0	?	150	3.0	400
(e)	100	?	300	200	2.0	400

5.3 Heating things up

FIRST THOUGHTS

How easily can different materials be heated or cooled? Investigations outlined in this section show you how to test different materials to answer this question.

Have you ever noticed how hot a piece of metal can get in sunlight? A stone of similar size gets a little warmer in the sun, but the metal can get quite hot. The metal panels of a parked car can get very hot in the sun. Some things heat up more easily than others. Hammering a piece of lead repeatedly makes it hot; do the same to a piece of steel of the same mass and it will warm up much less.

Investigating heating

To make something hotter, it must be given energy. Suppose two different objects of the same material are heated by giving them the same amount of energy. *Will they have the same temperature rise?*

For example, suppose an electric heater connected to a joulemeter is used to heat water in a plastic cup. The mass of water could be measured and the rise of temperature could be measured. The joulemeter could be used to measure the energy supplied. The test could then be repeated using twice as much water with the same amount of energy being supplied. Some results from the test are shown in the table.

Mass of water (kg)	0.10	0.20
Energy supplied (J)	5000	5000
Temperature rise (°C)	12	6

Investigations show that when energy is supplied to a given material to warm it up, the temperature rise depends on the following three factors:

1. **The amount of energy supplied**
   For any given object, the temperature rise is proportional to the energy supplied. In this test, if 0.1 kg of water had been supplied with 10 000 J of energy, its rise of temperature would have been 24 °C
2. **The mass of the material**
   For a given amount of energy supplied, the temperature rise is inversely proportional to the mass. In other words, doubling the mass halves the temperature rise, etc.
3. **The type of material being heated**
   Some materials heat up more easily than others.

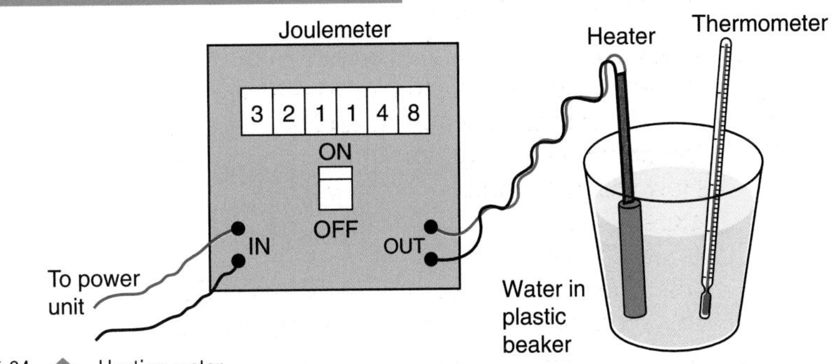

Figure 5.3A Heating water

The results show that the larger the mass, the smaller the temperature rise for the same amount of energy supplied. *Can you work out what temperature rise to expect for a mass of water of 0.30 kg supplied with 5000 J of energy?*

Specific heat capacity

The **specific heat capacity**, *c*, of a material is defined as the energy needed to raise the temperature of 1 kg of material by 1 °C. The unit of specific heat capacity is J/kg/°C.

Consider the results from the test using water.

- 5000 J of energy raises the temperature of 0.10 kg of water by 12 °C.
- For ten times the mass, ten times as much energy is needed. 50 000 J of energy raises the temperature of 1.0 kg of water by 12 °C.
- For a 1 °C rise, only $\frac{1}{12}$ as much energy is needed

$$\frac{50\,000\text{ J}}{12} \simeq 4200\text{ J}$$

4200 J of energy raises the temperature of 1.0 kg of water by 1 °C. Hence the specific heat capacity for water is approximately 4200 J/kg/°C.

The steps in this calculation can be summarised by the following equation

$$\text{Specific heat capacity} = \frac{\text{Energy supplied}}{\text{Mass} \times \text{Temperature rise}}$$

To work out the energy needed to raise the temperature of a given mass, the above equation can be rearranged to give

Energy supplied = Mass × Specific heat capacity × Temperature rise
(in joules) (in kg) (in J/kg/°C) (in °C)

Table 5.1 Some specific heat capacities

Material	Water	Oil	Aluminium	Iron	Copper	Lead	Concrete
Specific heat capacity (J/kg/°C)	4200	2100	900	390	490	130	850

SUMMARY

Specific heat capacity, c, is the energy to raise the temperature of 1 kg of material by 1 °C. Some materials heat up more easily than others. This is because materials that heat up easily have low specific heat capacities.

The **heat capacity, C,** of an object is the energy needed to raise its temperature by 1 °C. The unit of C is J/°C.

If the object is made of a single material.

Heat capacity = Mass × Specific heat capacity

For example, an aluminium kettle of mass 2.00kg, the specific heat capacity of aluminium is 900 J/kg/°C. Therefore, the heat capacity of the kettle is 1800 J/°C (= 2.00 = 900).

If the object is made of several materials, its heat capacity can be worked out by adding together the heat capacities of the individual materials. For example, the heat capacity of a 2.00 kg aluminium kettle containing 1 kg of water is 1800 + 4200 = 6000 J/°C. This means that 6000 J of energy must be supplied for each °C temperature rise.

Caloric and the count

About 200 years ago, there were two conflicting theories about the nature of heat. One school of thought considered heat as a form of motion of atoms, essentially as a form of energy although ideas about energy were not well-understood then. The fact that objects became hot through friction when rubbed was the basis of the theory.

Figure 5.3B Sir Humphry Davy

Figure 5.3C Count Rumford

Figure 5.3D ▲ James Prescott Joule

The other group, the Calorists, supposed that heat was a fluid which they called 'caloric'. When a hot object was placed in contact with a cold object, caloric transferred from the hot object to the cold object to make the hot object cooler and the cold object warmer. Experiments involving transfer of heat between hot and cold bodies supported the caloric theory. They explained heating due to friction by supposing that tiny particles created by the rubbing action released caloric. The **calorie** as a unit of heat (equal to the energy needed to heat 1 g of water by 1 °C) dates from this time.

The caloric theory was the accepted view about the nature of heat until it was challenged by Count Rumford in Munich in 1798. He was born Benjamin Thompson in North America and had to flee to Europe during the American Revolution because he remained loyal to Britain. He was a restless, energetic person who developed great enthusiasm and interest in science and founded the Royal Institution in Britain. He settled in Bavaria where he was made Minister of War and was given the title Count Rumford in recognition of his work there.

Rumford measured the temperature rise of a 50 kg brass cannon when a blunt steel drill was applied to it. After 30 minutes, the temperature had risen by 40°C and just 54 g of metal dust had been produced. The idea that heat from 54 g of metal dust could raise the temperature of a 50 kg brass cannon by 40 °C seemed unlikely to Rumford. More importantly, he realised that the supply of heat was inexhaustible – provided the drill kept turning.

The calorists insisted that the results could be explained if solid metal was much easier to heat than metal dust. The dispute was settled in 1799 by Sir Humphry Davy. He showed that two pieces of ice below freezing point melted when rubbed together. The fact that water at 2 °C was produced could only be explained in terms of work done.
In Manchester in 1840, James Joule took the investigations further. He measured the specific heat capacity of water by different methods, all involving doing work to heat the water. All the methods gave the result as 4.2 J/g/°C. This proved conclusively that heat is a form of energy. The scientific unit of energy is named after Joule in recognition of his work.

CHECKPOINT

1 Use the information in Table 5.1 to explain the following statements.
(a) A mass of lead heats up more easily than an equal mass of aluminium.
(b) A mass of water heats up less easily than an equal mass of oil.
(c) A metal seat gets much hotter than a bucket of water in the sun.

2 How much energy is needed for each of the following?
(a) To raise the temperature of 2.0 kg of aluminium by 30 °C.
(b) To raise the temperature of 0.05 m^3 of water in a water tank from 20 °C to 60 °C. (Assume the density of water is 1000 kg/m^3).
(c) To heat a 20 kg concrete block in a storage heater from 10 °C to 40 °C.
(d) To heat 1.50 kg of water in a copper kettle of mass 0.050 kg from 20 °C to boiling point.

3 An electric shower heater is capable of heating water from 20 °C to 45 °C when water passes through the shower unit at a rate of 0.012 kg per second.
(a) Calculate the energy required to raise the temperature of 0.012 kg of water from 20 to 45 °C.
(b) The shower heater is rated at 3.0 kW. Estimate its efficiency.

4 (a) A 3.0 kW electric kettle is capable of heating 2.0 kg of water from 20 °C to 100 °C in 300 seconds. How much electrical energy is supplied and how much of the energy is used to heat the water? Explain why more energy is supplied than is used to heat the water.
(b) Water in a plastic jug was put in a fridge where it cooled from 15 °C to 5 °C in 600 seconds. If the mass of water in the jug was 0.8 kg, how much energy was removed from it to reduce its temperature? How much energy was removed each second?

5.4 Latent heat

FIRST THOUGHTS

In a busy kitchen, water is boiled in a kettle and frozen in a freezer. What energy changes are involved in these processes? What is the cost of each of these processes?

See www.keyscience.co.uk for more about latent heat.

'Latent' means hidden. No change in temperature occurs when a substance changes state – even though heat transfer occurs.

In the kitchen

Heat some water in a saucepan. As the water warms up, water vapour comes off its surface. This is because water in the liquid state is changing into water vapour. This is the process of **vaporisation**. As the water becomes hotter, it reaches a temperature at which vaporisation takes place throughout the water. Bubbles of vapour rise to the surface – the water boils.

Suppose you put an egg into some water which is being heated. The temperature of the water rises until it reaches boiling point. No further increase in temperature occurs while the water boils, even though heat is still being supplied. On 'full heat', the water boils vigorously but doesn't cook the egg any faster. Turn the 'heat' down on the cooker so the water is just 'on the boil'. This stops the kitchen steaming up and saves fuel.

The energy supplied to vaporise water is called **latent heat**. It is needed to break the attractive forces between the molecules.

Condensation is the reverse of vaporisation. For example, when a lot of steam is produced in a kitchen, water may be seen running down window panes or wall tiles. This happens because the steam condenses on the cold surfaces. Latent heat is given out in condensation.

Investigating boiling

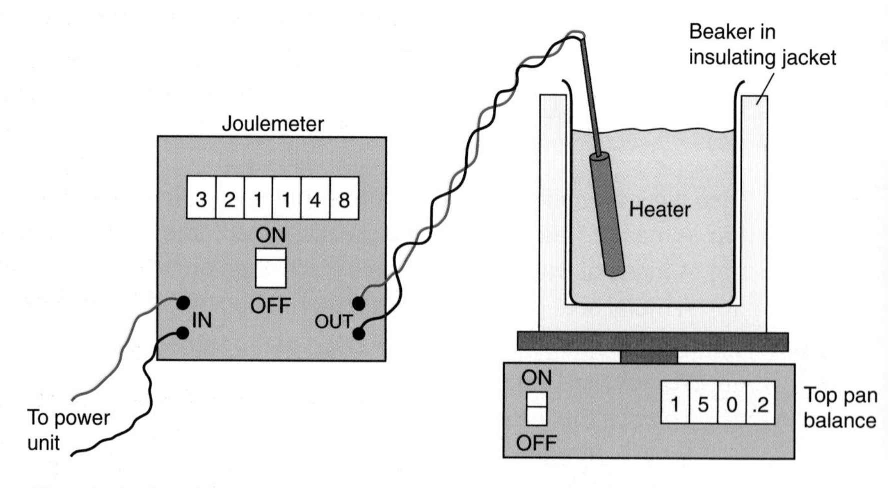

Figure 5.4A ▲ Measuring the specific latent heat of vaporisation of water

How much energy is needed to vaporise water? Figure 5.4A shows how this can be measured. Here are some measurements from this apparatus;

Joulemeter reading (J)	0	4600	9200	13 800	18 400
Top pan balance reading (kg)	0.152	0.150	0.148	0.146	0.144

As the water boiled away, the total mass on the balance went down. *Can you work out how much energy was needed to boil away 0.001 kg of water?* The results in the table above should give 2300 J for this calculation. To vaporise 1.00 kg of water, 2 300 000 J (= 2300 J × 1000) would be needed.

The **specific latent heat of vaporisation, *l***, of a pure liquid is the energy needed to vaporise 1 kg of pure liquid without change of temperature. The unit of l is J/kg (or MJ/kg).

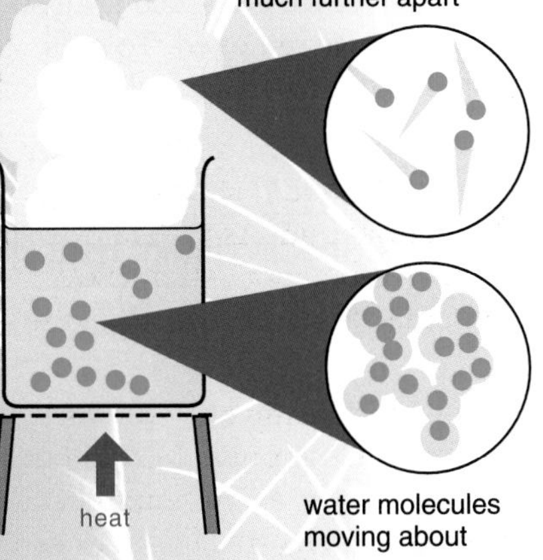
Molecules in the vapour state move about much faster than molecules in the liquid state and are much further apart
heat
water molecules moving about

Figure 5.4B From water to steam

Therefore, the energy needed to vaporise a certain mass m of pure liquid is equal to mass $m \times$ specific latent heat of vaporisation l.

Worked example How much energy is needed to change 0.50 kg of water at 20 °C to steam at 100 °C? The specific heat capacity of water is 4200 J/kg/°C and the specific latent heat of steam is 2 300 000 J/kg.

Solution Work out the energy needed to raise the temperature of the water to 100 °C and then the energy needed to boil the water away.

To raise the temperature of 0.50 kg of water from 20 °C to 100 °C

Energy needed = Mass × Specific heat capacity × Temperature rise
= 0.50 kg × 4200 J/kg/°C × 80 °C = 168 000 J = 0.168 MJ

To boil away 0.50 kg of water

Energy needed = Mass × Specific latent heat
= 0.50 kg × 2 300 000 J/kg = 1 150 000 J = 1.15 MJ

The total energy needed = 0.168 + 1.15 = 1.32 MJ

Using a freezer

EXTENSION FILE
ACTIVITY

Ice cubes from a freezer melt quickly in a kitchen. During melting, there is no change in the temperature of the substance. The energy needed to melt a substance is called **latent heat of melting**. It is used to enable the molecules of the substance to break free from each other.

Put a plastic beaker of water in a freezer and the reverse process happens. The water cools to freezing point then it turns to ice. The freezer takes energy from the water.

Food in a freezer is preserved because micro-organisms need water for growth. The water content of the food is frozen and micro-organisms can not multiply. However, freezing does not kill the micro-organisms and they revive when the food thaws out. Food from a freezer must be cooked thoroughly to kill any micro-organisms in it before being eaten.

Specific latent heat

When any liquid solidifies, latent heat is given out by the substance as the molecules bond more closely together. For example, a test tube of wax gives out heat when the liquid solidifies. The word latent means 'hidden'; when a liquid turns into a solid, energy 'hidden' in the bonds between the molecules is released. When food is put into a freezer, its water content turns into ice and latent heat is given out. This is taken away by the refrigeration unit of the freezer.

The energy removed from food put in a freezer depends on the mass of food and the type of food. Doubling the mass means that twice as much energy must be removed.

The **specific latent heat of fusion, l**, of a substance is the energy needed to melt 1 kg of solid (or given out when 1 kg of solid freezes) without change of temperature. The unit of l is J/kg.

£1.39
SAVE 60

Figure 5.4C A supermarket freezer. Freeze-drying to remove water is a very successful method of preserving food. Meat, fish, and vegetables are freeze-dried and vacuum-packed to give a long shelf-life

SUMMARY

Melting, vaporisation and sublimation are processes that take in energy. Condensation and solidification are processes that give out energy. The energy involved in these changes is called latent heat.

To work out the energy removed when freezing a substance, its mass must be known and its specific latent heat. For example, to freeze 5.00 kg of water at 0 °C, the energy to be taken away is $5.00 \times 340\,000$ J (= 1.7 MJ). This is because the specific latent heat of fusion of ice is 340 000 J/kg so to freeze each kg of ice, 340 000 J of energy must be removed. To melt 5.00 kg of water at 0 °C, $5.00 = 340\,000$ J of energy must be supplied.

Energy needed (or given out) (in joules)	=	Mass (in kg)	×	Specific latent heat (in J/kg)

CHECKPOINT

1 Describe the energy changes in each of the following situations.
- (a) Clearing a misted car window using an electric demister heater.
- (b) Freezing home-baked fruit pies in a deep freeze.
- (c) Distilling ethanol (alcohol).

2 The figure opposite shows an experiment to measure the specific latent heat of ice.

With the heater switched off, 0.024 kg of water was collected in the beaker in 5.00 minutes. The beaker was then emptied and replaced as shown. With the heater switched on for 5.00 minutes, the joulemeter reading increased from 3500 J to 18 500 J. During this time, 0.068 kg of water was collected in the beaker.

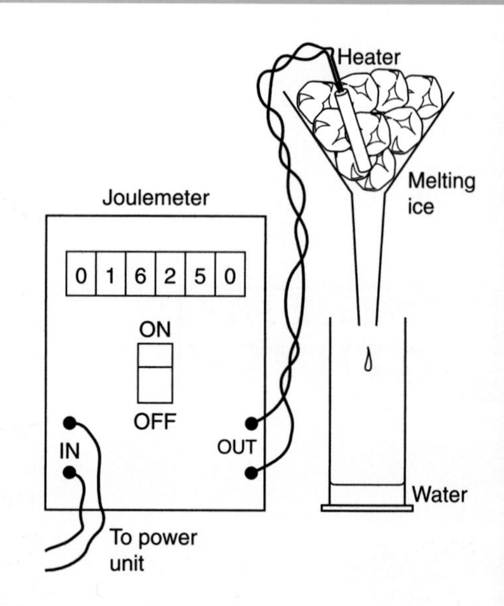

- (a) Some of the ice in the funnel melted when the heater was off. Why?
- (b) How much ice was melted in 5 minutes due to the heater being on?
- (c) How much energy was supplied by the heater in this time?
- (d) Work out the specific latent heat of ice from the above results.

3 Carlos has just completed a half-marathon and is hot, tired and sweaty. He wants to use an anti-perspirant to stop sweating but Louise tells him it's not such a good idea while he is still hot. Why?

4 Design an experiment to test whether a cup of hot water placed in a freezer freezes faster than cold water. If possible, try out your ideas. You may be surprised by the result.

5 A 3.0 kW electric kettle is fitted with a thermal cut-out designed to switch it off as soon as the water boils. Unfortunately, the cut-out does not operate correctly and allows the water to boil for 30 seconds longer than it is supposed to.
- (a) How much electrical energy is used in this time?
- (b) The specific latent heat of vaporisation of water is 2.3 MJ/kg. Use this value and your answer to (a) to work out the mass of water boiled away in this time.
- (c) If you used this kettle in your kitchen, what would happen to the steam it produced?

5.5 Heat transfer

FIRST THOUGHTS

Different methods of heat transfer are compared here with a brief look at some applications.

Look around your home and make a list of the types of heaters you find. Include central heating radiators in your list. Heaters keep your home warm by using energy in a particular form to supply heat.

Heat convectors

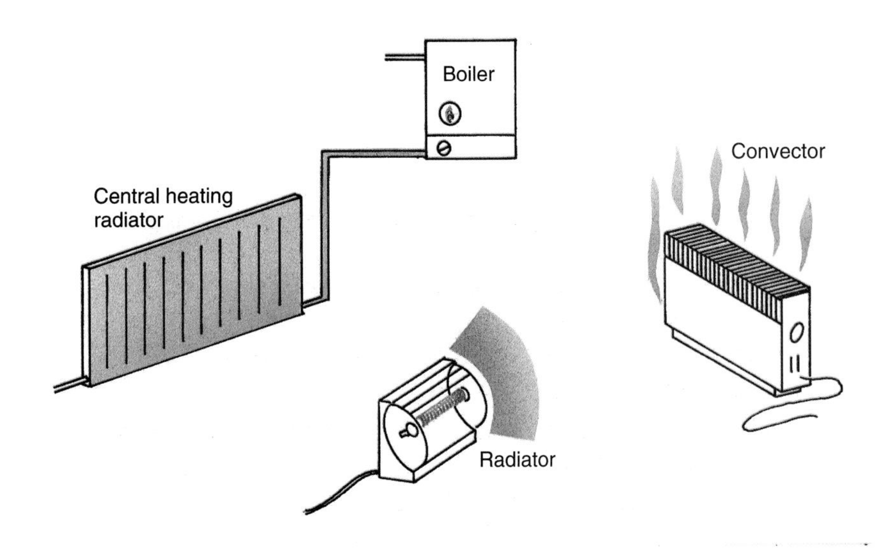

Figure 5.5A Heaters at home

Convectors are designed to heat the air in a room. They rely on the fact that hot air, being less dense than cold air, rises. In a convector, a heating element warms the air which rises. The warm air is replaced by cold air which in turn is heated. In this way the warm air circulates and the room heats up.

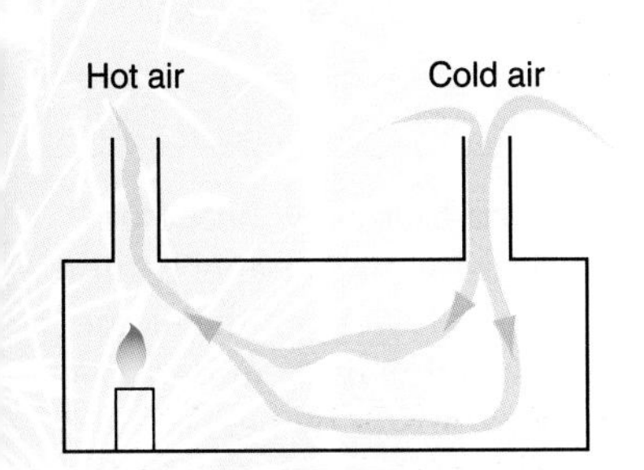

Figure 5.5B Convection

A simple demonstration of convection is shown in Figure 5.5B. The hot gases from the burning candle go straight up the chimney above the candle. Cold air is drawn down the other chimney to replace the air leaving the room.

Convection happens whenever a liquid or a gas is heated. Here are some examples.

Hot air balloons

The hot air from the burner rises and fills the balloon. Because the air in the balloon is warm, it is less dense than the surrounding air so the balloon rises. To keep the balloon floating, the burner must be operated at intervals. Otherwise the trapped air cools and the balloon descends. What would happen if the burner is left on for too long?

Figure 5.5C A hot air balloon

Thermals

Warm air rises, creating air currents known as thermals. These currents can keep birds and gliders high above the ground for long periods.

Domestic water heating

Water from a hot water tank is usually drawn off near the top of the tank. This is because the hot water in the tank rises to the top. If the feed pipe for the hot water tap was connected near the bottom of the tank, it would draw cold water.

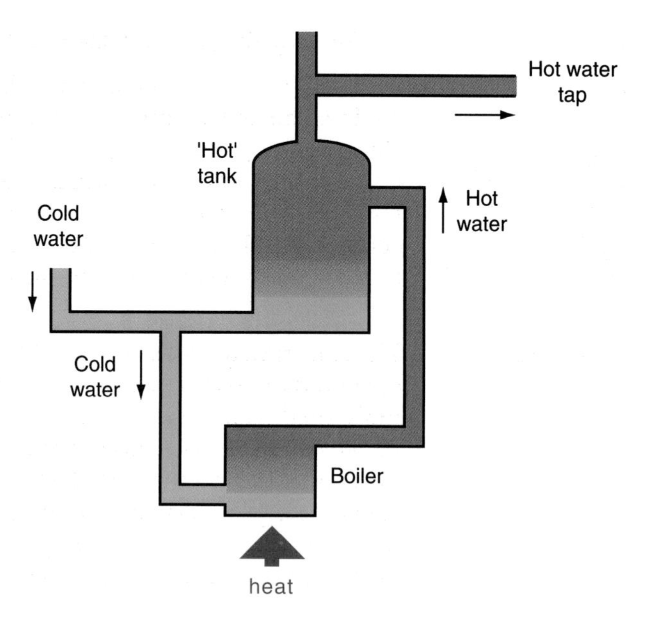

Figure 5.5D ▲ Convection

Heat radiators

Radiators like the one in Figure 5.5A are designed to 'throw' heat from the heating element into the room. The curved shiny metal panel behind the heating element is important because it reflects heat from the element into the room. Heat **radiation** is carried by **infra-red** waves emitted from the heating element. No substance is needed for transfer by heat radiation because infra-red waves can travel through a vacuum. Heat from the Sun is carried by this type of radiation.

SUMMARY

Heat transfer can be by conduction, convection or radiation.

- Radiation: no substance is needed to transfer heat.
- Convection: when a liquid or a gas is heated, the hot fluid rises, taking heat with it.
- Conduction: heat passes through a solid, liquid or gas.

Heat conductors

Central heating radiators transfer heat to the room by convection as well as by radiation. Each hot radiator panel warms the nearby layers of air which carry heat to other parts of the room by convection of air currents. *How does the panel get hot?* Water from the central heating boiler is pumped through the panel via pipes. The water is cooler when it leaves the panel than when it enters. This is because it gives up energy to heat the room. This energy passes through the metal walls of the radiator from the inside to the outside of the radiator. The process where heat passes through a solid is called **conduction**. Radiator panels made from plastic would be little use because plastic is a much poorer conductor of heat than metal is. Heat can pass through liquids and gases by conduction as well as by convection.

5.6 More about conduction

FIRST THOUGHTS

Thermal conductors and insulators are compared in this section and their uses outlined.

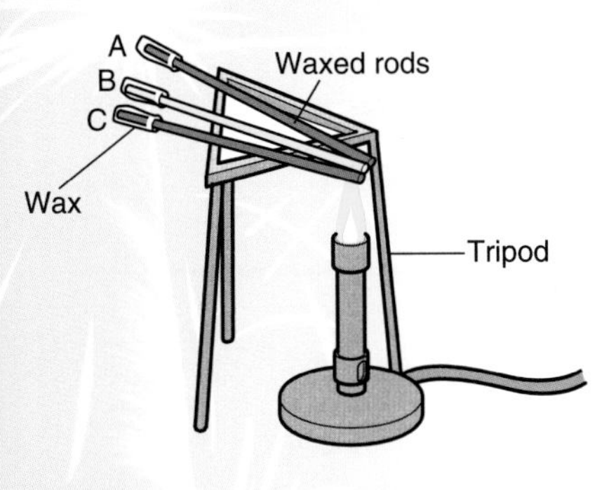

Figure 5.6A ▲ Comparing conductors

Conductors

Why do saucepans have plastic or wooden handles? If the handle was made from metal, what would happen when you took hold of a hot saucepan? Why must the handle be made from material that is a poor conductor of heat?

Testing thermal conductors

To test how well different materials conduct heat, rods coated with wax could be heated as shown in Figure 5.6A. The rods need to be the same size. As heat passes along each rod, the wax along the rod melts. The best conductor is the rod which 'de-waxes' fastest. Look carefully at Figure 5.6A to decide which material is the best and which is the poorest conductor of heat. Metals are much better **thermal conductors** (conductors of heat) than non-metals. Copper is one of the best thermal conductors.

The reason why metals are good thermal conductors is because they contain **free electrons**. Most of the electrons in a metal are attached to individual atoms. These electrons cannot move about inside the metal. However, some electrons can move about freely inside the metal. If a metal is heated in one part, the free electrons in this part gain kinetic energy. They spread out randomly and collide with other free electrons, transferring their kinetic energy in the process. This is the main mechanism for heat conduction in a metal.

The heat exchanger

Thermal energy can be transferred from a gas or liquid to another gas or liquid using a heat exchanger. One of the fluids is pumped through metal pipes which pass through the other fluid. Heat transfer through the pipe wall occurs without the fluids mixing. Figure 9.6A shows a heat exchanger in a nuclear reactor.

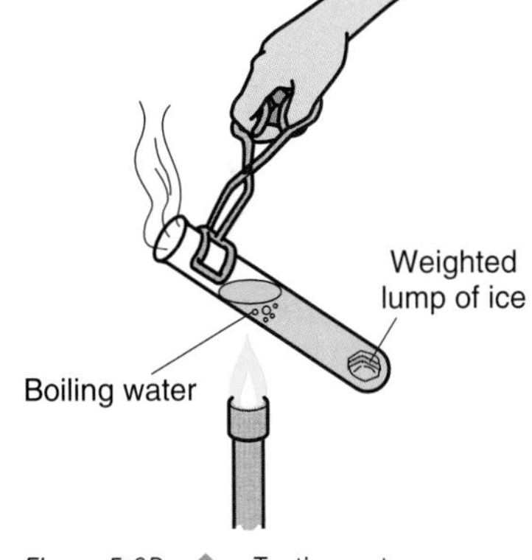

Figure 5.6B ▲ Testing water

How well does water conduct heat? Heat a test-tube of water near the top with a 'weighted' ice cube near the bottom. Even when the water at the top starts boiling, the ice cube does not melt. Water is a poor conductor of heat.

Figure 5.6C ▲ Loft insulation

Insulators

Insulators are used to lag pipes, lofts, hot water tanks and many other objects. Insulators are very poor conductors. Wrapped round hot objects, they act as a barrier, reducing the heat loss. Fibre glass insulation is a popular choice for insulating the lofts of homes. This material is manufactured in thick layers and sold in rolls. Trapped air is an effective insulator too. Fibre glass layers contain lots of tiny pockets of trapped air. Thermal clothing, designed to wear outdoors in winter, contains materials with lots of tiny air pockets.

SUMMARY

Metals are the best conductors of heat. Materials used for thermal insulation are very poor conductors of heat. The heat flow per square metre through a material increases as the temperature difference across the material increases. Making the material thicker reduces the heat flow.

Tests show that the heat flow per second passing through each square metre of a material depends on the following factors.

- The thickness of the material: using more layers cuts down the heat flow.
- The temperature difference across the material: the greater the difference from the hot side to the cool side, the greater the heat flow.
- The type of material chosen: some materials are better insulators than others.

CHECKPOINT

1 Explain why;

(a) an electric iron has a plastic handle and a metal base,
(b) fitting an insulation jacket to a domestic hot water tank saves money,
(c) wrapping a block of ice cream in paper helps to stop it melting.

2 In the ice cube experiment shown in Figure 6.6B, suppose the ice cube was at the top of the tube and the tube was heated at the bottom. Would the ice cube melt more easily this time? Explain your answer.

3 Keith's parents are planning to put insulation down in their loft. His mother works out that the insulation ought to cut the heat loss by 80 J/s for each °C of temperature difference.
(a) How much energy will be 'saved' in 1 day when the temperature difference is 10 °C?
(b) Keith's father works out that their heating costs 1.5p per million joules. How much money will be saved in (a)?
(c) The insulation material costs them £80. How many days like (a) will it take to pay for it?

4 In an experiment to test the effectiveness of insulation material, hot water was poured into a test-tube and the water temperature was measured at intervals while it cooled. The test was then repeated with the same volume of water, with the tube wrapped in the insulating material. The results are shown opposite.

	Temperature (°C)	
Time (min)	*Unwrapped tube*	*Wrapped tube*
0	70	70
1	64	67
2	60	64.5
3	57	62
4	55	60
5	53.5	58.5

(a) Plot graphs of temperature (on the vertical axis) against time for both the wrapped and the unwrapped tube on the same axis.
(b) How long did each tube take to cool from 70 °C to 65 °C?
(c) How effective was the insulating material?

5.7 Radiation

FIRST THOUGHTS

The nature and properties of radiant heat are outlined here through investigation and applications.

In winter at night, a clear starry sky usually means it's going to be very cold by the next morning. This happens because of heat radiation going from the ground out into space. During the daytime, heat radiation from the Sun reaches the Earth and warms the ground. At night, when the sky is clear, the ground temperature falls as the ground radiates energy into space.

Satellites fitted with special cameras sensitive to heat radiation give very interesting pictures. Towns and cities show up because they emit much more radiation than rural areas.

All objects emit heat radiation. Heating an object up makes it radiate more energy. Cooling it makes it radiate less. Have you ever walked into a refrigerated cold storage room? Don't get locked in because you radiate your energy away very quickly and you would soon freeze! Go into a hot room and you soon warm up. This is because the hot walls radiate energy which you absorb. So you get warm.

To detect radiation, you could use a thermometer with a blackened bulb. The bulb absorbs heat radiation much more effectively if it is blackened. Some other types of detectors are shown in Figure 5.7B.

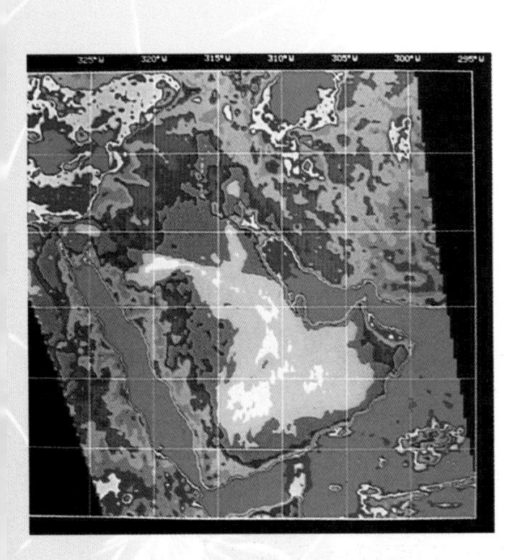

Figure 5.7A An infra-red satellite photograph of the Arabian peninsula

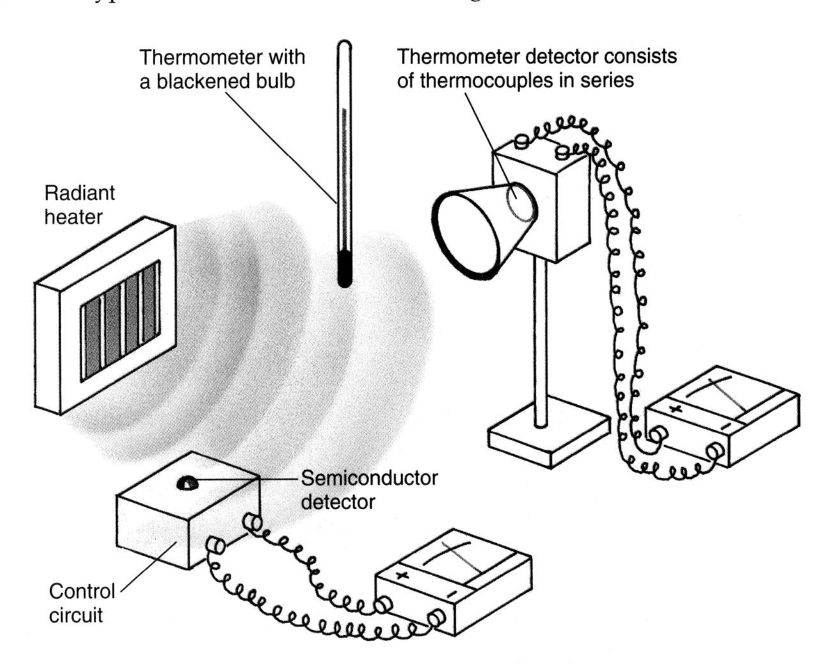

Figure 5.7B Detectors of heat radiation

Heat radiation consists of **electromagnetic waves**. There is a wide range of these waves known as the electromagnetic spectrum. Heat radiation is also known as **infra-red** radiation because it fits between red light and microwaves in the spectrum. Heat radiation travels at the same speed as light because both are electromagnetic waves. Like light, heat radiation can be reflected. However, infra-red waves have a longer wavelength than light and their absorption properties differ. For example, light passes through glass but heat radiation does not.

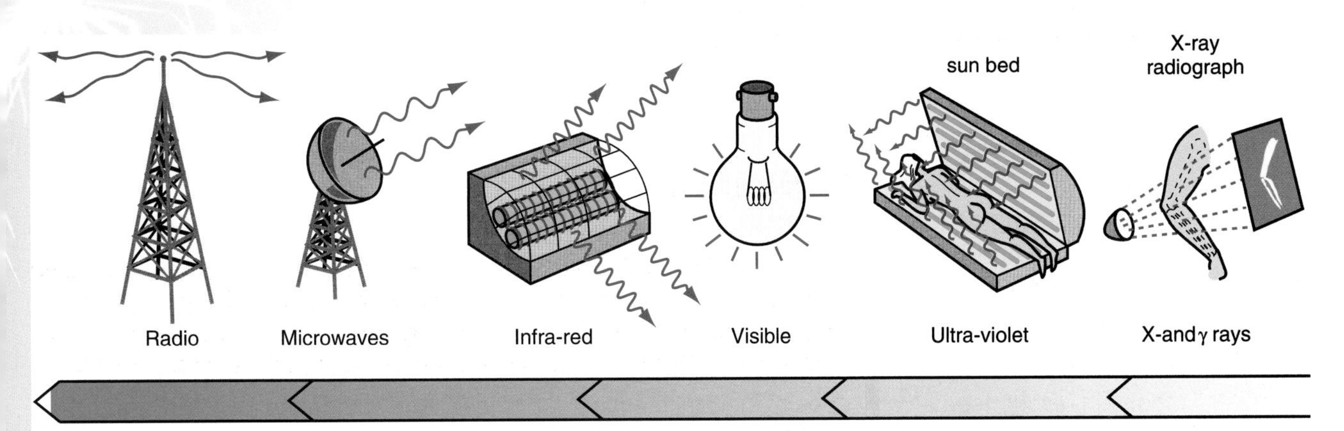

Figure 5.7C The electromagnetic spectrum

Investigating radiation from different surfaces

Figure 5.7D shows how you can test how well different surfaces radiate heat. A beaker containing water is tested first. Then the test is repeated with the beaker wrapped in aluminium foil, using the same volume of hot water. The test could also be tried with a similar beaker painted black.

The measurements could be plotted as graphs on the same axis of temperature against time. Results from tests like the ones above show that;

- a dull black surface is the best emitter of radiation
- a shiny silvery surface is the worst emitter of radiation,

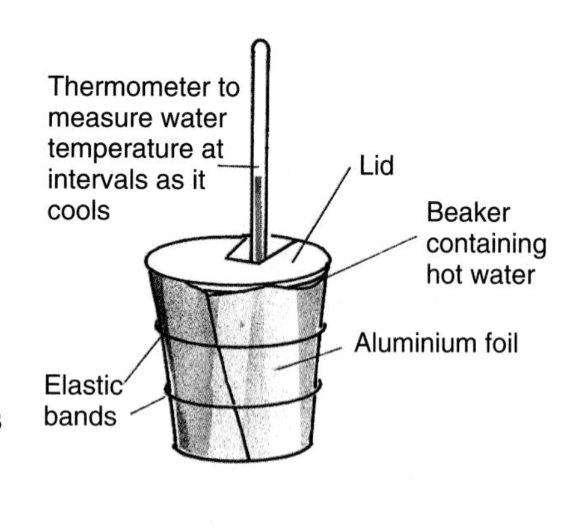

Figure 5.7D ▲ Testing different surfaces as radiators

A surface that is a good absorber of radiation is also a good emitter of radiation.

EXTENSION FILE ACTIVITY

What type of surface is the best absorber of radiation?

Figure 5.7E shows one way to test different surfaces. Results from this type of test show that,

- a dull black surface is the best absorber of radiation,
- a shiny silvery surface is the worst absorber or radiation.

Now you know why a kettle shines! It keeps hot longer than if it was black and dull. A car radiator is usually painted black because it needs to radiate heat effectively.

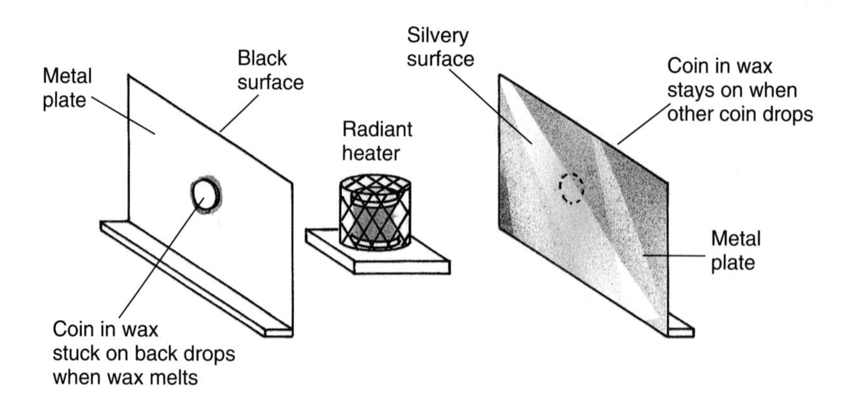

Figure 5.7E ▲ Testing different absorbers of radiation

SUMMARY

Heat radiation consists of electromagnetic waves from the infra-red range of the electromagnetic spectrum. The hotter an object becomes the more heat it radiates away. A black surface is the best absorber and emitter of radiation. A white or a silvered surface is much less effective.

CHECKPOINT

1. Explain the following;
   (a) Hot meals for delivery to the elderly are wrapped in aluminium foil.
   (b) Solar heating panels are painted black.
   (c) Houses in hot countries are usually painted white.
   (d) Infra-red photographs taken from satellites usually show rivers darker than land areas.
2. When the night sky in winter is clear, the ground often becomes frosty. This does not happen as often when the night sky is clouded. Why?
3. Mountain rescue teams carry special thermal blankets which are lined with metal foil. How do these keep accident victims warm?

5.8 Keeping warm

FIRST THOUGHTS

In this section you will find out which are the most effective methods for keeping warm in winter. You could help to reduce your home heating costs by studying this section carefully.

Keeping warm at home

Home heating costs money, not just to buy heaters but to pay for the fuel used. The amount of fuel we use as a nation each year to keep warm is equivalent to 30 million tonnes of coal! Better insulation in our homes cuts down on fuel bills and could reduce the need for more power stations.

Double glazing
Cuts down on outdoor noise as well as reducing heat losses. Sealed double-glazed units consist of two panes of glass with a layer of dry air between. Another effective way of cutting out heat losses through a window is to draw the curtains!

Aluminium foil
Placed between a radiator panel and the wall,the foil reflects heat radiation away from the wall. Radiator panels are often fixed on walls under windows, otherwise the air near the window would be cold.

Draught excluders
Fitted around the doors they help to keep us warm in winter. Cutting out the airflow can be very dangerous where gas and coal fires and paraffin heaters are used. They all use up the oxygen from the air and usually produce the harmless products, carbon dioxide and water. If there is insufficient air flow these fuels burn to form the poisonous gas, carbon monoxide. So never use draught excluders to block air vents.

Loft insulation
Is very cost effective. This means the initial costs for the material are soon repaid by the savings from the reduced fuel bills. A home without loft insulation loses heat through the roof almost as quickly as if there were no roof. Lagging the hot water tank saves money too. Insulation jackets are manufactured to fit around the hot water tanks.

Cavity wall insulation
Another popular method of reducing fuel bills. The outer walls of a house are double-brick with a cavity between the two layers of bricks. An air filled cavity cuts down heat losses: however, by filling the cavity with suitable insulation foam, it becomes an even more effective barrier against heat losses.

Figure 5.8A ▲ Keeping warm at home

U-values

Materials used to insulate homes are rated in terms of their **U-value**. This is the rate of heat flow passing through one square metre of material when the temperature difference across the material is 1 °C. For example, a wall with a U-value of 1.6 W/m^2/°C lets 1.6 J pass through each square metre each second when the temperature inside is 1 °C higher than outside.

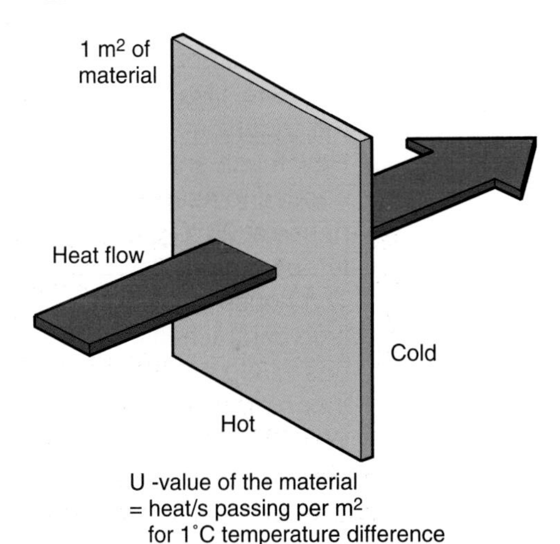

Figure 5.8B ▲ U-value

Keeping warm outdoors

What about when you go outdoors in winter? Outdoor clothing for winter use must let your body 'breathe' otherwise sweating can make you uncomfortable. Another problem is that your clothing must not absorb rain. If it did, you would become chilled and if there was any wind, you might even freeze. This is because the rain evaporates, a process that needs energy. The effect of the wind is to increase the rate of evaporation which means that energy is lost even faster. So there are a number of factors to be considered in the choice of materials for outdoor clothes.

Figure 5.8C ▲ The chill factor. The chill factor is the decrease in skin temperature due to adverse weather. Rain may cause a chill factor of 5–10°C but strong winds as well as rain may increase it to 20 °C

SUMMARY

There are a variety of measures that can be taken to reduce heat losses from homes. These measures reduce fuel bills. Suitable clothing is needed to keep warm outdoors in winter.

CHECKPOINT

1. Hot drinks can be kept warm in a vacuum flask. The figure opposite shows a cross-section of this type of flask.
   (a) Why does the inner flask have silvery walls?
   (b) Why is there a vacuum between the silvery walls of the inner flask?
   (c) Polythene is used to hold the inner flask in position in the can. Why is polythene chosen?
   (d) Why is it important to keep the lid on the flask when it is filled with hot liquid?

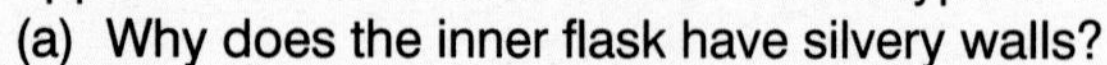

Stopper
Polythene
Can
Silvery walls
Hot liquid
Vacuum
Polythene

2. Keeping warm in winter is a major problem for elderly people. They can suffer from hypothermia, a condition which happens when the body gets too cold and the victim becomes unconscious. What advice would you give to help an elderly person at home to keep warm without running up big fuel bills?
3. Design an experiment to investigate the chill factor. Use a test-tube containing hot water as a 'body'. State what additional equipment you would use, what measurements you would make and how you would use your measurements.
4. (a) Work out the energy loss per second through the windows and wall of the house in the figure opposite. Use the information in the picture.
   (b) How much energy per second is needed to keep the inside of this house at 20 °C when the outdoor temperature is 0 °C?
   (c) The house in the figure has gas-fired central heating. The cost of gas is 36p per therm and one therm gives 106 MJ of energy. Work out the cost per hour of keeping the house at 20 °C when it is freezing outside.

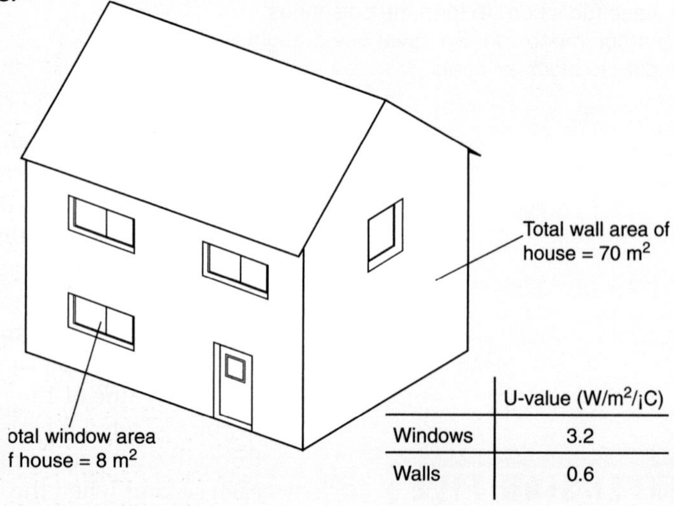

	U-value (W/m²/¡C)
Windows	3.2
Walls	0.6

Topic 6 Energy resources

6.1 Our energy needs

FIRST THOUGHTS

Find out in this section how much energy we as a nation use in a year and where it comes from.

If energy cannot be created or destroyed, why do we need to keep searching for new sources of energy? Jobs need energy: *what happens to the energy used after the job is done?* The problem with energy is that it spreads out when it is used.

When energy is used, it tends to spread out and become less useful.

Figure 6.1A ▲ Energy for fun

How much energy do you use in a day? To keep warm in winter, you could switch an electric heater on. Suppose your heater uses 1000 joules of electrical energy each second. If used for a whole day, it would use

$$1000 \times 24 \times 60 \times 60 = 86.4 \text{ MJ}$$

More energy would be needed to feed you, entertain you, carry you about and so on. In fact, on average, each person in the United Kingdom needs about 500 MJ each day. *Where does it all come from and where does it all go?*

- **Useful energy** is the energy that is used to do a particular task.
- **Primary energy** is the total energy needed to supply useful energy.

Energy is used to heat water in a kettle to make a hot drink. But much more energy is 'needed' than is used. This is because the energy used by the kettle is in a convenient form delivered to where we want to use it. The energy to heat the kettle may have come from an oil-fired power station. Energy is needed to change the energy in the oil to a convenient form and deliver it to the user. Suppose making a hot drink uses 1 MJ of electrical energy and 0.5 MJ of energy is used to produce and deliver the electrical energy to you. The total energy needed to make the drink (i.e. the primary energy) is therefore 1.5 MJ, just part of the 500 MJ of primary energy each of us needs per day on average.

See www.keyscience.co.uk for more about energy resources.

How much energy does the United Kingdom need in one year? Each person needs about 500 MJ in a day and there are about 50 million people in the UK. Reach for your calculator and work out the nation's primary energy need per year. More than 8 million, million, million joules!

Where do we get all this energy from? Life on Earth depends on the Sun because most forms of energy that we use come, directly or indirectly, from the Sun.

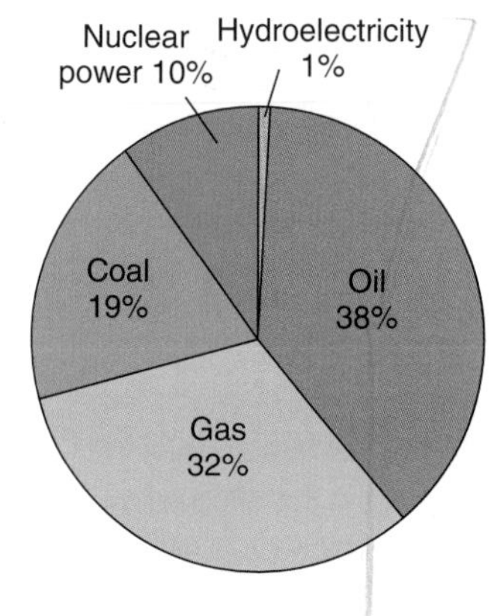

Figure 6.1B ▲ Britain's energy supplies

Fuels such as coal and oil were formed millions of years ago from prehistoric forests and marine organisms. Layer upon layer of decaying vegetation became compressed and eventually formed coal and other **fossil fuels**. Energy from the Sun made these forests grow so when we burn coal or oil we are releasing energy that arrived on Earth long ago. Uranium, the fuel used in nuclear reactors, has always been part of the Earth's crust. Coal and oil are examples of primary fuels because they can be used to produce secondary fuels such as electricity or petrol which must be made.

Renewable energy resources such as hydroelectricity and solar power use processes that are part of our natural environment. Energy from these resources is 'tapped off' and used. Coal, oil and uranium are not renewable resources because they were formed long ago; once used as fuel, they cannot be replaced. The term 'renewable' is used for energy from sources that are constantly replenished. For example, energy from hydroelectricity comes from rainwater running down hill, and this is part of the natural **water cycle**.

Figure 6.1B shows how the United Kingdom energy needs were met in the late 1990s. Fuel reserves are limited and, sooner or later, there will be no more oil or gas or coal or uranium. Some scientists think that gas reserves will be used up by the year 2025. *How old will you be then?*

SUMMARY

Energy always spreads out when it is used. Our present energy needs are met by using fuels such as gas, oil, coal and uranium. Little use is made of renewable energy resources at present.

CHECKPOINT

1. Make a list of the secondary fuels you use between getting up in the morning and arriving at school. What do you use each type of fuel for?
2. In Figure 6.1B, if oil was not available, how would this affect the usage of energy in terms of transport, heating, industry and so on?
3. World energy reserves are shown in the figure opposite.
   (a) Which fuels will run out first at the present rate of usage?
   (b) How will the absence of this fuel affect people in general?
   (c) Which type of fuel would appear to offer the best long-term supply of energy?
   (d) What factors make predictions about the use of energy uncertain?

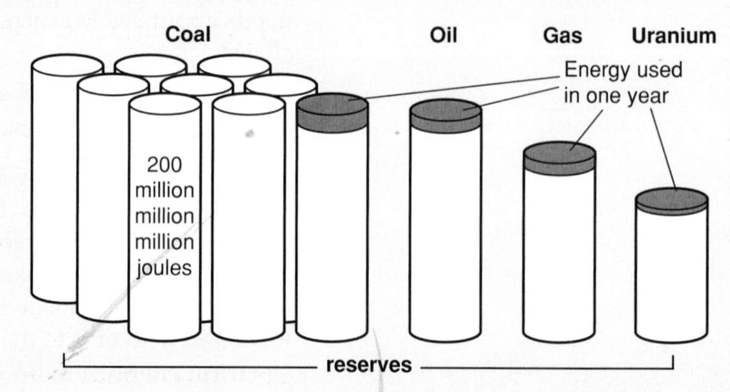

6.2 ▶ Fuel

FIRST THOUGHTS

Why is it important to use our fuel reserves carefully? What are the benefits of each type of fuel? Which fuels are likely to be in short supply in the future?

Fossil fuels such as coal, oil and gas lie underground and are often difficult to extract. Some coalfields in coastal areas extend beneath the sea-bed. Recovery of coal from such mines is much more complicated than recovery from open-cast mines. Open-cast mines are mines where the coal seams can easily be reached by tunnels rather than vertical shafts. Oil and gas fields under the sea bed are much more difficult to reach than fields under dry land. As the demand for fossil fuels increases, so the search for new fields is being extended to remote regions of the world. However, reserves of fossil fuels are not unlimited.

Figure 6.2A ▲ Open cast mining

Fuels release energy as they burn. Fossil fuels release about 30 MJ per kilogram.

Agriculture depends on petroleum; the **Petrochemicals industry** makes **pesticides** that farmers use to protect their crops from insects and weeds. Without the petrochemicals industry we should have fewer modern drugs and medicines. Most manufacturing industries depend on the oil industry. Without oil, our cars, trains, boats, planes would be useless. Without oil and gas, our present economy would collapse returning us to primitive living conditions. Unless scientists find substitutes for oil and gas, our future is going to be a low-technology future. This is what you hear described as the **world energy crisis**. It crops up constantly in newspaper articles and TV programmes.

Coal

Coal will continue to be an important fuel for many years to come. Britain is literally built on coal and new coalfields like Selby in Yorkshire have been developed to meet future energy needs. Coal-fired power stations could produce much of our electricity.

Figure 6.2B Inside a coal mine

Research scientists have found methods of making liquid fuels from coal.

Coal + Hydrogen $\xrightarrow[\text{pressure with a catalyst}]{\text{At high temperature and}}$ Liquid hydrocarbon fuel

This fuel can be used to power diesel engines. South Africa has invested in enormous coal **liquefaction** plants for making transport fuel. These factories are unhealthy places to work in because contact with the products can cause skin cancer.

Coal can also be used as a source of fuel gas. The gasification of coal is less risky and less expensive than liquefaction. It is done by heating coal and then passing steam over the residue. A mixture of gaseous fuels is formed.

Coal $\xrightarrow{\text{Heat then steam}}$ Fuel gas (Carbon monoxide + Hydrogen + Methane)

There are drawbacks, however, over increasing coal production. Open-cast mining ruins the landscape. Underground mining creates spoil heaps, great mounds of soil and rock that have been dug out of the mine. A spoil heap collapsed on the Welsh village of Aberfan in 1968, killing dozens of people.

Oil and natural gas

Oil and natural gas reserves throughout the world are likely to be used up by the middle of the twenty-first century unless major new fields are discovered. Searching for new oilfields takes scientists to remote parts of the world like Alaska. Oil rigs in the North Sea are designed to withstand extremely hostile weather conditions. One of the most important products from oil is petrol. Scientists in several countries are trying to develop alternative fuels for cars.

Natural gas is mostly methane and is found trapped in pockets underground, often below the sea-bed. Gas is used in homes for heating and cooking.

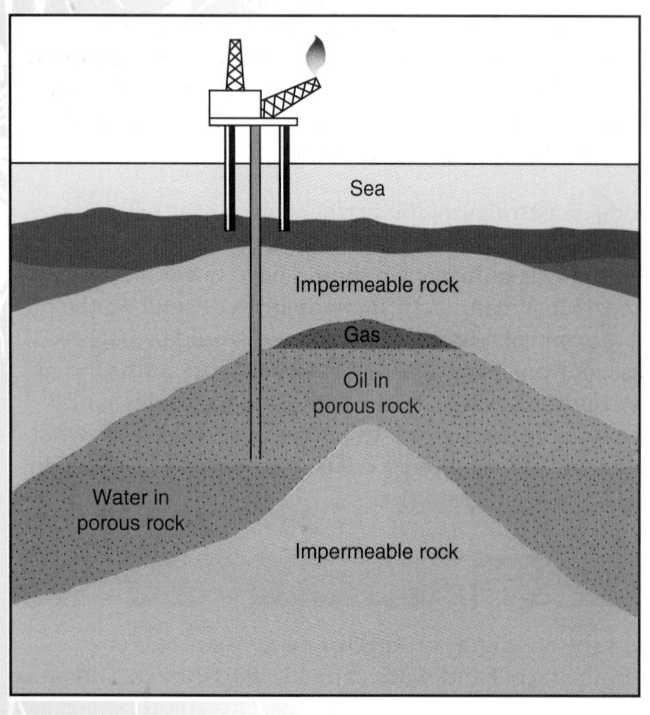

Figure 6.3C Drilling for oil

Figure 6.2D Oil rig

Tar sands and oil shales are sources of oil which have not yet been exploited. This is because it is too costly to extract oil from them at present. Tar sands are deposits of tar in sand. Oil shales are deposits of oil in porous shale rocks.

Nuclear fuels

Fission

One kilogram of uranium releases about 1 million times as much energy as one kilogram of fossil fuel.

Figure 6.2E ▲ Torness nuclear power station

Uranium is the fuel for nuclear power stations. Uranium ores were formed in the Earth's crust billions of years ago. Uranium ores must be refined before they are suitable for using as fuel. One tonne of uranium can give the same amount of energy as one million tonnes of coal. However, nuclear power stations have disadvantages. The waste products from nuclear fuel are highly radioactive and must be stored for thousands of years until they become harmless.

Nuclear reactors are designed to produce heat which drives steam turbines to generate electricity. Reactors must be continually monitored to ensure radioactive fuel does not escape. Inside a nuclear reactor, the temperature and pressure are very high and the materials used for the reactor must withstand these extreme conditions.

Figure 6.2F ▲ Storage of radioactive waste

Fusion

Fusion reactors are nuclear reactors that do not use uranium as the fuel. Instead, the two isotopes of hydrogen called **deuterium** and **tritium** will be used. Deuterium can be obtained from water and tritium can be made within the reactor from the light metal lithium. Although the reactor structure will become radioactive, the waste produced will be easier to deal with. Waste from fusion reactors needs to be stored for much shorter periods than waste from uranium reactors. Fusion reactors themselves are much safer than uranium reactors. Fusion reactors are still at the experimental stage, and many more years of research are needed to find out whether they will be able to produce electricity on a large scale.

Figure 6.2G JET – the Joint European Torus – is the world's largest and most powerful fusion experiment. It aims to produce fusion reactions similar to those which occur in the Sun. JET is a collaborative venture involving 14 European nations.

SUMMARY

Most of our present energy needs are met by using fossil fuels and uranium. Fossil fuels add to the 'greenhouse effect' and will result in a warming of the atmosphere. Uranium fuels produce radioactive waste products that must be stored for many years.

CHECKPOINT

1 Many homes were heated by coal fires before 1960. Smoke and fog together produce an unhealthy atmosphere called 'smog' which can last for days. Converting homes to gas or electric heating has eliminated the smog problem.
- (a) Why did the introduction of gas and electric heating in homes eliminate the smog problem?
- (b) Why is it important that waste gases from coal-fired power stations should be cleansed of smoke?
- (c) What sort of illnesses could be caused by breathing in smog?

2 France has few reserves of coal. Why has the French government built many more nuclear power stations than Britain has?

3 The demand for electricity in Britain in the year 2025 could be 15 000 MW greater than at present.
- (a) How many 1000 MW power stations need to be built to meet this demand?
- (b) What type of power stations should be built to meet this demand? Explain your answer.

4
- (a) Nuclear power stations are generally located far away from cities. Why?
- (b) Power stations need vast quantities of water in their cooling systems. How does this limit the number of possible locations?

5 The demand for electricity varies during each day, as shown in the figure opposite.
- (a) Why is demand highest in the evening?
- (b) Demand becomes very small after midnight. Why?
- (c) Why is demand in winter greater than in summer?

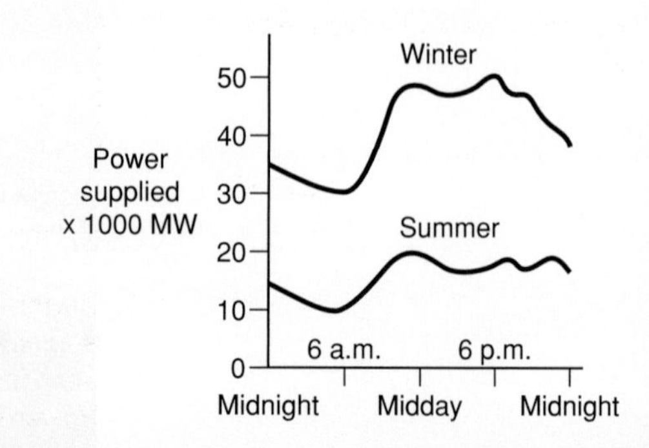

6.3 Renewable energy resources

FIRST THOUGHTS

The alternatives to present fuel supplies are outlined and compared in this section. Are these alternatives realistic? Read on to find out.

See www.keyscience.co.uk for more about renewable energy resources.

The Sun's energy drives the earth's atmosphere like a huge engine, creating rain, wind and waves. Plants absorb energy from sunlight. Scientists reckon that the amount of the sun's energy reaching the Earth each second is about 1 million million million ($= 10^{18}$) joules. Just a tiny fraction of this vast amount of energy would be enough to meet our energy needs. No fuel would be needed. We would be using energy from the Sun as it arrives and the Sun has 5000 million years to go before it runs short of its fuel.

Renewable energy is a term used to describe energy supplied by processes that are part of our natural environment. For example, windmills are machines that produce useful energy from the energy of the wind. Windmills have been around for many centuries but now there is renewed interest in using wind energy to meet our energy needs. The oil reserves will run out, the use of coal may need to be limited, nuclear reactors may be considered too risky. Can renewable energy resources meet our needs in the future?

All except two of the renewable energy resources described in the next few pages are driven by the Sun's energy. The two exceptions are geothermal energy and tidal power.

Hydroelectricity is produced from the potential energy given up by rainwater running off mountains. The water is channelled so it flows downhill through huge turbines which produce electricity. In the late 1990s, hydroelectric power stations supplied less than 0.25% of the UK's primary energy needs. Large hydroelectric schemes in Wales and the Scottish Highlands could supply more of our present energy needs. With no fuel needed, the cost of hydroelectricity is much less than electricity from coal or uranium powered stations.

(a) Hydroelectric dam

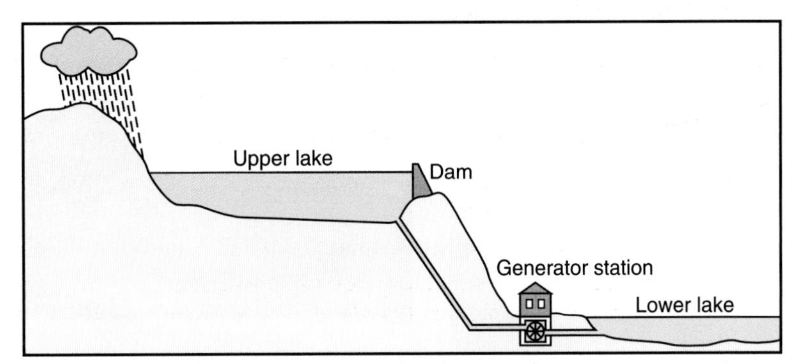

(b) Pumped storage

Figure 6.3A ▶

Oil and gas reserves are likely to be exhausted within the next century.

Pumped storage schemes, as at Dinorwic in North Wales, are designed to help engineers cope with surges in demand. At the end of popular TV programmes, electricity demand surges as people leave their TV sets and make hot drinks. Power stations cannot be switched on and off like light bulbs. When there is little demand for electricity, water from a low-level reservoir is pumped uphill to a reservoir up in the mountains. When demand suddenly reverses, the flow is reversed and electricity is produced (see Figure 6.4B on page 82).

Wavepower is another promising energy resource. Scientists reckon that each metre of suitable coastline could give about 20 kW of power. Just 50 km of coastline would give the same power output as a 1000 MW nuclear power station. Some scientists think that wavepower could supply 10% of the UK's energy needs.

Wave-powered electricity generators would need to withstand extreme weather conditions. They would not produce a constant supply of electricity but they could be used to feed energy into pumped storage schemes.

Wavepower and windpower are driven by the effects of solar heating on the atmosphere and oceans.

Figure 6.3B ▲ Wave power

Biomass is material from living organisms. Wood, charcoal and ethanol are examples of biomass fuels. Methane gas from rotting plant and animal matter is called **biogas**. Many farms have a methane generator which produces methane for heating. Household rubbish tipped into landfill sites also generates biogas. In the UK, refuse from homes and factories could provide sufficient biogas for heating to supply 10% of national energy needs.

Figure 6.3C ▲ Using biomass

Aerogenerators are driven by wind. They can be much larger than old-style windmills and much more powerful. However, the output of an aerogenerator is not steady because it varies with windspeed.

Clusters of aerogenerators in remote areas could make a big contribution to our energy needs. Imagine a cluster of 100 aerogenerators over an area of 10×10 km, each generator producing 5 MW of electricity. Two such clusters would give the same output as a 1000 MW power station.

Figure 6.3D ▲ Wind power

Figure 6.3E Geothermal energy

Figure 6.3F Tidal power

Geothermal energy from the hot interior of the Earth can raise the temperature of underground rocks in the Earth's crust to more than 200 °C. Water pumped down to these rocks is converted into steam which can then be used to drive turbines that make electricity.

Some geothermal wells produce their own hot water from water that has gradually seeped into sedimentary layers in the Earth's crust. The energy from a single geothermal well could heat several large blocks of flats or offices. Geologists reckon that geothermal energy could supply up to 3% of the UK's primary energy needs.

Tidal power stations in suitable coastal areas are designed to trap each rising tide behind a barrage. The high tide is then released into the sea through turbines in the barrage to make electricity. One of the most promising sites in Britain is the Severn estuary where the incoming tide is reflected by the estuary banks making it even higher.

Solar energy can be collected directly using solar heating panels or solar cells. Solar panels fitted to south-facing roofs can be very effective. Water trickling through the panels is heated directly by the Sun. Even on cloudy days in Britain, a solar panel of area 11 m^2 could collect 300 W of solar energy. A south-facing roof covered with solar panels could supply the hot-water needs of a house. Solar heating could produce the equivalent of ten 1000 MW power stations.

Solar energy is free once the panels and cells are paid for.

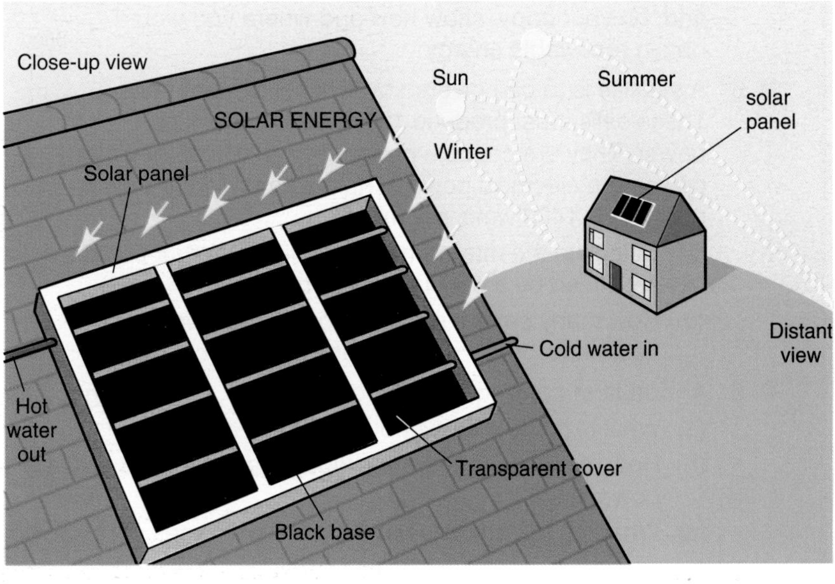

Figure 6.3G Solar energy

SUMMARY

Renewable energy resources are natural processes that can be 'tapped' to supply useful energy. Materials used in these processes are replaced by natural means. Renewables could provide up to 33% of our energy needs.

Solar cells convert solar energy directly into electricity. A solar cell contains solid substances such as silicon; solar radiation entering these substances frees electrons from the atoms of these substances to produce electricity. The best solar cells available at present convert no more than about 30% of the incident solar energy into electricity. Because they are expensive, they are used to provide electricity for specialist purposes such as calculators and satellites.

Figure 6.3H ▲ A solar powered vehicle

CHECKPOINT

1 (a) Why are tropical countries better able to produce biomass fuels than countries like Britain?
  (b) Use the figure opposite to find out how much solar energy per square metre falls on Britain each year on average.
  (c) What area of solar panels would each person in Britain need to obtain an annual energy supply of about 200 000 MJ from solar energy?

Annual average solar radiation

2—3,000 3—4,000 4—5,000 5—6,000 6—7,000 7—8,000 over 8,000
Megajoules/sq metre/year

2 (a) Tidal power is a much more reliable source of energy than wind power or wavepower. Why?
  (b) How would people be affected by a tidal power station being built in their locality?

3 Make a list of renewables that can supply energy for
  (a) running vehicles,
  (b) home heating,
  (c) running electric trains.

4 If you lived on a small isolated island with a population less than 100, which renewable resource would meet your community's needs best? Copy the map opposite and, on your copy, show how and where you would obtain renewable energy.

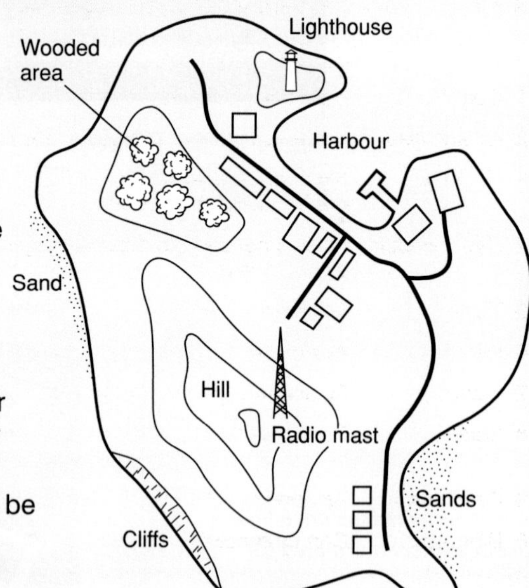

5 A satellite is to be powered by a panel of solar cells. These cells must produce 1000 watts of electrical power. They are capable of converting 25% of the solar power into electrical power. The maximum solar power possible is 1400 watts per m^2.
  (a) Calculate the maximum electrical power that could be produced by a panel of area 1 m^2.
  (b) How many square metres of panel area would be needed to produce 1000 watts of electrical power?

6 A wind farm consists of 10 aerogenerators, each designed to produce 4 MW of electricity.
  (a) What is the total power output of the windfarm?
  (b) How many aerogenerators would be needed to give the same power output as a 1200 MW power station?
  (c) What are the advantages and disadvantages of replacing power stations with wind farms?

6.4 Energy for the future

FIRST THOUGHTS

What options are available for meeting future energy demands? What are the advantages and disadvantages of each option? Choices made now will affect you throughout your lifetime!

It's a fact!

How much coal must be burnt in a power station to keep a single light bulb alight for its entire operating life? The answer is a staggering 150 kg or more. How much carbon dioxide gas goes into the atmosphere as a result of burning 150 kg of coal? Almost half a tonne!

Alternative supplies

The demand for energy has risen throughout the twentieth century as more and more labour-saving machines have come into use. As living standards have gone up, so too has the demand for energy. Fuels such as oil will soon be in short supply. *Can alternative supplies of energy be developed?*

Nuclear power could be developed to supply much of our electricity but many people are unhappy about the disposal of nuclear waste. Also, electric vehicles would need to be developed to take the place of petrol-powered vehicles – unless coal or biomass can be used to produce an alternative fuel to petrol.

Coal is in plentiful supply and known reserves would last several centuries at the present rate of usage. Coal could fill the gap left by dwindling oil and gas reserves. However, the greenhouse effect is partly due to burning fossil fuels so this may limit use of coal as a fuel. In addition, acid rain is caused by gases released into the atmosphere from coal-fired power stations. Plant growth in remote areas can be severely affected by acid rain from distant power stations.

Renewable energy resources could meet much more of Britain's energy needs than at present. This is attractive to many people but money and new skills are needed to develop and build 'renewable energy' power stations. Politicians need to be made more aware of the importance of renewable energy resources. Environmental problems such as aerogenerator noise may arise as a result of renewable resources developments. Some of the advantages and disadvantages of each type of energy resource are summarised in Table 6.1.

Table 6.1 ▾ Advantages and disadvantages of different energy resources

Type	Advantages	Disadvantages
1 Coal	Abundant reserves	Air pollution, acid rain, greenhouse effect, spoil heaps
2 Oil and gas	Undersea reserves, pipeline supply	Reserves limited, air pollution, greenhouse effect
3 Nuclear	Much more energy released per kg of fuel than above	Reactor core is highly radioactive, radioactive waste must be stored for centuries
4 Hydroelectric	No fuel needed, no disadvantages as 1, 2 and 3	Large upland reservoir necessary
5 Tidal	As above	Ecology of tidal estuaries affected, variable power
6 Wind	As above	Aerogenerator noise, large areas needed, landscape, variable power
7 Waves	As above	Variable power, floating generators needed
8 Solar	As above	Variable power if ground-based, high cost of solar cells, large areas needed
9 Geothermal	As above	Deep boreholes needed, limited power output
10 Biomass	Can be renewed	Water and large areas in a hot climate needed

Efficiency savings

What happens to all the energy we get from fuel and renewable resources? Industry, transport and homes use about 50% of the total and about 30% is exported. At present, the other 20% is wasted. Finding ways of doing jobs more efficiently would reduce this colossal wastage. So too would switching off unwanted lights! Better insulation in our homes is another big saver.

Combined heat and power (CHP) stations would save fuel by supplying 'waste' heat to local factories and offices. Any power station needs huge quantities of water to keep its turbines cool. In non-CHP stations, the hot water from the cooling system is discharged into a reservoir or river. A CHP station is designed to supply this hot water to local buildings. Figure 6.4A shows the overall savings made by replacing a 100 MW non-CHP station with a CHP station.

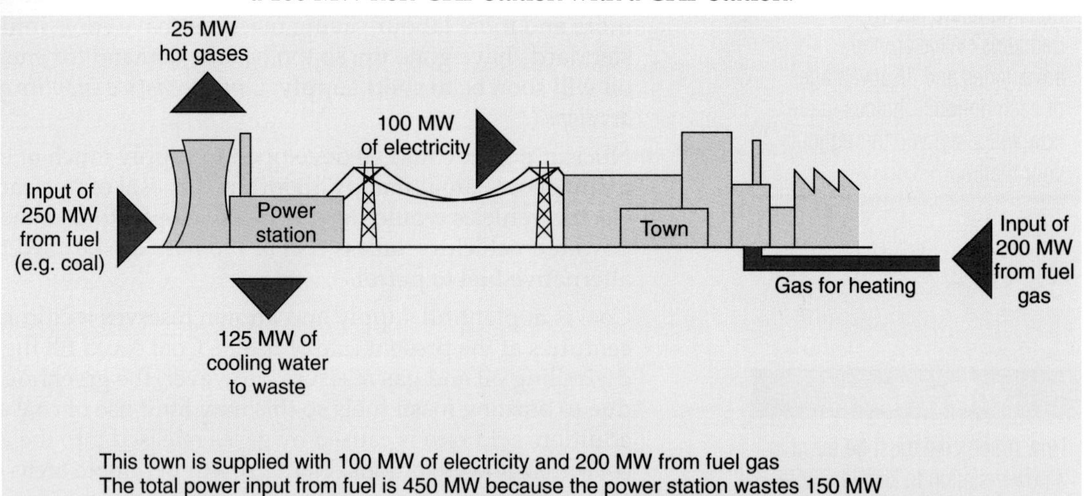

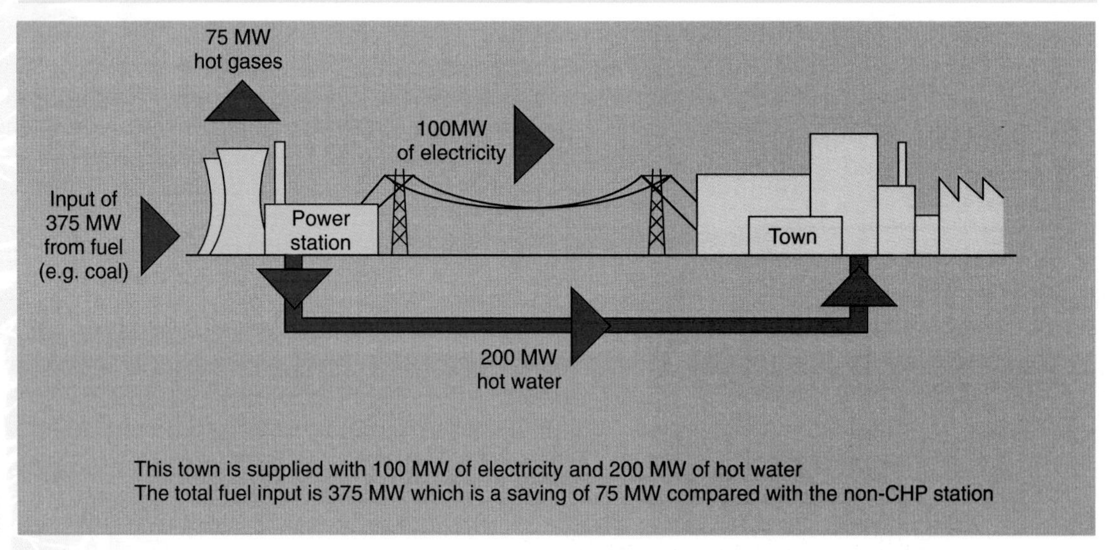

Figure 6.4A Saving fuel (a) A non-CHP power station (b) A CHP power station

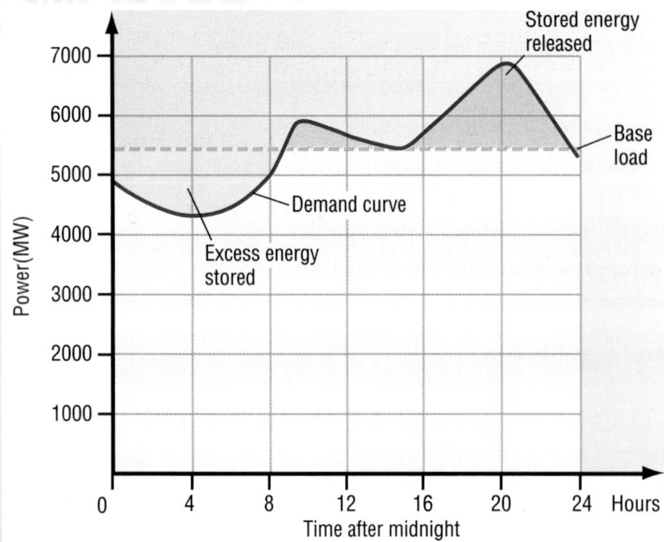

Figure 6.4B Meeting demand

Pumped storage power stations are designed to use electricity from nuclear and fossil power stations when demand is low. The electricity is used to drive turbines which pump water from a lake to an uphill reservoir. When demand is high, the flow is reversed and the turbines drive electricity generators which supply electricity to the grid system. Conventional power stations cannot be switched on and off to meet sudden changes of demand; average demand each day is met using conventional power stations. This is referred to as **base load** demand. When the demand is below average, the excess energy is stored in pumped storage schemes; this is released when the demand is above average.

In 2001, the UK government announced the construction of 18 off-shore 'windfarms' round Britain's coasts. Each windfarm will consist of 30 2MW aerogenerators) or 'wind turbines'

Could the energy gap caused by dwindling oil and gas reserves be filled by renewable energy resources and more efficient usage of energy? The answer is yes – if people are willing to accept a slower increase in their standard of living. If not, nuclear and coal-fired power stations will continue to be needed.

See www.keyscience.co.uk for more about energy resources.

SUMMARY

Energy for the future will depend on coal, nuclear power, renewable resources and more efficient energy usage. Planning is essential to ensure continued energy supplies. Otherwise, living standards will not continue to rise and may even fall.

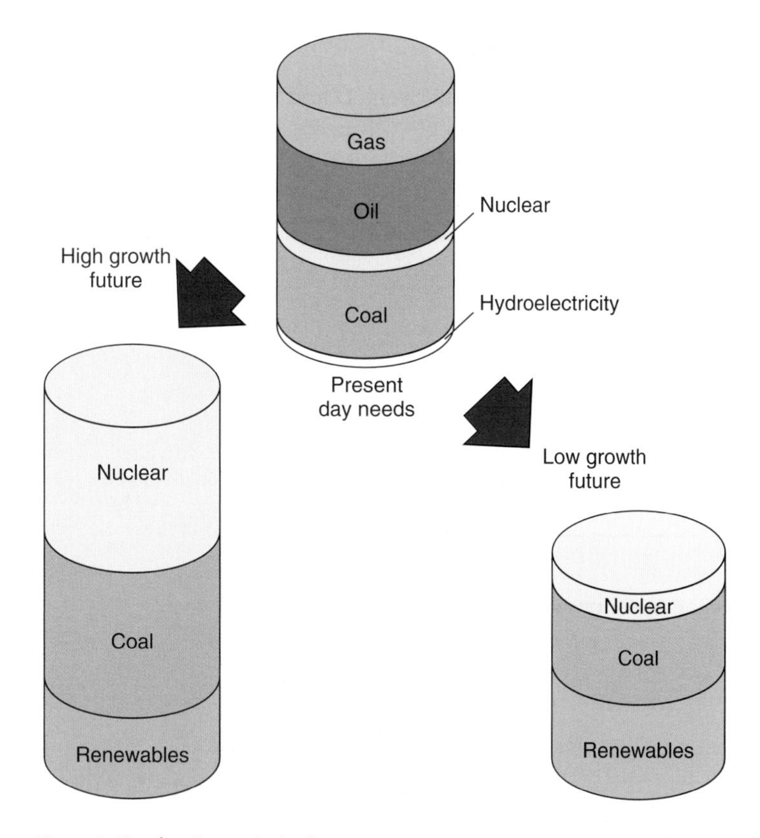

Figure 6.4C ▲ Energy in the future

CHECKPOINT

1. (a) Supplies of energy from renewable sources such as windpower and wavepower are not as reliable as fuels such as coal. Why?
   (b) Hydroelectric power stations are designed to use electricity to pump water uphill as well as being able to produce electricity. The water flow can be reversed within minutes. Why is this useful?
2. The increasing use of computers in the future will allow many people to work from their homes. How will this affect the demand for energy?
3. Suppose the Government proposed to build a cluster of aerogenerators a few miles from your home in place of a planned nuclear power station far away. Some residents in your area have formed an action group to oppose the windfarm. However, the local Green party welcome the proposals.
   (a) List the points in favour of the windfarm.
   (b) List the points against the windfarm.
   (c) Which side would you support? Perhaps you and your friends would like to form two opposing groups to discuss and debate the proposal.
4. How should our future needs be met? Outline the choices you support, giving reasons for your choices. If possible, form a group to discuss the choices available.

Theme Questions

TOPIC 3

1 The graph shows how the temperature of a substance rises as it is heated. At A, the substance is a solid.

(a) Say what happens:
(i) between A and B,
(ii) between B and C,
(iii) between C and D,
(iv) between D and E,
(v) between E and F.

(b) Name the temperatures T_1 and T_2.

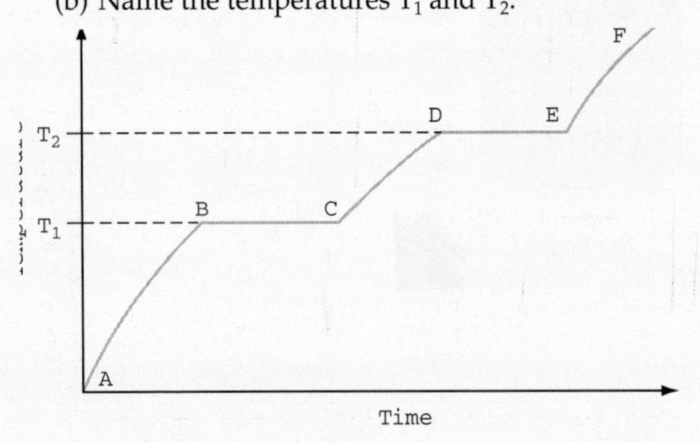

2 To answer the question you may find this equation useful:

$$\text{Density} = \frac{\text{Mass}}{\text{Volume}}$$

Steve measured the mass and volume of some rectangular blocks. The table shows the results he obtained.

Block	*Mass* (g)	*Volume* (cm³)
Aluminium	160	60
Brass	170	20
Iron	475	60
Lead	110	10
X	320	120

(a) Which block had the largest volume?
(b) Which block was the heaviest?
(c) Block X had the same density as one of the others. Which one? Explain your answer.
(d) Describe how you would measure the **volume** of a rectangular block of iron. (MEG)

TOPIC 4

3 Sally and Jane decide to compare their work rates. They each lift as many 10 newton sandbags as they can on to a table 1 metre high in 1 minute. Here are their results:

	Sally	*Jane*
Weight of sandbag (N)	10	10
Distance lifted (m)	1	1
Number of sandbags lifted	30	36
Time of exercise (min)	1	1

(a) What total weight does Sally lift in 1 minute?
(b) Work done = Weight × Distance
Jane lifts a total of 360 N. Calculate how much work she does. (NEAB)

4 A group of fifth year students planning an expedition to the Lake District decide to do some scientific research before they go. They decide to look at what activity they are fit for and how far they can walk in a day. Their findings are shown below.

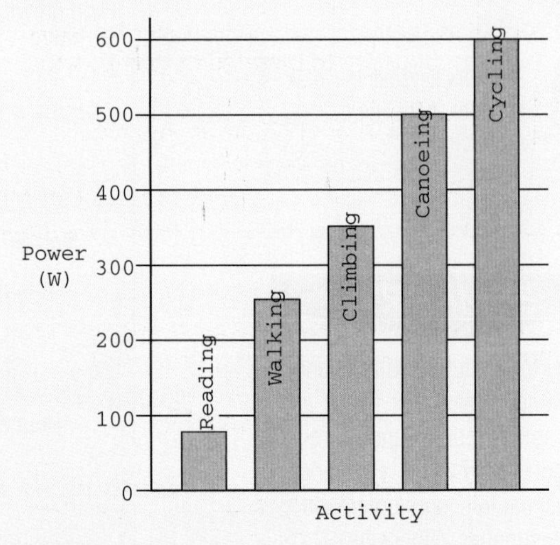

(a) What power is required for canoeing?
(b) What pattern connects the power and the type of activity in the bar chart?
(c) The group decides on a walking expedition. One of the students is interested in how much energy he will use up in one minute. He reads the power rating at 250 W (joules/second). How much energy does he use in one minute? (NEAB)

TOPIC 5

5 (a) Explain why the smoke coming out of chimneys continues to rise up through the surrounding air. Why will the smoke eventually stop rising?

(b) An apparatus set up to investigate the effect of heating water is shown below.
Draw a sketch graph showing what readings you would expect to take from the two thermometers. How is heat transmitted to the thermometer labelled B, other than by conduction, convection and radiation?

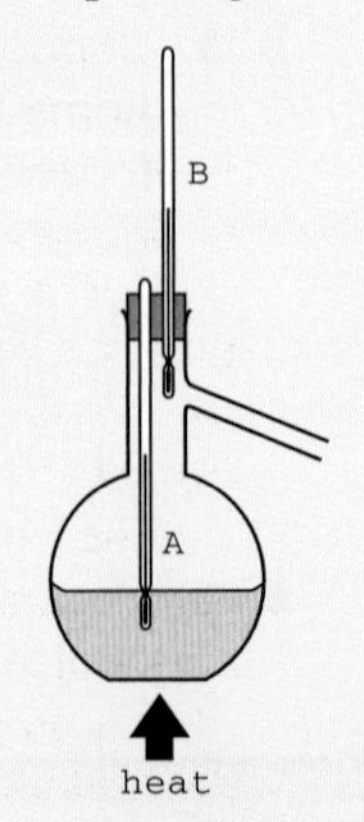

(c) In an experiment with this apparatus it was found that the temperature of 500 cm^3 of water was raised from room temperature to boiling point in 8 minutes exactly. The bunsen burner had a heat output of 875 J/s. What percentage of the heat from the bunsen passed into the water? (It takes 4.2 J to heat 1 g of water through 1 °C. Water has a density of 1 g/cm^3)

(MEG)

6 1 kilogram of rock will lose or gain 800 joules of heat energy as its temperature changes by one degree Celsius.
1 kilogram of rock will lose or gain 4200 joules of heat energy as its temperature changes by one degree Celsius.

(a) Which **one** of these two substances will need the most heat energy to increase the temperature of 1 kg by 10 °C?

(b) 1 kg rock and 1 kg of water, at the same temperature, lose energy at the same rate, until their temperatures fall by 10 °C. Which substance will take longest for this to happen? Explain your answer.

(SEG)

7 The diagram shows a heater used to warm a large building.

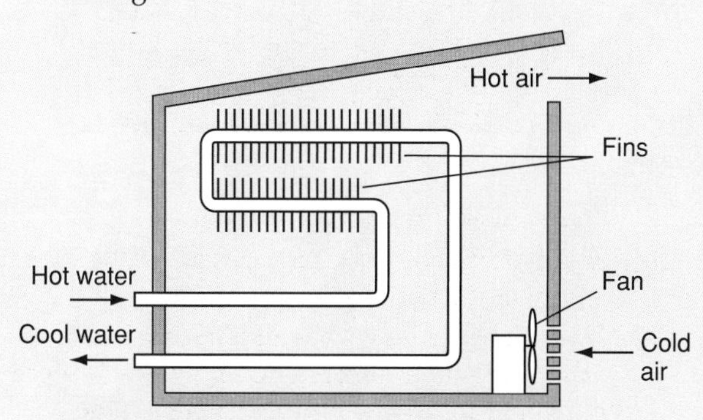

The heater consists of pipes through which hot water flows. Fins are attached to the pipes. An electric motor operates a fan which draws air into the heater and blows it out after it passes through the fins.

(a) State how heat is transferred from the water to the fins.

(b) (i) Name a suitable material for the fins, and give **one** reason for your choice of material.
(ii) Explain why fins are used.

(c) Water flows through the heater at a rate of 2.5 kg each minute. It enters at an average temperature of 70 °C and leaves at an average temperature of 67 °C. The specific heat capacity of water is 4200 J/kg K. Calculate
(i) the temperature fall of the water,
(ii) the quantity of heat energy removed from the water each minute.

(d) Air of specific heat capacity 180 J/kg K enters the heater at 18 °C and leaves at a temperature of 43 °C. Calculate the mass of air flowing through the heater each minute.

(e) The room heated by the heater has the dimensions 5m × 4m × 3m.
(i) Calculate the volume of air in the room.
(ii) If the density of air is 1.25 kg/m^3 how long will it take to heat all the air in the room?

(f) After this time will the temperature of the room be 43 °C? Explain your answer.

(g) Give **one** advantage that this type of heater has over an ordinary radiator.

(NEAB)

TOPIC 6

8 (a) Electrical power could be obtained from the tides by building dams. The table lists some of the places which are being considered and gives information about tidal power schemes in these places.

Location	*Barrage length* (m)	*Power output* (MW)	*Cost of energy* (p/kWh)
Loch Broom	500	29	13.9
Mersey	1750	620	3.6
Morecambe Bay	16 600	3040	4.3
Severn	17 000	7200	3.7
Thames	9000	1120	8.3
Wash	19 600	2760	7.2

(i) Electricity from coal-fired power stations costs about 4 pence per kilowatt hour. Which places in the table would produce cheaper electricity than this?
(ii) Many people think the Morecambe Bay scheme is worthwhile. Suggest a reason why.

(b) (i) A tidal scheme is more expensive to build than a coal-fired power station. But it is cheaper to run. Explain this.
(ii) Morecambe Bay would have a power output of 3040 MW but Loch Broom could produce only 29 MW. What does the power output depend on?

(c) The map below shows Morecambe Bay. Explain how people living in Silverdale might feel about:

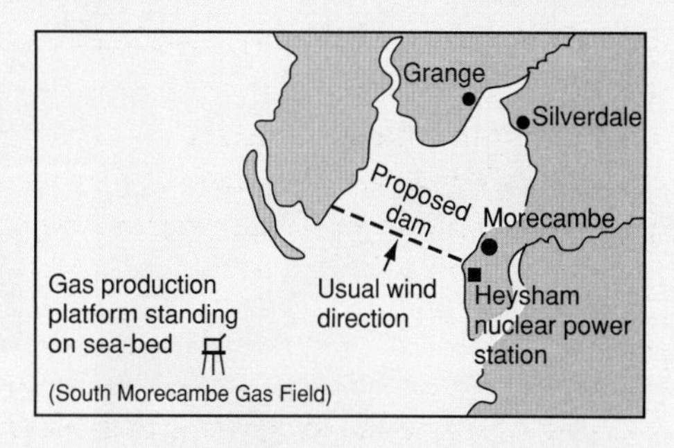

(i) natural gas from the South Morecambe Gas Field,
(ii) nuclear energy from Heysham power station,
(iii) the Morecambe Bay tidal scheme, as alternative energy sources.

In your answer consider facts already given in this question and anything you already know about these types of energy source.

(LEAG)

Theme C

The Atom

'Matter is composed of atoms.'
This is what John Dalton said in 1808. What a simple statement this appears to be! Yet the complex developments that followed from this statement fill the whole of physics and chemistry.

What are atoms?

How many different kinds are there?

How do atoms differ from even smaller particles?

You will find some of the answers to these questions in Theme C.

Topic 7 Particles

7.1 The atomic theory

FIRST THOUGHTS

What exactly is an atom, and why did the atomic theory take nearly 2000 years to catch on? Combine your imagination and your experimental skills in this section to find out why.

The idea that matter consists of tiny particles is very, very old. It was first put forward by the Greek thinker Democritus in 500 BC. For centuries the theory met with little success. People were not prepared to believe in particles which they could not see. The theory was revived by a British chemist called John Dalton in 1808. Dalton called the particles **atoms** from the Greek word for 'cannot be split'. According to Dalton's **atomic theory**, all forms of matter consist of atoms.

The atomic theory explained many observations which had puzzled scientists. Why are some substances solid, some liquid and others gaseous? When you heat them, why do solids melt and liquids change into gases? Why are gases so easy to compress? How can gases diffuse so easily? In this topic, you will see how the atomic theory provides answers to these questions and many others.

Elements and compounds

There are two kinds of pure substances: **elements** and **compounds**. An element is a simple substance which cannot be split up into simpler substances. Iron is an element. Whatever you do with iron, you cannot split it up and obtain simpler substances from it. All you can do is to build up more complex substances from it. You can make it combine with other elements. You can make iron combine with the element sulphur to form iron sulphide. Iron sulphide is made of two elements chemically combined: it is a compound.

The smallest particle of an element is an **atom**. In some elements, atoms do not exist on their own: they join up to form groups of atoms called **molecules**. Figure 7.1A shows models of the molecules of some elements.

Matter is made up of particles. Pure substances can be classified as elements and compounds. Elements are substances which cannot be split up into simpler substances. The smallest particle of an element is an atom. In many elements, groups of atoms join to form molecules.

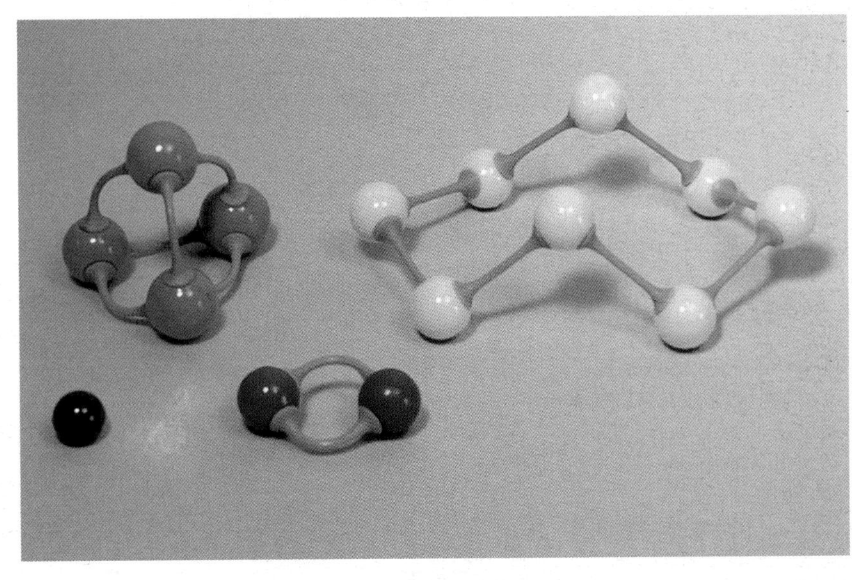

Figure 7.1A ▲ Models of molecules (helium, He; oxygen, O_2; phosphorus, P_4;sulphur, S_8)

A compound is a pure substance that contains two or more elements. The elements are not just mixed together: they are chemically combined. Many compounds consist of molecules, groups of atoms which are joined together by **chemical bonds**. All gases, whether they are elements or compounds, consist of molecules. Figure 7.1B shows models of molecules of some compounds.

Compounds are pure substances that contain two or more elements chemically combined. Many compounds are made up of molecules; others are made up of ions. All gases consist of molecules.

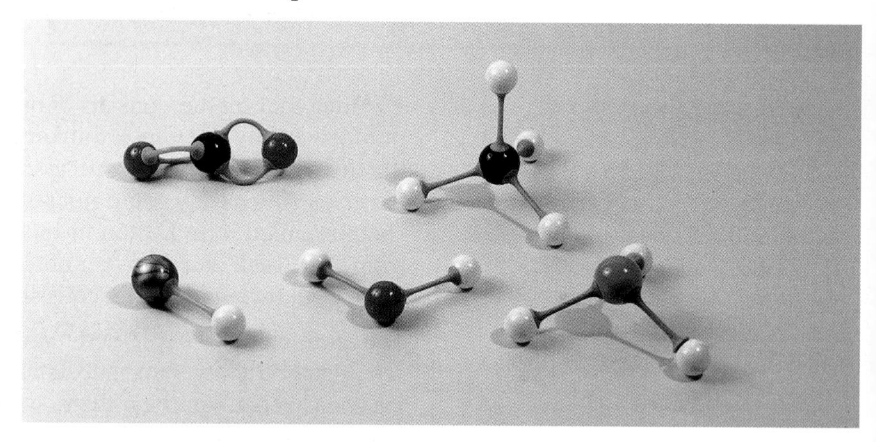

Figure 7.1B ▲ Models of molecules of some compounds (carbon dioxide, CO_2; methane, CH_4; ammonia, NH_3; water, H_2O; hydrogen chloride, HCl)

Some compounds do not consist of molecules; they are made up of electrically charged particles called **ions.** The word particle can be used for an atom, a molecule and an ion. Gaseous compounds and compounds which are liquid at room temperature consist of molecules.

How big is a molecule?

You can get an idea of the size of a molecule by trying the experiment shown in Figure 7.1C. This experiment uses olive oil, but a drop of detergent will also work. You can try this experiment at home.

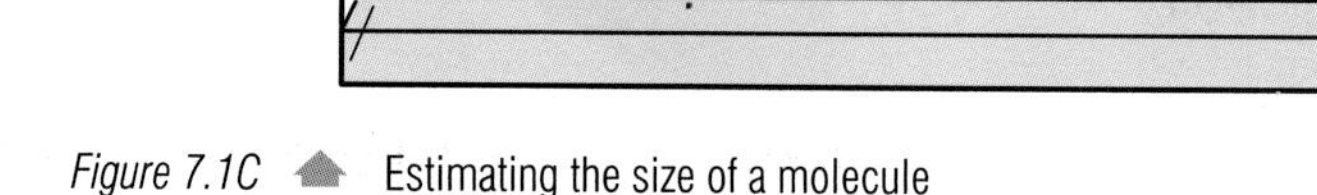
1. Fill a clean tea tray with clean water
2. Sprinkle fine talcum powder on the surface
3. Dip a fine piece of wire into the olive oil. Lift out a tiny drop of oil. Aim to get a drop about 0.5mm in diameter
4. Dip the wire into the water. The droplet of oil spreads out and pushes back the talcum powder. Measure as well as you can the area of the patch of olive oil
Olive oil

Figure 7.1C ▲ Estimating the size of a molecule

SUMMARY

The smallest particle of an element is an atom. Atoms are tiny! A pinhead would hold 5×10^{12} atoms. A molecule consists of two or more atoms. Olive oil molecules are about 2×10^{-6} mm in diameter.

Sample results

Diameter of drop = 0.5 mm
Volume of drop = $(0.5\ \text{mm})^3 = 0.125\ \text{mm}^3$
Area of patch = $(25\ \text{cm})^2 = (250\ \text{mm})^2 = 6.25 \times 10^4\ \text{mm}^2$

Volume of patch = area × depth (d)
$0.125\ \text{mm}^3 = 6.25 \times 10^4\ \text{mm}^2 \times d$

$$d = \frac{0.125\ \text{mm}^3}{6.25 \times 10^4\ \text{mm}^2} \qquad d = 2.5 \times 10^{-6}\ \text{mm}$$

The layer is only 2.5×10^{-6} mm deep (two millonths of a millimetre). We assume that it is one molecule thick.

7.2 ▶ The kinetic theory of matter

FIRST THOUGHTS

Particles in motion: what does this idea explain? The difference between solids, liquids and gases for a start, and the beauty of crystalline solids.

The **kinetic theory of matter** states that matter is made up of small particles which are constantly in motion. (Kinetic comes from the Greek word for 'moving'.) The higher the temperature, the faster they move. In a solid, the particles are close together and attract one another strongly. In a liquid the particles are further apart and the forces of attraction are weaker than in a solid. Most of a gas is space, and the particles shoot through the space at high speed. There are almost no forces of attraction between the particles in a gas.

Scientists have been able to explain many things with the aid of the kinetic theory.

Solid, liquid and gaseous states

The differences between the solid, liquid and gaseous states of matter can be explained on the basis of the kinetic theory.

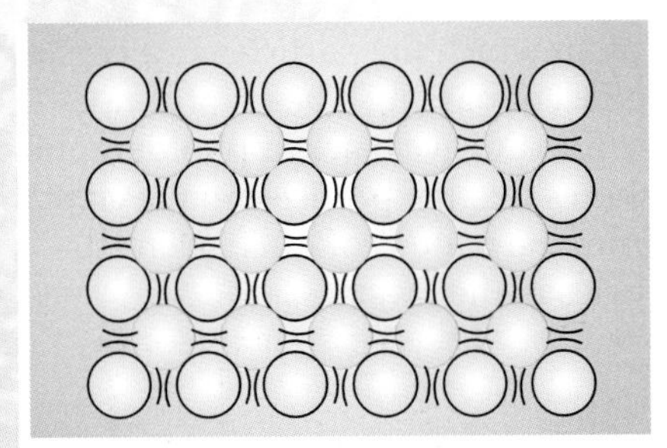

A solid is made up of particles arranged in a regular 3-dimensional structure. There are strong forces of attraction between the particles. Although the particles can vibrate, they cannot move out of their positions in the structure.

When a solid is heated, the particles gain energy and vibrate more and more vigorously. Eventually they may break away from the solid structure and become free to move around. When this happens, the solid has turned into liquid: it has melted.

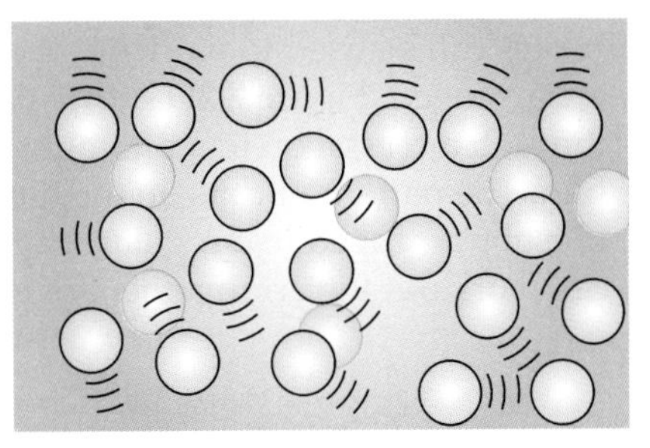

In a liquid the particles are free to move around. A liquid therefore flows easily and has no fixed shape. There are still forces of attraction between the particles.

When a liquid is heated, some of the particles gain enough energy to break away from the other particles. The particles which escape from the body of the liquid become a gas.

In a gas, the particles are far apart. There are almost no forces of attraction between them. The particles move about at high speed. Because the particles are so far apart, a gas occupies a very much larger volume than the same mass of liquid.

The molecules collide with the container. These collisions are responsible for the pressure which a gas exerts on its container.

Figure 7.2A ▲ The arrangement of particles in a solid, a liquid and a gas

Crystals

Crystals are a very beautiful form of solid matter. A crystal is a piece of solid which has a regular shape and smooth faces (surfaces) which reflect light (see Figure 7.2B). Different salts have differently shaped crystals. *Why are many solids crystalline?* Viewing a crystal with an electron microscope, scientists can actually see individual particles arranged in a regular pattern (see Figure 7.2C). It is this regular pattern of particles which gives the crystal a regular shape.

The kinetic theory of matter can explain the differences between the solid, liquid and gaseous states, and also how matter can change state. X-ray photographs show that crystals consist of a regular arrangement of particles.

Figure 7.2B ▲ Crystals of copper (II) sulphate

X-rays can be used to work out the way in which the particles in a crystal are arranged. Figure 7.2D shows the effect of passing a beam of X-rays through a crystal on to a photographic film. X-rays blacken photographic film. The pattern of dots on the film shows that the particles in the crystal must be arranged in a regular way. From the pattern of dots, scientists can work out the arrangement of particles in the crystal.

Metals are composed of lots of tiny crystals called grains.

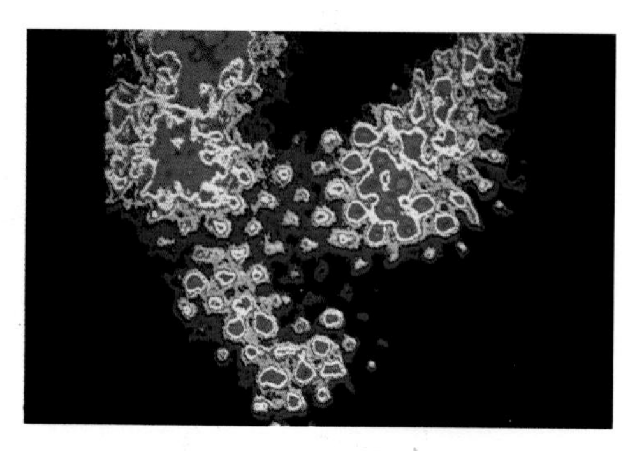

Figure 7.2C ▲ An electron microscope picture of uranyl acetate: each spot represents a single uranium atom

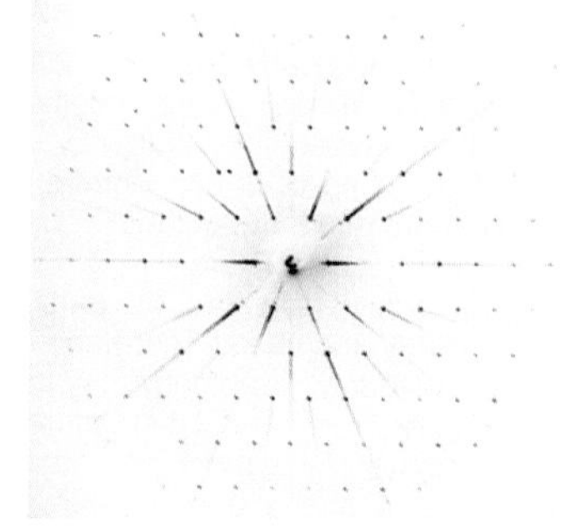

Figure 7.2D ▲ X-ray pattern from crystals of the metal palladium

How do solids dissolve, how do liquids vaporise and how do gases diffuse? Imagine particles in motion, and you will be able to explain all these changes.

Dissolving

Crystals of many substances dissolve in water. You can explain how this happens if you imagine particles splitting off from the crystal and spreading out through the water. See Figure 7.2E.

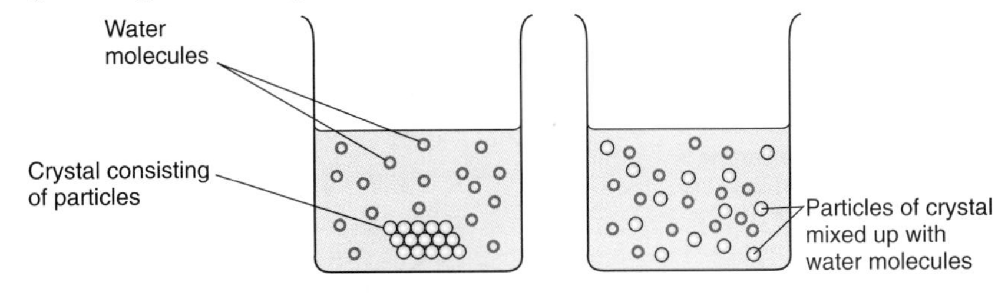

Figure 7.2E ▲ A coloured crystal dissolving

Diffusion

Gases diffuse: they spread out to occupy all the space available to them. The kinetic theory of matter explains gaseous diffusion.

What evidence have we that the particles of a gas are moving? The diffusion of gases can be explained. **Diffusion** is the way gases spread out to occupy all the space available to them. Figure 7.2F shows what happens when a jar of the dense green gas, chlorine, is put underneath a jar of air.

On the theory that gases consist of fast-moving particles, it is easy to explain how diffusion happens. Moving molecules of air and chlorine spread themselves between the two gas jars. The molecules repeatedly collide with each other at random and gradually achieve a uniform distribution throughout the containers.

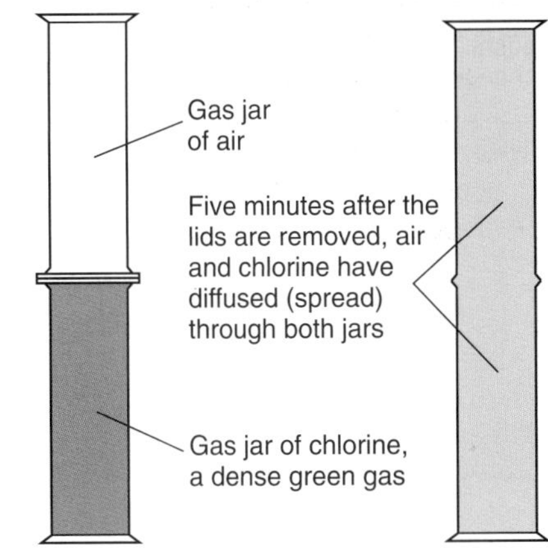

Figure 7.2F ▲ Gaseous diffusion

Brownian motion

This section describes Brownian motion. You will see how neatly the kinetic theory can explain it.

Figure 7.2G shows a smoke cell and the erratic path followed by a particle of smoke.

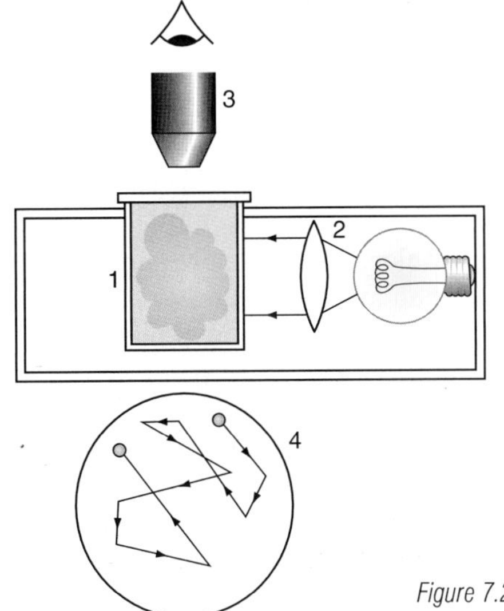

Figure 7.2G ▲ A smoke cell

Brownian motion puzzled scientists until the kinetic theory of matter offered an explanation.

We call this kind of motion **Brownian motion** after the botanist Robert Brown, who first observed it in 1785. He used a microscope to observe pollen grains floating on water. He was amazed to see that the pollen grains were constantly moving about and changing direction. It was as if they had a life of their own. Brown could not explain what he saw. Brownian motion puzzled scientists until the kinetic theory of matter offered an explanation.

The smoke particle is much larger than the air particles

The cell contains air particles which are in constant erratic motion. As they collide with the smoke particle they give it a push. The direction of the push changes at random

Figure 7.2H ▲ Brownian motion

Figure 7.2H shows how the Brownian motion of smoke particles in air is caused. Air molecules must be moving very fast to cause this effect because they are far too small to see. The results of their random collisions with much bigger smoke particles are easily observed. Can you explain, with the aid of a diagram, what makes pollen grains move like this in water?

Evaporation

Why can you cool a hot cup of tea by blowing on it? Read this section to see if your idea is correct.

When a liquid evaporates, it becomes cooler (see Figure 7.2I).

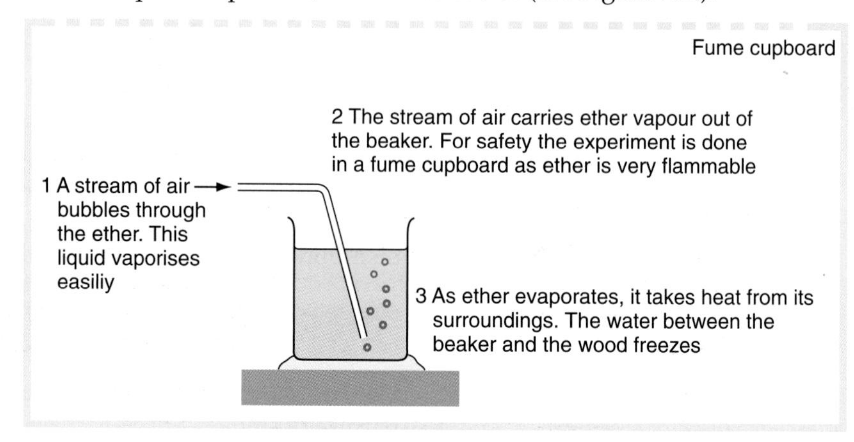
Fume cupboard
2 The stream of air carries ether vapour out of the beaker. For safety the experiment is done in a fume cupboard as ether is very flammable
1 A stream of air bubbles through the ether. This liquid vaporises easiliy
3 As ether evaporates, it takes heat from its surroundings. The water between the beaker and the wood freezes

Figure 7.2I ▲ The cooling effect produced when a liquid evaporates

The kinetic theory can explain this cooling effect. Attractive forces exist between the molecules in a liquid (see Figure 7.2J). Molecules with more energy than average can break away from the attraction of other molecules and escape from the liquid. After the high energy molecules have escaped, the average energy of the molecules which remain is lower than before: the liquid has become cooler.

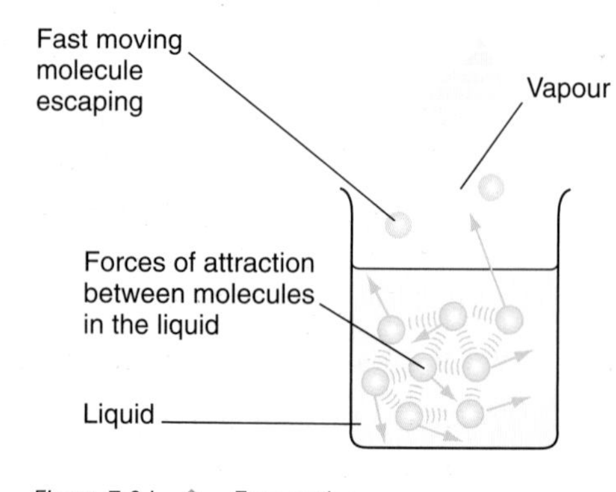
Fast moving molecule escaping
Vapour
Forces of attraction between molecules in the liquid
Liquid

Figure 7.2J ▲ Evaporation

EXTENSION FILE
ACTIVITY

SUMMARY

The kinetic theory of matter can be used to explain the differences between the solid, liquid and gaseous states of matter, change of state (including evaporation), dissolving, diffusion and Brownian motion.

What happens when you raise the temperature? More molecules have enough energy to break away from the other molecules in the liquid. The rate of evaporation increases.

What happens if you pass a stream of dry air across the surface of the liquid? The dry air carries vapour away. The particles in the vapour are prevented from re-entering the liquid, that is, condensing. The liquid therefore evaporates more quickly.

CHECKPOINT

1. Explain the following statements in terms of the kinetic theory.
   (a) Water freezes when it is cooled sufficiently.
   (b) Heating a liquid makes it evaporate more quickly.
   (c) Heating a solid makes it melt.
2. The solid X does not melt until it is heated to a very high temperature. What can you deduce about the forces which exist between particles of X?
3. Of the five substances listed in the table below, which is/are (a) solid (b) liquid (c) gaseous (d) unlikely to exist?

Substance	*Distance between particles*	*Arrangement of particles*	*Movement of particles*
A	Close together	Regular	Move in straight lines
B	Far apart	Regular	Random
C	Close together	Random	Random
D	Far apart	Random	Move in straight lines
E	Close together	Regular	Vibrate a little

4. Supply words to fill in the blanks in this passage.
   A solid has a fixed _____ and a fixed _____. A liquid has a fixed _____, but a liquid can change its _____ to fit its container. A gas has neither a fixed _____ nor a fixed _____. Liquids and gases flow easily; they are called _____. There are forces of attraction between particles. In a solid, these forces are _____, in a liquid they are _____ and in a gas they are _____.
5. Imagine that you are one of the millions of particles in a crystal. Describe from your point of view as a particle what happens when your crystal is heated until it melts.
6. Which of the two beakers in the figure opposite represents (a) evaporation, (b) boiling? Explain your answers.
7. When a stink bomb is left off in one corner of a room, it can soon be smelt everywhere. Why?
8. Beaker A and dish B contain the same volume of the same liquid. Will the liquid evaporate faster in A or B? Explain your answer.
9. Why can you cool a cup of hot tea by blowing on it?
10. A doctor dabs some ethanol (alcohol) on your arm before giving you an injection. The ethanol makes your arm feel cold. How does it do this?

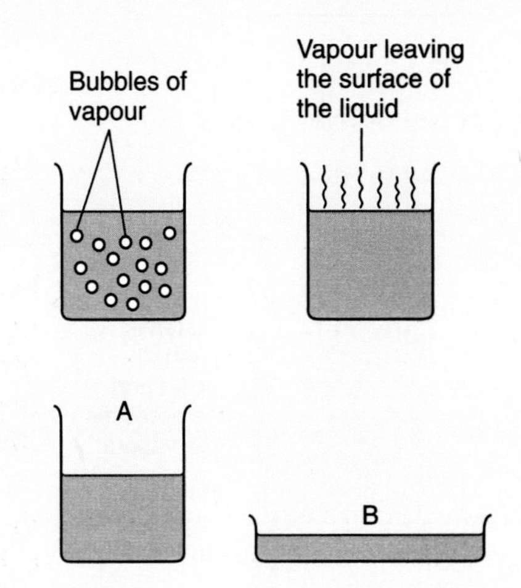

7.3 Saturated and unsaturated vapours

FIRST THOUGHTS

Meteorologists, anaesthetists, mining engineers and chemical engineers are just some of those who need to know about vapours. Read on to find out why.

In the steamroom

Any liquid in contact with air vaporises as molecules move from the liquid into the air space above the liquid. *What do you think happens if you leave a saucer of water in a living room?* Gradually, the water in the saucer will vaporise. This happens because the water molecules that move from the liquid into the air space above spread out through the room. Put some more water in the saucer and that too will evaporate. The air in the room is said to be **unsaturated** because it is capable of containing more water vapour.

Now suppose you left a saucer of water in a steamy room. The water in the saucer would still be present as long as the room stays steamy. *Why doesn't the water evaporate in this situation?* In fact, the air in the room already contains so much water vapour that it cannot contain any more water molecules. Molecules do transfer from the liquid water to the air space but they also transfer in the reverse direction at an equal rate. The air in the room is said to be **saturated** because it is incapable of containing any more water vapour.

(a)

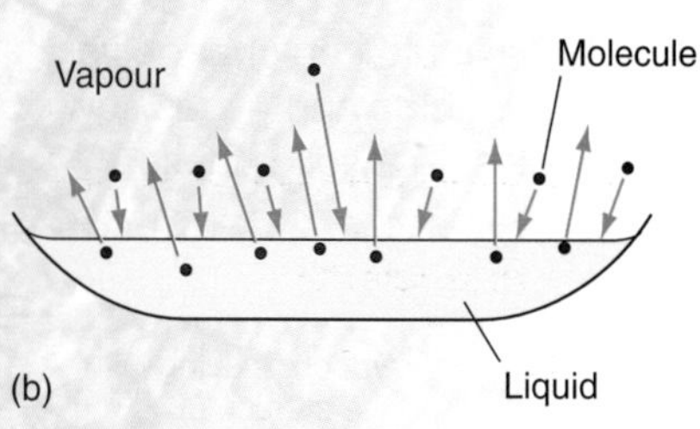

(b)

Figure 7.3A A saturated vapour

Under pressure

In a refrigerator, a volatile liquid vaporises because it absorbs energy from the objects put into the refrigerator. The vapour is then compressed by a pump, causing it to condense and release energy as latent heat. The condenser unit is outside the refrigerator so the heat released warms the room. In this way, heat is transferred from objects in the refrigerator to the room, making the objects colder than the room.

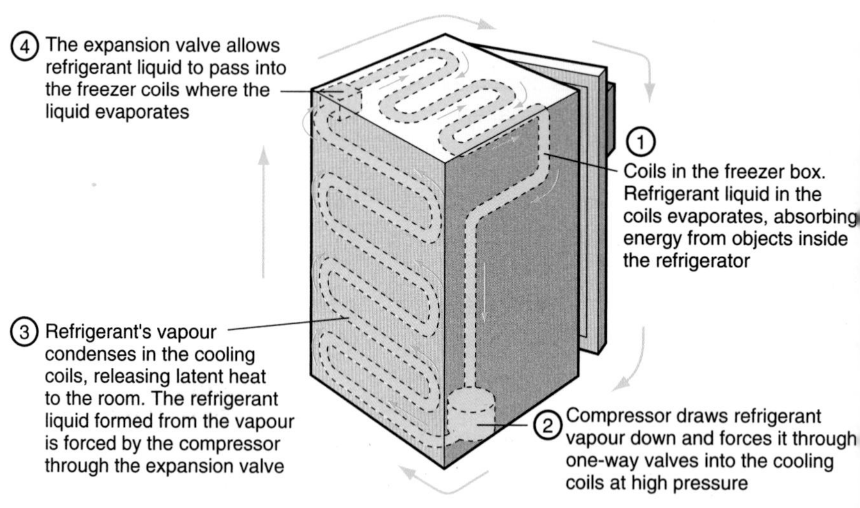

4 The expansion valve allows refrigerant liquid to pass into the freezer coils where the liquid evaporates
1 Coils in the freezer box. Refrigerant liquid in the coils evaporates, absorbing energy from objects inside the refrigerator
3 Refrigerant's vapour condenses in the cooling coils, releasing latent heat to the room. The refrigerant liquid formed from the vapour is forced by the compressor through the expansion valve
2 Compressor draws refrigerant vapour down and forces it through one-way valves into the cooling coils at high pressure

Figure 7.3B How a refrigerator works

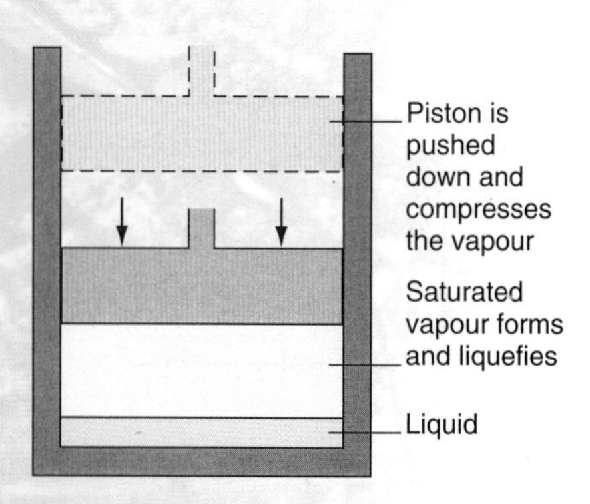

Figure 7.3C Compression of a vapour

Why does a vapour condense when it is compressed? Compression reduces the volume of the vapour and forces the vapour molecules closer together. When they are forced closely together, they attract each other and form the liquid state. In other words, the vapour is initially unsaturated and when it is compressed it becomes saturated and turns into liquid. Figure 7.3C shows the process.

Table 7.1 ▼ Critical temperature

Gas	Critical Temperature
Helium	−268
Hydrogen	−240
Nitrogen	−146
Oxygen	−122
Carbon dioxide	31
Ether	197

A vapour can be converted into a liquid by compression only if its temperature is sufficiently low. For any vapour, there is a **critical temperature** above which it cannot be liquefied no matter how much pressure is applied. Table 7.1 gives the critical temperatures of some common gases. The term *vapour* is used for a gas below its critical temperature.

It's a fact!

A gas can be liquefied by repeatedly compressing it, cooling it and then allowing it to expand. Liquid nitrogen is used in processes such as the freeze-drying of food.

Saturated vapour pressure

The pressure of a saturated vapour does not depend on the volume occupied by the vapour. Reducing this volume does not increase the pressure as it would with an unsaturated vapour or a gas. This is because when the volume of a saturated vapour is reduced, there is an overall transfer of molecules from the vapour state to the liquid state.

The pressure of a saturated vapour increases with temperature. Figure 7.3D shows how to measure the boiling point of a liquid at different pressures below atmospheric pressure. At a given liquid temperature, the vacuum pump is used to reduce the pressure in the flask until the liquid boils. At this point, the pressure of the saturated vapour is equal to the pressure in the flask. In this way, the saturated vapour pressure of the liquid can be determined at different temperatures.

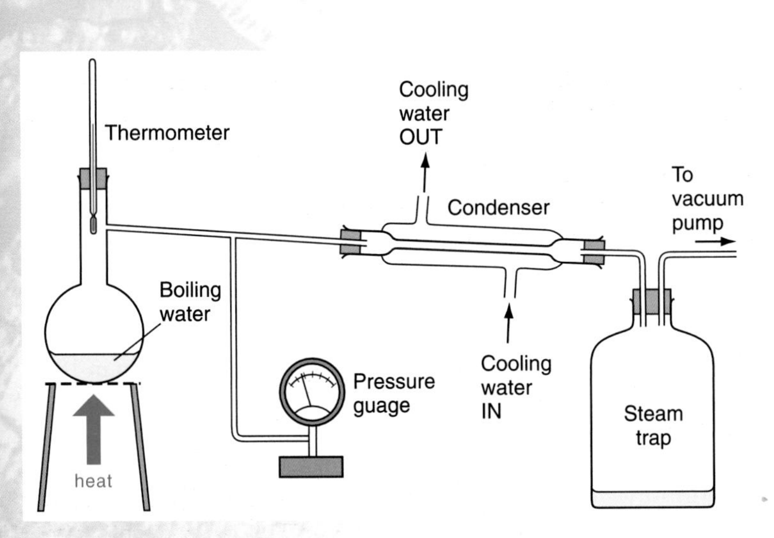

Figure 7.3D ▲ (a) Measuring the saturated vapour pressure of water at reduced pressure

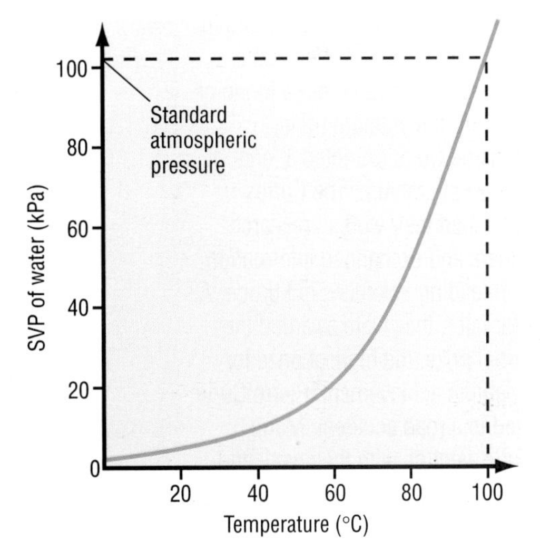

(b) SVP against temperature for water

CHECKPOINT

1. Why is humid weather uncomfortable?
2. (a) Wet clothes on a washing line dry faster on a windy day than on a still day. Why?
   (b) A tumble drier is designed to dry damp clothing by blowing a current of warm air through a rotating drum containing the clothing. Why should the air from the drum then be 'vented' to outdoors?
3. Simon is collecting a suit from the dry cleaners. A notice in the shop warns customers using cars to open the car windows when they have put dry cleaned clothes in the car. Why is this necessary?
4. Why is the boiling point of water reduced at high altitude?
5. Describe and explain what happens if a sealed bottle of a fizzy drink is shaken and then opened.

Topic 8 Inside the atom

8.1 Becquerel's key

FIRST THOUGHTS

This topic will give you a glimpse of the fascinating story of the discovery of radioactivity.

KEY SCIENTISTS

The Curies worked for four years in a cold, ill-equipped shed at the University of Paris. From a tonne of ore from the uranium mine, Madame Curie extracted a tenth of a gram of uranium. The Curies published their work in research papers and exchanged information with leading scientists in Europe. A year later, they were awarded the Nobel prize, the highest prize for scientific achievement. Pierre Curie died in a road accident. Marie Curie went on with their work and won a second Nobel prize. She died in middle-age from leukaemia, a disease of the blood cells. This was caused by the radioactive materials she worked with.

In 1896, a French physicist, Henri Becquerel, left some wrapped photographic plates in a drawer. When he developed the plates, he found the image of a key. The plates were 'fogged' (partly exposed). The areas of the plates which had not been exposed were in the shape of a key. Looking in the drawer, Becquerel found a key and a packet containing some uranium compounds. He did some further tests before coming to a strange conclusion. He argued that some unknown rays, of a type never met before, were coming from the uranium compounds. The mysterious rays passed through the wrapper and fogged the photographic plates. Where the key lay over the plates, the rays could not penetrate, and the image formed on the plates.

Figure 8.1A ▲ Becquerel's key

A young research worker called Marie Curie took up the problem in 1898. She found that this strange effect happened with all uranium compounds. It depended only on the amount of uranium present in the compound and not on which compound she used. Madame Curie realised that this ability to give off rays must belong to the 'atoms' of uranium. It must be a completely new type of change, different from the chemical reactions of uranium salts. This was a revolutionary new idea. Marie Curie called the ability of uranium atoms to give off rays **radioactivity**.

Marie Curie's husband, Pierre, joined in her research into this brand new branch of science. Together, they discovered two new radioactive elements. They called one **polonium**, after Madame Curie's native country, Poland. They called the second **radium**, meaning 'giver of rays'. Its salts glowed in the dark.

Many scientists puzzled over the question of why the atoms of these elements, uranium, polonium and radium, give off the rays which Marie Curie named radioactivity. The person who came up with an explanation was the British physicist, Lord Rutherford. In 1902 he suggested that radioactivity is caused by atoms splitting up. This was another revolutionary idea. The word 'atom' comes from the Greek word for 'cannot be divided'. When the British chemist John Dalton put forward his Atomic Theory in 1808, he said that atoms cannot be created or destroyed or split. Lord Rutherford's idea was proved by experiment to be correct. We know now that many elements have atoms which are unstable and split up into smaller atoms.

8.2 Protons, neutrons and electrons

FIRST THOUGHTS

Atoms are made up of even smaller particles: protons, neutrons and electrons. As you read this section try to visualise how these particles are arranged inside the atom.

The work of Marie and Pierre Curie, Rutherford and other scientists showed that atoms are made up of smaller particles. These **subatomic particles** differ in mass and in electrical charge. They are called **protons, neutrons** and **electrons** (see Table 9.1).

Table 8.1 Sub-atomic particles

Particle	Mass (in atomic mass units)	Charge
Proton	1	$+e$
Neutron	1	0
Electron	0.0005	$-e$

Protons and neutrons both have the same mass. We call this mass one **atomic mass unit**, one u ($1.000\ \text{u} = 1.67 \times 10^{-27}$ kg). The mass of an atom depends on the number of protons and neutrons it contains. The electrons in an atom contribute very little to its mass. The number of protons and neutrons together is called the **mass number**.

Electrons carry a fixed quantity of negative electric charge. This quantity is usually written as $-e$. A proton carries a fixed charge equal and opposite to that of the electron. The charge on a proton can be written as $+e$. Neutrons are uncharged particles. Whole atoms are uncharged because the number of electrons in an atom is the same as the number of protons. The number of protons (which is also the number of electrons) is called either the **atomic number** or the **proton number**. You can see that

$$\text{Number of neutrons} = \text{Mass number} - \text{Atomic (proton) number}$$

For example, an atom of potassium has a mass of 39 u and an atomic (proton) number of 19. The number of electrons is 19, the same as the number of protons. The number of neutrons in the atom is

$$39 - 19 = 20$$

Rutherford had an easier life in science than the Curies. He arrived in Britain from New Zealand in 1895. By the age of 28, he was a professor. He worked in the universities of Montreal in Canada, Manchester and Cambridge. He was knighted in 1914 and made Lord Rutherford of Nelson in 1931. His co-worker Otto Hahn described him as 'a very jolly man'. Many stories were told about Rutherford. He would whistle 'Onward, Christian soldiers' when the research work was going well and 'Fight the good fight' when difficulties had to be overcome.

Relative atomic mass

The lightest of atoms is an atom of hydrogen. The mass of other atoms may be compared with the mass of a hydrogen atom. This gives the **relative atomic mass** of an atom. The relative atomic mass, A_r, of calcium is 40. This means that one calcium atom is 40 times as heavy as one hydrogen atom. Because some hydrogen atoms are heavier than normal hydrogen atoms, scientists now use the carbon-12 scale. On this scale, 1 unit of atomic mass is defined as exactly $\frac{1}{12}$th of the mass of a carbon-12 atom.

$$\text{Relative atomic mass of an element} = \frac{\text{Mass of one atom of the element}}{\frac{1}{12}\text{th of the mass of one atom of carbon-12}}$$

CHECKPOINT

1 Some relative atomic masses are:
$A_r(H) = 1$, $A_r(He) = 4$, $A_r(C) = 12$, $A_r(O) = 16$, $A_r(Ca) = 40$.
Copy and complete the following sentences.
(a) A calcium atom is ______ times as heavy as an atom of hydrogen.
(b) A calcium atom is ______ times as heavy as an atom of helium.
(c) One carbon atom has the same mass as ______ helium atoms.
(d) ______ helium atoms have the same mass as two oxygen atoms.
(e) Two calcium atoms have the same mass as ______ oxygen atoms.

2 Element E has atomic number 9 and mass number 19. Say how many protons, neutrons and electrons are present in one atom of E.

3 State (i) the atomic number and (ii) the mass number of:
(a) an atom with 17 protons and 18 neutrons,
(b) an atom with 27 protons and 32 neutrons,
(c) an atom with 50 protons and 69 neutrons.

8.3 The arrangement of particles in the atom

It's a fact!

If the nucleus of an atom was the size of a cricket ball, the nearest electron would be in the stands.

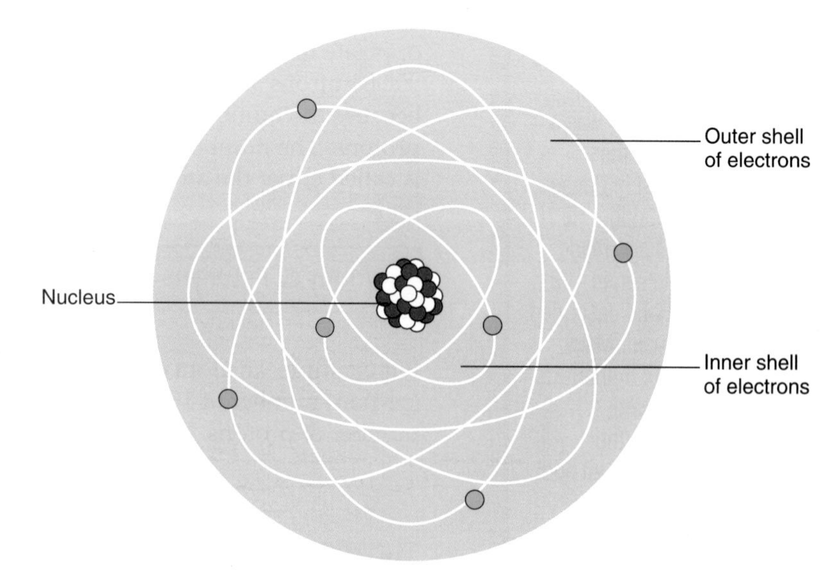

Figure 8.3A The structure of an atom

Lord Rutherford showed, in 1914, that most of the volume of an atom is space. Only protons and electrons were known in 1914; the neutron had not yet been discovered. Rutherford pictured the massive particles, the protons, occupying a tiny volume in the centre of the atom. Rutherford called this the **nucleus**. We now know that the nucleus contains neutrons as well as protons. The electrons occupy the space outside the nucleus. The nucleus is minute in volume compared with the volume of the atom.

The electrons of an atom are in constant motion. They move round and round the nucleus in paths called **orbits**. The electrons in orbits close to the nucleus have less energy than electrons in orbits distant from the nucleus.

8.4 How are the electrons arranged?

Figure 8.4A illustrates the electrons of an atom in their orbits. The orbits are grouped together in **shells**. A shell is a group of orbits with similar energy. The shells distant from the nucleus have more energy than those close to the nucleus. Each shell can hold up to a certain number of electrons.

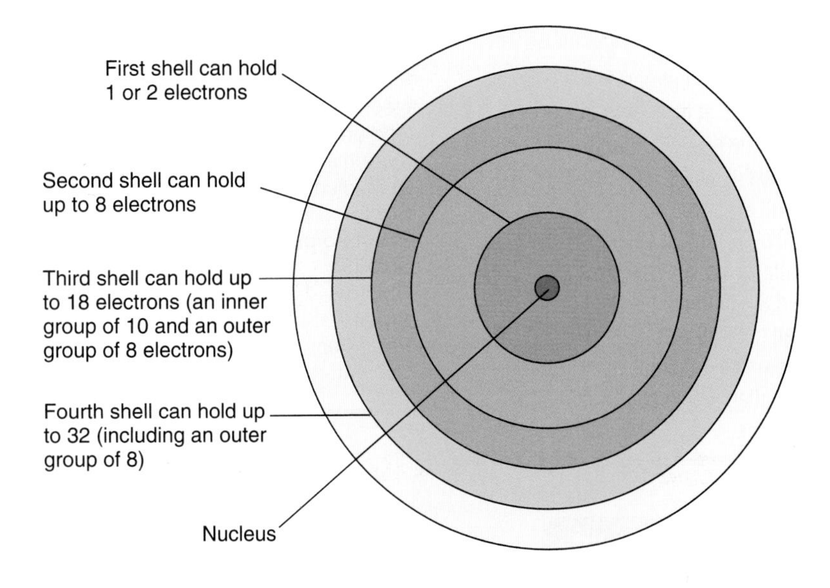
First shell can hold 1 or 2 electrons
Second shell can hold up to 8 electrons
Third shell can hold up to 18 electrons (an inner group of 10 and an outer group of 8 electrons)
Fourth shell can hold up to 32 (including an outer group of 8)
Nucleus

Figure 8.4A Shells of electron orbits in an atom

The chemical properties of an element are determined by the arrangement of the electrons in the atoms of the element.

See www.keyscience.co.uk for more about atoms, molecules and elements.

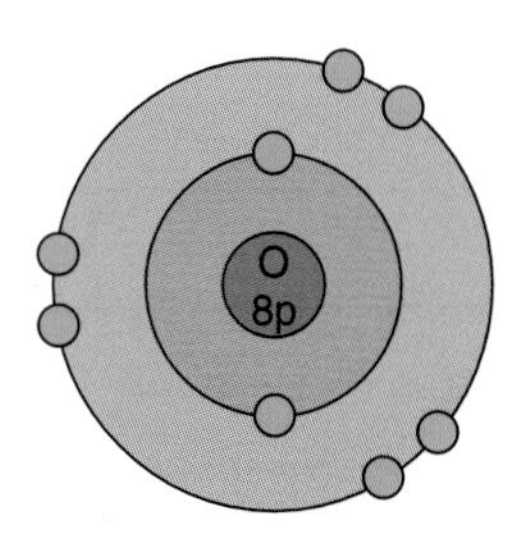
O
8p

Figure 8.4B The arrangement of electrons in the oxygen atom

The atomic number of an element tells you the number of electrons in an atom of the element. The electrons fill the innermost orbits in the atom first. An atom of oxygen has 8 electrons. Two electrons enter the first shell, which is then full. The other 6 electrons go into the second shell (see Figure 8.4B).

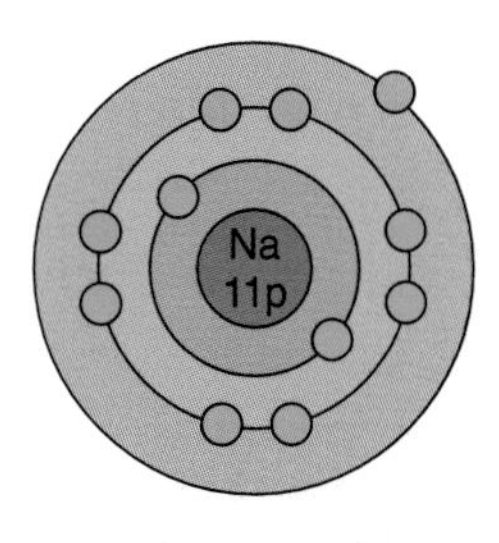
Na
11p

Figure 8.4C The electron configuration of sodium

An atom of sodium has atomic number 11. The first shell is filled by 2 electrons, the second shell is filled by 8 electrons, and 1 electron occupies the third shell. The arrangement of electrons can be written as (2.8.1). It is called the **electron configuration** of sodium. Table 8.2 gives the electron configurations of the first 20 elements.

SUMMARY

The atoms of all elements are made up of three kinds of particles. These are:

- protons, of mass 1 u and electric charge $+e$
- neutrons, of mass 1 u, uncharged
- electrons, of mass 0.0005 u and electric charge $-e$.

The protons and neutrons make up the nucleus at the centre of the atom. The electrons circle the nucleus in orbits. Groups of orbits with the same energy are called shells. The 1st shell can hold 2 electrons; the 2nd shell can hold 8 electrons; the 3rd shell can hold 18 electrons. The arrangement of electrons in an atom is called the electron configuration.

Table 8.2 Electron configurations of the atoms of the first twenty elements

Element	Symbol	Atomic (proton) number	1st shell	2nd shell	3rd shell	4th shell	Electron configuration
Hydrogen	H	1	1				1
Helium	He	2	2				2
Lithium	Li	3	2	1			2.1
Beryllium	Be	4	2	2			2.2
Boron	B	5	2	3			2.3
Carbon	C	6	2	4			2.4
Nitrogen	N	7	2	5			2.5
Oxygen	O	8	2	6			2.6
Fluorine	F	9	2	7			2.7
Neon	Ne	10	2	8			2.8
Sodium	Na	11	2	8	1		2.8.1
Magnesium	Mg	12	2	8	2		2.8.2
Aluminium	Al	13	2	8	3		2.8.3
Silicon	Si	14	2	8	4		2.8.4
Phosphorus	P	15	2	8	5		2.8.5
Sulphur	S	16	2	8	6		2.8.6
Chlorine	Cl	17	2	8	7		2.8.7
Argon	Ar	18	2	8	8		2.8.8
Potassium	K	19	2	8	8	1	2.8.8.1
Calcium	Ca	20	2	8	8	2	2.8.8.2

CHECKPOINT

1. Silicon has the electron configuration (2.8.4). What does this tell you about the arrangement of electrons in the atom? Sketch the arrangement. (See Figures 8.4B and 8.4C for help.)
2. Sketch the arrangement of electrons in the atoms of (a) He (b) C (c) F (d) Al (e) Mg. (See Table 8.2 for atomic numbers.)

8.5 Isotopes

FIRST THOUGHTS

There are two sorts of chlorine atom and three sorts of hydrogen atom. You can find out the difference between them in this section.

Atoms of the same element all contain the same number of protons, but the number of neutrons may be different. Forms of an element which differ in the number of neutrons in the atom are called **isotopes**. For example, the element chlorine, with relative atomic mass 35.5, consists of two kinds of atom with different mass numbers.

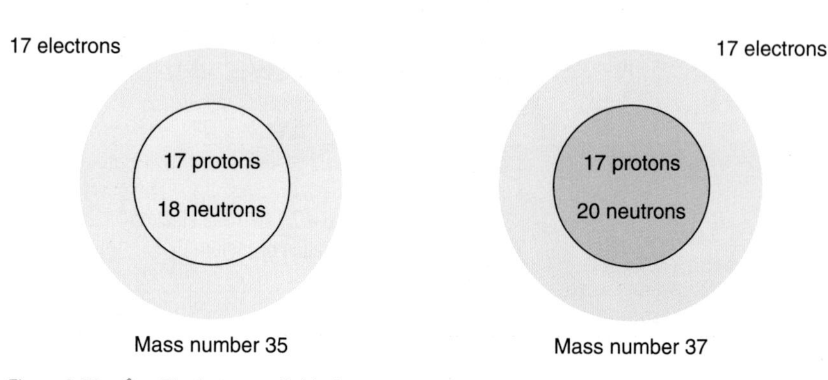

Figure 8.5A The isotopes of chlorine

Since the chemical reactions of an atom depend on its electrons, all chlorine atoms react in the same way. The number of neutrons in the nucleus does not affect chemical reactions. The different forms of chlorine are isotopes. Their chemical reactions are the same. In any sample of chlorine, there are three chlorine atoms with mass 35 u for each chlorine atom with mass 37 u so the average atomic mass is

$$\frac{(3 \times 35) + 37}{4} = 35.5 \text{ u}$$

Isotopes are shown as ${}^{A}_{Z}\mathbf{X}$, where X is the chemical symbol, *A* is the mass number (i.e. the number of neutrons and protons) and Z is the atomic number (i.e. the number of protons).

The isotopes of chlorine are written as ${}^{35}_{17}Cl$ and ${}^{37}_{17}Cl$. The isotopes of hydrogen are ${}^{1}_{1}H$, ${}^{2}_{1}H$ and ${}^{3}_{1}H$. They are often referred to as hydrogen-1, hydrogen-2 and hydrogen-3. Hydrogen-2 is also called deuterium, and hydrogen-3 is also called tritium.

SUMMARY

Isotopes are atoms of the same element which differ in the number of neutrons. They contain the same number of protons (and therefore the same number of electrons).

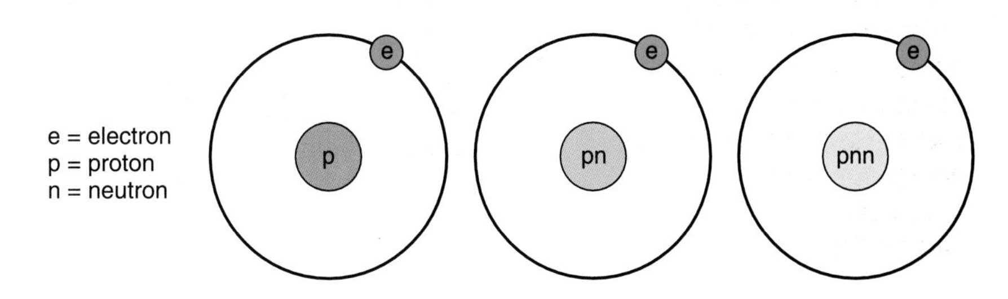

Figure 8.5B The isotopes of hydrogen

CHECKPOINT

1 Hydrogen, deuterium and tritium are isotopes.

(a) Copy and complete this sentence. Isotopes are ______ of an element which contain the same number of ______ and ______ but different numbers of ______.

(b) Copy and complete the table.

	Hydrogen	*Deuterium*	*Tritium*
Atomic number			
Mass number			

(c) Write the formula of the compound formed when deuterium reacts with oxygen.

(d) Explain why isotopes have the same chemical reactions.

2 Write the symbol with mass number and atomic number (as above) for each of the following isotopes.

(a) oxygen with 8 protons and 8 neutrons

(b) argon with 18 protons and 22 neutrons

(c) bromine (symbol Br) with 35 protons and 45 neutrons

(d) chromium (symbol Cr) with 24 protons and 32 neutrons

3 Copy this table, and fill in the missing numbers.

Particle	*Mass number*	*Atomic number*	*Number of... protons*	*neutrons*	*electrons*
Nitrogen atom	14	7	–	–	–
Sodium atom	23	–	–	–	11
Potassium atom	39	–	19	–	–
Uranium atom	235	92	–	–	–

Topic 9 Radioactivity

9.1 What is radioactivity?

FIRST THOUGHTS

Radioactivity: what is it? How was it discovered? What are α-, β-, γ-radiation? These are some of the things you will find out in this section.

The changes which Marie Curie described as radioactivity are **nuclear changes**. A nuclear change is quite different from a chemical change. In a chemical change, bonds between atoms are broken, and new bonds are made, but the nuclei of the atoms stay the same. In a nuclear change, new atoms of different elements are formed because the nuclei change. Marie Curie and her fellow scientists knew their work was revolutionary, but none of them could have imagined how it would change the world.

Isotopes which give off radioactivity are said to be **radioactive**. They are **radioisotopes**. The atoms of radioactive isotopes each have an unstable nucleus. The nucleus becomes stable by emitting small particles and energy. The particles and energy are called **radioactivity**; the breaking-up process is **radioactive decay**.

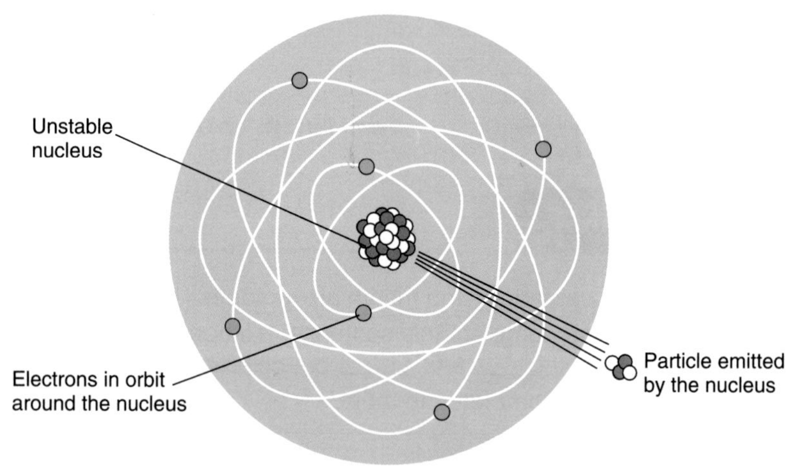

Figure 9.1A ▲ Radioactive decay

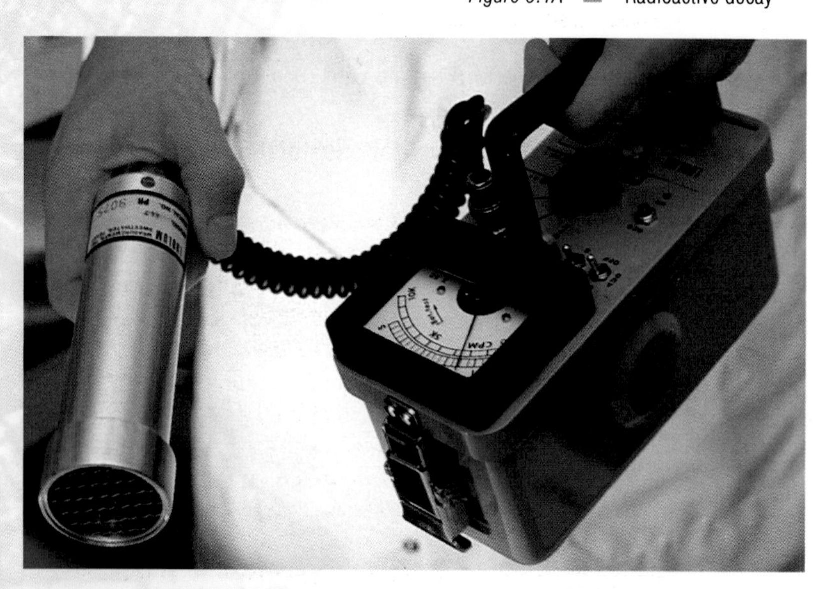

Figure 9.1B ▲ Using a Geiger–Müller tube.
(NOTE: experiments on radioactivity must only be done by a teacher.)

Investigating radioactivity

One way to detect radioactivity is to use a Geiger–Müller counter (named after its inventors). When radioactive particles enter it, a Geiger–Müller counter gives out a clicking sound. The number of clicks per minute is called the count rate. Even when there is no radioactive source nearby, the counter still clicks occasionally. This is due to **background radioactivity** from certain building materials and from cosmic radiation (from space). When you measure the count rate of a radioactive source, you have to subtract the background count.

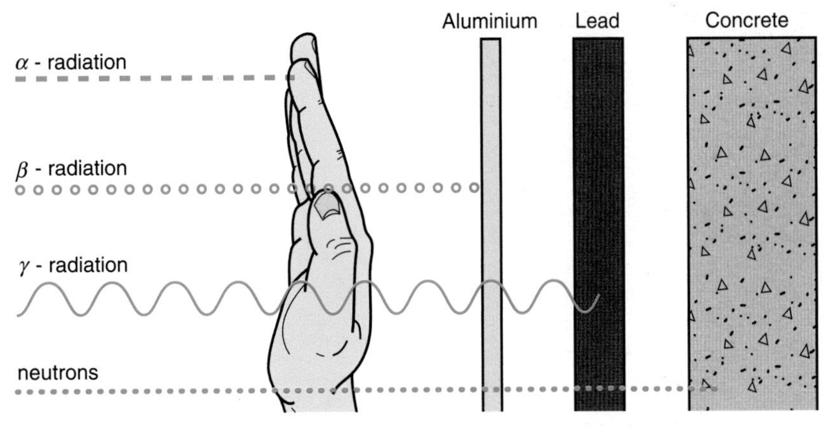
Aluminium
Lead
Concrete
α - radiation
β - radiation
γ - radiation
neutrons

Figure 9.1C ▲ The penetrating power of radiation

There are three types of radioactivity: **alpha radiation** (α-radiation), **beta radiation** (β-radiation) and **gamma radiation** (γ-radiation). They differ in penetrating power (see Figure 9.1C). A thin metal foil will stop α-radiation, but a metal plate about 5 mm thick is needed to stop β-radiation, and γ-radiation will penetrate several centimetres of lead, γ-radiation passes through the skin and can penetrate bone. It can cause burns and cancer. People who work with sources of γ-radiation protect themselves by building a wall of lead bricks between themselves and the source (see Figure 9.1D).

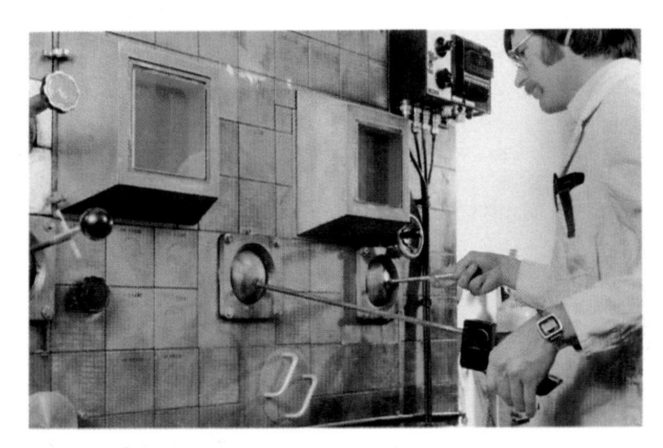

Figure 9.1D ▲ Lead bricks shield workers from radiation

The nature of α, β and γ-radiations

What are these mysterious radiations? This was the question that Lord Rutherford and other scientists tackled early in this century. Within ten years of Henri Becquerel's discovery of radioactivity, scientists had unravelled the mystery. They discovered three new types of radiation by using a magnetic field to separate them (see Figure 9.1E). Scientists invented new instruments, like the Geiger–Müller counter and the cloud chamber (see Figure 9.1F) to use in their study.

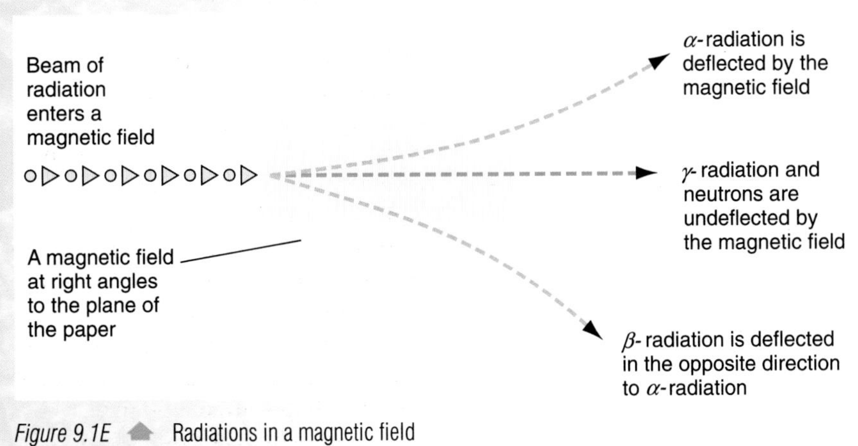
Beam of radiation enters a magnetic field
A magnetic field at right angles to the plane of the paper
α-radiation is deflected by the magnetic field
γ-radiation and neutrons are undeflected by the magnetic field
β-radiation is deflected in the opposite direction to α-radiation

Figure 9.1E ▲ Radiations in a magnetic field

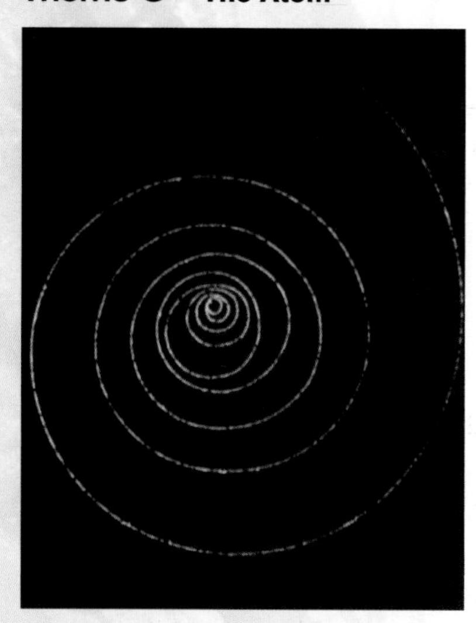

Figure 9.1G ▲ A β-particle track in a cloud chamber with a strong magnetic field

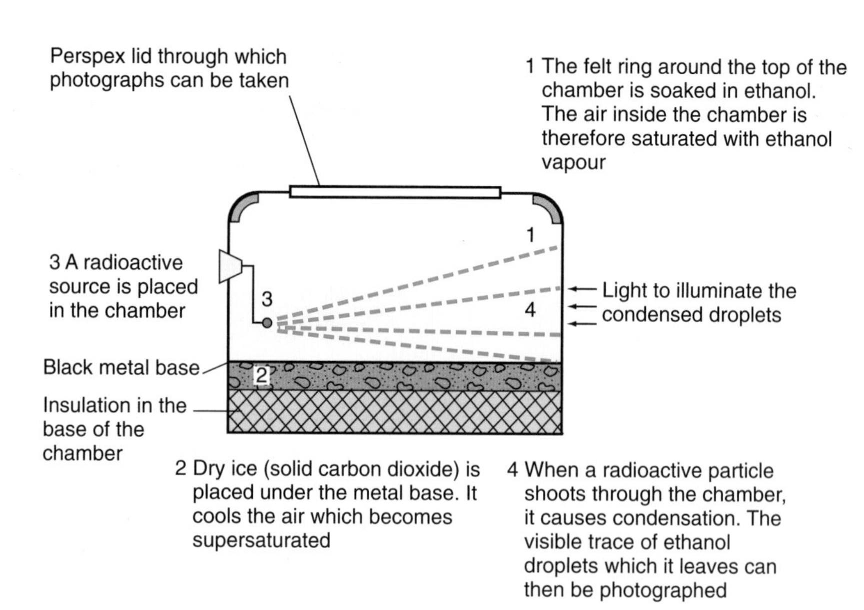
Perspex lid through which photographs can be taken
1 The felt ring around the top of the chamber is soaked in ethanol. The air inside the chamber is therefore saturated with ethanol vapour
1
3 A radioactive source is placed in the chamber
3
4
Light to illuminate the condensed droplets
Black metal base
2
Insulation in the base of the chamber
2 Dry ice (solid carbon dioxide) is placed under the metal base. It cools the air which becomes supersaturated
4 When a radioactive particle shoots through the chamber, it causes condensation. The visible trace of ethanol droplets which it leaves can then be photographed

Figure 9.1F ▲ A cloud chamber

α-radiation

α-radiation consists of α-particles. These are the nuclei of helium atoms; they consist of 2 protons and 2 neutrons and carry a positive charge. A nucleus of an α-emitting isotope changes from an unstable nucleus to a more stable nucleus by ejecting 2 protons and 2 neutrons as a single particle. The stable nucleus has 2 protons less: it has a different atomic number: it is a nucleus of an atom of a different element. An example of α-emission is

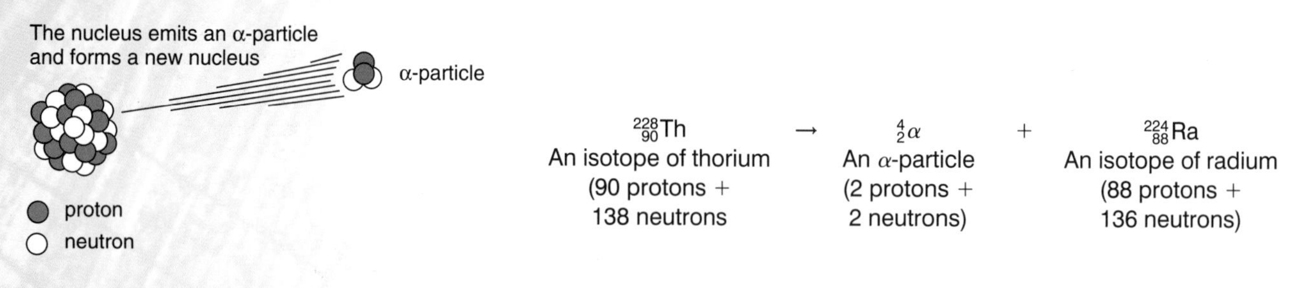
The nucleus emits an α-particle and forms a new nucleus
α-particle
proton
neutron
$^{228}_{90}$Th → $^{4}_{2}\alpha$ + $^{224}_{88}$Ra
An isotope of thorium (90 protons + 138 neutrons)
An α-particle (2 protons + 2 neutrons)
An isotope of radium (88 protons + 136 neutrons)

Figure 9.1H ▲ α-emission

β-radiation

β-radiation consists of electrons. A nucleus of a β-emitting isotope changes from an unstable nucleus to a stable nucleus by changing a neutron into a proton plus an electron and ejecting the electron from the nucleus. The new nucleus has one more proton than before and a different atomic number. It is a nucleus of an atom of a different element. An example of β-emission is

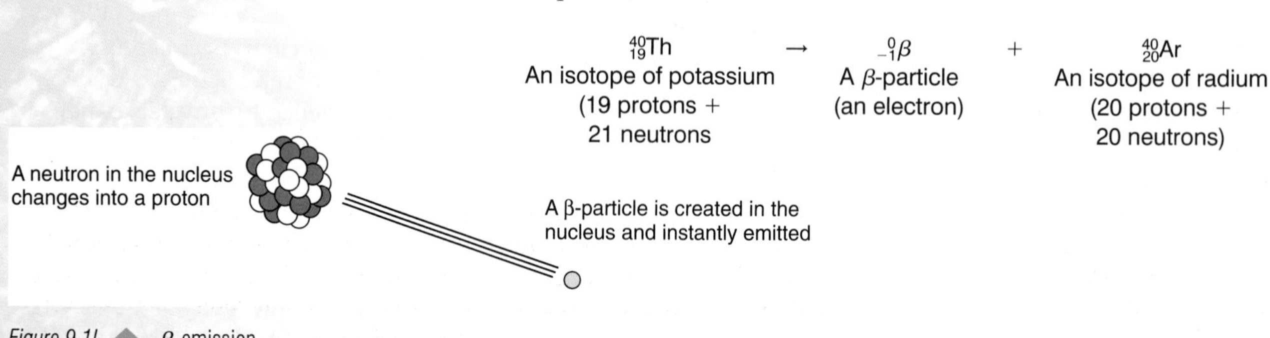
$^{40}_{19}$Th → $^{0}_{-1}\beta$ + $^{40}_{20}$Ar
An isotope of potassium (19 protons + 21 neutrons
A β-particle (an electron)
An isotope of radium (20 protons + 20 neutrons)
A neutron in the nucleus changes into a proton
A β-particle is created in the nucleus and instantly emitted

Figure 9.1I ▲ β-emission

Unstable nuclei become stable by emitting an α-particle or a β-particle or γ-radiation.

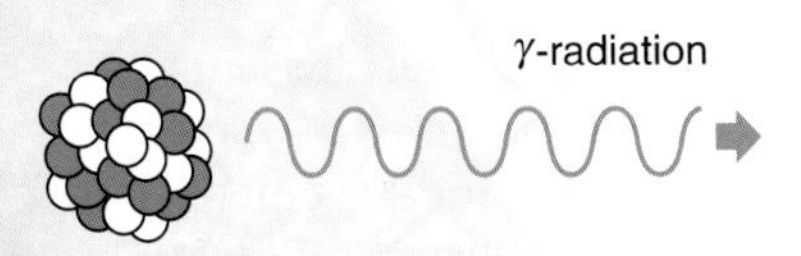

γ-radiation
The nucleus emits γ-radiation so that it can lose surplus energy

Figure 9.1J ▲ γ-emission

γ-radiation

γ-radiation is high-energy electromagnetic radiation. After an unstable nucleus has emitted an α- or a β-particle, it sometimes has surplus energy. It emits this energy as γ-radiation.

Probing the nucleus

Lord Rutherford showed that most of the atom is empty space. He proved conclusively that

- most of its mass is concentrated in a tiny volume which he called the nucleus of the atom
- the nucleus carries a positive charge equal to Ze, where Z is the atomic number of the element the atom belongs to and e is the magnitude of the charge of the electron.

The diameter of an atom is about 10^{-10} m.

The diameter of its nucleus is about 10^{-15} m.

How did Rutherford arrive at these conclusions? Two of his research workers, Hans Geiger and Ernest Marsden, investigated the scattering of alpha particles in a beam by thin metal foils, as outlined in Figure 9.1K.

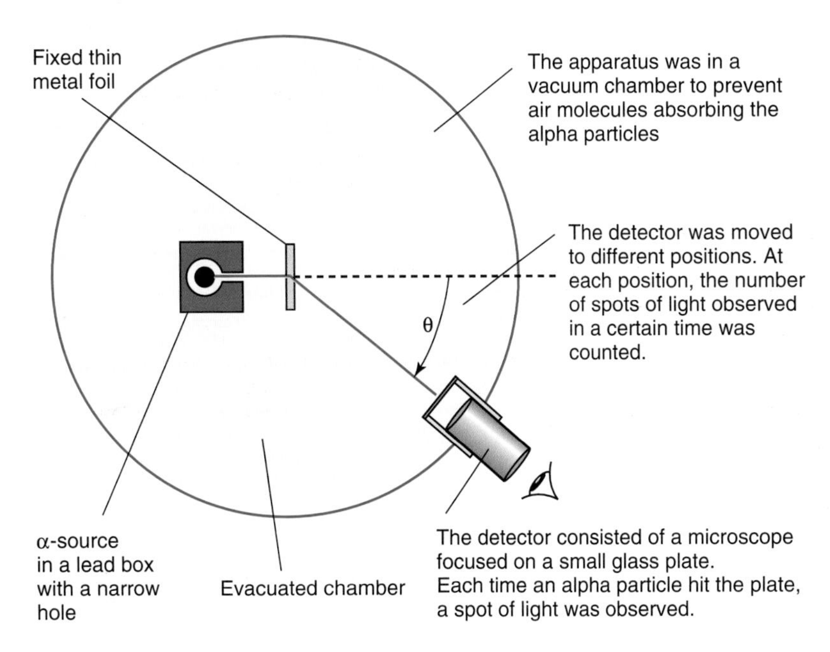

Fixed thin metal foil
The apparatus was in a vacuum chamber to prevent air molecules absorbing the alpha particles
The detector was moved to different positions. At each position, the number of spots of light observed in a certain time was counted.
θ
α-source in a lead box with a narrow hole
Evacuated chamber
The detector consisted of a microscope focused on a small glass plate. Each time an alpha particle hit the plate, a spot of light was observed.

Figure 9.1K ▲ Alpha particle scattering

Geiger and Marsden's results showed that

- most of the alpha particles were undeflected
- the proportion of alpha particles scattered decreased as the angle θ was increased.
- about 1 in 10000 of the incident alpha particles were deflected by more than 90°.
- some of the alpha particles bounced off the foil back towards the source.

Rutherford was astonished by these results. He likened the situation to shooting pellets at a sheet of paper and discovering that an occasional pellet rebounds. He realized that the atom must be mostly empty space with most of its mass concentrated in a tiny volume which he referred to as the nucleus. He assumed that the nucleus is positively charged so it

It's a fact!

Ideas about the atom

Before Rutherford's great discovery, J.J. Thompson who discovered the electron in 1895 reckoned that the atom was like a miniature plum pudding with its electrons like small plums dotted about inside a ball of positive charge. A currant bun is a better comparison (the currants being the electrons) but Thompson seemed to prefer a plum pudding. However, the Thompson atom has no concentration of mass and positive charge inside it so could not be used to explain why alpha particles passing through an atom are sometimes strongly deflected.

SUMMARY

Radioactivity is caused by nuclear changes. Some isotopes have unstable nuclei. An unstable nucleus changes into a stable nucleus by emitting an α-particle or a β-particle or γ-radiation or a combination of these. A Geiger–Müller counter is used to detect and measure radioactivity.

repels alpha particles because they carry positive charge too. Based on this model, he then worked out

- how the path of each alpha particle depends on the least distance between the nucleus and the line of approach of the alpha particle. Figure 9.1 shows some of the possible paths.
- the proportion of alpha particles deflected at each position. His results agreed exactly with the measurements obtained by Geiger and Marsden. The very close agreement between the measurements and his theoretical calculations demonstrated his nuclear model of the atom conclusively.

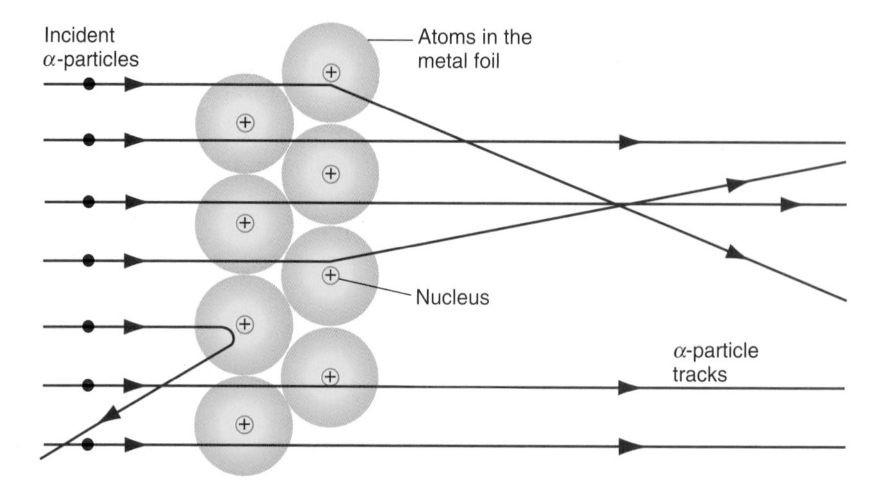

Figure 9.1L Alpha particle paths

CHECKPOINT

1. Henri Becquerel discovered radioactivity when he developed an unused photographic plate and found an image of a key on it. What type of radioactivity was emitted by the uranium salts?

2. In an investigation to find out what type of radioactivity was emitted from a given source, the following measurements were made with a Geiger–Müller counter.

Source	Average count rate (counts per minute)
No source present	29.5
Source 20 mm from the tube and...	
1 no absorber present	385.2
2 a sheet of metal foil between the source and the tube	387.2
3 a 10 mm thick aluminium plate between the source and the tube	32.4

(a) What was the count rate due to the background activity?
(b) What was the count rate due to the source?
(c) What type of radioactivity did the source emit?

3. Atomic nuclei are composed of protons and neutrons. Work out the number of protons and the number of neutrons in each of the following atoms.
(a) $^{238}_{92}U$ (b) $^{234}_{91}Pa$ (c) $^{227}_{89}Ac$

4. (a) $^{238}_{92}U$ is an α-emitter. When a uranium-238 nucleus emits an α-particle, it forms a new nucleus. Which of the following list is the new nucleus?
$^{234}_{92}U$ $^{233}_{90}Th$ $^{234}_{90}Th$ $^{228}_{90}Th$ $^{231}_{91}Th$
Write an equation for the nuclear change.
(b) $^{234}_{91}Pa$ is a β-emitter. From the list in part (a), identify the nucleus formed when this change happens. Write the equation for the change.

9.2 Radioactive decay

FIRST THOUGHTS

- You will find a computer simulation of radioactive decay interesting.
- You can also use IT for data capture and display. A radioactive source, a VELA and an oscilloscope will show a graph of radioactivity against time.

Radioactivity is caused by unstable nuclei emitting particles and energy as they change into more stable nuclei. Each unstable nucleus is said to **decay** or **disintegrate** when it emits a radioactive particle.

The **activity** of a radioactive source is the number of its nuclei that disintegrate per second. The unit of activity is the **becquerel**, Bq.

1 Bq = disintegration per second

Suppose a radioactive isotope decays to form a non-radioactive product. The number of radioactive atoms decreases with time as the unstable nuclei of the radioactive isotope decay to form stable nuclei. *How does the activity of the radio-isotope change with time?* You can refer back to Figure 9.1B to see how a Geiger–Müller counter can be used to find out. The results of the investigation can be plotted as a graph of count rate against time.

The activity of a radioactive isotope halves every half-life.

You can see from the graph that:

- the time taken for the count to fall from 600 to 300 c.p.m is 45 minutes,
- the time taken for the count to fall from 300 to 150 c.p.m is 45 minutes,
- the time taken for the count to fall from 150 to 75 c.p.m is 45 minutes.

The time for the activity to fall to half its value is the same, no matter what the original activity is. This time is called the **half-life**. The half-life of the isotope in this example is 45 minutes.

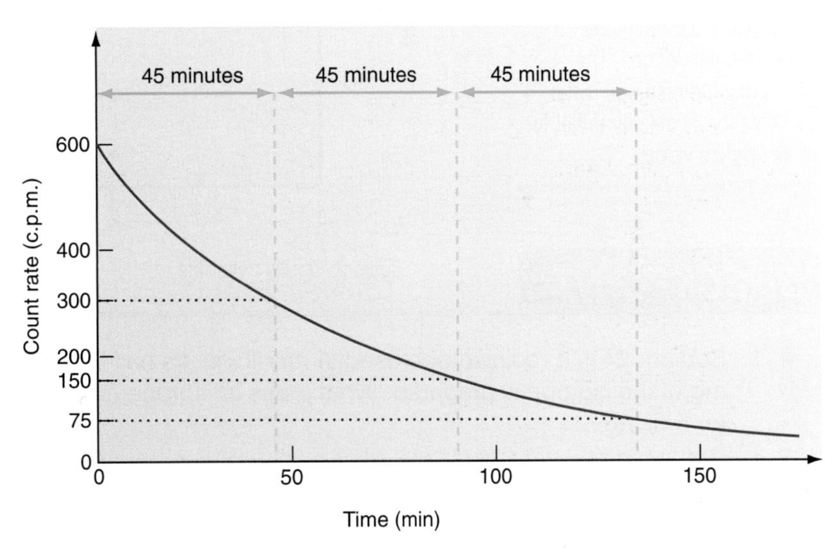

Figure 9.2A ▲ Radioactive decay: a graph of count rate against time

The number of atoms of a radioactive isotope halves every half-life.

The disintegration of a radioactive nucleus is a **random** process. No-one can say when an individual nucleus will suddenly split up. However, for a large number of nuclei, you can predict how many nuclei will disintegrate in a certain time. This is a bit like throwing dice. You can't predict what number you will get with a single throw, but if you threw 1000 dice you would expect one-sixth of them to come up with the number 4 and so on.

Suppose you have a sample of material which contains 1000 radioactive nuclei and that 10 per cent disintegrate per hour. After one hour, 100 nuclei will have disintegrated, leaving 900. During the next hour, 90 (10% of 900) will disintegrate to leave 810. The table shows the number of radioactive nuclei remaining as each hour passes.

Table 9.1 The number of radioactive nuclei that decay every hour

Time from the start (hours)	0	1	2	3	4	5	6	7
No. of radioactive nuclei present	1000	900	810	729	656	590	530	477
No. of nuclei that decay in the next hour	100	90	81	73	66	59	53	48

Figure 9.2B shows how the number of radioactive nuclei changes with time. The rate of decay is proportional to the number of radioactive nuclei left. The half-life can be read from the graph. It is the time taken for the number of nuclei to fall from, say, 1000 to 500.

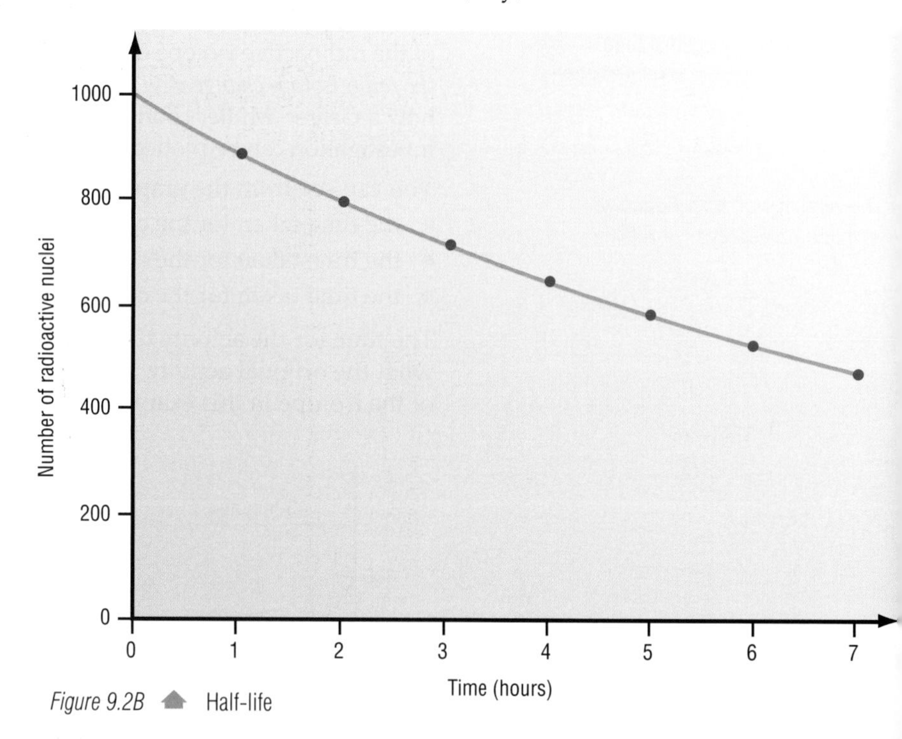

Figure 9.2B Half-life

SUMMARY

The activity of a radioisotope is the number of its nuclei that disintegrate per second. The half-life of a radioisotope is the time taken for its activity to fall to half its original value.

CHECKPOINT

1 Sodium-24 is a radioisotope used in medicine. Its half-life is 15 hours. A solution containing 8.0 mg of the isotope is prepared. What mass of isotope remains after (a) 15 hours (b) 30 hours (c) 5 days?

2 Cobalt-60 is a radioisotope made by placing cobalt in a nuclear reactor. It has a half-life of 5 years. The activity of a piece of cobalt-60 is 32.0 kBq. How long would it take for its activity to fall to (a) 16.0 kBq (b) 1.0 kBq?

3 In Figure 9.2B, how long would it take for the count rate to fall from 800 c.p.m to 100 c.p.m?

4 The following measurements were made in an experiment using a Geiger-Müller tube near a radioactive source.

Time (hours)	0	0.5	1.0	1.5	2.0	2.5
Count (c.p.m)	510	414	337	276	227	188

The background count was 30 c.p.m.

(a) The initial count rate from the source alone was 480 c.p.m (510–30). This is the corrected count rate. Work out the corrected count rates for the other readings.

(b) Plot a graph of the count rate (on the vertical axis) against time.

(c) Use your graph to find the half-life of the source.

9.3 Making use of radioactivity

FIRST THOUGHTS

When a new discovery is made, scientists are keen to find ways of using the discovery both in research and in everyday life. Radioactivity is a good example of a fascinating discovery for which many useful applications have been found.

Radioactive dating

Carbon-14 dating

Carbon is made of the isotopes carbon-12, carbon-13 and carbon-14. The isotope carbon-14 is radioactive. It has a half-life of 5700 years. Carbon-14 is present in the carbon dioxide which living trees use in photosynthesis. After a tree dies, it can take in no more carbon-14. The carbon-14 already present decays slowly; carbon-12 does not change. The ratio of the amount of carbon-14 left in the wood to the amount of carbon-14 in living trees can be used to tell the age of the wood. Animals take in carbon-14 in their food while they are alive. After their death, the proportion of carbon-14 in their bones tells how long it is since they died. A carbon-14 decay curve is shown in Figure 9.3A.

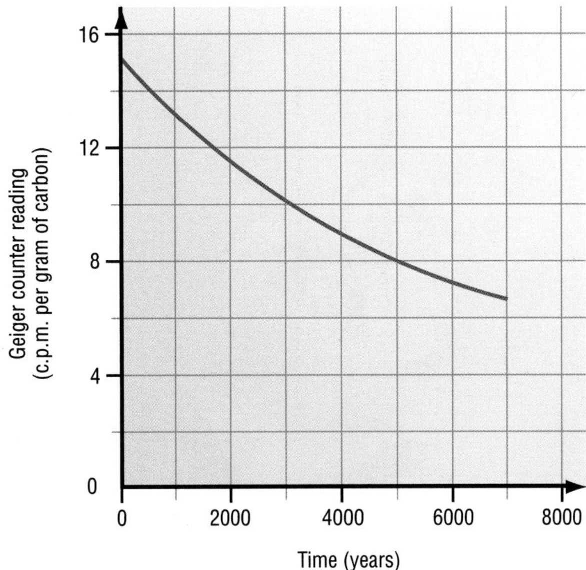

Figure 9.3A ▲ Carbon-14 dating

Argon dating

Rocks containing trapped argon gas can be dated by measuring the proportion of the gas to the radioactive isotope of potassium, K-40, which produces the gas as a result of radioactive decay. K-40 is an unstable isotope producing the argon isotope, Ar-40, as a result of the nucleus capturing an inner electron, turning one of its protons into a neutron. This gas is trapped in molten rock containing K-40 after the rock solidifies. K-40 also decays by β-emission to form an isotope of calcium Ca-40, a process 8.3 times more probable than electron capture. The effective half life of K-40 is 1250 million years.

The age of the rock can be determined by measuring the relative proportions of the two isotopes Ar-40 to K-40. Suppose a rock is analysed and found to contain 1 atom of Ar-40 for every 5 atoms of K-40. For every Ar-atom, there must be 8.3 atoms of Ca-40 and so there must have been 14.3 K-40 atoms (8.3 + 1 + 5) originally present for every 5 atoms of K-40 now present. In other words, the number of K-40 atoms is now 35% ($= \frac{5}{14.3} \times 100\%$) of the original number present. Use the half-life curve like Figure 9.2A to estimate how long it takes for the proportion of the 'parent' isotope to decrease to 35%.

Using radio-isotopes

Radioactive isotopes differ enormously in their half-lives. Some are listed in Table 9.2.

Table 9.2 States of Matter

Isotope	Radiation	Half-life	Isotope	Radiation	Half-life
Uranium-238	α	5000 million years	Iodine-131	β	8 days
Uranium-235	α	700 million years	Sodium-24	β	15 hours
Plutonium-239	α, γ	24 000 years	Bromine-82	β, γ	36 hours
Carbon-14	β	5700 years	Uranium-239	β	24 minutes
Strontium-90	β	59 years	Strontium-93	β, γ	8 minutes
Caesium-137	β, γ	30 years	Barium-143	β	12 seconds
Cobalt-60	β, γ	5 years	Polonium-213	α	4×10^{-6}s

Scientists have found many uses for radioactive isotopes. Figures 9.3 B–E show some of them.

Figure 9.3B Separating radioactive isotopes for medical use

Uses of radioactivity are:
- radioactive dating
- medical treatment
- tracers
- food preservation.

Medical uses

Radioactivity can be used to penetrate the body and kill cancerous tissue. Cobalt-60 and caesium-137 are often used for this purpose. They emit γ-rays. The dose of radiation must be carefully calculated to destroy cancerous tissue and leave healthy tissue alone. *Why is a γ-emitter used, rather than an α-or a β-emitter for this job?*

Radiation is used to destroy germs on medical instruments. It is more convenient than boiling, and it is also more efficient. Cobalt-60 is often used. *Why is a γ-emitter better than an emitter of α- or β-radiation for this job?*

The thyroid gland in the throat takes iodine from food and stores it. To find out whether the thyroid is working correctly, a patient is given food containing iodine-131, a β-emitter. The radioactive iodine can be detected as it passes through the body. The half-life of iodine-131 is 8 days. After a few weeks, there will be little left in the patient's body. *Can you explain why a β-emitter is better than a γ-emitter for this purpose?*

Figure 9.3C Radioactivity in agriculture

Agricultural research

This research worker is studying the uptake of fertilisers by plants. He has used a fertiliser containing phosphorus-32. This isotope is a β-emitter with a half-life of 14 days. By measuring the radioactivity of the leaves, the scientist can find out how much fertiliser has reached them.

Industrial uses

This research worker is measuring engine wear. The pistons of this engine have been in a nuclear reactor. Some of the metal atoms have become radioactive. As the engine runs, the pistons wear away and radioactive atoms enter the lubricating oil. The more the engine wears, the more radioactive the oil becomes. You can tell how well a lubricating oil reduces engine wear by timing the uptake of radioactivity.

Figure 9.3E shows the production of metal foil (e.g. aluminium baking foil). The detector measures the amount of radiation passing through the foil. If the foil is too thick, the detector reading drops. The detector sends a message to the rollers, which move closer together to make the foil thinner. *Why must the source be a β-emitter, not an α- or a γ-emitter? Why does it need to have a long half-life? Can you suggest a suitable isotope (see Table 9.2)?*

Figure 9.3D ▲ Measuring engine wear

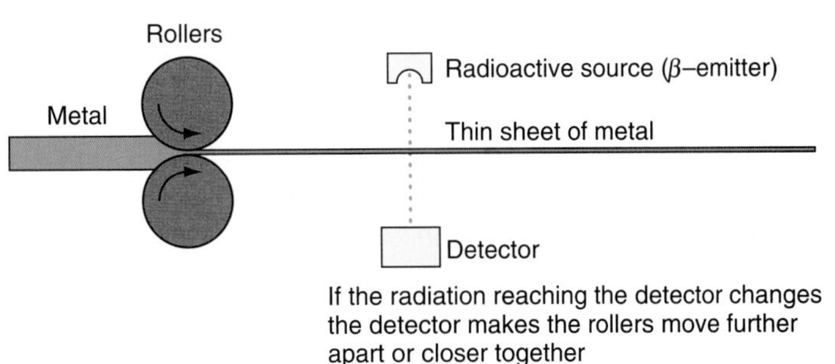

Figure 9.3E ▲ The manufacture of metal foil

Figure 9.3F ▲ A pool used for research into the irradiation of food

SUMMARY

Radioactive isotopes contained in naturally-occurring substances can be used to date these substances. Natural and artificial radioactive isotopes are used for a wide range of purposes in medicine, research and industry. Gamma radiation is used to destroy disease-carrying organisms in food.

Radioactivity and food

Food spoilage is a serious problem. About 20% of the world's food is lost through spoilage. The major cause is the bacteria, moulds and yeasts which grow on food. Some bacteria produce waste products which are toxic to people and cause the symptoms of food poisoning (sickness and diarrhoea). There are thousands of cases of food poisoning every year and some are fatal. Now there is an answer.

Irradiation of food with γ-rays kills 99% of disease-carrying organisms. These include *Salmonella*, which infects a lot of poultry, and *Clostridium*, the cause of botulism, which is often fatal. Spices, which are likely to contain micro-organisms as they are imported from tropical countries, can be irradiated with no loss of flavour. Irradiating potatoes is useful because it stops them sprouting without affecting the taste. The treatment is not suitable for all foods. Red meats turn brown and develop an unpleasant taste, eggs develop a smell, shrimps turn black and tomatoes go soft.

Some people fear that irradiated food will be radioactive. In fact, foods contain a natural low level of radioactivity, and the treatment increases this level only slightly. The dose of radiation which the food receives is carefully calculated. By the time the food is eaten, the extra radioactivity has decayed. The best proof that irradiation is safe is that you can not detect it.

CHECKPOINT

1 In 1985, a body was found in a peat bog in Norfolk. The reading which scientists obtained for the carbon-14 radioactivity of the body was 9 c.p.m. per gram of carbon.
   (a) Archaeologists called the body Pete Marsh. Where do you think they got the idea for the name?
   (b) Refer to Figure 9.3A, which shows a decay curve for carbon-14. How long ago did Pete Marsh die?

2 Testing the filling of cans on a production line. Can A is full, but something has gone wrong on the production line, and Can B is only partly filled. Can you design a scheme for using a radioactive source and a detector to tell whether the cans coming off the production line are completely filled? Say whether you would use a source of a or b or g radiation and say what kind of half-life would be suitable. Choose an isotope from Table 9.2 which could be used.

3 Detecting leaks. Imagine that you are a scientist working for a water company. An underground pipe 2 km long is leaking. You want to find out where the leak is without digging up the pipe. You decide to add a radioactive isotope to the water at one end of the pipe and then drive slowly along the route which the pipe takes, testing the ground with a Geiger–Müller counter. If you detect radioactivity you will know where the underground leak is. Which radioisotope will you choose from Table 9.2? You will need:

   - an element which forms a water-soluble compound,
   - radiation which can penetrate the soil,
   - a half-life which will give you time to drive slowly along the route,
   - to avoid contaminating the water for long.

4 A firm which manufactures plastic syringes gets an order from a hospital. The hospital wants the syringes to be sterile. The firm cannot sterilise the syringes by heating because this would soften the plastic. What can the firm do? (See Table 9.2 for details of radioactive isotopes.)

5 An engineer is planning a method of checking the stability of an oil rig which is to be used in the North Sea. She decides to mix a radioactive isotope with the concrete which fixes the legs of the oil rig to the sea bed. Then she can install a detector to find out whether there is any movement in the concrete. Which isotope from Table 9.2 should she choose? Concrete contains calcium compounds so the isotope should be chemically similar to calcium. It should have a suitable half-life.

6 An ancient rock is analysed and is found to contain 1 argon-40 (Ar-40) atom for every 9 atoms of the isotope potassium-40 (K-40) now present. Potassium-40 is a radioactive isotope with a half life of 1250 million years. Its decay products are the stable isotopes argon-40 and calcium-40 which it forms in the proportions 1:8.
   (a) For every atom of argon-40 now present, how many atoms of
      (i) calcium-40 are now present in the rock?
      (ii) potassium-40 were originally present in the rock?
   (b) The scientist who obtained these measurements deduced that the rock is 1250 million years old. Explain how she arrived at this conclusion from the above measurements.

9.4 The dangers of radioactivity

FIRST THOUGHTS

As well as being useful, radioactivity can be dangerous. People who work with it must understand the science in what they are doing. This sign is the radioactivity hazard sign.

We make good use of radioactivity. However, large doses of radiation are dangerous. Exposure to a high dose of radiation burns the skin. Delayed effects are damage to the bones and the blood. People who are exposed to a lower level of radiation for a long time may develop leukaemia (a disease of the blood cells) and cancer. When radioactive elements get inside the body, they are very dangerous. They irradiate the body organs near them, and the risk of cancer is very great. Even an emitter of α-rays, the rays with least penetrating power, can do immense damage if it gets inside the body. Fairly low doses of radioactivity can damage human **genes**. This may result in the birth of deformed babies. People who work with radioactive sources take precautions to protect themselves.

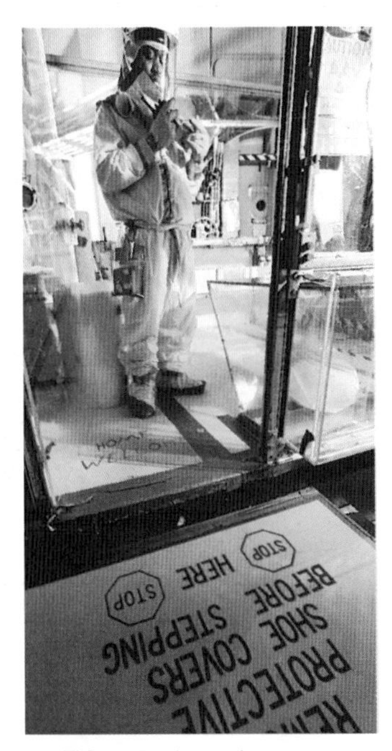

(a) This worker is wearing protective clothing so that radioactive material does not get onto his hands or clothes

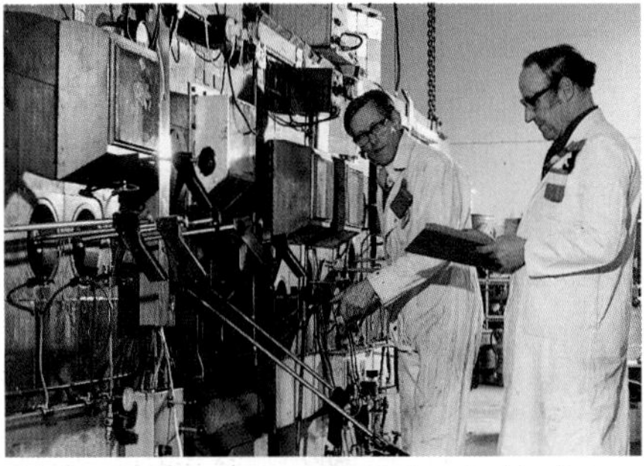

(b) These workers are using long handled tools to distance themselves from radioactive sources

(c) A film badge measures the dose of radiation its wearer receives

Figure 9.4A ▲ Precautions

They didn't know what the symbol meant

In the city of Goiania in Brazil in 1987, two junk collectors broke into a disused medical clinic and stole a heavy cylindrical object, which they took to be lead. They sold the metal cylinder to a junk yard. It emitted an eerie blue light from narrow slits. The manager of the junk yard hammered the head off the cylinder. Inside lay a capsule containing a powdery blue substance which stuck to the skin and glowed. It was caesium-137 a radioactive isotope which is used in cancer therapy. Massive doses of the γ-radiation from caesium-137 cause leukaemia, bleeding, sterility and cataracts. The dealer did not know this: he thought the bluish powder was so pretty and shiny that he gave away bits of it to his neighbours. Almost immediately, those who touched it became sick and feverish. One man stored some under his bed because he wanted to see it glow in the dark. Several children rubbed it on their bodies, and their skin became blistered and burnt.

SUMMARY

Radioactive isotopes are used in medicine, in research and in industry. Workers handling radioactive materials take precautions to shield themselves from radiation.

As a result of handling caesium-137, 20 people were taken to hospital, and 50 more people were put under medical observation, 25 homes were evacuated, and a large area of the city was cordonned off while the authorities cleared the area of radioactivity.

CHECKPOINT

1. Why is it essential for people who handle radioactive material to wear gloves?
2. Why do people build lead walls to protect them from γ-emitters but not from α-emitters?
3. Why are a-emitters dangerous if they get on to your hands?
4. The radioisotope iodine-131 is used for medical diagnosis. It has a half-life of 8 days. On 1 January, a doctor injects a patient with a solution containing 0.04 mg of iodine-131. What is the maximum mass of iodine-131 that could be left in the patient's body on 2 January? Why is the actual mass less than this?

9.5 The nuclear bomb

FIRST THOUGHTS

Splitting the atom was a marvellous achievement. Who could have foreseen the horrifying use that would be made of this discovery?

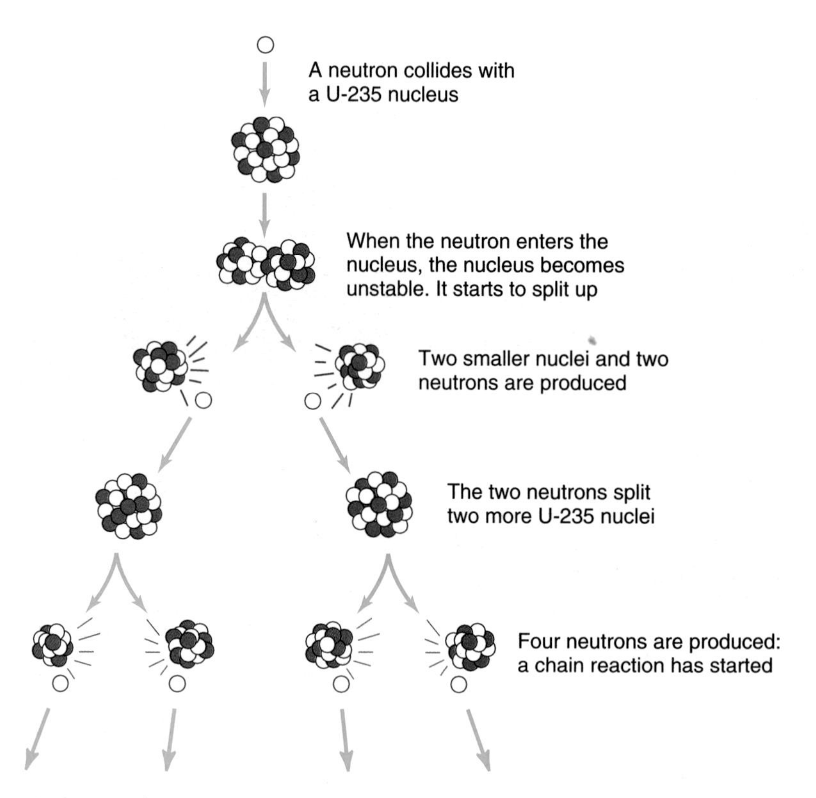
A neutron collides with a U-235 nucleus
When the neutron enters the nucleus, the nucleus becomes unstable. It starts to split up
Two smaller nuclei and two neutrons are produced
The two neutrons split two more U-235 nuclei
Four neutrons are produced: a chain reaction has started

Figure 9.5A The chain reaction in uranium fission

A German scientist called Otto Hahn made history in 1939. He split the atom! Hahn had been experimenting on firing neutrons at different types of nuclei. When neutrons struck uranium nuclei, some nuclei of uranium-235 split into two new nuclei and two neutrons. **Nuclear fission** (nucleus splitting) had occurred. The electrons of the original uranium-235 atom divided themselves between the two nuclei and two new atoms were formed. At the same time, an enormous amount of energy was released.

The energy given out was much greater than the energy given out in a chemical reaction. The energy which is released in nuclear fission is called **nuclear energy** (and also called **atomic energy**). The discovery that energy on a grand sale could be obtained from nuclear fission was made in 1939. Soon afterwards, the Second World War broke out. Scientists on both sides began trying to invent an 'atom bomb', a bomb which would release nuclear energy.

Figure 9.5A shows what happens in a block of uranium-235. The fission of one U-235 nucleus produces two neutrons. These two neutrons split two more uranium nuclei, producing four neutrons. A chain reaction is set off. In a large block of uranium-235, it results in an explosion. In a small block of uranium-235, an explosion does not occur because many neutrons escape from the surface before producing fission. A nuclear bomb consists of two blocks of uranium-235, each smaller than the critical mass. The bomb is detonated by firing one block into the other to make a single block which is larger than the critical mass. The detonation is followed by an atomic explosion.

On 6 August, 1945, a uranium bomb was dropped on the Japanese city of Hiroshima. It destroyed 60 percent of the city, and killed 140 000 people.

A huge mushroom-shaped cloud rose to a height of 10 km. A plutonium bomb was dropped on the city of Nagasaki three days later. Half the city was destroyed and thousands of people were killed in the blast. There had been air raids before which had destroyed large parts of cities and killed thousands of people. This time, the damage had been done in seconds by a single bomb. One nuclear bomb had done as much damage as 20 000 tonnes of TNT. There was a new horror as the unknown danger of nuclear radiation and fall-out revealed itself.

Figure 9.5B ▲ Hiroshima after the bomb

KEY SCIENTIST

Where was Otto Hahn, who first split the uranium atom, when the bomb fell on Hiroshima? He was one of a group of German scientists who were interned in Britain. He was distressed by the thought of the great misery which the bomb had caused. Hahn told the officer in charge of the internment camp that when he first saw that fission might lead to an atomic bomb he had thought of suicide.

Figure 9.5C ▲ A CND rally

Thousands of people received a large dose of radiation, from which they never recovered. The damage was so terrible that no nation has used nuclear weapons since. Many nations have huge stockpiles of nuclear weapons. The Campaign for Nuclear Disarmament (CND) wants nations to destroy their stockpiles in case an accident brings about another nuclear explosion.

Some of the scientists who worked on nuclear fission foresaw a new age, the 'atomic age', in which the energy of the atomic nucleus could be used for great benefit to mankind. They did not want to usher in the atomic age with a bomb. One group of scientists proposed that the USA should demonstrate the new bomb on a desert or on a barren island and then tell Japan to surrender or face atomic bombs. The military decision to drop the bombs was the result of the huge casualties suffered in the war against Japan. It was estimated that a million more lives would be lost in an invasion of Japan. Even after the first bomb was dropped, Japan refused to surrender. After the second bomb, the emperor himself ended the war, inspite of an attempt by a group of army officers to depose him and continue the fight.

This section shows how nuclear power stations utilise the energy of fission (splitting atomic nuclei) and explores the hope of harnessing the energy of fusion (joining atomic nuclei) in the future.

9.6 Nuclear reactors

FIRST THOUGHTS

This section shows how nuclear power stations utilise the energy of fission (splitting atomic nuclei) and explores the hope of harnessing the energy of fusion (joining atomic nuclei) in the future.

Nuclear reactors obtain energy from the same reaction as the nuclear bomb: the fission of uranium-235. In reactors, fission is carried out in a controlled way. Reactors use naturally occurring uranium, which is a mixture of uranium-235 and uranium-238. Only uranium-235 undergoes fission with slow neutrons. Figure 9.6A shows a nuclear reactor. Neutrons from the fuel rods go into the graphite core, where they collide with graphite atoms and lose kinetic energy. The graphite is called a 'moderator' because it slows down the neutrons. The neutrons then pass into the fuel rods and cause fission. The boron steel rods control the rate of fission by absorbing some of the neutrons. The heat generated by nuclear fission warms a coolant fluid which circulates through the moderator.

The coolant may be water (as in pressurised water reactors) or a gas, e.g. carbon dioxide (as in gas cooled reactors). The heat is used to turn water into steam. The steam drives a turbine and generates electricity.

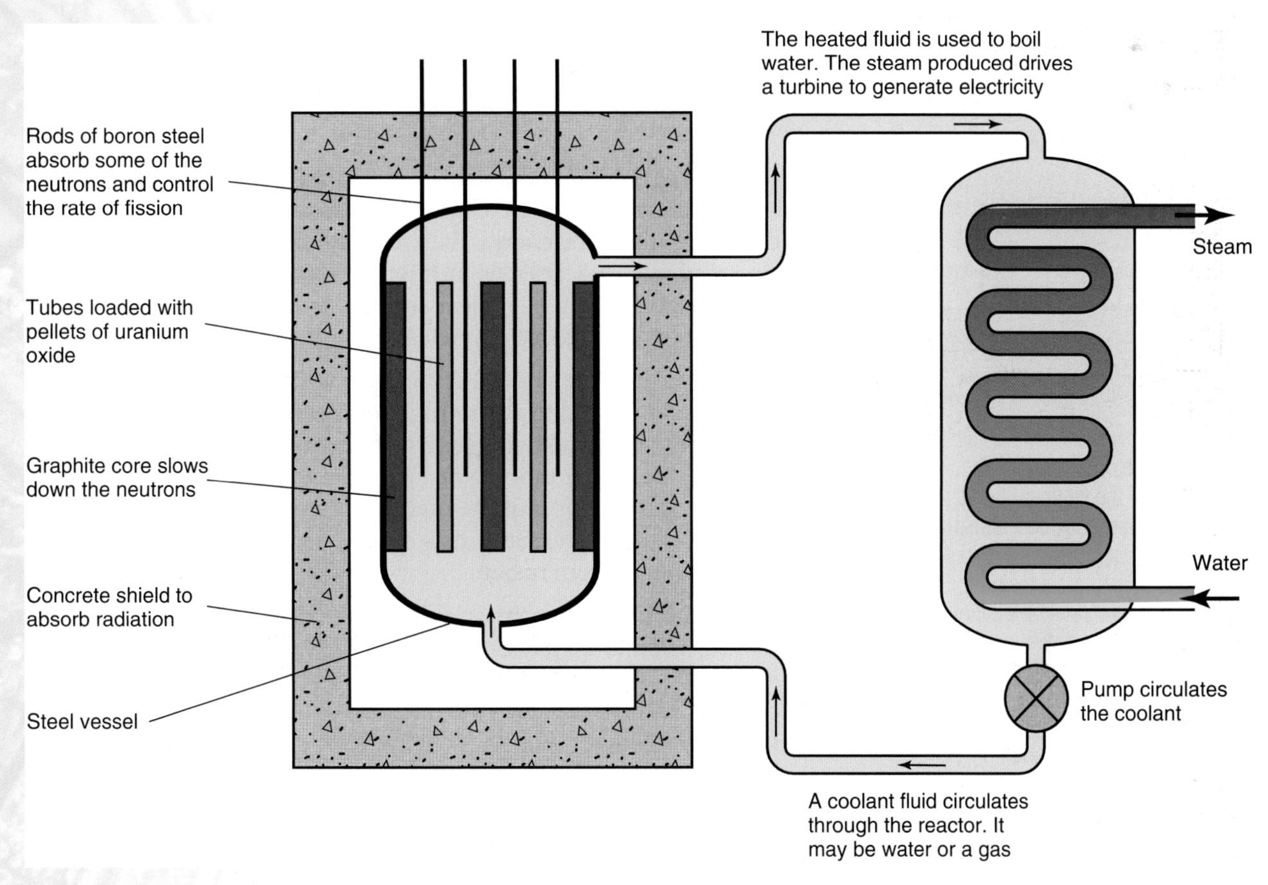
The heated fluid is used to boil water. The steam produced drives a turbine to generate electricity
Rods of boron steel absorb some of the neutrons and control the rate of fission
Tubes loaded with pellets of uranium oxide
Graphite core slows down the neutrons
Concrete shield to absorb radiation
Steel vessel
Steam
Water
Pump circulates the coolant
A coolant fluid circulates through the reactor. It may be water or a gas

Figure 9.6A ▲ A nuclear reactor

It's a fact!

The first nuclear reactor was built in an old squash court under a football stadium in Chicago in 1942. In charge was an Italian, Enrico Fermi, who had fled from Fascism in his native country to the USA in 1938. He assured the scientists present that the chain reaction would not get out of hand. It was held in check by a control rod. The control rod was pulled further and further out of the reactor. The neutron count rate rose and rose. Finally, Fermi ordered the reactor to be shut down. The scientists and engineers who saw the demonstration were impressed.

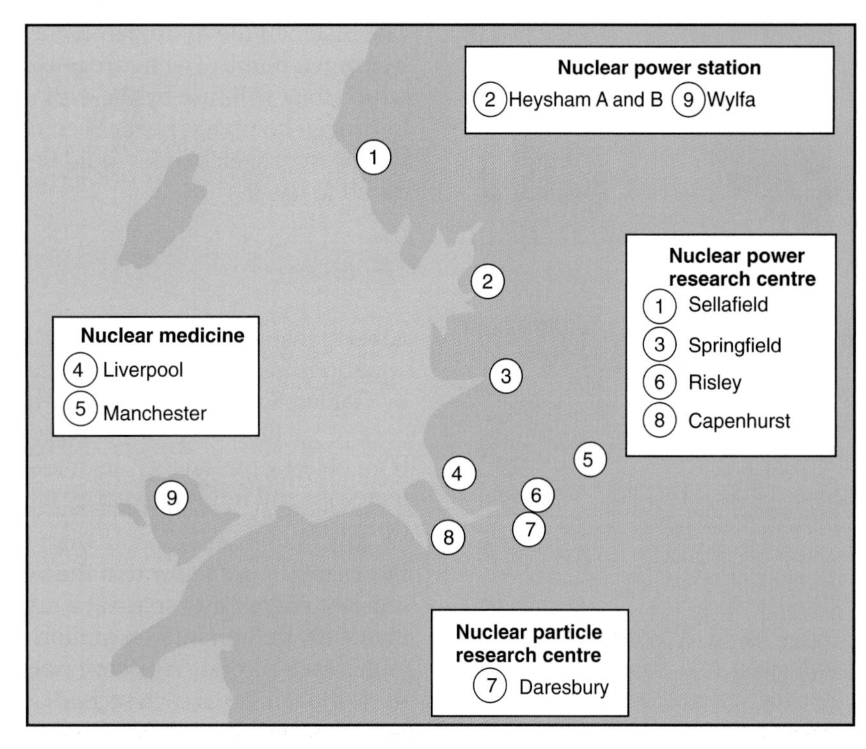
Nuclear power station
2 Heysham A and B 9 Wylfa
Nuclear power research centre
1 Sellafield
3 Springfield
6 Risley
8 Capenhurst
Nuclear medicine
4 Liverpool
5 Manchester
Nuclear particle research centre
7 Daresbury

Figure 9.6B ▲ The nuclear industry in North-west England and North Wales

9.7 Nuclear fusion

It's a fact!

The first successful controlled fusion experiment took place on 13 August 1978 in Princeton University, USA. A temperature of 60 million °C was reached for 0.10 seconds.

KEY SCIENTIST

The fusion bomb which Russia detonated in 1953 was made possible by the work of Andrei Sakharov. Later Sakharov became anxious about radioactive fallout. In 1957 he began to campaign for a stop to tests on nuclear weapons. In 1968, he was sacked from his post in a research institute and later interned as a dissident. After his release in 1986, he continued to be active in politics until his death in 1989.

SUMMARY

Energy is released in nuclear reactions. This nuclear energy can be released in an explosive manner in a nuclear (atomic) bomb. The fission (splitting) of uranium-235 was used in nuclear bombs. The nuclear energy from the fission of uranium-235 is released in a safe, regulated manner in a nuclear power station. The fusion (joining together) of small nuclei also releases energy. The fusion of hydrogen-2 nuclei is the source of energy in the hydrogen bomb.

One glass of water could provide the same amount of energy as 1 tonne of petrol! The source of this energy is nuclear fusion. When two atoms of hydrogen-2 (deuterium) collide at high speed, the nuclei fuse together.

$$^{2}_{1}H + ^{2}_{1}H \rightarrow ^{3}_{2}He + ^{1}_{0}n$$

An isotope of helium and a neutron are produced. A large amount of energy is released when two light nuclei fuse together. It gives more hydrogen-2 atoms the energy they need to fuse with other atoms. Thus a chain reaction starts. There is no shortage of hydrogen-2 because it occurs in water as $^{2}H_2O$. Nuclear reactors of the future may use the fusion process as a source of nuclear energy.

There are enormous technical difficulties with fusion. The hydrogen-2 atoms must be heated to a very high temperature before they will fuse. Repulsion between the two positive charges sets in when the nuclei get close. If the atoms are moving fast enough, this repulsion can be overcome. The reactor must be made of materials which will withstand very high temperatures. Scientists are still working on the problems of obtaining energy from fusion.

The products of fusion are not radioactive. However, the metal structure in which fusion takes places does become radioactive by interaction with the neutrons produced in fusion. Fusion reactors of the future would, like fission reactors, involve the disposal of both high-level and low-level radioactive waste.

The Sun obtains its energy from the fusion of hydrogen-2 atoms. In the Sun, the temperature is about 10 million °C, and the hydrogen-2 atoms have enough energy to fuse.

The hydrogen bomb

The fusion of the hydrogen-2 nuclei is the source of energy in the **hydrogen bomb**. The hydrogen-2 atoms are raised to the temperature at which they will fuse by the explosion of the uranium-235 bomb. The hydrogen bomb has never been used in warfare. The destruction caused by one hydrogen bomb would be so widespread that no nation has dared to use it.

$E = mc^2$

Albert Einstein was a genius. He derived one of the most important equations in physics by reconsidering the laws of motion first established by Galileo. Einstein proved that mass and energy are linked by the speed of light, *c*. If an object gains energy, its mass increases and if it loses energy its mass decreases.

For example, we know that the Sun radiates energy into space at a rate of about 400 million million million million joules each second. Work out how much mass the Sun loses each second, given that the speed of light is 300 million metres a second.

Figure 9.7A ▲ Albert Einstein

9.8 Disposal of radioactive waste

FIRST THOUGHTS

Everyone who uses radioactive materials has to find a solution to the problem of disposing safely of radioactive waste.

Radioactive waste comes from uranium mines, nuclear power stations, hospitals and research laboratories. It must be disposed of in some place where it is not a health hazard. The method used for waste disposal depends on whether the radioactivity is low-level, intermediate-level or high-level.

Low-level waste

Power stations produce a lot of slightly radioactive cooling water. This passes through long pipes out to sea. It is discharged 1-2 km from the shore.

Laboratory equipment and protective clothing are placed in metal containers. These are buried (see Figure 9.8A).

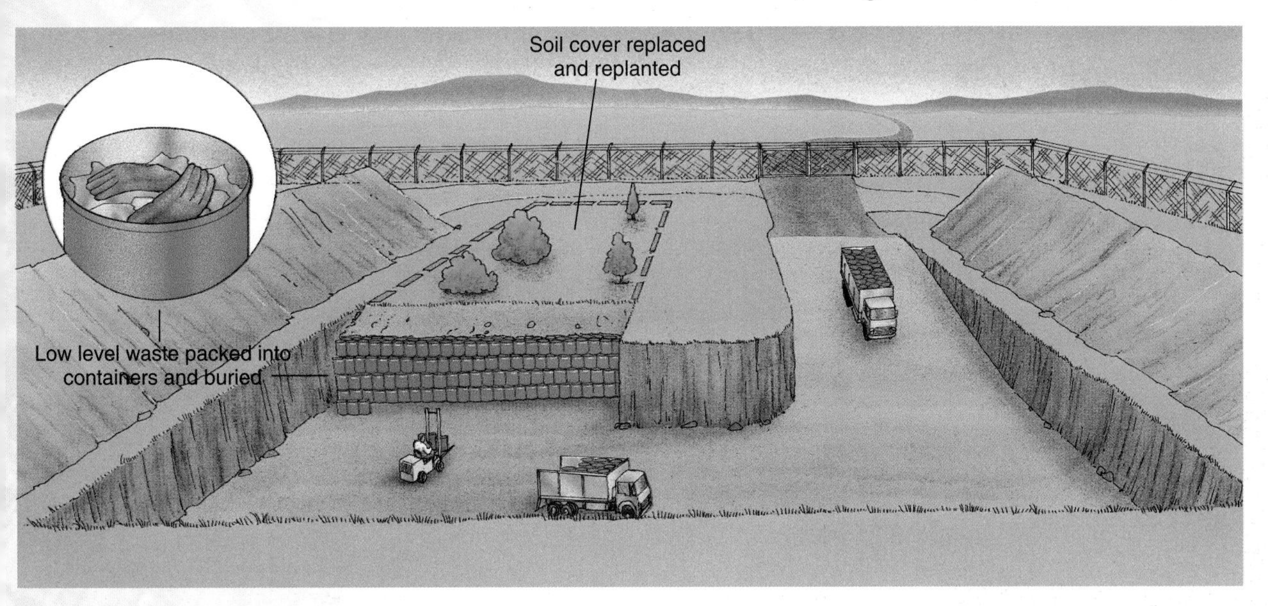

Figure 9.8A Burial of low-level solid waste

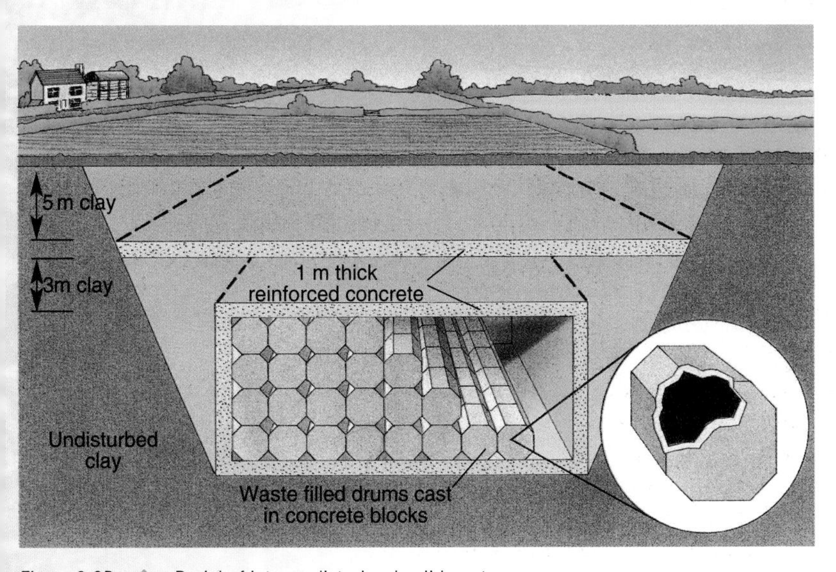

Figure 9.8B Burial of intermediate-level solid waste

Intermediate-level waste

Nuclear power stations produce large quantities of intermediate waste. Figure 9.8B shows one method of disposal. Drums containing waste are cast in concrete and buried under 8 m of clay. Another method is to bury drums of waste deep underground in disused mines.

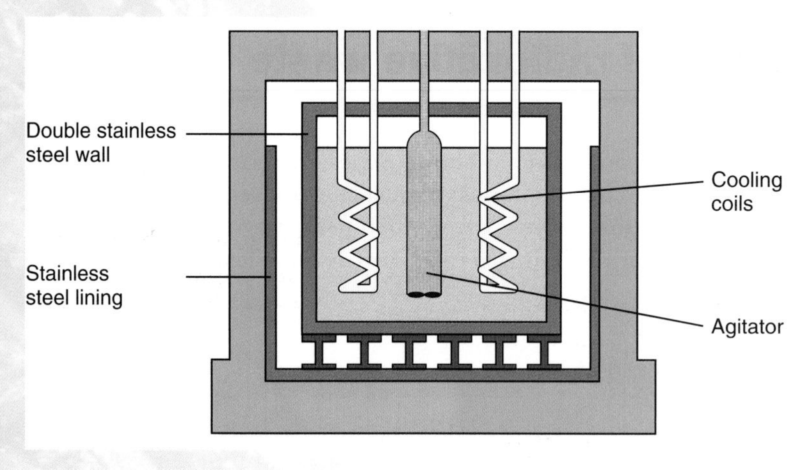

Figure 9.8C A storage tank for high-level liquid waste

High-level waste

Nuclear reactors contain fuel rods (see Figure 9.6A). After a time, the fuel rods must be replaced. The spent fuel rods are highly radioactive. They are stored in cooling ponds of water until they are less radioactive. Then they are removed to a special treatment plant where they are dissolved in acid. The solution must be stored and cooled until it becomes less radioactive (see Figure 9.8C).

Liquid waste can be **vitrified** (turned into glass). France stores vitrified high-level waste, and Britain plans to start using this method. Figure 9.8D shows deep underground burial of steel canisters containing vitrified waste. There is a shortage of land sites for this purpose. British Nuclear Fuels Ltd is planning to store radioactive waste in tunnels under the sea.

See www.keyscience.co.uk for more about radioactivity.

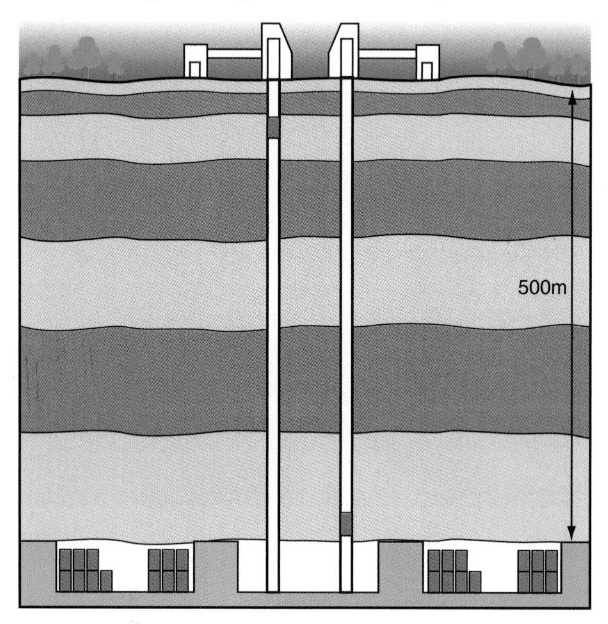

Figure 9.8D Deep underground burial of high level radioactive waste

CHECKPOINT

1 Why is a clay site chosen for the burial of intermediate-level waste (see Figure 9.8B) rather than a sandy soil?

2 Some countries use disused salt mines for the storage of radioactive waste. Why do they feel sure that such sites will be dry? Why is it necessary to store the waste in dry conditions?

3 When British Nuclear Fuels planned to build a site like that in Figure 9.8D in Humberside, people all over the county protested strongly. They said they did not want a 'nuclear dump in our back yard'. Posters read, 'Say no to a shallow grave'. The plan was dropped.
   (a) Why do you think that people did not want a site of this kind in the county?
   (b) Is it fair to describe the site as 'a shallow grave'?
   (c) What will happen if every county refuses to store radioactive waste?
   (d) Waste from nuclear power stations contains plutonium. How long does plutonium have to be stored before it has lost (i) half its radioactivity (ii) three quarters of its radioactivity? (See Table 9.2)
   (e) Compose a letter to your local newspaper. Explain *either* why you think your county should accept a burial site for radioactive waste *or* why you think your county should resist such a plan.

9.9 Safety

FIRST THOUGHTS

As you read this section, try to weigh up the advantages of nuclear power against the dangers.

Background radioactivity

Radioactive materials occur naturally in rocks, in soil and in the air. The low level of radioactivity which they give out is called **background radioactivity**. Figure 10.9A shows the radiation we receive from natural sources and from artificial sources.

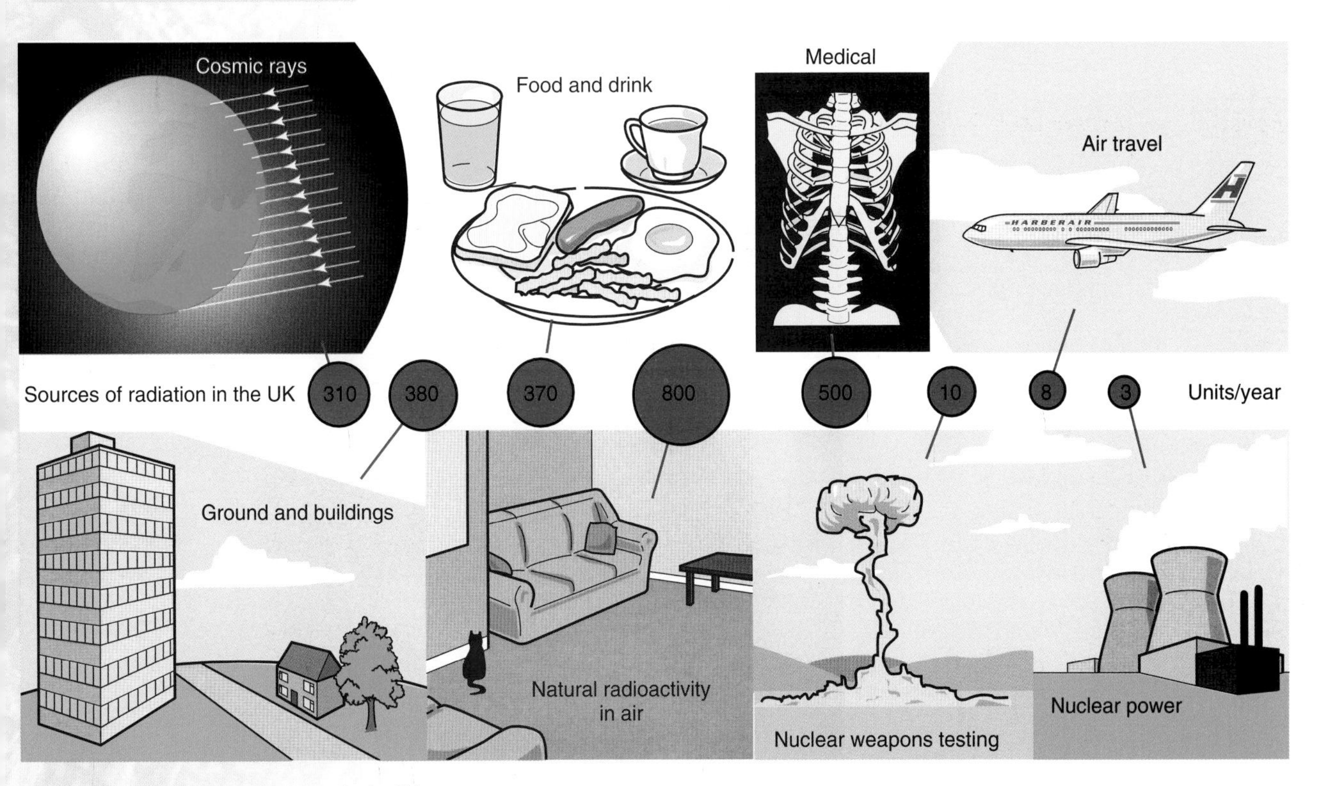

Figure 9.9A ▲ Sources of radiation in the UK. The unit is the microsievert, a measure of the effect of radioactivity on cells

Accidents

Windscale

They call it 'the day the reactor caught fire'. It happened at Windscale in Cumbria on 11 October 1957. The power station is now called Sellafield. The reactor overheated and graphite rods caught fire. They managed to put out the fire by flooding the reactor with water. Radioactive isotopes were blown from the reactor over the Lake District. In fact, no-one was injured in the accident.

Three Mile Island

A bigger accident happened in Three Mile Island in the USA in 1979. The reactor was a pressurised water reactor. The pumps which fed cold water into the reactor stopped and the temperature of the reactor shot up. After two hours, the operators rectified the fault. It took a week for the temperature to fall. The reactor was crippled. A cloud of radioactive substances fell over the island. Over three million litres of cooling water were radioactive. Luckily, no-one was injured in the accident. There may be long-term effects on health, however. In 1985, permission was given to reopen the plant. The public have staged such huge demonstrations against the reopening of the power station that work has not yet begun.

Chernobyl

The worst nuclear accident happened in 1986 at the Chernobyl power station in the USSR. It was a pressurised water reactor with a graphite moderator, and a loss of cooling water caused the reactor to overheat. Steam reacted with graphite to produce hydrogen, which exploded. The explosion set the building on fire. Firemen battled heroically to put out the blaze. Many of those who fought the fire have since died of leukaemia. The reactor also caught fire. It burned for days while people tried to bring it under control. When the core of the reactor reached 5000 °C, they feared that it might melt the container and burn its way down into the earth. Helicopters managed to get near enough to drop sand and lead on the burning reactor to cool it.

Chernobyl and the surrounding area will remain radioactive for many years.

Figure 9.9B The Chernobyl power station taken from a helicopter after the explosion

The roof of the reactor blew off in the explosion, and a cloud of radioactive material spread over the Ukraine in the USSR and drifted over other European countries. A week later, the cloud of radioactive material reached the UK.

From the area round Chernobyl, 135 000 people were evacuated. Thirty or more people died in the accident, and hundreds suffered from radiation sickness. People who received a smaller dose of radiation may develop leukaemia or cancer in the future. The final toll will not be known for many years.

Could Chernobyl happen here? The Chairman of the then Central Electricity Generating Board answered this question in 1986. These are some of the points he made.

- The Chernobyl design is not used outside the USSR.
- The design is poor: the reactor is unstable when operated at low power.
- The operators at Chernobyl ignored some safety instructions.
- Chernobyl had no automatic fast-acting shutdown system to close the reactor if it became unsafe.
- The UK's reactors all have computer-controlled safety systems which can shut down reactors if faults are detected.

CHECKPOINT

1. Look at Figure 9.9A.
   (a) Background radiation is made up of sources A, B, C and D. What dose of radiation do these sources add up to?
   (b) Radiation from sources E, F, G and H is under our control. What dose of radiation do these sources add up to?
   (c) Which of the sources under our control adds the most to our total dose of radiation? What is the difficulty in reducing the dose from this source of radiation?
   (d) What is the total dose of radiation we receive in a year? What fraction of this comes from (i) nuclear power stations and (ii) nuclear weapons testing?
2. Russia did not announce the accident at Chernobyl until after Sweden had detected an increase in atmospheric radioactivity. What do you think that a country which has a nuclear accident should do? If the neighbouring countries had known that a cloud of radioactive fallout was on the way, what precautions could they have taken?
3. Strontium-90 is produced in nuclear reactors. It is radioactive. Why are babies especially at risk if an accident releases strontium-90 into the environment? (The clue is in the Periodic Table. Which elements resemble strontium? Which of these elements is present in milk?)
4. (a) Explain what is meant by background radioactivity.
   (b) Figure 9.9A shows the sources of radiation in the UK.
      (i) Which is the most significant source of radiation in Figure 9.9A?
      (ii) Which is the least significant source of radiation in Figure 9.9A?
      (iii) Which sources of radiation have always been present?

Topic 10 Inside the nucleus

10.1 Stable and unstable nuclei

FIRST THOUGHTS

The nucleus of an atom consists of protons and neutrons packed together in a tiny space. Some nuclei are unstable and disintegrate. Most nuclei are stable which is just as well otherwise we wouldn't exist!

The strong nuclear force

Hydrogen is the lightest element. An atom of hydrogen consists of one proton as its nucleus and one electron. An atom of any other element contains two or more protons and neutrons in its nucleus. An atom of a heavy element such as lead contains many protons and neutrons in its nucleus. Each proton carries a positive charge and therefore tries to repel any other proton. There must be a very strong attractive force between the neutrons and protons in a stable nucleus, otherwise they would fly apart because of the electrostatic repulsion of the protons.

The force holding the protons and neutrons together in a nucleus is called the **strong nuclear force**. Unstable nuclei disintegrate because this force is not strong enough to prevent the particles in the nucleus flying apart.

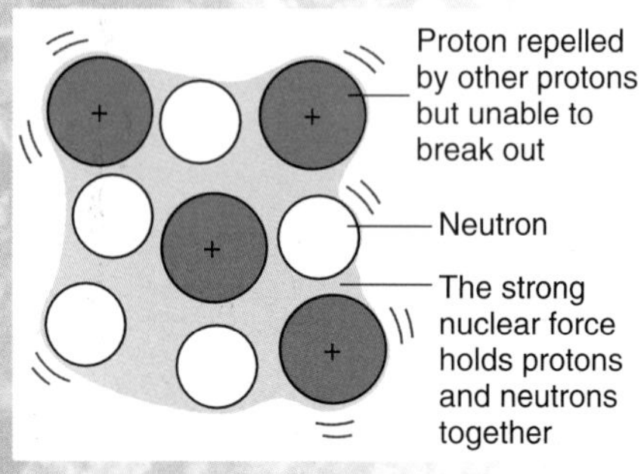

Figure 10.1A The strong nuclear force

The *N–Z* plot

Every known type of nucleus can be plotted on a graph of neutron number *N* on the *y*-axis against proton number *Z* on the *X*-axis, as shown in Figure 10.1B.

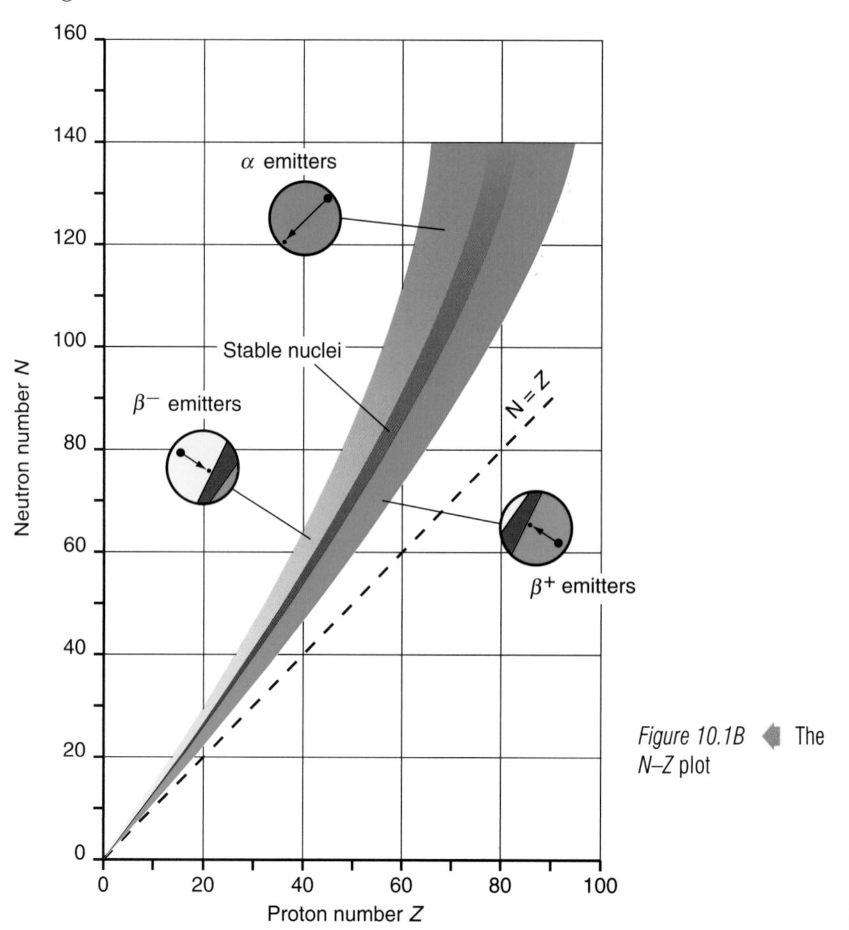

Figure 10.1B The *N–Z* plot

- An atom of the isotope ${}^{A}_{Z}X$ contains Z protons and $(A - Z)$ neutrons in its nucleus. Its position on the graph is therefore at $N = A - Z$ on the vertical axis and Z on the horizontal axis. For example, atoms of the radioactive isotope ${}^{14}_{6}C$ each have 6 protons and 8 neutrons in the nucleus. This isotope is therefore plotted at $N = 8$, $Z = 6$.
- All the isotopes of an element would appear on this graph at the same horizontal position (i.e. same Z) and at different vertical positions, corresponding to different neutron numbers.
- Stable nuclei form a well-defined line on the N–Z plot. This line is straight from $Z = 0$ to about $Z = 20$ then it curves up towards increasing N:
  1. Stable nuclei up to about $Z = 20$ have equal numbers of neutrons and protons in their nuclei.
  2. Stable nuclei beyond $Z = 20$ have more neutrons than protons. For the largest stable nuclei, the number of neutrons, N, is approximately equal to $1.5Z$.

Radioactive changes on the *N–Z* plot

1. Alpha decay

α emitters lie near the top of the N–Z curve.

Very heavy nuclei are unstable α emitters. In such a heavy nucleus, the strong nuclear force is not strong enough to prevent the nucleus from disintegrating by emitting an alpha particle. As an alpha particle consists of two protons and two neutrons bound tightly together, the nucleus formed has two fewer protons and two fewer neutrons than the original nucleus. This change is shown on the N–Z plot and is represented by the equation below.

$$ {}^{A}_{Z}X \rightarrow {}^{4}_{2}\alpha + {}^{A-4}_{Z-2}Y $$

Note: the change results in a daughter nucleus Y which is further from the stability line of the N–Z *plot. To become stable, the nucleus emits a series of α and β^- particles, ending on the stability line at* Z = *82 which is a lead nucleus.*

2. Beta minus decay

β^- emitters lie above the N–Z curve.

Neutron-rich nuclei have too many neutrons to be stable. Such a nucleus becomes stable by changing one of its neutrons into a proton. A β^- particle is created in the nucleus and emitted at the instant of change. Because a β^- particle carries a negative charge equal and opposite to that of the proton and is much lighter than a proton or neutron, the nucleus must have become more positive as a result of one of its neutrons changing into a proton. The nucleus formed therefore has 1 more proton and 1 less neutron than the original nucleus. This change is shown on the N–Z plot and is represented by the equation below.

$$ {}^{A}_{Z}X \rightarrow {}^{0}_{-1}\beta + {}^{A}_{Z+1}Y $$

As neutron-rich nuclei are above the stability line on the N–Z plot, the change is always towards the stability belt from above.

For example, a nucleus of the β^- emitting isotope ${}^{60}_{27}Co$ consists of 27 protons and 33 neutrons. As a result of emitting a β^- particle, it changes to a nucleus of the isotope ${}^{60}_{28}Ni$ which consists of 28 protons and 32 neutrons.

$$ {}^{60}_{27}Co \rightarrow {}^{0}_{-1}\beta + {}^{60}_{28}Ni $$

3. Beta-plus decay

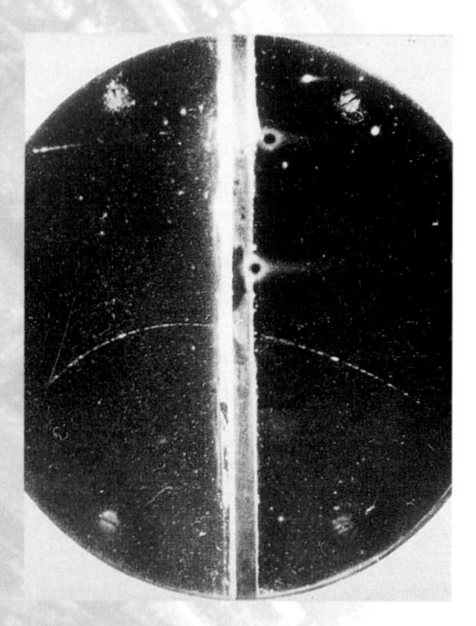

Figure 10.1C Discovery of the positron

A β^+ particle, also called a positron, is the antiparticle of the electron. Antimatter was predicted in 1928 by a British physicist called Paul Dirac. He put forward the theory that there is an antiparticle for every particle and that a particle and its antiparticle annihilate each other and turn into radiation when they meet. The positron was discovered in 1932 as a result of observing cosmic ray tracks in a cloud chamber. Now we know that positrons are produced by unstable nuclei that have too many protons.

Proton-rich nuclei have too many protons to be stable. Such a nucleus becomes stable by changing one of its protons into a neutron. A β^+ particle is created in the nucleus and emitted at the instant of change. Because a β^+ particle carries a positive charge equal to that of the proton and is much lighter than a proton or neutron, the nucleus must have become more negative as a result of one of its protons changing into a neutron. The nucleus formed therefore has 1 more neutron and 1 less proton than the original nucleus. This change is shown on the *N*–*Z* plot and is represented by the equation below.

$$^{A}_{Z}X \rightarrow {}^{0}_{+1}\beta + {}^{A}_{Z-1}Y$$

As proton-rich nuclei are below the stability line on the *N*–*Z* plot, the change is always towards the stability belt from below.

β^+ emitters lie below the *N*–*Z* curve.

For example, a nucleus of the β^+ emitting isotope $^{11}_{6}C$ consists of 6 protons and 5 neutrons. As a result of emitting a β^+ particle, it changes to a nucleus of the isotope $^{11}_{5}B$ which consists of 5 protons and 6 neutrons.

$$^{11}_{6}C \rightarrow {}^{0}_{+1}\beta + {}^{11}_{5}B$$

The PET scanner

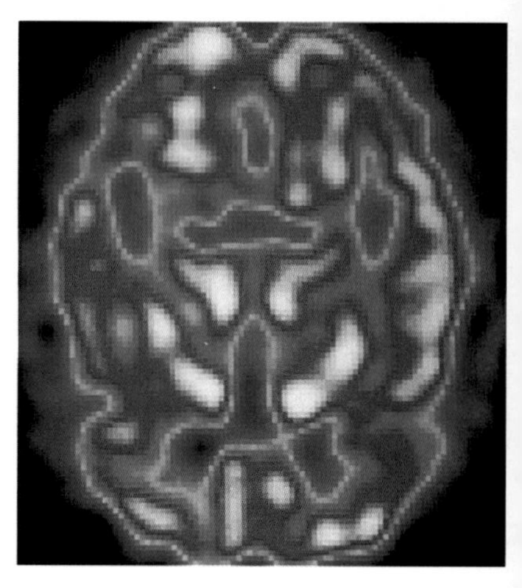

Figure 10.1D (a) PET scanner in use (b) PET image

Antimatter in the form of positrons is used in medicine to scan the brain. Before the scan is carried out, the patient is given a drink of water containing a tiny amount of a positron-emitting isotope. Each positron travels no more than about a millimetre before it is annihilated by an electron, causing the emission of two 'bursts' of gamma radiation travelling in opposite directions. A ring of detectors connected to a computer registers a positron emission when opposite detectors respond at the same time. The computer is programmed to map out the location of the positron-emitting isotope in the body to give a three-dimensional image of the brain. The PET scanner is also used in industry to detect internal flaws in metals and to monitor fluid flow.

SUMMARY

- Stable nuclei up to $Z = 20$ contain equal numbers of protons and neutrons. Heavier nuclei have more neutrons than protons. The heaviest nuclei contain 50% more neutrons than protons.
- Unstable nuclei that contain too many neutrons become stable by changing a neutron into a proton and emitting a β^- particle.
- Unstable nuclei that contain too many protons become stable by changing a proton into a neutron and emitting a β^+ particle.
- Very heavy unstable nuclei become stable by emitting a series of α and β^- particles.

CHECKPOINT

1 (a) The carbon isotope $^{12}_{6}C$ is stable. How many protons and how many neutrons are in each nucleus?
(b) The carbon isotope $^{14}_{6}C$ is unstable.
(i) How many neutrons and how many protons are in each nucleus of this isotope?
(ii) What type of radiation is emitted by this isotope?
(iii) Write down the equation that represents this change, given that the isotope formed is $^{14}_{7}N$.

2 How many protons and how many neutrons are present in a nucleus of each of the following isotopes of oxygen?
(a) $^{14}_{8}O$ (b) $^{16}_{8}O$ (c) $^{19}_{8}O$

3 (a) Which one of the isotopes in question 2 is stable?
(b) The other two isotopes are unstable. What type of radiation would you expect from each of these unstable isotopes?

4 A sample of copper is placed in a nuclear reactor. After removal, it is radioactive and emits β^- particles.
(a) Why does it become radioactive as a result of being placed in a nuclear reactor?
(b) Why does it become a β^- emitter?

10.2 Quarks

FIRST THOUGHTS

Is an atom like a box which has a box inside which has another box inside and so on? The nucleus of the atom consists of protons and neutrons. Are these two particles made up of even smaller particles?

Probing the nucleus

Figure 10.2A An aerial view of the Stanford Linear Accelerator

The **Stanford Linear Accelerator (SLAC)** is the longest electron gun in the world, built in 1968 to accelerate electrons through 20 000 million volts almost to the speed of light. The electron gun in a TV tube accelerates electrons in a beam through a voltage of about 5000 volts to hit a TV screen and produce light. SLAC accelerates electrons to much higher kinetic energies and makes them collide with nuclei in a solid target. At such high energies, an electron is capable of penetrating a nucleus and burrowing into a proton or a neutron. These experiments were first carried out in 1968. The measurements showed that most electrons passed straight through but some were scattered by very large angles. This proved that there are three tiny 'hard' particles inside each proton or neutron. Further measurements showed that these particles carried a charge of either $+\frac{2}{3}e$ or $-\frac{1}{3}e$, where e is the magnitude of the charge of an electron.

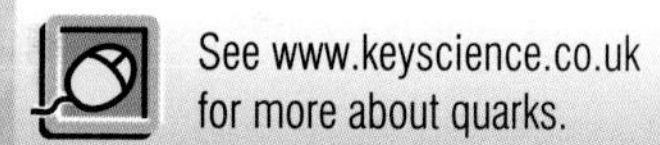

See www.keyscience.co.uk for more about quarks.

These particles inside a neutron or proton are called **quarks.** They were predicted earlier to explain the patterns of particles and antiparticles created in high-energy particle collisions. SLAC provided direct experimental evidence that they exist. Half a century earlier, Rutherford probed the atom using alpha particles and discovered the nucleus. The SLAC experiment probed the nucleus and discovered three quarks in each proton and in each neutron. Prove for yourself that a proton must contain one negative and two positive quarks and a neutron must contain one positive and two negative quarks.

Up and down quarks

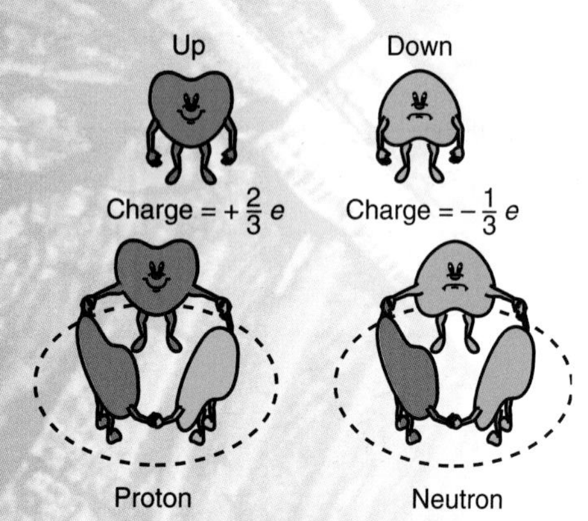

Figure 10.2B ▲ *Making protons and neutrons*

The quark model of the proton and the neutron is based on two types of quark, referred to as the **up** quark and the **down** quark. An up quark carries a charge of $+\frac{2}{3}e$ and a down quark carries a charge of $-\frac{1}{3}e$, where e is the magnitude of the charge carried by an electron.

- A proton consists of two up quarks and a down quark (uud).
- A neutron consists of one up quark and two down quarks (udd).

Quark flavours

Hundreds of different short-lived particles have been discovered as a result of colliding beam experiments in high-energy accelerators. No more than six different types of quarks are needed to account for all these short-lived particles. The six different types, referred to as quark **flavours**, are shown in Figure 10.2C. Only the up quark and the down quark are stable in ordinary matter. The other flavours can only be produced in high-energy collisions and they are short-lived.

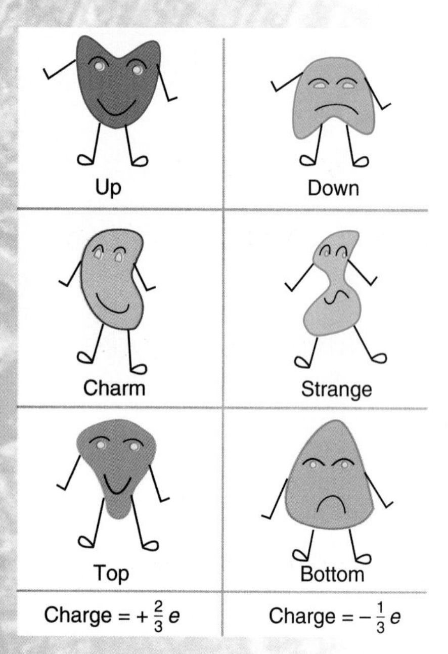

Figure 10.2C ▲ The quark family

Back to beta decay

- A neutron-rich nucleus becomes stable when one of its neutrons changes into a proton. This happens because a down quark in a neutron suddenly changes into an up quark. A β^- particle is created when this change occurs to conserve the total charge. The beta particle is instantly emitted from the nucleus.

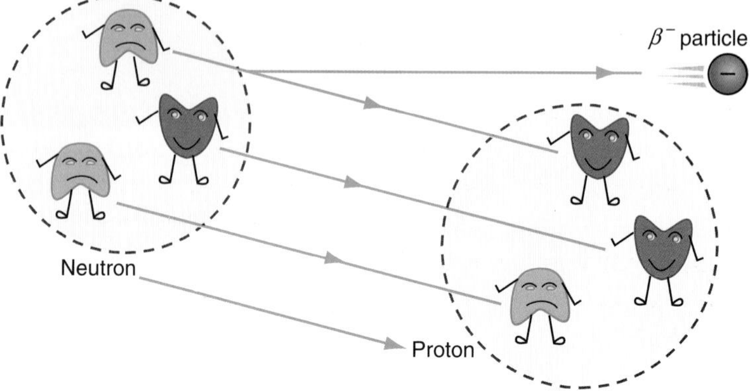

Figure 10.2D ▲ β^- decay

- A proton-rich nucleus becomes stable when one of its protons changes into a neutron. This happens because an up quark in a proton suddenly changes into a down quark. A β^+ particle is created when this change occurs to conserve the total charge. The beta particle is instantly emitted from the nucleus.

Quark slavery

The quarks in a neutron or proton cannot be freed from each other. Attempts to free quarks by slamming protons or neutrons into each other at huge energies do not work. This is because quark–antiquark pairs are created by the collision. These new quarks and antiquarks and the existing quarks form new particles which are often short-lived.

Production of a quark–antiquark pair, known as **pair production**, is the reverse of annihilation. Astronomers think that the Big Bang created quarks and antiquarks which then annihilated each other. There were more quarks than antiquarks in our part of the Universe. The remaining quarks joined together in trios to form protons and neutrons which joined together to form nuclei. The nuclei attracted electrons to form atoms which eventually formed all the substances in existence.

The quark model takes you to the frontiers of physics. Research continues to find out more about the fundamental nature of matter and energy. Science really is stranger than anyone could imagine!

SUMMARY

- A proton consists of two up quarks and a down quark.
- A neutron consists of two down quarks and an up quark.
- An up quark carries a charge of $+\frac{2}{3}e$.
- A down quark carries a charge of $-\frac{1}{3}e$.

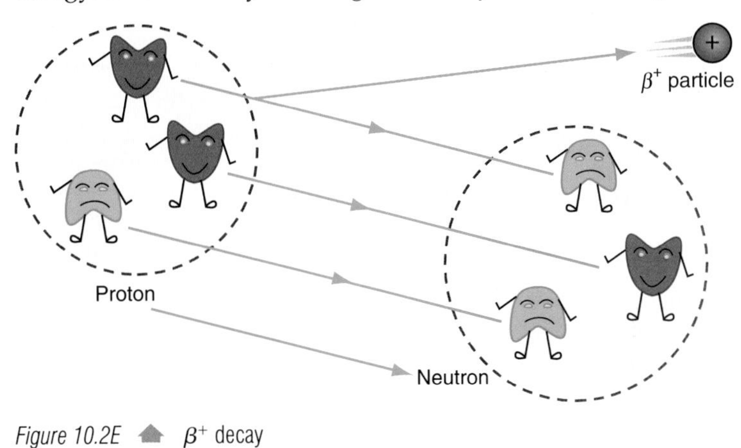

Figure 10.2E ▲ β^+ decay

CHECKPOINT

1. The cobalt isotope $^{60}_{27}Co$ emits β^- radiation.
   (a) How many protons and how many neutrons are there in the nucleus of an atom of this isotope?
   (b) How many protons and how many neutrons are present in the nucleus after a β^- particle has been emitted?
   (c) Use the quark model to describe this change.
2. How many up quarks and how many down quarks are present in (a) the nucleus of a hydrogen atom, (b) an alpha particle?
3. (a) How many protons could be formed from 4 up quarks and 2 down quarks?
   (b) How many neutrons could be formed from 4 up quarks and 8 down quarks?
4. How many protons and how many neutrons could be formed from (a) 3 up quarks and 3 down quarks, (b) 4 up quarks and 5 down quarks?
5. What is the total charge of (a) 3 up quarks and 3 down quarks, (b) 3 up quarks and 6 down quarks?

Theme Questions

TOPIC 7

1 (a) Sue is inflating a weather balloon, which will rise 10 km into the atmosphere. Pat advises Sue to fill the balloon only partially when it is on the ground. Why does Pat say the balloon should not be filled completely?

(b) Michael pumps up his car tyres before driving from Britain to sunny Spain. When he arrives, his friend Jose advises him to let some of the air out of his tyres. Why does Jose advise doing this?

2 The diagram shows what happens when you breathe.

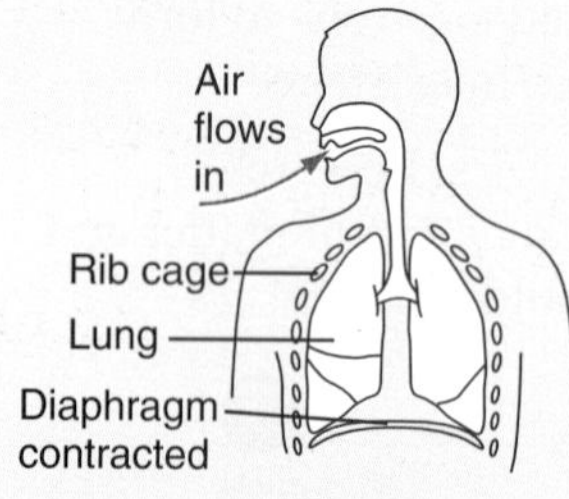

The diaphragm contracts and flattens, so increasing the volume of the chest cavity. What does this do to the air pressure in the cavity? Why does it cause air to flow into the lungs?

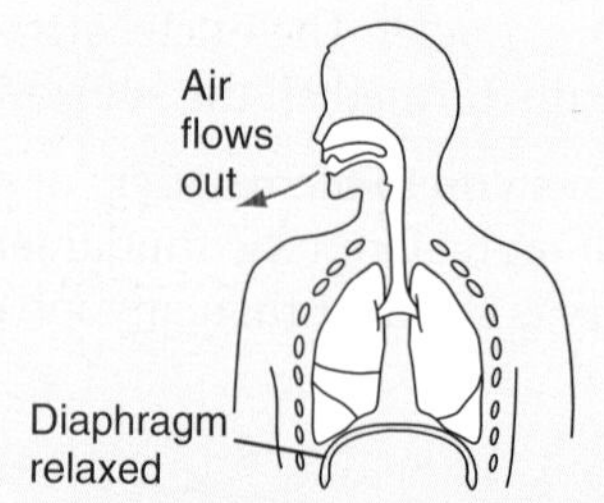

The diaphragm relaxes and pushes up into the chest cavity. What does this do to the air pressure in the cavity? Why does it cause air to flow out of the lungs?

3 (a) The particles that make up solids, liquids and gases have different spacing and forces of attraction.

(i) Copy and complete the table below:

Form of matter	Forces of attraction	Spacing
Solid	Strong	
Liquid		Close
Gas		

(ii) Using the idea of particles, explain why gases are easier to compress than liquids.

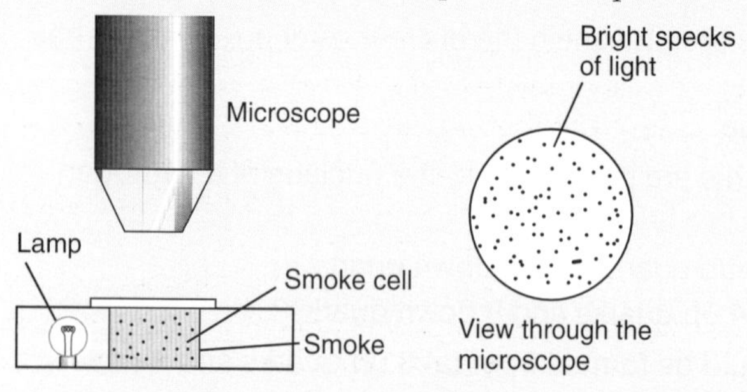

(b) A smoke cell is illuminated with a lamp and viewed using a microscope.

(i) Explain why you see bright specks of light.

(ii) Describe the motion of the specks.

(iii) This is called Brownian motion. Write down the correct ending to the following sentence.

The observed motion is caused by:
convection of the air;
Random impacts of air molecules with the smoke particles;
smoke particles being pushed by the heat of the lamp.

(c) A soluble deep purple crystal is placed carefully at the bottom of a beaker of water as shown in the diagram.

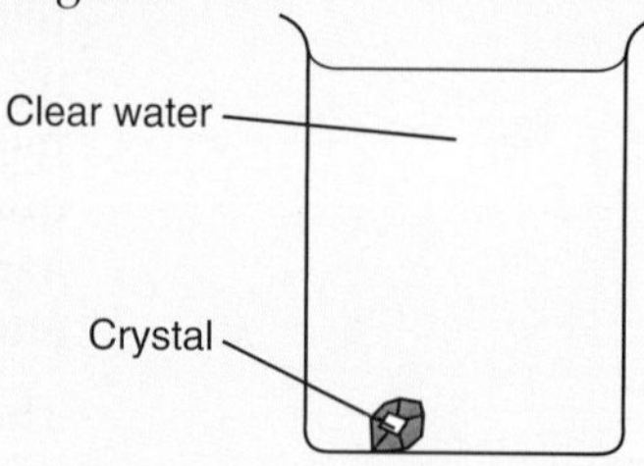

After a few days, the purple colour has spread through the water.

(i) What is this process called?

(ii) Using the ideas from part (b), explain why this happens. (NEAB)

TOPIC 8

4 The element X has atomic number 11 and mass number 23. State how many protons and neutrons are present in the nucleus. Sketch the arrangement of electrons in an atom of X.

5 Atom A has atomic number 82 and mass number 204. Atom B has atomic number 80 and mass number 204. How many protons has atom A? How many neutrons has atom B? Are atoms A and B isotopes of the same element? Explain your answer.

6 The three isotopes of hydrogen are ${}^{1}_{1}H$, ${}^{2}_{1}H$ and ${}^{3}_{1}H$.

(a) How many protons and how many neutrons are present in a nucleus of each type of atom?

(b) State the number of electrons present in each type of atom.

(c) How many times heavier is the heaviest atom in comparison with the lightest atom?

7 There are 82 protons and 126 neutrons in each nucleus of the most abundant isotope of lead.

(a) What is the atomic number and the mass number of this type of atom?

(b) What is the ratio of the mass of this type of atom to the mass of a carbon-12 atom?

(c) Why is the diameter of an atom of lead larger than that of an atom of carbon?

(d) Why is the density of lead more than the density of carbon?

TOPIC 9

8 Phosphorus-32 is radioactive, with a half life of 14 days. A solution of sodium phosphate, containing phosphorus-32, gives a count rate of 6000 c.p.m in a Geiger-Müller counter. What will be the count rate after (a) 56 days (b) 140 days?

9 An engineer needs to put a radioactive isotope into an oil pipeline to investigate a leak. The isotope chosen must have a suitable half-life and emit a suitable type of radiation. Which of the following list would be suitable?

Isotope	*Emission*	*Half-life*
Copper-29	β	13 hours
Iodine-131	β, γ	8 days
Nitrogen-16	β, γ	7 seconds
Phosphorus-32	β	14 days
Sulphur-35	β	87 days
Sodium-24	β, γ	15 hours
Thallium-208	β, γ	3 minutes

Give reasons for your choice.

10 Why are people more worried about the accident risk posed by nuclear power stations than the accident risk in using cars?
This is not an easy question, and perhaps you would like to form a group to discuss it.

11 Argon-44 is a radioactive isotope. The amount of argon-44 in a sample was measured every ten minutes over a period of one hour. The results are shown in the table.

Radioactive argon remaining in the sample (%)	100	58	33	19	10	6	4
Time (min)	0	10	20	30	40	50	60

(a) Explain the meaning of 'radioactive' when used to describe an atom.
(b) On graph paper, draw a graph to show how the percentage of argon-44 in the sample changed during the hour.
(c) Use your graph to find the time when only 50% of the argon-44 was left in the sample.
(d) Some of the waste from nuclear power stations contains radioactive isotopes with very long half-lives. These isotopes gives out large amounts of radiation.
(i) Explain why getting rid of this waste is a problem.
(ii) How could this waste be stored safely?
(LEAG)

12 (a) The diagram shows three types of container which could be used to store radioactive material.

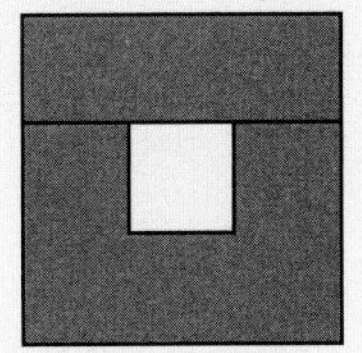
Lead container
2 cm thick walls

Aluminium container
5 mm thick walls

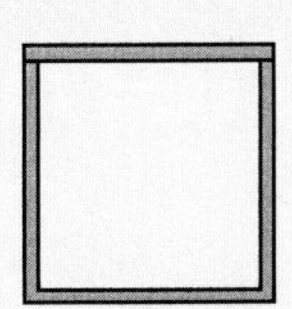
Cardboard container
2mm thick walls

(i) Which one of the containers should be used to store a radioactive source of gamma rays?
(ii) Which of the other two containers should be used to store a radioactive source of beta particles?
(b) In a radioactivity experiment, a detector recorded 45 counts in 100 seconds with no source of radiation obviously present.
(i) What is the count rate in this experiment?
(ii) What causes this count rate?
(iii) With a radiation source present, the detector recorded 295 counts in 100 seconds. What is the true count rate of the source?
(c) Give **three** safety precautions that should be taken when experiments with radioactive sources are carried out in the laboratory.
(NEAB)

13 (a) When a ^{235}U atom captures a neutron it undergoes fission, producing two or three high-speed neutrons as well as other radioactive fission products and energy. Where does the energy come from?
(b) The uranium is most likely to capture low energy (thermal) neutrons. In a nuclear reactor the high energy neutrons are showed by a material called the ______
A suitable material would be ______
(c) For energy to be produced continuously, a chain reaction must be maintained in the reactor core. Explain what is meant by a 'chain reaction'.
(NEAB)

TOPIC 10

14 (a) A uranium-235 nucleus contains 92 protons. How many neutrons does it contain?
(b) When this nucleus is struck by a neutron it splits into two daughter nuclei and releases two neutrons. One of the daughter nuclei contains 44 protons and 67 neutrons. How many protons and neutrons are in the other daughter nucleus?
(c) The two daughter nuclei are unstable because they are neutron-rich. What type of radiation do they emit?

15 (a) What is the quark composition of (i) a proton, (ii) a neutron?
(b) Describe the changes that occur in an unstable proton-rich nucleus when it becomes stable?

Using Waves

Wherever you are, you are probably using waves in one form or another. In the photograph, the surfrider is using waves for enjoyment. Sunbathers use ultra-violet waves from the Sun to become tanned, lifeguards use light waves to watch out for sharks and sirens can emit sound waves as a warning to swimmers.

Topic 11 Waves

11.1 ▶ Making and using waves

FIRST THOUGHTS

This section introduces you to some different types of wave. What uses do you make of each type of wave you read about? Which devices that you use depend on waves?

Every day we use waves in one form or another. Waves carry energy. Figure 11.1A shows some 'snapshots' from a typical morning routine. An alarm clock wakes you up because it emits sound waves when it 'goes off'. The television set picks up signals in the form of radio waves and transforms them into a picture, which is carried to your eye by light waves. In these examples waves are being used to send information.

Figure 11.1A ▲ Making waves (a) Alarm clock (b) Breakfast television

It is easy to make water waves. Drop a stone into a pond and you will see waves spread out in expanding circles. Anything floating on the water, like a duck or a toy boat, will bob up and down as the waves pass. At any point on the water surface, the water rises as each wave crest passes and then falls as each crest is followed by a trough.

To make waves in a rope, stretch the rope out with one end fixed, and move the free end from side to side repeatedly to send waves along it (Figure 11.1B). The strings of musical instruments, such as the piano, guitar and violin, vibrate to produce sound waves when they are plucked, struck or bowed. The human voice box, the larynx, works in the same way. It contains strings, the vocal cords, which vibrate to produce sound waves.

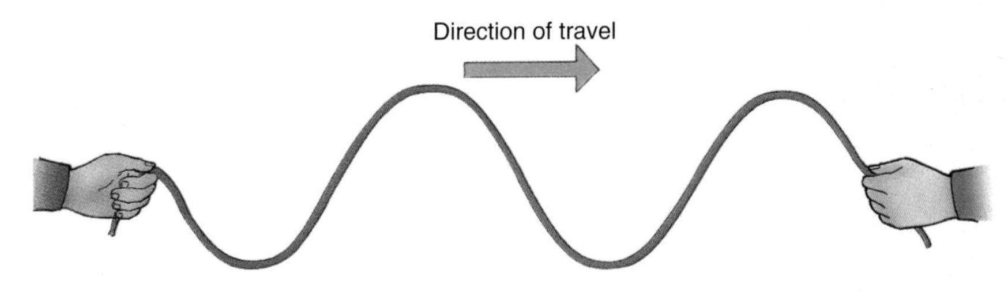

Figure 11.1B ▲ Making waves in a rope

Radio waves are produced by a radio transmitter which forces electrons to move up and down the transmitter aerial. The motion of electrons causes radio waves to be sent out from the aerial. The radio receiver aerial can pick up the radio waves.

Electromagnetic waves pass through a vacuum and do not need to be carried by a substance.

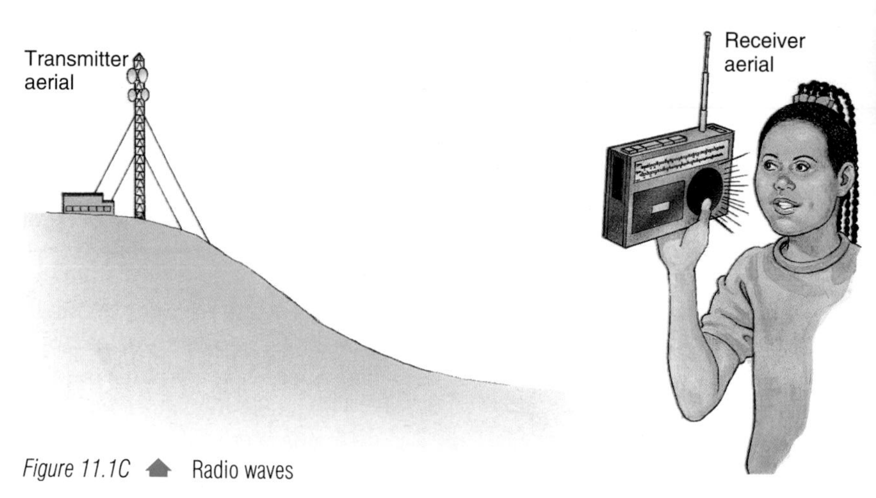

Figure 11.1C ▲ Radio waves

Light waves and radio waves are part of the **electromagnetic spectrum** of waves. All electromagnetic waves travel through space at a speed of 300 000 km/s. Light bulbs convert electrical energy into light energy.

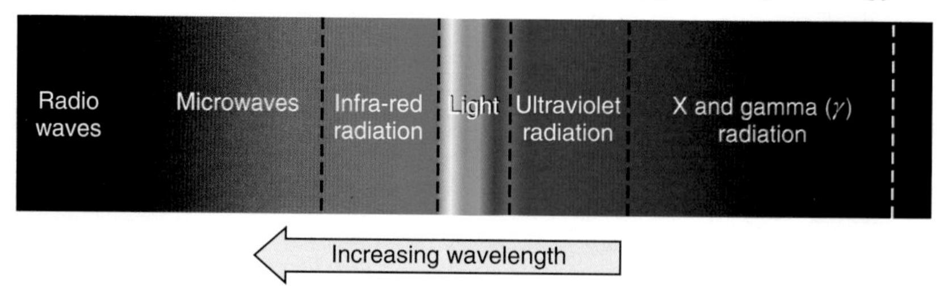

Figure 11.1D ▲ The electromagnetic spectrum

Figure 11.1E ▲ A natural buzzer

We make **sound waves** every time we speak. Our vocal cords vibrate, causing the air around them to vibrate and producing sound waves in the air. Anything that vibrates creates sound waves. A buzzing bee makes sound waves by moving its wings to and fro repeatedly very fast. A loudspeaker makes sound waves because it has a diaphragm that vibrates, pushing the air back and forth and making **compression waves** pass through the air. These waves can be detected by the human ear.

SUMMARY

Waves carry energy and they can be used to transmit information. Water waves, waves in a string, electromagnetic waves and sound waves are all examples of different forms of waves.

CHECKPOINT

1. In each of the following situations, write down whether the waves are being used to transfer energy or information.
   (a) Microwaves used to cook a meal.
   (b) Light from the headlamps of a car.
   (c) Light from a flashing indicator of a car.
   (d) Sound from a loudspeaker.
   (e) Light from a laser used to destroy unwanted tissue in the human body.
2. When a wave travels across the surface of water, the water does not travel with the wave. How would you prove this to a friend who thinks the water moves along with the wave?
3. A 'tsunami' is a tidal wave created by an earthquake under the ocean. Such a wave can travel for thousands of kilometres and is capable of devastating coastal areas. Yet ships at sea 'ride out' these waves. Imagine you are on a ship at sea that rides out a tidal wave. Write a brief account of the events for the ship's log book.
4. For each of the following sound producers, describe the sound and explain how it is produced.

   (a) Bluebottle.
   (b) Grasshopper.
   (c) Cuckoo.
   (d) Drum.
   (e) Whistle.

11.2 Investigating waves

FIRST THOUGHTS

All waves can be reflected, refracted and diffracted. This section explains their behaviour.

All types of wave have certain properties. For example, they can all be reflected. Sea waves reflect off sea walls; sound waves reflect to cause echoes; light waves reflect off mirrors. We can study the properties of water waves in carefully controlled conditions by using a **ripple tank**, as shown in Figure 11.2A.

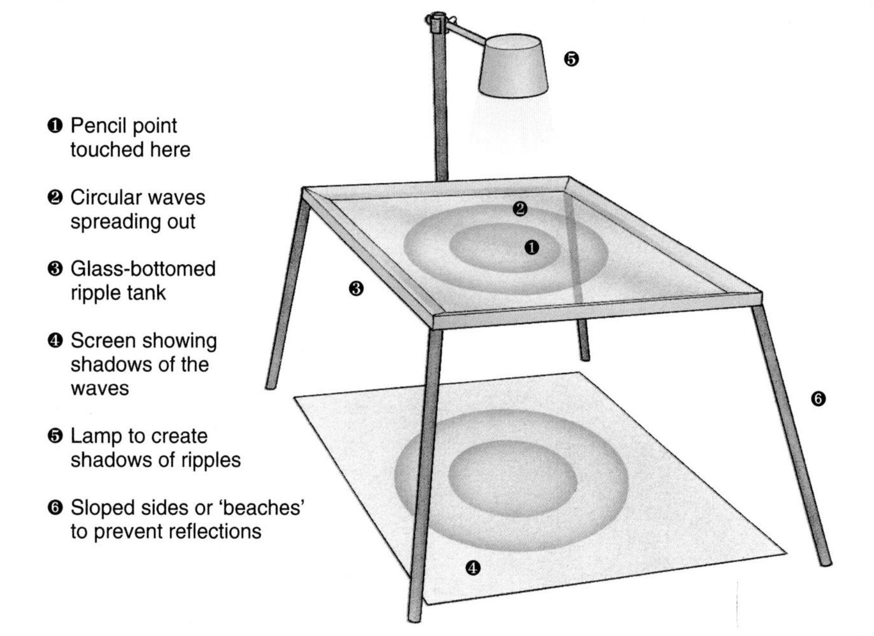

Figure 11.2A ▲ The ripple rank

Reflection

The reflection of waves can be studied by putting differently shaped reflecting walls in the ripple tank and seeing how waves are reflected off them. Figure 11.2B shows the shape of some reflectors.

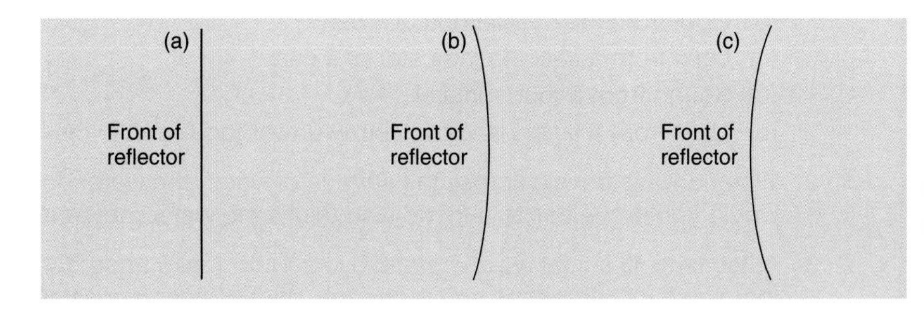

Figure 11.2B Some reflector shapes
(a) Plane or straight reflector (b) Concave reflector (c) Convex reflector

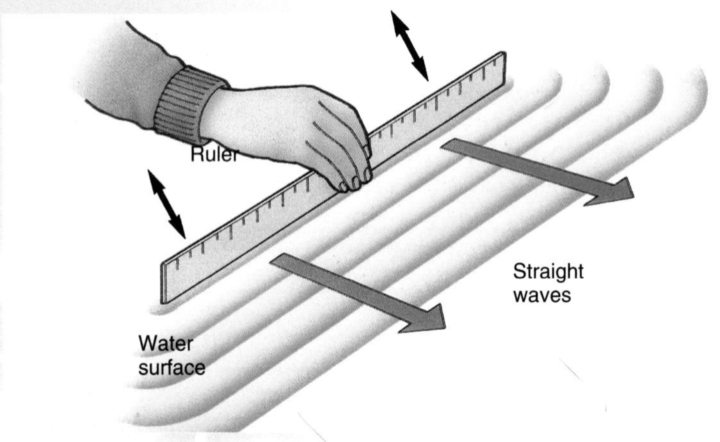

Figure 11.2C Making straight waves

In Figure 11.2C plane waves (straight waves) are created by moving a ruler up and down in the water. The waves travel across the water and are reflected off a straight or plane reflector (Figure 11.2D). Notice that the reflected waves are at the same angle to the reflector as the incident waves. This can be demonstrated easily with light waves too.

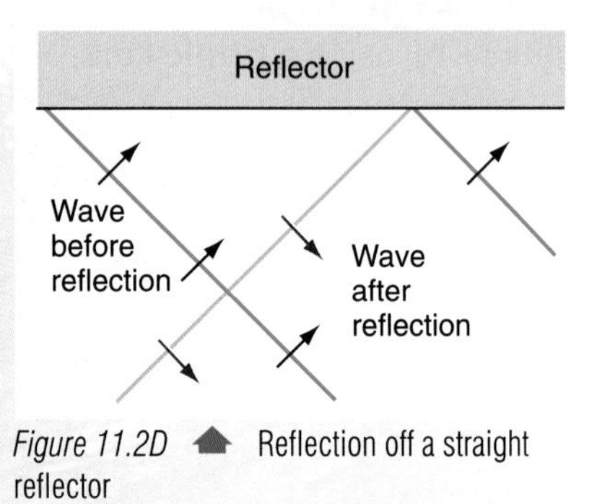

Figure 11.2D Reflection off a straight reflector

In Figure 11.2E plane waves are being reflected by a concave reflector. Again, the reflected waves are at the same angle to the reflector as the incident waves. The reflected waves all meet at a point, called the **focus** of the reflector. Radio waves are reflected from concave reflectors in just the same way, which is why concave dishes are used to 'pick up' television programmes broadcast via satellites. The detector is actually at the focus of the dish, so the dish concentrates the signal by reflecting all the incoming radio waves to the detector.

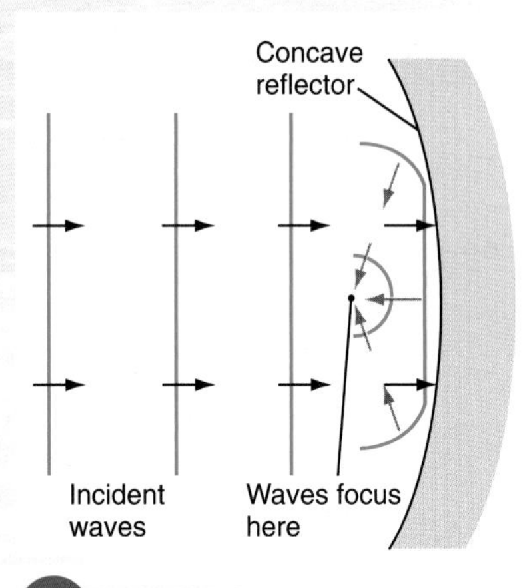

Figure 11.2E (a) Reflection off a concave reflector (b) Television satellite dish with detector at the focus of the concave reflector

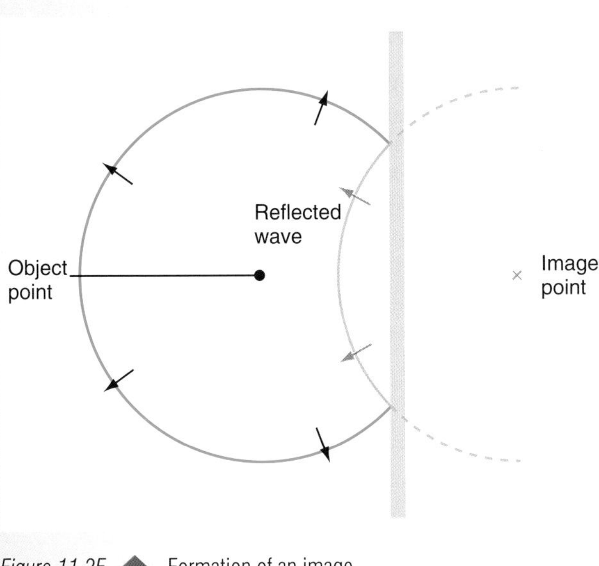

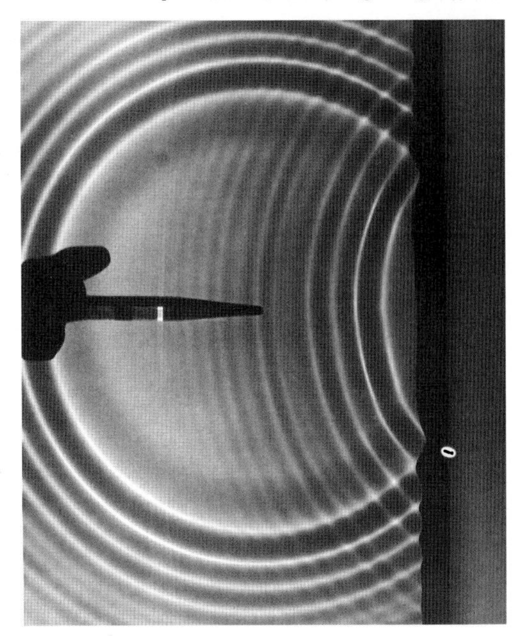

Figure 11.2F ▲ Formation of an image

Circular waves can be made by dipping a pencil in the tank. Figure 11.2F shows circular waves being reflected off a straight reflector. The reflected waves appear to come from an image point behind the reflector. Notice that the image point and the object point (where the waves come from) are both at the same distance from the reflector. The same happens with light waves. This is why your image in a plane mirror is the same distance behind the mirror as you are in front of it.

Refraction

All types of wave can be refracted. Refraction is a change of direction due to a change of speed. For example, water waves travel more slowly in shallow water than in deep water. Figure 11.2G shows how water waves change direction when they pass from deep to shallow water.

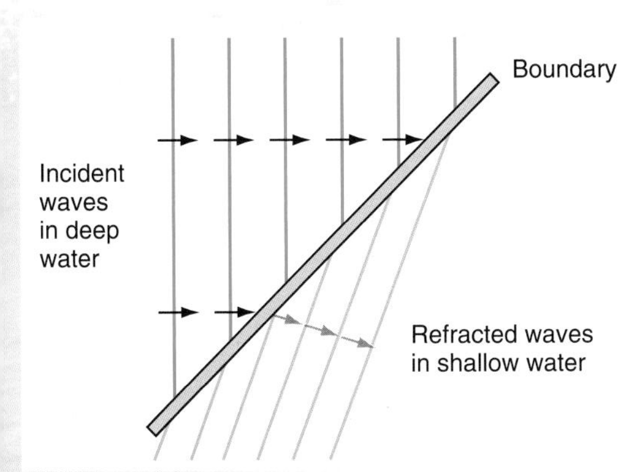

Figure 11.2G ▲ Refraction of waves

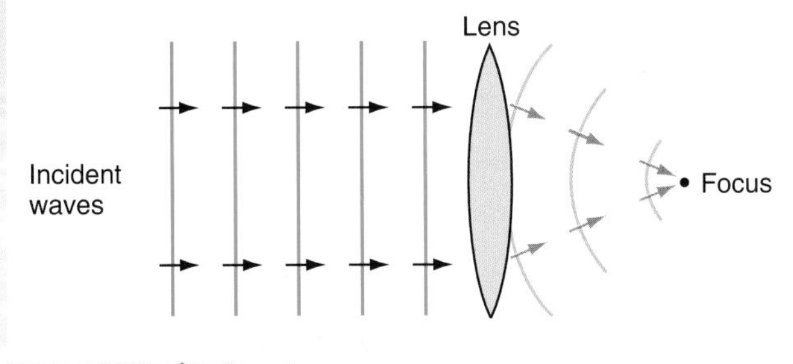

Figure 11.2H ▲ Focusing waves

Light waves are refracted when they pass between air and glass. Figure 11.2H shows how light waves are refracted (change direction) as they enter and leave a convex glass lens. They meet at a point. If a screen is put at this point, the light will be focused on the screen.

SUMMARY

All types of wave can be reflected, refracted and diffracted. Reflection is when waves bounce off a suitable reflector. Refraction of waves is when waves change direction due to change of speed. Diffraction happens when waves pass through a gap or round an obstacle and spread out.

Diffraction

Diffraction is the name given to the way waves spread out when they pass through a gap or round an obstacle. The narrower the gap, the more the waves spread out (see Figure 11.2I). Sea waves passing into a harbour entrance spread out behind the harbour walls. You can demonstrate this for water waves using a ripple tank. An important application is where long wavelength radio waves passing over a hill spread out to the ground behind the hill whereas short wavelength radio waves do not. This is because diffraction is greater the greater the ratio of the wavelength to the obstacle size. The same effect is caused by large buildings.

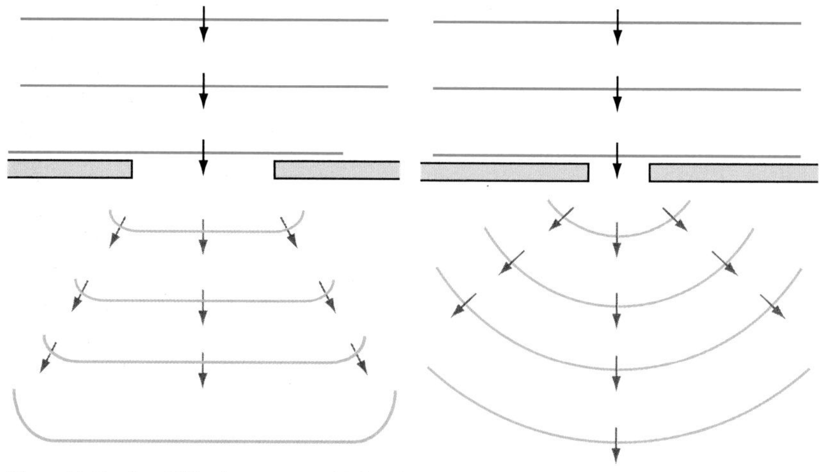

Figure 11.2I ▲ Diffraction at a gap (a) A wide gap (b) A narrow gap

CHECKPOINT

1. Sea waves rolling up a sandy beach are not reflected. Why are 'beaches' used to line the sides of a ripple tank? What would happen if the 'beaches' were left out?
2. Copy the sketches below showing circular waves produced near a concave reflector. In each case sketch in some reflected waves, indicating their direction. (Figure 11.2E will help you.)

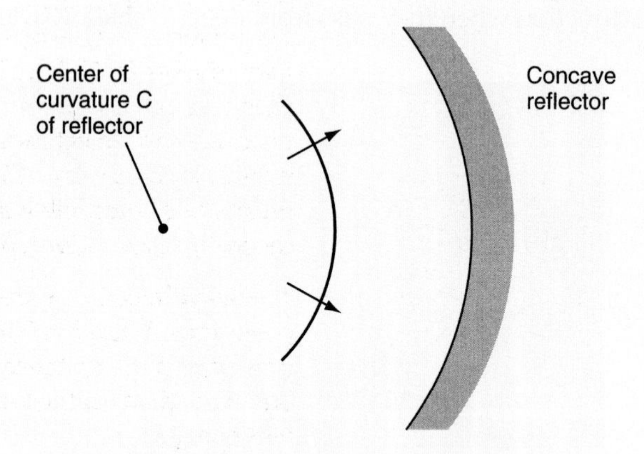

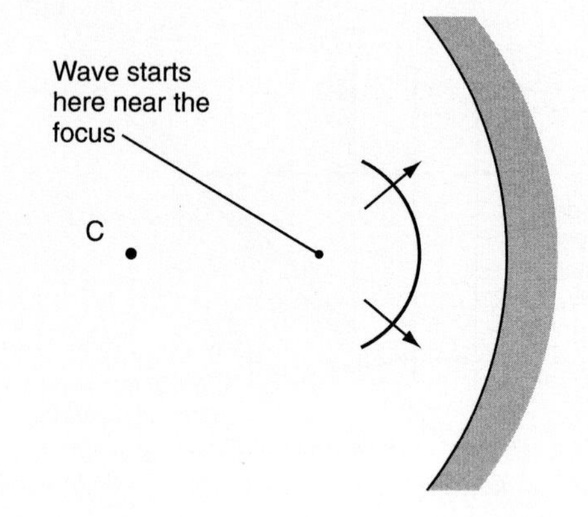

3. (a) Sketch what you see when you look through a pinhole in metal foil at a distant light source.
   (b) Try two more pinholes, one smaller and one larger than in (a). Is there any difference in what you see?
4. Copy the sketch opposite, which shows plane waves passing from shallow to deep water. Draw in some refracted waves, indicating their direction. Remember that waves travel faster in deep water than in shallow water.

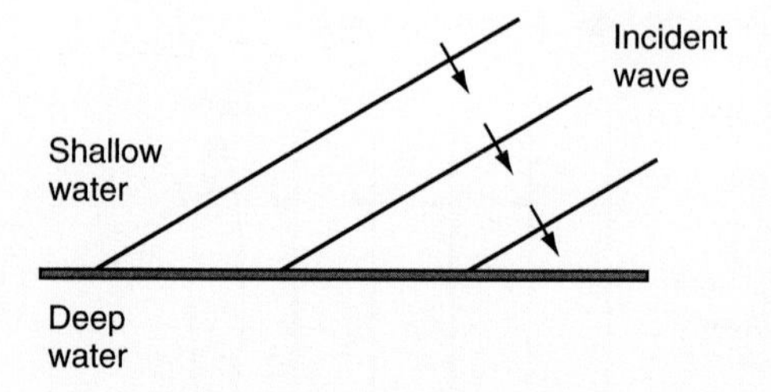

11.3 Measuring waves

This section explains how waves are measured and the uses we make of these measurements.

When the *Voyager 1* space probe flew past the planet Jupiter, it sent back amazing pictures of the planet and its moons. These pictures took about 40 minutes to reach Earth, even though they were carried by radio waves travelling at a speed of 300 000 km/s through space. All electromagnetic waves travel at this speed through space. Thus light from the Sun takes about eight minutes to reach Earth, a distance of about 150 million kilometres.

The **wavespeed**, *c*, of a wave is the distance travelled by the wavepeak per second. Sound waves in air travel at a speed of about 340 m/s. An aeroplane like Concorde can break the 'sound barrier' because it can travel faster than sound. However, nothing can travel faster than the speed of light, the 'cosmic speed limit'.

Water waves travel much more slowly than sound or light waves and they give a useful picture of measurements that can be made on any type of wave. Imagine you are fishing from a boat on a lake when a speedboat passes near. The speedboat creates waves that spread across the lake, and your boat goes up with each wavecrest (or wavepeak) and down with each wavetrough (Figure 11.3A). Each successive wave takes you through a cycle of motion.

Figure 11.3A ▲ Waves from a speedboat

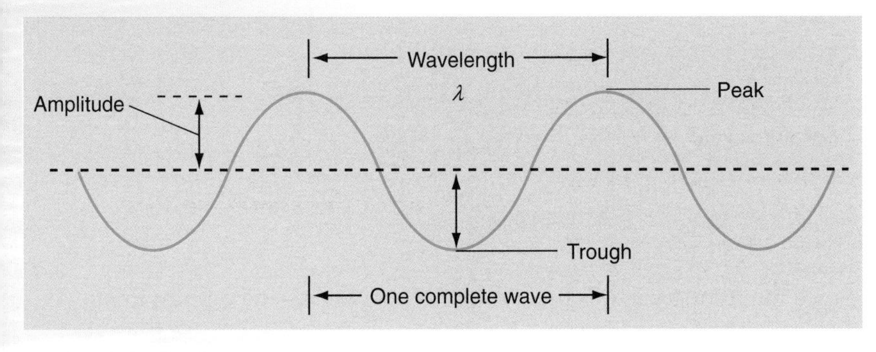

Figure 11.3B ▲ Parts of a wave

The following terms that we use to describe waves are illustrated in Figure 11.3B.

- One **complete wave** is from one wavepeak to the next wavepeak.
- The **frequency**, *f*, is the number of complete waves passing a point in one second. The unit of frequency is the hertz, Hz, named after Heinrich Hertz who discovered how to produce and detect radio waves. A frequency of 1 hertz means one complete wave per second passes a given point.
- The **wavelength**, λ (Greek: pronounced 'lambda'), of a set of waves is the distance from one wavepeak to the next wavepeak.
- The **amplitude** of a wave is the height of its wavepeak or the depth of its wavetrough from the middle. The bigger the amplitude, the more energy the waves carry. Water waves decrease in height as they spread out. This is because the same amount of energy is being carried across a wider area.

It's a fact!

It is possible to tune in to radio broadcasts from distant countries at night in summer. The Sun's radiation causes a layer of ions to form in the ionosphere (upper atmosphere). At night this layer cools and becomes dense enough to reflect radio waves. By being bounced off this layer, radio waves can reach distant continents.

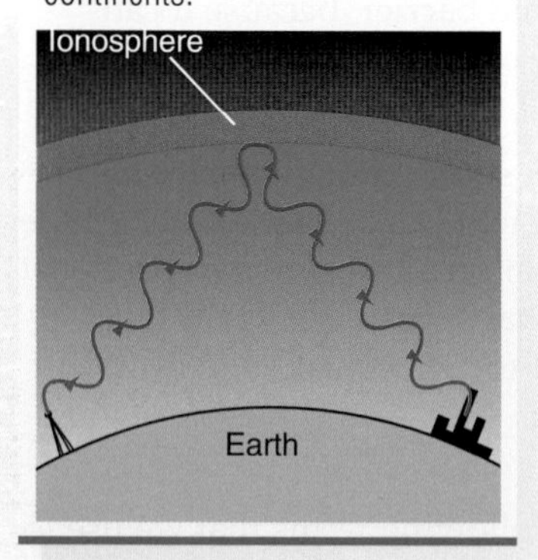

Tuning in

Now we will look at how these ideas apply to waves that cannot be seen. For example, radio waves from a transmitter aerial spread out as they travel away from the aerial. The amplitude of the waves becomes smaller as they spread out. This is why it is difficult to receive programmes from your local radio station when you travel away from your own area.

Each radio station has its own broadcasting frequency. *What is your favourite radio station? What frequency does it broadcast on?* To listen to it, you need to tune your radio in to that particular frequency. Radio programmes are broadcast using waves with frequencies ranging from about 1000 Hz (= 1 kHz) up to about 1 000 000 Hz (= 1 MHz). Television programmes are carried by radio waves with much higher frequencies, usually around 100 MHz. This is sometimes called the ultra high frequency (or UHF) range.

Suppose a new radio station has started in your neighbourhood. You have read that it broadcasts at a frequency of 983 kHz, but your radio dial is marked in wavelengths in metres. You can work out the station wavelength using this formula.

$$\underset{f}{\text{Frequency}} \times \underset{\lambda}{\text{Wavelength}} = \underset{c}{\text{Wavespeed}}$$

Radio waves travel at a speed of 3.00×10^8 m/s through air. Hence the wavelength for a frequency of 983 kHz is given by

$$983 \times 10^3 \times \text{Wavelength} = 3 \times 10^8$$

thus $$\text{Wavelength} = \frac{3.00 \times 10^8}{983 \times 10^3} = 305 \text{ metres}$$

The formula $c = f\lambda$ applies to all waves

To prove the formula consider Figure 11.3C, which shows two stages of a ripple tank experiment in which straight waves are produced at a constant frequency, f.

Wave W moves forward from X to Y in time t.
Hence distance XY = Wavespeed × Time = $c \times t$.

See www.keyscience.co.uk for more about waves and vibrations.

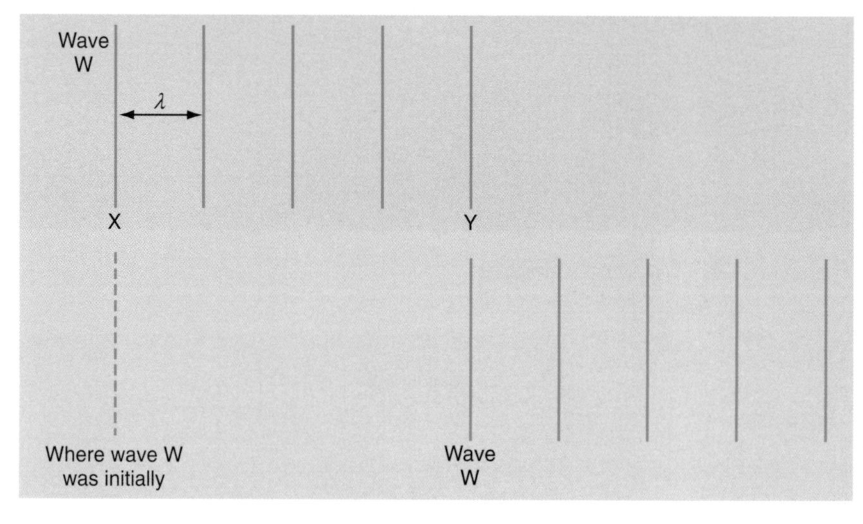

Figure 11.3C ▲ $c = f\lambda$ (a) Initially (b) After time t

Since the frequency, f, is the number of waves passing a point in one second, then $f \times t$ waves pass Y in time t. Therefore there are ft waves between X and Y.

SUMMARY

A complete wave is from one wavepeak to the next. One wavelength is the distance from one wavepeak to the next. The frequency is the number of complete waves passing a point per second. The amplitude of a wave is the height of the wavepeak above the middle.

So another way of finding the distance XY is

$$XY = \text{Number of waves from X to Y} \times \text{Wavelength}$$
$$= f \times t \times \lambda$$

Combining [1] and [2] we have

$$f \times t \times \lambda = c \times t$$

and cancelling t on both sides gives

$$f \times \lambda = c$$

CHECKPOINT

1 (a) Explain why you hear the sound of distant thunder some time after seeing the lightning flash that produced the thunder.

(b) In a thunderstorm, Katie observes that the interval between a lightning flash and its thunder is 6 seconds. She knows that sound travels at 340 m/s. How far away was the lightning strike?

2 Radio waves travel through air at a speed of 300 000 km/s. Work out the frequency of the radio waves from each of the following radio stations.

(a) BBC Radio 4; λ = 1500 m

(b) BBC Radio 1; λ = 3.07 m

(c) BBC Radio 3; λ = 3.25 m

(d) World Service; λ = 463 m

3 Two fishermen in different boats are 30 m apart when a speedboat passes.

(a) The waves from the speedboat travel along the line between the two boats as in the illustration, causing the fishermen to bob up and down once every two seconds for a few minutes. What is the frequency of the waves?

(b) When one fisherman is on a wavepeak, the other is in a wavetrough as shown. Work out the wavelength and the wavespeed of the waves.

4 Work out the wavelength of the sound waves produced in each of the following situations, given the speed of sound in air is 340 m/s.

(a) A tuning fork vibrating at a frequency of 512 Hz in air.

(b) A siren operating at 3000 Hz.

(c) A dog whistle emitting sound at 25 kHz.

5 The diagram shows a snapshot of a wave travelling from left to right along a rope.

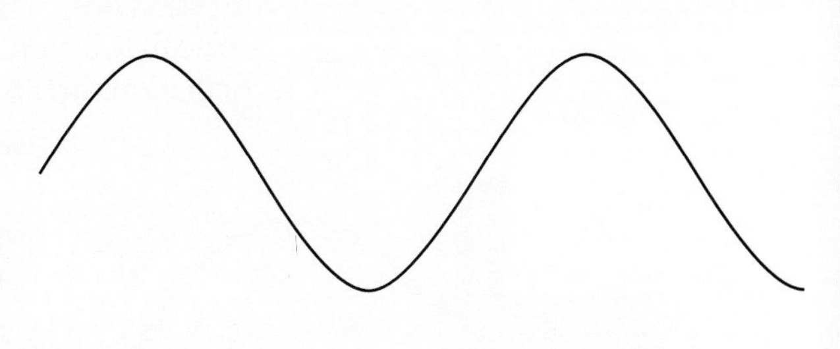

(a) What is the amplitude and the wavelength of this wave?

(b) Each point on the rope vibrates up and down at a frequency of 2.0 Hz. Calculate the speed of the waves along the rope.

(c) Copy the diagram opposite and sketch on it how the rope would appear exactly one second later.

11.4 Transverse and longitudinal waves

FIRST THOUGHTS

When you think about waves you probably picture water waves. This picture is reasonable for light waves and electromagnetic waves, but not for sound waves. Read on to find out why not.

In the previous sections, water waves have been used to explain some of the basic properties of all types of wave. When waves travel across water, objects floating on the water surface bob up and down. The objects are said to oscillate because they repeatedly move up and down.

You can use a long rope to demonstrate wave properties, as shown in Figure 11.4A. By moving one end of the rope from side to side repeatedly, you can make waves travel along the rope towards the other end. Each point along the rope oscillates from side to side as the waves pass.

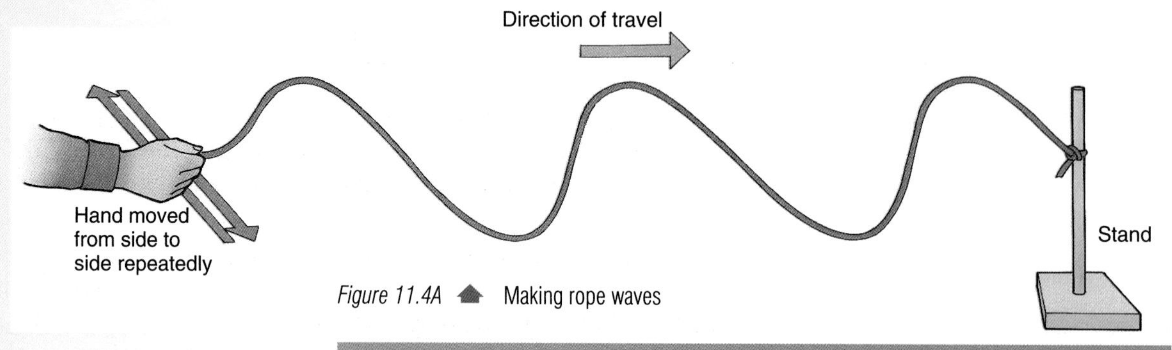

Figure 11.4A ▲ Making rope waves

Transverse waves

Waves on a rope and light waves are examples of **transverse waves**. Transverse means 'across' and transverse waves are waves where the oscillations are at right angles to (i.e. across) the direction in which the waves travel.

In waves travelling along a rope, the oscillations can be along any line at right angles to the rope, not just up and down or side to side. To see this, try the demonstration shown in Figure 11.4B.

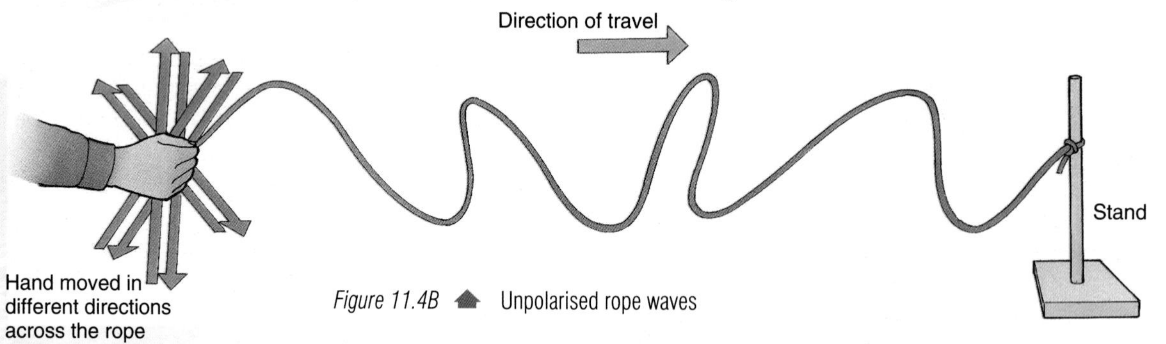

Figure 11.4B ▲ Unpolarised rope waves

Transverse waves are said to be **polarised** if they oscillate in one fixed direction only. Figure 11.4C shows polarised waves oscillating in a vertical direction. The waves in Figure 11.4B are unpolarised waves – the oscillations are in several different directions.

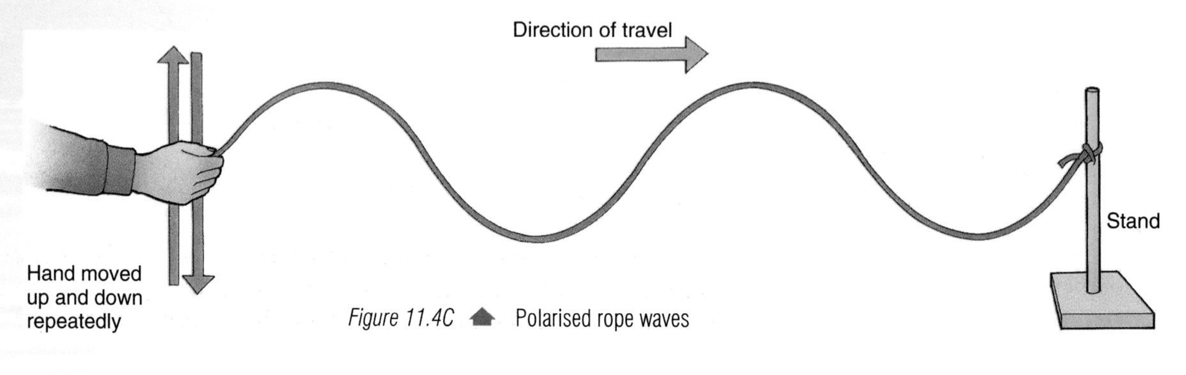

Figure 11.4C ▲ Polarised rope waves

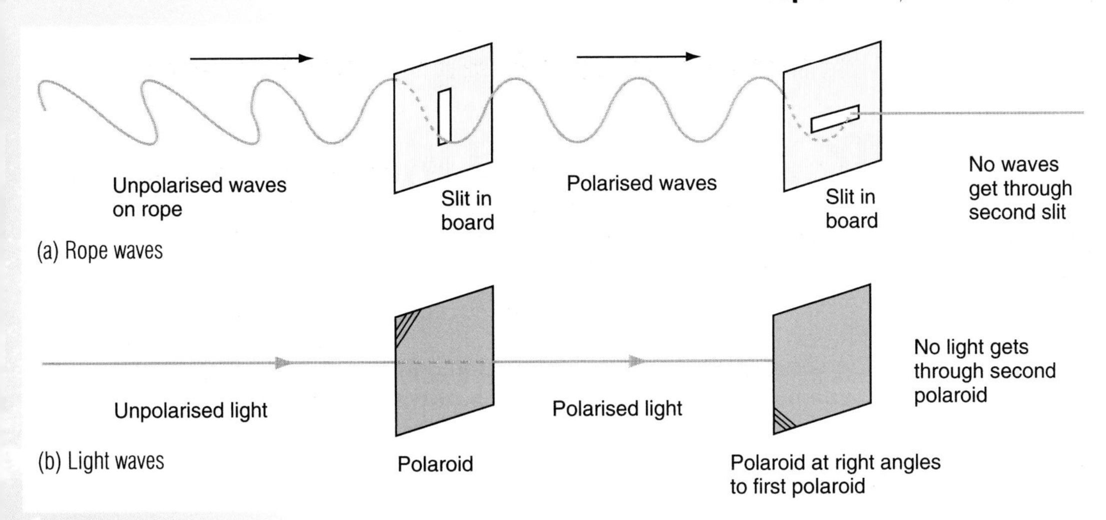

Figure 11.4D ▲ Polarisers

Light waves from a lamp bulb are unpolarised transverse waves. They can be polarised by being passed through special material called Polaroid®. This material only allows light waves through that are oscillating in a certain plane.

Polarisation is a property of transverse waves only.

The particular plane depends on the Polaroid molecules. Figure 11.4D shows how waves on a rope are polarised by being passed through a vertical slit that allows only vertical oscillations through. If the vertically polarised waves are then passed through a horizontal slit, no waves will come through at all. In the same way, light waves polarised in one direction by a piece of Polaroid cannot pass through a piece of the same Polaroid held at right angles to the first.

Polaroid filters are used in liquid crystal displays (LCDs) in calculators, and in Polaroid sunglasses. Look at an LCD through Polaroid sunglasses and rotate the display as you look. You should find that the display disappears then reappears as it is turned. This is because light from the LCD is polarised. Therefore it can only pass through the Polaroid sunglasses if the light is polarised in the 'correct' plane for the Polaroid molecules.

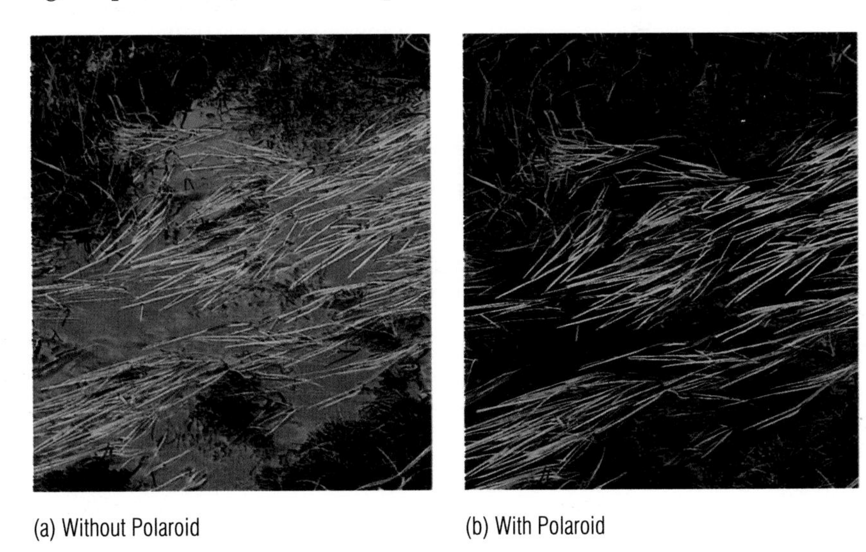

(a) Without Polaroid (b) With Polaroid

Figure 11.4E ▲ Polaroid sunglasses eliminate glare

An LCD display

Scientists in different countries are racing to develop flat screen televisions. These are likely to prove very popular because they can be wall-mounted. One promising line of research uses liquid crystals. However, present LCDs are too slow to respond for television signals and scientists are hoping to develop crystals with faster responses.

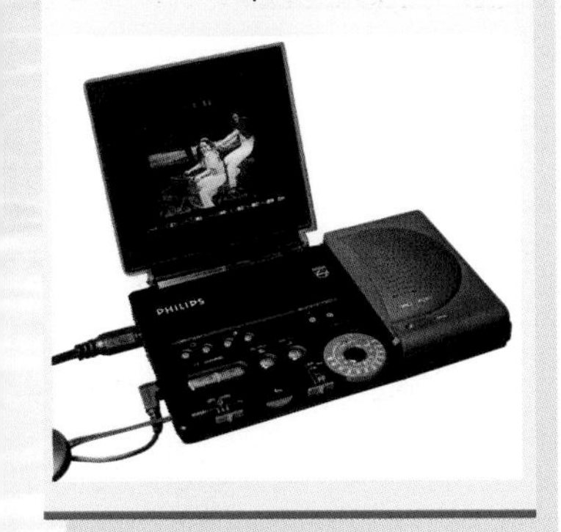

Longitudinal waves

Sound waves are created by surfaces vibrating in air. The motion of a vibrating surface in air sends pressure waves through the air as the surface pushes and pulls repeatedly on the air surrounding it.

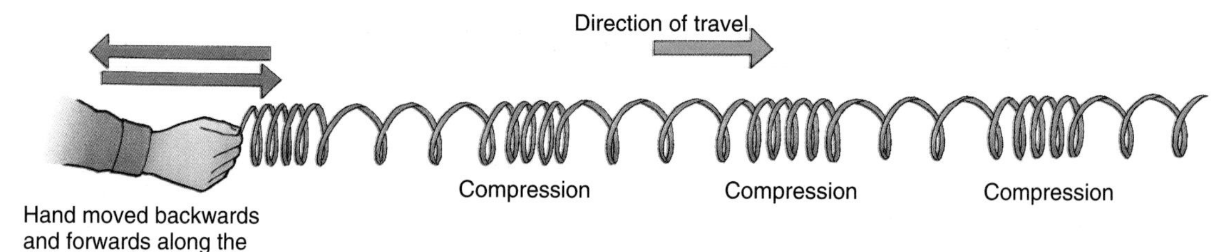

Figure 11.4F ▲ Making longitudinal waves in a slinky spring

A slinky spring is useful for demonstrating how sound waves travel. If you move one end of the spring backwards and forwards repeatedly along the direction of the spring, you will see waves of compression travelling along it (Figure 11.4F). Each part of the slinky oscillates along the line of the spring. In the same way, when sound travels through air, each layer of air oscillates along the direction in which the sound travels (Figure 11.4G).

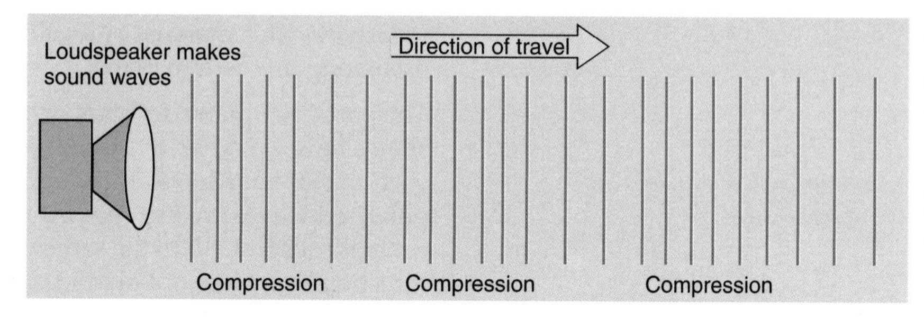

Figure 11.4G ▲ Making sound waves

SUMMARY

Transverse waves oscillate at right angles to the direction of travel of the waves. Longitudinal waves oscillate along the direction of travel of the waves. Transverse waves can be polarised.

Sound waves and compression waves along a slinky are examples of longitudinal waves, where the direction of oscillation is along the same line as the direction of travel of the wave itself. Like transverse waves, longitudinal waves can be reflected, refracted and diffracted. However, longitudinal waves cannot be polarised. Polarisation is a property of transverse waves only.

CHECKPOINT

1. Describe how you would use a slinky to demonstrate to a friend the difference between longitudinal and transverse waves.
2. If a beam of light is passed through two polaroid filters, as in the diagram, the beam can be stopped by turning one of the filters. Explain why this happens.

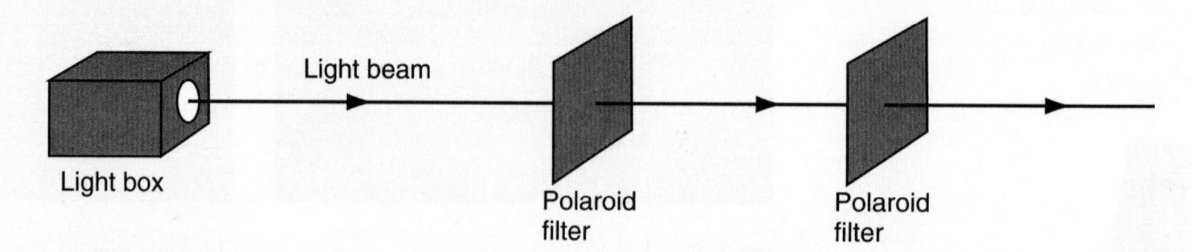

3. (a) Susan's mother finds driving to work difficult because of the 'glare' of the Sun on the road. She thinks Polaroid sunglasses might be useful. Test a piece of Polaroid to find out if it cuts out glare.
   (b) It is not advisable to wear Polaroid sunglasses for driving because the toughened zones on the windscreen may be polarised. Why does this make Polaroid sunglasses unsuitable?

11.5 ▶ Interference of waves

FIRST THOUGHTS

How do we know that light is a waveform? In this section you will learn about one of the most important experiments in the history of physics. It was first carried out in 1803 and proved that light is a waveform.

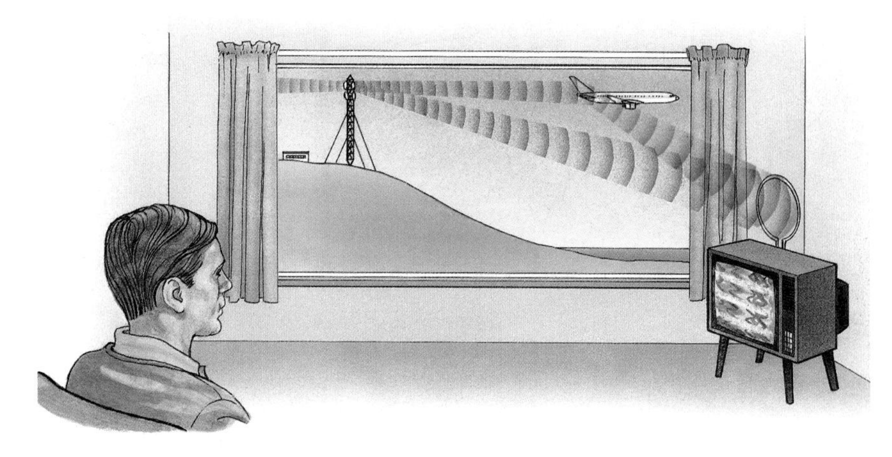

Figure 11.5A ▲ Television interference

Television programmes can sometimes be disturbed by poor reception when a low-flying aircraft passes near. Radio waves reflected from the aircraft arrive at the television aerial at the same time as radio waves direct from the transmitter (Figure 11.5A). The two sets of waves are said to **interfere** where they meet.

Figure 11.5B(a) shows two waves travelling along a rope towards each other. One wave is a crest and the other is **trough**. They meet and pass through each other. When they meet, they **cancel** each other out at that instant. Figures 11.5B(b) and (c) show what happens when a crest meets a crest or a trough meets a trough. In these cases the waves are 'added together' to make a larger crest or trough. These are examples of **interference**.

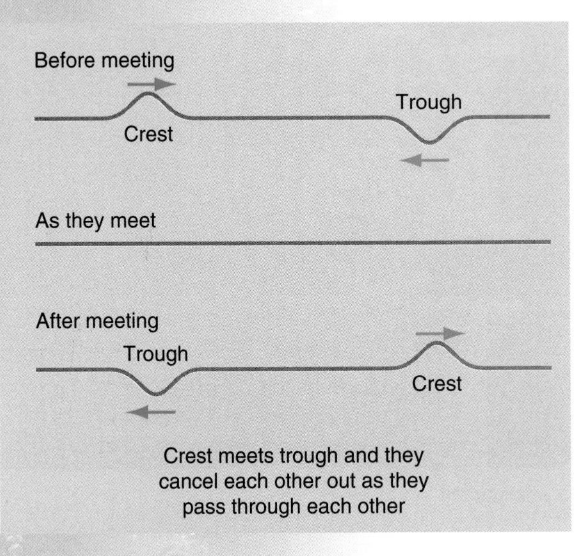

(a) Crest meets trough

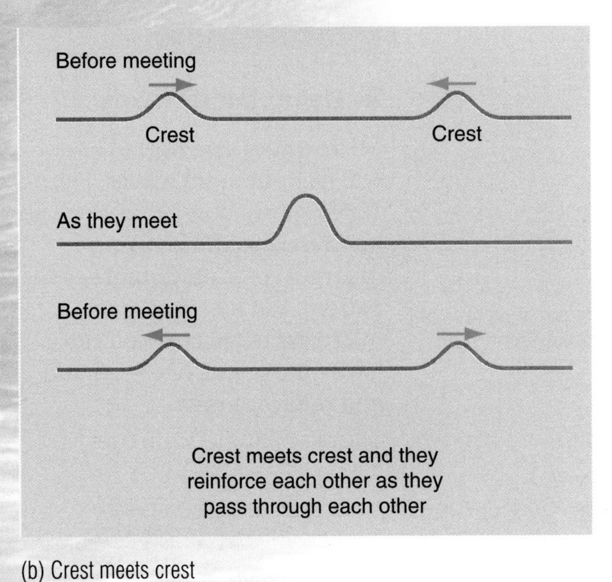

(b) Crest meets crest

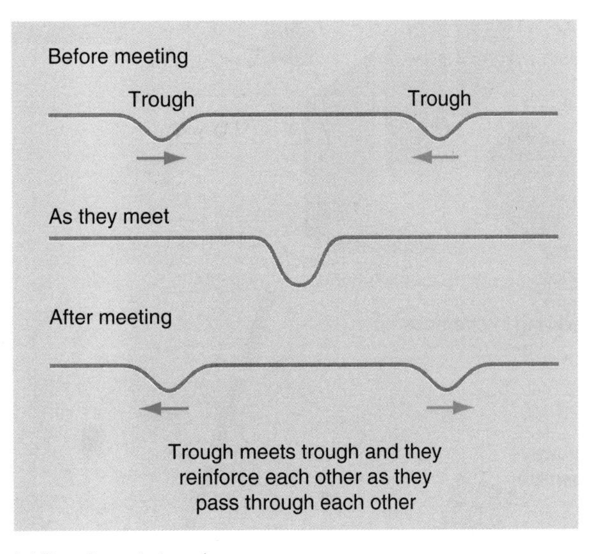

(c) Trough meets trough

Figure 11.5B ▲ Waves meeting

❶ Lamp casts shadows of ripples on to screen ❹
❷ Beam and dippers suspended by elastic bands. The electric motor on the beam makes it vibrate, hence creating waves where the dippers ❸ touch the water.

(a) Using the ripple tank

Crest spreading out
Dipper •
Line of reinforcements
Dipper •
Line of gaps
Trough spreading out

(b) Pattern on the screen

Figure 11.5C ▲ Interference of water waves

All types of wave can be made to interfere. Figure 11.5C shows what happens in a ripple tank when two sets of water waves overlap. Gaps are seen where crests from one dipper cancel troughs from the other dipper. The gaps are **points of cancellation.**

Between these gaps, there are **points of reinforcement** where crests (or troughs) from one dipper arrive at the same time as crests (or troughs) from the other dipper. The pattern of cancellations and reinforcements is called an **interference pattern**.

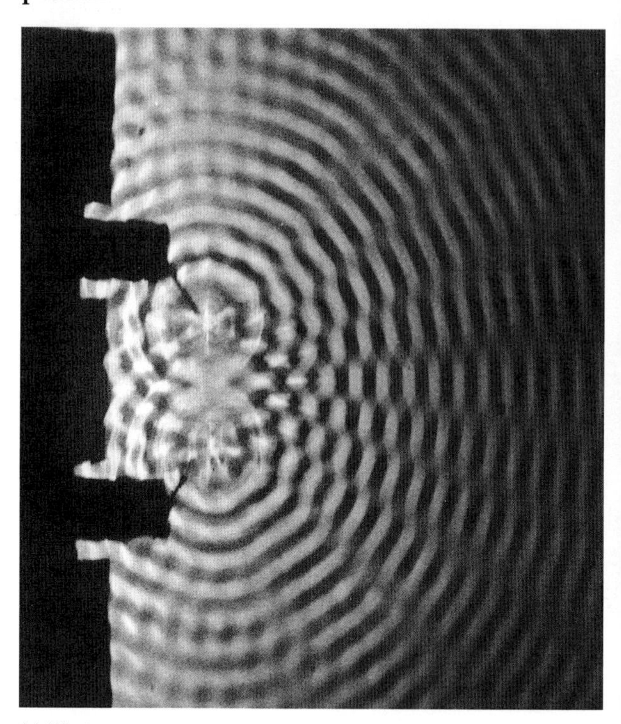

(c) What you see

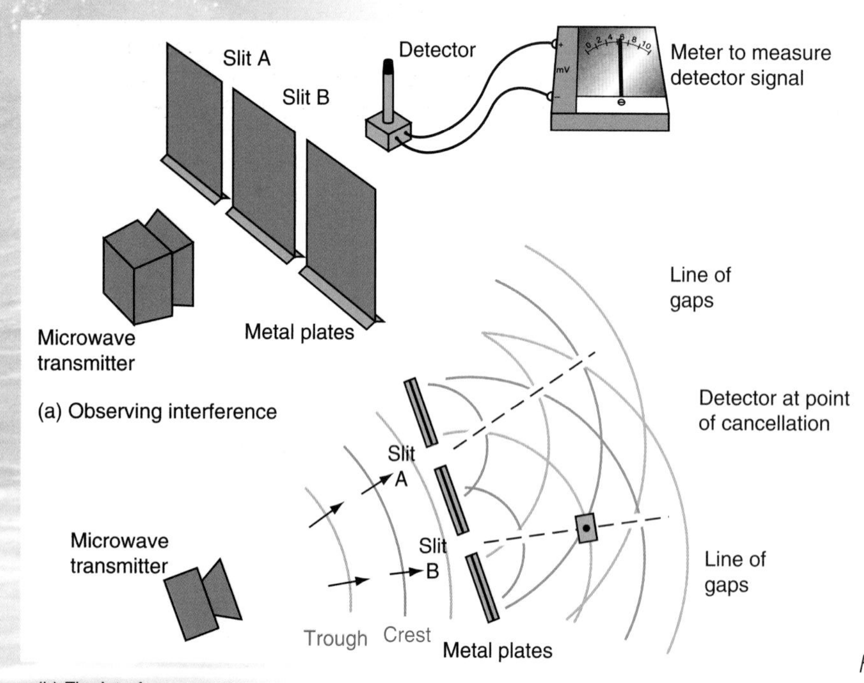

(b) The interference pattern

Investigating interference

Using microwaves

Microwaves are short wavelength radio waves. Figure 11.5D shows how to produce an interference pattern using microwaves. You cannot see the pattern, but by moving a microwave detector you can locate the points of cancellation and reinforcement.

Figure 11.5D ◀ Using microwaves to investigate interference

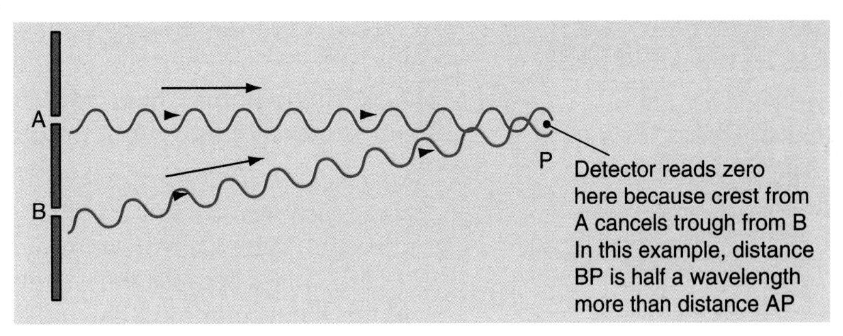

Figure 11.5E ▲ Interference conditions

Suppose the detector is moved to a point P where the meter reads zero (Figure 11.5E). This is a point of cancellation. At this point, waves from slit A cancel out waves from slit B. This is because a crest from one slit arrives at P at the same time as a trough from the other slit.

In fact, cancellation happens wherever the distances AP and BP differ by one half-wavelength or three half-wavelengths or five half- wavelengths, etc. Then crests from one slit arrive at P half a cycle later than crests from the other slit. Hence the two sets of waves cancel.

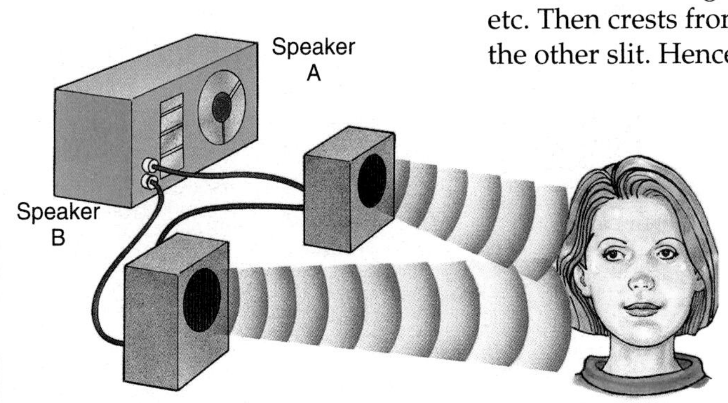

Figure 11.5F ▲ Interference of sound waves

Using sound waves

Use two loudspeakers working from the same signal generator and walk about in front of the speakers (Figure 11.5F). You will hear the sound intensity vary from place to place. At each position of minimum intensity, crests from one speaker are arriving at the same time as troughs from the other speaker, and cancellation is taking place.

How does the interference pattern change if the wavelength of the sound waves is increased? The points of cancellation and reinforcement are now further apart.

Using light

Observe the light from a narrow slit through a pair of slits, as shown in Figure 11.5G. A pattern of alternate bright and dark fringes will be seen stretching outwards from the double slits.

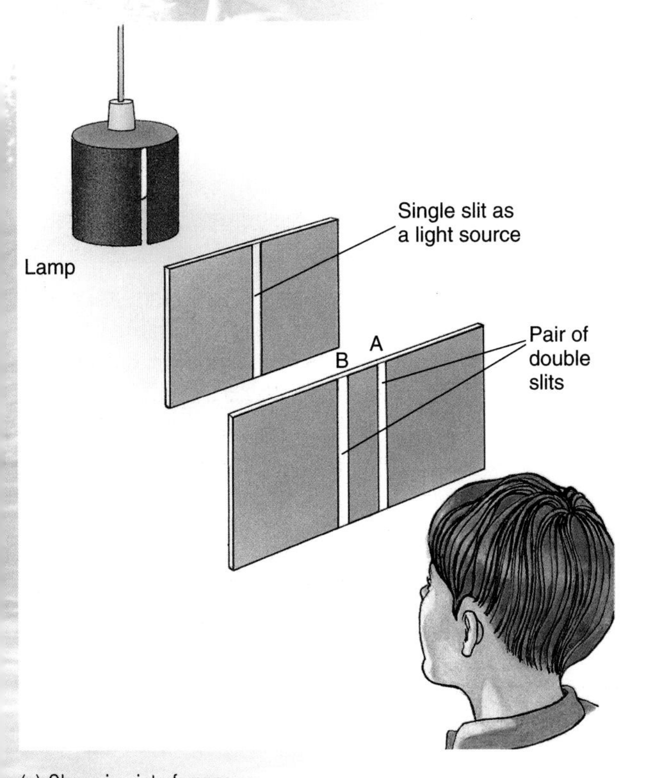

(a) Observing interference

(b) Interference pattern

Figure 11.5G ▲ Interference of light

It's a fact!

This experiment was first performed by Thomas Young in 1803. It provided the first direct evidence that light consists of waves. Measurements from this experiment can be used to estimate the wavelength of light. About 2000 light wavelengths fit into 1 mm!

Light waves from the single slit source pass through each of the double slits. Where these two sets of waves overlap, interference takes place. Dark fringes occur where crests from one of the double slits arrive at the same time as troughs from the other double slit.

What happens if a different colour of light is used? Alternate bright and dark fringes are seen whatever colour is used. However, the spacing between adjacent fringes is greatest for red light and least for blue light (Figure 11.5H). This is because the wavelength of light waves differs for each colour; longest for red light and shortest for blue light.

Alternate bright and dark red fringes

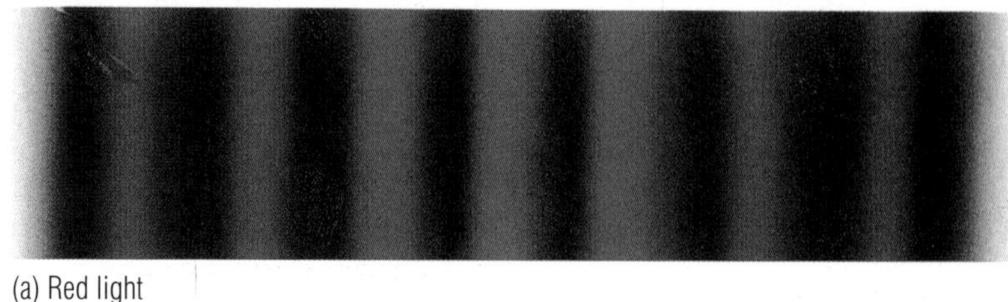

(a) Red light

Alternate bright and dark blue fringes

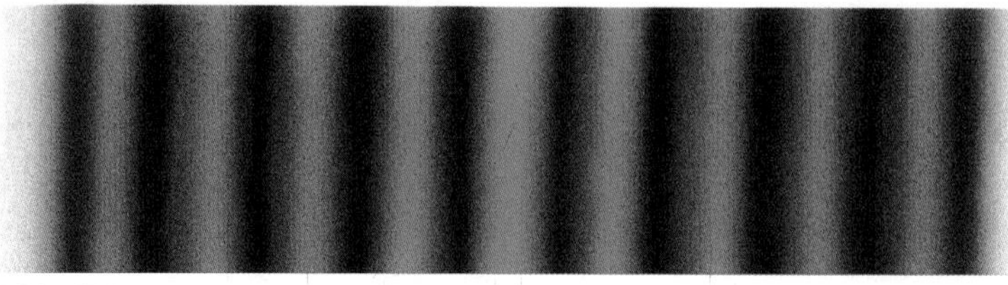

(b) Blue light

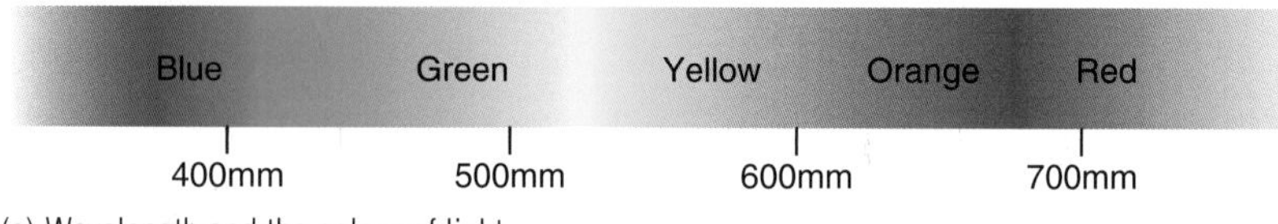

(c) Wavelength and the colour of light

Figure 11.5H Interference patterns for different wavelengths of light

SUMMARY

When two waves meet, they pass through each other. Where a crest meets a trough, they cancel as they pass through each other. Where a crest meets a crest or a trough meets a trough, they reinforce as they pass through each other.

In the shadows

If you hold a coin in a beam of light from a torch in a darkened room, a sharp shadow of the coin is formed on the wall. In 1704 Sir Isaac Newton, the famous British scientist, put forward a theory of light to explain how shadows are formed.

Newton supposed that a ray of light consisted of tiny particles streaming out from the source of light. He called the particles corpuscles. Newton's theory was that a coin in the path of a beam of light casts a shadow because it stops the light particles. Newton also used his theory to explain other properties of light, such as reflection by mirrors. He imagined the corpuscles bouncing off the mirror like balls bouncing off a wall.

A different theory of light had been proposed by Christian Huygens in 1678 in the Netherlands. He imagined light as a waveform, spreading out from a point source like water waves spreading out from a stone dropped into water. He used his theory to explain reflection but was unable to explain how shadows are formed. According to wave theory, light passing the edge of a sharp object should be diffracted and spread out behind the object, making the edge of the shadow fuzzy rather than sharp. The sharpness of shadows was the main reason why Newton's theory was accepted, rather than Huygen's theory.

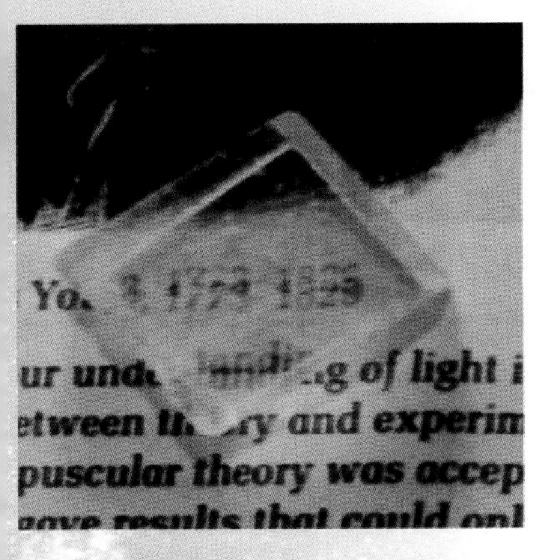

Figure 11.5I ▲ Double refraction

The corpuscular theory of light was unchallenged for a century until another British scientist, Thomas Young, discovered interference of light. He observed that light passing through two closely spaced slits produced a pattern of bright and dark fringes. This can be explained by wave theory but not by corpuscular theory.

Young presented his discovery to other scientists at a lecture to the Royal Society in 1803. However, although his discovery of interference of light was accepted, his fellow scientists would not accept that light was a waveform. They could not use Young's wave theory to explain why light passing through a calcite crystal formed a double image, as shown in Figure 11.5I. This was because they thought in terms of longitudinal waves, not transverse waves. Double refraction happens when light passes through calcite because the speed of light in calcite depends on the directions of vibration of the light waves.

The solution did not occur to Young until 1818, when he realised that light waves are transverse not longitudinal. Augustin Fresnel, a French scientist, used this idea to give a complete explanation of double refraction. He then developed Young's theory to explain how fringe patterns are formed when light passes through any gap or round any obstacle. He also used wave theory to show that there should be a bright spot at the centre of the shadow of a coin in a beam of light. Scientists looked again at the sharpness of shadows. They used microscopes that had not been available to Huygens and observed the fringe patterns predicted by Fresnel. They found that there is indeed a bright spot at the centre of the shadow of coin. Huygens had had the right idea after all!

Why did it take fifteen years for Thomas Young to realise that light waves are transverse, not longitudinal? At this time the ruling classes in Britain were worried that there could be a revolution as there had been in France. Britain was almost a police state and people were transported or imprisoned for opposition to the repressive government policies.

Shortly before his 1803 lecture to the Royal Society, Young resigned from his post as assistant lecturer at the Royal Institution. Benjamin Thompson (who had appointed Young) had proposed radical plans to provide technical education for working people. The Royal Institution turned this down, since education could make the 'masses' more powerful. Thompson took his plans to Bavaria and Young, discouraged by the Institution's attitude, returned to his full-time medical practice.

It's a fact!

The short-sighted policies of the ruling class in Britain restricted the development of science and technology in Britain for almost two decades. Pressure for change built up and the Reform Act of 1832 introduced democracy to Britain. The balance of power shifted from the landowning classes to the industrialists, and the importance of scientific and technical education to promote industrial and economic growth was recognised. As a result, institutes and colleges were established in many cities and towns.

CHECKPOINT

1. Simon notices that when he moves a metal tea tray near the aerial of a portable television, the television reception is affected. Why do you think this happens?
2. In the microwave experiment to demonstrate interference (Figure 11.5D), the detector is placed at a point where its signal strength is zero. What would you expect to observe if each slit were blocked in turn?
3. (a) In Figure 11.5D, the detector is moved to a point equidistant from the slits. The detector signal strength is then a maximum. Why?
   (b) The detector is then moved to one side until the signal becomes zero. When it is moved further in the same direction the signal rises again. Explain these observations.
4. In a microwave oven, microwaves are reflected from the inside surface of the oven. A metal 'paddle wheel' inside the oven spreads the reflected microwaves around. How would food being cooked in the microwave be affected if the wheel stopped turning during the cooking process?
5. In Figure 11.5G, how would the pattern of light fringes change if:
   (a) one of the double slits was covered completely,
   (b) blue light was used instead of red light?

Topic 12 Sound

12.1 Sound patterns

FIRST THOUGHTS

In this section you will find out how characteristic sounds are produced.

Can you recognise people by the sound of their voices? How would you describe different voices? Loud, soft, deep, high-pitched; these are all terms we use to describe the human voice. Voice recognition locks are designed to open when commanded by the correct voice. Provided you don't have a cold, electronic voice recognition devices can use your voice to check your identity.

Investigating different sounds

Sound waves can be displayed using a microphone connected to an oscilloscope, which is an instrument designed to display waveforms on its screen (Figure 12.1A). The microphone converts sound waves into electrical waves, which are then supplied to the oscilloscope. The pattern on the oscilloscope screen shows how the amplitude of the sound waves changes with time.

Figure 12.1A ▲ Displaying sound waves

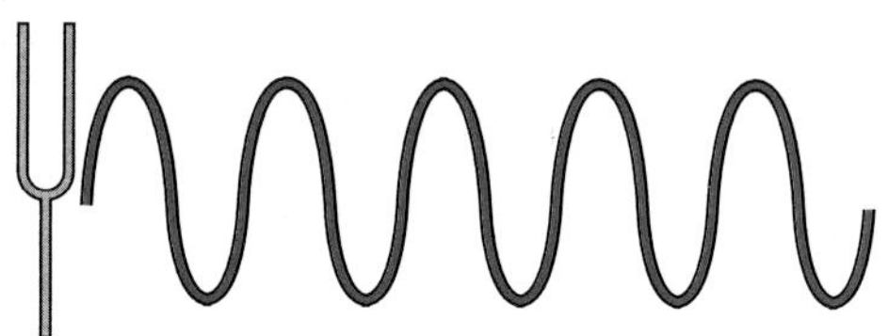

Figure 12.1B ▲ Tuning fork pattern

The pattern for a tuning fork sounded near the microphone shows waves of constant frequency, as in Figure 12.1B. Whistling at a constant pitch at the microphone gives a pattern like that in Figure 12.1C. The way in which changing the pitch or the loudness alters the pattern is also shown in Figure 12.1C.

Playing a flute produces a pattern like that in Figure 12.1D. Unlike the previous examples, this pattern is a mixture of frequencies rather than a single frequency.

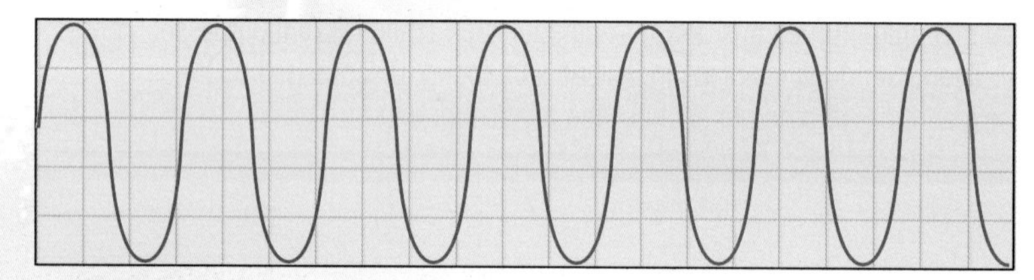

(a) Loud and high-pitched

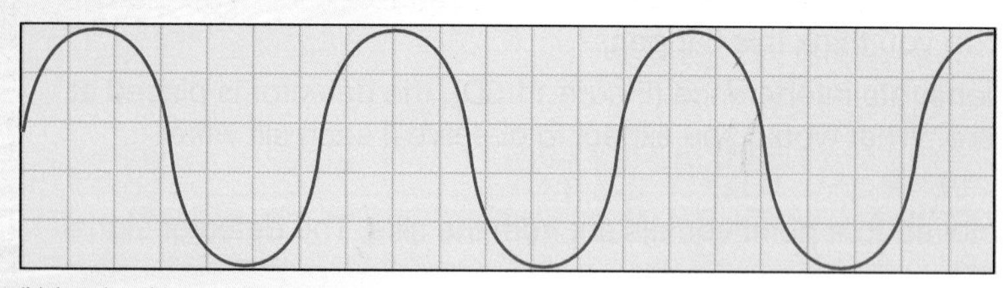

(b) Loud and low-pitched

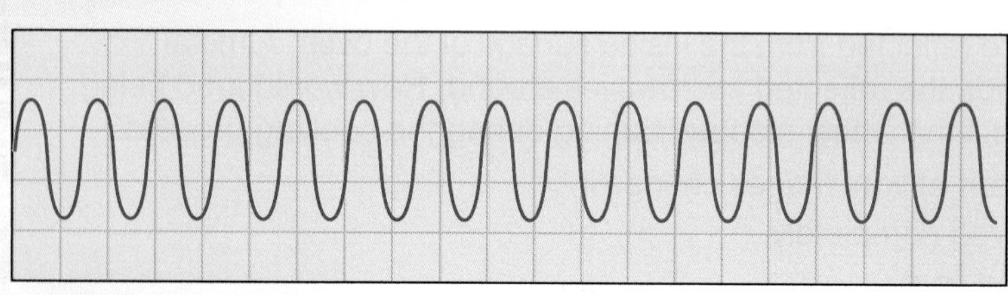

(c) Quiet and high-pitched

Figure 12.1C ▲ Whistling wave patterns

Figure 12.1D ▲ Flute wave pattern

SUMMARY

Sounds can be described in terms of loudness, pitch and quality. Increasing the amplitude of sound waves makes them louder. Increasing the frequency makes the pitch higher.

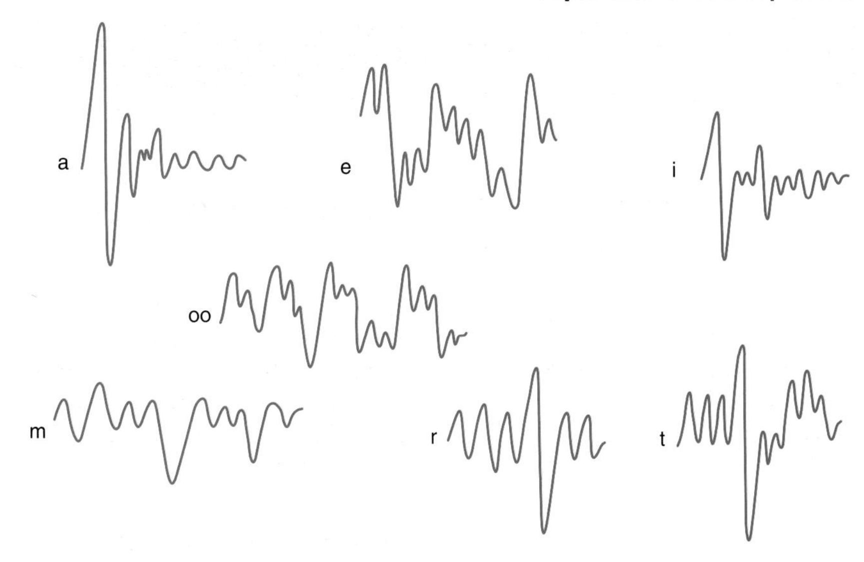

Figure 12.1E ▲ Making sounds

Speaking into the microphone produces changing patterns. Each voice sound produces a distinctive pattern. Some examples of patterns for different speech sounds are shown in Figure 12.1E.

- The **loudness** of a sound depends on the amplitude of the sound waves. Increasing the 'volume' of a radio increases the amplitude of the sound waves produced by the radio's speaker, so the sound from the radio is louder. Changing the loudness does not alter the frequency of the sound.
- The **pitch** depends on the frequency of the sound waves. Increasing the frequency makes the pitch higher. A signal generator is an electrical instrument that can be connected to a loudspeaker to produce sounds at different pitches.
- The sound **quality** depends on how close the sound wave is to a wave with constant frequency. A high quality note is said to be more 'pure' than a low quality note. For example, in Figure 12.1E, the 'r' sound is a higher quality note than the 'i' sound.

CHECKPOINT

▶ **1** Use the patterns in Figure 12.1E to sketch the oscilloscope pattern produced when Marie says her name at the microphone.

▶ **2** The patterns shown below are words made up of the sounds shown in Figure 12.1E. Work out what the words are.

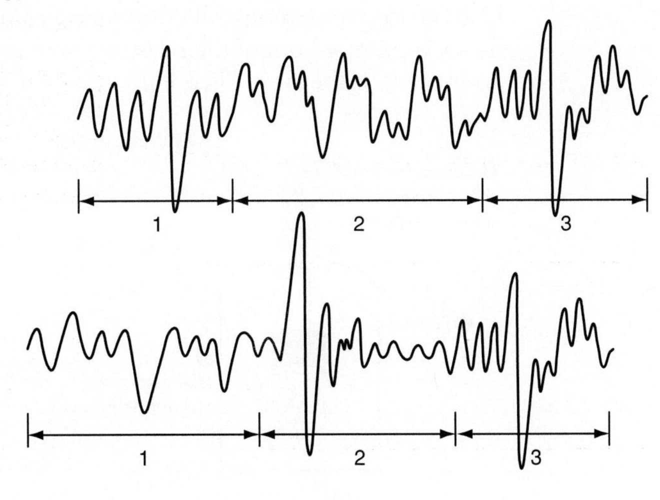

3 The pattern produced by a tuning fork is shown in Figure 12.1B. Copy this pattern and sketch further patterns produced by:
(a) the same tuning fork making a louder sound,
(b) the same tuning fork making a softer sound,
(c) a higher pitched tuning fork making a louder sound,
(d) a lower pitched tuning fork making a sound of the same loudness as the one in Figure 12.1B.

4 If you strike a tuning fork and hold it near your ear, you can hear it faintly for a long time. If you strike the same tuning fork and hold its base on a worktop, making the worktop vibrate as well, the sound is much louder but doesn't last as long. Try this for yourself and then explain the differences.

5 Devise a voice recognition test, using a cassette recorder, and use your friends as 'guinea pigs'. Explain how you would test to find out if their voices can be recognised easily. If possible, conduct the test and present your findings to the group.

12.2 The properties of sound

FIRST THOUGHTS

If you live near an airport the properties of sound affect you considerably. Read on to find out about the effects of different materials on sound.

Sound consists of vibrations transmitted through a substance. For example, an electric bell creates sound in the surrounding air because the bell vibrates as a result of being repeatedly struck by the clapper. The vibrating bell pushes and pulls on the surrounding air, sending pressure waves through the surrounding air. Sound waves are longitudinal because the substance carrying the waves vibrates to and fro along the direction of travel of the waves.

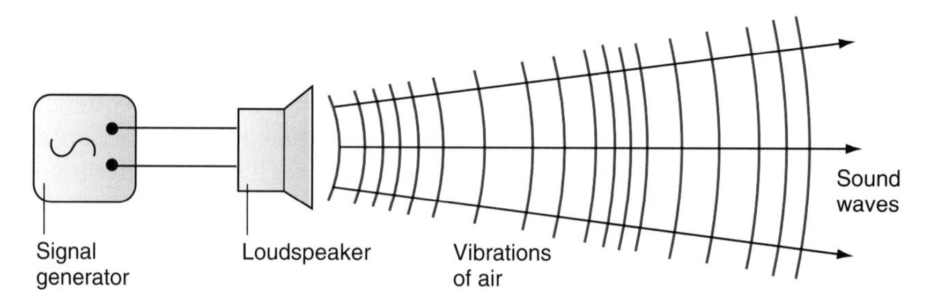

Figure 12.2A Creating sound waves in air

Sound cannot travel through a vacuum. The apparatus shown in Figure 12.2B is used to demonstrate that sound cannot travel through a vacuum. As air is pumped out of the flask, the sound of the ringing bell disappears. If the jar is filled with air again, the sound returns.

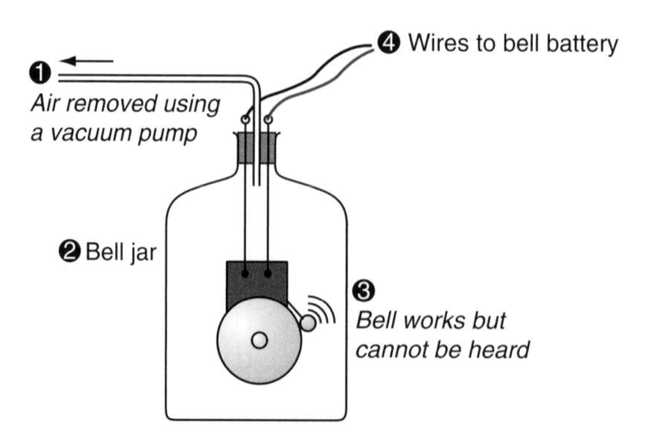

Figure 12.2B Sound does not pass through a vacuum

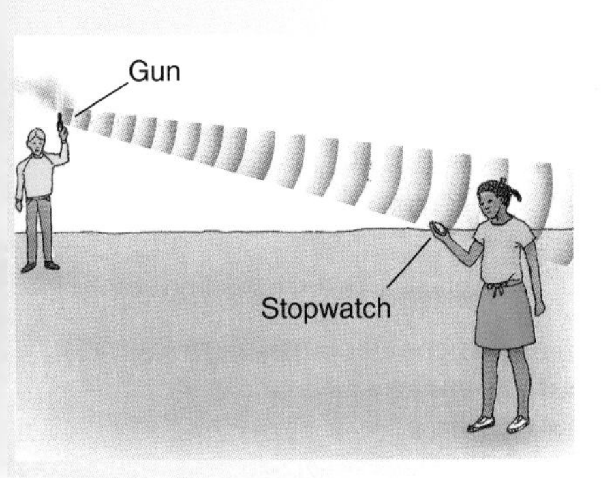

Figure 12.2C ▲ Measuring the speed of sound

Sound waves travel through solids and liquids as well as through gases. The speed of sound depends on the substance the sound is travelling through. The speed of sound in water is greater than the speed of sound in air. The speed of sound in a gas depends on the temperature of the gas; the higher the temperature, the faster the sound travels.

Measuring the speed of sound in air

You need two people for this. You and a friend should stand on opposite sides of a field as far apart as possible, but within sight of each other. If your friend fires a starting pistol in the air you will see the smoke from the pistol almost straight away. However, the 'bang' will be delayed because sound travels much more slowly than light. Use a stopwatch to time the interval between the smoke and the bang. This is the time taken for the sound to travel from your friend to you. Then measure the distance from your friend to where you did the timing. You can then work out the speed of sound using the formula

$$\text{Speed} = \frac{\text{Distance}}{\text{Time}}$$

Echoes

A hard, smooth wall is a good reflector of sound.

Echoes are sounds reflected off hard surfaces, such as bare walls and cliff faces. Shouting in a sports hall produces lots of echoes. Stand at the centre of a quiet hall facing a bare wall and clap your hands. You should be able to hear the echo of the clap. You may even be able to clap in time with the echoes.

To produce echoes the reflecting surface must be hard and smooth. If the surface is soft, the sound waves are absorbed by the surface instead of being reflected. If the surface is bumpy the waves are all reflected in different directions and the reflection is 'broken up', so no echo is heard. The sound is said to reverberate if it can still be heard after each clap. For example, in a curtain-lined hall no echoes are produced, so sound does not reverberate.

See www.keyscience.co.uk for more about light and sound.

You can estimate the speed of sound by clapping near a bare wall. Suppose you stand 30 m from the bare wall and you clap in time with the echoes at a rate of six claps per second, timed by a friend.

Each clap travels a distance of $2 \times 30 = 60$ m from you to the wall and back again. Since each clap returns at the instant the next clap is being produced, it takes $\frac{1}{6}$th second to travel to the wall and back.

Figure 12.2D ▲ The Whispering Gallery, St Paul's Cathedral

Hence Distance travelled $= 60$ m

Time taken $= \frac{1}{6}$ s

Thus Speed of sound $= \frac{\text{Distance}}{\text{Time}} = \frac{60}{\frac{1}{6}} = 360$ m/s

Noise

A school canteen at lunchtime is no place for anyone who wants a quiet rest. All the chatting and clattering can be very noisy. Noise is unwanted sound.

How could the din in a busy canteen be cut down? Eating in silence is not a practical proposition in a school. Wearing ear muffs is unrealistic too in this situation. A much more effective approach is to redesign the building to keep noise levels down.

Noisy sounds consist of sound waves that vary irregularly.

Noise can be a problem in many situations. For example, people living near airports or busy motorways need protection in their homes from excessive noise. Noise can be cut down by:

- Reducing it at source; for example, cutting down on the number of flights allowed in and out of an airport or fitting quieter engines to jets.
- Absorbing it after it has been produced; for example, using fences or walls to shield homes in urban areas from traffic noise.

Figure 12.2E ▲ A jumbo jet landing over houses

Ultrasonics

The human ear can detect sound waves in the frequency range from about 20 Hz to about 18 000 Hz. Sound waves above the frequency limit of the human ear are called **ultrasonic** waves. Ultrasonics are used in underwater sonar systems, in industry and in medicine.

- **Echo sounders** are used at sea to measure the depth of the sea bed. Ultrasonic pulses from the ship are beamed at the sea bed which reflects the pulses back to the surface. The pulses are timed and the timing is used to work out the depth of the sea bed.

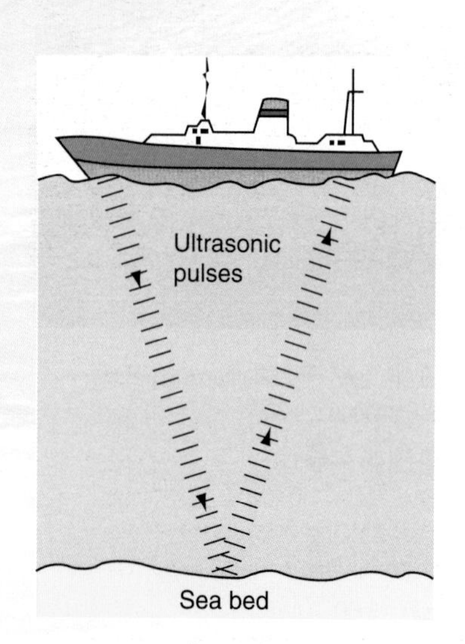

Figure 12.2F ▲ Depth finding

Worked example An echo sounder on a ship is tested where the sea bed is known to be 60 m deep. The time between sending each ultrasonic pulse and receiving its echo was measured at 80 milliseconds.

(a) Calculate the time taken for each pulse to travel to the sea bed.

(b) Calculate the speed of the pulses through the water.

(c) The ship moved to another part of the sea. Each pulse now took 60 milliseconds. Calculate the depth of the sea bed at its new location.

Solution (a) The time taken for each pulse to travel from the ship to the sea bed is the same as the time it takes to travel back from the sea bed to the ship. Hence each pulse takes 40 ms to travel to the sea bed.

(b) Speed = distance/time = 60 metres/0.040 seconds = 1500 m/s.

(c) Each pulse took 30 ms to reach the sea bed from the ship.
Depth of the sea bed = distance travelled by each pulse in 30 ms = speed × time
= 1500 m/s × 0.030 s = 45 m.

Ultrasonics are sound waves above the frequency range of the human ear.

- **Ultrasonic cleaners** are used to clean street lights. The lighting unit is immersed in a tank of water and cleaned by passing 40 kHz ultrasonic waves through the water. The ultrasonic waves dislodge particles from inaccessible surfaces of the lighting unit. The lighting unit is designed so it does not need to be removed from its support when it is cleaned in this way.

Figure 12.2G ▲ Ultrasonic cleaning

- **Ultrasonic scanners** are used to produce images of organs in the body or of babies in the womb. A scanner consists of a probe, a control unit and a display screen. The probe produces and detects ultrasonic waves at a frequency of about 1 MHz. As the probe is moved over the body surface, its signals are used to build up an image of internal tissue boundaries on the screen. Ultrasonic waves are non-ionizing unlike X-rays and therefore ultrasonic scans are safer than X-ray scans. Ultrasonic scanners are also used to detect cracks inside metals.

SUMMARY

Sound cannot travel through a vacuum. The speed of sound depends on the substance through which it passes. Echoes are caused by sound waves being reflected off hard surfaces. Ultrasonic waves are sound waves above the upper frequency limit of the human ear. Ultrasonics are used in echo sounders at sea, in industrial cleaning, for medical scanning, and for detecting cracks inside metals.

❶ The probe is used to direct pulses of ultrasonic waves in the body.

❷ These pulses are reflected by **internal tissue boundaries** in the body. After each pulse is emitted, the probe is programmed to detect reflected pulses until the next pulse is emitted.

❸ The reflected pulses are displayed as a line of spots on the display screen.

❹ Electrical sensors are attached to the probe control arm. The sensors are connected to the display screen to make the direction of the line of spots the same as the direction in which the probe points.

❺ As the probe is moved over the body, an image of the internal tissue boundaries is built up on the screen.

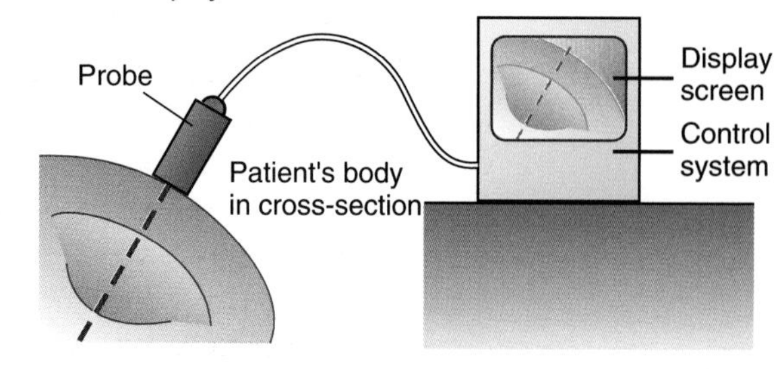

(a) An ultrasonic scanner system

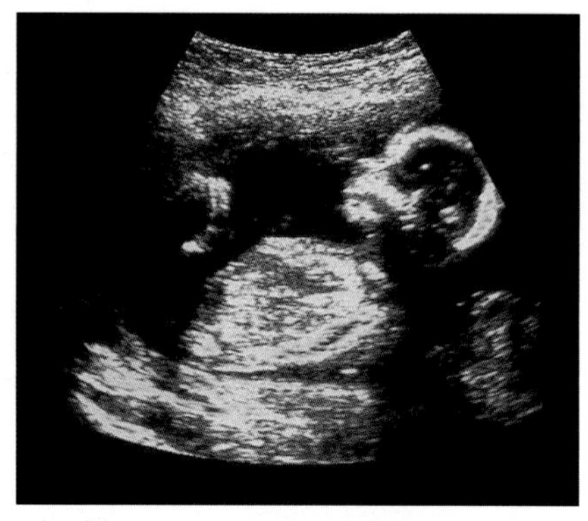

(b) An ultrasonic image of a baby in the uterus

Figure 12.2H

CHECKPOINT

1. (a) In an experiment to measure the speed of sound, a student fires a starting pistol. Another student 600 m away times the interval between the flash and the bang at 1.8 s. Work out the speed of sound.
   (b) Does sound travel faster downwind than upwind? Design an outdoor experiment to find out if it does.
2. Why would your voice echo more in an empty house than in a house that is carpeted and contains furniture?
3. Concert halls are designed very carefully to eliminate unwanted echoes. However, the hall walls must not absorb the sound completely. How do you think a concert in a bare hall would sound compared with one in a hall with totally absorbing walls?
4. Ultrasonics are used in hospitals to scan babies in the uterus. Figure 12.2H shows an image produced by an ultrasonic scan. A special device sends ultrasonic pulses into the mother's body. Tissue and bone boundaries reflect the pulses, which are detected and used to build up a picture of the baby on a monitor.
   (a) Why is this method preferable to an internal examination of the mother?
   (b) Why is it important to be able to see the placenta on the scan?
   (c) What other things can a scan warn the doctor about?
5. In a test to measure the depth of the sea bed, ultrasonic pulses took 0.4 s to travel from the surface to the sea bed and back. Given that the speed of sound in sea water is 1350 m/s, work out the depth of the sea bed.

12.3 The ear

FIRST THOUGHTS

Your ears are vital organs that enable you to receive information from other people. In this section you will find out how the ear works and how to test your own hearing response.

Have you ever listened to the sound of your own voice played back on a tape recorder? Try it and you will hear yourself as others hear you. To you, your voice will sound different on the recording. When you hear yourself speak, sound waves from your voice travel through your head as well as through the air to reach your ears. Other people only receive the sound of you speaking through the air.

The human ear is a remarkable organ that can detect an enormous range of sound waves. The loudest sounds it can withstand carry over one million million times more energy than the quietest sounds it can hear. It can detect frequencies from about 20 Hz (a bee buzzing) to about 18 000 Hz (a very high-pitched whistle).

The ear sends signals to the brain in response to sound waves arriving at the **eardrum**. Sound waves arriving at the ear make the eardrum vibrate. These vibrations are passed through the middle ear by **three tiny bones**, the **hammer**, **anvil** and **stirrup**, to reach the **oval window** of the inner ear. The vibrations of the oval window are transmitted through the fluid of the **cochlea**, making the **basilar membrane** vibrate. Tiny hair cells, which are sound-sensitive receptors, are lined up on the basilar membrane. The vibrating membrane activates the hair cells, which fire off nerve impulses to the brain along the **auditory nerve**.

It's a fact!

The fleshy lobe of the outer ear is called the **pinna**. It funnels sound waves down the ear tube to the ear drum. Cats, dogs and other mammals can adjust the pinna and cock it towards sources of sound. In most humans it is fixed. The walls of the ear tube produce wax which keeps the ear drum soft and supple.

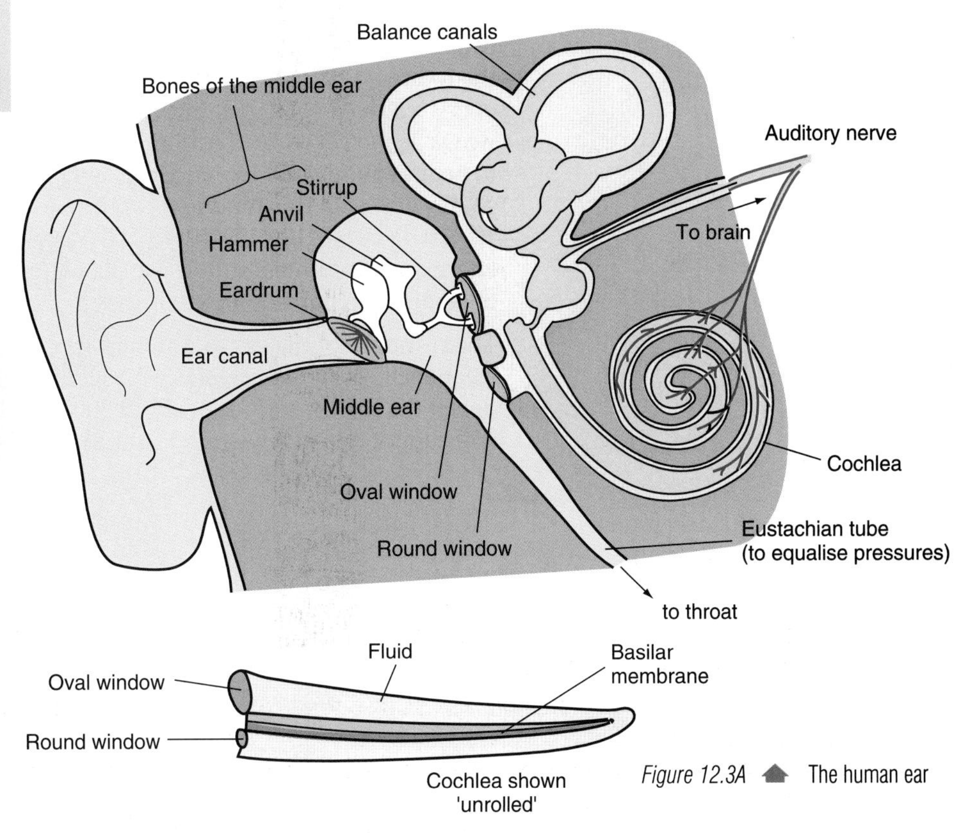

Figure 12.3A ▲ The human ear

What happens to the ear if very loud sound falls on it? If this happens too often, the ear becomes less and less sensitive and deafness can occur. One reason for this is that the hair cells in the cochlea became damaged. Another reason is that excessive vibrations make the bones of the middle ear less effective at passing the vibrations from the ear drum to the oval window. Operators of noisy machines must wear ear pads or they suffer permanent loss of hearing. Protect your ears at noisy discos by putting cotton wool in your ears.

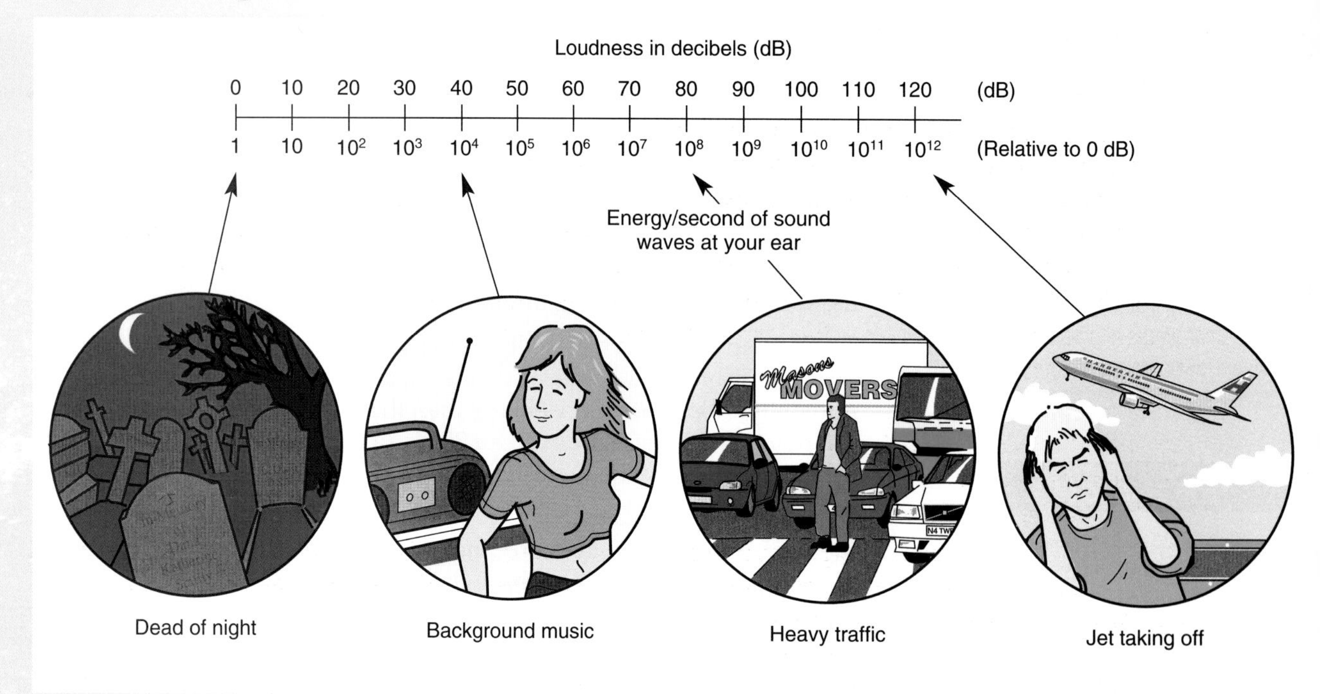

Figure 12.3B ▲ Decibel levels of everyday sounds

The faintest sound that can be heard is 0 dB by definition.

Loudness is measured in **decibels** (dB). The faintest sound that the ear can hear is defined as zero decibels (0 dB). Imagine steadily increasing the loudness of a radio from zero until it becomes too loud to bear. For every ten decibel (10 dB) increase in loudness, the energy of the sound waves is increased by a factor of 10. The sound would become too loud to bear at about 120 dB. Since this is 12 steps at 10 dB for each step, sound waves at this loudness carry $10 \times 10 \times 10 \times 10 \times 10 \times 10 \times 10 \times 10 \times 10 \times 10 \times 10 \times 10 = 10^{12}$ times as much energy as the faintest sound waves. Figure 12.3B shows the decibel levels of some everyday sounds.

The response of the ear to different levels of loudness varies with frequency, as shown in Figure 12.3C. The ear is most sensitive and can detect the softest sounds at about 3000 Hz. It is completely insensitive and cannot detect any sound over 18 000 Hz.

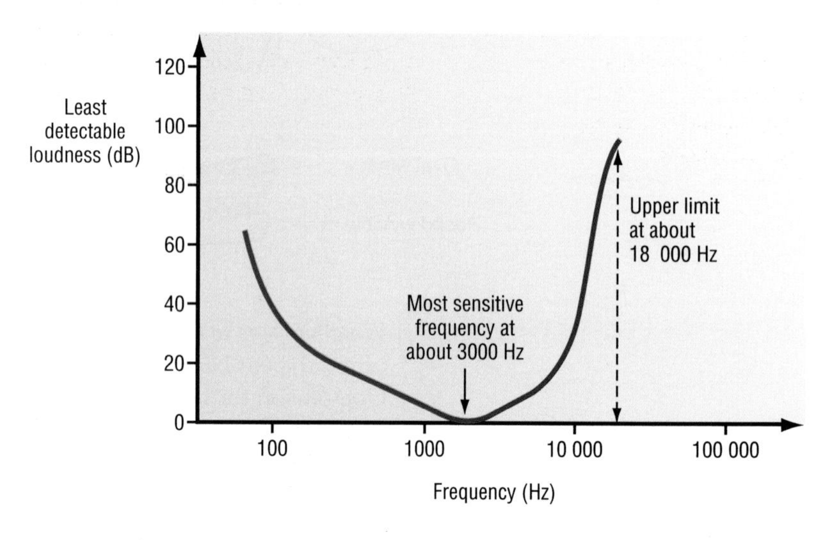

Figure 12.3C ▲ Frequency response of the ear

The ear converts signals carried by sound waves into nerve impulses that it sends to the brain. The ear cannot detect frequencies above 18 000 Hz. Loudness levels above 120 decibels can cause deafness if the ear is not protected.

CHECKPOINT

1. (a) One of the first hearing aids was the 'ear trumpet', which was a large hollow horn held to the ear. Why do you think this device improves hearing ability?
   (b) Modern hearing aids are so small that they can be worn behind the ear. Such a device contains an electronic amplifier, a tiny microphone and an earpiece speaker. The amplifier makes electrical signals bigger without changing the frequency of the signal. What is the purpose of the microphone and what is the earpiece speaker for?
2. What is the purpose of each of the following parts of the ear:
   (a) the eardrum,
   (b) the pinna,
   (c) the bones of the middle ear,
   (d) the oval window,
   (e) the hair cells?
3. Play back your own voice using a tape recorder. How does it differ from what you hear when you speak? Compare the voices of your friends on a tape recorder. Do their voices seem different from when they speak directly to you?
4. Here is a passage from Claire's diary describing part of an evening out with her friends. 'It was very noisy and hot in the disco. We could only hear each other when the music stopped. I got a lift home with Michelle and her dad. There was a thunderstorm on the way home. When I got home, the TV was on very loud so I went to my bedroom for some peace and quiet.'
   (a) When was the loudness level greatest and when was it least?
   (b) Estimate the loudness when it was greatest.
   (c) Which do you think was most damaging to the ears: the thunderclap or the disco noise?

12.4 ▶ Making music

FIRST THOUGHTS

How does music differ from other sounds? How is it produced? As you work through this section, think about your own experiences of making music, and perhaps try out some instruments.

Notes of music are sounds that are easy to listen to because they are rhythmic. The waves that carry these notes change smoothly and the wave pattern repeats itself regularly. Figure 12.1D shows the wave pattern for a note played on a flute, displayed on an oscilloscope connected to a microphone. Compare this note pattern with the pattern shown in Figure 12.1B produced by a tuning fork, which gives a note of constant frequency, sometimes called a **pure note**.

Figure 12.4A ▲ Making a swing resonate

Wind instruments such as the trombone or the flute produce notes because the air inside the instrument is made to **resonate**. Anything that moves to and fro repeatedly can be made to resonate. For example, pushing a child on a swing every time the swing descends makes the swing resonate and the child go higher and higher (Figure 12.4A).

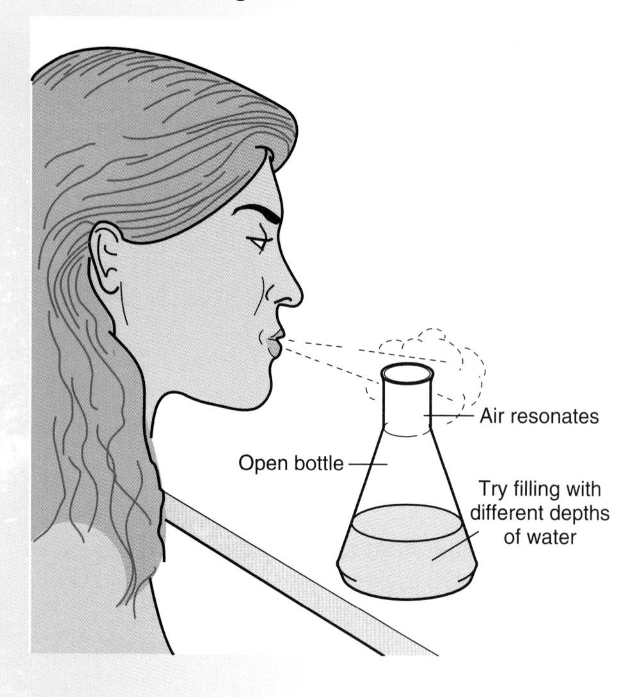

Figure 12.4B ▲ Making air resonate

Figure 12.4C ▲ Vibrating guitar strings

Wind instruments are designed to make layers of air in the instrument oscillate so much that the instrument resonates with sound. You can make an empty bottle resonate by blowing gently across the top (Figure 12.4B). At the right speed, a booming sound is produced as the air inside resonates.

In general, the longer a wind instrument is, the deeper the note it produces. A deeper note is produced when the trombone is lengthened. This is because the instrument resonates at longer wavelengths when it is lengthened. A pipe organ has pipes of different lengths; the longer a pipe is, the deeper the note it produces.

Percussion instruments such as the drum or the handbell are struck to produce sound. A drum has a tightly stretched membrane, the drum skin, which vibrates when it is struck. The vibrating drum skin pushes the surrounding air to and fro, making sound waves that travel outwards from the drum.

String instruments such as the guitar and the violin produce sound when the strings vibrate. A guitar string is plucked and vibrates, whereas a violin string vibrates when a bow is drawn across it. In both cases, the vibrating strings make the body of the instrument vibrate at the same frequency, which makes the surrounding air vibrate, so sound waves are produced.

A guitar has six strings, each of a different thickness. The pitch of the note produced by a plucked string is changed by making the string tighter or shorter. Pressing the string on the frets along the neck of the guitar makes the vibrating length shorter. Tuning the appropriate key changes the tension (i.e. tightness) of the string. *How would you make the pitch of a string higher? How does the pitch differ between a thin string and a thick string?*

It's a fact!

The most famous make of violin is the Stradivarius. This was made by Antonio Stradivari in the seventeenth century in Italy. He perfected the design to such an extent that no one since has ever surpassed it. Even attempts to reproduce the design have failed.

Investigating the vibrations of a stretched string

Fix one end of a string to a vibrator driven by a signal generator and pass the other end over a pulley. Fix a weight on this end, as shown in Figure 12.4D. The string is then under constant tension and its length is fixed.

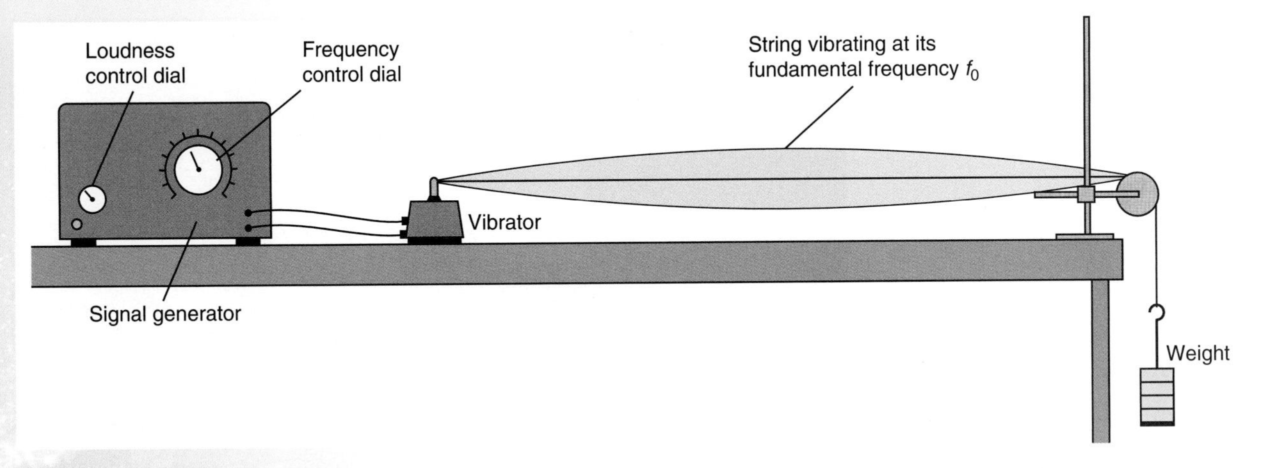

Figure 12.4D ▲ Fundamental resonance

The output frequency of the signal generator can be changed using the control dial. When the frequency is increased from zero, different patterns of vibration are produced in the string when it is made to resonate by the vibrator.

- The simplest pattern is called the fundamental pattern (Figure 12.4D). The vibrations are strongest at the middle of the string. The vibrating string produces sound waves at the same frequency as it is vibrating.
- Increasing the frequency produces more complicated patterns. These are called overtones. They occur at frequencies equal to two, three, four, etc. Times the fundamental frequency. Figure 12.4E shows the patterns of the 1st and 2nd overtones.

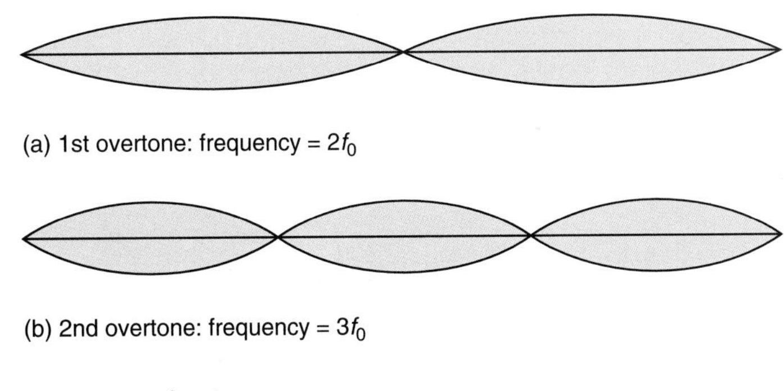

Figure 12.4E ▲ Overtones

When a wire or a string is plucked it produces sound waves composed of a mixture of the fundamental note and the overtones. In comparison, a tuning fork produces a pure note (just the fundamental). The note from a wire sounds slightly different because of the presence of the overtones.

Synthesisers are electronic instruments that can make any type of sound (Figure 12.4F). Recording studios use them to make the sounds of other musical instruments and interesting sound effects for television, radio and films. A synthesiser has many channels, each of which can produce a pure note at a different frequency. The loudness of the note can be altered. To make a particular sound, different channels are selected, each feeding a note into a mixer circuit. The mixer adds the notes together and its output is then supplied to a loudspeaker. The loudness level of each channel can be adjusted to allow the user to produce the desired sound. The synthesiser can be linked up with a computer and programmed to produce different sounds in sequence.

SUMMARY

Musical notes are wave patterns that repeat themselves. Musical instruments are designed to vibrate, thus creating sound waves in the surrounding air. A synthesiser can produce the same note as any musical instrument.

Figure 12.4F ▲ Synthesiser in use

CHECKPOINT

1. (a) List ten musical instruments, stating whether each one is a percussion instrument, a wind instrument or a string instrument.
   (b) Why do you think big instruments usually produce deeper notes than small instruments?
2. A guitar has six strings, each of a different thickness. The vibrating length of each string can be changed by pressing it against the frets. The string tension can be changed by using the tension keys.
   (a) How does making the string (i) tighter (ii) shorter affect the pitch of the note produced by the string?
   (b) How does the pitch of a note produced by a thin string compare with that of a note produced by a thick string at the same length and tension?
3. (a) Design an experiment to test how the fundamental frequency of a vibrating string under constant tension varies with the length of the string.
   (b) Here are some results from such an experiment. Can you see the connection between the frequency and the length? Plot a graph to show the relationship between these quantities.

Length (mm)	1000	800	600	400	200
Frequency (Hz)	96	120	160	240	480

4. (a) Some wind instruments have a mouthpiece that contains a reed. Why?
   (b) You can make some interesting sounds using a comb and paper. Try it and then explain how it works.
   (c) Explain why a cracked handbell will not ring.
   (d) Why do pianos need to be tuned regularly?
5. Most microcomputers can be used as synthesisers. Here is a program that turns a PC into a keyboard using the ABCDEFG keys to represent the same musical notes. Do not worry about how it works. Click on 'File' then 'Run' then type in QBASIC and return. Then key the program in and see if you can recognise the tune FFGEFG AA+BAGFGFEF when you press f5.

```
PRINT : PRINT "KEY IN FFGEGFAA+BAGFGFEF": PRINT "KEY IN Z TO ESCAPE"
DO: DO: S$ = INKEY$: PRINT S$:
LOOP UNTIL S$ = "Z" OR S$ = "A" OR S$ = "B" OR S$ = "+B" OR S$ = "C" OR S$ = "D" OR
S$ = "E" OR S$ = "F" OR S$ = "G"
IF S$ = "Z" THEN END
PLAY "L8": PLAY "O" + STR$(2): PLAY S$
LOOP UNTIL S$ = "Z": END
```

Topic 13 Seismic waves

13.1 The structure of the Earth

FIRST THOUGHTS

What is Earth like deep down, many kilometres below the surface? It sounds an impossible question to answer, but scientists have found methods of investigating the deep structure of the Earth.

The study of the Earth is called **geology**, and a person who works in this branch of science is called a geologist. The research work of geologists has enabled them to construct a model of Earth's structure (see Figure 13.1A).

How was Earth formed? A molten mass cooled down over millions of years. Dense materials sank deeper into the centre to form a core of dense molten rock. Less dense material remained on the surface to form a crust of solid rock (50 km thick). Gaseous matter outside the crust is the atmosphere. Earth's atmosphere is chiefly oxygen and nitrogen.

As Earth cooled, water vapour condensed to form rivers, lakes and oceans on the surface of Earth. No other planet has oceans and lakes, though Mars has some water vapour and polar ice caps.

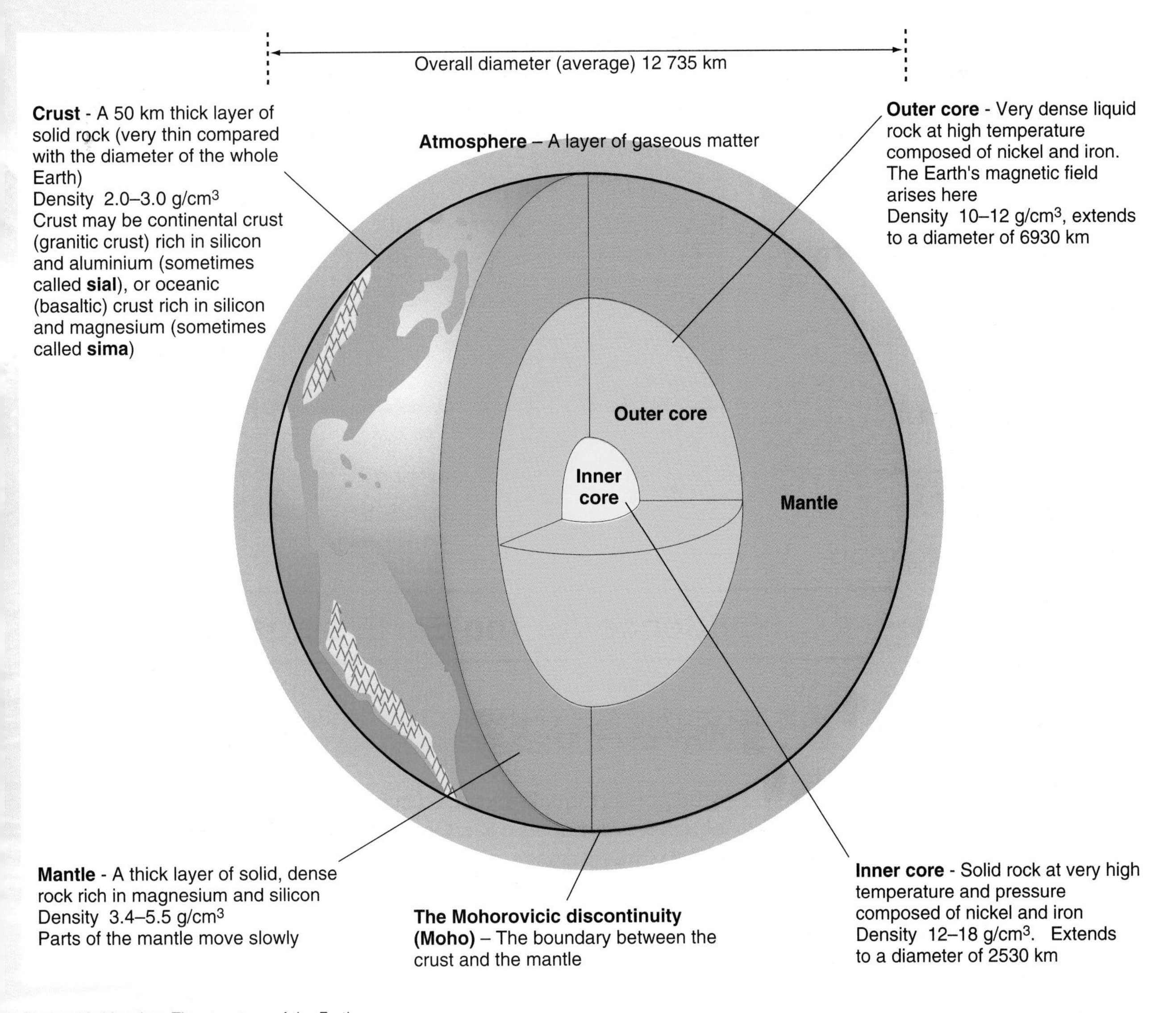
Overall diameter (average) 12 735 km
Crust - A 50 km thick layer of solid rock (very thin compared with the diameter of the whole Earth)
Density 2.0–3.0 g/cm3
Crust may be continental crust (granitic crust) rich in silicon and aluminium (sometimes called sial), or oceanic (basaltic) crust rich in silicon and magnesium (sometimes called sima)
Atmosphere – A layer of gaseous matter
Outer core - Very dense liquid rock at high temperature composed of nickel and iron. The Earth's magnetic field arises here
Density 10–12 g/cm3, extends to a diameter of 6930 km
Outer core
Inner core
Mantle
Mantle - A thick layer of solid, dense rock rich in magnesium and silicon
Density 3.4–5.5 g/cm3
Parts of the mantle move slowly
The Mohorovicic discontinuity (Moho) – The boundary between the crust and the mantle
Inner core - Solid rock at very high temperature and pressure composed of nickel and iron
Density 12–18 g/cm3. Extends to a diameter of 2530 km

Figure 13.1A ▲ The structure of the Earth

Earth's crust

Earth's crust is composed of rocks and soils. Soils have been formed by the breakdown of rocks and vegetation. The crust is divided into continental and oceanic crust. Earth's overall diameter is 12 735 km.

Continental crust	Oceanic crust
• Forms continents and their shelves • Up to 70km thick in mountain ranges • Density ~ 2.7 g/cm^3 • Age: up to 3700 million years • Same composition as granite rock • Often called granitic crust • Rich in silicon and aluminium • The deeper parts of continental crust are of a denser material similar to oceanic crust.	• Beneath deep sea floors • Average thickness 6 km • Density ~ 3.0 g/cm^3 • Age: up to 220 million years • Same composition as basalt rock • Often called basaltic crust • Rich in silicon and magnesium • Material similar to oceanic crust is thought to lie beneath the continents.

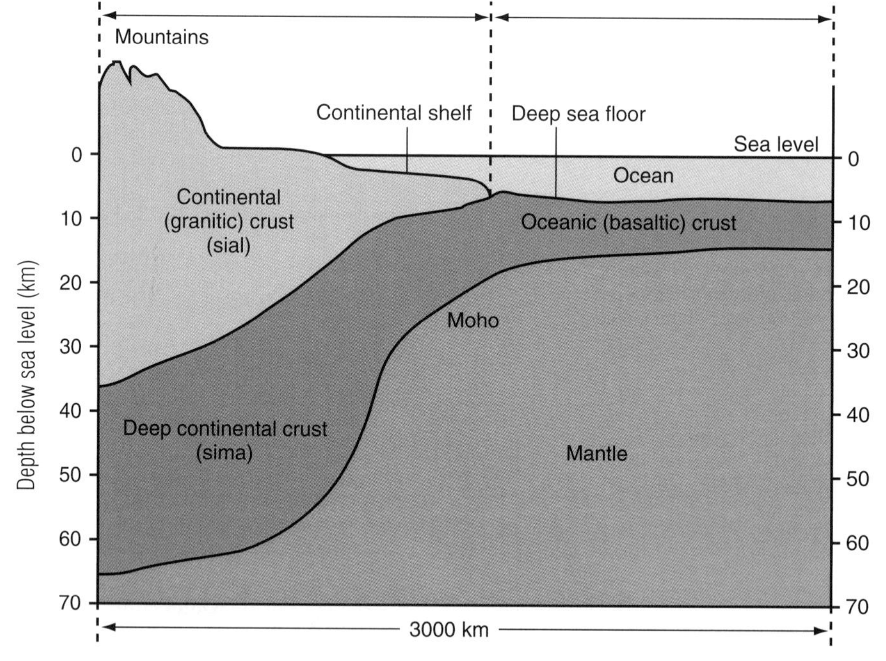

Figure 13.1B ▲ Continental crust and oceanic crust

SUMMARY

The Earth has a layered structure: inner core, outer core, mantle, crust and atmosphere. The crust is composed of oceanic (basaltic) crust beneath the ocean floors and continental (granitic) crust, which forms the Earth's land masses. The boundary between the crust and the mantle is called the Moho. The study of the Earth is called geology.

13.2 Evidence for the Earth's structure

FIRST THOUGHTS

We cannot obtain information about the deep structure of Earth by mining or by drilling. The world's deepest mine is 3.5 km in depth, and the radius of Earth is over 6000 km. Evidence about the interior of Earth comes from the study of earthquakes and volcanoes.

Earthquakes

The study of earthquakes is called **seismology**. About 500 000 earthquakes occur every year. Only about 1000 of these are strong enough to cause damage, and only a few are serious. An earthquake occurs when forces inside Earth become strong enough to fracture large masses of rock and make them move. The energy which is released travels through the Earth as a series of shock waves, referred to as **seismic waves**. Earthquakes are limited to the rigid part of the crust. They cannot occur in the molten part of the mantle. Most earthquakes are generated within 600 km of Earth's surface. The point where an earthquake originates is called the **focus**. The nearest point on Earth's surface directly above it is the **epicentre.** Shock waves are felt most strongly at the epicentre and then spread out from it. Earthquake shocks are recorded by an instrument called a **seismometer** (see Figure 13.2A).

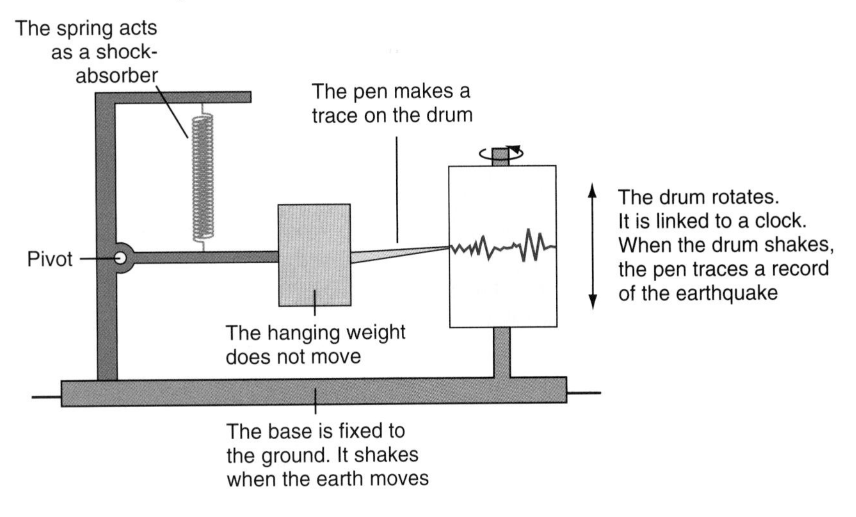

Figure 13.2A ▲ A seismometer

The recording is called a seismogram. The energy of the earthquake is measured on the Richter scale. Each point on the scale means an increase by a factor of ten: a scale of 5 is ten times as powerful as a scale of 4.

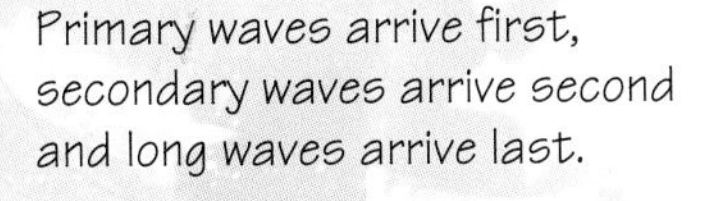

Primary waves arrive first, secondary waves arrive second and long waves arrive last.

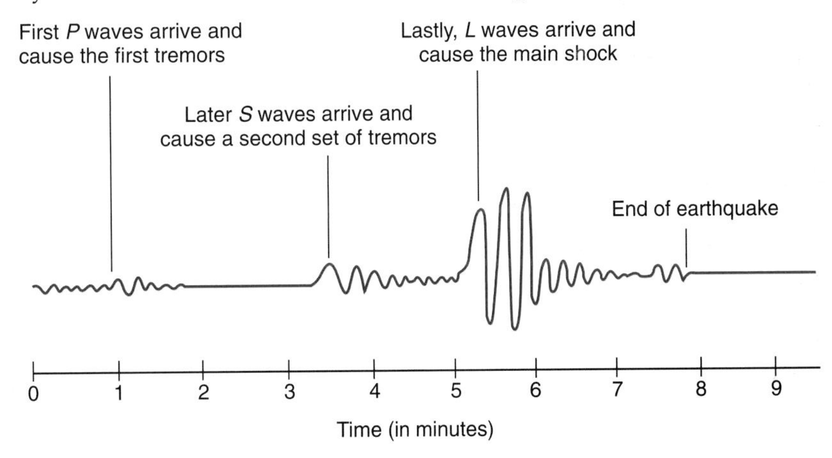

Figure 13.2B ▲ A seismogram

See www.keyscience.co.uk for more about waves, vibrations and seismic waves

Analysing seismic waves

A typical seismomenter recording of an earthquake is shown in Figure 13.2B.

- **Primary (P)** waves cause the first tremors. These are longitudinal waves which make the material which they pass through vibrate to and fro as they push and pull on it. They are faster than S or L waves.
- **Secondary (S)** waves arrive a few minutes later, causing more tremors. They travel slower than P waves. These are transverse waves that shake the material they travel through from side to side.
- **Long (L)** waves arrive last to cause the main shock. These travel along the crust only, making the surface move violently up and down as well as to and fro. They are slower than P or S waves.

Handy Hint: **P** waves push and pull. **S** waves shake side to side. **L** waves go the long way round.

The Earth's structure has been deduced from seismic waves. The pattern of waves received by a seismometer depends on what the waves have passed through inside the Earth. Waves are either reflected or refracted when they travel from one type of material into another.

- When an earthquake occurs, its epicentre can be located by comparing seismometer readings from different parts of the world.
- The structure of the Earth has been deduced as a result of analysing seismometer readings.

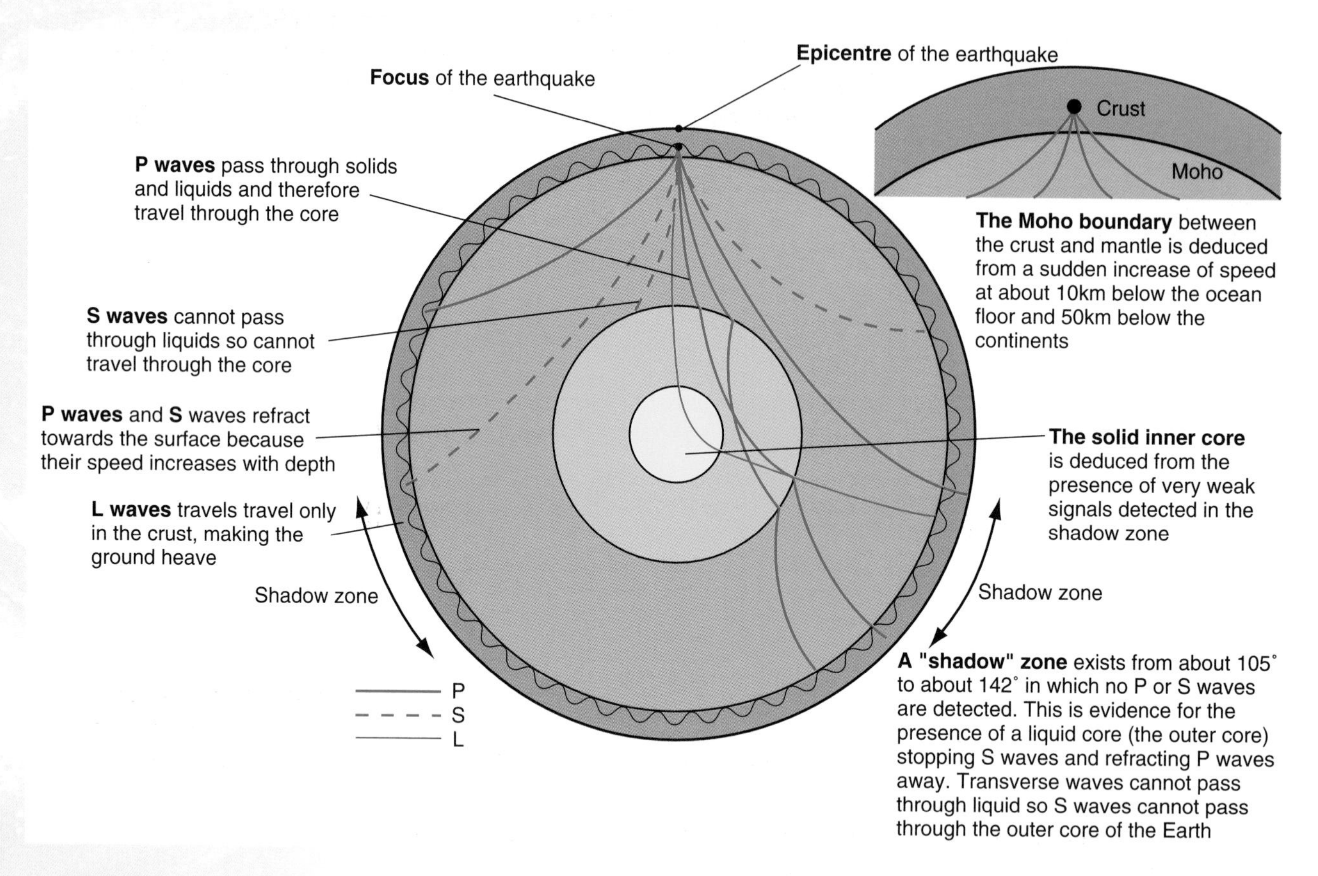
Epicentre of the earthquake
Focus of the earthquake
Crust
Moho
P waves pass through solids and liquids and therefore travel through the core
The Moho boundary between the crust and mantle is deduced from a sudden increase of speed at about 10km below the ocean floor and 50km below the continents
S waves cannot pass through liquids so cannot travel through the core
P waves and S waves refract towards the surface because their speed increases with depth
The solid inner core is deduced from the presence of very weak signals detected in the shadow zone
L waves travels travel only in the crust, making the ground heave
Shadow zone
Shadow zone
P
S
L
A "shadow" zone exists from about 105° to about 142° in which no P or S waves are detected. This is evidence for the presence of a liquid core (the outer core) stopping S waves and refracting P waves away. Transverse waves cannot pass through liquid so S waves cannot pass through the outer core of the Earth

Figure 13.2C ▲ Shock waves from an earthquake

Figure 13.2D ▲ Searching for oil

Oil prospectors produce seismic waves by detonating small explosive charges. Then they record the wave patterns on seismometers and feed the information into computers. The computers record the position, thickness and density of the rock layers. This is very much more convenient than drilling holes to find out!

Volcanoes

The lava erupted by volcanoes gives information about the crust and upper mantle, where lava is produced, but not about deeper layers. Plotting on a map the places where volcanoes have occurred gives information about the regions of Earth where heat is being generated and causing volcanic activity.

Meteorites

Meteorites reach Earth from space. They are pieces of rock and dust which have been attracted towards Earth by Earth's gravity. Most meteorites burn up when they reach Earth's atmosphere, but some fall to Earth's surface. Geologists believe that meteorites may be samples of planetary material dating from the time of formation of the Solar System.

Iron meteorites are thought to have come from the cores of objects in the Solar System long since destroyed by collisions. In 1996, scientists in America discovered fossil evidence for very primitive life forms in a meteorite that fell to Earth long ago in Antartica. This meteorite is thought to have been thrown into space from Mars as a result of a very large object hitting the martian surface in the early Solar System. If confirmed, this discovery would imply that life is not unique to Earth!

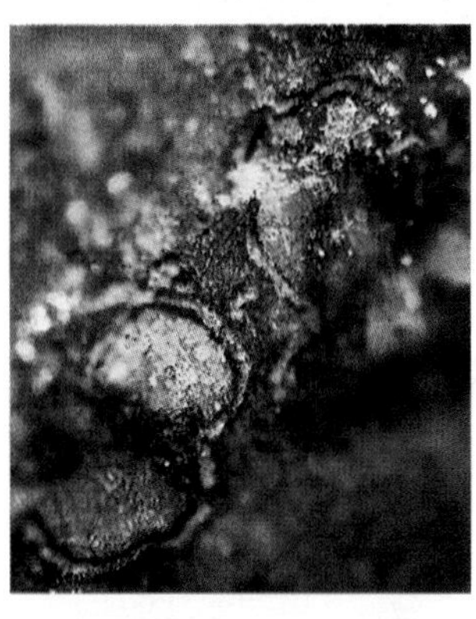

Figure 13.2E ▲ Life on Mars?

Magnetism

Earth's magnetic field is evidence for the presence of iron in the core. Because the outer core is liquid iron, it is thought that convection in the core creates electric currents which produce the Earth's magnetic field.

Evidence for the structure of the Earth comes from:

- the patterns of shock waves produced by earthquakes
- material erupted by volcanoes
- the positions of earthquakes and volcanoes on the map
- meteorites
- the Earth's magnetic field.

CHECKPOINT

1 (a) List the three types of seismic waves in order of increasing speed.
(b) Which type is (i) longitudinal only, (ii) transverse only?
(c) Why do seismic waves bend towards the surface?

2 (a) Sketch the Earth in cross-section to show its structure. Label the crust, the mantle, the inner core and the outer core.
(b) (i) Mark a point on your sketch which could be the focus of an earthquake.
(ii) Sketch the path of a seismic wave from this focus which reaches the surface 90° further round.
(iii) Mark the shadow zone from about 105° to about 142° from this focus.
(iv) Explain why this is referred to as a shadow zone.

3 The overall density of Earth is 5.5 g/cm^3. The rocks in the Earth's crust have densities of 2.5 to 3.0 g/cm^3. How can you explain the difference between these values and the much higher density of the whole Earth?

4 Take a piece of string 3 m long. Imagine that this length represents the 3000 million years that have passed since the first living things appeared on Earth. Mark on the string the length that represents the 2 million years since the human race appeared.

5 Meteorites are thought to be debris of objects in orbit round the Sun that collided with each other. Some meteorites are iron, some are stony and some are stone and iron. What does this suggest about the structure of the objects which they originated from?

13.3 Earthquake belts

FIRST THOUGHTS

Why do many parts of Earth experience earthquakes and volcanoes? Why do some parts of Earth have neither? Geologists have found answers to this puzzle.

Earthquakes and volcanoes occur in certain parts of Earth's crust but not in others. Geologists speculated for many years on reasons for the difference. Some patterns emerge from Figure 13.3A.

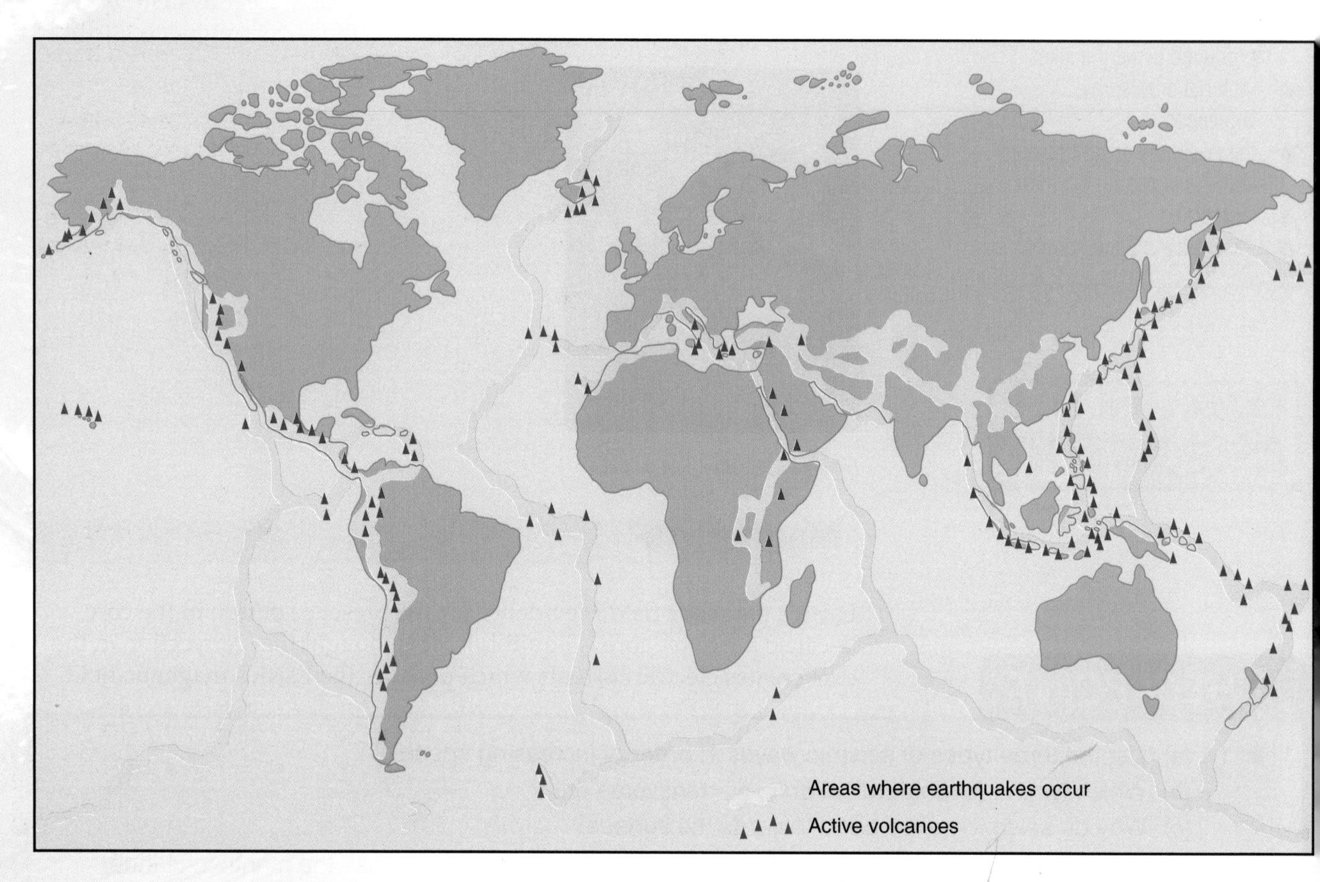

Figure 13.3A ▲ Areas of earthquake and volcanic activity

See www.keyscience.co.uk for more about volcanoes.

- Earthquakes and volcanoes occur in belts of activity. The belts are hundreds of kilometres wide and thousands of kilometres long. In places, belts join up.
- On land, belts occur along chains of high mountains like the Alps.
- Beneath the sea, belts pass through the centres of oceans and through chains of volcanic islands like the Philippines.
- Surveys of the mid-ocean belts show that the sea floor rises to form chains of huge mountains beneath the sea. These mountain chains are called oceanic ridges. The mid-Atlantic ridge rises above sea-level to form Iceland.
- Surveys show that belts which pass through chains of islands are close to deep oceanic trenches on the sea floor. The same is true of mountain ranges near the edges of continents. A trench in the Pacific called the Peru-Chile trench runs parallel to the Andes.

Plate tectonics

Figure 13.3B ▲ The result of an earthquake in Los Angeles in 1994

Geologists believe that the outer layer of Earth is made up of separate pieces called **plates**. Each plate is a piece of lithosphere (crust and uppermost layer of mantle) of 80–120 km in thickness. Movements in the mantle beneath make the plates move very slowly, a few centimetres per year. As a result, plates sometimes rub against each other. If stress builds up to a large extent, the plates may bend. When they spring back into shape, the ground shakes violently: there is an earthquake. There has to be a source of energy to produce the movement of plates. Many geologists believe that it is the heat given out when radioactive elements decay.

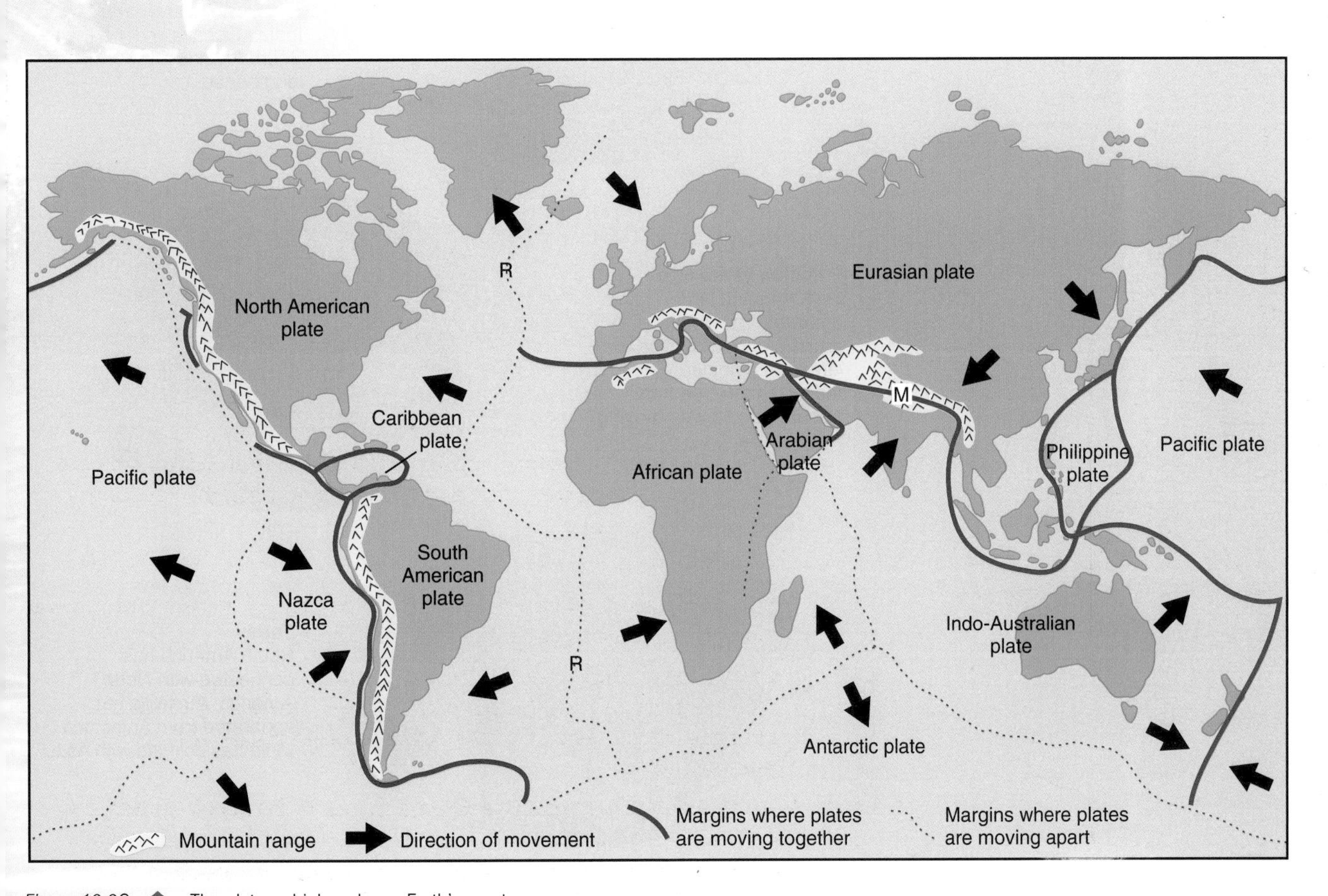

Figure 13.3C ▲ The plates which make up Earth's crust

Continental drift

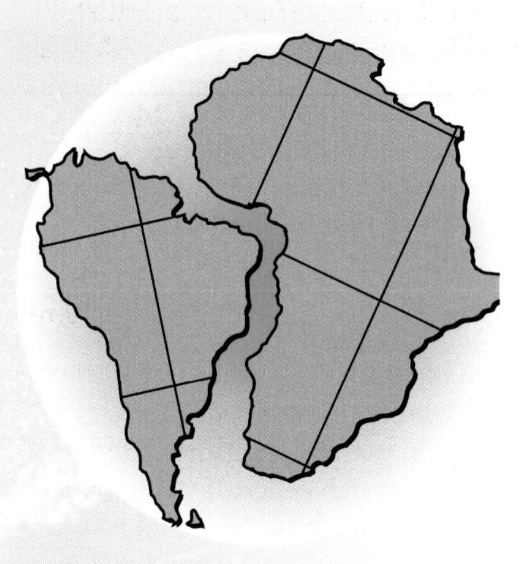

Figure 13.3D ▲ The 'jigsaw' fit of South America and Africa

When studying a map of the World, have you noticed the similarity between the east coastline of South America and the west coastline of Africa? This was first noticed in 1915 by a German scientist called Alfred Wegener. He promoted the theory that these continents were joined together long ago and have drifted apart since then. He found many similarities in fossils, plants and animals from the two continents. He believed that all the continents were once joined as a single land mass called **Pangaea** which split into separate parts that drifted apart. He based his theory on the idea that the continents float on a dense fluid crust. However, he was unable to come up with a convincing source of power to cause the drift.

180 million years ago
The original land mass,
Pangaea, had split into two
najor parts. Gondwanaland
nad started to break up

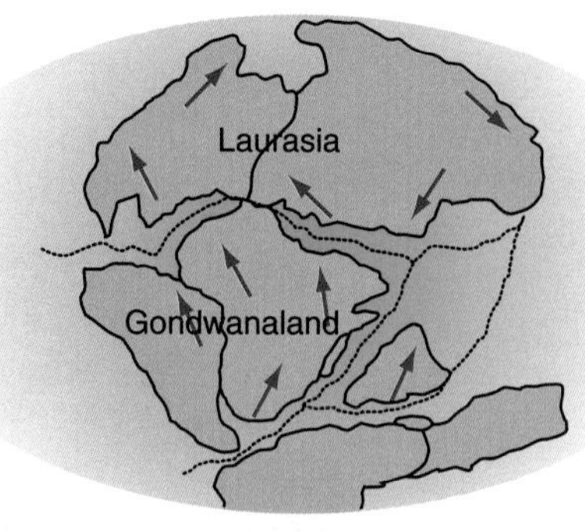

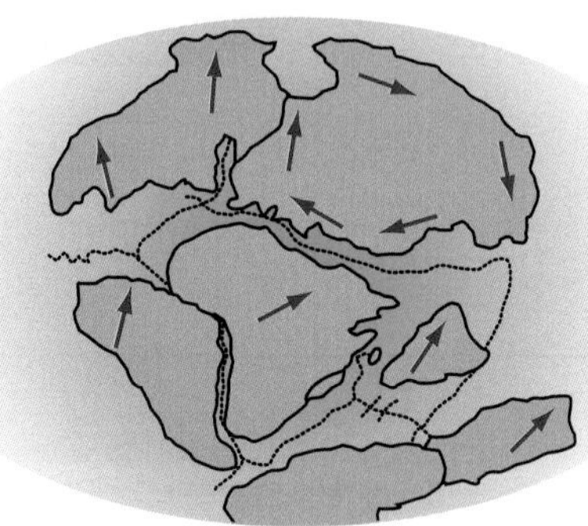

135 million years ago
Gondwanaland and Laurasia drifted northwards. The North Atlantic and Indian Oceans widened. The South Atlantic rift lengthened

EXTENSION FILE
ACTIVITY

5 million years ago
outh America had
eparated from Africa.
ustralia and Antarctica
ere still combined. The
lediterranean Sea had
ppeared. India was moving
owards Asia

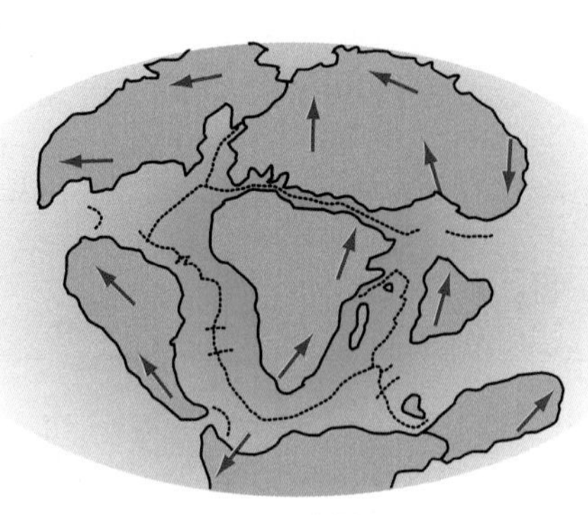

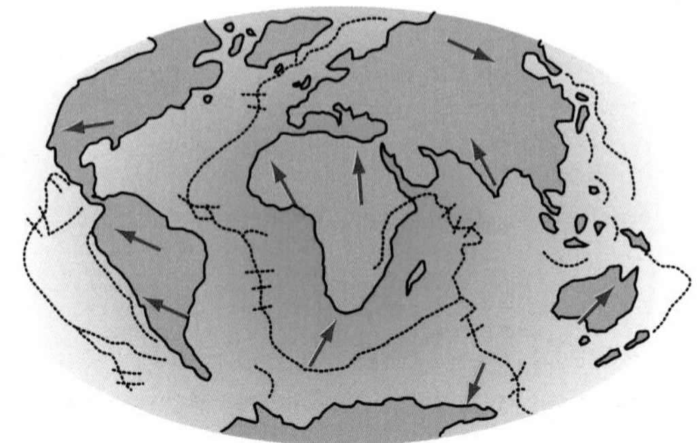

Today
South America has connected with North America. Australia has separated from Antarctica. India has collided with Asia.

Figure 13.3E ▲ The formation of the continents

SUMMARY

Earth's crust and the upper part of the mantle are together called the lithosphere. Geologists believe the lithosphere consists of separate plates and that these plates are moving slowly. Plate movement has given rise to mountain ranges and causes earthquakes and volcanoes. Earthquakes and volcanoes occur in belts of activity. These belts run:

- along mountain ranges
- along oceanic ridges (mountains on the sea floor)
- along oceanic trenches (deep channels in the sea floor).

Wegener's theory was revived in 1931 when Arthur Holmes, a British geologist, suggested that heat generated by radioactive decay inside the Earth provides the source of power. This heat causes convection currents of molten rock called **magma** in the mantle. These slow-moving currents move the outer layers of the Earth. By the mid-1970s, these ideas had been developed into the theory of plate tectonics. The continents on the plates are carried by the plates and therefore drift on the surface. Figure 13.3E shows how the continents are thought to have arrived at their present positions. Earthquakes and volcanoes are thought to arise where the plates are colliding or separating.

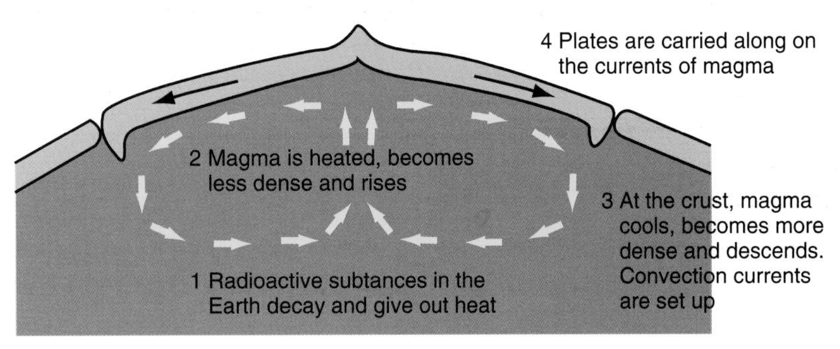
4 Plates are carried along on the currents of magma
2 Magma is heated, becomes less dense and rises
3 At the crust, magma cools, becomes more dense and descends. Convection currents are set up
1 Radioactive subtances in the Earth decay and give out heat

Figure 13.3F

Direct evidence for the theory of continental drift was found in the 1960s when ocean ridges were studied.

The age of the oceanic crust either side of a ridge was found to increase with distance from the ridge. New crust forms at a ridge due to magma rising and solidifying; this new crust pushes previously formed crust away from the ridge along the sea floor in either direction. This effect is known as **sea-floor spreading** and varies with location from 2 to 10 cm every year.

Bands of magnetised crust either side of such ridges were discovered. The Earth's magnetic field reverses its polarity abruptly every few thousand years. The newly formed crust contains iron particles and therefore becomes magnetised on solidification. Each time the Earth's magnetic field reverses its polarity, the magnetisation of the newly formed crust reverses. Sea floor spreading gradually carries these magnetic bands away from the ridge.

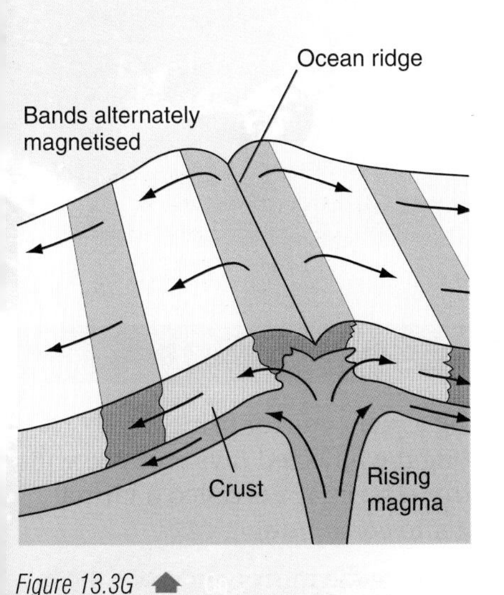
Ocean ridge
Bands alternately magnetised
Crust
Rising magma

Figure 13.3G

CHECKPOINT

1. The North Atlantic is spreading at a rate of about 5 cm/year.
   (a) How far will it spread during your lifetime (say 80 years)?
   (b) How tall are you?
   (c) How does your answer to (a) compare with your answer to (b)?
2. (a) Why does the UK experience few earthquakes?
   (b) Rocks on the Isle of Skye in Scotland show that volcanoes erupted there about 50 million years ago.
   Explain how this could have happened.
3. (a) What is meant by continental drift?
   (b) How does the theory of plate tectonics explain continental drift?
   (c) What is the source of power that makes the continents drift?

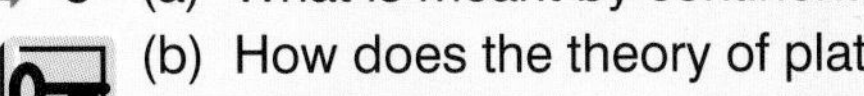

Topic 14 Light

14.1 Mirror images

FIRST THOUGHTS

Compare a photograph of yourself with your image in a mirror. Why do they differ? Images can be very deceptive.

Look in a mirror and you will see an image of yourself created by the reflection of light by the mirror. Your mirror image is reversed – use your left hand to wave at it and you will see that your image's right hand waves at you. However, provided your mirror is a plane (i.e. flat) mirror, your image has the same shape as you. In the 'Hall of Mirrors' at the funfair, using curved mirrors instead of plane mirrors gives very amusing images.

To understand how mirrors form images, we can reflect light rays off mirrors and then make scale drawings called ray diagrams to explain what we see. The light rays are drawn to show the direction in which the light waves travel (Figure 14.1A).

Light rays show the direction of travel of light waves.

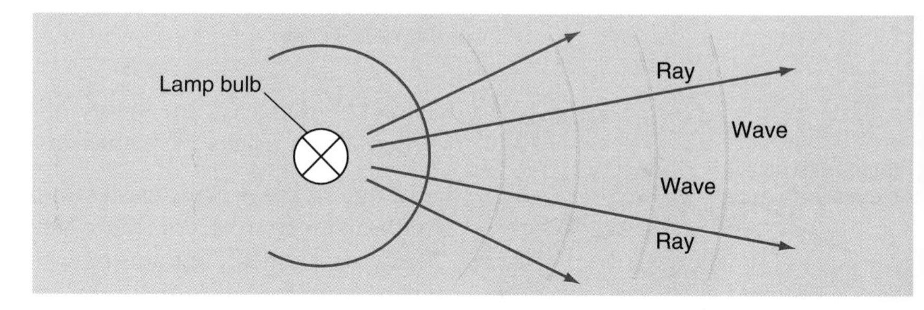

Figure 14.1A ▲ Rays and waves

Investigating reflection by plane mirrors

You can see how an image is formed by a plane mirror using the arrangement in Figure 14.1B. Look along the reflected rays and you will see an image of the ray box in the mirror. The image is called a **virtual image** because the reflected rays appear to come from it.

How far is the image behind the mirror? Remove the mirror and trace the ray paths on the paper to find out where they meet behind the mirror. This is where the image is formed. Compare the distance from the ray box to the mirror with the distance from the image to the mirror. Try the experiment again with the ray box in a different position. You should find that the image is always the same distance from the mirror as the object.

Figure 14.1B ▲ (a) Equal distances; the candle and its image are at equal distances from the mirror

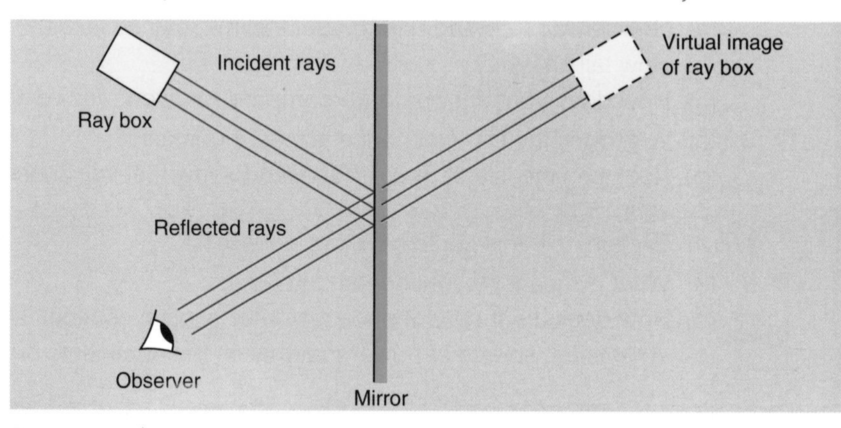

Figure 14.1B ▲ (b) Investigating reflection

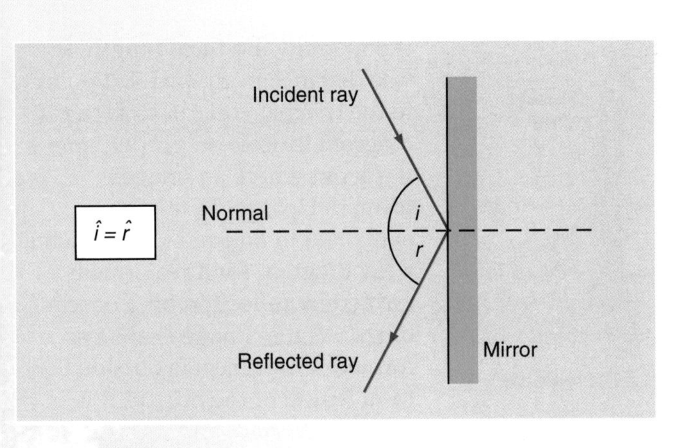

Figure 14.1C ▲ (a) The law of reflection

To understand why this always happens, look carefully at Figure 14.1C(a) *Can you see that each ray reflects off the mirror at the same angle as it strikes the mirror?* To check this, use a protractor to measure the angles shown. Each angle is always measured from the light ray to the normal, which is the line at right angles to the mirror.

The **Law of Reflection** states that the angle between the incident ray and the normal is always equal to the angle between the reflected ray and the normal.

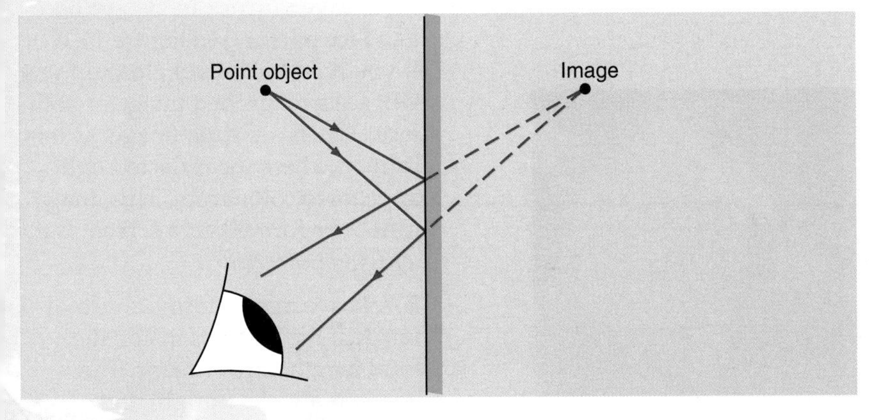

Figure 14.1C ▲ (b) Image formation

Light rays from the point object in Figure 14.1C(b) reflect off the mirror in accordance with the law of reflection. The reflected rays appear to originate from a point image behind the mirror. The perpendicular distance from this image to the mirror is the same as the corresponding distance from the object to the mirror.

Investigating the concave mirror

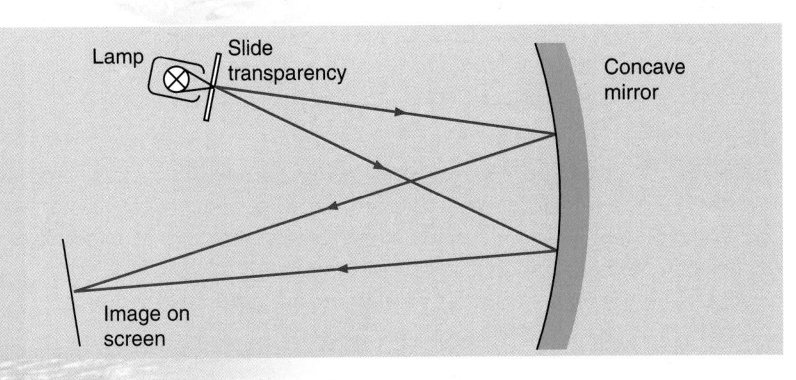

Figure 14.1D ▲ Investigating image ormation

You can use a concave mirror to form an image of a distant object on a screen. The ray diagram in Figure 14.1D shows how the light from the lamp shines through the slide on to the mirror and is reflected off the mirror on to the screen. An inverted image of the slide is formed on the screen. This image is called a real image, because it can be formed on a screen.

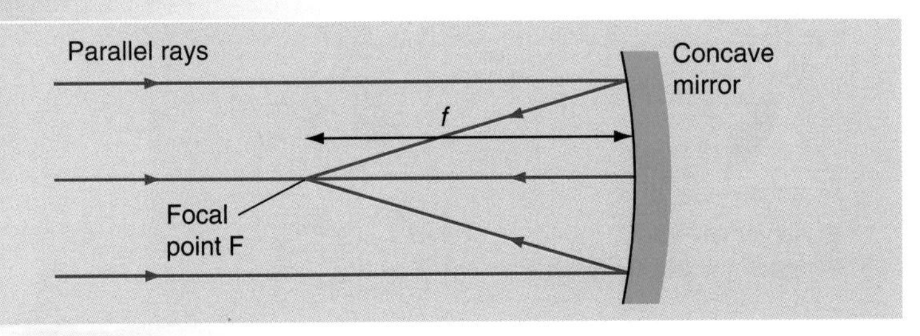

Figure 14.1E ▲ Focal length

Parallel rays of light always meet at the **focal point**, F, of the concave mirror after reflection (Figure 14.1E). Rays of light from a distant object reach the mirror as a parallel beam, so the image of a distant object is always formed at the focal point.

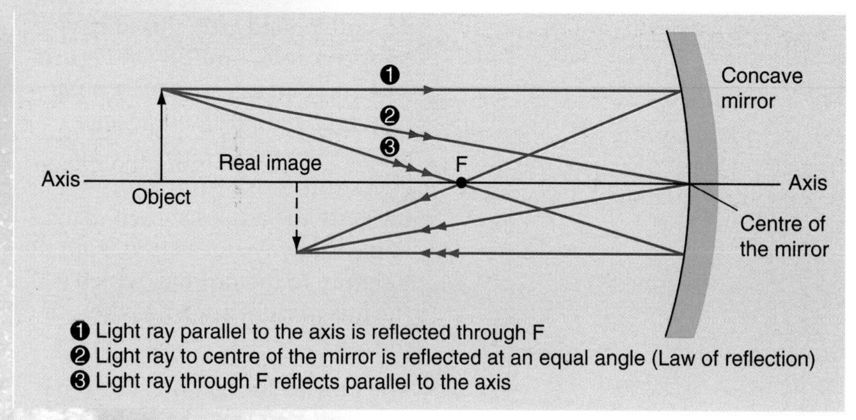

Figure 14.1F Formation of a real image

If you know the **focal length**, f, (the distance from the focal point to the mirror) you can draw a ray diagram to scale on graph paper to work out where an image is formed. How to do this is explained in Figure 14.1F, which is a ray diagram for a real image formed by reflection off a concave mirror. A real image is always formed if the object is outside the focal length of the mirror. It is always inverted, and its size depends on the distance of the object from the mirror.

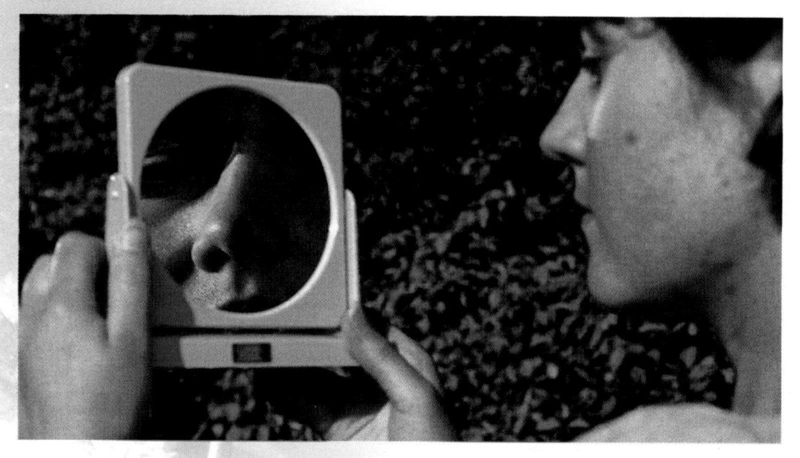

Figure 14.1G Forming a magnified image

A concave mirror can also be used as a face mirror (see Figure 14.1G). If you hold the mirror close up you will see a magnified image of your face. This is a virtual image, as it is formed where the reflected light appears to come from. This image cannot be formed on a screen. *Is it upright or inverted?*

A concave mirror forms a virtual image if the object is inside the focal length of the mirror. The image is always upright and enlarged. The ray diagram in Figure 14.1H shows how the image is formed.

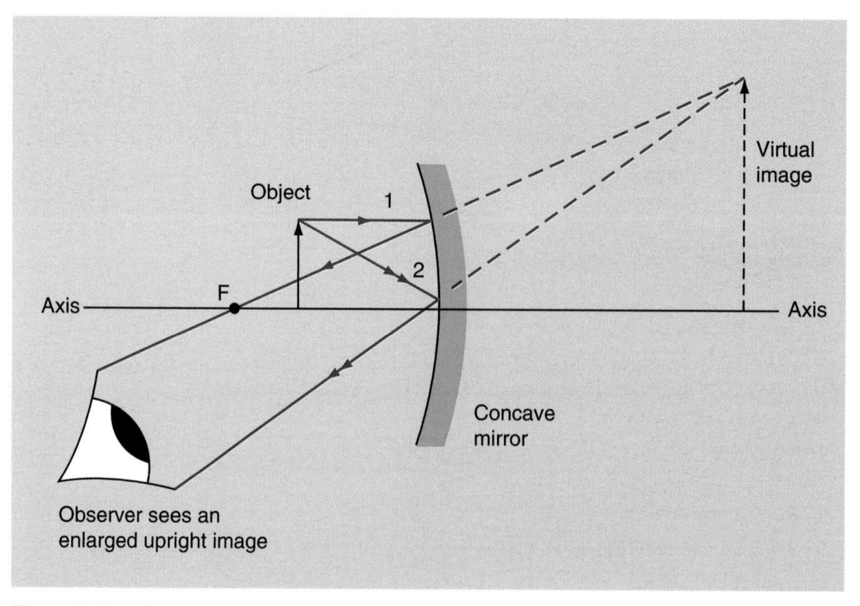

Figure 14.1H Formation of a virtual image

Uses of mirrors

Plane mirrors are used to see round awkward corners as well as to check personal appearance. They are also used in meters where a pointer has to be read against a scale (Figure 14.1I). For an accurate reading the image of the pointer must be under the pointer.

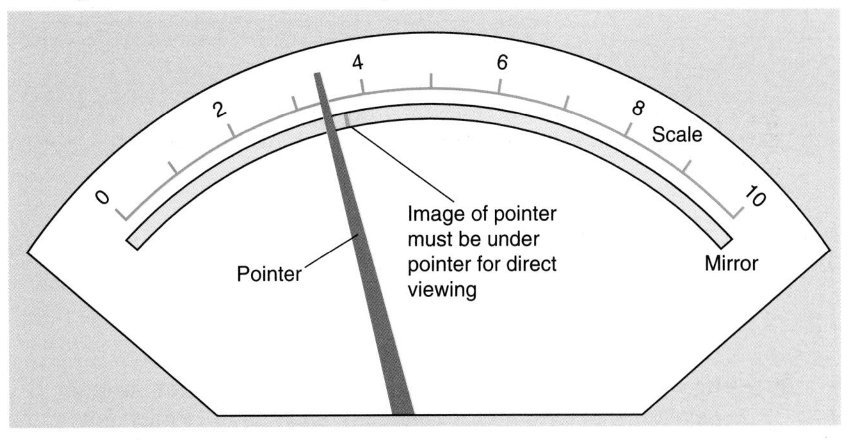

Figure 14.1I ▲ Using a plane mirror

All mirrors obey the law of reflection

$i = r$

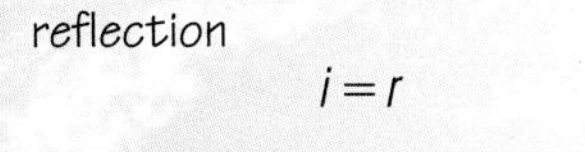

A **periscope** can be constructed using two plane mirrors. You can use this if your view is obstructed when you are in a crowd. Figure 14.1J shows the construction of a periscope using two plane mirrors.

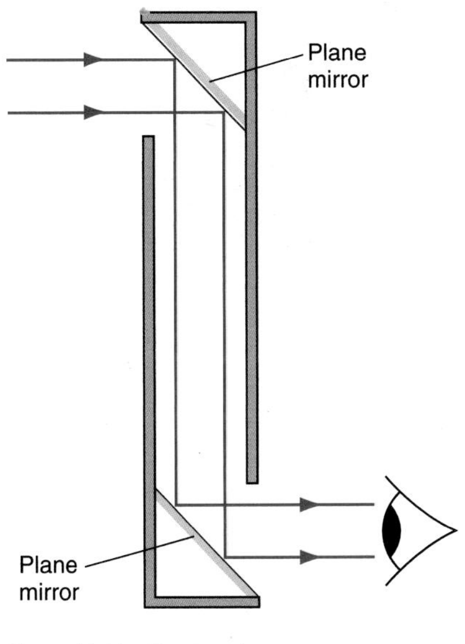

Figure 14.1J ▲ A periscope

Concave mirrors can be used as face mirrors, or to collect light or to project it. Sunlight can be focused to a tiny 'hot spot' by a concave mirror, as in a solar energy collector. To project light, a light bulb is placed at the focus of the mirror, which gives a parallel beam of reflected light, as shown in Figure 14.1K.

Convex mirrors are used as driving mirrors, as security mirrors in shops and to see around corners. Convex mirrors have a wide field of view. Figure 14.1L. shows how a convex mirror reflects light falling on it.

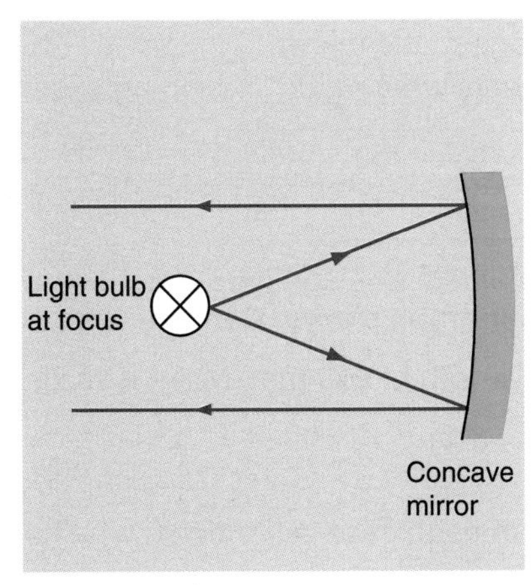

Figure 14.1K ▲ Projecting a parallel beam

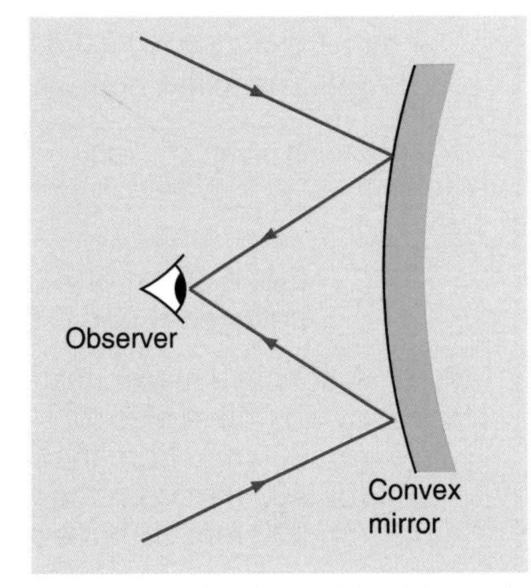

Figure 14.1L ▲ A wide field of view

SUMMARY

A real image can be formed on a screen. A virtual image cannot be formed on a screen. A plane mirror always gives a virtual image that is the same size as the object. A concave mirror can give a real or virtual image depending on the distance of the object from the mirror.

CHECKPOINT

1 Test your drawing skills. A point object O is placed 50 mm from a plane mirror, as shown. Copy the figure and on your copy show the path of two rays from O that reflect off the mirror. Use the reflected rays to mark the image position 1 and show that the image is 50 mm behind the mirror.

Plane mirror

Object O

2 Kim notices that letter 'p' held near a mirror appears as letter 'b' in the mirror. Are there any other pairs of letters that form mirror images of each other? Which letters appear the same in a plane mirror?

3 The diagram shows a person standing in front of a wall-mounted flat mirror.

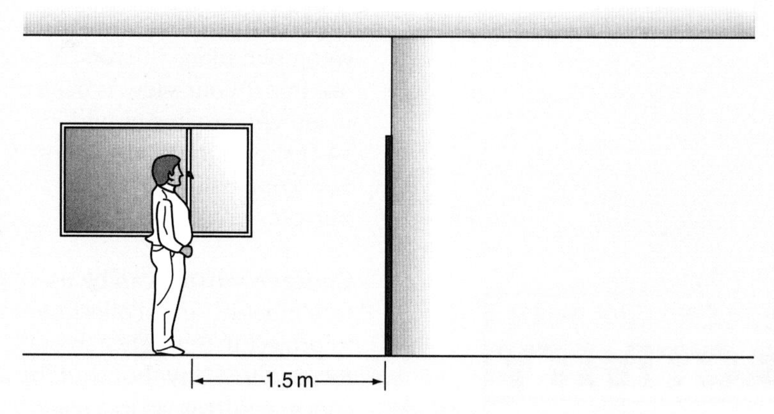

(a) Copy the diagram and sketch the image of the person on your diagram.
(b) The person stands 1.5 m from the mirror. What is the distance from the person to her image?
(c) The person is 1.8 m tall. Which part of the mirror does she use to observe her image from head to toe?

4 (a) Describe how you would use two small plane mirrors to observe the back of your head.
(b) A dentist uses small mirrors to observe teeth inside the mouth. Why is a small concave mirror better than a small flat mirror for this purpose?

5 Make your own 'Hall of Mirrors'. Suppose you are given a flexible plastic sheet with a mirror finish on one side. How would you bend it to make the image:
(a) thinner than normal,
(b) shorter than normal,
(c) longer and thinner than normal,
(d) wide at the bottom and thin at the top.

6 (a) A concave mirror can be used to form a real image of an object. Design an experiment, using a pair of crosswires as an object, to find out how the height (H) of the image varies with the distance (D) from the image to the mirror.
(b) The following measurements were made in such an experiment.

D (mm)	1000	850	700	550	400
H (mm)	46	37	26	17	6

Plot a graph of H on the vertical axis against D on the horizontal axis. Describe the connection between H and D that your graph shows. Can you write an equation to show this connection?

7 A concave mirror has a focal length of 250 mm. Make a scale drawing to locate the image formed by an object placed:
(a) 600 mm from the mirror,
(b) 400 mm from the mirror,
(c) 200 mm from the mirror. In each case, write down whether the image is real or virtual.

14.2 ▶ Fibre optics

FIRST THOUGHTS

Surgeons and security firms use fibre optics for very different purposes. Read on to find out how fibre optics work.

Mr Green is in terrible pain. The surgeon thinks the cause may be a stomach ulcer. Fortunately for Mr Green the surgeon can find out without cutting into him, by using an endoscope. An endoscope is a thin, flexible tube, which the surgeon can pass down Mr Green's throat into his stomach. Light is sent down part of the tube and is reflected back along another part of the tube. It shows there is an angry-looking ulcer on the lining of Mr Green's stomach (Figure 14.2A).

Total internal reflection can occur when light in a medium is incident on a boundary with air.

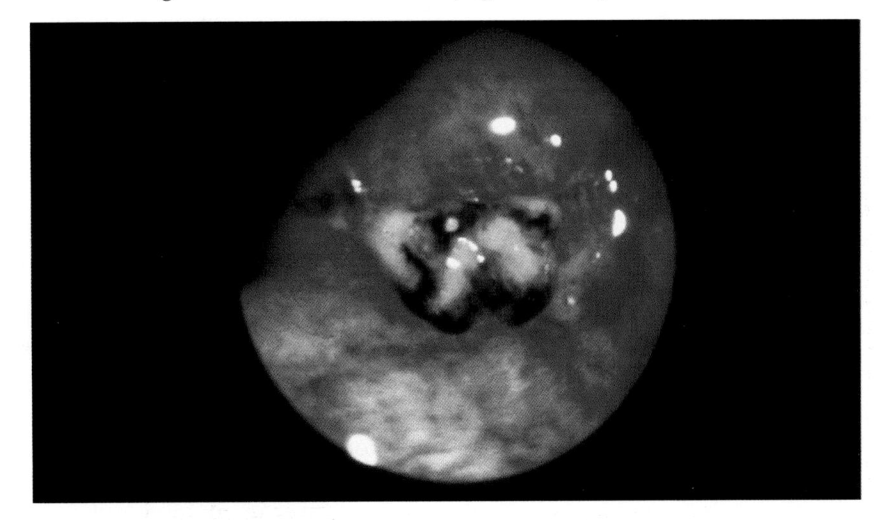

Figure 14.2A ▲ Stomach ulcer viewed through an endoscope

An endoscope contains a bundle of optical fibres, which are very thin strands of glass. Light enters the fibre at one end, passes along it, and comes out at the other end, even if the fibre is bent or twisted. Light passes along an optical fibre by **total internal reflection**.

Investigating total internal reflection

You can use a semicircular glass block to investigate this. In Figure 14.2B the light ray is **refracted** by the glass and comes out through the straight edge of the block. Here the angle of incidence, i, is less than the **critical angle**, c. The critical angle for glass is about 42°, but varies for different types. If a light ray enters the block at the critical angle (Figure 14.2C) the refracted ray emerges along the straight edge of the block. If the angle of incidence of the light ray is greater than the critical angle, the light ray is totally internally reflected, as shown in Figure 14.2D. *What measurements would you make to show that the angle of reflection is equal to the angle of incidence?*

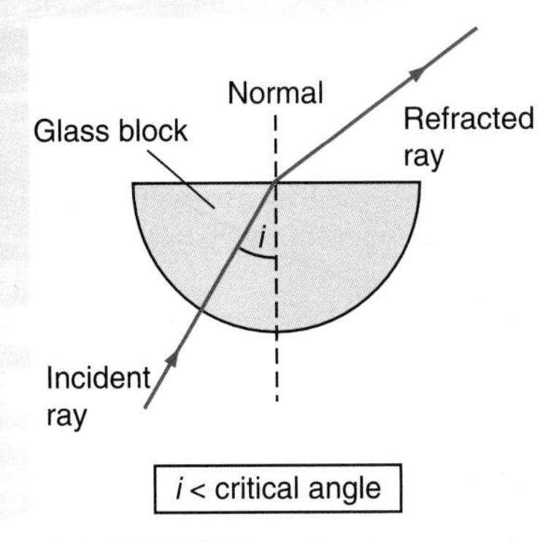

Figure 14.2B ▲ Refraction of light

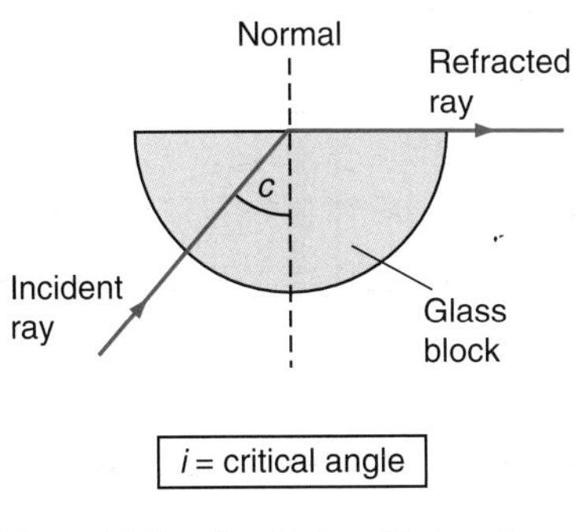

Figure 14.2C ▲ At the critical angle

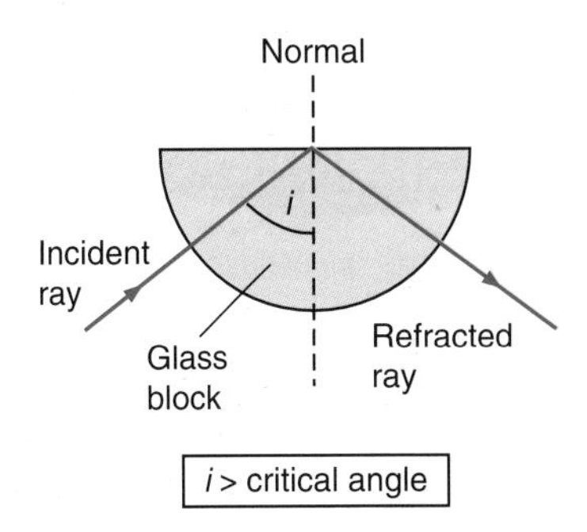

Figure 14.2D ▲ Total internal reflection

Light rays enter an optical fibre and are totally internally reflected from the sides of the fibre each time they hit the glass boundary at the edge of the fibre (Figure 14.2E). The fibres must be very thin. If the fibres were too thick, the angles of incidence of the light rays inside hitting the boundary would be smaller on average (see Figure 14.2F). If they were smaller than the critical angle, the light would not all be totally internally reflected, and some would be lost through the sides of the fibre. The thinner the fibre, the more it can be bent without light being lost.

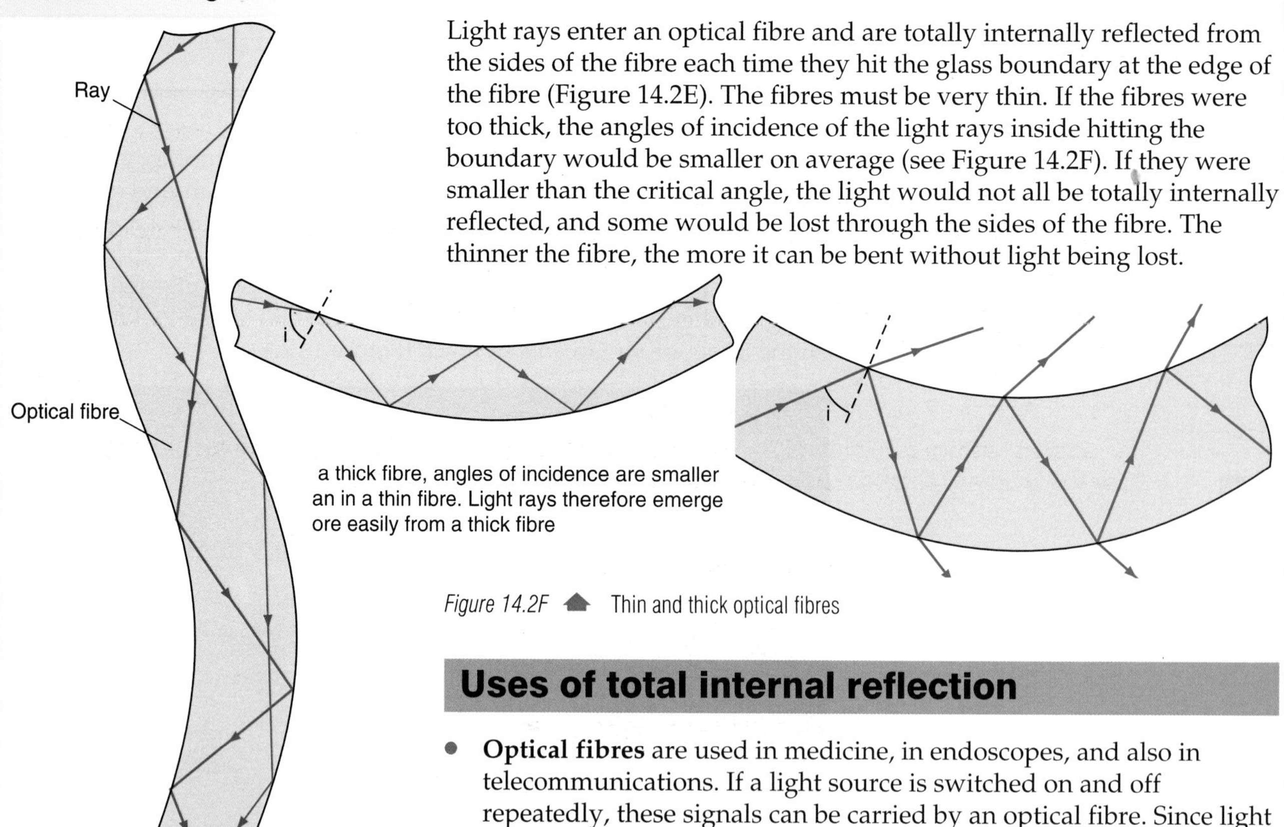

Figure 14.2E ▲ Fibre optics

Figure 14.2F ▲ Thin and thick optical fibres

Uses of total internal reflection

- **Optical fibres** are used in medicine, in endoscopes, and also in telecommunications. If a light source is switched on and off repeatedly, these signals can be carried by an optical fibre. Since light waves have a much higher frequency than radio waves, optical fibres can carry many more pulses than radio beams. Also, the light cannot be seen from outside the fibre, only by the receiver at the other end, so no-one can spy on the signals.
- **Prisms** are used in periscopes, binoculars and expensive cameras to reflect light instead of plane mirrors. The reflective surface of a mirror tarnishes gradually and becomes non-reflecting. This does not happen when a prism is used as a reflector because the light is totally internally reflected. Figure 14.2G shows the idea. Note that right-angled prisms are used.
- **A cycle reflector** is designed to reflect light back in the same direction it came from. Its rear surface consists of lots of tiny prisms which reflect light internally. Figure 14.2H shows how it works.

SUMMARY

Optical fibres are used to guide light rays. The fibres must be thin to ensure light does not escape through the sides of the fibre. They are used in medicine and in telecommunications.

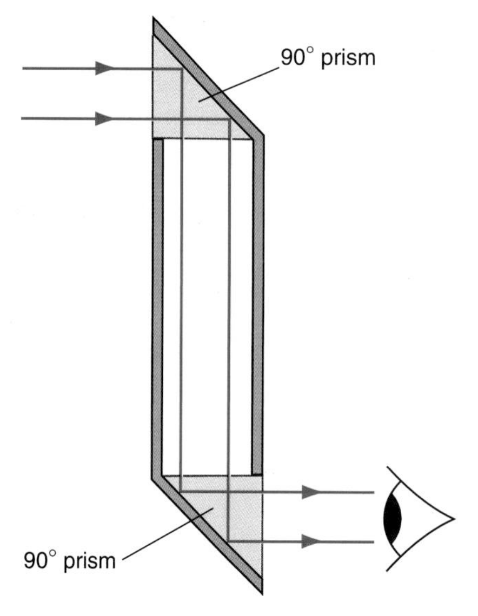

Figure 14.2G ▲ Prisms in a persicope

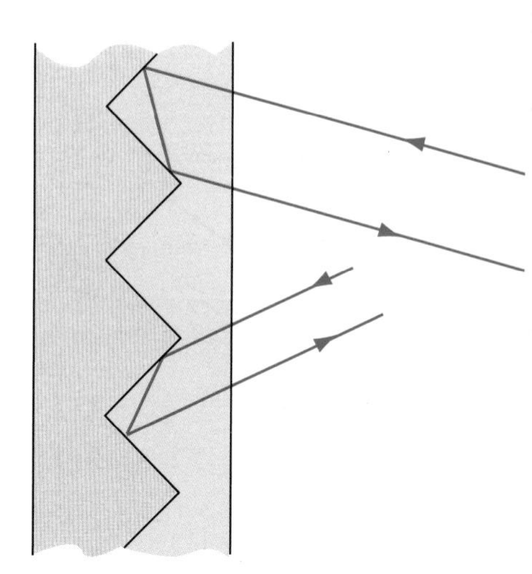

Figure 14.2H ▲ A prismatic reflector

CHECKPOINT

➧ **1** In medicine, separate 'bundles' of optical fibres are used to light up cavities in the body to enable doctors to see what is there.
   (a) Why is it necessary to light up the cavity?
   (b) Give two reasons why lots of thin fibres are used to make a bundle, rather than a few thick fibres.

➧ **2** In telecommunications, messages carried by pulses of light can be sent along individual fibres.
   (a) Why is it important that light does not leak between the fibres?
   (b) Each fibre is coated with a special 'cladding' to prevent light leakage where fibres touch. What would happen if this cladding got rubbed away?

➧ **3** The diagram shows a section of an optical fibre made of a material with a critical angle of 40°. Copy the diagram and complete the paths of the two rays marked X and Y.

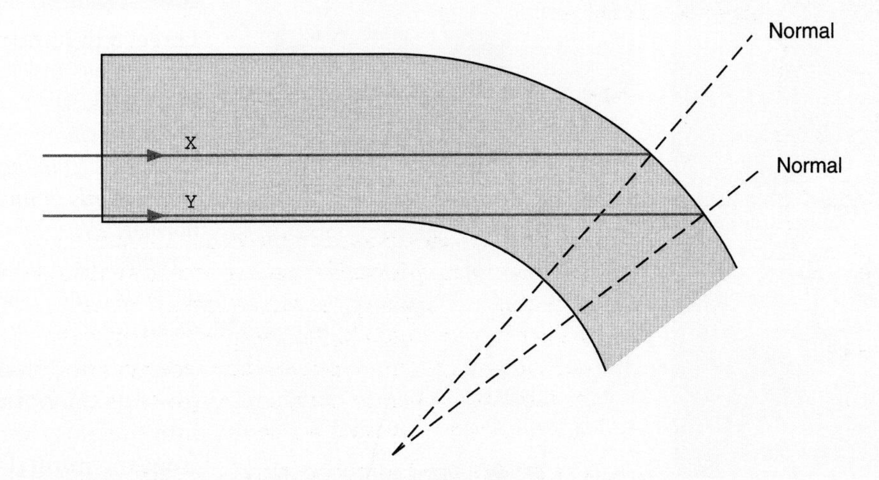

➧ **4** A pulse of light is sent down an optical fibre as shown.
   (a) The light rays that are reflected from side to side take longer to travel the length of the fibre than those that pass straight along. Why?
   (b) Why does this time lag cause the light pulse reaching the far end to be 'smeared out'?

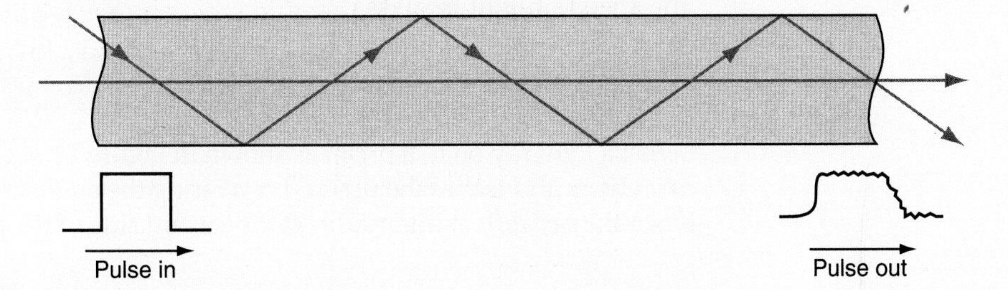

➧ **5** (a) With the aid of a diagram, explain the construction of a periscope containing two right-angled prisms.
   (b) Mark the path of a light ray through the periscope, showing clearly where it changes direction.

14.3 Refraction

FIRST THOUGHTS

Refraction is a property of every type of wave. In this section you will study refraction of light to give you the understanding necessary for work on lenses and optical instruments.

When you visit the swimming pool, take care near the deep end. It might seem shallow, but if you fall in you may be surprised. The water is deeper than it looks, because light from the bottom of the pool is bent (refracted) when it leaves the water, so the image of the bottom of the pool appears to be closer than it really is (Figure 14.3A).

Light is also refracted when it passes between air and glass. Light waves travel more slowly in glass than in air, because glass is more dense than air.

Figure 14.3A ▲ Refraction makes the water look shallower

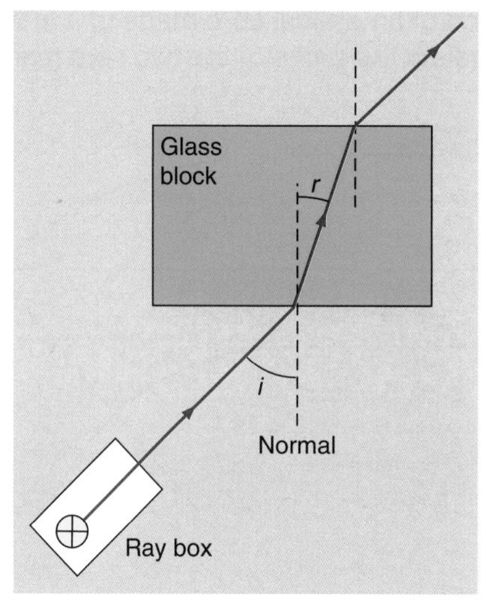

Figure 14.3B ▲ Refraction by a glass block

Refraction in a glass block

Direct a light ray from a ray box into a glass block, as shown in Figure 14.3B. *What can you say about the change of direction when the ray passes from the air into the glass? Is it towards or away from the normal? What about the change when it comes out of the glass? What happens if you change the angle of incidence, i?*

You will find that **the ray bends towards the normal when it goes into the glass and bends away from the normal when it leaves the glass.**

The ratio $\frac{\sin i}{\sin r}$ is constant, independent of the angle of incidence. This ratio is called the **refractive index** of glass. Theory shows it is equal to

$$\frac{\text{the speed of light in air}}{\text{the speed of light in glass}}$$

A light ray bends towards the normal on passing from air into glass.

Refraction in a prism

Direct a light ray on to a prism as shown in Figure 14.3C. The ray is refracted as it enters and leaves the prism. Try altering the angle of incidence, *i*, to make the ray **reflect internally** off the second side of the prism.

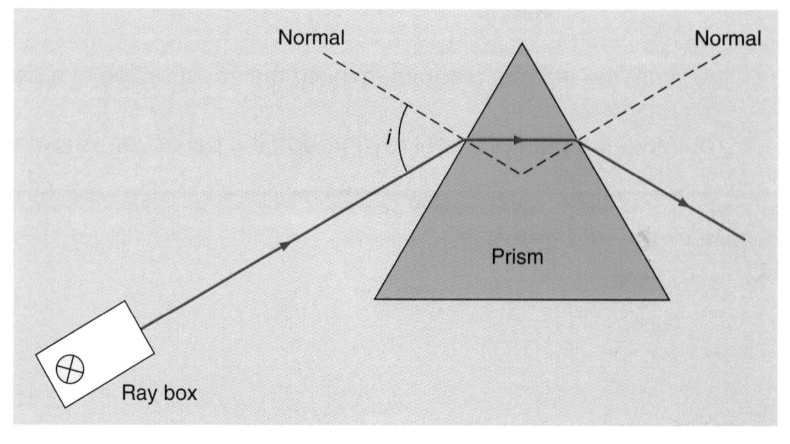

Figure 14.3C ▲ Refraction by a prism

Refraction in lenses

If you use a ray box plate with several slits you can produce several light rays at once. Aim the light rays at different shaped lenses and sketch the ray patterns for each one.

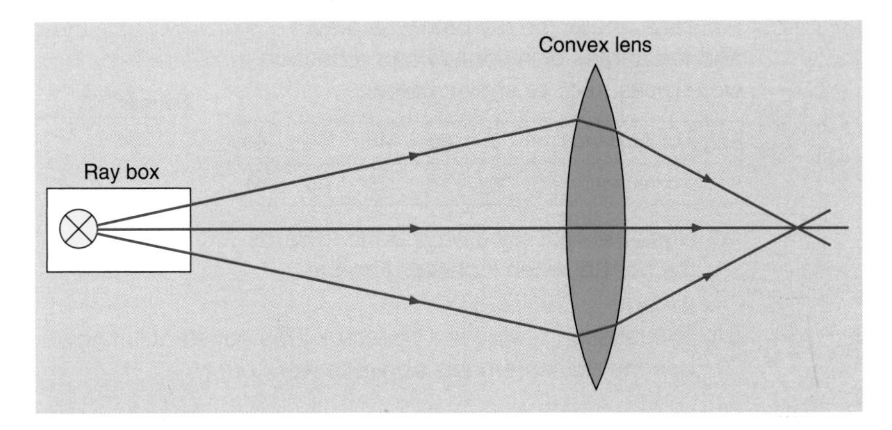

Figure 14.3D ▲ The convex lens

A convex lens makes a beam of light converge, or diverge less.

A concave lens makes a beam of light diverge, or converge less.

Figure 14.3D shows a ray diagram for a convex lens, which makes the light rays converge (i.e. meet). However, if the lens is too close to the ray box, the lens just reduces the spreading of the rays, without making them actually meet. Imagine that the lens is divided into small sections (Figure 14.3E). Each section behaves like a small prism, refracting the rays at each glass-air boundary.

See www.keyscience.co.uk for more about light and sound.

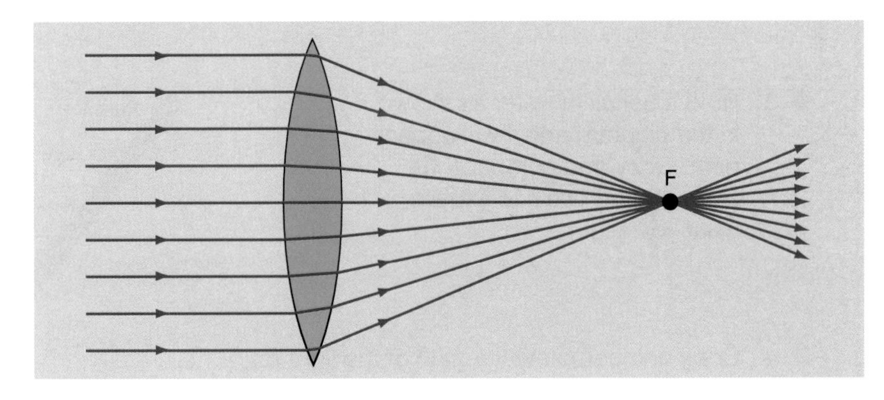

Figure 14.3E ▲

A concave lens makes the light rays diverge (i.e. spread out), as in Figure 14.3F.

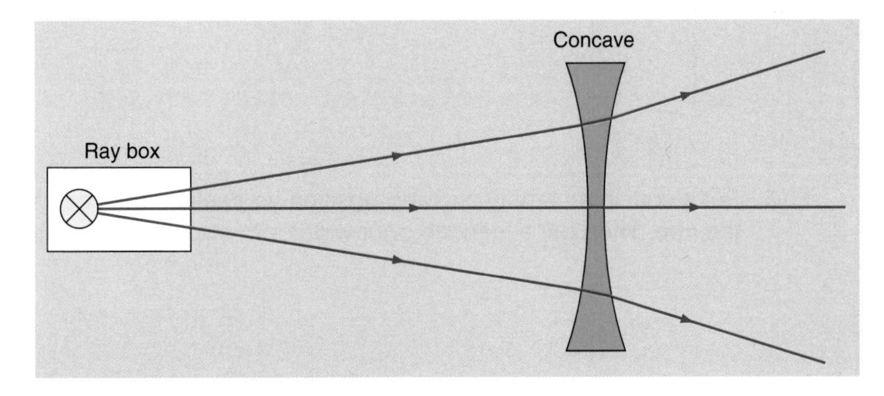

Figure 14.3F ▲ The concave lens

SUMMARY

When a light ray passes from air into glass, it bends towards the normal. A convex lens can make light rays converge.

CHECKPOINT

1 In an investigation of the refraction of light, a light ray was aimed at an air-glass boundary at different angles of incidence. For each angle, the ray path was drawn and the angles of incidence and refraction were measured, as shown below.

Angle of incidence *i* (°)	0	20	40	60	80
Angle of refraction *r* (°)	0	13	25	35	41

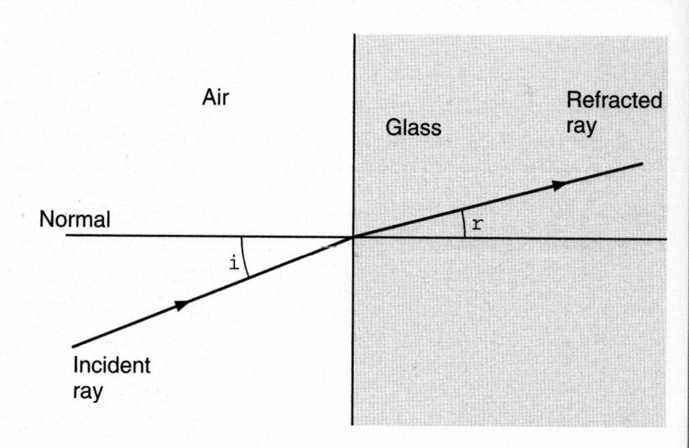

(a) Does the light ray always bend towards the normal when it passes into the glass?

(b) The ratio sin i / sin r is a constant. This constant is known as the refractive index, n, of glass. Use the measurements above to work out n.

2 Light bends when it passes from air into water. Copy the sketch and complete the ray path into air.

(a) Explain why the pond appears shallower than it really is.

(b) Would the fish be able to see the fly above the water?

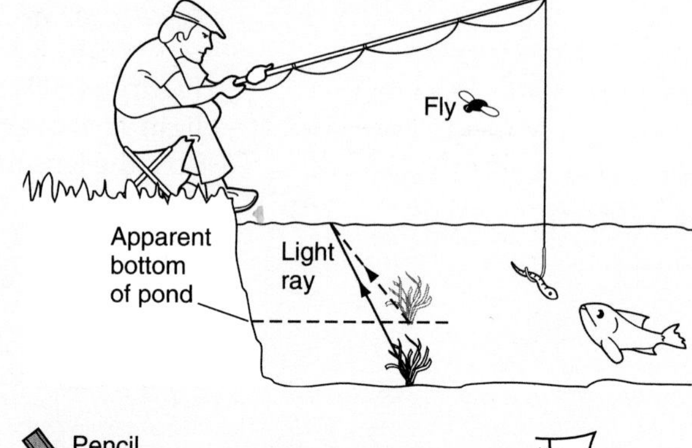

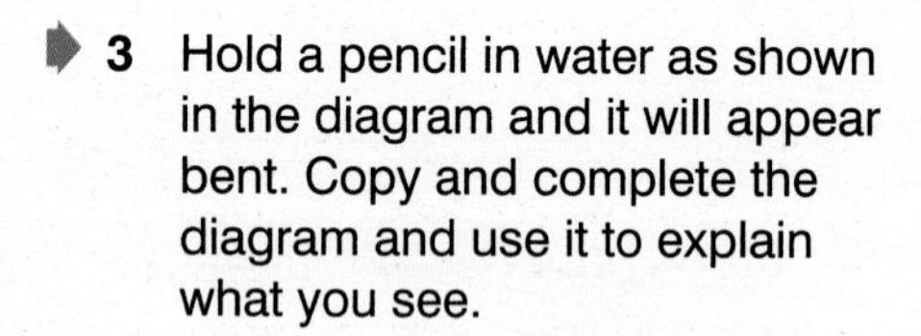

3 Hold a pencil in water as shown in the diagram and it will appear bent. Copy and complete the diagram and use it to explain what you see.

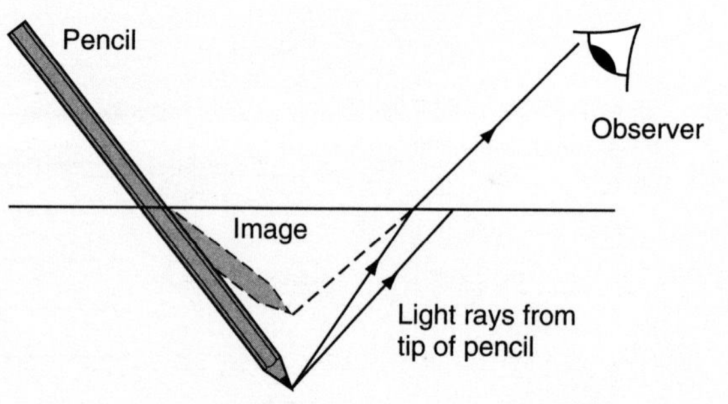

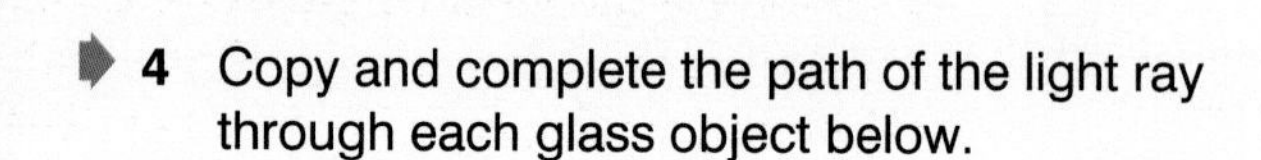

4 Copy and complete the path of the light ray through each glass object below.

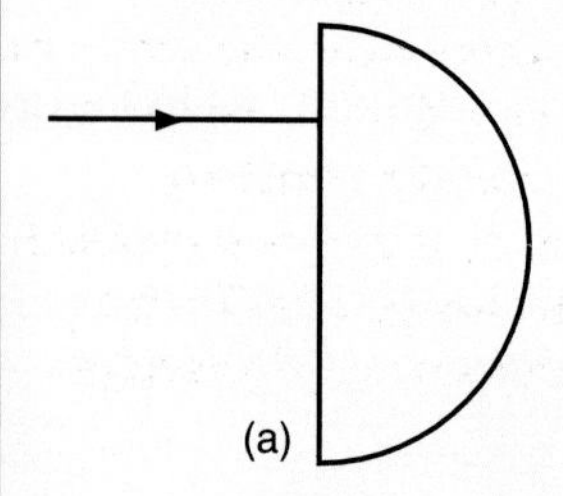

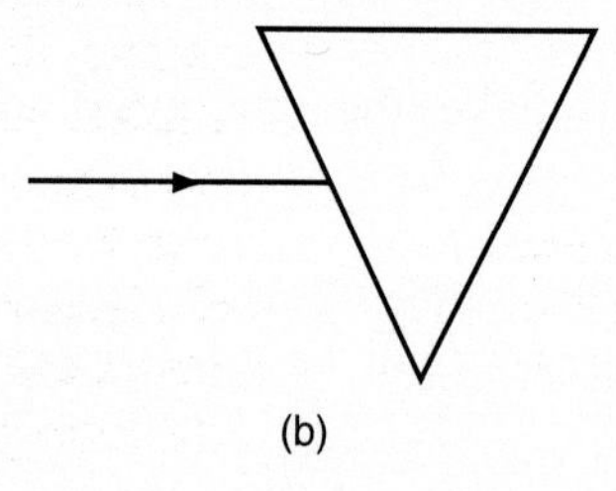

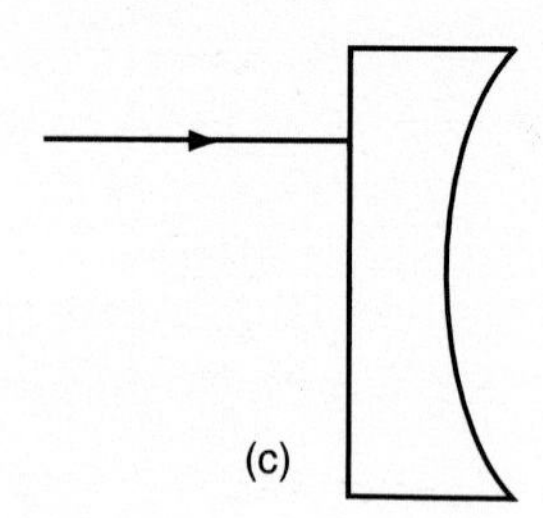

5 Put a coin in an empty cup and position yourself so the coin is just out of sight below the rim of the cup. Then get a friend to pour water into the cup. Now you can see the coin. Why?

14.4 Lenses

FIRST THOUGHTS

The eye, the camera and the microscope are examples of optical instruments that use convex lenses. In this section you will find out about the action of convex lenses before studying optical instruments.

Our eyes contain lenses to enable us to see clearly. Cameras contain lenses to capture images on film. Microscopes contain lenses to magnify tiny objects. Film projectors use lenses to throw images onto screens. Telescopes and binoculars use lenses to enable us to see distant objects.

Figure 14.4A ▲ A convex magnifying lens

A **convex lens** can be used to project or to magnify images. Hold a convex lens over this page and you will see that it magnifies the print (see Figure 14.4A). The image of the print is a virtual image because it cannot be formed on a screen. It is formed at the point where the rays appear to come from.

Figure 14.4B ▲ Forming an image of a distant object

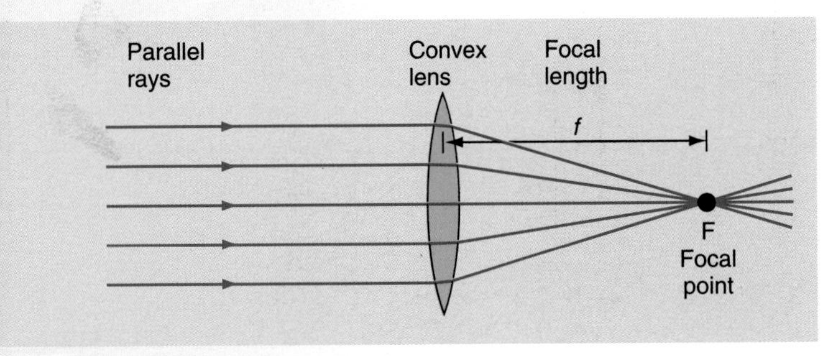

Figure 14.4C ▲ Focal length of a convex lens

Now hold the same lens up near a window and use it to form an image of a distant object on a sheet of white paper (Figure 14.4B). If you shine a torch bulb through the lens you should be able to project an image of the bulb onto a nearby screen or wall. This is a real image because it is formed on a screen.

The type of image formed by the lens depends on the distance of the object from the lens. The lens has a **focal point**, F, which is where parallel rays shining on the lens are brought to a focus. The distance from the focal point to the lens is called the **focal length**, f (Figure 14.4C).

To measure the focal length of a convex lens, use it to form an image of a distant object on a screen, as in Figure 14.4B. The focal length of the lens is the perpendicular distance from the centre of the lens to the screen. This is because the rays from a distant point object are effectively parallel by the time they reach the lens. Hence the image of a distant object is formed in focus on a screen at the focal point.

Investigating the convex lens

Put the lens in a holder and place it facing an open window with a white screen behind it, as shown in Figure 14.4B. Move the screen until you can see a clear image of a distant object on the screen. The light rays from the distant object are almost exactly parallel by the time they reach the lens, and so are focused at the focal point. Thus the screen is at the focal point and you can measure the focal length, f, of the lens.

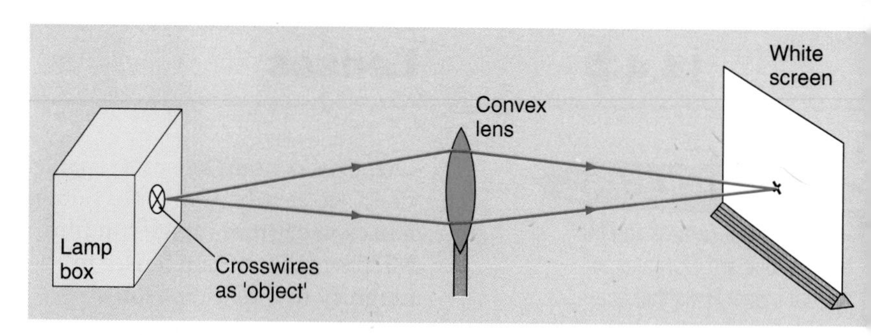

Figure 14.4D ▲ Investigating the convex lens

Now use a lamp box with a pair of crosswires as an object (Figure 14.4D) Set the object at distance $3f$ from the lens, and move the screen until you can see a clear image of the crosswires on it. Measure the image distance (the distance from the lens to the screen). Repeat this for different object distances. In each case, observe the image formed and measure the object and image distances. You should find that the screen needs to be placed nearer the lens as the object is moved further away.

Ray diagrams are useful to understand how a convex lens forms an image of an object. Three key rays from the object are used to locate the image. Figures 14.4E–G use these three key rays to show how a convex lens forms the different images you will have seen in your investigations.

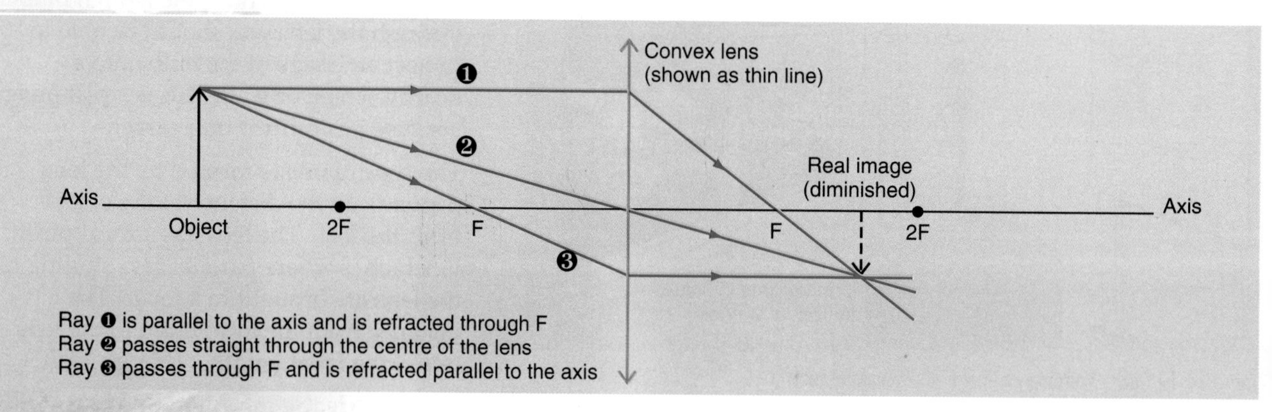

Figure 14.4E ▲ Object outside 2F

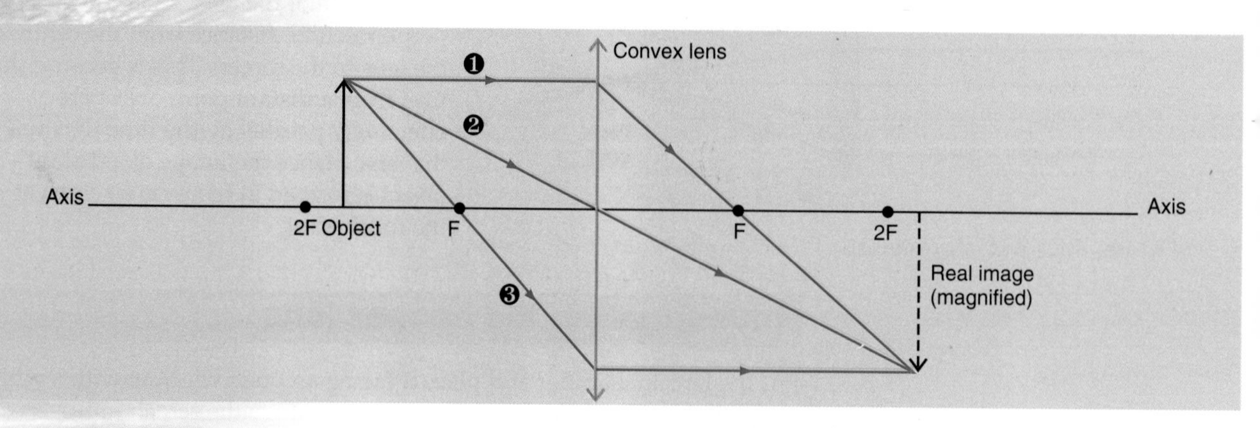

Figure 14.4F ▲ Object between F and 2F

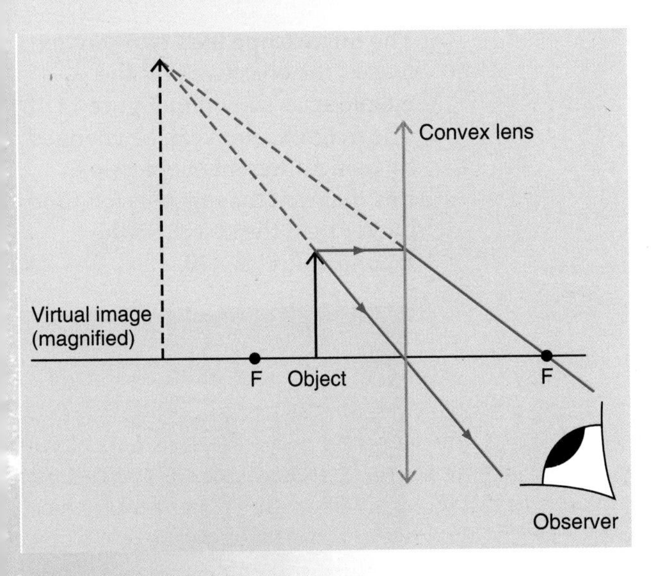

Figure 14.4G ▲ Object inside F

If the image is larger than the object, it is said to be **magnified**.

If the image is smaller than the object, it is said to be **diminished**.

If the object distance exceeds $2f$, the image is diminished; if the object distance is less than $2f$, the image is magnified.

A **real image** is formed where the rays from a point object meet. A **virtual image** is formed where the rays from a point object appear to come from. For a real image to be formed, the object must lie beyond the focal point F of the lens. For a virtual image to be formed, the object must lie between the focal point and the lens. Table 14.4A summarises the information shown in Figures 14.4E–G and also lists applications for each situation.

Table 14.4A ▼ Using a convex lens

Object distance	Image distance	Nature of the image	Applications
Beyond 2F	Between F and the lens	Real, inverted, diminished	The camera, the eye
At 2F	At 2F	Real, inverted, same size	Inverter lens
Between F and 2F	Beyond 2F	Real, inverted, magnified	Projector
At F	At infinity	No image formed	Producing a parallel beam
Between F and the lens	Same side as the object	Virtual, upright, magnified	Magnifying glass

Convex lens applications

The **camera** uses a convex lens to form a real image on a photographic film. 'Focusing' the camera adjusts the distance from the lens to the film, so that objects at different distances can be focused on to the film. Figure 14.4H explains how the camera works.

The image is real, inverted and diminished in the camera.

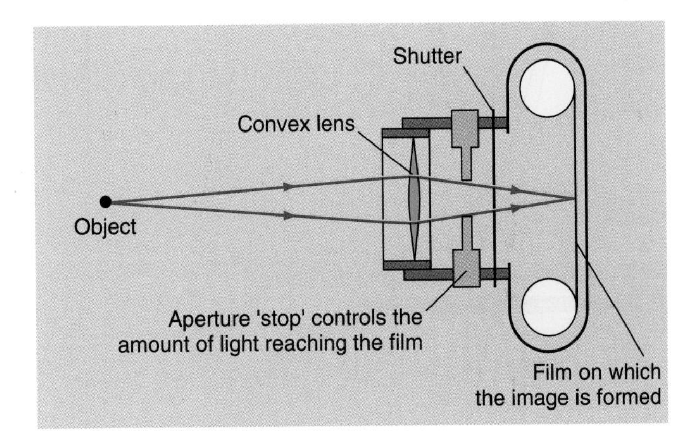

Figure 14.4H ▲ The camera

The **magnifying glass** is shown in Figure 14.4G. The object is closer than the observer's near point so it appears larger than when it is at the near point without the lens present.

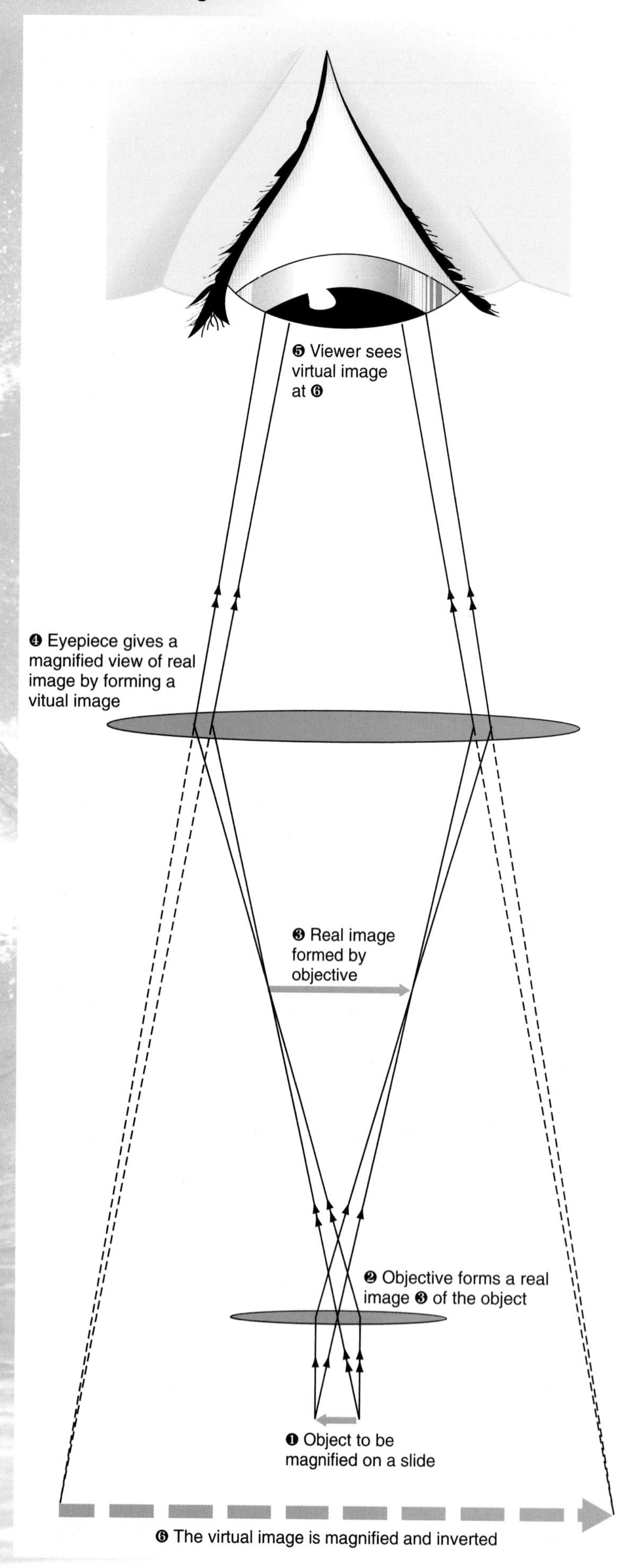
❺ Viewer sees virtual image at ❻
❹ Eyepiece gives a magnified view of real image by forming a vitual image
❸ Real image formed by objective
❷ Objective forms a real image ❸ of the object
❶ Object to be magnified on a slide
❻ The virtual image is magnified and inverted

The **microscope** uses two convex lenses, the objective and the eyepiece, as shown in Figure 14.4I. The magnification can be changed by using different eyepieces. In this diagram the image is ten times bigger than the object, so the magnification is $\times 10$.

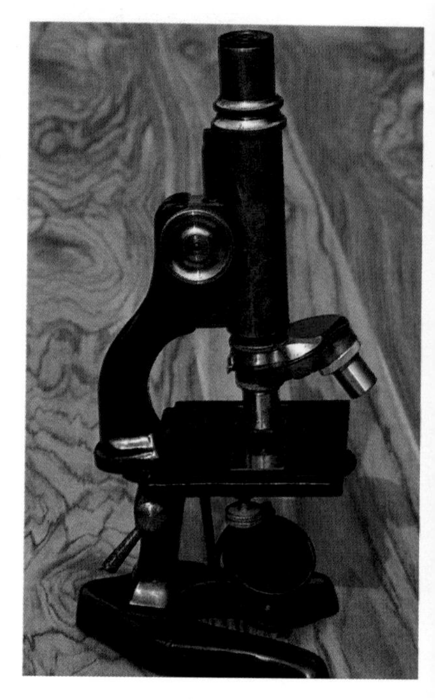

Figure 14.4I (a) The microscope (b) Ray diagram of the microscope

$$\text{Objective magnification} = \frac{\text{Height of ❸}}{\text{Height of ❶}} = 3$$

$$\text{Eyepiece magnification} = \frac{\text{Height of ❻}}{\text{Height of ❸}} = \frac{10}{3}$$

$$\text{Overall magnification} = 3 \times \frac{10}{3} = 10$$

The projector uses a convex lens to cast an image of a slide transparency onto a screen. The slide is illuminated by a beam of light, created using a condenser lens and a compact light bulb. The distance from the lens to the screen is adjusted to produce a clear enlarged image of the slide on the screen. Because the image is inverted, the slide must be placed in position upside down so the image is the correct way up.

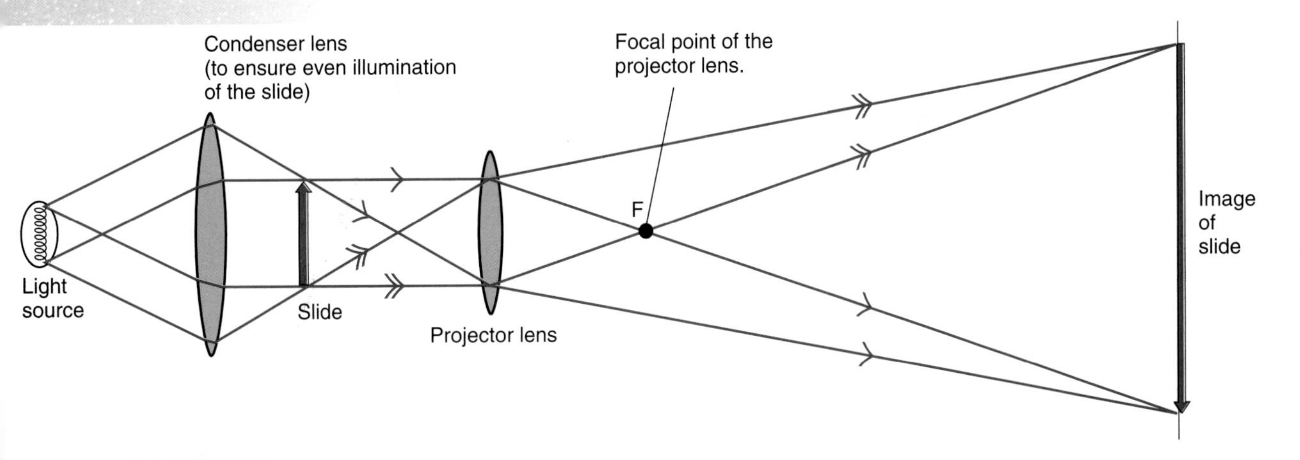

Fig 14.4J ▲ The slide projector

The image is real, inverted and magnified in the projector.

The lens formula

For an object at distance u from a lens of focal length f, the distance v from the lens to the image formed is given by the formula below.

$$\frac{1}{u} + \frac{1}{v} = \frac{1}{f}$$

To use the formula, real objects and images are given positive values for u and v, and virtual objects and images are given negative values for u and v. This rule is called the **'real is positive'** sign convention.

Also, for a convex lens, the focal length is given a positive value and for a concave lens, the focal length is given a negative value.

Worked example 1 A slide transparency of height 20 mm is placed 170 mm from a convex lens of focal length 150 mm used to project an image of the slide onto a screen. Work out (a) the image distance and (b) the height of the image on the screen.

Solution

(a) To work out the image distance v, use the lens formula $1/u + 1/v = 1/f$ where $u = +\,0.170$ m and $f = +\,0.150$ m.
Hence

$$\frac{1}{0.170} + \frac{1}{v} = \frac{1}{0.150}$$

$$\frac{1}{v} = \frac{1}{0.150} - \frac{1}{0.170} = 6.667 - 5.882 = 0.785 \text{ m}^{-1}$$

giving $$v = \frac{1}{0.785} = 1.274 \text{ m}$$

A real image is formed 1274 mm from the lens on the opposite side to the object (see Figure 14.4F)

(b) The linear magnification

$$= \frac{v}{u} = \frac{1.274}{0.170} = 7.5$$

Hence the image height = 7.5 × the object height = 7.5 × 20 mm = 150 mm

SUMMARY

The focal length, f, of a convex lens is the distance from the focal point, F, to the lens. A real image is always formed if the object is outside the focal length. A virtual image is formed if the object is inside the focal length.

Worked example 2 A convex lens of focal length 100 mm is used as a magnifying glass to inspect a postage stamp. The lens is held close to the eye and the stamp is positioned 75 mm from the lens. Work out (a) the image distance and (b) the linear magnification.

Solution

(a) To work out the image distance v, use the lens formula $1/u + 1/v = 1/f$ where $u = +\ 0.170$ m and $f = +\ 0.100$ m.

Hence

$$\frac{1}{0.075} + \frac{1}{v} = \frac{1}{0.100}$$

$$\frac{1}{v} = \frac{1}{0.100} - \frac{1}{0.075} = 10.00 - 13.33 = -3.33\ \text{m}^{-1}$$

giving $$v = \frac{1}{-3.33} = -0.300\ \text{m}$$

The image is a virtual image 300 mm from the lens (see Figure 14.4G)

(b) The linear magnification

$$= \frac{v}{u} = \frac{-0.300}{0.100} = -3$$

The image height is 3 × the height of the object. Note that the minus sign denotes a virtual image.

CHECKPOINT

1 Copy and complete the ray paths in each of the following sketches.

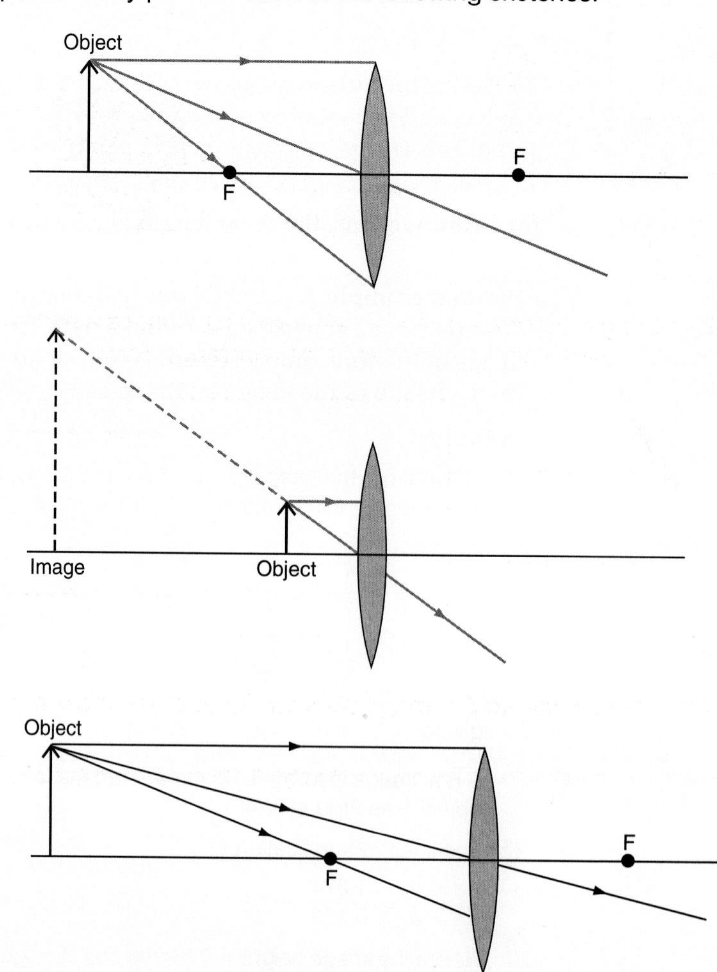

2 With the aid of a ray diagram, explain how a single convex lens may be used as a magnifying glass.

3 (a) With the aid of a diagram, explain the construction of a simple lens camera.
(b) State the function of (i) the shutter (ii) the aperture stop in a simple lens camera.

4 Draw ray diagrams to find the position of the image formed by a convex lens of focal length 15 cm when an object is placed:
(a) 30 cm from the lens,
(b) 20 cm from the lens,
(c) 10 cm from the lens. In each case, describe the image formed.

5 Ahmed has made a simple camera using a convex lens, a cardboard box and some greaseproof paper, as shown. He points the camera at a distant tree.
(a) Make a sketch of what he ought to see on the screen.
(b) Keith walks in front of the camera but his image is out of focus on the screen. What adjustment should be made to the camera to bring Keith's image into focus?
(c) Why should the inside of the camera be painted black?

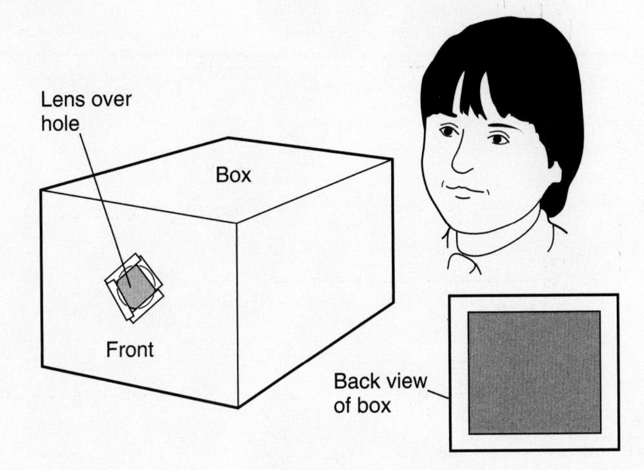

6 Design a system using a convex lens and a concave lens to make a narrow parallel beam of light into a wider parallel beam of light, as shown in the diagram below.

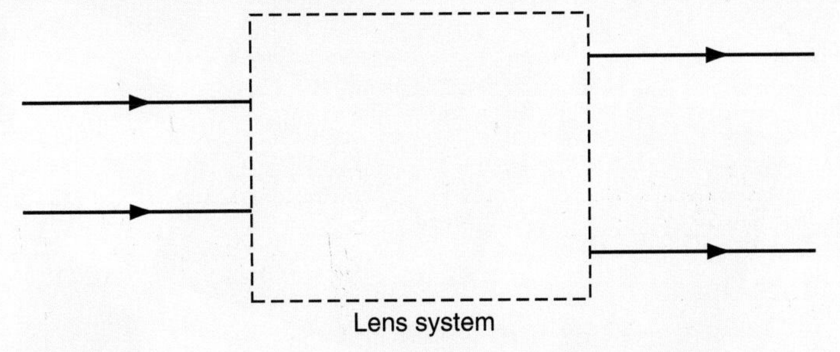

7 Use the lens formula to work out the image distance for each example in question 4 above.

14.5 The eye

FIRST THOUGHTS

This section concentrates on the eye as an optical instrument rather than on its biological details. Read on to find out about sight defects and how they are corrected.

Have you ever been told to 'use your eyes' when you complain that you cannot find something or other? Think about what we use our eyes for. They tell us about colour, shape, position and movement. When you look at an object, each eye forms an image of it and sends signals to your brain. Your brain 'reads' these signals and you 'see' the object. Figure 14.5A shows four unusual views of everyday objects. *Can you recognise them?*

Figure 14.5A ▲ Recognising things

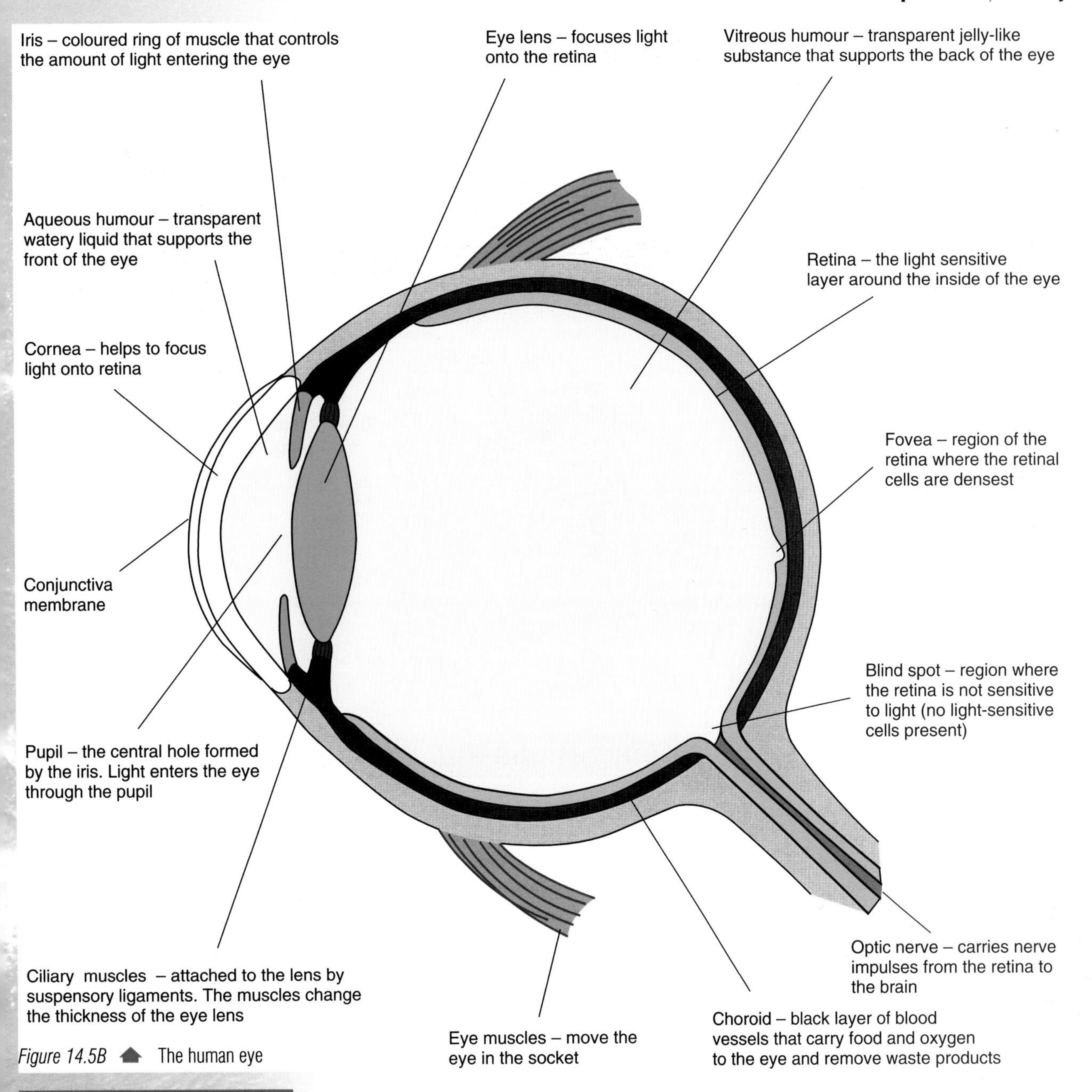
Iris – coloured ring of muscle that controls the amount of light entering the eye
Eye lens – focuses light onto the retina
Vitreous humour – transparent jelly-like substance that supports the back of the eye
Aqueous humour – transparent watery liquid that supports the front of the eye
Retina – the light sensitive layer around the inside of the eye
Cornea – helps to focus light onto retina
Fovea – region of the retina where the retinal cells are densest
Conjunctiva membrane
Blind spot – region where the retina is not sensitive to light (no light-sensitive cells present)
Pupil – the central hole formed by the iris. Light enters the eye through the pupil
Optic nerve – carries nerve impulses from the retina to the brain
Ciliary muscles – attached to the lens by suspensory ligaments. The muscles change the thickness of the eye lens
Choroid – black layer of blood vessels that carry food and oxygen to the eye and remove waste products
Eye muscles – move the eye in the socket

Figure 14.5B ▲ The human eye

It's a fact!

The front of the eye is covered by a thin transparent membrane called the conjunctiva. Dust particles that collect on the conjunctiva are washed away by a watery fluid from the tear glands, which are under the eyelids. This fluid contains lyosyme – a chemical that destroys germs. Blinking helps to spread the fluid across the conjunctiva. When the fluid reaches the lower part of the eye, it drains into a tube and goes down into the nose.

Figure 14.5B explains how the parts of the eye work. Light enters the eye through a tough transparent layer called the **cornea**. This protects the eye and it helps to focus the light onto the **retina**, the layer of light-sensitive cells at the back of the inside of the eye. The amount of light entering the eye is controlled by the **iris**, which adjusts the size of the **pupil** – the circular opening in its centre. The **eye lens** focuses the light to give a sharp image on the retina. Although the image on the retina is inverted, the brain interprets it so you see it the right way up.

Focusing

How does the eye focus on objects at different distances? If you look up from this book and gaze out of the window, your eye lens becomes thinner to keep your vision in focus. This is called **accommodation**. Your eye muscles alter the thickness of your eye lens. The muscles fibres run round the eye lens, so when they contract they shorten and squeeze the eye lens, making it thicker.

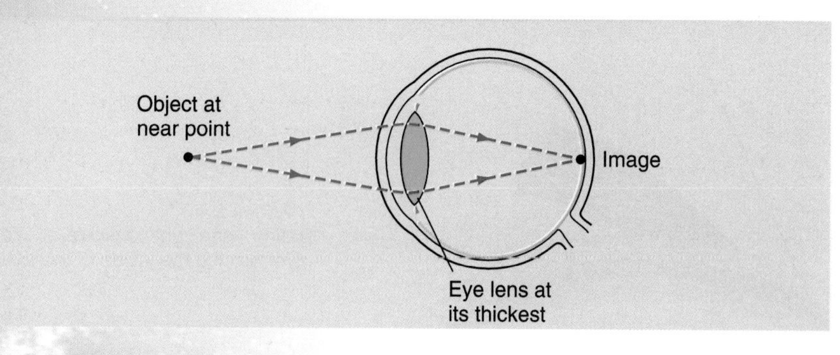

Figure 14.5C The near point

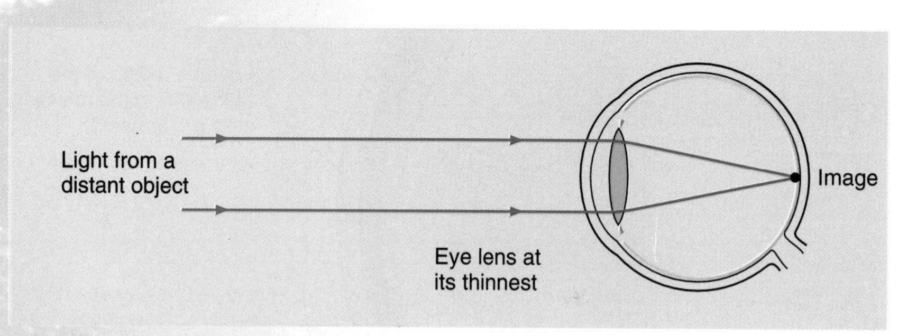

Figure 14.5D The far point

What is your range of vision? A normal eye can see clearly any object from far away to 25 cm from the eye.

- The **near point** of the eye is the closest point to the eye at which an object can be seen clearly. The eye lens is then at its thickest (Figure 14.5C).
- The **far point** of the eye is the furthest point from the eye at which an object can be seen clearly. The eye lens is then at its thinnest (Figure 14.5D).

Seeing shape and colour

How do we recognise the shape of an object? An image of the object is formed on the retina, which consists of lots of light-sensitive cells. When light falls on a cell, the cell sends an electrical impulse as a signal to the brain. The brain recognises the pattern of the signals from the cells covered by the image, and so recognises the object's shape.

How do we tell the colour of an object? There are two types of cells on the retina – **rods** and **cones** (Figure 14.5E). Rods occur mostly near the edges of the retina. They are not sensitive to colour and only respond to the brightness of light. Ask a friend to test you to see if you can tell the colour of something at the edge of your field of vision.

Cones are packed densely together at the middle of the retina. This area is called the **fovea**. Each cone is sensitive to red or blue or green light. For example, when red light falls on the retina, it 'activates' the red-sensitive cones, so you see red. Other colours activate more than one type of cone. For example, yellow light activates the red and the green cones, so they send messages to the brain. When the brain receives signals from adjacent red and green cones, it knows yellow light is on that part of the retina.

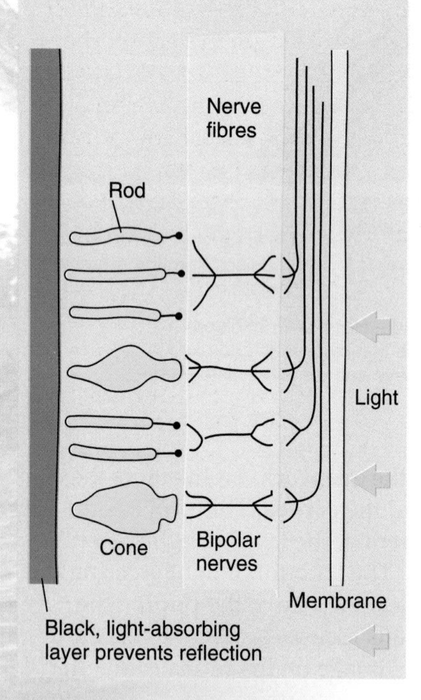

Figure 14.5E Rods and cones

Cones are colour-sensitive.
Rods are not.

It's a fact!

There are more rods than cones. Reliable estimates are about 130 million rods and 7 million cones in a pair of human eyes.

It's a fact!

Some scientists think that reading in poor light harms your eyes. Poor light causes the retinal cells to expand a little, effectively making the retina a bit closer to the eye lens, so the lens needs to be stronger to see near objects clearly. If the eye lens can not become strong enough, the objects become blurred. Make sure you don't read in poor light!

Sight defects

Sight defects occur when the eye lens cannot form a sharp image on the retina. Spectacles contain lenses that compensate for sight defects.

Short sight is caused by over-strong eye muscles. A short-sighted eye cannot see far away objects clearly because the eye muscle cannot relax enough to make the eye lens thin enough. A suitable concave lens in front of the eye counteracts the effects of the over-strong eye lens (Figure 14.5F).

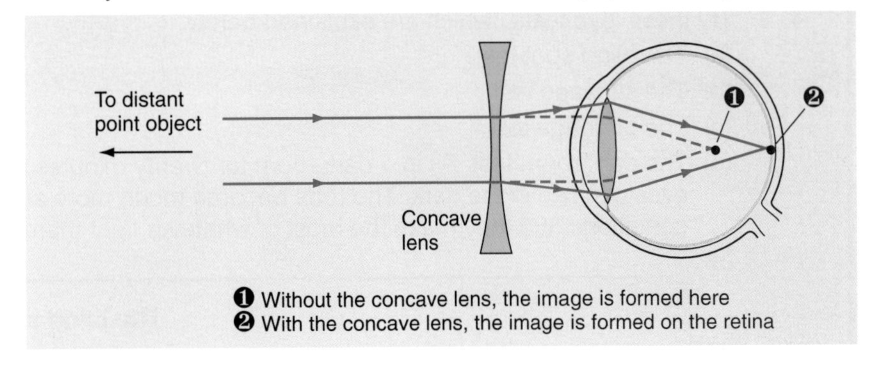
To distant point object
Concave lens
❶
❷
❶ Without the concave lens, the image is formed here
❷ With the concave lens, the image is formed on the retina

Figure 14.5F ▲ Short sight

A short-sighted person is unable to see distant objects.

A long-sighted person is unable to read a book without the aid of spectacles.

Long sight is caused by weak eye muscles, and often develops as the muscles weaken with age. The muscles are unable to contract enough around the lens to make it thick enough to focus near objects. A suitable convex lens in front of the eye helps the eye lens to form a clear image on the retina (Figure 14.5G).

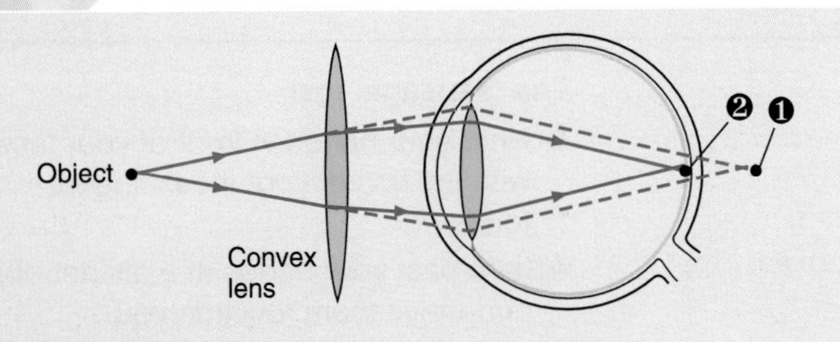
Object
Convex lens
❷ ❶
❶ Without the convex lens, the image is formed here
❷ With the convex lens, the image is formed on the retina

Figure 14.5G ▲ Long sight

The **power** of a lens is defined as 1/its focal length in metres. The unit of lens power is the dioptre (D). The power is positive for a convex lens and negative for a concave lens. For example, a concave lens of focal length 0.5 m has a power of -2.0 D ($= 1/0.5$ m).

It's a fact!

Contact lenses are a popular alternative to spectacles. The photograph shows a man fitting a contact lens to his eye.

SUMMARY

The eye lens forms a clear image on the retina. The iris controls the amount of light entering the eye. The thickness of the eye lens changes to accommodate (focus) objects at different distances. Spectacle lenses compensate for defects in the eye muscles controlling the eye lens, to give the wearer normal vision.

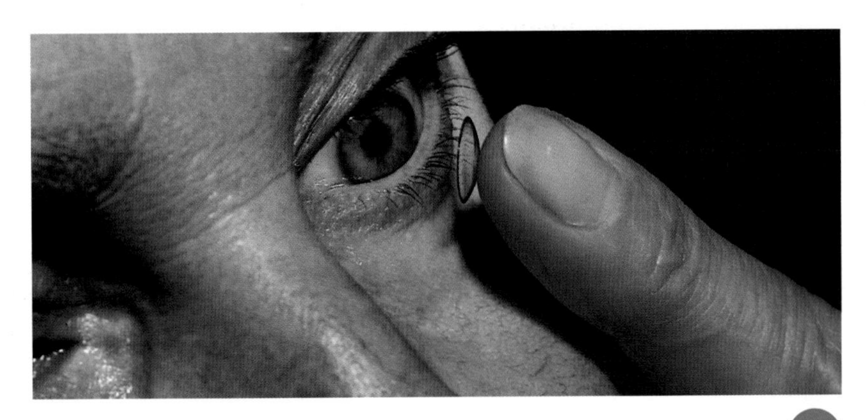

CHECKPOINT

1 (a) Why do your eye muscles relax when you look at a distant object and contract when you look at something close to you?

(b) Why is it easier to study an object in detail if you look straight at it?

(c) Why does the eye pupil dilate (i.e. widen) in dim light?

2 Explain how (a) short sight and (b) long sight are caused and how they are corrected.

3 Try these 'eye tests', which are explained below.

(a) The blind spot test.

(b) The sausage test.

(c) The birdcage test.

(d) The dark room test. Sit in a dark room for twenty minutes or more and you will discover your eyes can see in the dark. The rods become much more sensitive than normal in dark conditions and they make the most of whatever light there is.

(a)

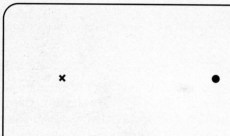

The blind spot test

❶ Position the black spot in front of your left eye. Cover your right eye

❷ Move the book closer and keep staring at the spot

❸ The X disappears when its image falls on the blind spot of your left eye

(b)

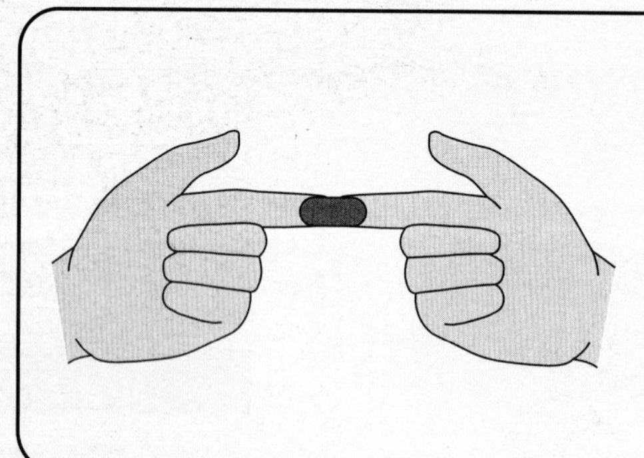

The 'sausage' test

❶ Hold your hands in front of your face with the tips of your index fingers touching

❷Stare past your hands at a distant object and move them towards you

❸ You should see a 'sausage' between the tips of your index fingers caused by overlapping images from each eye

(c)

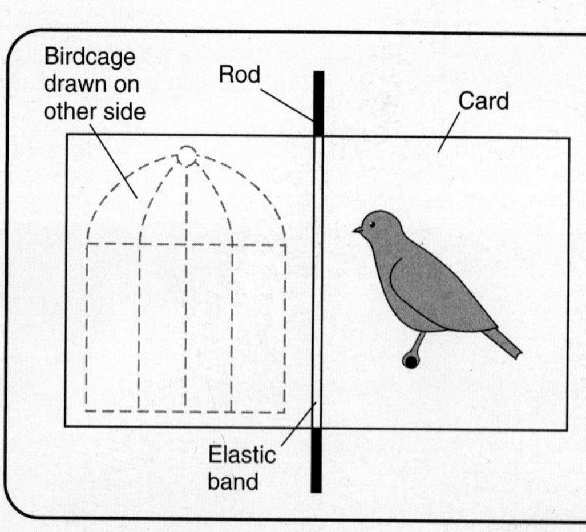

The birdcage test

❶ Make the test card as shown

❷ Spin the card about the rod

❸ The bird appears in the cage because the images last for a fraction of a second and hence overlap. This is called 'persistence of vision'

❹ Try spinning the card at different speeds. What difference does it make?

Topic 15 Electromagnetic waves

15.1 The spectrum

Figure 15.1A ▲ Splitting white light

FIRST THOUGHTS

In this section you will find out what determines the colour of a surface. Objects on a stage change their appearance when the stage lights change colour. Read on to find out why.

Can you recall the colours of the rainbow? You can make these colours by shining a ray of white light through a prism. White light is made up of a mixture of colours. The prism splits white light into these colours because it refracts each colour by a different amount. Splitting white light into colours in this way is called **dispersion**. Seen on a white screen, the colours form a pattern called the **white light spectrum**, as shown in Figure 15.1A. Dispersion occurs because the speed of light in glass depends on the colour of the light. Blue light travels slower in glass than red light so it bends more than red light does when it passes through a prism.

Investigating colour

You can use colour filters to obtain different colours of light. For example, if you direct a beam of white light at a red filter, the beam that emerges is red (Figure 15.1B). This is because the filter absorbs all the other colours that make up white light

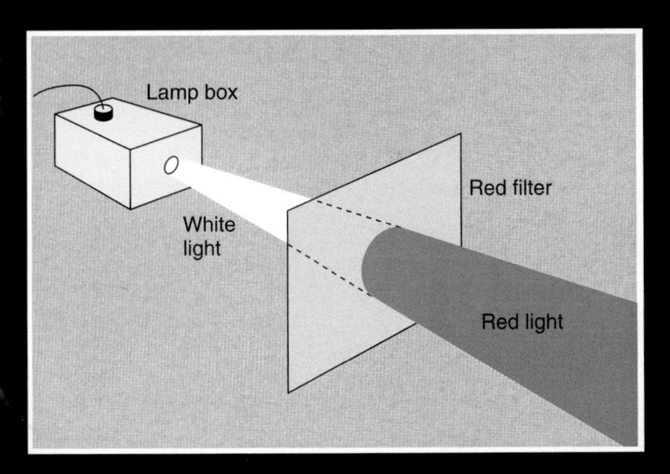

Figure 15.1B ▲ Investigating colour

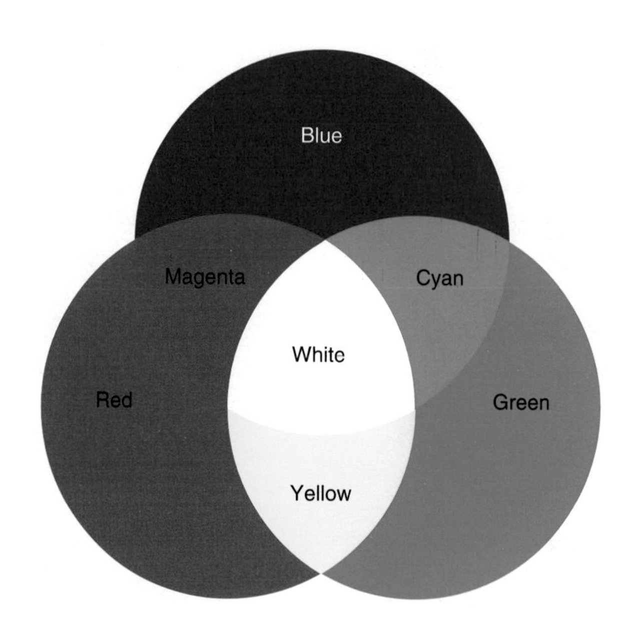

Figure 15.1C ▲ Overlapping colours

White light contains all the colours of the spectrum.

Red, blue and green are called the **primary** colours of light because they can be mixed to produce any other colour of the spectrum. Figure 15.1C shows the colours produced when light of the three primary colours is shone on a white screen. White light is seen where all three primary colours overlap. Two primary colours overlapping give a **secondary** colour. For example, red and green light give yellow, so yellow is one of the three secondary colours. Blue light added to yellow light gives white light, because yellow is made up of red and green.

The colour of a surface depends on the colour of the light falling on it, as well as the surface pigments. A book that is red in normal white light has a surface that absorbs all the colours of white light except red. When white light falls on its surface, only the red component of the incident white light is reflected (Figure 15.1D(a)).

What colour does the book appear if it is viewed in blue light? Since the incident light does not contain any red component, the surface absorbs all the light falling on it and reflects none (Figure 15.1D(b)), so the book appears black.

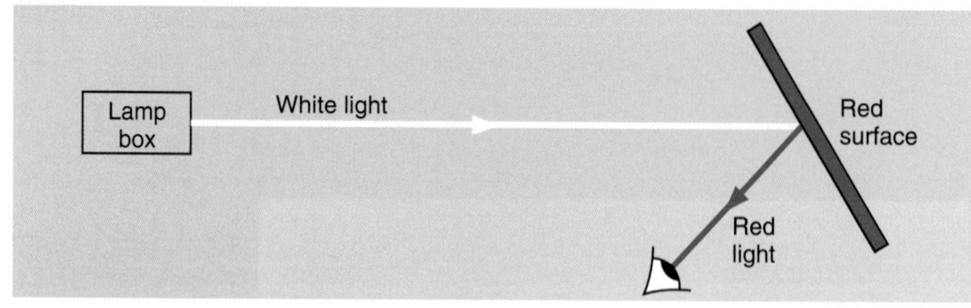

(a) Observing in white light (a red surface is seen)

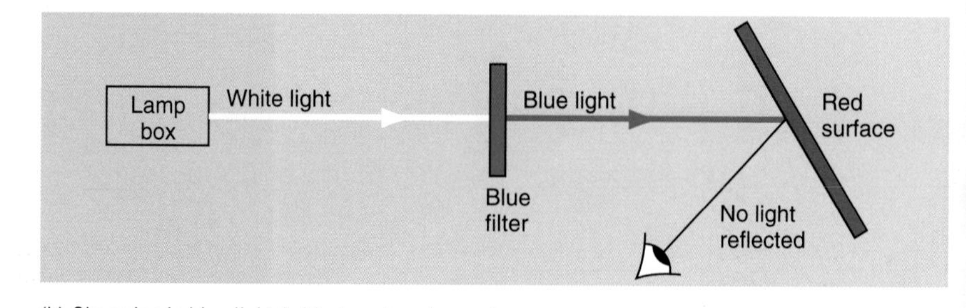

(b) Observing in blue light (a black surface is seen)

Figure 15.1D ▲ Observing in different coloured lights

SUMMARY

White light consists of all the colours of the spectrum. Any colour of the spectrum can be formed by mixing different proportions of red, blue and green light. These three colours are called the primary colours of light.

CHECKPOINT

1. In the school's drama production, the first scene is a sunny day outdoors. The stage lights are red, green, blue and white. Each colour is controlled separately.
   (a) Which light would you use for the first scene?
   (b) The scene then changes to a red sunset. How should the light be altered?
   (c) The leading actress wears a green dress in the first scene. What colour will it appear in the second scene?
2. Explain why a blue book appears black in red light. What colour would it appear in green light?
3. Top Gear Clothes Ltd want to advertise their name in discos so they have asked your school for some ideas. One suggestion is that the poster should have a red background with the words TOP GEAR in blue. How will this look in a disco where the lights flash red and blue alternately? Have you a better suggestion? If so, try it out.
4. Suppose that in the experiment shown in Figure 14.1A the white light beam was shone through a beaker of chlorophyll before passing into the prism. Describe and explain what you would expect to see on the screen.

15.2 ▶ The electromagnetic spectrum

FIRST THOUGHTS

Our eyes are sensitive to only a small part of the electromagnetic spectrum. This section explains the properties and uses of other electromagnetic waves.

Light is just a small part of the spectrum of electromagnetic waves. Our eyes cannot detect the other parts. The world would appear very different to us if they could. The wavelengths of electromagnetic waves range from more than 1000 m (radio) to less than 10^{-12} m (X and γ (gamma) radiation).

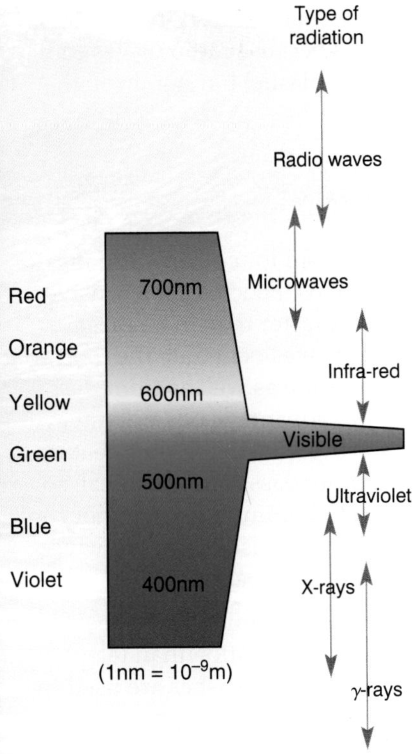

Table 15.1 ▼ The electromagnetic spectrum

Wavelength (m)	Sources	Detectors	Applications
10^3	Radio TV Microwave transmitters	Receivers fitted with aerials	Long wave radio Medium wave radio VHF and CB radio TV (called UHF) Microwave signals
10^0	Microwave ovens		Radar, microwave cookers
	Hot objects	Blackened thermometer bulb	Infra-red cookers Infra-red heaters
10^{-3}		Phototransistor Special film	Infra-red photography
10^{-6}	Glowing objects	The eye Film	Optical instruments Photography
	UV lamps The Sun	Film Fluorescent chemicals	UV sun-tan lamps UV ink driers
10^{-9}			
	X-ray tubes	Film Geiger–Müller tube	X-radiography
10^{-12}	Radioactive isotopes	Film Geiger–Müller tube	Crack detection X-and γ-ray therapy
			Radioactive tracers Sterilising equipment
10^{-15}			

Type of radiation
Radio waves
Microwaves
Infra-red
Visible
Ultraviolet
X-rays
γ-rays
Red
Orange
Yellow
Green
Blue
Violet
700nm
600nm
500nm
400nm
(1nm = 10^{-9}m)

All electromagnetic waves travel at the same speed through a vacuum. However, the behaviour of electromagnetic waves in different materials depends on the wavelength and the type of material.

Radio waves

Radio waves are emitted when electrons are forced to move up and down an aerial by a specially designed electronic circuit called a radio transmitter connected to it. Lightning also produces radio waves, which is why radio programmes 'crackle' when there is a thunderstorm.

Figure 15.2A ▲ Local radio transmission

When radio waves pass across a wire aerial, the electrons in the wire are pushed to and fro along the wire by the radio waves. The motion of the electrons produces a tiny electrical voltage, which can be used to detect the radio waves (Figure 15.2A).

Portable radios have plastic cases with the aerial inside, but a car radio needs an aerial fitted to the car body. This is because radio waves can pass through non-conducting materials (e.g. plastic) but not through electrical conductors (e.g. metal).

Figure 15.2B ▲ Concave microwave receiving dish

Microwaves

Microwaves are used for cooking and in telecommunications. Food in a microwave oven cooks much faster than in a normal oven. The microwaves penetrate the food so it cooks from the inside as well as from the outside. Another advantage of a microwave oven is that the oven itself does not heat up. Microwaves cannot pass through metal, so food wrapped in aluminum foil will not cook in a microwave oven. Also the microwaves make the foil produce sparks.

In telecommunications, microwave beams can carry much more information than telephone wires can. Huge concave dishes are used to send and receive microwave signals (Figure 15.2B).

Infra-red radiation

All objects emit infra-red radiation. The hotter an object is, the more energy it emits. Your body emits infra-red radiation. In medicine, infra-red scanners are used to detect 'hot spots' under the body's surface. Figure 15.2C shows an infra-red image of a man's head. Hot spots can often mean that the underlying tissue is unhealthy.

Infra-red radiation is just beyond the red part of the visible spectrum. A thermometer with a blackened bulb absorbs infra-red radiation. If the bulb is placed just beyond the red part of a visible spectrum in an experiment, the thermometer reading rises, because the bulb is absorbing infra-red energy.

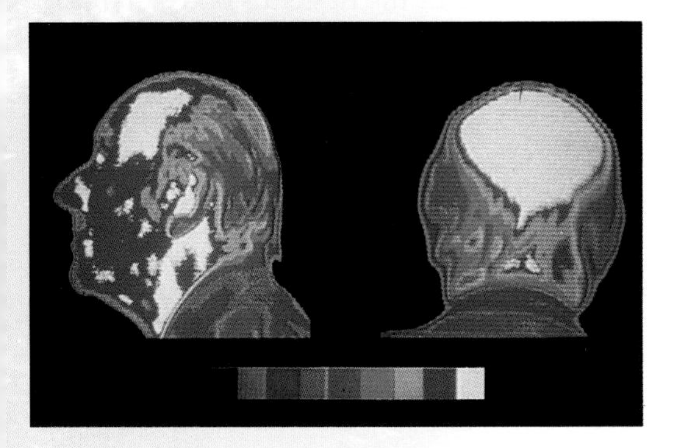

Figure 15.2C ▲ Infra-red imaging

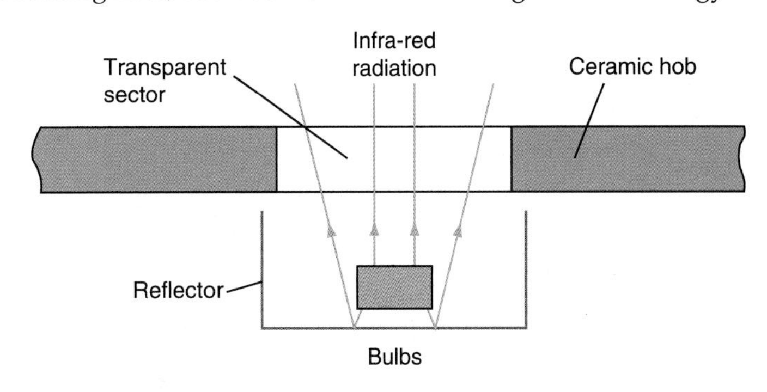

Figure 15.2D ▲ A ceramic hob

Remember the wavelength range of each part of the electromagnetic spectrum.

Cookers with ceramic hobs use special halogen lamp bulbs that emit mainly infra-red radiation. The radiation passes through the transparent part of the hob to heat up a saucepan placed on it (Figure 15.2D). A ceramic cooker ring has four 500 W bulbs above a silvered reflector. Most of the radiation from the bulbs is then directed upwards through the transparent hob.

A ceramic hob cooker cools quickly when the ring is switched off because the bulbs cool down fast when they are switched off. Ordinary cooker rings stay hot for much longer because they contain much more material than halogen lamps.

Infra-red cameras and sensors are used for security purposes to detect the infra-red radiation emitted by people and animals in darkness. Infra-red beams are also used to send signals from remote control handsets to televisions and video recorders.

Ultraviolet radiation

Figure 15.2E ▲ Goggles protect the woman's eyes from UV radiation while she uses a sunbed

Ultraviolet (or UV) radiation is harmful to human eyes and can cause blindness. Solar radiation contains UV radiation as well as light. UV radiation tans the skin, but does not heat it like infra-red radiation does. Too much UV radiation causes sunburn. It damages skin tissue, which then becomes painful and may blister. It can also cause skin cancer. To prevent sunburn, sunbathers should use skin creams that block UV radiation, preventing it from reaching the skin. Sunbed users should never exceed the exposure times recommended by the manufacturer.

Ozone in the atmosphere absorbs much of the ultraviolet radiation from the Sun, but there is much concern about damage to the ozone layer caused by aerosol propellants and other chemicals. Some scientists think that this is why skin cancer has become more common in recent years.

Ultraviolet radiation makes certain chemicals glow. Such chemicals in washing powder make clothes 'whiter than white'. The disco lighting that makes white clothes glow uses ultraviolet radiation.

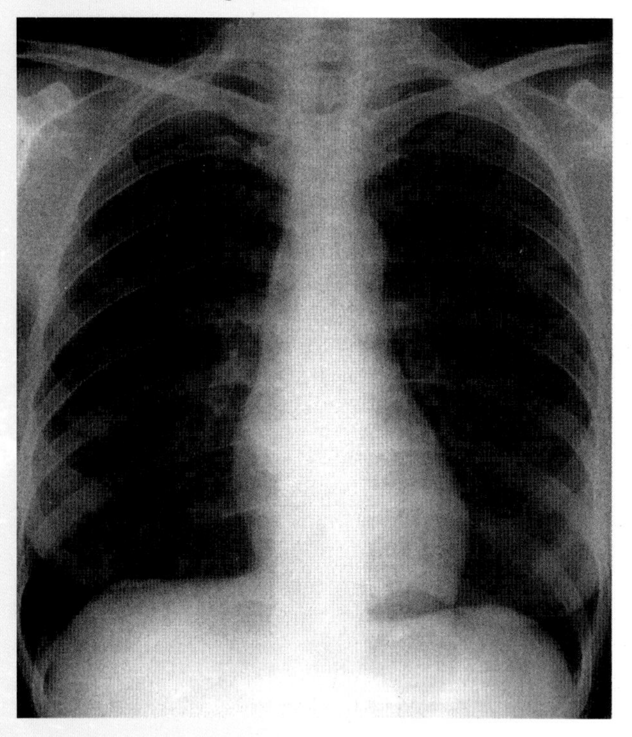

Figure 15.2F ▲ A chest X-ray

X-rays

X-radiation is used in hospitals to take X-ray pictures of limbs and organs (Figure 15.2F). X-rays pass through tissue but they are stopped by bones. To make a **radiograph** or X-ray picture, the radiographer places the patient between the X-ray tube and a photographic film in a cassette (Figure 15.2G). When the X-ray tube is switched on, X-rays pass through the patient's body and affect the film (Figure 15.2H). Too much X-radiation can induce cancer, so the radiographer and parts of the patient's body that do not need to be X-rayed are protected by thick lead screens, which stop the X-rays. X-rays and gamma rays can be used to destroy tumours inside the body.

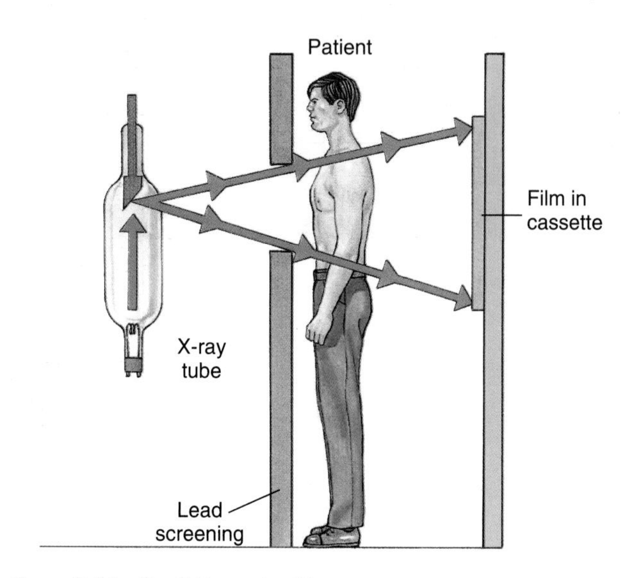
Patient
Film in cassette
X-ray tube
Lead screening

Figure 15.2G ▲ Taking a chest X-ray

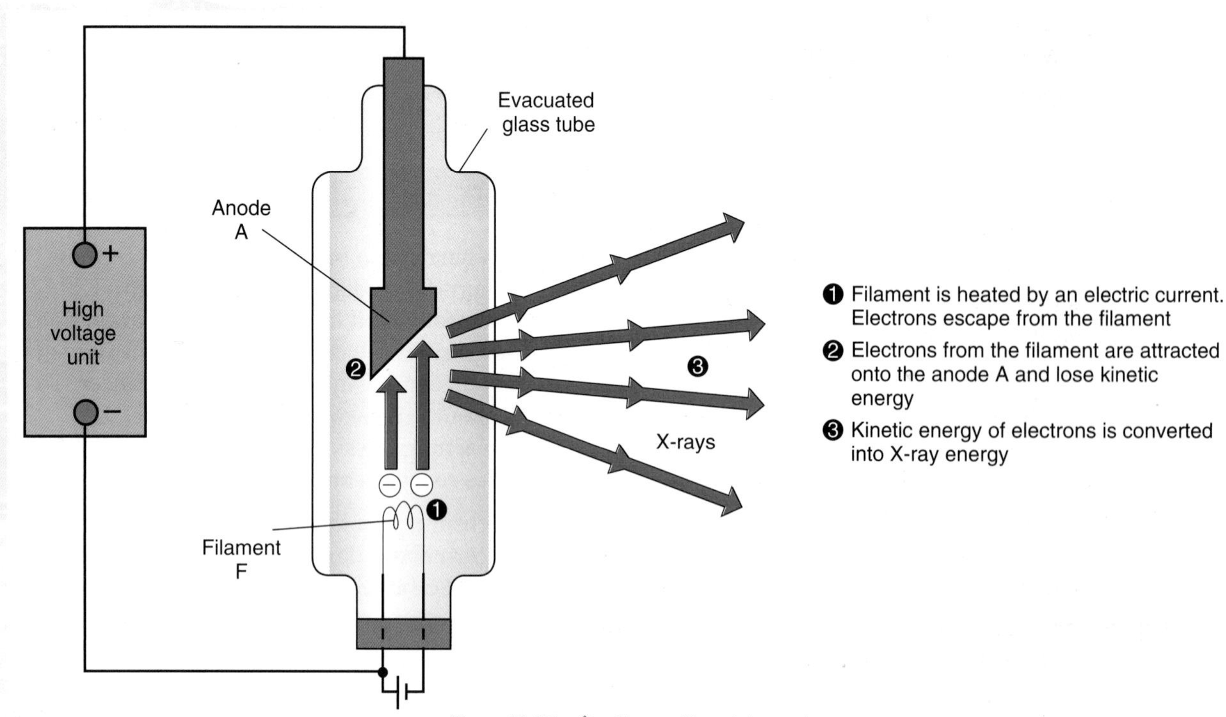
Evacuated glass tube
Anode A
High voltage unit
+
–
Filament F
X-rays
❶ Filament is heated by an electric current. Electrons escape from the filament
❷ Electrons from the filament are attracted onto the anode A and lose kinetic energy
❸ Kinetic energy of electrons is converted into X-ray energy

Figure 15.2H ▲ How an X-ray tube works

Electromagnetic waves and living cells

1. **Microwaves** penetrate tissues and are absorbed by the water content of living cells, causing internal heating which may damage or even kill the cell.
2. **Infra-red radiation** is absorbed by skin cells, causing internal heating which may damage skin tissue.
3. **Ultraviolet radiation** damages the cells of the retina of the eye. Special protective goggles must be worn by users of ultraviolet lamps. UV radiation also damages cells below the skin because it can penetrate more than infra-red radiation. Users of sunbeds must **not** exceed recommended exposure times or severe burns will result.
4. **X-rays and gamma radiation** create ions in substances they pass through. Other forms of ionizing radiation include alpha particles, beta particles and proton and neutron beams. Living cells are damaged by ionizing radiation. High doses kill living cells and low doses can cause cell mutation and cancerous growth. For this reason, a film badge as shown in Figure 9.4A must be carried by anyone using X-ray and gamma equipment or equipment that produces any other form of ionizing radiation. If the badge is overexposed, the wearer is not allowed to continue working with the equipment. The equipment should be shielded by means of thick lead plates (or very thick concrete walls in the case of a nuclear reactor) and operating personnel should be present for as little time as necessary and kept as far from the source as possible. There is no evidence for a lower limit below which living cells would not be damaged. The maximum permissible dose limit is a legal limit decided on the basis of acceptable risk.

SUMMARY

Light is part of the spectrum of electromagnetic waves. All electromagnetic waves travel at the same speed through a vacuum but their properties in materials depend on their wavelength and the type of material.

CHECKPOINT

▸ 1 (a) Microwave ovens should not be used to cook food wrapped in aluminium foil. Why?
(b) Why is it important that microwave ovens do not 'leak' microwaves when in use?
(c) A microwave oven operates at a frequency of 2500 MHz. Work out the wavelength of the microwaves. (The wavespeed in air is 3×10^8 m/s.)

▸ 2 The manager of a local cotton mill intends to use a 25 kW microwave oven to replace a 5 kW oven to dry large cotton reels at the end of the manufacturing process. This will cut the drying time from 10 hours to half an hour and so reduce the cost.
(a) How many kilowatt-hours of electricity does the 5 kW oven use to dry each batch?
(b) How many kilowatt-hours of electricity does the microwave oven use to dry each batch?
(c) How many kilowatt-hours per batch will be saved by using this process?

▸ 3 High power ultraviolet lamps can be used to dry printing ink very quickly. Special UV sensitive ink is used so the ink dries without heating the paper.
(a) Test ordinary ink to see how long it takes to dry at room temperature.
(b) A continuous printing press delivers pages at a rate of 10 per second. Why is a UV drier better than drying the print using radiant heaters?

▸ 4 Ceramic hobs for cooking use special bulbs.
(a) Why is a reflector fitted under each bulb?
(b) Why is the hob safer than an ordinary electric ring or a gas ring?
(c) Why is the hob easier to clean than an ordinary electric ring or a gas ring?

▸ 5 Certain pens contain ink that is invisible in ordinary light but shows up in ultraviolet radiation. Suggest some uses for such a pen.

15.3 Radio and television communications

FIRST THOUGHTS

Are you familiar with technical terms such as UHF waves and DBS systems? Read on to find out what they mean.

Where would you be without your favourite radio and television programmes? By using **satellites**, we can watch events on the opposite side of the Earth as they happen. Yet radio broadcasting only began in the 1920s and television was not widely available until the 1960s.

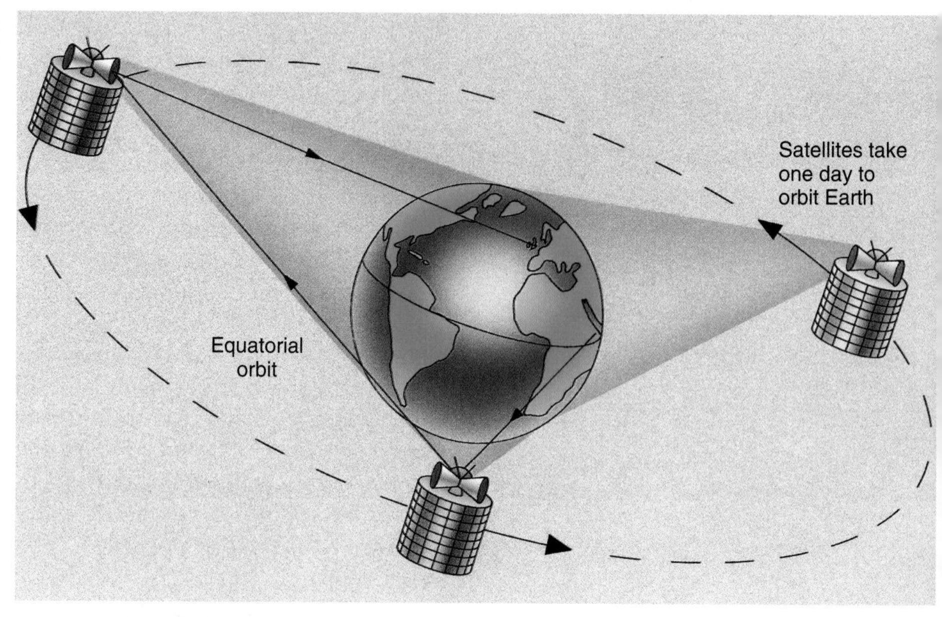

Figure 15.3A ▲ Global links

Sending information using radio waves

Morse code is the simplest way to send a message by radio. The operator uses a tapping key to switch the radio transmitter on and off repeatedly. Each burst of radio energy is made either short or long, corresponding to the dots and dashes of Morse code (Figure 14.3B).

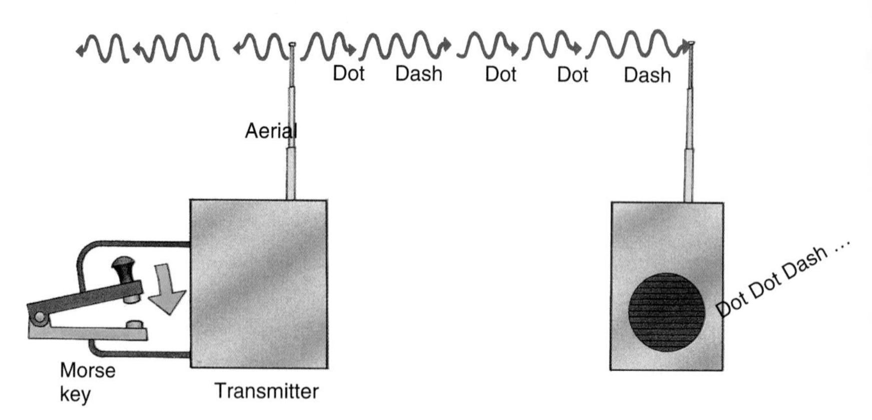

Figure 15.3B ▲ Morse code

- **Amplitude modulation** (AM) is one way to send speech and music over the radio. A microphone and amplifier are used to convert the sound waves into electrical signals called **audio waves**. These signals are used to **modulate** or vary the amplitude of radio waves, as shown in Figure 15.3C. In this way, the radio waves 'carry' the audio waves, and are sometimes called **carrier waves**.

It's a fact!

Radio caught the public interest in 1910 when a murderer called Dr Crippen slipped through the police net in Britain by catching a ship to Canada. The police sent radio signals across the Atlantic Ocean and Crippen was apprehended on his arrival in Quebec. He was sent back to Britain where he was convicted and executed.

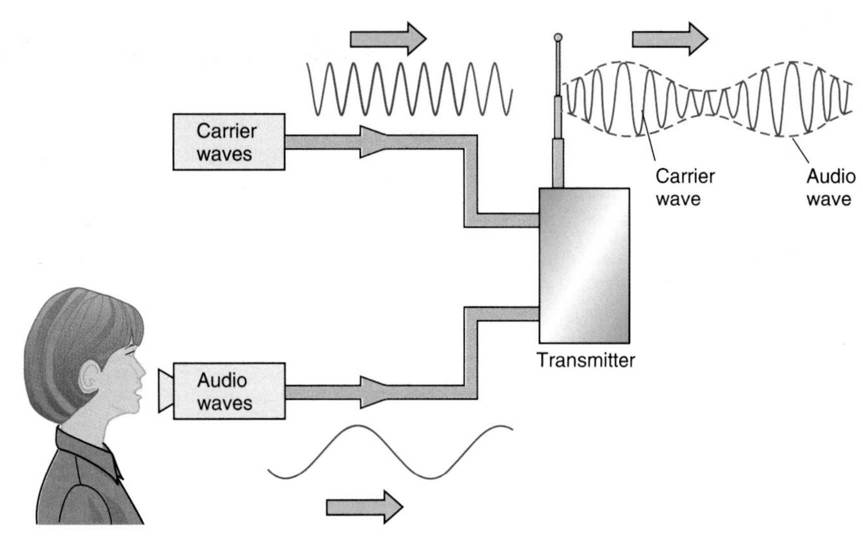

Figure 15.3C ▲ Amplitude modulation

Radio waves from a transmitter become weaker as they spread out. This decrease of the wave amplitude the further it travels is known as **attenuation**.

The radio waves are picked up by a receiver and the carrier waves are filtered out. The audio waves are then amplified and converted back to sound, using speakers.

The **bandwidth** of a signal is the frequency range covered by the signal. For example, a radio signal with a carrier wave of frequency 1000 kHz carrying an audio frequency range of 4 kHz could occupy a frequency channel from 1000 to 1004 kHz. The bandwidth of the channel is thus 4 kHz. Electrical **noise** is generated by components in radio and TV circuits. This can be heard as a hissing sound on a radio which is not tuned in to a radio station. Weak AM signals can be masked by noise.

FM needs to be at a higher frequency than AM because the bandwidth needed for FM is much greater.

- **Frequency modulation** (FM) is used to broadcast high-quality music programmes. Noise is reduced considerably using FM. The carrier frequency is changed or modulated as a result of changes of the amplitude of the audio signal. Figure 15.3D shows the idea. Noise does not affect the carrier frequency which is why FM broadcasts are of much higher quality than AM broadcasts. However, the bandwidth is much broader than for AM signals.

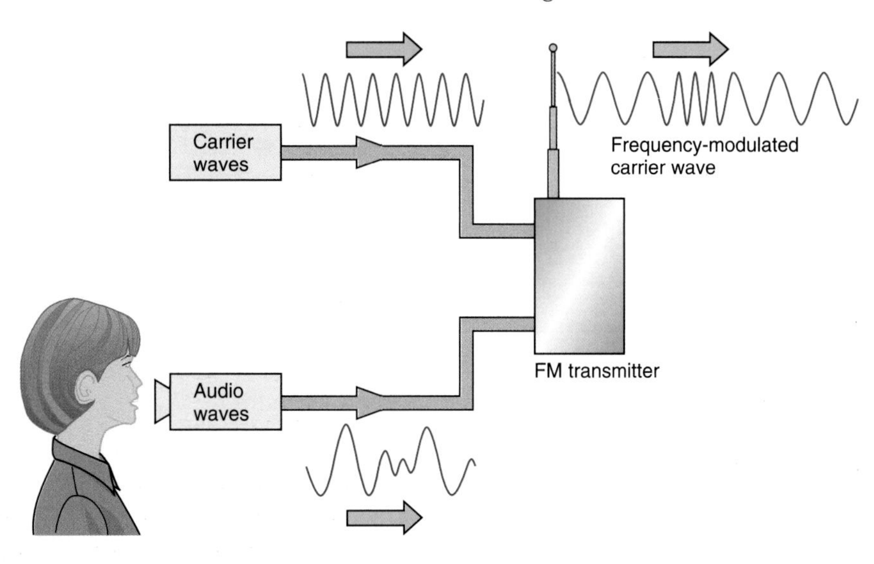

Figure 15.3D ▲ Frequency modulation

Noise is eliminated in digital transmission

- **Digital transmission** is used to send telephone calls and computer data along cables and to send television signals and telephone calls to and from satellites along microwave beams. A digital signal is a series of **bits**, which are 1's and 0's corresponding to the carrier wave being switched off (for a 0) or on (for a 1). Audio and video signals need to be converted from analogue to digital form for digital transmission, then reconverted at the receiver. Computer systems consist of digital circuits so no converter circuits are needed for transmission. Figure 15.3E shows an outline of how a digital audio link works.

It's a fact!

Mobile phones transmit and receive microwave radiation. When you make a mobile phone call, the handset emits a microwave signal which is detected by a receiver 'mast' serving the area. The receiver is linked by a 'landline' cable to a telephone exchange so your call is then routed to its destination. Your local mast emits a microwave signal to you which carries the audio signal from the phone at the other end. WAP mobile phones provide direct access to an Internet Service Provider. Mobile phone systems using digital transmission are replacing analogue systems. Is the radiation from a mobile phone hazardous? At the present time, there is no conclusive scientific evidence one way or the other. See Extension File activity 'You and Your Mobile Phone'. Advice from the UK Government recommends caution, particularly for young people, until more is known.

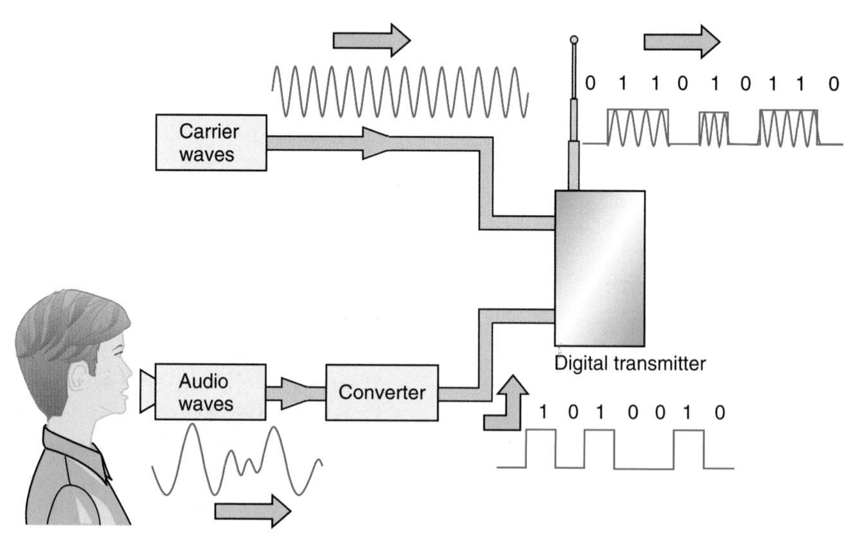

Figure 15.3E ▲ Digital transmission

Noise is eliminated in a digital link because the signals are 'cleaned' by regenerators. Each pulse entering a regenerator switches its output signal on then off so producing a sharp output pulse, free of noise. Phone calls to distant countries are usually very clear because they are carried by digital signals along satellite links. See Figure 23.8G for more information.

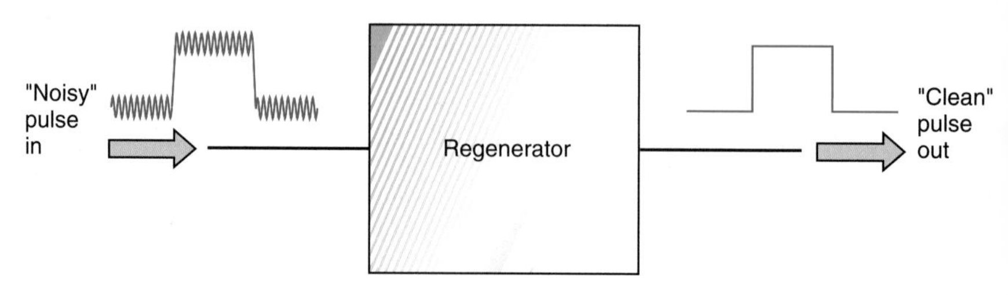

Figure 15.3F ▲ Cleaning a noisy pulse

Terrestial television signals carrying the national TV channels are broadcast from tall television masts. These signals are carried by ultra high frequency (UHF) radio waves. High frequency waves have a smaller wavelength than low frequency waves, so they can carry more information. UHF waves are used for television because much more information is needed for television pictures than for radio waves. The bandwidth for a TV channel needs to be 8 MHz which is much wider than the bandwidth needed for an AM radio programme.

Satellite television programmes are carried by microwaves. These are electromagnetic waves with a much higher frequency than UHF radio waves. Microwaves are not used for terrestial broadcasting because microwaves become much weaker as they spread out, in comparison

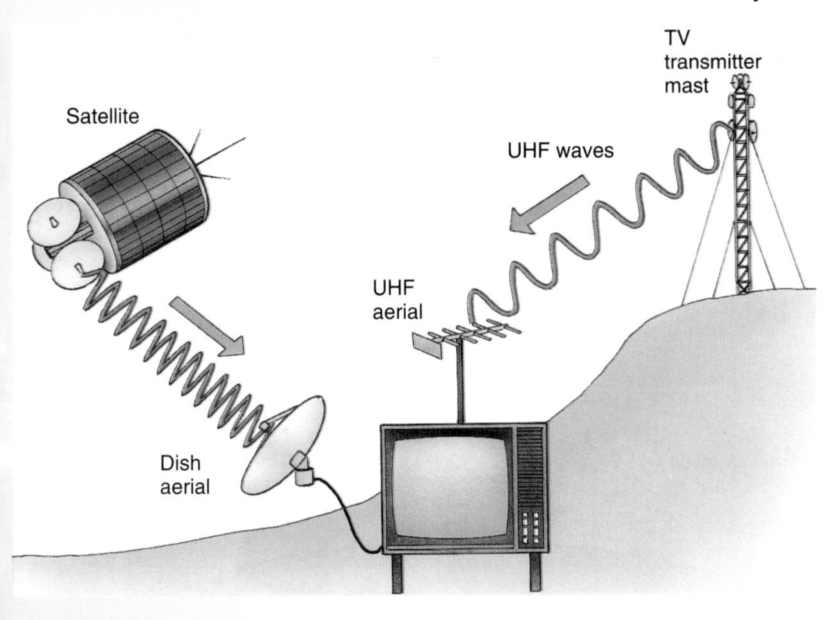

Figure 15.3G ▲ Television signals

with UHF waves. Most satellite broadcasts are now carried by digital signals because digital signals are less affected by noise and therefore are easier to detect than analogue signals.

Each satellite receiver must be connected to a dish aerial, which is necessary to collect sufficient power from the microwaves. (See Figures 15.4G and H.) The dish must point directly towards the satellite which is always above the equator in a geosynchronous orbit. The size of the dish is important. If it is too small, it does not focus sufficient microwave power on the aerial to give a satisfactory picture; if it is too large, it is difficult to align because diffraction by the dish is very small.

Digital terrestial TV broadcasting carried by UHF radio waves is likely to be introduced within the next few years. At present, TV channels must be separated by unused channels otherwise they interfere with each other. This happens with analogue signals in adjacent channels but not with digital signals. Switching to digital transmission would release about 10 channels for every channel presently in use. Every TV would need to be fitted with a decoder.

(a) A DB satellite

(b) A DB control room

Figure 15.3H ▲ (c) A DB dish aerial

It's a fact!

Summertime is the best time to tune in to distant radio stations. Solar radiation is more effective at producing ions in summer than in winter. Hence the ionosphere is a better reflector of radio waves in summer than in winter.

The ionosphere

Solar radiation creates ions from gas atoms in the Earth's upper atmosphere. These ions become trapped by the Earth's magnetic field in a layer about 150 km above the ground. This layer, known as the **ionosphere**, will reflect radio waves of frequencies up to about 30 MHz. This is because the presence of free ions makes the ionosphere a conducting layer. It acts like a metal plate and reflects incident radio waves of frequencies no more than about 30 MHz.

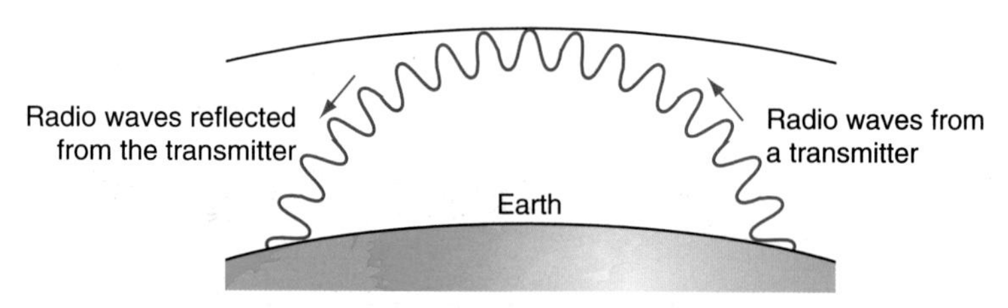

Figure 15.3l ▲ The ionosphere

Radio waves that are reflected from the ionosphere back to the ground are called sky waves. The reason you can listen to medium wave (MW) and high frequency (HF) radio programmes from other countries is that the carrier waves from the transmitter spread out and are reflected off the ionosphere to reach the ground far from the transmitter.

Microwaves and UHF waves pass through the ionosphere. This is why microwave beams can pass between the ground and satellites in orbit about the Earth. UHF waves used to carry TV programmes pass through the ionosphere too but they are not used for satellite-based communication links as microwaves carry much more information than UHF waves. At frequencies more than about 1000 GHz, electromagnetic waves are absorbed by uncharged gas molecules in the atmosphere. The **radio window** in the atmosphere stretches from frequencies of about 30 MHz to about 1000 GHz. This is why astronomers can use huge radio telescopes to map radio sources in the galaxy and beyond.

SUMMARY

Radio waves carry radio and television programmes. The carrier frequency for television broadcasting needs to be much higher than for radio broadcasts. Microwaves are used to carry satellite television channels.

CHECKPOINT

1. Electromagnetic waves travel through the atmosphere at 300 000 km/s, the same speed as through space. How long would a radio signal take to travel:
   (a) across the Atlantic Ocean, a distance of 5000 km,
   (b) to the Moon and back (Earth to Moon = 380000 km),
   (c) to the nearest star, Proxima Centauri, which is 4.2 light years away?
2. Radio waves at frequencies up to about 30 MHz are reflected from the ionosphere, whereas higher frequencies pass out into space. Use this information to explain why people in Britain can listen to foreign radio programmes but cannot watch foreign television stations.
3. Why is it important that each radio or television station is allocated a frequency (or 'channel') to broadcast on, which no other station can use?
4. Direct broadcasting satellites must travel round the Earth over the equator, taking 24 hours for each orbit.
   (a) A DB satellite in such an orbit always remains over the same point on the Earth. Why?
   (b) Why is a dish needed to receive DBS signals?
   (c) Why must the dish always be pointed in a certain direction to receive the signals?

15.4 Photons

FIRST THOUGHTS

Imagine taking a photograph using a camera with a shutter that opens very very briefly. The photo would show an image composed of a huge number of dots. Why?

Sir Isaac Newton was a famous 17th century scientist who made many discoveries, including the fact that white light consists of a continuous spectrum of colours. He imagined that light is a stream of tiny particles which he called *corpuscles*. In 1805, Thomas Young demonstrated **interference** of light (see Topic 11.5) and concluded that light is a *waveform*. By the end of the 19th century, the wave theory of light was in question because scientists were unable to explain why light above a certain frequency was capable of causing electrons to be emitted from metals. Fortunately, an imaginative young scientist called Albert Einstein turned his mind to the problem of light.

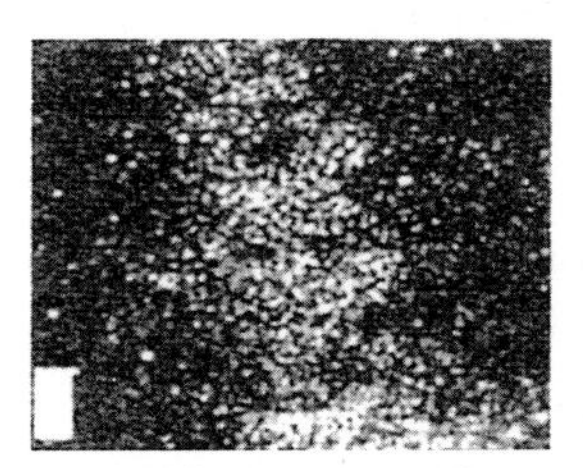

Figure 15.4A ▶ Photons at work. This sequence of photographs of a girl's face shows that photography is a photon process. The probabilistic nature of the process is evident from the first photograph, in which the number of photons is very small. As the number of photons increases the photograph becomes more and more distinct until the optimum exposure is reached.

Einstein put forward a revolutionary new theory, based on the idea that light consists of *wavepackets* which he called **photons**. When an atom emits light, Einstein imagined the atom producing a short 'burst' of electromagnetic waves in one direction. More importantly, he put forward an equation based on assuming that the energy of a photon is directly proportional to the frequency of the electromagnetic waves.

$$\text{Energy} = \text{constant} \times \text{frequency}$$

The constant is known as the Planck constant after a scientist whose ideas helped Einstein to develop the photon theory.

Figure 15.4B ▲ Emitting photons

Identifying elements

Neon tubes used for advertising signs produce a bright red glow. In comparison, sodium lamps used for street lighting emit yellow light. Different chemical elements, if heated strongly enough, emit light of different colours. Figure 15.1A shows how to produce a spectrum from a beam of white light produced by a filament lamp. If the lamp is replaced by a light source such as a mercury discharge tube or a cadmium lamp, a spectrum of coloured lines is seen rather than a continuous spectrum. The pattern of lines, called a line spectrum, is unique to the chemical element that produces the light in the light source and may be used to identify the element (Figure 15.4C).

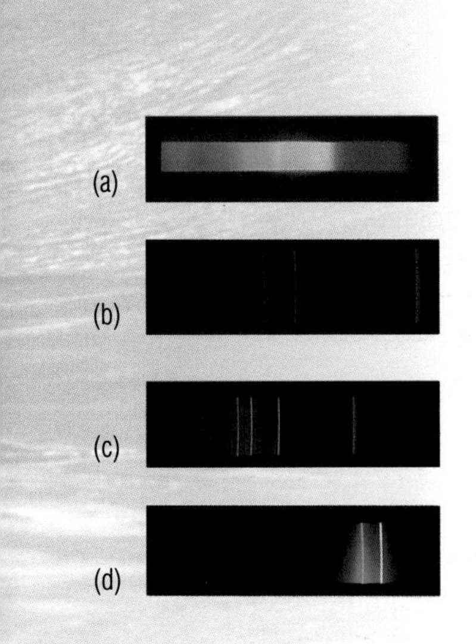

Figure 15.4C ▲ Spectra of (a) a filament lamp, (b) hydrogen, (c) cadmium and (d) mercury

Inside the atom

Electrons in an atom orbit the nucleus along certain paths only. An electron in orbit round the nucleus of an atom possesses constant energy. The closer an orbit is to the nucleus, the lower the energy of an electron in that orbit.

If two atoms of a gaseous element are in violent collision, an electron may gain sufficient energy from the collision to jump to a higher energy orbit. This will leave a vacancy for an electron in the lower orbit. An electron from a higher orbit will drop into this vacancy, emitting a photon in the process. This is how an atom emits light.

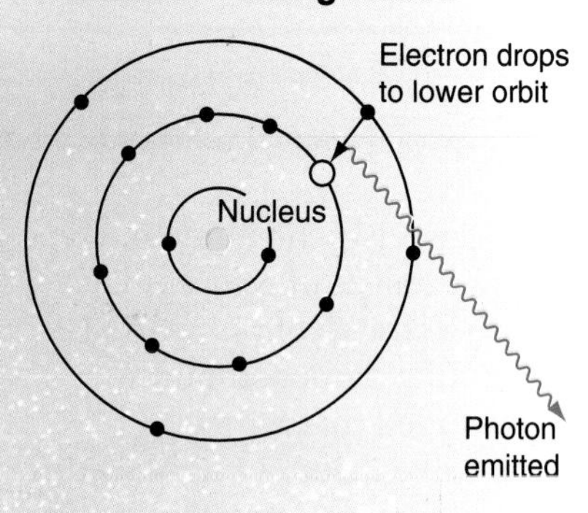

Figure 15.4D An electron transition

It's a fact!

Red shift

A star moving away from Earth produces a line spectrum in which all the lines are shifted by a constant amount towards the red part of the spectrum. This is known by astronomers as the red shift. It happens because the receding motion of the star makes all the wavelengths of emitted photons longer. Stars moving towards Earth produce light with a blue shift (See p.21).

SUMMARY

Electromagnetic radiation consists of photons. Each photon carries a certain amount of energy. Photons are created or absorbed when electrons move from one orbit to another.

The energy carried by a photon emitted in this way is equal to the change of energy of the electron that emitted the photon. Since each orbit has a certain energy, the photon emitted also has a certain energy and hence a certain frequency. Each coloured line in the spectrum of a chemical element is produced as a result of a vacancy in an atomic orbit being filled by an electron from a higher energy orbit. The pattern of lines in the spectrum of an element is characteristic of the element because the energy levels of the electrons are characteristic of the element.

Ionising radiation

If a sufficiently energetic photon hits an electron in an atom, the electron can absorb the photon and move to a higher energy orbit. This is a bit like a fairground coconut shy when a ball thrown at a coconut on a shelf hits its target – however you would be a little surprised if the coconut then jumped up to a higher shelf and the ball completely disappeared. That's what happens when a photon hits an orbital electron (Figure 15.4E).

The photon must carry sufficient energy to enable the electron to make the jump to a higher energy orbit. **Ionisation occurs if the electron is completely ejected from the atom by the impact**.

X-ray photons and gamma particles ionise atoms because they are high energy photons and so they have enough energy to knock electrons out of atoms. High-energy photons penetrate living tissues and create ions inside the cells which are then damaged or killed.

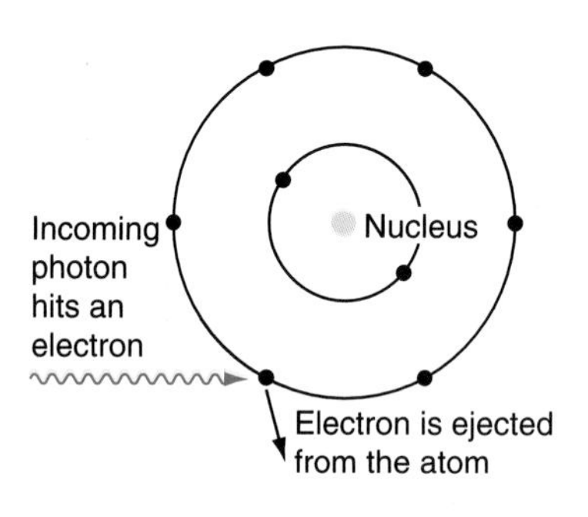

Figure 15.4E Ionisation

CHECKPOINT

1. Why is the photograph of a very faint object likely to be speckled?
2. Consider the following types of electromagnetic radiation: infra-red radiation, X-rays, blue light, microwaves.
   (a) Which has the longest wavelength?
   (b) Which has the highest frequency?
   (c) Which types cannot produce ionisation?
3. Alpha, beta and gamma radiation and X-rays are types of ionising radiation. Which of these are not electromagnetic waves?
4. Ultra-violet radiation is harmful to the eye. Explain this by comparing the energy of a light photon with that of an ultra-violet photon.
5. The lines of the light spectrum from a certain star are all shifted towards the red part of the spectrum. What does this tell you about the motion of the star relative to Earth?

Theme Questions

TOPIC 11

1 The figure below shows some water waves made by a wave machine in a swimming pool, seen from the side.

Waves from wave machine
x
w
B
A
z
y
Beach ball

(a) Which letter represents (i) the wavelength (ii) the amplitude?
(b) How many **complete** waves are shown in the diagram?
(c) How will the beach ball move as the waves reach it?
(d) If the wave moves from A to B in two seconds, what is its frequency?
(e) How many waves does the wave machine make in one minute?
(f) The figure below shows the position of the wave machine at the side of the pool.

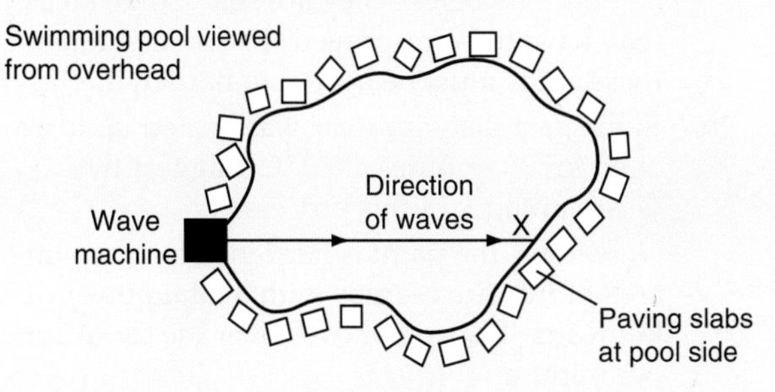

Draw a line on a copy of the diagram to show the direction in which the waves will move after they hit the side of the pool at X.

(SEG)

2 Instruments on a ship send out an ultrasonic sound wave and detect an echo from the sea bed exactly six seconds later. The water is 4500 metres deep.

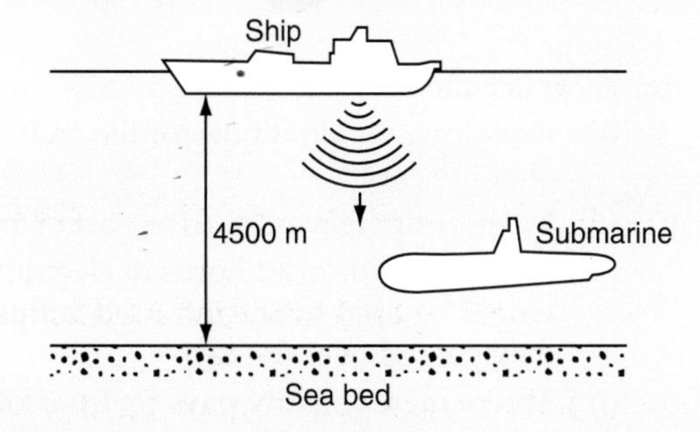

(a) What is meant by 'ultrasonic'?
(b) How far, in metres, has the sound travelled in 6 seconds?
(c) Calculate the speed of sound in water.
(d) An echo is detected after 4 seconds as the ship passes over a submarine. How deep is the submarine?
(e) Suggest why ultrasonic echoes are sometimes preferred to X-rays for showing structures inside the human body.

(MEG)

TOPIC 12

3 The bar chart below shows the loudness of various sounds.

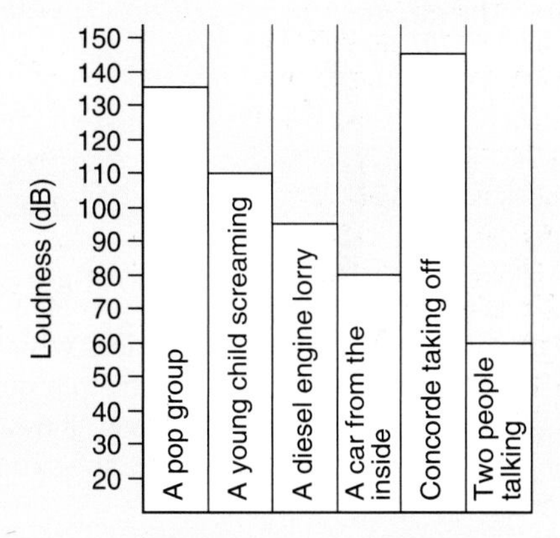

Sound can damage hearing and there are government regulations regarding the length of time that workers can be exposed to certain sounds.
Two examples are given below.
1 A loudness of 90 decibels for up to 8 hours per day.
2 A loudness of 100 decibels for up to 15 minutes per day.

(a) Use the bar chart to help you answer the following questions:
  (i) What is the maximum loudness of the pop group?
  (ii) Which **two** sounds would a worker not be allowed to work in for a quarter of an hour?
  (iii) Explain why it is likely that people who regularly attend pop concerts will become deaf.
(b) What can be done to protect workers from ear damage?
(c) A man was driving a car with a broken exhaust. He was breaking the law because he was guilty of noise pollution. Explain this statement.

(WJEC)

TOPIC 13

4 (a) With the aid of a labelled diagram, describe the structure of the Earth. On your diagram, you should label the crust, the mantle, the inner core and the outer core.

(b) Explain why secondary waves cannot travel through the Earth from the focus of an earthquake to a point on the surface directly opposite the focus.

(c) Describe one piece of evidence that was used by scientists to establish the theory of plate tectonics.

(d) Use the theory of plate tectonics to explain why earthquakes rarely occur in the United Kingdom.

TOPIC 14

5 (a) (i) The figure below shows a lamp bulb placed at the principal focus of a concave mirror. Draw on a copy of the diagram two rays of light which leave the lamp and reflect from the mirror.

(ii) Give **two** examples of devices which use concave reflectors for electromagnetic waves other than light. In each case, state which part of the electromagnetic spectrum is involved.

(b) The figure below shows a section of a rear reflector of a bicycle.

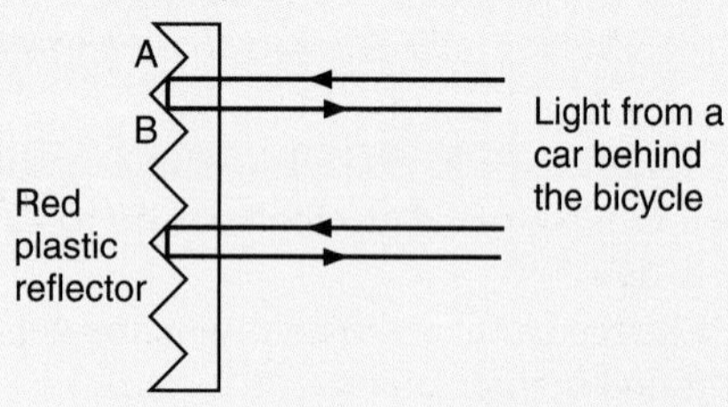

(i) What is the name of the effect at points A and B?

(ii) What are the **two** essential conditions for this effect to take place?

(iii) The figure below shows a section of an optical fibre which uses the above effect. State one use of optical fibres.

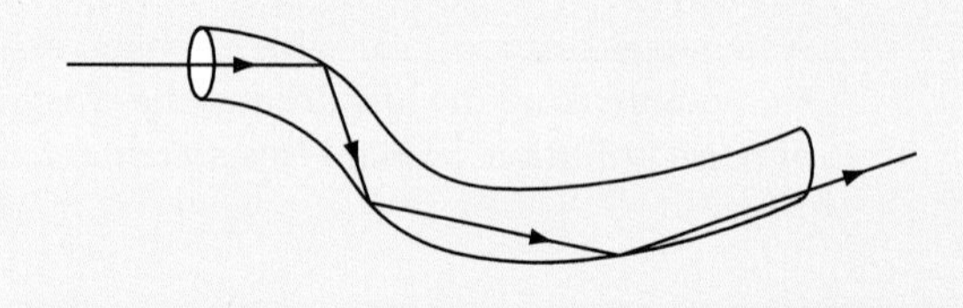

(NEAB)

6 A camera forms on the film a well-focused image of an object. The diagram shows how the camera's lens produces this image.

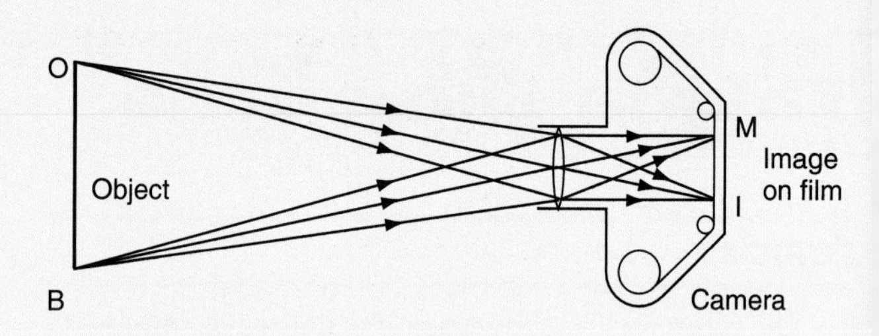

(a) Explain the following, referring to the diagram.

(i) The image which the camera produces is upside-down.

(ii) The size of the image remains unchanged if the size of the hole in front of the lens is decreased.

(b) The object moves further away from the camera. What adjustment will be needed to the lens to keep the image in sharp focus on the film?

(MEG)

TOPIC 15

7 Infra-red radiation forms part of the electromagnetic spectrum.

(a) Name **two** other types of electromagnetic radiation.

(b) The Acme Car Company checks the paintwork on new cars before they are sold. Some areas may have to be resprayed. The new paint on these areas must be heated to harden it.

In the past the whole car was heated up in an oven for 35 minutes at 80 °C. Suggest **two** disadvantages of this process.

Nowadays the paint is hardened by a moving arch of infra-red lamps with a total power of 72 kilowatts. The arch moves over the car at 1 metre per minute.

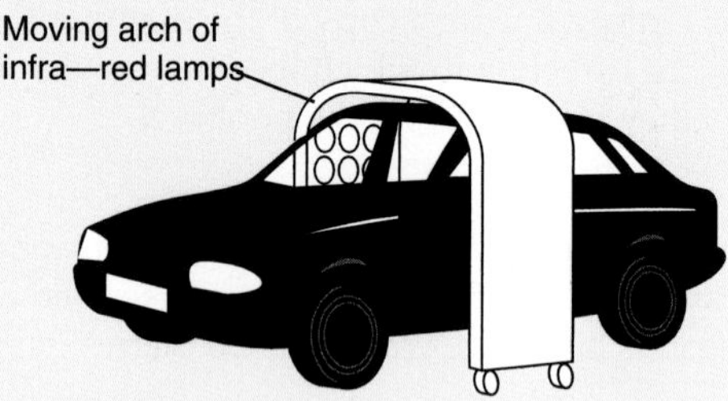

(c) A car is 6 metres long.

(i) How long would it take for the arch to pass over the car?

(ii) Write your answers to (i) as part of an hour.

(iii) How many kilowatt hours of electric energy would be used by the infra-red lamps as they passed over the car?

(iv) The Acme Company pays 5 p for a kilowatt hour of electrical energy. How much will it cost them to harden the paint on the car?

(LEAG)

8 (a) A national radio station broadcasts from a single transmitter at a wavelength of 1500 m.

(i) Calculate the frequency of the carrier waves from this radio station. The speed of radio waves in air is $3.0 \times 10^8\,m\,s^{-1}$.

(ii) The audio signal from this radio station modulates the amplitude of the carrier waves. With the aid of a diagram, explain what is meant by 'amplitude modulation'.

(b) A local radio station broadcasts on FM at a frequency of 104 MHz.

(i) FM is an abbreviation for 'frequency modulation'. With the aid of a diagram, explain what is meant by the term 'frequency modulation'.

(ii) Give one reason why FM transmission is better than AM transmission for local radio stations.

(iii) National radio programmes are transmitted on FM using a network of local transmitters. Why is it important that adjacent transmitters should broadcast at different frequencies.

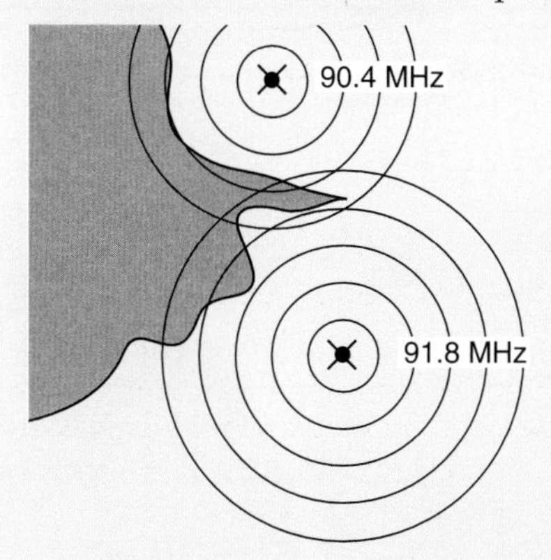

9 X-rays are electromagnetic waves that are used to photograph broken bones in limbs. The diagram shows an X-ray source which is used to direct X-rays at a broken leg. A photographic film in a light-proof wrapper is placed under the leg. When the film is developed, an image of the broken bone is observed on the film.

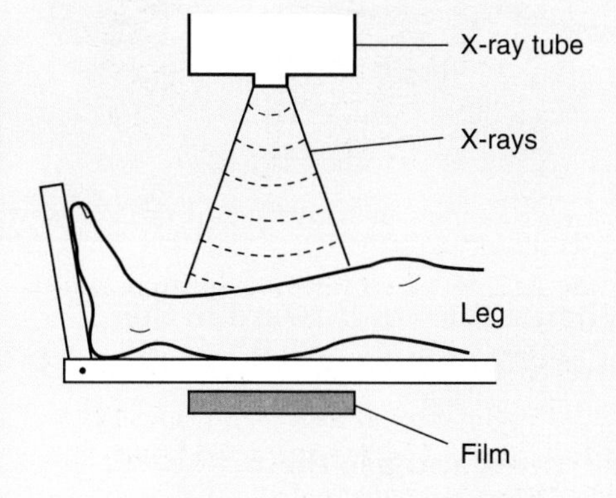

(a) (i) Explain why an image of the bone is formed on the film?

(ii) Why is it possible to see the fracture on the image?

(b) When an X-ray photograph of the stomach is taken, the patient is given food containing barium before the photograph is taken.

(i) Why is it necessary for the patient to be given this food before the photograph is taken?

(ii) The exposure time for a stomach X-ray must be shorter than for an X-ray of a limb. Why?

(c) An ultrasonic scanner is used to observe an unborn baby. Why is ultrasound instead of X-rays used to observe an unborn baby?

10 (a) Ultraviolet (UV) radiation is electromagnetic radiation just beyond the violet end of the visible spectrum.

(i) Copy and complete the chart below to show the other main parts of the electromagnetic spectrum.

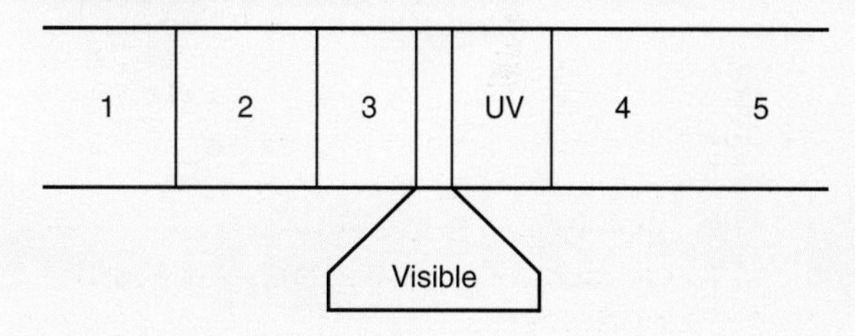

(ii) State two similarities between ultraviolet radiation and visible radiation.

(b) (i) Why is it advisable to cover exposed skin with a suitable cream if you stay outdoors in summer on a sunny day?

(ii) Describe one application of ultraviolet radiation.

(c) An atom emits a photon of electromagnetic radiation when an electron moves nearer the nucleus to fill a vacancy in an electron shell. The energy carried away by the photon is equal to the energy lost by the electron. The wavelength of a photon depends on its energy.

(i) Explain why the spectrum of light from a glowing gas consists of well-defined lines of different colours.

(ii) Why is it possible to identify an element from the spectrum of light its atoms emit when heated?

Theme E

Forces

Car seat belts are designed to prevent car passengers being thrown forward in the event of a crash. The seat belts and fittings in a car are tested to ensure they can withstand the huge forces caused by an impact.

There are many situations where scientists need to measure forces and use their measurements. The test car in the photograph is fitted with sensors that measure the dummy's movement and the forces on it. Read on to find out how to measure motion and force.

Topic 16 Forces in balance

16.1 What forces do

FIRST THOUGHTS

How many different types of force can you think of? For each type of force, think about a situation where the force acts.

Figure 16.1A ▲ Using force

Forces change the motion or the shape of an object.

Have you ever tried to push a piece of furniture across a room? Your push is opposed by the friction between the furniture and the floor. Not much push is needed if the floor is slippery, but if the floor is rough you may need help to move the furniture.

Work must be done to move a piece of furniture. Anything that can do work is called a **force**. Pushing on something is an example of applying a force.

Some more forces are illustrated in Figure 16.1B. *What is each force doing to the object it acts on?* A force acting on an object can change the shape or change the motion of the object. To bring about these changes, work is done by the force. Two or more forces acting on an object can keep it stationary. Since there is no movement, the forces do no work.

Two magnets do not need to be in contact to attract or repel each other. For this reason, magnetic forces are sometimes called action-at-a-distance forces. *Which other forces in Figure 16.1B act between objects that are not in direct contact?*

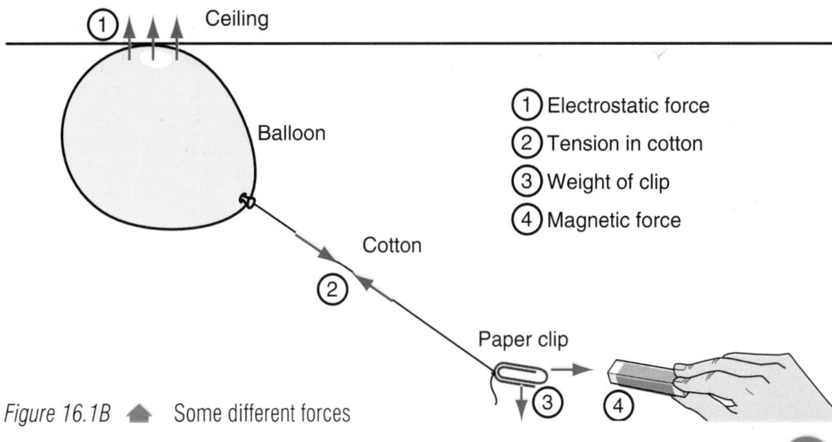

Figure 16.1B ▲ Some different forces

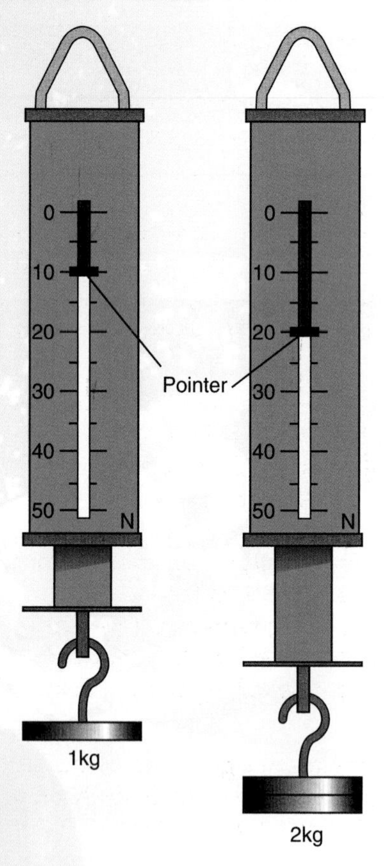

Figure 16.1C ▲ Using a newton balance

The **newton**, N, is the unit of force. It is used to measure every type of force. It is defined as the force necessary to give a mass of 1 kg an acceleration of 1 m/s^2.

Weight

If you release any object above the ground it falls, because it is attracted towards the Earth. This force of attraction is called **gravity**. The Earth's gravity acts on every object near the Earth. The force of gravity on an object is its **weight**.

Weight is measured using a newton balance, which is a spring balance with a scale marked in newtons. Figure 16.1C shows different masses being weighed. A 1 kg mass has a weight of 10 N. A 2 kg mass has a weight of 20 N. *What is the weight of a 3 kg mass?* These results show that the strength of gravity on the Earth's surface is 10 N/kg. More accurate measurements give a result of 9.8 N/kg. This is usually denoted by the symbol g.

To work out the weight of an object from its mass, you can use the equation

$$\text{Weight (in newtons)} = \text{Mass (in kg)} \times g$$

where $g \approx 10$ N/kg.

You can reduce your weight by going to the equator. The strength of gravity varies slightly over the Earth's surface, and is least at the equator. However, your mass is constant, so you would not lose mass by going to the equator.

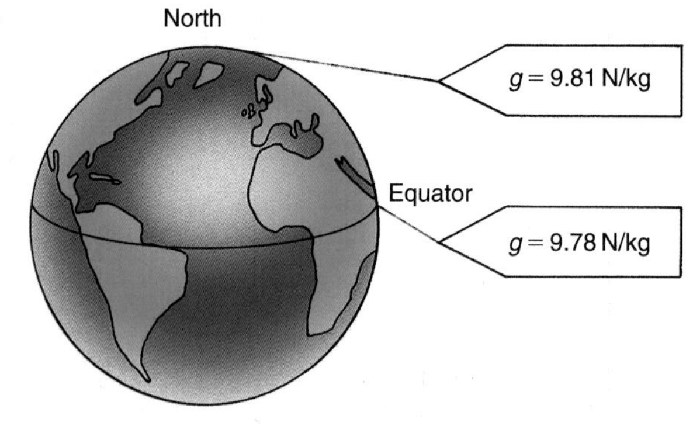

Figure 16.1D ▲ Variation of g with latitude

Figure 16.1E ▲ Astronauts weigh less on the Moon

The weight of astronauts is reduced when they go into space. Imagine you are an astronaut on a journey to the Moon. As your rocket moves away from the Earth, your weight becomes less because you are moving away from the Earth's gravity. Approaching the Moon, you begin to feel the effect of the Moon's gravity. On the Moon's surface your weight would be about one sixth of your weight on Earth, because lunar gravity is about one sixth of Earth's gravity at the surface.

Force and shape

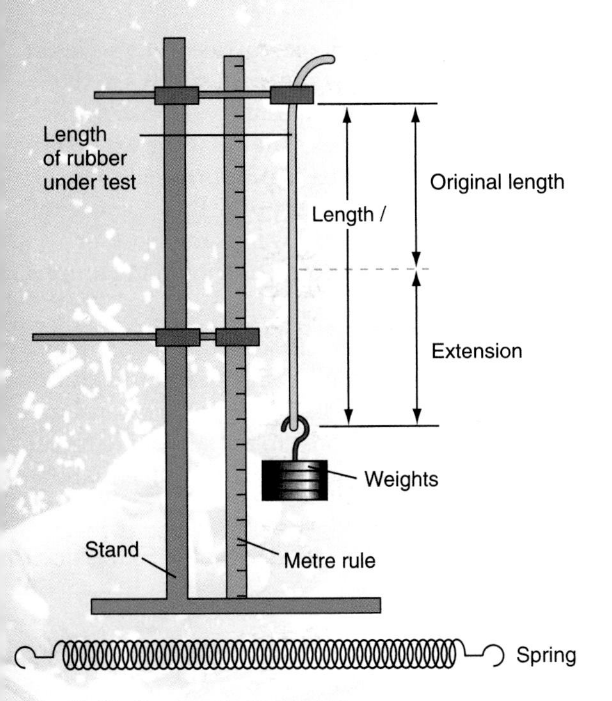

Figure 16.1F ▲ Investigating stretching

Squash players know that hitting a squash ball hard changes its motion and can even change its shape. Forces can change the motion or shape of an object. An object that regains its original shape when the force is removed is said to be elastic. For example, a rubber band that is stretched and then released usually returns to its original length. Rubber is an example of a material that possesses elasticity.

To investigate how easily a material stretches, you can hang weights from it as in Figure 16.1F. First, measure the initial length of the material. A small weight attached to the material will keep it straight. Then increase the amount of weight in steps, measuring and recording the length at each step. To check your measurements you can measure the material again as the weights are removed. For each step calculate the extension of the material – the increase in length from the initial length.

Plot your measurements on a graph, as extension (on the vertical axis) against weight. Some results for different materials are shown in Figure 16.1G.

EXTENSION FILE ASSIGNMENT

The extension of a steel spring is proportional to the weight it supports.

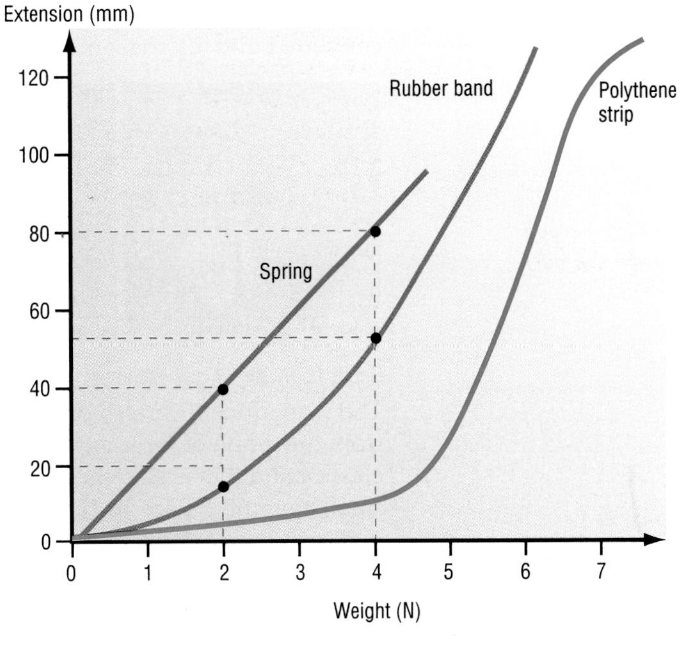

Figure 16.1G ▲ Graph of extension against weight for some different materials

The experiments show that the steel spring and the rubber band behave elastically. That is to say they regain their initial length after the weights are unloaded. The extension of the steel spring is proportional to the weight suspended on it. For example, doubling the weight from 2.0 N to 4.0 N doubles the extension of the spring. This is not true for the rubber band. When the weight on the band doubles from 2.0 N to 4.0 N, the extension of the rubber band more than doubles.

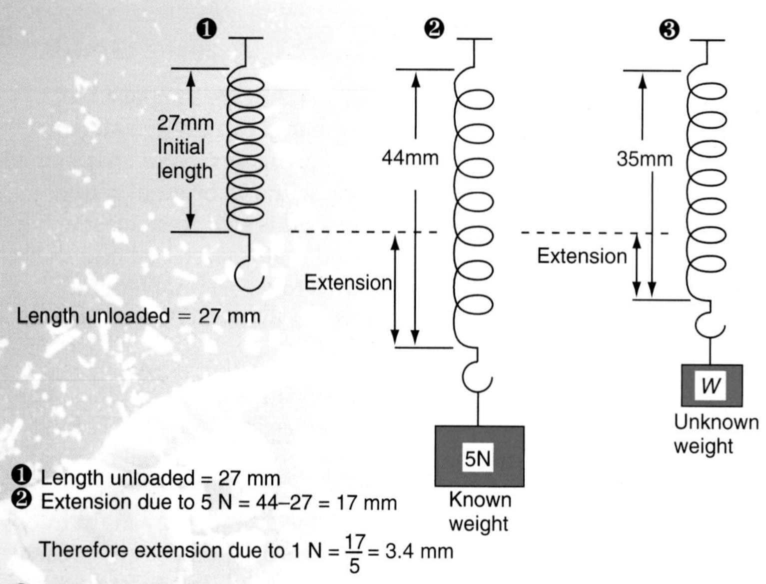

❶ Length unloaded = 27 mm
❷ Extension due to 5 N = 44–27 = 17 mm

Therefore extension due to 1 N = $\frac{17}{5}$ = 3.4 mm

❸ Extension due to W = 35–27 = 8 mm

Therefore $W = \frac{8}{3.4}$ = 2.4 N

Figure 16.1H Using Hooke's Law to find an unknown weight

Robert Hooke carried out a series of investigations with springs in the seventeenth century. Hooke's Law for springs states that **the extension of a spring is proportional to the weight it supports**. By applying this law you can weigh an object, or measure force, as shown in Figure 16.1H.

Although Hooke's Law refers only to springs, other materials are said to obey Hooke's Law if the extension is proportional to the stretching force. This may be written as an equation

$$\text{Force} = \text{Constant} \times \text{Extension}$$

Elastic energy

When a golf club hits a golf ball forcefully, the ball is deformed by the impact but regains its shape afterwards. In the first half of the impact, as the ball is becoming more and more deformed, the kinetic energy of the club is converted into **elastic energy** of the ball; in the second half, the elastic energy is converted back into kinetic energy as the ball leaves the club at high speed. The golf ball will be slightly warmer after the impact than before because some of the initial kinetic energy of the club is converted into internal energy of the molecules of the ball.

The area under a force versus distance graph gives the work done.

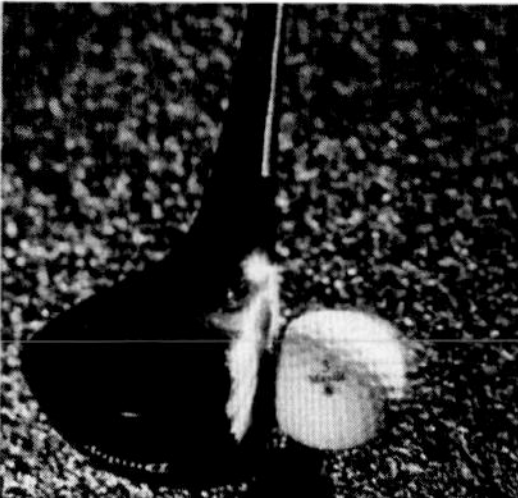

Figure 16.1I A golf ball impact

The work done to stretch or compress a solid object can be estimated from the graph of force against change of length of the object. Imagine an elastic band that is stretched in steps, each step increasing its length by a small amount. Figure 16.1J shows how the force needed to stretch it increases with extension.

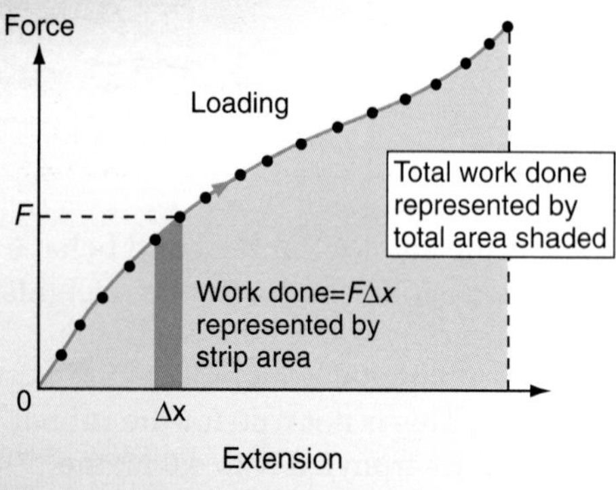

Figure 16.1J (a) Loading curve for an elastic band

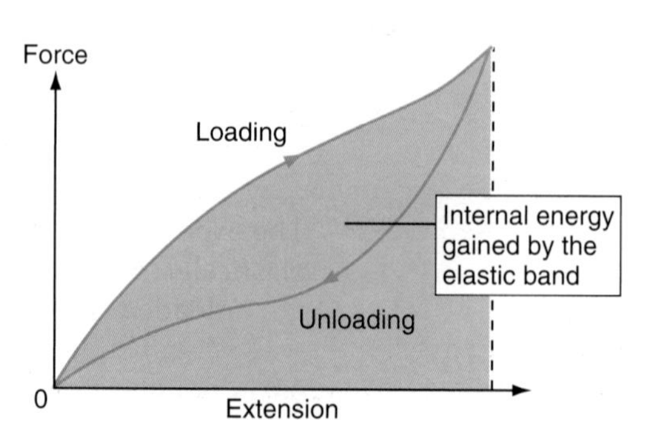

(b) Loading and unloading an elastic band

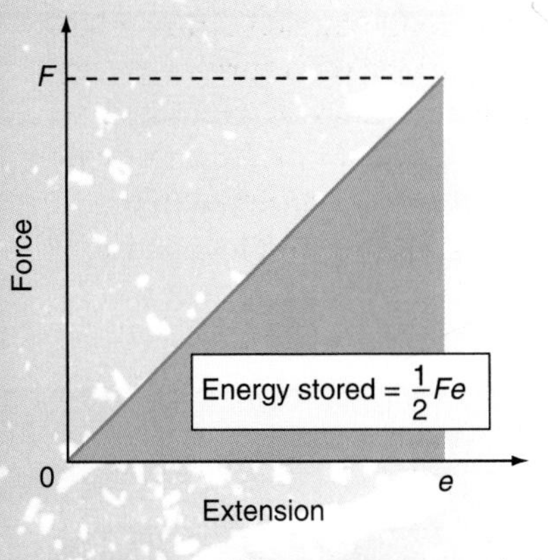

Figure 16.1K ▲ Energy stored in a stretched spring

SUMMARY

Forces can change the shape or change the motion of objects. The newton is the unit of force. Weight is the force of gravity on an object. A newton balance is used to weigh objects.

The work done in each step is given by

Force needed (F) × Increase in length (Δx)

This is represented on the graph by the area of the strip between the curve and the 'Extension' axis. The total work done in all the steps is therefore the total area of all the strips. Thus the total area between the curve and the 'Extension' axis represents the total work done to stretch the object.

What happens when a stretched elastic band is loaded then unloaded? Figure 16.1J(b) shows how the extension of an elastic band varies as it is loaded then unloaded. The area under the loading curve represents work done to stretch the elastic band; the area under the unloading curve represents recovery of elastic energy; the area between the two curves represents internal energy gained by the molecules of the elastic band.

What happens when a steel spring is loaded then unloaded? The graph of force against extension is a single straight line, as shown in Figure 16.1K. No internal energy is gained by the spring. The work done to stretch the spring is stored as elastic energy in the spring and all the elastic energy is recovered from the spring. Since the graph is a straight line through the origin, the area under the line representing the elastic energy stored in the spring = $\frac{1}{2}$ × force × extension (corresponding to $\frac{1}{2}$ × height × base).

CHECKPOINT

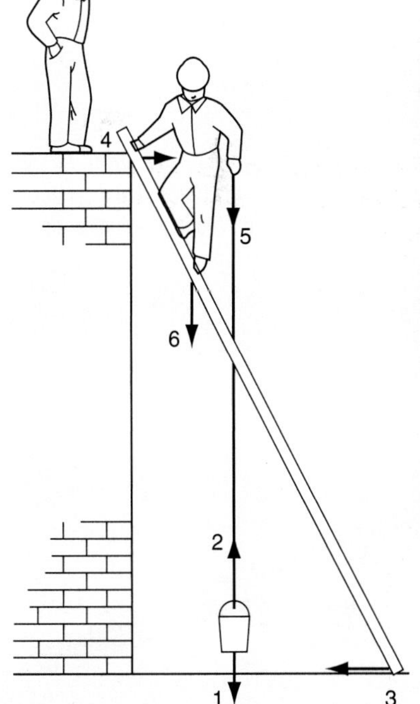

1. (a) Figure 16.1B shows different forces acting. Write down what each force is doing to the object it acts on.
   (b) Name the forces acting in the diagram opposite. What is each force doing?

2. Copy and complete the following table.

Mass (kg)	1.0	0.1	?	60	?
Weight (N)	10	?	25	?	0.2

3. Suppose you were given several 1 kg masses, some string and a 'suspect' newton balance. Explain how you would check the accuracy of the balance.

4. The strength of the Earth's gravity varies slightly over the Earth from 9.81 N/kg at the poles to 9.78 N/kg at the equator (Figure 16.1D). What would be the change in weight of a person of mass 60 kg who went from the equator to the North Pole?

5. In a Hooke's Law test on a spring, the following results were obtained.

Weight (N)	0.0	1.0	2.0	3.0	4.0	5.0	6.0
Length (mm)	245	285	324	366	405	446	484

   (a) Use these results to plot a graph of weight (on the vertical axis) against extension.
   (b) How much weight is needed to extend the spring by (i) 100 mm, (ii) 1 mm?
   (c) If an object produces an extension of 230 mm, what is its weight?
   (d) Work out the mass of the object from its weight.
   (e) Work out the energy stored when the extension is 230 mm.

16.2 Centre of gravity

FIRST THOUGHTS

A tightrope artiste, a cyclist and a waiter carrying a stack of plates know all about the importance of the centre of gravity. Imagine carrying a stack of plates while cycling along a tightrope. It has been done!

The design of racing cars has changed considerably since the first models. More powerful engines, better streamlining and less dense materials have all contributed to the improved performance of the modern racing car.

Figure 16.2A
(a) 1920s racing car design
(b) Modern racing car design

However, one feature that has not changed is the need to keep the car low. The weight of the car must be as near to the ground as possible, otherwise the car would overturn when cornering at high speeds.

The weight of a tractor also needs to be as low as possible, or it would tip over on rough ground. The top of the tractor cab may well be two or three metres above the ground. However, mounting the heavy parts like the engine as low as possible keeps the weight low and makes the tractor stable.

We can think of the weight of an object as if it acts at a single point. This point is called the **centre of gravity** of the object.

If you balance a ruler on the tip of your finger (Figure 16.2B) the point of balance is the centre of gravity of the ruler. The easiest way to carry a ladder is to lift it at its centre of gravity.

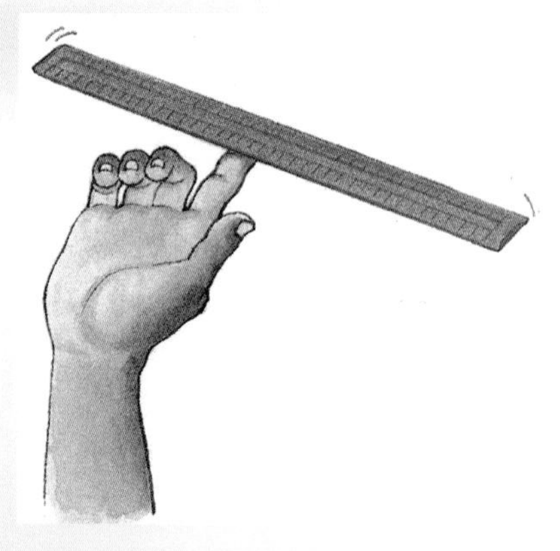

Figure 16.2B A balancing act

> The centre of gravity of an object is the point where its entire weight seems to act.

A centre of gravity test

To find the centre of gravity of a flat card, hang the card from a pin so it is free to rotate, as in the diagram and let it come to rest. Use a plumbline to draw a vertical line from the pin downwards. The centre of gravity is somewhere along this line. Now repeat this step with the card hung at a different point. The centre of gravity is where the two lines drawn on the card meet. Test your results by seeing if you can balance the card at this point.

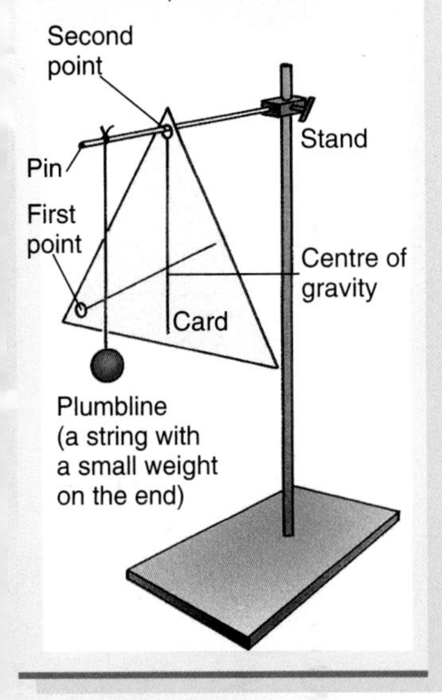

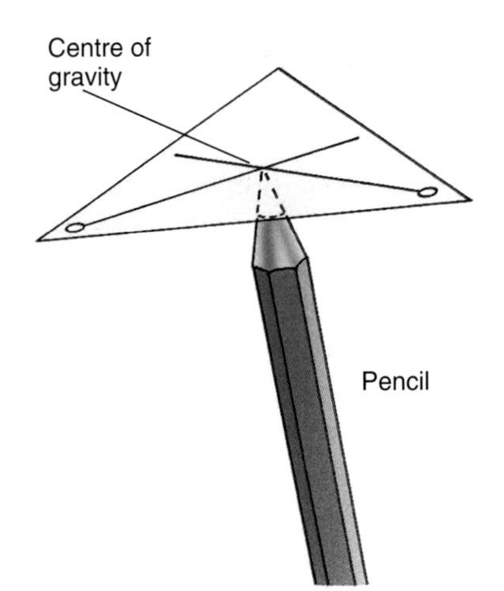

Figure 16.2C ▲ Unstable equilibrium

Stability

A card that hangs from a pin without moving is said to be in **equilibrium**. This means the forces acting on the object are balanced.

If the card is pushed to one side and then released, it swings back to its equilibrium position. This is an example of **stable equilibrium** as in the test diagram. The same card balanced on the tip of a pencil (Figure 16.2C) is in **unstable equilibrium**. A small push to one side will make the card fall off.

Tilting and toppling

How far can a tractor tilt before it topples over? How far can a pram with a bouncing baby tilt safely? Stand a brick on its end and then tilt it until it just balances on one edge. Figure 16.2D shows the idea. The brick will balance when its centre of gravity is directly above the edge on which it balances. The brick topples over if it is released when its centre of gravity is 'outside' this edge. The same applies to a vehicle when it tilts to one side. If the centre of gravity goes outside the wheel base, the vehicle topples over.

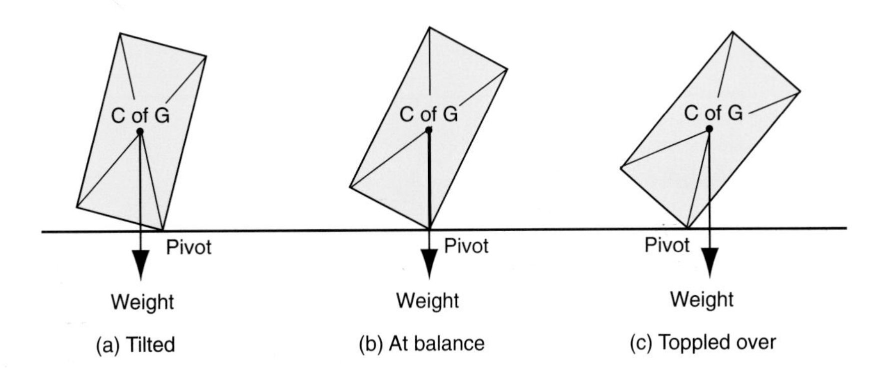

Figure 16.2D ▲ Tilting and toppling

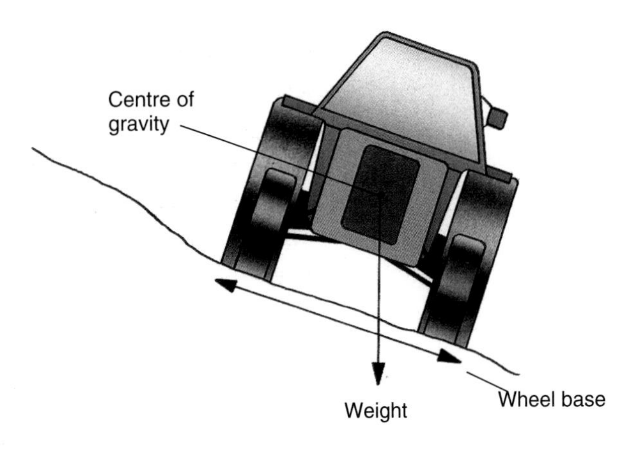

Figure 16.2E ▲ Forces on a tilting tractor

SUMMARY

The centre of gravity of an object is the point where its weight seems to act. An object in stable equilibrium returns to equilibrium when it is pushed and released.

CHECKPOINT

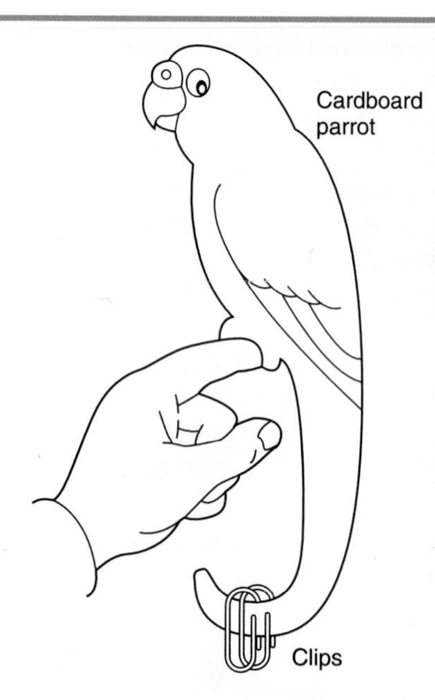

1. (a) A tightrope walker often carries a long thin pole. How does this help the walker to balance on the rope?
   (b) Balancing a spinning dinner plate on the end of a vertical pole is another circus trick. Why must the pole support the plate at its centre?
2. Balance a parrot on your finger. Cut the parrot shape shown from a piece of card. Fix some paper clips to the parrot's tail. Why does this help to improve its stability?
3. Babies and young infants are often top-heavy. This is because a baby's head is usually large compared to its body. Why should a well-designed baby chair have a wide base? If the base was too narrow, what might happen if the baby leaned over the side of the chair?
4. (a) Filing cabinets are designed so that only one drawer at a time can be pulled out. What might happen if all the drawers, loaded with documents, were pulled out at the same time?
   (b) High-sided lorries are at risk in strong winds. Why?
5. A Bunsen burner has a wide heavy base so it cannot be knocked over easily. How many other objects can you think of that are designed to be difficult to knock over? Make a list. Can you think of any object that needs to be redesigned because it is knocked over too easily?

16.3 Push and pull forces

FIRST THOUGHTS

In this section you will find out how to work out the effect of two forces acting on an object in opposite directions.

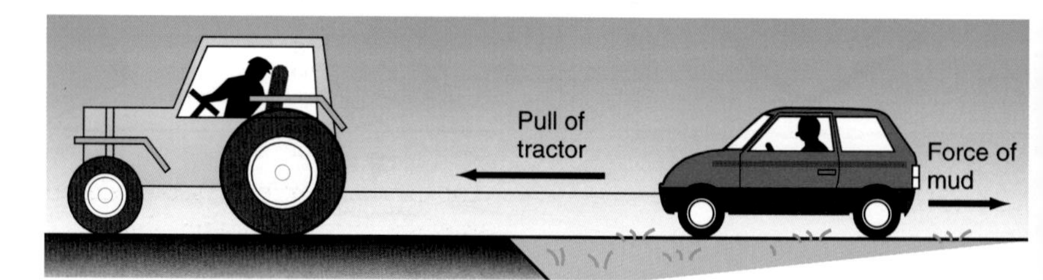

Figure 16.3A A muddy problem

John has just passed his driving test but he still has much to learn. He has reversed his car into a very muddy cart track and is stuck. He has had to ask a farmer with a tractor to pull his car out of the mud (Figure 16.3A).

One end of a rope is tied to the back of the tractor and the other end to the front of the car. At first, the car stays stuck in the mud. The pull of the tractor is not enough to overcome the force of the mud. As the tractor uses more power, the force it exerts becomes greater than the force of the mud and the car is pulled out.

The situation is like a tug-of-war; the winning team is the one that pulls with most force. If the two teams pull with equal force, there is stalemate. The rope is in equilibrium because the force at one end is equal and opposite to the force at the other end.

The size and direction of a force can be represented by a vector. A vector is an arrow whose length represents the magnitude (size) of the force and whose direction gives the direction of the force. Any force has magnitude and direction and so can be drawn as a vector.

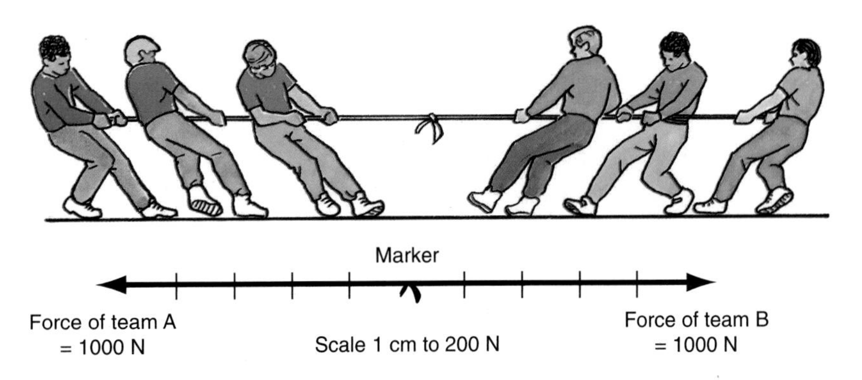

Figure 16.3B ▲ A tug-of-war

Any physical quantity with magnitude and direction is a vector.

Figure 16.3B shows the pull of two tug-of-war teams as vectors. A scale of 1 cm to 200 N is used here, so the force of 1000 N is represented by a vector 5 cm long. The two vectors are the same length because the magnitudes of the forces are the same. Because the forces are in opposite directions, the two vectors point in opposite directions.

If one force were larger than the other, what would the overall effect be? In a tug-of-war, the team that exerts the larger force wins. Suppose one team pulls with a force of 1000 N and the other team with a force of 950 N (see Figure 16.3C). The smaller force nearly cancels out the other force, but not quite. The combined effect of these two forces is called the resultant. *What is the resultant here?* The stronger team can exert a force 50 N greater than the other team. So the resultant is 50 N.

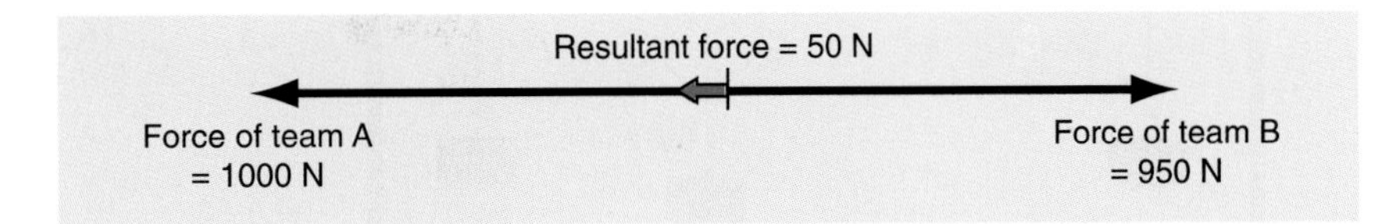

Figure 16.3C ▲ Unequal forces

Another example of two forces in equilibrium is when an object is supported by a newton balance. The forces acting on the object are the force of gravity (i.e. its weight) and the pull of the spring balance. The weight of the object is equal and opposite to the pull of the balance.

What is the weight of the iron bar in Figure 16.3D(a)? Now suppose a magnet is brought near the iron bar as in Figure 16.3D(b). *If the newton balance reading increases to 5.5 N, what is the force on the bar due to the magnet?* The bar's weight is less than 5.5 N. The force due to the magnet must make up the difference.

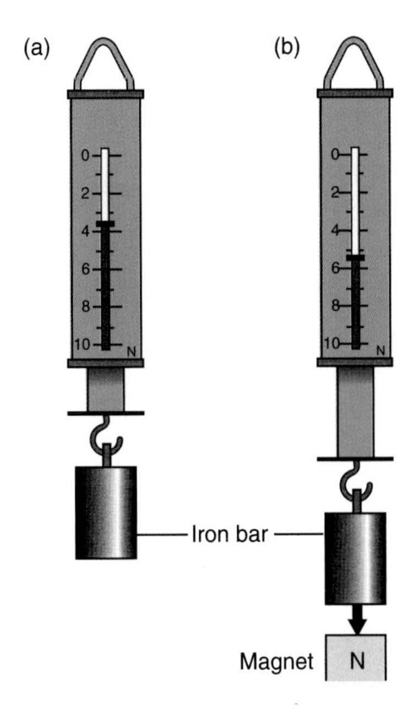

Figure 16.3D ▲ Measuring forces

SUMMARY

The magnitude and direction of any force can be represented on a diagram by a vector. An object is in equilibrium if the forces acting on it balance each other. If the forces do not balance, they do not cancel each other out. Their combined effect is called the resultant.

Figure 16.3E Under tow

The parallelogram of forces

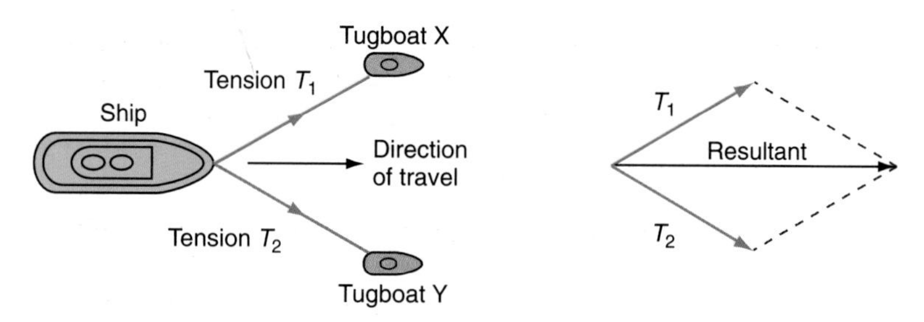

Figure 16.3F Combining forces

Consider a vessel being towed by two tugboats, as shown in Figure 16.3E. The combined effect of the pull forces from the towing cables is to pull the vessel forwards. The combined effect is called the **resultant**. Figure 16.3F shows how the pull forces (i.e. tensions) in the two cables, represented as vectors, combine to produce the resultant. The tension vectors T_1 and T_2 are shown as adjacent sides of a parallelogram; the resultant is the diagonal from the origin of T_1 and T_2. This method of determining the resultant is called the **parallelogram of forces**.

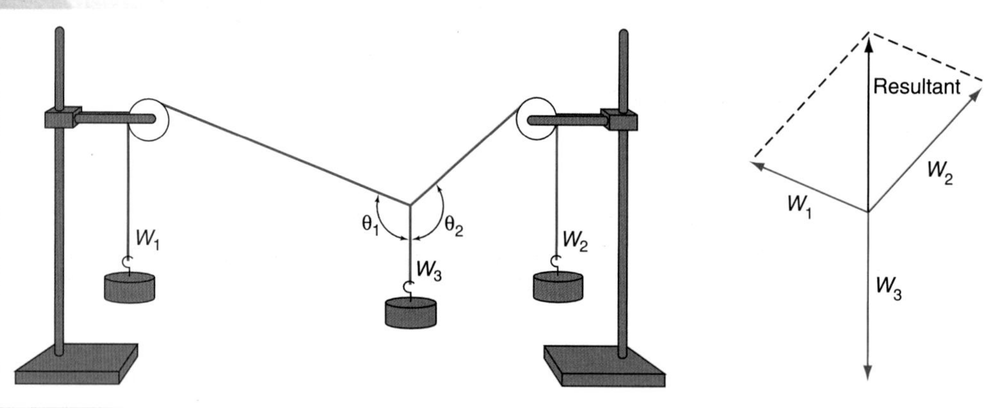

Figure 16.3G (a) Using pulleys (b) Using the parallelogram

Figure 16.3G shows how to demonstrate the parallelogram of forces in the laboratory using two pulleys and three known weights. The tension in each string is due to the weight it supports, either directly or over a pulley. The angles between the strings are measured using a protractor. A scale diagram of a parallelogram is then drawn, with the vectors representing weights W_1 and W_2 as adjacent sides of the parallelogram. The resultant, represented by the diagonal as shown, should be equal and opposite in direction to the vector representing W_3.

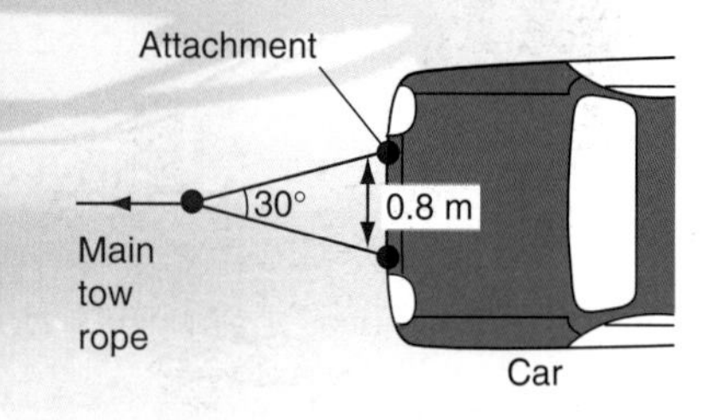

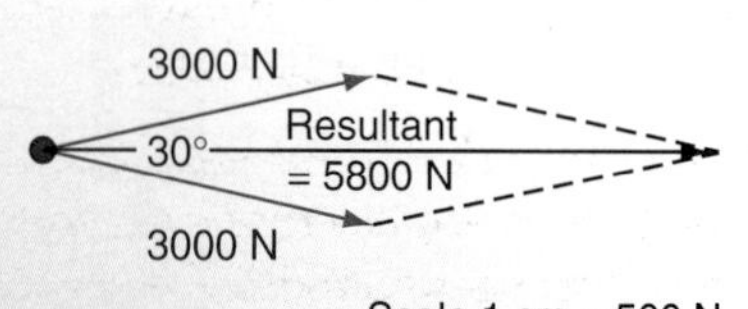

Figure 16.3H

Worked example A tow rope is attached to a car at two points 0.80 m apart, as in Figure 16.3H. The two sections of rope joined to the car are of the same length and are at 30 ° to each other. The pull on each attachment should not exceed 3000 N. Use the parallelogram of forces to determine the maximum tension in the main tow rope.

Solution The maximum tension T in the main tow rope is obtained when the tension in each of the two rope sections attached to the car is 3000 N. Hence T is the resultant of two forces of 3000 N at 30° to each other. The parallelogram of forces may therefore be used to determine the resultant, giving T = 5800 N.

CHECKPOINT

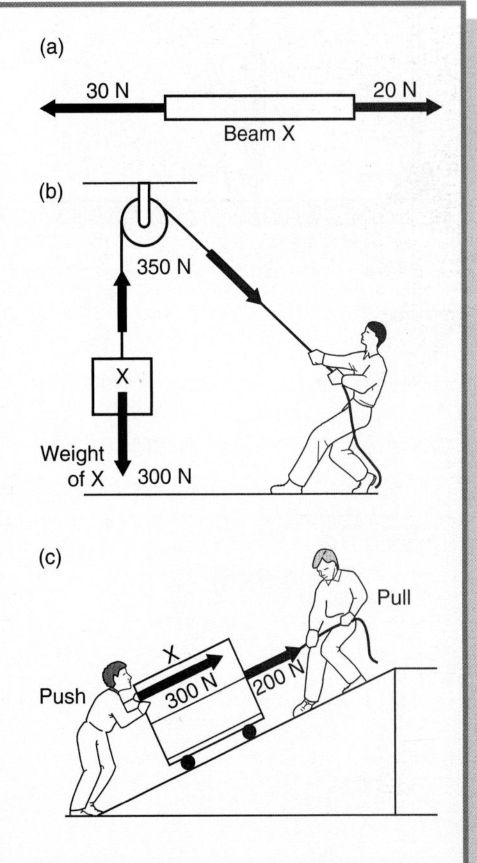

1. A trailer is attached to a car by a tow bar. The tow bar must be strong enough to withstand being pulled or pushed. Is the tow bar under tension or in compression when:
   (a) the car and trailer drive off from rest,
   (b) the car and trailer halt?

   Make a sketch showing the trailer and the forces on it when the car drives off from rest.
2. The diagrams opposite show several situations where two forces act on an object. In each case, work out the magnitude and direction of the resultant.
3. In the village of Much Watering, the highlight of the annual fair is when the tug-of-war team challenges the team from Little Steeping, the next village. The challenge takes place with the teams on opposite banks of the River Steeping. Each team is determined to pull the other team into the river. The team captains toss a coin to decide which side of the river to take. This is important because one bank is slightly higher than the other. Which bank would you take? Explain why.
4. A magnet of weight 2.4 N was hung from a newton balance. An identical magnet was hung from the first magnet. The second magnet was pulled down until it broke free from the first magnet. The spring balance reading just before the break was 9.6 N.
   (a) What was the reading on the balance when it supported the first magnet only?
   (b) What was the reading when the second magnet was hung on?
   (c) What was the force of attraction between the two magnets just before they broke apart?
5. An empty car has a weight of 5500 N. Its driver has a weight of 600 N. The weight on each wheel should not be greater than 2500 N. In addition to the driver, how much extra weight can the car support safely?

16.4 Turning forces

FIRST THOUGHTS

Your skeleton is a walking example of turning forces in action. Read on to find out about more turning forces.

To undo a very tight wheel nut on a bicycle you need a spanner. The force you apply to the spanner is called a **turning force**. Its effect is to turn the nut to undo it. *If you had the choice between a long-handled spanner and a short-handled one, which would you choose?*

Figure 16.4A ▲ A turning force

You cannot turn a tight nut with your fingers, but the spanner can turn it. Therefore the spanner must be exerting a bigger force on the nut than your fingers could. The spanner is an example of a **force multiplier**.

Figure 16.4B shows a crowbar being used to shift a heavy weight. The weight is called the **load**, and the force the person applies to the crowbar is called the **effort**. Using the crowbar, the effort needed to lift the safe is only a small fraction of its weight. A turning force is at work. The point about which the crowbar turns is called the **pivot** or the **fulcrum**. The crowbar is another example of a force multiplier.

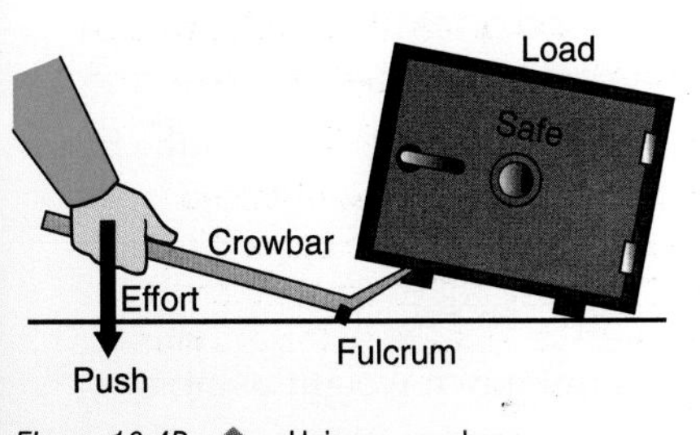

Figure 16.4B ▲ Using a crowbar

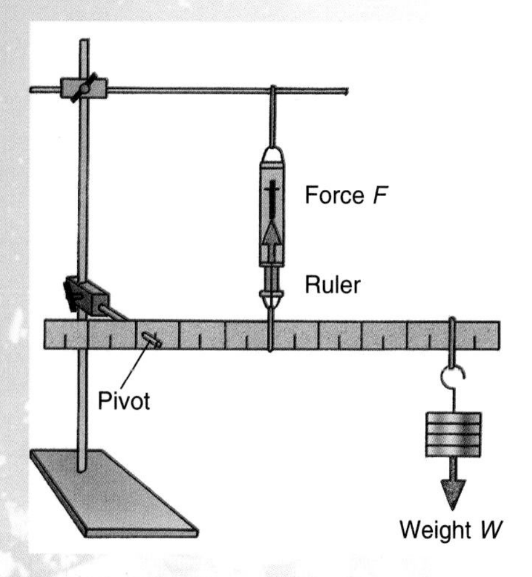

Figure 16.4C Investigating turning forces

Investigating the effect of a turning force

Figure 16.4C shows one way of investigating turning forces. The weight W is moved along the metre rule. *How do you think the reading on the balance compares with the weight? How does this reading change as the weight is moved away from the pivot?* Record the balance reading for different distances d from the pivot to the weight. The balance readings show that the force F needed to support the weight becomes larger as the weight is moved away from the pivot. The turning effect of the weight is called its **moment**.

The moment of a turning force is defined as

> Moment = Force × Perpendicular distance from the pivot to the line of action of the force

In Figure 16.4C, the moment of the weight is $W \times d$ where d is the distance from the pivot to the point on the rule where W is suspended. The greater the distance, d, the bigger the moment. The weight tries to turn the bar clockwise. Force F tries to turn the bar anticlockwise. The moment of the weight is cancelled out by the moment of force F.

Figure 16.4D The seesaw

The Principle of Moments can be applied about any point of an object in equilibrium.

A seesaw is another example where clockwise and anticlockwise moments balance an object. Figure 16.4D shows Vida sitting near the fulcrum to balance her younger brother, Tariq, at the far end of the seesaw. Vida's larger weight acting at a short distance from the fulcrum gives an anticlockwise moment. Tariq is not as heavy as his big sister and sits further away from the fulcrum on the other side. His smaller weight at a greater distance from the fulcrum gives a clockwise moment that balances Vida's anticlockwise moment.

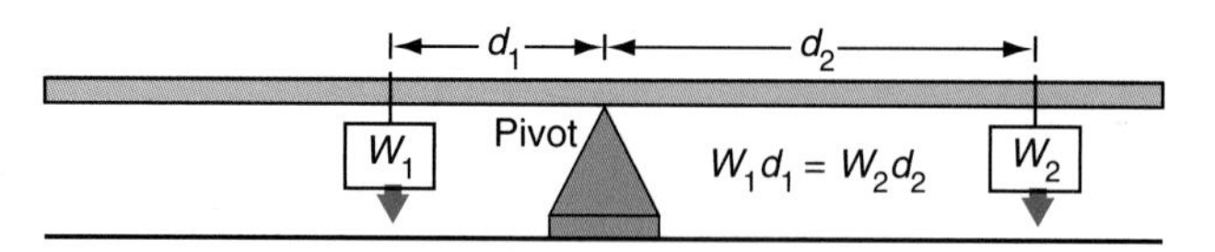

Figure 16.4E The principle of moments

This is an example of the **principle of moments**, which states that for an object in equilibrium,

> Total clockwise moment = Total anticlockwise moment

Figure 16.4E shows how to test the principle. You will find that the beam is balanced when

$$\underset{\text{(anticlockwise)}}{W_1 \times d_1} = \underset{\text{(clockwise)}}{W_2 \times d_2}$$

This principle can be used to measure an unknown weight using a known weight.

In Figures 16.4D and E the beam is balanced at its centre of gravity. Its own weight acts at the pivot so has no turning effect. However, if the beam is balanced at another point then its weight does have a turning effect. Figure 16.4F shows how this can be used to find the weight of the beam using a known weight.

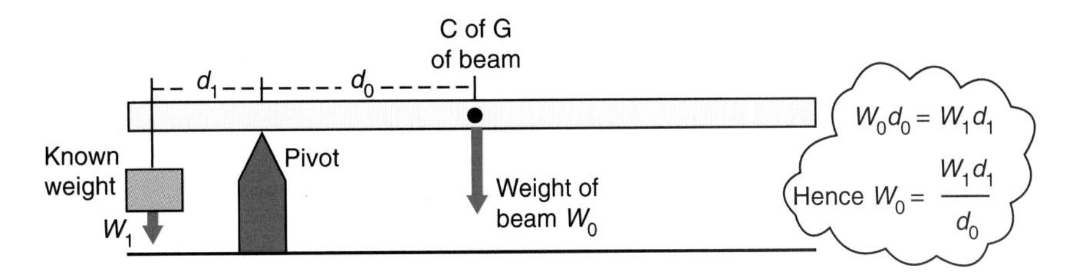

Figure 16.4F ▲ Finding the weight of a beam

The principle of moments explains why levers are so useful. Figure 16.4G shows some more examples that you may be familiar with. In each case, the effort provides a moment ($F_1 \times d_1$) that equals the moment of the load ($W_0 \times d_0$).

Note that the marked distances in Figures 16.4E, F and G are the perpendicular distances from the pivot to the line of action of each force. The moment of a force about a point is defined as the force x the perpendicular distance from the point to the line of action of the force.

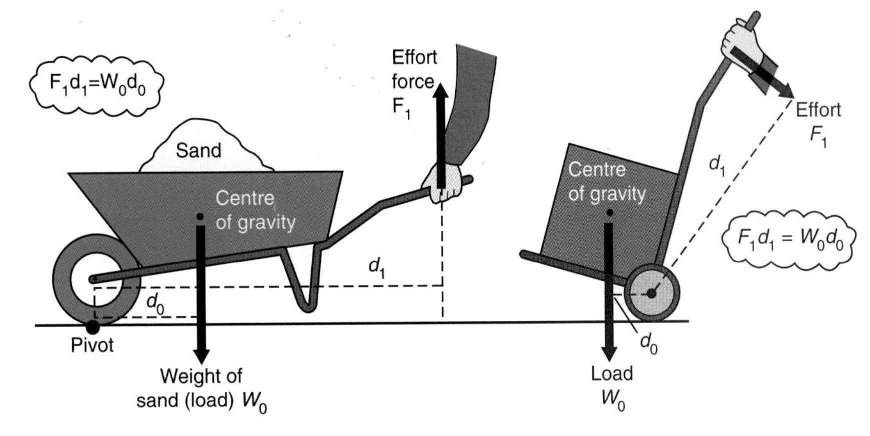

Figure 16.4G ▲ Using moments

SUMMARY

The moment of a turning force is the product of the force and the distance to the pivot from the point where the force is applied. For any object in balance, the total clockwise moment is equal to the total anticlockwise moment.

Consider the wheelbarrow where $d_1 = 0.80$ m, $d_0 = 0.20$ m, $W_0 = 300$ N

Since $F_1 \times d_1 = W_0 \times d_0$
then $F_1 \times 0.80 = 300 \times 0.20$
so $F_1 = \frac{300 \times 0.20}{0.80} = 75\text{N}$

Thus the wheelbarrow allows a load of 300 N weight to be shifted by applying an effort of 75 N.

The joints in the human body act as pivots. Each joint allows two bones to swivel about each other. The elbow allows the lower arm to swivel about the upper arm. The biceps muscle controls the movement of the lower arm. Compare Figures 16.4H and 16.4C. The muscle supports the lower arm in the same way as the balance supports the rule. In both cases, the force needed to support the weight is much larger than the weight itself.

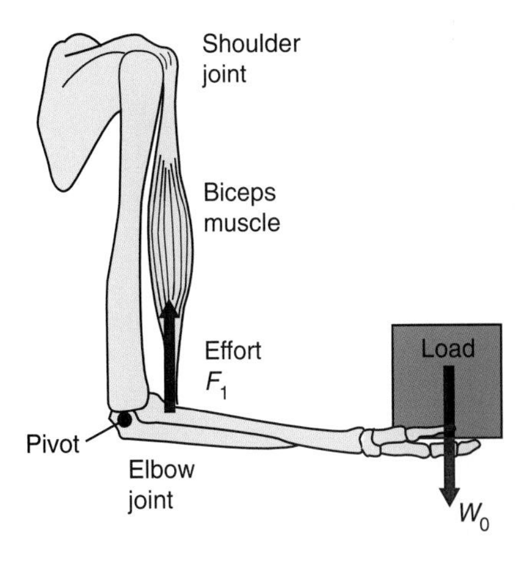

Figure 16.4H ▲ Using your muscles

CHECKPOINT

1 Explain each of the following statements.
   (a) Trapping your finger in a door near the hinge can be very painful.
   (b) A claw hammer is easier to use if it has a long handle.
   (c) Using a wheelbarrow saves effort.

2 For each of the balanced beams in the diagram opposite work out the unknown weight W.

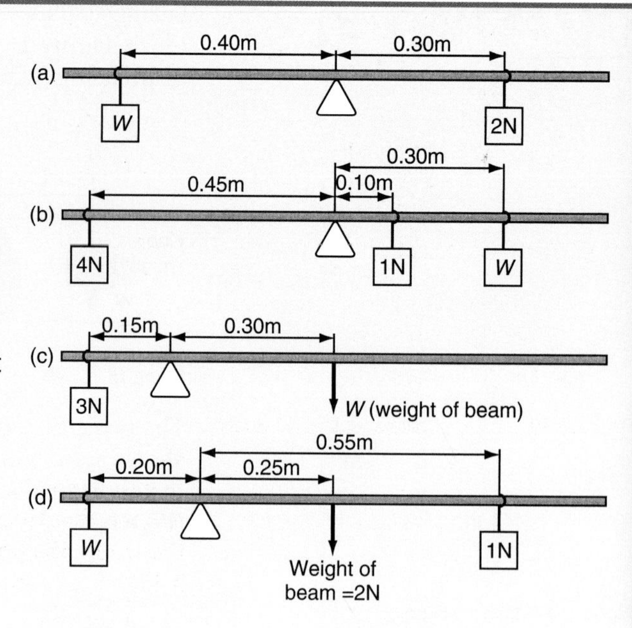

3 Jill weighs 425 N and thinks she is the lightest girl in her class. Dawn claims she is lighter than Jill. They go to the local park to find out who weighs least. Dawn sits on the seesaw 2.50 m from the fulcrum. Jill balances the seesaw by sitting 2.00 m from the fulcrum on the other side of the pivot.
   (a) Who is the lighter?
   (b) What is Dawn's weight?
   (c) Dawn gets off the seesaw so John can sit on to balance Jill. His weight is 450 N. How far from the fulcrum should he sit?

4 The diagram opposite shows a bottle opener being used to remove a bottle top. Copy the diagram and add to it to show:
   (a) where the effort is applied,
   (b) where the bottle opener acts on the bottle top,
   (c) where the fulcrum is.

   Explain how the bottle opener acts as a force multiplier.

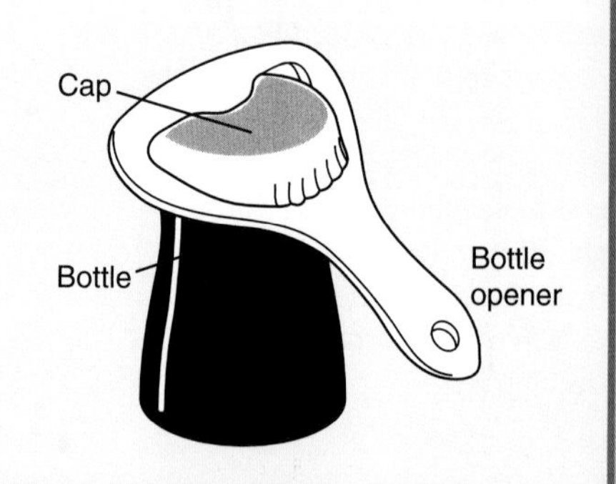

16.5 Forces and frames

FIRST THOUGHTS

Architects, civil engineers and builders need to work out at the design stage what forces to expect in buildings. Miscalculating the forces could be disastrous.

Figure 16.5A The Eiffel Tower

Towers and bridges

The Eiffel Tower in Paris is one of the most famous buildings in the world. The iron tower, 300 m high, was the tallest building in the world when it was completed in 1889. If the upper section had been solid iron rather than a lattice of girders, perhaps it could have been made taller. *Why do you think a lattice of girders was used?* Modern skyscraper towers like the World Trade Centre in New York, which is 412 m high, are made from steel and concrete.

Figure 16.5B ▲ (a) Coalbrookedale Iron Bridge (b) Modern motorway bridge

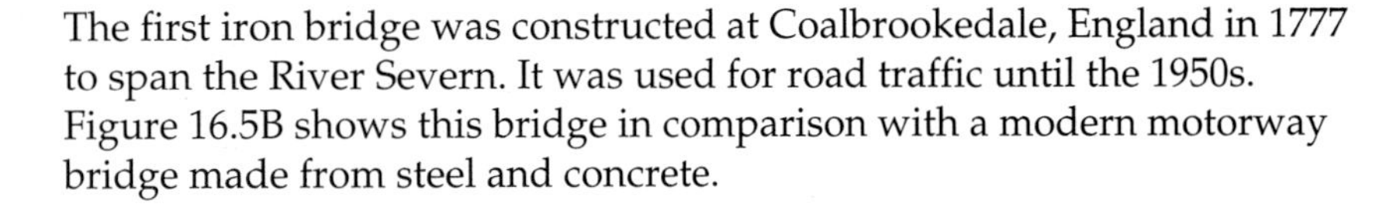

The first iron bridge was constructed at Coalbrookedale, England in 1777 to span the River Severn. It was used for road traffic until the 1950s. Figure 16.5B shows this bridge in comparison with a modern motorway bridge made from steel and concrete.

Metre rule

'Load'

Figure 16.5C ▲ Testing a bridge

Investigating bridges

Use a metre rule to make a simple beam bridge, as shown in Figure 16.5C. Load the rule at its centre with different weights and measure how much it sags.

Devise some means of reducing the 'sag' when the rule is loaded. Repeat your measurements and see how effective your design is. Some ideas are shown in Figure 16.5D.

- The arch bridge is strengthened by the arch under the span. The weight of the span and its load compresses (i.e. squeezes) the arch. The material of the arch must withstand huge compression forces.
- The suspension bridge roadway is supported by steel cables slung between towers. The weight of the roadway and traffic pulls on the cables. The cables must withstand very large tension forces.
- The cantilever bridge roadway consists of two beams that press against each other end-to-end. The upper part of each beam is in tension and the lower part is in compression. The beams are made from concrete reinforced with steel cables.

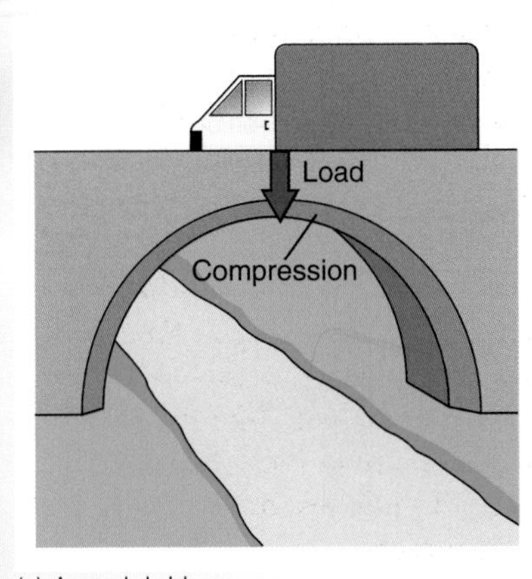

(a) An arch bridge

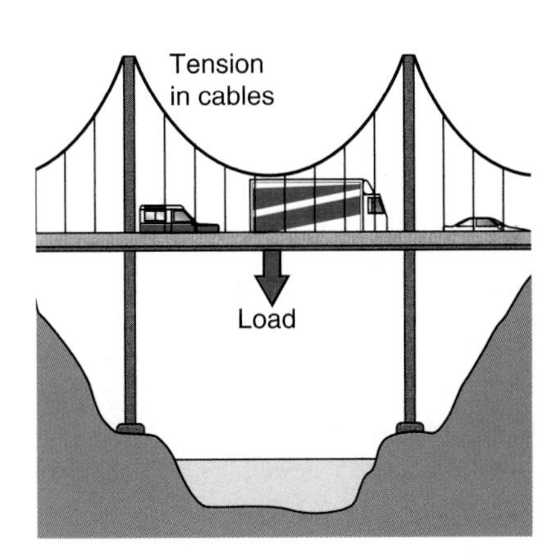

(b) A suspension bridge

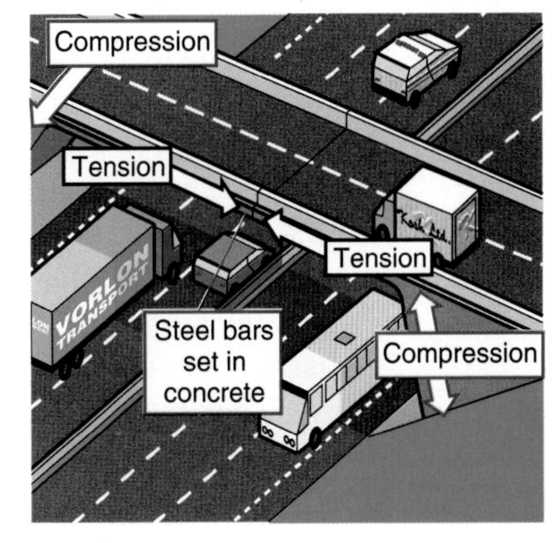

(c) A cantilever bridge

Figure 16.5D ▲ Types of bridge

It's a fact!

Motorway flyovers are made from reinforced concrete. Unfortunately, the combined action of rain and salt (used in winter to prevent ice on roads) has weakened the concrete on some flyovers. Construction programmes are underway to strengthen them. Traffic delays are inevitable where flyovers are being repaired.

SUMMARY

The design of any structure must take account of:

- the forces likely to act on the structure,
- how to strengthen the structure to withstand the forces on it,
- the best materials to use to make the structure, considering factors such as strength, weight and cost.

On the scaffold

To build a high wall, builders stand on a scaffold platform erected alongside it. This is much safer than using a ladder. The scaffold needs to be strong to support the weight of the bricklayers and the bricks.

The scaffold is made of metal tubes clamped together. *Why are metal tubes used rather than metal bars?*

Some of the tubes are vertical, some horizontal and some are 'diagonal'. The horizontal tubes support the weight of the people and bricks on the platform. The vertical tubes support the horizontal tubes. *What are the diagonal tubes for?* If they were not included, the scaffold could collapse sideways. Their effect is to spread the weight of the people and bricks on the scaffold evenly across the structure.

Figure 16.5E ▲ Scaffolding

CHECKPOINT

1 Make a sketch of a well-known bridge and describe how the 'span' of the bridge is supported.

2 (a) Explain why an arch bridge is much stronger than a simple beam bridge of the same length.
(b) Cardboard storage drums are much stronger than cardboard boxes. Why?

3 (a) The diagram shows how reinforced concrete is used in the construction of a building. What is the purpose of the steel bars in the concrete?
(b) In the diagram, why is the steel bar in the top part of the concrete in the balcony section and in the lower part of the floor section?

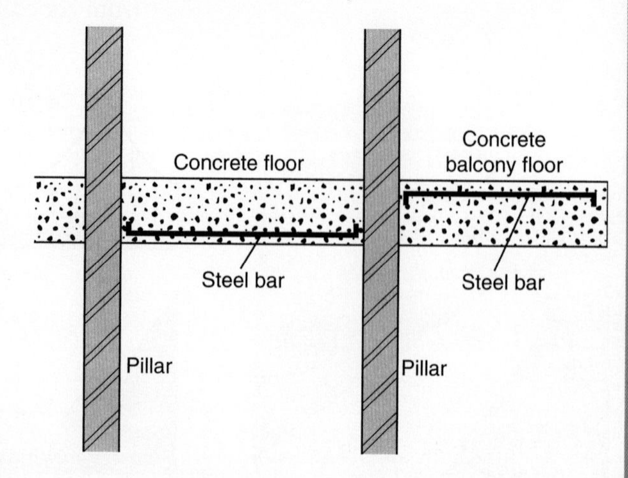

4 Elderly people often have difficulty in getting in and out of the bath. A 'bath seat' could be useful to an elderly person.
(a) Why can taking a bath be difficult for an elderly person?
(b) Design a bath seat to make it easier. Explain your choice of materials.

Topic 17 Pressure

17.1 What is pressure?

FIRST THOUGHTS

Pressure is important wherever you are and whatever you do! In this topic you will find out why.

If you have stood barefoot on a sharp object, you will have found out about **pressure** in a very painful way. All your weight acts on the tip of the object, so there is a huge pressure on your foot at the area of contact.

Pressure is caused when objects exert forces on each other. The pressure caused by any force depends on the area of contact where the force acts, as well as on the size of the force.

Ellen weighs 500 N. She stands barefoot on the kitchen floor. Then she puts on her stiletto heels and walks across the kitchen. Heelmarks appear on the vinyl floor wherever she treads in her high-heeled shoes. *Why has this happened*? Her weight has not changed, so the force she exerts on the floor has not changed. However, the area over which her weight acts has changed, changing the pressure that she exerts on the floor.

Snowshoes like those shown in Figure 17.1A are useful for walking across soft snow without sinking into it. Each showshoe has a much bigger area than the average human foot. The weight of the wearer is spread over a much bigger area of contact than in ordinary shoes. Therefore the pressure exerted on the snow by someone wearing snowshoes is much less than that exerted by the same person wearing ordinary shoes.

Figure 17.1A ▶ Spreading the weight

Foot pressure

To estimate the pressure on your feet first draw around your shoes on centimetre squared paper. Count the number of squares in each footprint (ignoring any square less than half filled) to find the area of contact in square centimetres (cm^2). Convert this to square metres (m^2) using the conversion 10 000 cm^2 = 1 m^2.

Use bathroom scales to find your weight. If the scales read mass in kg, work out your weight in newtons using $g = 10$ N/kg.

Work out the pressure on your feet using

$$\text{Pressure} = \frac{\text{Weight}}{\text{Area}}$$

Pressure is defined as force per unit area. The unit of pressure is the pascal (Pa), which is equal to one newton per square metre (N/m^2).

$$P = \frac{F}{A}$$

where P = pressure in pascals
A = area of contact in square metres
F = force in newtons acting at right angles to the surface

How much has the pressure Ellen exerts on the floor increased as a result of her wearing stiletto heels? Her weight is 500 N. The area of her bare foot in contact with the floor is 50 cm^2, which equals 0.0050 m^2. If she stands barefoot in the kitchen, the pressure she exerts on the floor with both feet is given by

$$P_1 = \frac{\text{Weight}}{\text{Contact area}} = \frac{500}{2 \times 0.0050} = 50\,000 \text{ Pa}$$

The area of each stiletto heel is 1.0 cm^2 = 0.0001 m^2. Work out the pressure she exerts on the floor now. You should obtain a value equal to 50P_1. In other words, the pressure is 50 times greater as a result of wearing the stiletto heels. No wonder the stiletto heels mark the floor.

SUMMARY

Pressure is force per unit area. The unit of pressure is the pascal, which is equal to 1 N/m^2.

CHECKPOINT

1. Explain each of the following.
   (a) When you do a handstand, the pressure on your hands is greater than the pressure on your feet when you stand upright.
   (b) Caterpillar tracks are essential for vehicles used in the Arctic.
   (c) A sharp knife cuts meat more easily than a blunt one.
2. Hospital patients confined to bed need to be moved to stop bed sores forming. These occur where the body presses on the bed for long periods. Estimate your area of contact when you lie on a bed and work out the pressure of your body on the bed.
3. A concrete paving slab of weight 1400 N has dimensions $1.00\ m \times 0.80\ m \times 0.050\ m$. What pressure does it exert when (a) it is laid flat on a bed of sand, (b) it stands upright on its shorter side?
4. Skis are designed to allow the wearer to travel fast across snow. Why are skis made so much longer than even the largest human foot?

17.2 Pressure at work

FIRST THOUGHTS

Huge forces can be produced and huge loads shifted using hydraulic pressure systems. Read on to find out how hydraulic machines work.

Figure 17.2A Mechanical digger

Roadworks are an everyday feature of life in any town or city. Mains services such as water and gas reach our homes through underground pipes, usually beneath the road. Mechanical diggers are usually used to reach broken pipes. The 'grab' of the digger, designed to remove earth, operates by a hydraulic pressure system. Most machines that shift or lift things operate hydraulically. The hydraulic system of a machine is its 'muscle power'.

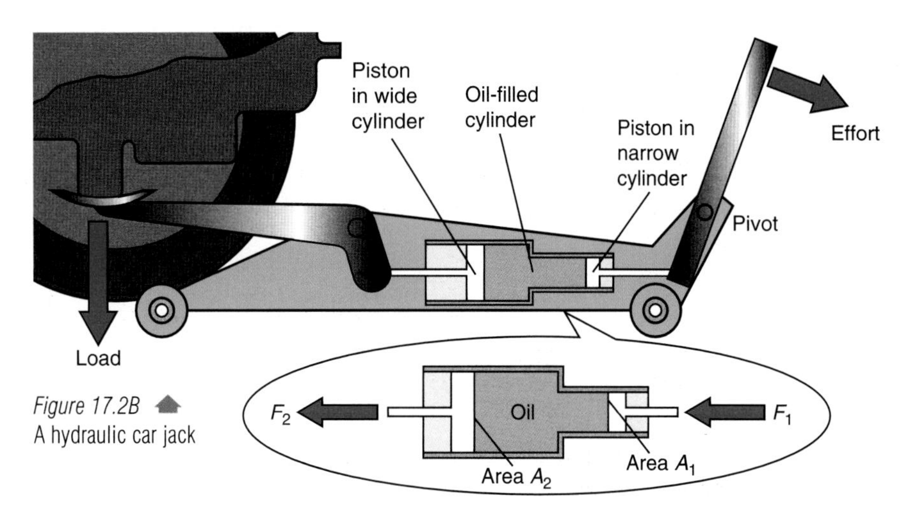

Figure 17.2B ▲
A hydraulic car jack

See www.keyscience.co.uk for more about forces.

A car can be lifted 'by hand' using a hydraulic car jack (Figure 17.2B). When you press the handle down, a narrow piston is forced along an oil-filled cylinder. The oil is forced out of this cylinder along a pipe and into a wider cylinder. The pressure of the oil forces the piston in this cylinder outwards. The second piston acts on a lever to raise the car. The force applied to the narrow piston (the master piston) causes a large pressure on the oil. This pressure then acts on the wide piston (the slave). In Figure 17.2B the pressure on the narrow piston $= F_1/A_1$ so

Force F_2 on the large piston = Pressure × Area of large piston

$$F_2 = \frac{F_1}{A_1} \times A_2$$

Since area A_2 is much greater than area A_1, then F_2 is much greater than F_1. A force on the first piston supplies a bigger force to lift the car. The jack therefore acts as a 'force multiplier'.

Hydraulic brakes use pressure. When the driver presses the brake pedal of a car, pressure is exerted on the oil (or brake fluid) in the master cylinder (Figure 17.2C). This pressure is transmitted along oil-filled brake pipes to a slave cylinder at each wheel. The oil pressure forces the piston in each slave cylinder to push the brake disc pads on to the wheel disc.

The larger the area of a piston, the greater the force it can exert.

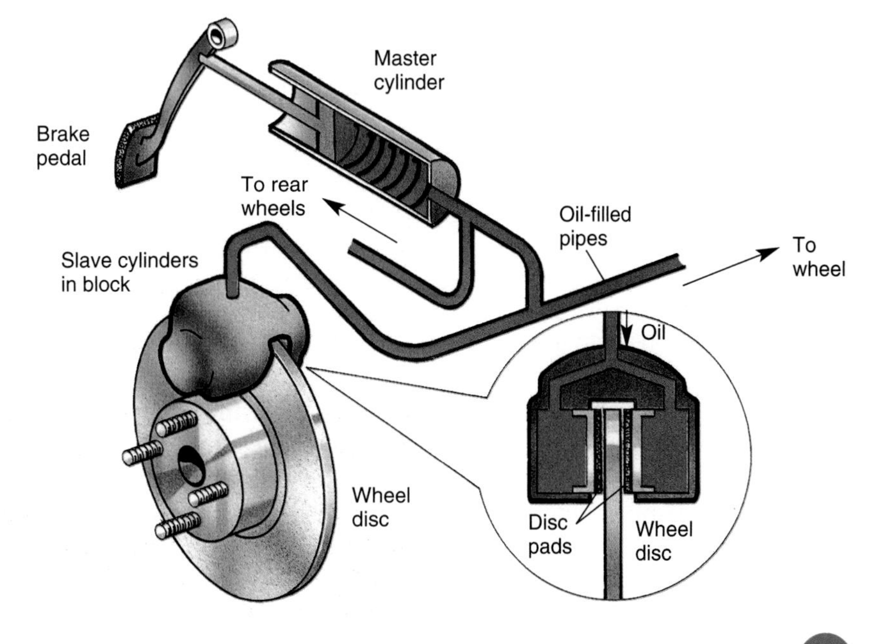

Figure 17.2C ▸
Disc brakes

Power-assisted brakes fitted to heavy goods vehicles and coaches use compressed air (air brakes). When the driver applies the brakes, compressed air at very high pressure is released to push on the piston in the master cylinder. The compressed air is used instead of the force of the brake pedal. That is why such vehicles hiss when the brakes are released.

Robots use hydraulics for 'muscle power'. Robots are used more and more in factories. Do not let your imagination run away with you though. The robot of science fiction is a long way off. Factory robots are fixed machines that operate such things as welding gear or paint sprays on assembly lines. They work non-stop without the need for a human operator. Robot muscles use compressed air.

Figure 17.2D ▲ Remote controlled robot used by bomb disposal teams for handling suspect packages

SUMMARY

Hydraulics are used to shift or lift heavy objects. The pressure in a hydraulic system is transmitted through the oil or air in the system. The force on the slave piston due to the oil pressure is much greater than the force applied to the master piston.

CHECKPOINT

1 Write down as many things as you can think of that are operated hydraulically.

2 A force is applied to the brake pedal of a vehicle, creating a pressure of 30 kPa in the brake fluid pipes. The area of the piston of the master cylinder is 0.0010 m^2.
   (a) Calculate the force exerted on this piston.
   (b) In Figure 17.2C, explain whether the force applied to the brake pedal is more or less than the force exerted on the piston of the master cylinder.
   (c) The pistons in the slave cylinders each have an area of 0.010 m^2. Calculate the force exerted on the wheel disc by each slave cylinder piston.

3 Why is it important to ensure that air does not leak into the brake fluid of a car?

4 The hydraulic lift shown has four pistons, each of area 0.01 m^2, to lift the platform. The pressure in the system must not be greater than 500 kPa. The platform weight is 2 kN. What is the maximum load that can be lifted on the platform?

Ramp
Ground level
Oil
Area 0.01m^2

5 Figure 17.2B shows a hydraulic car jack being used to raise one side of a car off the ground.
   (a) If the wide piston has an area which is 8 times greater than the area of the narrow piston, calculate the force exerted by the oil on the wide piston when a force of 100 N is applied to the narrow piston.
   (b) Explain why the effort needed to apply this 100 N force to the narrow piston is less than 100 N.
   (c) State one further reason why the force applied to raise the load is much greater than the effort.

6 Figure 17.2A shows the arm of a mechanical digger. It is controlled by three hydraulic pistons, referred to as 'rams', labelled X, Y and Z.
   (a) Explain why the arm is raised when compressed air is released into ram X.
   (b) What is the purpose of each of the other two rams, Y and Z?

17.3 ▶ Pressure in liquids

FIRST THOUGHTS

In a two-storey house, why is the water pressure upstairs less than the pressure downstairs? Find out in this section.

Imagine you are swimming underwater with a snorkel tube. Provided the top of the tube is above water, you ought to be able to breathe air safely. *Why not give deep-sea divers very long snorkel tubes*? They would not work because of the huge pressure on the diver. The diver's chest muscles would not be strong enough to expand his or her chest muscles against the water pressure on his or her body. So the diver would be unable to draw air down the tube.

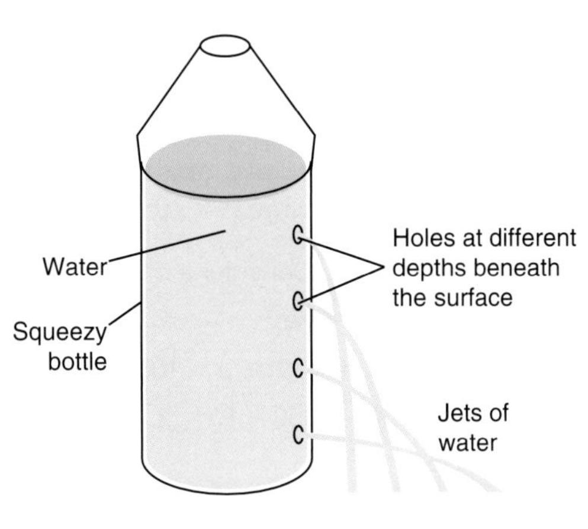

Figure 17.3A ▲ Pressure increases with depth

The pressure in a liquid increases with depth. A simple way to show this is to make small holes down the side of an empty plastic bottle. Then fill the bottle with water and place it in a bowl. A jet of water will emerge from each hole, as shown in Figure 17.3A. The deeper the hole is beneath the water level in the bottle, the greater the pressure of the jet.

The pressure in a liquid depends on depth and density.

The pressure along a fixed level is constant. Use the same bottle and make several holes around the bottle at the same level. The jets from these holes should be at the same pressure.

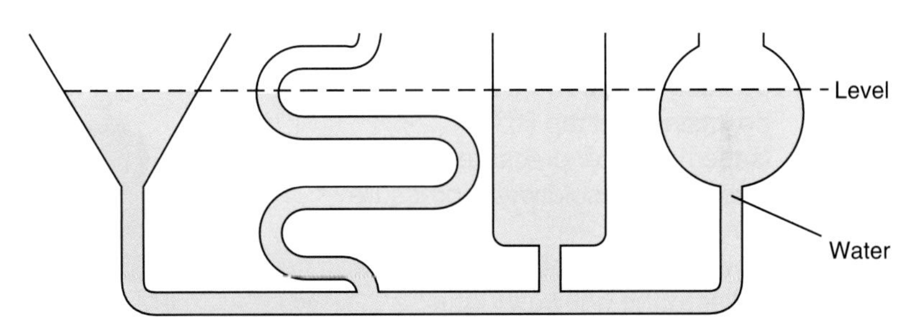

Figure 17.3B ▲ Pascal's vases

Pascal's vases, shown in Figure 17.3B, consist of several containers linked so that water can flow between them. When water is poured into one of them, the water level in each one rises until it is the same in all the containers. This is because the water will not come to rest until all the pressures on it are equalised. The pressure of the water in the container depends only on its depth, so if the depths are equal, the pressures are equal.

The pressure depends on density. Suppose water is poured into a U-shaped tube, as in Figure 17.3C, until the level is about one third of the way up each side. Then oil is poured carefully down one side to form a column on top of the water. The oil level is higher than the water level because oil is less dense than water.

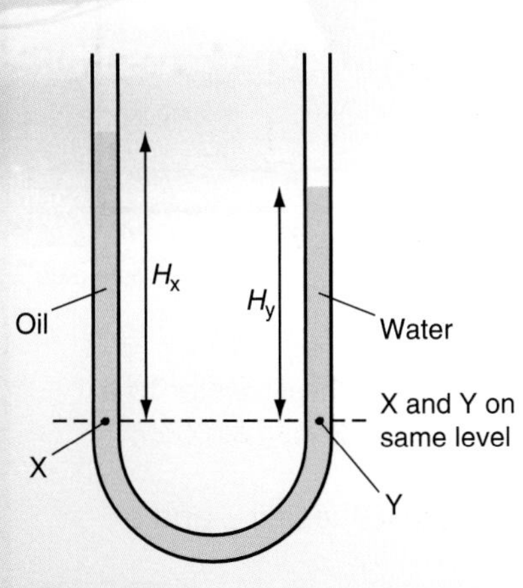

Figure 17.3C ▲ Comparing densities of liquids

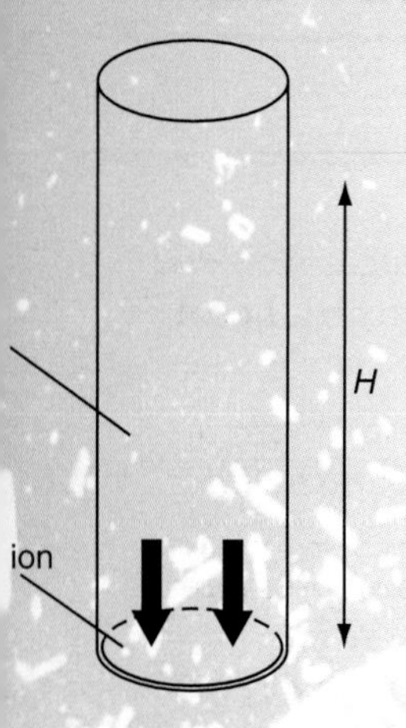

Figure 17.3D ⬆ Calculating liquid pressure

SUMMARY

The pressure in a liquid increases with depth. Along a fixed level, the pressure does not change.

Pressure of a liquid column

How much pressure is exerted by a liquid column? Consider the column in Figure 17.3D.

Volume of liquid in the column = Area of cross-section $A \times$ Height H
Mass of liquid = Volume $\times$ Density of liquid $\rho = A \times H \times \rho$
Weight of liquid = Mass $\times g = A \times H \times \rho \times g$ (where g = 10 N/kg)

$$\text{Pressure at base of column} = \frac{\text{Weight of liquid}}{\text{Area of cross-section } A} = \frac{AH\rho g}{A} = H\rho g$$

Pressure of a liquid column = $H\rho g$

Worked example Calculate the pressure due to sea water on the floor of the sea bed at a depth of 200 m. The density of sea water is 1050 kg/m^3. Assume g = 10 N/kg.

Solution Consider a column of sea water of height 200 m above the sea bed.

Using $p = H\rho g$,

where H = 200 m, ρ = 1050 kg/m^3,

$p = 200 \times 1050 \times 10 = 2.10 \times 10^6$ Pa

CHECKPOINT

1 In most homes, the water pressure upstairs is less than the pressure downstairs. The diagram shows a house with a water tank in the loft to supply the hot water boiler.

(a) Why is the cold water pressure upstairs (at tap A) less than downstairs (at tap B)?
(b) Is the hot water pressure at C less than the cold water pressure at A?
(c) How does hot water get from the boiler to the hot water tank?
(d) What is the function of the expansion pipe?

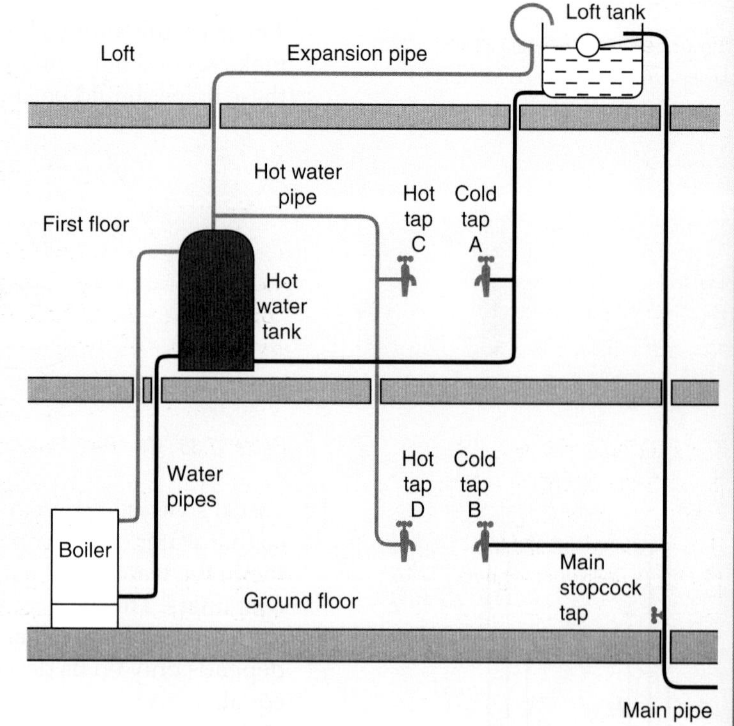

2 Why is the wall of a dam thicker at the base than at the top?

3 In a hydro-electric scheme, water is piped from an upland reservoir down to a generator station, 700 m below the reservoir. Work out the pressure in the water pipe from the reservoir when at the station. The density of water is 1000 kg/m^3. Assume g = 10 N/kg.

4 A sink plug has an area of 0.0006 m^2. It is used to block the outlet of a sink filled with water to a depth of 0.09 m.

(a) What is the pressure on the plug due to the water?
(b) How much force is needed to remove the plug from the outlet? Use the values for ρ and g given in question 3.

17.4 Measuring pressure

FIRST THOUGHTS

In this section you will learn how pressure is measured and why such measurements are important.

Keeping the recommended pressure in a system means measuring the pressure regularly, so that action can be taken if the pressure is not at the correct level.

What do nurses, car mechanics, gas board officials and submariners have in common? They all need to measure pressure. Nurses measure blood pressure, car mechanics measure tyre pressure, gas board officials measure pressure to detect gas leaks and submariners use a pressure indicator as a depth gauge.

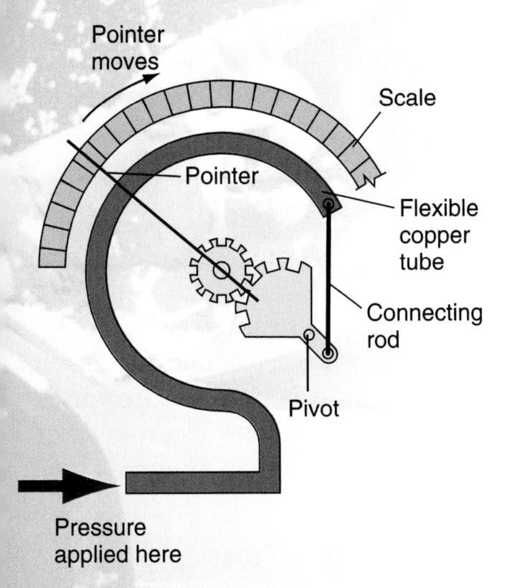

Figure 17.4A ▲ Bourdon gauge

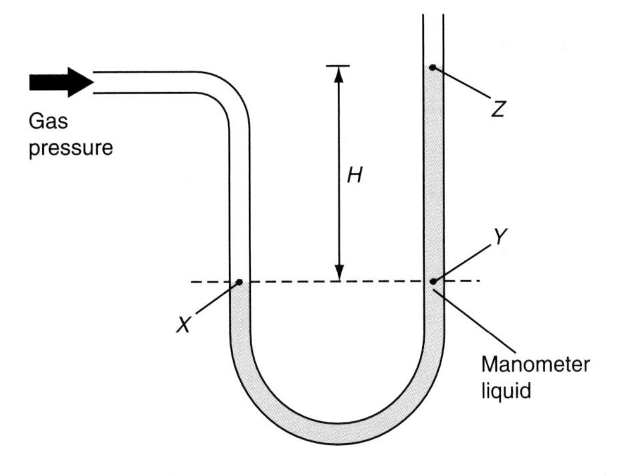

Figure 17.4B ▲ The U-tube manometer

The **Bourdon gauge** contains a flexible copper tube (Figure 17.4A). When pressure is applied at the inlet, the tube uncurls a little. This movement is magnified by a lever to make a pointer move across a scale. Tyre pressure is measured using a Bourdon gauge. The reading is the excess pressure in the tyre above atmospheric pressure.

The **U-tube manometer** (Figure 17.4B) is a much simpler instrument than the Bourdon gauge. The gas pressure forces the manometer liquid up the open side of the U-tube until it remains steady when the difference in levels balances the gas pressure. The height difference between the levels, H, is measured. The equation $p = H\rho g$ gives the excess pressure of the gas (i.e. its pressure above atmospheric pressure) where ρ is the density of the manometer liquid.

Electronic pressure gauges use special crystals that generate a voltage when squeezed. This is called the **piezoelectric effect**. You may have used a piezoelectric gas lighter: press the trigger of the lighter and the high voltage generated produces sparks. In this type of pressure gauge, the pressure to be measured squeezes the crystal. A voltmeter connected across the crystal measures the voltage generated and thus the pressure.

SUMMARY

There are three main types of pressure-measuring instruments. They are the Bourdon gauge, the U-tube manometer and the electronic pressure gauge.

CHECKPOINT

1. In Figure 17.4B, atmospheric pressure acts at point Z. The gas pressure to be measured acts at X. How does the pressure at X compare with:
   (a) the pressure at Z,
   (b) the pressure at Y?
   (c) The manometer liquid is water. Its density is 1000 kg/m^3. If the difference in levels is 0.25 m, what is the difference between the pressure at X and at Z? Assume g = 10 N/kg.

2 Typical values of car tyre pressures are in the range 150 to 200 kPa. What would be the height of a column of water that produced a pressure of 200 kPa at its base? Use the values of ρ and g in question 1. Why is a Bourdon gauge more suitable for measuring tyre pressure than a U-tube manometer?

3 Blood pressure is usually expressed in millimetres of mercury rather than in pascals. This unit is used because the most common type of blood pressure gauge contains a tube of mercury. The blood pressure of a healthy human is 120 mm of mercury, on average. Work out what this pressure is in pascals, given the density of mercury is 13 600 kg/m^3 and g = 10 N/kg. (See KS: Biology. Topic 14.4 for more about blood pressure.)

4 State one advantage and one disadvantage of an electronic pressure gauge compared with (a) a Bourdon pressure gauge, (b) a U-tube manometer.

17.5 Atmospheric pressure

FIRST THOUGHTS

The fact that the atmosphere exerts pressure was not proved until the seventeenth century. Changes in the weather can be predicted by measuring atmospheric pressure.

The Earth's atmosphere extends 100 km or more into space. It becomes less dense the higher it is above sea level, so the pressure falls faster with height than it does in a liquid. At sea level, atmospheric pressure varies from day to day, changing with weather conditions. Fine, clear weather usually occurs when atmospheric pressure is higher than average.

Investigating atmospheric pressure

- You can use atmospheric pressure to siphon water from a tank (Figure 17.5A). Use a clean tube and suck water into it from the tank. Keep the end of the tube in the water. Take the other end from your mouth and hold it below the tank. *Why does water flow down the tube from the upper tank?*
- Drinking through a straw makes use of atmospheric pressure (Figure 17.5B). Sucking on the straw reduces the pressure in the straw. The pressure of the atmosphere on the liquid in the glass then forces the liquid up the straw from outside.
- Estimate atmospheric pressure using a suction cap (Figure 17.5C). Measure the area of the cap and then estimate the force needed to pull it off a wall. You can then work out the atmospheric pressure by Pressure = Force/Area. Its value is about 100 kPa.

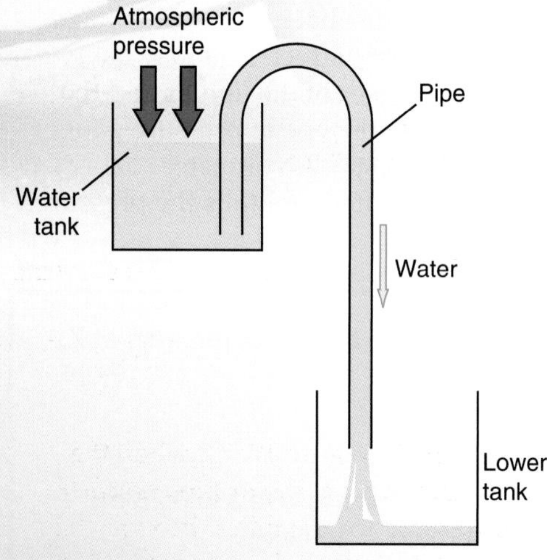

Figure 17.5A ▲ A siphon

Figure 17.5B ▲ Sucking through a straw

Atmospheric pressure
Pull
Suction cap

Figure 17.5C ▲ A suction cap

EXTENSION FILE ACTIVITY

KEY SCIENTIST

Figure 17.5D ▲ Blaise Pascal (1623-62)

Pascal and pressure

Fill a milk bottle in a bowl of water and then hold the bottle upside down with its open end under the water. *Why doesn't the water fall out? What would remain in the bottle if the water did fall out?* The answer is nothing – a vacuum. Some people would say the water does not fall out because 'nature dislikes a vacuum'. Do you think this is a satisfactory explanation?

Blaise Pascal was a seventeenth century French scientist who did not think much of such vague statements. He carried out experiments to explain them scientifically. Here are some simplified versions of his experiments.

1. The inverted bottle experiment uses a tube instead of a milk bottle. The tube is sealed at one end with a cork (Figure 17.5E). When the tube filled with water is inverted in a bowl of water, the water remains in the tube. *What happens if the tube is raised so its lower end is lifted out of the water?*
2. Suppose the cork is removed from the inverted tube in Figure 17.5E. *What happens now?*

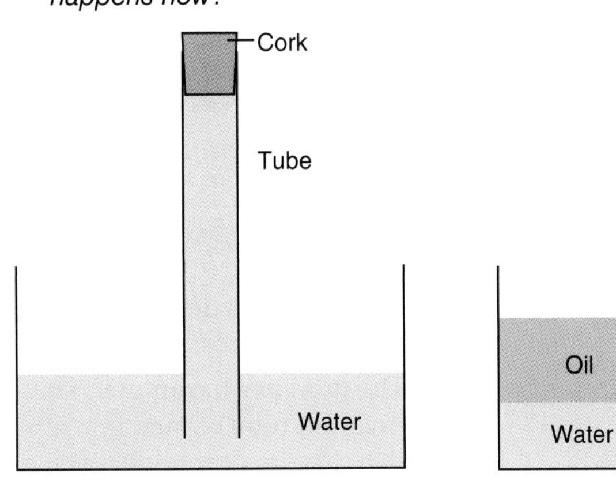

Figure 17.5E ▲

Figure 17.5F ▲

3. Now suppose oil is poured onto the water outside the tube. Oil is less dense than water so it floats on the water surface. Water goes up the inside of the tube (Figure 17.5E). *Why?*

Pascal realised that the weight of the oil pushing on the water outside the tube forced water up the tube. *What do you think would happen if oil was poured on to the water inside the tube as well?*

Can you use these observations to explain why the water remains in the tube in Figure 17.5E until the cork is removed? In Figure 17.5F the weight of the oil on the water forces the water up the tube. *So why does the water drop out when the cork is removed?* Letting air into the tube at the top is like pouring oil on to the water inside the tube. The water is pushed down the tube until it is at the same level as the water outside the tube.

Pascal realised that air has weight and the weight of the atmosphere creates pressure. Atmospheric pressure acting on the water outside the tube in Figure 17.5E pushes the water up inside the tube. When the cork is removed, atmospheric pressure now acts on the water in the tube as well as on the water outside. So the water in the tube is pushed down to the same level as the water outside the tube.

In recognition of Pascal's work on pressure, the unit of pressure is named after him.

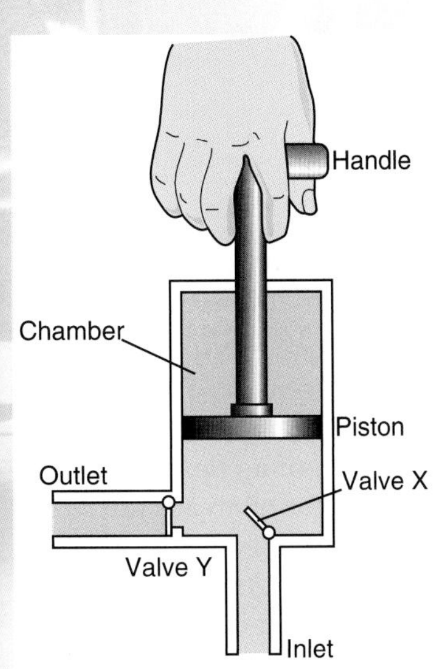

❶ Handle raised
❷ Piston pulled up
❸ Valve X opens, valve Y closes and chamber fills from inlet with water
❹ Handle pushed down, piston moves down
❺ Y opens, X closes and water forced through outlet

Figure 17.5G ▲ The force pump

Instruments that use atmospheric pressure

Pumps make use of atmospheric pressure. In Figure 17.5G, when the handle is pulled up, the piston is withdrawn from the pump chamber. Atmospheric pressure then forces water into the chamber through the inlet valve X. When the piston is pushed back in again, this valve closes and the outlet valve opens, so water leaves the chamber.

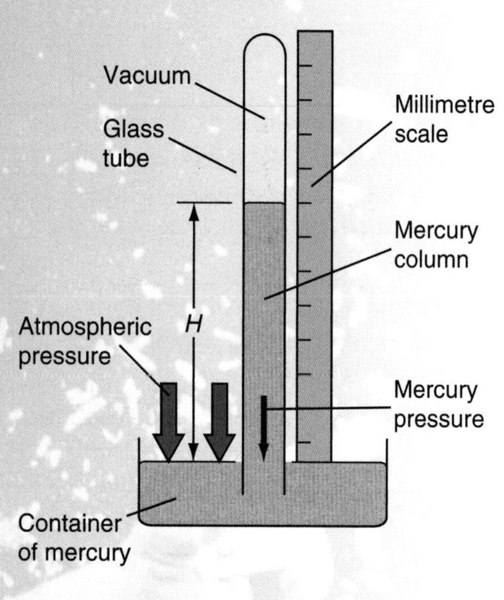

Figure 17.5H ▲ The mercury barometer

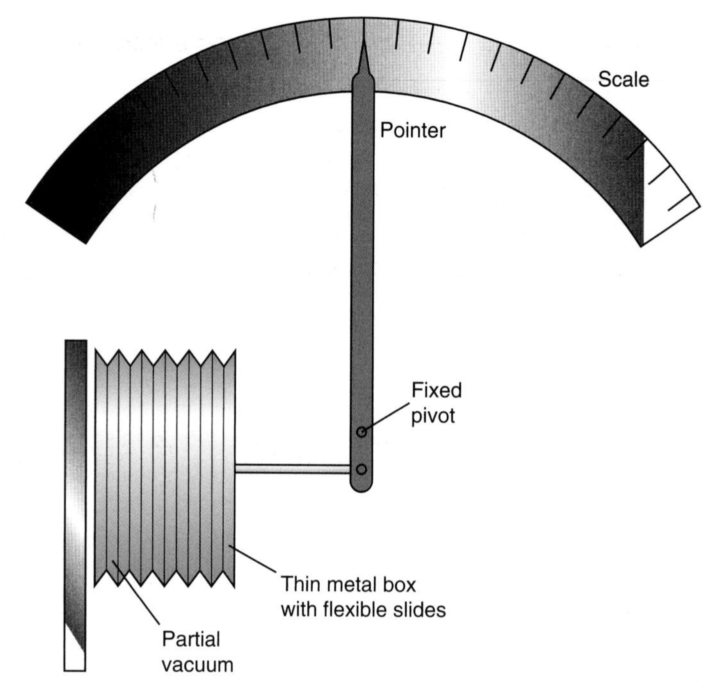

Figure 17.5I ▲ The aneroid barometer

It's a fact!

A spectacular demonstration of the force of atmospheric pressure took place in Magdeburg, Germany in 1651. Otto von Guericke made two thick copper hemispheres that fitted together with an airtight leather seal to form a sphere of diameter about 30 cm. The air inside was pumped out. The two hemispheres could not be pulled apart, even by two teams of eight horses.

SUMMARY

Atmospheric pressure at sea level changes slightly from day to day. Its average value is 760 mm of mercury. It can be measured using an aneroid barometer or a mercury barometer.

The **barometer** is designed to measure atmospheric pressure. There are two common types:

- The **mercury barometer**, shown in Figure 17.5H, consists of an inverted tube of mercury with its lower end under the surface of mercury in a container. The top end of the tube is sealed and there is a vacuum above the mercury in the tube. *Why doesn't the mercury drop out of the tube?* The atmospheric pressure acting on the mercury in the container holds it up by balancing the pressure due to the column of mercury in the tube.

  Atmospheric pressure $= H\rho g$

  where $H =$ height of the mercury column
  $\rho =$ density of mercury,
  $g = 10$ N/kg

  The average value of barometric height is 760 mm of mercury, often referred to as standard pressure. You can show that this is equal to 101 kPa, given that the density of mercury, ρ is 13 600 kg/m^3.

- The **aneroid barometer**, shown in Figure 17.5I, contains a sealed metal chamber with flexible sides. When atmospheric pressure increases, the box is squeezed in. The movement of its flexible sides makes a pointer move across a scale. The altimeter of an aircraft, used to measure its height, is an aneroid barometer with its dial marked in metres above sea level. It makes use of the fact that atmospheric pressure falls with increasing height above the ground.

CHECKPOINT

1. Explain how atmospheric pressure acts:
   (a) to keep a rubber sucker on a wall,
   (b) when you drink through a straw,
   (c) when water is siphoned from a tank.
2. Atmospheric pressure is about 100 kPa. What depth of water will give the same pressure? The density of water is 1000 kg/m^3. Assume $g = 10$ N/kg

3 How will the reading of a mercury barometer be affected if:
   (a) some air leaks into the top of the tube,
   (b) the tube and the scale tilt slightly?

4 The tyres of a car are at a pressure of 180 kPa above atmospheric pressure. Each of the four tyres has an area of contact with the ground of 0.015 m^2. Work out the weight of the car in newtons.

5 How would you find out if atmospheric pressure can be used to predict weather changes? Assume you have a barometer. What records would you keep each day?

17.6 Floating and sinking

FIRST THOUGHTS

Any object in a fluid is acted on by the pressure of the fluid. Find out why an object weighs less in water than in air.

Figure 17.6A ▲ Disabled people find it easier to move in water

The difference between an object's weight in air and its apparent weight in water is the upthrust of the water.

When you go swimming, have you noticed that you feel lighter in the water? Disabled people often find it much easier to move in water than in air. Water exerts an upward force on a body in it. This force is called the **upthrust** of the water.

Eureka!

Archimedes was one of the most famous scientists in Ancient Greece. His king asked him to find out if his new crown was made of pure gold, presumably to check that the royal crownmaker was not cheating. However, the king would not allow the crown to be damaged in any way.

After much thought Archimedes was no nearer solving the problem, so he decided to take a bath. In the bath he had a flash of inspiration. He realised that by weighing the crown in water and in air and making similar measurements on a piece of pure gold, he could tell whether the crown was pure gold. History records that he celebrated this discovery by running through the streets shouting 'Eureka!' (Greek for 'I've found it').

Archimedes realised that the upthrust on an object in water depended on how much water is **displaced** (i.e. pushed aside) by the object. He measured the amount of water displaced as illustrated in the exercise on the next page, and discovered the following principle

EXTENSION FILE ASSIGNMENT

The upthrust is always equal to the weight of fluid displaced. This is called Archimedes' Principle.

Measuring upthrust

The diagram shows a brick being weighed in air and then in water. The reading on the spring balance is less when the brick is in water. The difference is caused by the upthrust of the water on the brick. Work out the upthrust on the brick in the diagram.

Why do you think the reading of the spring balance changes as the brick is lowered into the water? Try it and you will find that the upthrust increases as more of the brick enters the water.

The upthrust is due to the upward pressure of the water on the underside of the brick. As the brick is lowered into the water, the upthrust increases, because the pressure on the underside increases with depth.

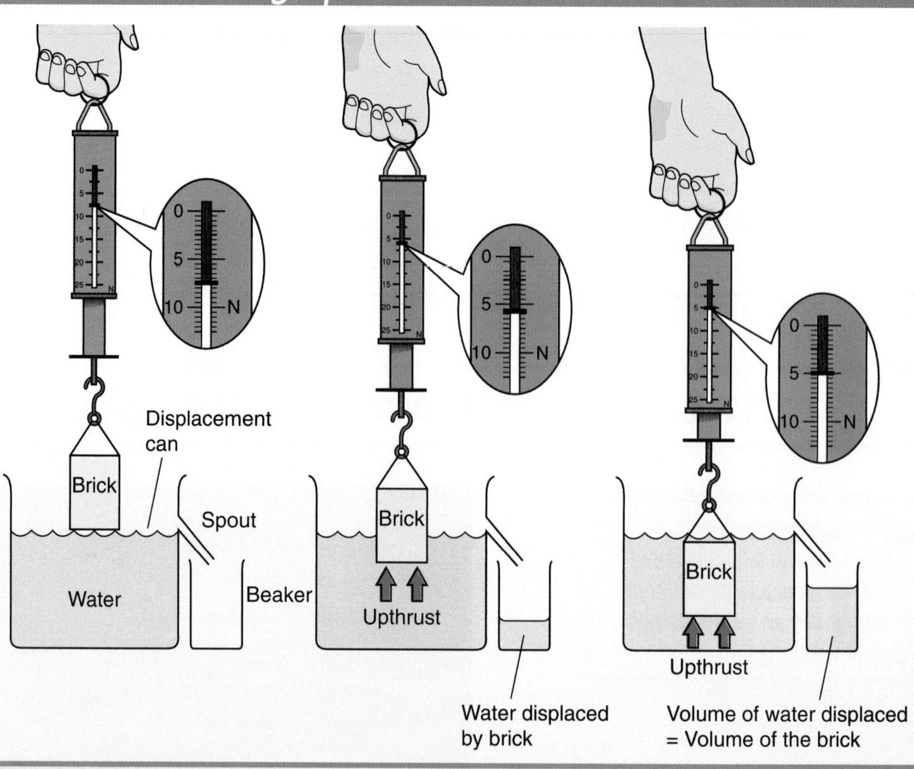

EXTENSION FILE
ACTIVITY

Will it float?

A busy waterway is fascinating to watch. Boats and ships laden with cargo float low in the water. A boat carrying cargo stays afloat provided the upthrust due to the water is equal to the total weight of the boat and its cargo.

A ship being loaded will float lower and lower in the water as the load increases. At any stage in the loading operation, the weight of water displaced by the ship is equal to the upthrust, which is greater than or equal to the total weight of the ship. If the ship is loaded too much, it sinks because the upthrust is unable to support the total weight of the ship.

Every ship has a horizontal line painted on its hull to show how low it can float safely in the water when loaded (see Figure 17.6C)-and also

Figure 17.6B ▲ Busy waterway

Figure 17.6C ▲ Loading lines and draft marks

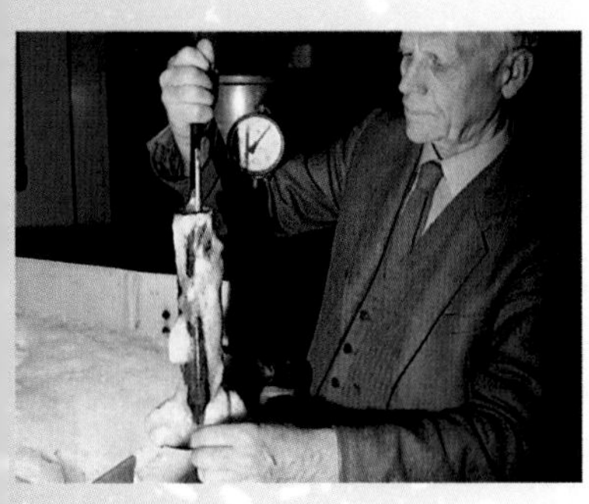

Figure 17.6D ▲ Using a hydrometer

the **Plimsoll line** after Samuel Plimsoll who guided the legislation through Parliament in 1875 to make this line compulsory. Before this time, many ships and crews were lost at sea due to overloading.

The hydrometer

In a brewery, the final product is tested by using a hydrometer to measure its density. If the beer's density is too low, it contains too much water; if the density is too high, the beer is too strong.

Figure 17.6D shows a hydrometer in use. The density of the liquid is given by the level of the liquid on the float. The density can be read off the scale.

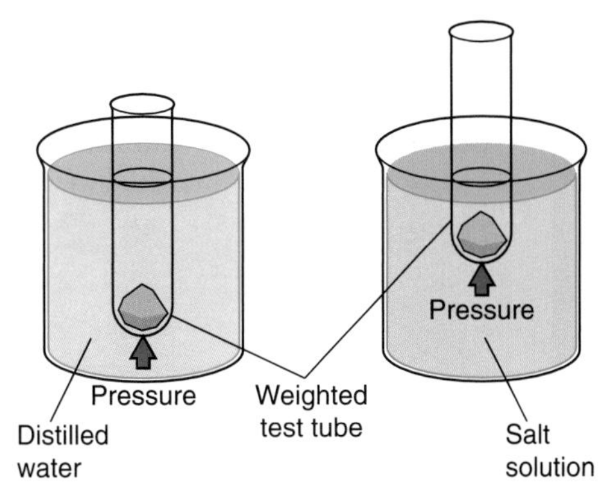

Figure 17.6E ▲ Using a weighted test tube as a hydrometer

A weighted test tube can be used as a hydrometer as shown in Figure 17.6E. The test tube floats higher in a salt solution than in pure water. Therefore the volume of salt solution displaced is less than the volume of pure water displaced. The weight of liquid displaced is the same in both cases, equal to the weight of the tube. Therefore the density of salt solution is greater than that of pure water.

SUMMARY

A body in water experiences an upthrust equal to the weight of the water it displaces. If the upthrust equals the weight of the body, the object does not sink.

CHECKPOINT

1. (a) Why is it difficult to hold an inflated plastic ball under water?
   (b) Why is cork a suitable material for filling life belts?
   (c) When a submarine surfaces, it uses compressed air to push water out of its 'ballast' tanks. Why does this allow it to rise?
2. In the diagram at the top of the previous page work out:
   (a) the mass of the brick, assuming a mass of 1 kg has a weight of 10 N,
   (b) the upthrust on the brick when it is totally submerged,
   (c) the density of the brick, assuming the density of water 5 1000 kg/m^3.
3. (a) A block of wood of weight 5.0 N floats in water, as shown. What is the upthrust on the block?
   (b) What weight of water is displaced by the block?
   (c) Work out the mass of the water displaced, assuming a mass of 1 kg has a weight of 10 N.
   (d) What can you say about the density of the block compared with the density of water?
   (e) A solid object is released in water and it sinks. What does this tell you about (i) the upthrust on the object in relation to its weight, (ii) the density of the object in relation to the density of water?

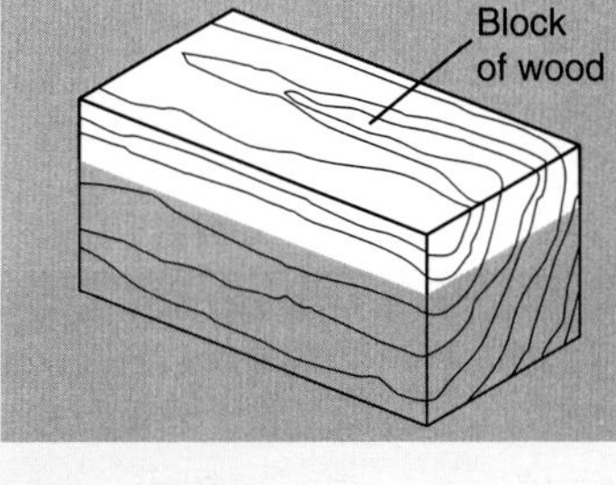

4. (a) Ice floats on water. What does this tell you about the density of ice compared with that of water?
   (b) In winter, why is it possible for fish to remain alive in ponds that are covered with ice?

17.7 Fluid flow

FIRST THOUGHTS

Birds and aeroplanes have similar body shapes and so do fish and submarines. In this topic, you will find out why.

If you are a keen cyclist, you will know about air resistance and its effects. To attain high speed, you need to keep your head down to reduce air resistance. Most forms of transport are shaped to cut down air resistance in order to reduce fuel consumption. If you think cyclists don't need to worry about fuel consumption, think again! Muscles use energy from food to do work.

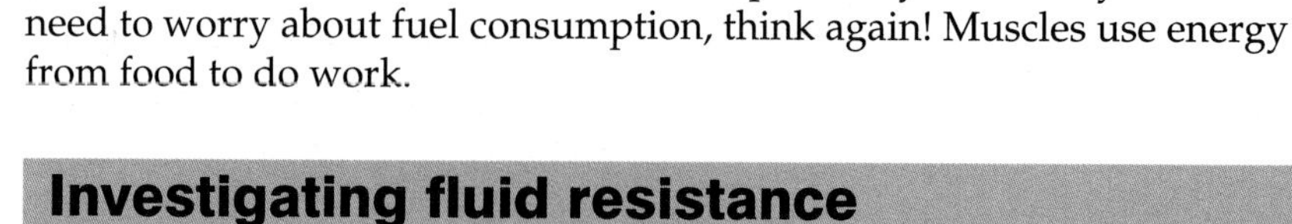

Figure 17.7A ▲ Reducing air resistance

Investigating fluid resistance

Make differently shaped boat hulls and compare the forces needed to pull each one through water. The force is needed to overcome the resistance of water to motion. This is called **fluid resistance**.

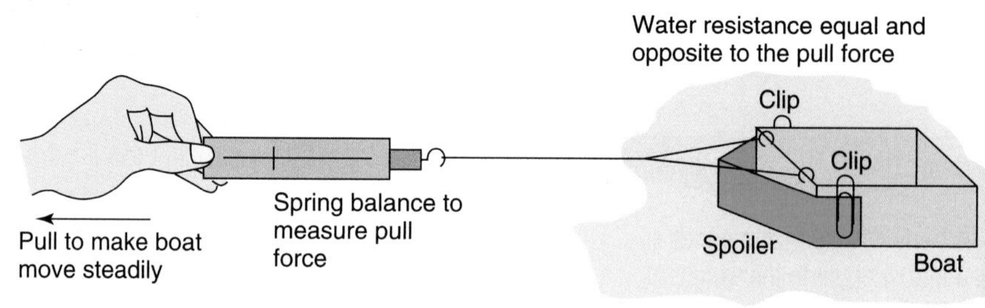

Figure 17.7B ▲ Measuring fluid resistance

You could start with an empty margarine tub, suitably weighted, and fit differently shaped 'spoilers' to it to test the effect of shape on fluid resistance. *Is the fluid resistance less if the bow is tapered or round? Should the stern be tapered or round?* Use your knowledge about the shape of fish to predict the best shape and then see if the measurements fit your prediction.

Streamlining

The designers of boats, planes and vehicles test body shapes to ensure minimum fluid resistance. They use flow tanks and wind tunnels. The flow of a fluid past a surface can be visualised by injecting coloured dye into the flow stream and observing the path of the dye. The path of the dye is called a **flowline**.

Flowlines that are stable are called **streamlines** and indicate steady flow. The streamlined shape of a fish is such that water passes over its surface steadily and the fish moves through the water effortlessly.

If the flowlines are unstable and move about unpredictably, the flow is said to be **turbulent**. If you have been on an aeroplane affected by turbulence, you will know that the motion is very uncomfortable. Turbulent flow creates unpredictable currents in the fluid and energy is wasted as a result.

The shape of an object can cause turbulent flow. For example, a boat with a box-shaped hull would leave turbulence in its wake as it moves through water. Even if the shape of the bow is rounded, a box-shaped stern would create turbulence in the wake. However, if the stern is tapered, the flow remains streamlined and so the resistance to motion is less.

Figure 17.7C ▲ Streamlines

It's a fact!

Painting and decorating Paint is a **thixotropic** fluid because it becomes less viscous as it is stirred. Wallpaper paste is a fluid that becomes more viscous as it is stirred. Fluids that become more or less viscous with stirring are referred to as non-newtonian fluids in comparison with newtonian fluids which do not change viscosity when stirred.

Understanding fluid resistance

When a fluid flows in a pipe or channel or a body moves through a fluid, the resistance to motion increases with speed. However, fluid resistance also depends on the fluid itself. Imagine pulling your test 'boat' in Figure 17.7B through syrup instead of water. Syrup is said to be more **viscous** than water because it flows less easily.

The **viscosity** of a fluid is a measure of its resistance to flow. The more easily a fluid flows, the lower is its viscosity. Gases are much less viscous than liquids and some liquids such as syrup and treacle are much more viscous than others.

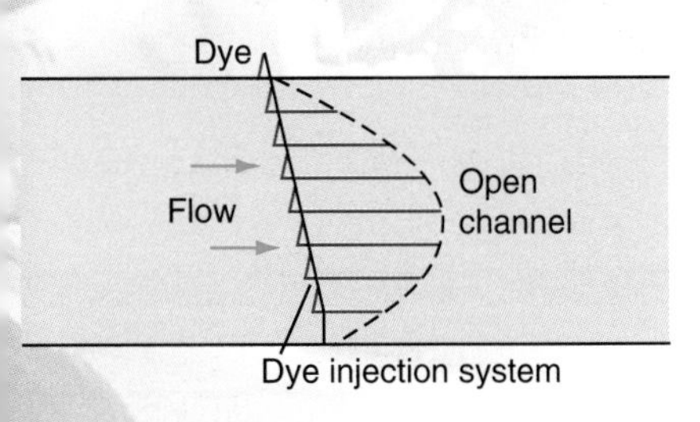

Figure 17.7D ▲ Flow in a pipe

In the pipeline

Gas, water and oil are just three of many fluids that are pumped through pipes. Pressure from a pump is needed to force a fluid through a pipe or along a channel because internal friction due to viscosity acts against the flow. The pipe surface drags on the adjacent fluid which drags on the fluid further away. Fig 17.7D shows how the speed of flow in a pipe varies across the pipe.

The flow rate through a pipe is the amount of fluid that flows per second in the pipe. The flow rate depends on

- the pressure gradient along the pipe. This is the pressure drop per metre along the pipe
- the pipe width
- the type of fluid and its temperature.

The flow rate in a pipe can be measured using a **Venturi tube** fitted to the pipe, as shown in Figure 17.7E. The speed of the fluid in the pipe is greatest where the tube is narrowest. This is where the pressure in the fluid is least, in accordance with the **Bernouilli principle**.

The pressure in a fluid decreases where the speed increases.

The lift force on an aircraft wing is explained by this principle, (see p. 284). Some further applications of the principle are shown in Figure 17.7F.

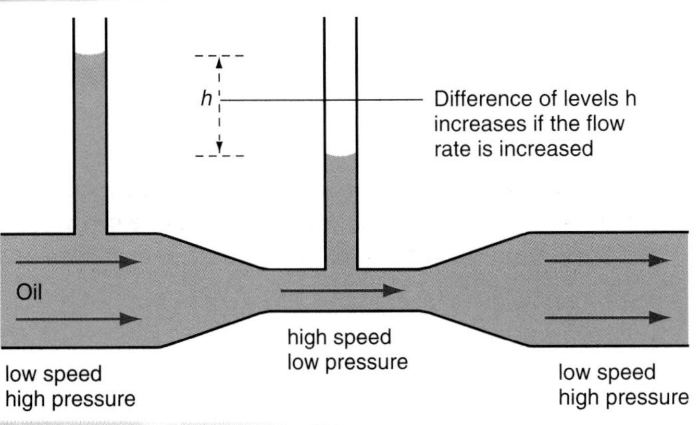

Figure 17.7E ▲ The Venturi tube

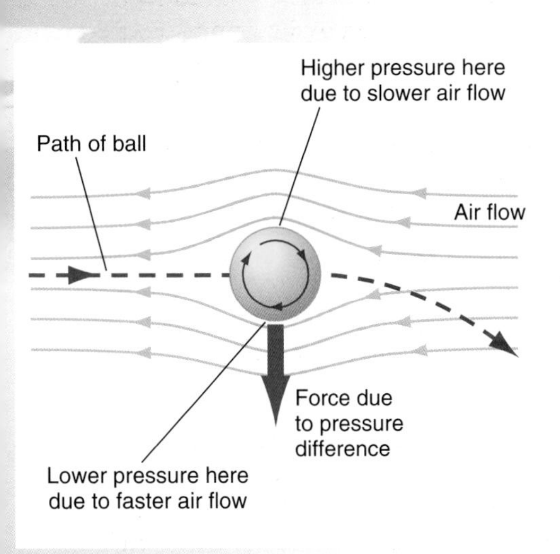

(a) Making a ball swerve

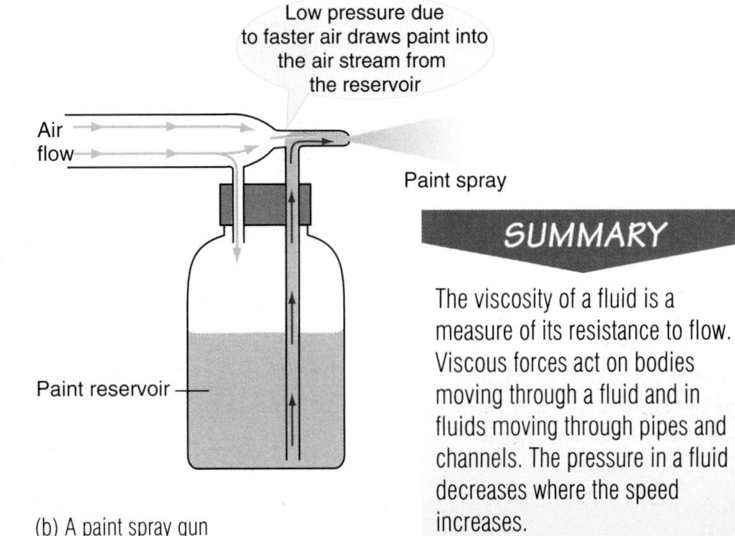

(b) A paint spray gun

Figure 17.7F ▲ Bernouilli effects

SUMMARY

The viscosity of a fluid is a measure of its resistance to flow. Viscous forces act on bodies moving through a fluid and in fluids moving through pipes and channels. The pressure in a fluid decreases where the speed increases.

CHECKPOINT

1. Explain why:
   (a) a motor cyclist crouches behind the handle bars at high speed,
   (b) a high speed locomotive is sloped at the front,
   (c) a table tennis ball falls more slowly than a solid rubber ball of the same size.
2. List the following newtonian fluids in order of increasing viscosity: oil, air, water, treacle, milk.
3. In a central heating system, a pump is used to force water round a closed system of pipes and radiators. Air in the system can create a blockage. Is it better to use narrow-bore pipes or wide-bore pipes in a central heating system? Explain your answer.
4. Blood is driven round the body by the pumping action of the heart. What would be the likely effect on blood pressure of:
   (a) narrowing of the arteries,
   (b) increased viscosity of the blood?
5. Natural gas reaches the consumer through a network of underground pipes.
   (a) Why are pumps necessary to force the gas through the pipes?
   (b) Why must the pump pressure not be too high?

17.8 Surface tension

FIRST THOUGHTS

Cleaning agents, waterproof clothing and pond life are all affected by the surface tension of water.

Place a clean needle on a piece of absorbent paper and then lay the paper on the surface of some water in a bowl. The paper gradually sinks leaving the needle floating on the water surface. The surface of the water acts like a stretched skin, preventing the needle from sinking.

The surface is in a state of tension because there are forces of attraction between the water molecules at the surface. The term **surface tension** is used to describe the fact that the surface of a liquid is in a state of tension.

The surface tension of a liquid is evident when a beaker or cup is overfilled. As liquid is poured into the cup, the liquid rises slightly above the rim of the cup before overflowing. Just before it overflows, the liquid is prevented from overflowing by its surface tension. However, as the liquid level rises, the surface tension is unable to withstand the pressure of the liquid trying to overflow.

It's a fact!

Mosquito larvae live in stagnant water, supported by the surface of the water. A small quantity of oil poured onto the surface reduces the surface tension of water considerably and causes the larvae to sink and die.

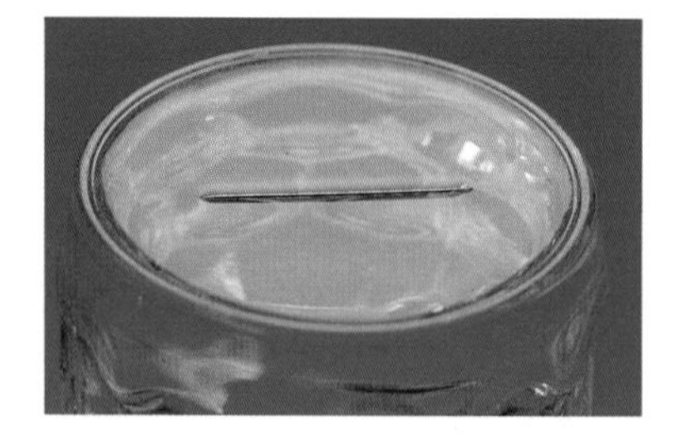

Figure 17.8A ▲ A floating needle

Figure 17.8B ▲ A water wall

Investigating droplets

Falling droplets

Use a magnifying glass or video camera to watch water drops forming at the outlet of a dripping tap. As a droplet forms at the outlet, its increasing weight causes its shape to change until it breaks free and falls.

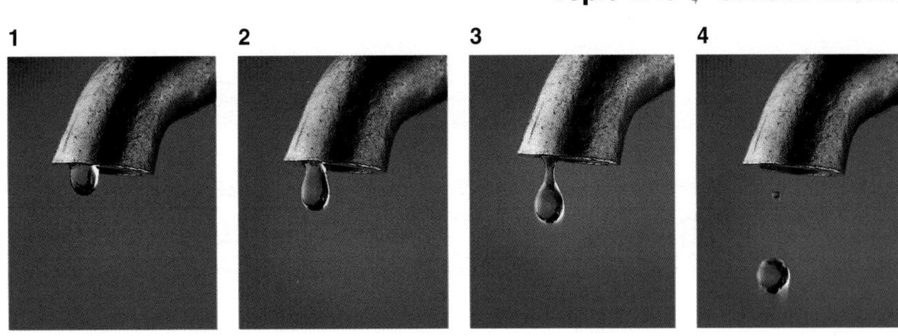

Figure 17.8C ▲ Water drop forming at a tap

Small droplets are spherical in shape because the surface tension makes the surface area of the droplet as small as possible. The water molecules at the surface attract each other and make the droplet spherical. A larger droplet would also be spherical but the force of gravity distorts it. The tension in the surface of water is **cohesive** because it tries to keep the water in.

Droplets on a flat surface

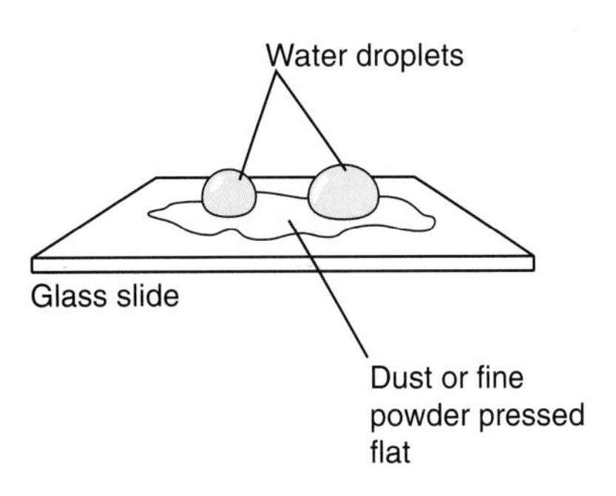

Figure 17.8D ▲ Droplet shapes

Use a magnifying glass to observe water droplets of different sizes on a clean glass slide. Then repeat the test using a dry slide covered with a layer of very fine powder.

In the first case, the water spreads but in the second case it forms spherical droplets. This is because the powder stops the water touching the glass so preventing the water molecules from being attracted by the glass molecules. In the absence of the powder, water molecules at the air-glass boundary are attracted by glass molecules, making the droplet lose its shape. The force between the water and the glass is **adhesive** because it tries to make the water stick to the glass. Water is said to *wet* a clean glass surface because it spreads out on the surface.

SUMMARY

The surface tension of a liquid is due to the molecules at the surface exerting attractive forces on each other.

CHECKPOINT

1. The fabric used for a raincoat consists of resin-coated fibres. Why does this prevent water penetrating the fabric?
2. Try to float a needle on soapy water. Describe what happens and explain your observations.
3. (a) The surface of the water in a tube is called a meniscus. Observe a water meniscus and explain its shape.
   (b) This diagram shows the shape of a mercury meniscus. What does this tell you about the force between mercury molecules in comparison with the force between mercury and glass molecules?

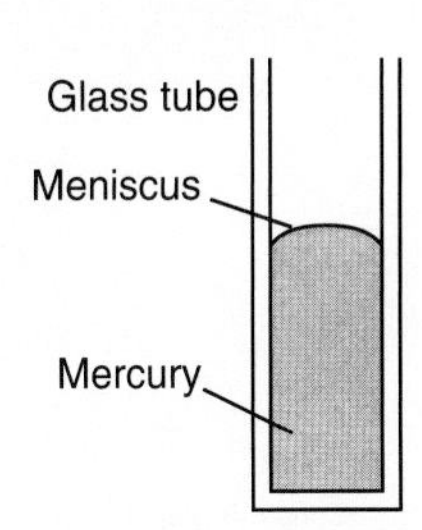

4. (a) The pressure inside a soap bubble is greater than the pressure outside. Do you think the pressure in a small soap bubble is more than the pressure in a large soap bubble? Explain your answer.
   (b) Devise an experiment to test your prediction.

Topic 18 Force and motion

18.1 Maps and routes

FIRST THOUGHTS

After you have worked through this section you will be able to plan journeys with confidence.

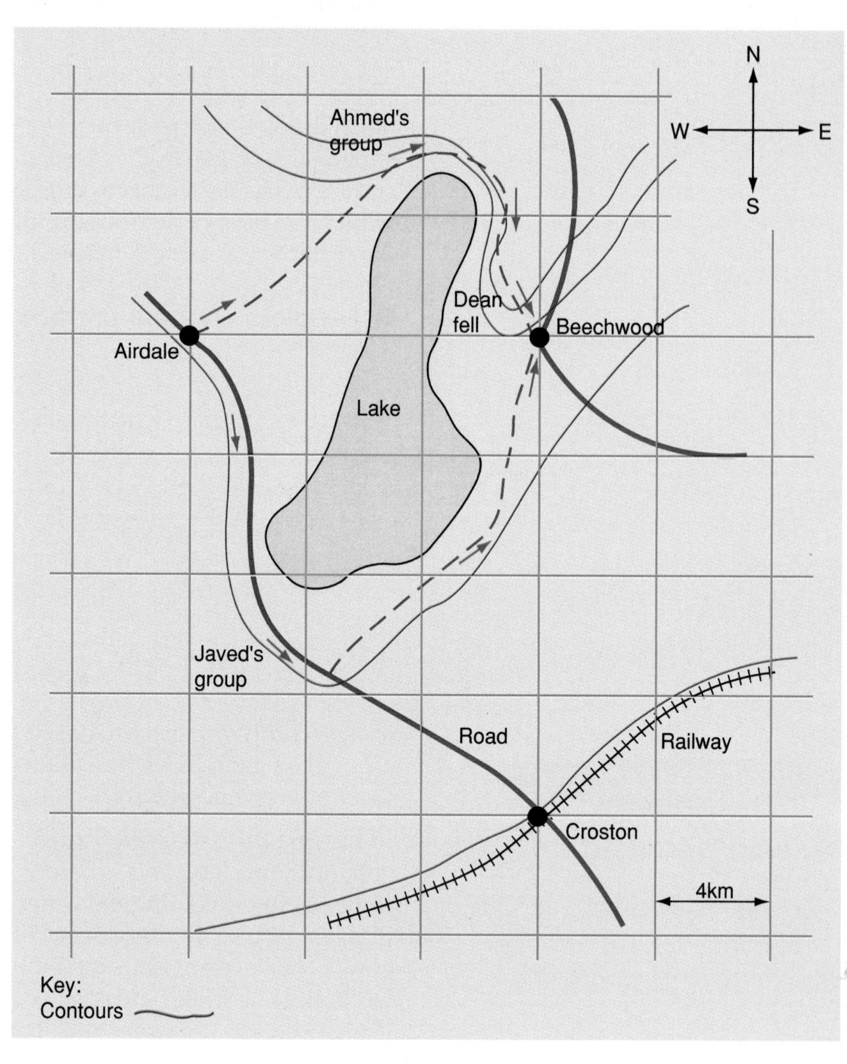

Figure 18.1A ▲ Routes

1 km = 1000 m
1 hour = 3600 s
3.6 km per hour = 1 m/s

A group of friends are planning a walking holiday, stopping overnight at youth hostels. Figure 18.1A shows part of the route. They decide to stop for two nights at Airdale and then walk to Beechwood. The shortest distance from Airdale to Beechwood is 12 km but this route is not very sensible. *Can you see why?*

There are two possible routes, one following the map contours and a shorter route over Dean Fell. Ahmed wants to take the shorter route and measures the distance at 20 km. Javed wants to take the other route, even though it is considerably longer than 20 km. So the group decide to split into two for this part of the walk.

The two groups set off at 10 a.m. from Airdale. Javed and his group arrive at Beechwood at 4 p.m. only to find that the other group have arrived one hour earlier. Ahmed's group travelled 4 km per hour. Javed reckons his group walked faster, even though they arrived one hour after Ahmed's group. *Which group do you think walked at the greatest speed?*

Speed is defined as distance travelled per unit time. Its scientific unit is metres per second (m/s). Other units such as cm/s or km/h (kilometres per hour) are used, but they may need to be converted into m/s for calculations. For example, a speed of 20 cm/s is 0.20 m/s, since 20 cm = 0.20 m. Car speeds are often given in km/h. A speed of 110 km/h is approximately 31 m/s.

Velocity is speed in a given direction. The map shows that Ahmed's route changes direction. Even if speed is constant, where the direction alters, the velocity changes.

The speed of each walker would have varied during the journey, going downhill faster than uphill perhaps. The average speed of each group can be worked out from

$$\text{Average speed} = \frac{\text{Total distance travelled}}{\text{Total time taken}}$$

Ahmed's group travelled a total distance of 20 km (= 20 000 m) in 5 hours (= $5 \times 60 \times 60$ seconds). The average speed was 4 km/h, which is the same as

$$\frac{20\,000}{5 \times 60 \times 60} = 1.11 \text{ m/s}$$

To work out the average speed of Javed's group, use the map to measure the distance they walked. One way to do this is to lay a length of cotton on the map along the route and then straighten it out to measure the distance from the map scale. Then divide this distance by the time taken to give the average speed.

$$\text{Speed} = \frac{\text{distance}}{\text{time}}$$

$$\text{Distance} = \text{speed} \times \text{time}$$

$$\text{Time} = \frac{\text{distance}}{\text{speed}}$$

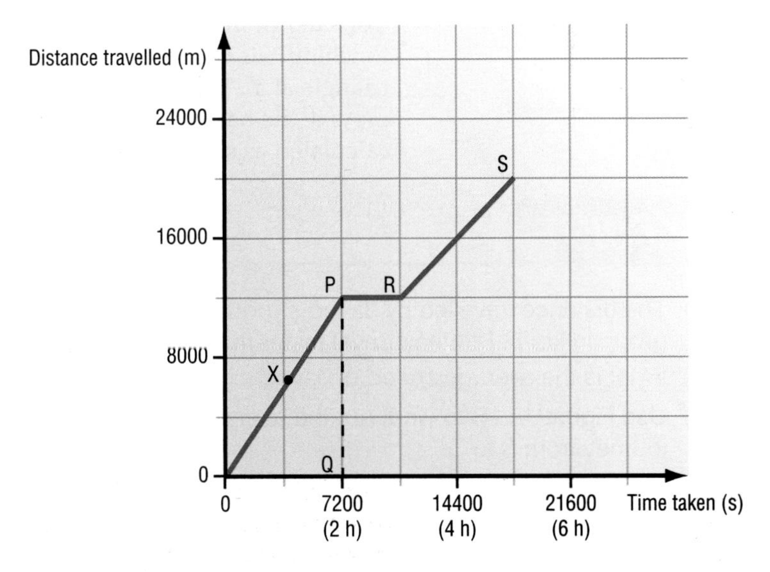

Figure 18.1B ▲ Ahmed's progress

The gradient of a distance–time graph is constant if the speed is constant.

Distance–time graphs are useful to illustrate a journey. From Figure 18.1B you can see that Ahmed's group travelled 12 km in the first two hours and then rested for an hour. Then they completed the remaining 8 km in two hours.

Uniform or constant speed is where the speed does not change. In Figure 18.1B the speed is uniform from O to P and from R to S. Uniform speed is shown on the graph by a constant gradient.

The speed at any point can be calculated from the **gradient** of the graph at that point. To find the gradient at any point, draw a 'gradient triangle'.

Consider point X in Figure 18.1B. The gradient triangle at X is labelled OPQ. Its height, PQ, represents distance travelled; its base, OQ, represents time taken.

$$\text{Speed at X} = \frac{\text{Distance}}{\text{Time taken}} = \frac{PQ}{OQ} = \frac{12\,000}{(7200)} = 1.67 \text{ m/s}$$

Where the speed changes, the gradient of the distance-time graph changes. Figure 18.1C is a distance-time graph for a cyclist going downhill. The increase in the cyclist's speed is shown by the increasing gradient. To find the speed at any point, a gradient triangle must be drawn at that point.

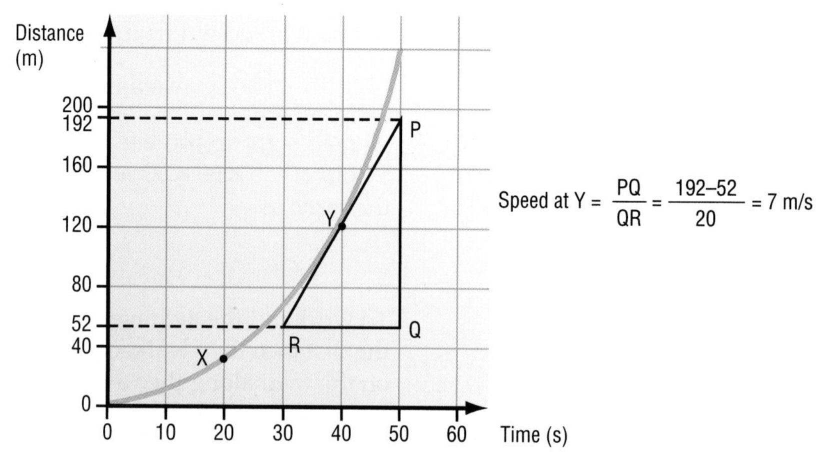

Figure 18.1C Using a distance-time graph

Consider point Y on Figure 18.1C. The graph is curved. The straight line touching Y is called the **tangent**. This is used to form the gradient triangle at Y. The height PQ of the triangle represents the distance moved; the base QR represents the time taken. Hence the speed can be calculated as shown.

SUMMARY

Speed is distance travelled per unit time. Velocity is speed in a given direction. The gradient of a distance–time graph gives the speed.

CHECKPOINT

1 (a) The distance travelled by Javed's group was 28 km. Work out the average speed of Javed's group in km/h. Did they travel faster than Ahmed's group?
(b) What is the average speed of Javed's group in m/s?

2 (a) Use Figure 18.1B to work out the average speed of Ahmed's group in the two hours of their journey from R to S.
(b) Why is their average speed for the whole journey, 1.1 m/s (see previous page), less than their speed for the first part of the journey?

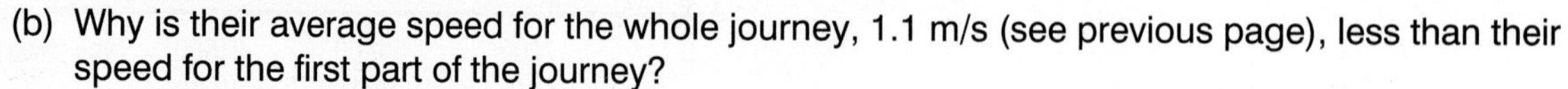

3 Javed's group walked 16 km from 10 a.m. until 12.30 p.m., when they took a break for an hour. Then they completed their journey without stopping. Sketch a distance-time graph for Javed and his friends.

4 (a) Imagine you are a cyclist. How could you measure your average speed between two points on your route?
(b) Work out the speed at point X of Figure 18.1C.

5 The next part of our friends' journey is from Beechwood to the railway station at Croston which is 16.0 km due south of Beechwood.
(a) What is the direct distance from Airdale to Croston?
(b) Ahmed reckons the best route to Croston from Beechwood is 24 km. How long would it take the group to make this journey at an average speed of 1.00 m/s?

18.2 ▶ Acceleration

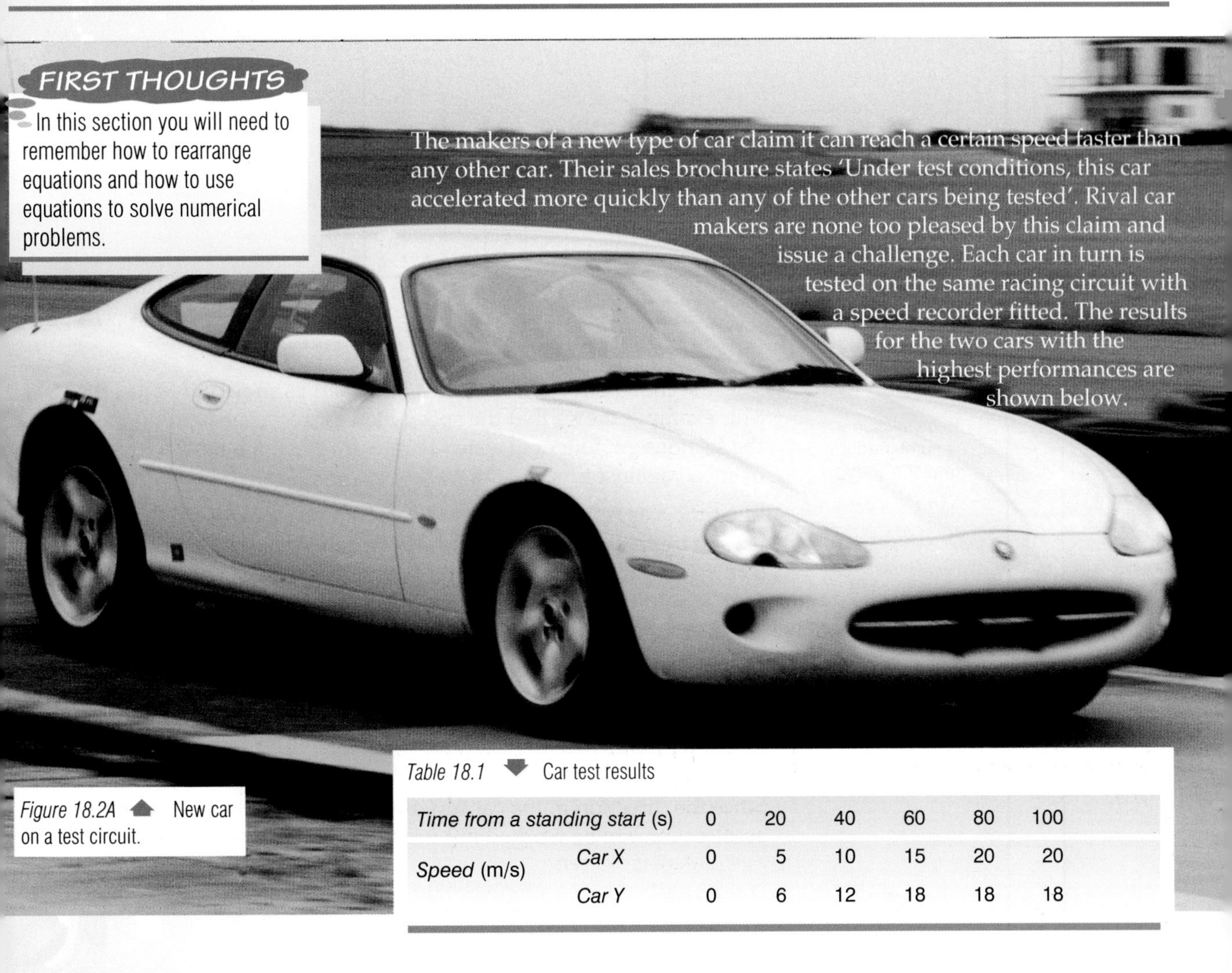

FIRST THOUGHTS

In this section you will need to remember how to rearrange equations and how to use equations to solve numerical problems.

The makers of a new type of car claim it can reach a certain speed faster than any other car. Their sales brochure states 'Under test conditions, this car accelerated more quickly than any of the other cars being tested'. Rival car makers are none too pleased by this claim and issue a challenge. Each car in turn is tested on the same racing circuit with a speed recorder fitted. The results for the two cars with the highest performances are shown below.

Figure 18.2A ▲ New car on a test circuit.

Table 18.1 ▼ Car test results

Time from a standing start (s)		0	20	40	60	80	100
Speed (m/s)	*Car X*	0	5	10	15	20	20
	Car Y	0	6	12	18	18	18

Which car accelerates more? The speed of X increases 5 m/s every 20 s, compared with 6 m/s every 20 s for Y. So Y **accelerates** more because its increase of speed in the same time is greater.

Investigating acceleration using a ticker timer

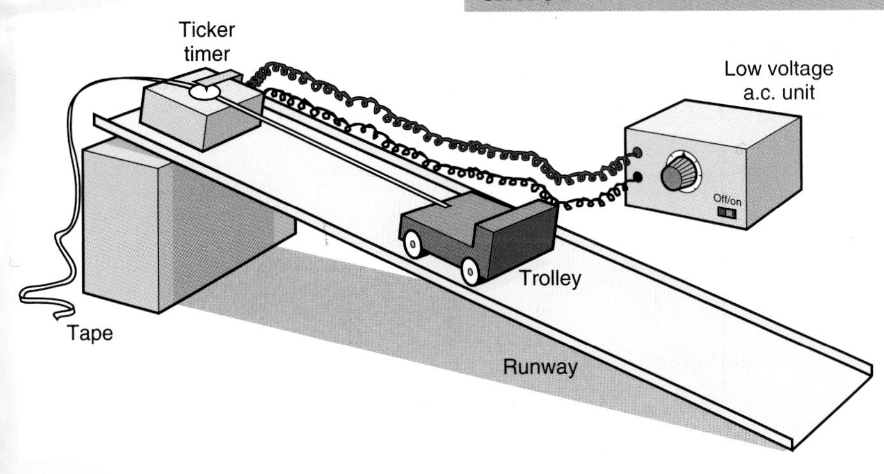

Figure 18.2B ▲ Investigating acceleration

The ticker timer is designed to print dots on to tape at a steady rate of 50 per second. Figure 18.2B shows how to use the ticker timer to record the motion of a trolley down a runway. As the trolley accelerates down the runway, it pulls the tape through the timer at a faster and faster rate. Figure 18.2C shows the result.

The dots become more widely spaced because the tape travels faster and faster. The time interval between successive dots is 0.02 s (= 1/50 s).

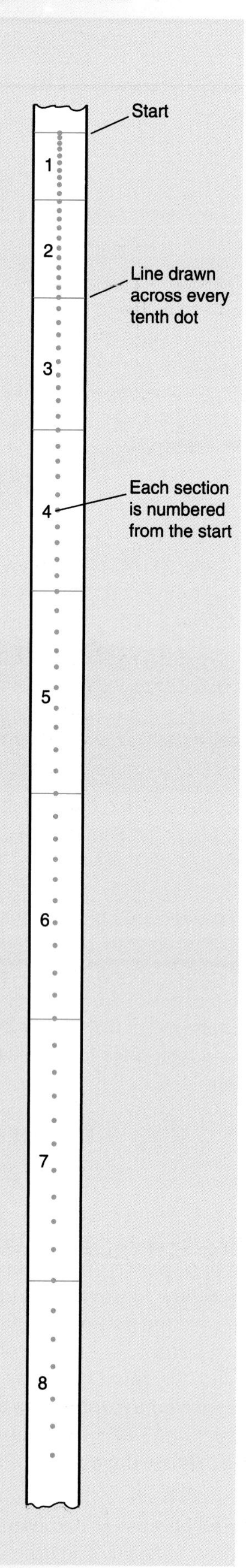

Figure 18.2C A ticker tape

The tape can be marked into 10-dot sections, each section taking 0.20 s (= 10/50s) to pass through the machine. The sections get longer as the tape goes through faster and faster.

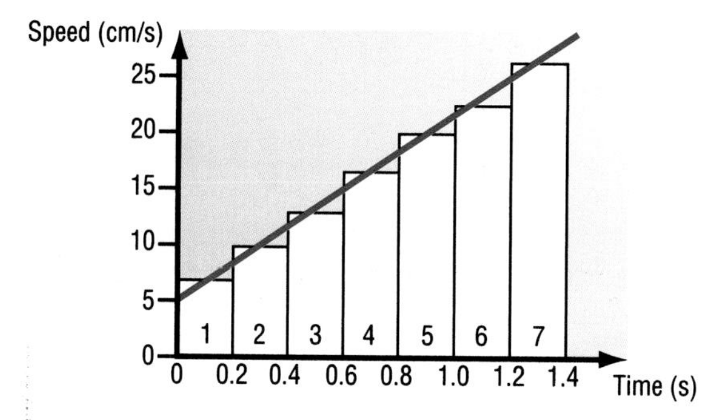

Figure 18.2D A tapechart

These sections can be made into a **tapechart** as shown in Figure 18.2D. This example shows that the speed increases at a constant rate, because the sections get longer at a steady rate. In effect, the tapechart is a graph of speed against time for the trolley.

The time axis is marked in intervals of 0.2 s because each section took 0.2 s to travel through the timer. The speed axis is marked in intervals of 5 cm/s for each centimetre of the scale. This is because a section of length 1 cm would pass through the timer in 0.2 s. Hence its speed would be $1/0.2 = 5$ cm/s.

The line through the tops of the sections has a constant gradient. The speed rises steadily. This shows the acceleration is **uniform.**

Acceleration is defined as the change of velocity per unit time. The unit of acceleration is the metre per second per second (m/s^2).

Since the direction in the car test example opposite is constant, the acceleration can be worked out from the change of speed per second.

$$\text{Acceleration} = \frac{\text{Change of speed}}{\text{Time taken}}$$

For example, for car X,

$$\text{Acceleration} = \frac{\text{Change of speed}}{\text{Time taken}} = \frac{20}{80} = 0.25 \text{ m/s}^2$$

When a moving object slows down, it is said to **decelerate**. This is represented by a negative-valued acceleration.

Speed-time graphs are used to show changes in motion. Figure 18.2E shows the results for cars X and Y plotted on the same graph. Line OY is steeper than line OX because Y accelerates more than X does.

The acceleration of each car can be worked out from the gradient of its line on the speed-time graph. Figure 18.2E shows how to do this for X.

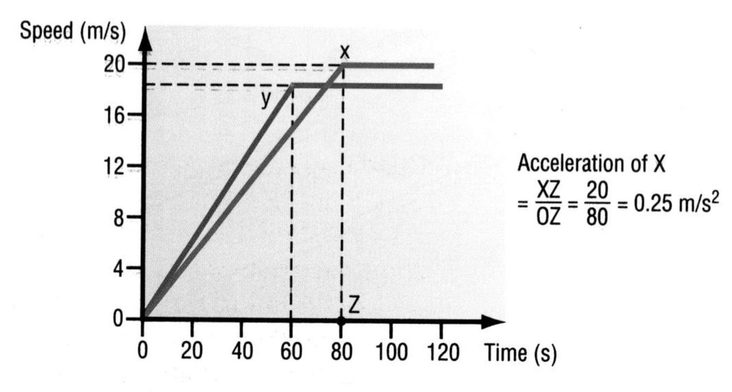

Figure 18.2E Speed-time comparison

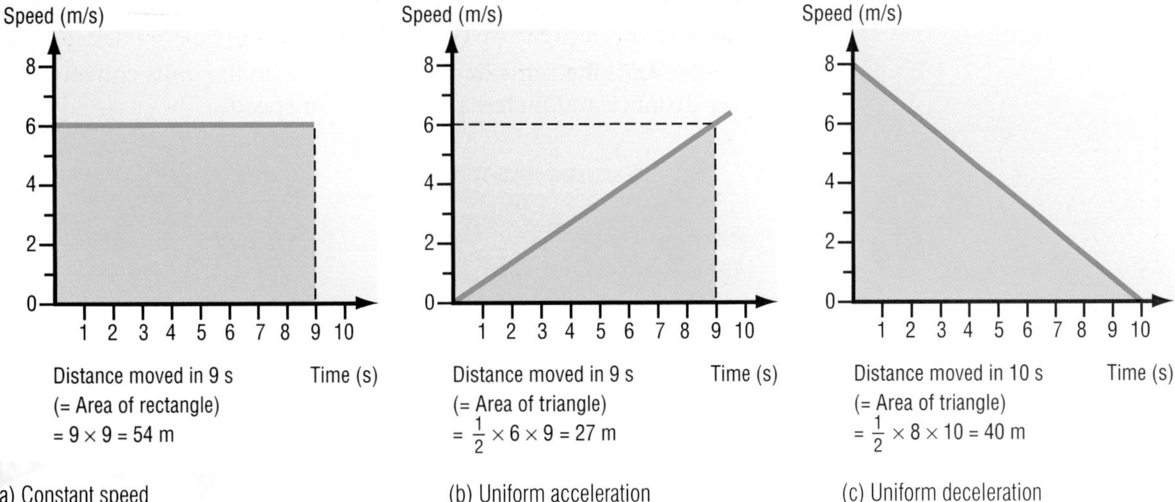

Figure 18.2F ▲ Speed-time graphs

SUMMARY

Acceleration is the change of velocity per unit time, which is equal to the change of speed per second for any object moving along a straight line. The gradient of a speed-time graph gives the acceleration; the area under the line gives distance travelled.

Graph (a) in Figure 18.2F is for an object moving at steady speed (acceleration = 0). The distance travelled in a certain time is given by Speed × Time. This is represented on the graph by Height × Base of the shaded rectangle, i.e. the area under the line.

For any speed-time graph the distance travelled by an object in a given time can be worked out from the area under the line. Graph (b) represents uniform acceleration and graph (c) represents uniform deceleration. In each case the area under the line is the area of the shaded triangle, which represents Average speed × Time taken = Distance moved.

Consider Figure 18.2E once again. *How far does car X move in 80 s from rest?* The area under the line OX is 1/2 × Height × Base of the triangle OXZ = 1/2 × 20 × 80 = 800 m.

Stopping distances

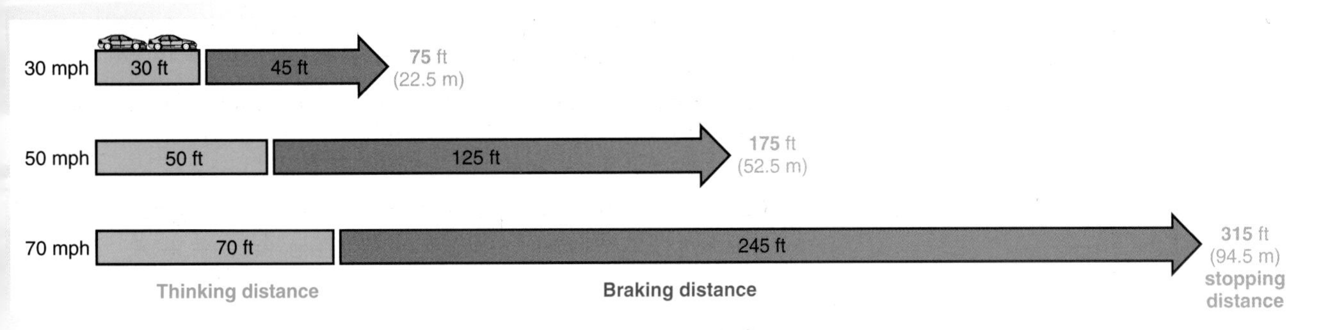

Figure 18.2G ▲ Stopping distances

Road users are advised to maintain safe distances to cut down the risk of accidents. The shortest stopping distance of a vehicle depends on its speed and on the road conditions. The chart above gives the shortest stopping distance on a dry road for different speeds.

- **The thinking distance** is the distance travelled by the car in the time it takes the driver to react.
- **The braking distance** is the distance travelled by the car from the point where the brakes are applied to where it comes to rest.

- **The stopping distance** is the thinking distance added to the braking distance.

Table 18.2 presents the same data (and more) with the units converted to metres for distance and metres per second for speed.

Table 18.2 Recommended stopping distances

Speed in mph	Speed in m/s	Thinking distance (m)	Braking distance (m)	Stopping distance (m)
10	4.5	3	1.5	4.5
20	9	6	6	12
30	13.5	9	13.5	22.5
40	18	12	24	36
50	22.5	15	37.5	52.5
60	27	18	54	72
70	31.5	21	73.5	94.5

Stopping distance = thinking distance + braking distance

The graphs in Figure 18.2H show how each of these distances increases with speed.

1 **The thinking distance is proportional to the speed.** This is because the vehicle travels at constant speed during the 'reaction' period before the brakes are applied. For constant speed, the distance moved = speed × time. The reaction time for the data in the table is 0.67 s. This is the average reaction time for a driver in an alert state of mind.

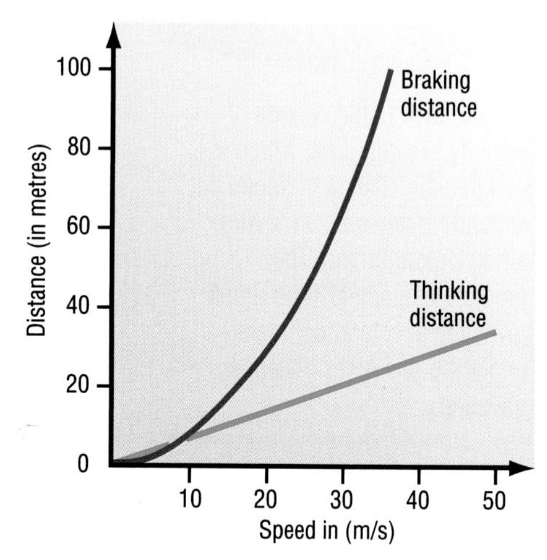

Figure 18.2H Safe distances

2 **The braking distance is proportional to the square of the speed.** This assumes the vehicle's deceleration is constant during braking. Therefore, the braking time t = initial speed u ÷ deceleration a. Since the braking distance = average speed × time taken and the average speed = 1/2 × initial speed, the braking distance is therefore equal to $u^2/2a$.

The data above give an acceleration of – 6.75 m/s^2 (i.e. a deceleration of 6.75 m/s^2).

See www.keyscience.co.uk for more about motion video clips.

Memorable numbers

The Highway Code carries a chart of stopping distances in feet at different speeds. The pattern of numbers makes memorising easy.

Speed in mph	20	30	40	50	60	70
Stopping distance in feet	40	75	120	175	240	315

Can you work out a pattern here?
$20 \times 2 = 40$; $30 \times 2.5 = 75$; $40 \times 3 = 120$ gives the pattern.

Remember this pattern if you intend to take a driving test!

SUMMARY

The stopping distance of a vehicle depends on road conditions as well as speed. The stopping distance is the thinking distance (the distance travelled by the car in the time it takes the driver to react) added to the braking distance (the distance travelled by the car with the brakes applied).

Factors affecting stopping distances

1 **Reaction time** The driver's reaction time depends on alertness. A driver in a drowsy state is a danger as the thinking distance would increase. For example, a reaction time of 1.0 s would result in a thinking distance of 31.5 m at 70 mph (= 31.5 m/s).

2 **Road conditions** When the brakes are applied to a vehicle, friction between the brakes and the wheels reduces the vehicle's speed, provided the tyres don't skid on the road. Skidding occurs if the braking force exceeds the grip (i.e. maximum amount of friction possible) between the tyres and the road. Grip depends on the road conditions as well as the tyre conditions and is reduced by water on the road and is almost eliminated by oil or ice. Worn tyres are dangerous because there is less road grip. Braking distance is increased where grip is reduced. This is why all drivers must reduce speeds where road conditions are poor.

CHECKPOINT

1 The diagram shows a tapechart made by Anna walking away from the timer while holding the end of the tape. Each section is a 10-dot length. The timer operated at 50 dots per second.
   (a) Describe how Anna's speed changed as she walked away.
   (b) How long did she take to reach her top speed?
   (c) What was her top speed in (i) cm/s, (ii) m/s?

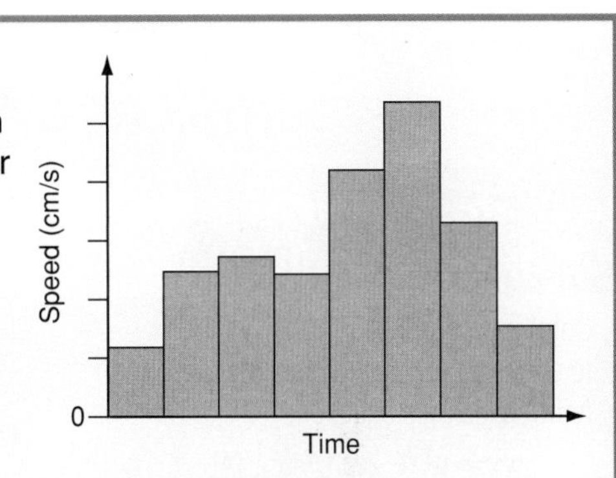

2 (a) From Figure 18.2E work out the acceleration of car Y.
   (b) How far did car Y travel in the first 60 s from rest?

3 A ticker timer is designed to print dots on a tape at a rate of 50 per second. How would you test that it is printing at the correct rate?

4 In a motor cycle test, the speed from rest was recorded at intervals.

Time (s)	0	5	10	15	20	25	30
Speed (m/s)	0	10	20	30	40	40	40

   (a) Plot a speed-time graph of these results.
   (b) What was the initial acceleration?
   (c) How far did it move in (i) the first 20 s, (ii) the next 10 s?

5 A rocket under test reaches a speed of 210 m/s from rest in 30 s before its fuel is used up. Assuming it accelerates uniformly, sketch a speed-time graph of its motion. Use the graph to work out how far it travelled in that time and what its acceleration was.

6 A car travelling at 30 m/s is being driven by a driver whose reaction time is 0.60 s.
   (a) Calculate the distance travelled by the car in this time.
   (b) The maximum possible deceleration of the car without skidding is 6.0 m/s^2. Calculate the braking time for the car.
   (c) Sketch a speed–time graph for the car, showing the driver's thinking time as well as the braking time.
   (d) Calculate the braking distance for the car at this speed.
   (e) Show that the stopping distance is 93 m.

18.3 Equations for uniform acceleration

FIRST THOUGHTS

The study of motion is called dynamics. In this section you will study motion in a straight line.

Consider an object accelerating uniformly in a straight line from initial speed u to final speed v in time t. Its motion is shown by the graph in Figure 18.3A.

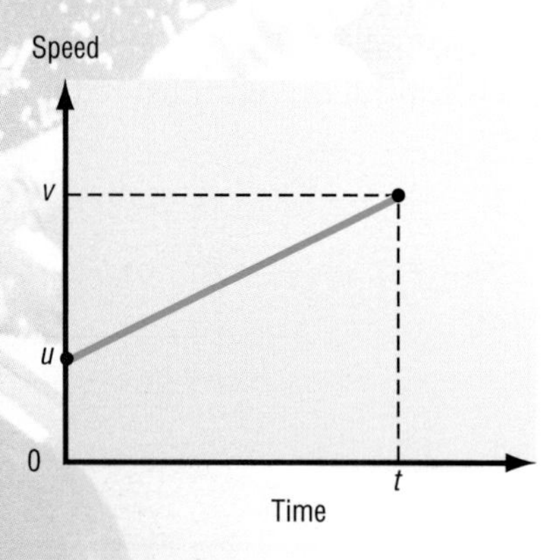

Figure 18.3A ▲ Uniform acceleration

Its change of speed is $(v - u)$ and since its acceleration is given by Change of speed/Time taken, then

$$\text{Acceleration, } a = \frac{(v-u)}{t} \qquad [1]$$

Because the acceleration is uniform the speed increases steadily from u to v. The average speed is therefore $\frac{1}{2}(u + v)$.

Since $\quad \text{Average speed} = \dfrac{\text{Distance}}{\text{Time}}$

Rearranging gives $\quad \text{Distance} = \text{Average speed} \times \text{Time}$

So $\quad \text{Distance travelled, } s = \frac{1}{2}(u + v)t \qquad [2]$

We can combine equations [1] and [2] to eliminate v.

Rearranging $\quad a = \dfrac{(v-u)}{t}$ gives $at = (v - u)$

Hence $\quad v = u + at$

Substituting this value for v into equation [2] gives

$$\text{Distance, } s = \tfrac{1}{2}(u + u + at)t$$

Hence $\quad s = ut + \frac{1}{2}at^2 \qquad [3]$

Test your maths skills

Combine equations [1] and [2] to eliminate t. Hence obtain a fourth dynamics equation,

$$v^2 = u^2 + 2as \qquad [4]$$

This equation is useful in situations where three of the quantities v, u, a and s are known and you have to calculate the fourth quantity.

Worked example 1 The speed of a train travelling between two stations changes with time as shown in the table.

Time (s)	0	50	100	150	200	250	300	350	400
Speed (m/s)	0	3.0	6.0	9.0	9.0	9.0	9.0	4.5	0

(a) Plot a speed-time graph of the journey.

(b) Work out the acceleration and distance travelled in each part of the journey.

(c) Work out the average speed for the whole journey.

Solution

(a) The speed-time graph is shown in Figure 18.3B.

(b) The journey is in three parts, as shown in Figure 18.3B.

OA: $\quad \text{Acceleration} = \dfrac{(v-u)}{t} = \dfrac{9.0 - 0.0}{150} = 0.06 \text{ m/s}^2$

$\quad \text{Distance} = \frac{1}{2}(u + v)t = \frac{1}{2}(9.0 + 0.0) \times 150 = 675 \text{ m}$

AB: $\quad$ Acceleration $= 0$ (since the speed does not change here)

$\quad \text{Distance} = \text{Speed} \times \text{Time} = 9.0 \times 150 = 1350 \text{ m}$

BC: $\quad \text{Acceleration} = \dfrac{(v-u)}{t} = \dfrac{0.0 - 9.0}{100} = -0.09 \text{ ms}^2$ (a deceleration)

$\quad \text{Distance} = \frac{1}{2}(u + v)t = \frac{1}{2}(0.0 + 9.0) \times 100 = 450 \text{ m}$

(c) Total distance travelled $= 675 + 1350 + 450 = 2475$ m

Total time taken $= 400$ s

Hence $\quad \text{Average speed for the whole journey} = \dfrac{\text{Distance}}{\text{Time}} = \dfrac{2475}{400} = 6.19 \text{ m/s}$

Speed (m/s)
10, 8, 6, 4, 2, 0
A, B, C
0, 100, 200, 300, 400
Time (s)

Figure 18.3B ▲ Speed–time graph for the train

SUMMARY

The four equations

$$a = \frac{v - u}{t}$$

$$s = \tfrac{1}{2}(u + v)t$$

$$s = ut + \tfrac{1}{2}at^2$$

$$v^2 = u^2 + 2as$$

can be used to solve problems where the acceleration is uniform.

Worked example 2 A bullet travelling at a speed of 120 m/s hits a large piece of wood and penetrates it to a depth of 60 mm. Work out the time taken to bring the bullet to rest and the deceleration.

Solution Initial speed u = 120 m/s, Final speed v = 0 m/s, Distance travelled
s = 60 mm = 0.060 m

(a) To find the time taken, t, use

$$s = \tfrac{1}{2}(u + v)t$$

$$0.060 = \tfrac{1}{2}(120 + 0)\,t = 60t$$

$$\text{Hence } t = \frac{0.060}{60} = 0.0010\text{s}$$

(b) To find the deceleration, use

$$a = \frac{(v - u)}{t}$$

$$a = \frac{(0 - 120)}{0.0010} = -120\,000 \text{ m/s}^2$$

CHECKPOINT

1 A sprinter is capable of accelerating from rest to a speed of 10 m/s in 1.5 s. Work out her acceleration and the distance she travels in this time.

2 The diagram shows a tapechart for a toy car released at the top of a ramp.

(a) How far did the car travel in 1 s from rest?

(b) Work out its average speed in the first second.

(c) Work out of the increase of speed in 1 second by measuring the length of two 10-dot sections 1 s apart. Hence work out the acceleration.

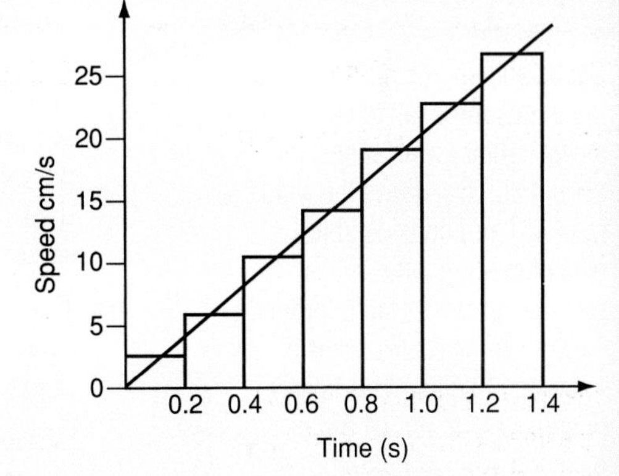

3 The speed of a car between two sets of traffic lights changes as shown in the table.

Time (s)	0	20	40	60	80	100	120
Speed (m/s)	0	2.5	5.0	7.5	10.0	5.0	0

(a) Plot a speed-time graph of the motion.

(b) Work out the acceleration and distance travelled in (i) the first 80 seconds, (ii) the last 40 seconds.

(c) Work out the average speed of the car between the two sets of lights.

4 In a test drive of a car on a dry road, the driver was instructed to travel at 30 m/s and to apply the brakes to stop the car when he passed a roadside marker X. The car stopped 75 m beyond X. This point was marked Y.

(a) What was (i) the initial speed, and (ii) the final speed of the car between X and Y.

(b) What was the average speed between X and Y?

(c) How long did the car take to stop and what was its deceleration?

5 Simon is going on holiday with his parents. He knows that the runway at the airport is 3500 m in length. While waiting in the departure lounge, he times a jet taking off. The timing was 32.0 s from when the plane started accelerating along the runway to when its wheels lifted off the ground.

(a) What was the average speed of the plane during take-off?

(b) Assuming its initial speed was zero, what was its final speed?

(c) What was its acceleration?

18.4 Free fall

Now that you know how to use the dynamics equations for motion in a straight line, let's see how they apply to falling objects.

Figure 18.4A Free fall

It's a fact!

Seeds and spores released into the atmosphere fall to earth very slowly. Their terminal speeds are small. This allows the wind to blow them considerable distances. Cigarette smoke particles in the air fall to ground slowly, so the smoke takes a long time to clear. This is why people who do not smoke can be at risk from people who do smoke.

Galileo Galilei was a famous scientist who lived in Italy in the seventeenth century. One of his best known experiments involved dropping objects from the top of the Leaning Tower of Pisa. He showed that different weights released at the same time reach the ground at the same time.

The **acceleration due to gravity**, g, is the same for all falling objects, provided there is no air resistance. The value of g near the Earth's surface is approximately 10 m/s².

Where there is air resistance the speed of a falling object builds up to a constant value. This value is called the **terminal speed** of the object. For example, a weight released under water falls at steady speed; so does a feather released in air. Because there is resistance acting against the motion these are not examples of free fall.

Multiflash Photography

This is a method for investigating motion (Figure 18.4B). The aperture of a camera is kept open in a darkened room. A light flashes at a constant rate as the object under test moves. Each time the light flashes, an image of the object is recorded on the camera film. Figure 18.4C overleaf shows the result for two objects falling freely.

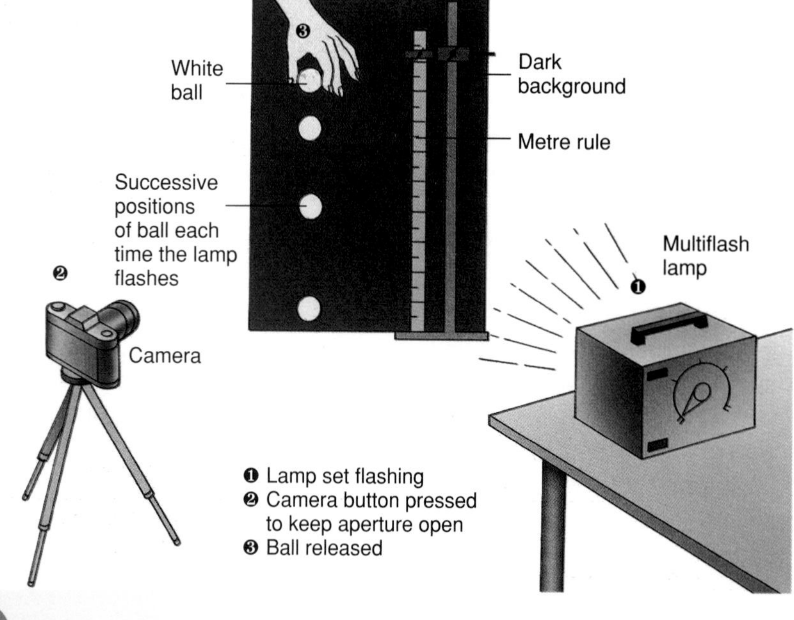

Figure 18.4B Multiflash photography

A projectile moves equal distances horizontally in equal times. Its vertical motion is at constant acceleration.

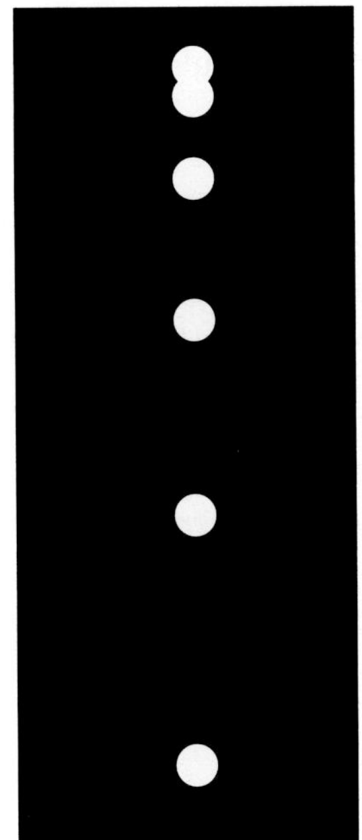

Figure 18.4C ▲ (a) Object A (b) Object B

Object A falls vertically after being released from rest. Object B, released at the same time, is given a push sideways. Any object acted on by gravity alone after being given a push is called a **projectile**.

What do you notice about the motion of B compared with A? Both fall at the same rate; this is because gravity acts downwards on every object. However, B moves across as well. *What can you say about the horizontal motion of B?* The photograph shows that B moves equal distances horizontally in equal times.

Famine relief

In countries hit by famine, transport planes are used to drop food sacks to remote villages. Since there are no runways to land on, each plane drops its sacks as it approaches the target.

What factors should the pilot take into account to make sure the sacks reach the target? Assume there is no wind. The height and speed of the aircraft must be considered. Suppose the plane approaches the target area in level flight at a speed of 80 m/s at 500 m height. Each sack released from the plane follows a path like that shown in Figure 18.4D.

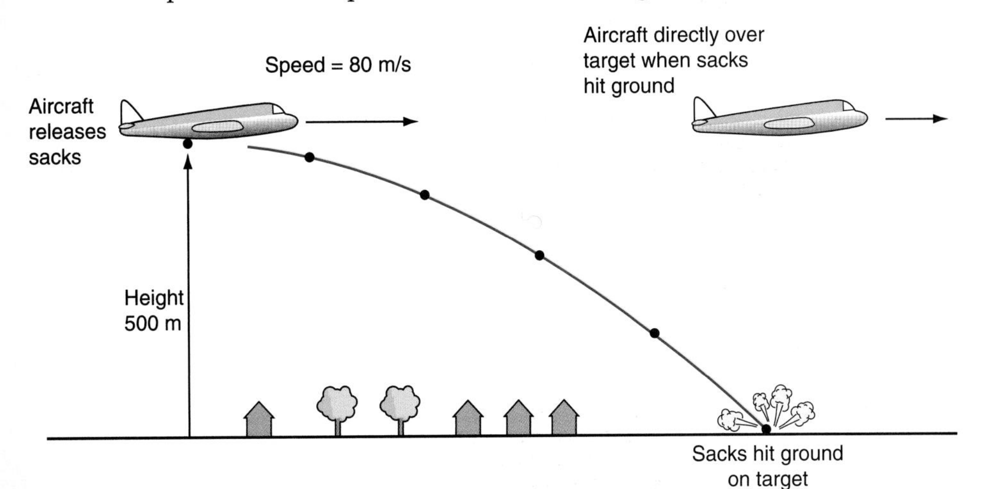

Figure 18.4D ▲ On target

How long does a sack take to reach the ground? A stationary sack released at the same height would take the same time. This time can be worked out as follows

Distance fallen, $s = 500$ m
Acceleration, $g = 10$ m/s^2
Initial **vertical** speed $u = 0$ (since the plane is moving horizontally).

To work out the time taken, t, use the equation $s = ut + \frac{1}{2}at^2$ to give

$500 = (0 \times t) + (\frac{1}{2} \times 10 \times t^2)$
$500 = 5t^2$
$t^2 = 100$
Therefore $t = 10$ s

How far does the sack travel horizontally in this time? As Figure 18.4C shows, the horizontal motion of a projectile is not affected by its vertical motion. The sack continues to move horizontally at the same speed as the plane (i.e. 80 m/s). Therefore in 10 seconds, the sack travel $80 \times 10 = 800$ m horizontally. The pilot must therefore release the sacks from the aircraft 800 m before the target area.

SUMMARY

All falling objects accelerate at the same rate, g, provided air resistance is negligible. The horizontal motion of a projectile is unaffected by its vertical motion.

CHECKPOINT

1 A feather and a penny are released from the same height at the same time. Which one reaches the ground first? Explain why they do not reach the ground at the same time.

2 (a) Test your reaction time. Ask a friend to help, as shown in the diagram.

(b) Rubinda and Karen try the test. The rule drops 0.20 m for Rubinda and 0.22 m for Karen. In both cases, the initial speed is zero. The girls use the equation $s = ut + \frac{1}{2}at^2$ to work out their reaction times.

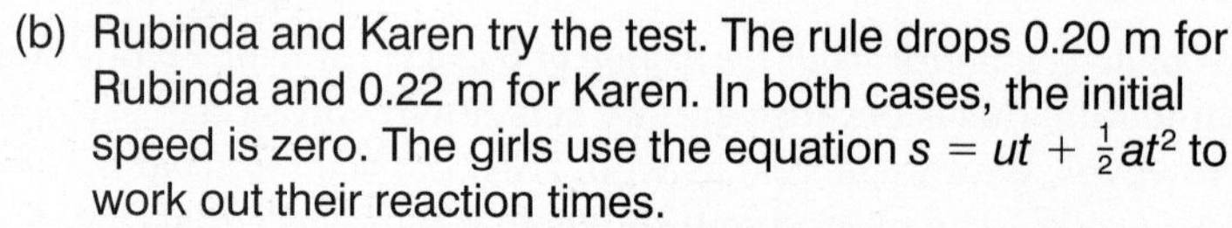

(i) Rubinda reckons her reaction time works out at 0.20 s. Do you agree?

(ii) Work out Karen's reaction time.

3 A stone released at the top of a well took 1.5 s to hit the water in the well. Use the equations of uniform acceleration to work out

(a) the speed at which the stone hit the water, (b) the distance from the top of the well to the water. Assume $g = 10$ m/s^2.

4 How fast can you throw a ball? One way to find out is to throw a ball directly up into the air and time its flight from launch to return. Suppose your timing is 4.2 s.

(a) How long did the ball take to reach maximum height from launch?

(b) At the instant it reaches maximum height, what was its speed?

(c) Use your values from (a) and (b) to work out its initial speed. Assume $g = 10$ m/s^2.

5 Alan and Simon are on a cliff top overlooking an empty beach. Alan throws a stone horizontally from the cliff top while Simon times it. It takes 3.0 s to hit the beach. Then Simon throws another stone horizontally. It hits the beach further away than the first one.

(a) How do the flight times of the two stones compare?

(b) Which stone was given the greatest initial speed?

(c) Work out the height of the cliff top above the beach.

(d) If Simon's stone landed 60 m away from the foot of the cliff, work out the initial speed of the stone.

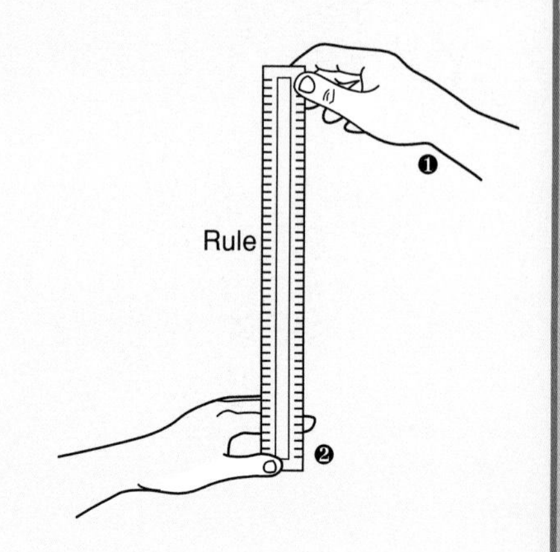

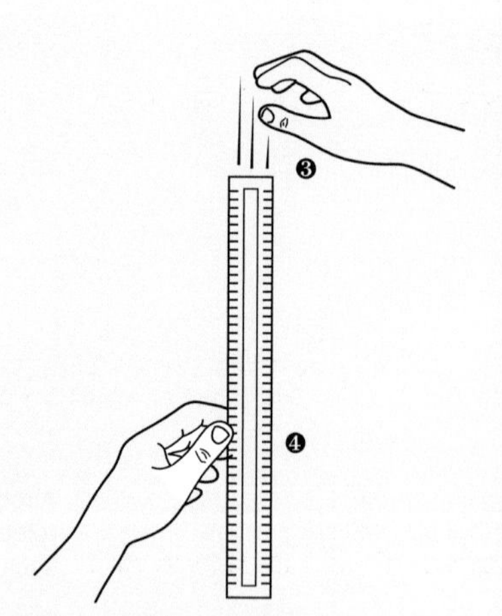

❶ Karen holds a metre rule
❷ Rubinda positions her hand at the zero mark, ready to catch the rule
❸ Karen lets go of the rule
❹ Rubinda catches the rule at the 0.20 m mark

18.5 Force and acceleration

FIRST THOUGHTS

The link between force and motion is not easy to discover, because of the effects of friction as a hidden force. Read on to find out what happens when friction is absent.

Friction, the hidden force

In winter, slides on icy playgrounds can be great fun if you manage to stay upright. Throw a stone across an icy pond and the stone will skid across the ice. Ice hockey players are experts at making pucks slide very fast across ice. In all these examples, friction is too small to affect the motion.

If you have ever tried to push a heavy crate across a rough concrete floor (Figure 18.5B) you will know about friction. The push force is opposed by friction and as soon as you stop pushing, friction stops the crate moving. If you do not know about friction, you might well think that a force is needed to keep an object moving.

What if the floor is smooth? Friction is almost absent and so you do not have to push as hard to move the crate. If the floor is very smooth, the crate will continue to slide when you stop pushing. As friction is almost absent the crate keeps moving.

Figure 18.5A ▲ Low friction motion

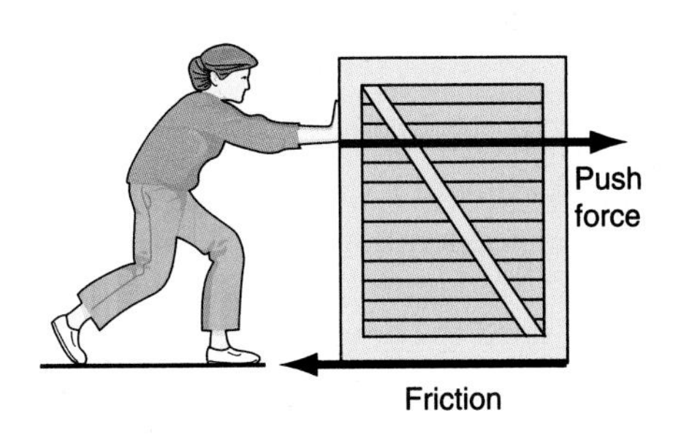

Figure 18.5B ▲ Overcoming friction

Figure 18.5C shows a linear air track where a glider floats on a cushion of air. Provided the track is level, the glider moves at constant speed along the track, because friction is absent.

Sir Isaac Newton was the first person to realise that objects either stay at rest or move with constant velocity unless acted on by a force. This is known as **Newton's First Law of Motion**. His discoveries made him the most famous scientist of his generation. He was able to show exactly how force affects motion and he showed that his theories apply everywhere.

Figure 18.5C ▲ The linear air track

Investigating the link between force and motion

The way in which force affects motion can be seen by using the apparatus shown in Figure 18.5D. The runway is sloped to compensate for friction. This means that if the trolley is given a push, it will move at constant speed down the runway.

A constant force is applied to the trolley to pull it down the runway. The ticker timer records the motion of the trolley. The tape is cut into 10-dot lengths to make a tapechart.

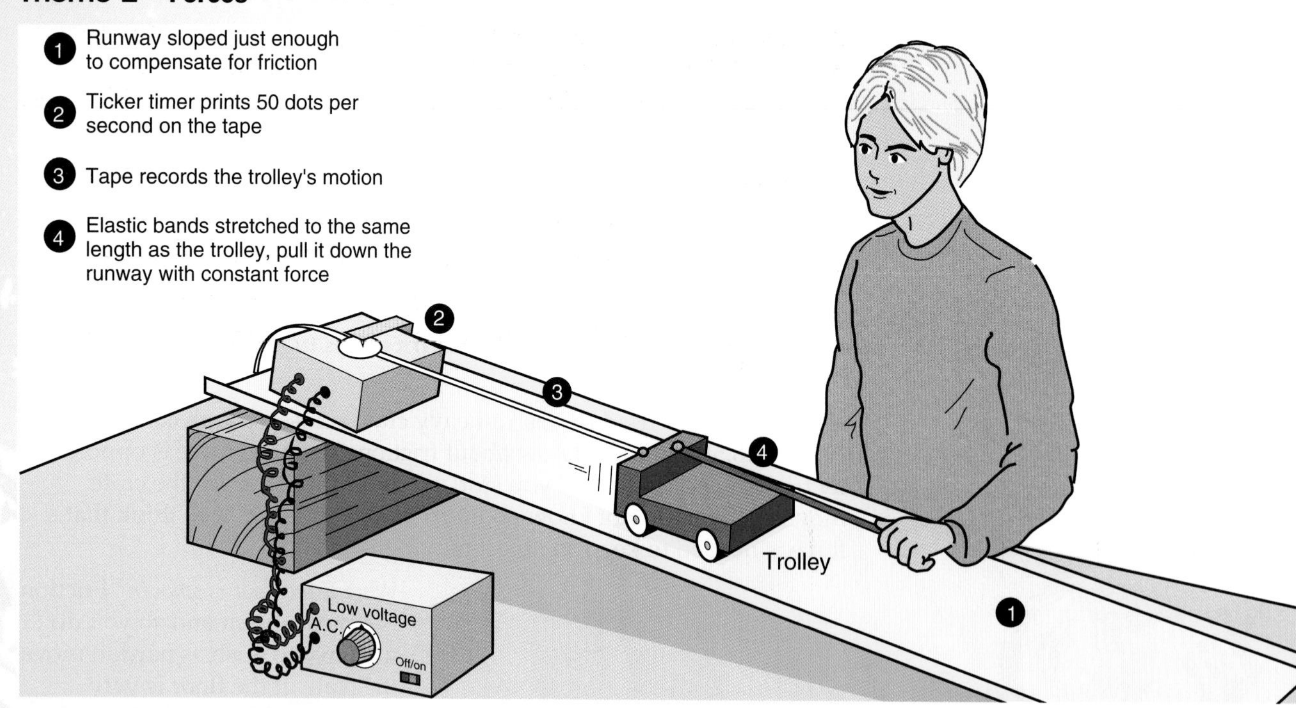

Figure 18.5D ▲ Investigating the link between force and motion

Using one or more elastic bands, as in Figure 18.5D, different forces are applied to the trolley and a tapechart is made for each test. The tests can be repeated using a 'double decker' trolley. Figure 18.5E shows some of the tapecharts produced by these tests.

The tapecharts are speed-time graphs. Each has a constant gradient, which shows that the acceleration of the trolley is uniform. Thus the experiment shows that a constant force produces a uniform acceleration.

The acceleration in each test can be worked out. Each gradient triangle OPQ in Figure 18.5E has the same base length. So the triangle heights can be used to compare the accelerations. The results are as follows.

Table 18.2 ▼ Investigating force and motion

Force (no. of elastic bands)	1	2	3	1	2	3
Mass (no. of trolleys)	1	1	1	2	2	2
Acceleration	12	23	37	6	12	18
Mass × *Acceleration*	12	23	37	12	24	36

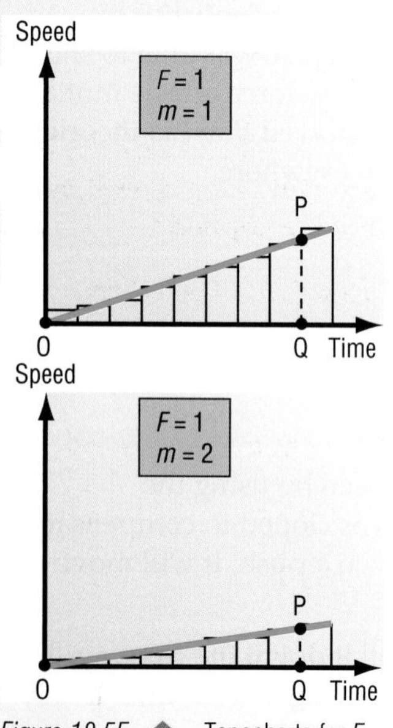

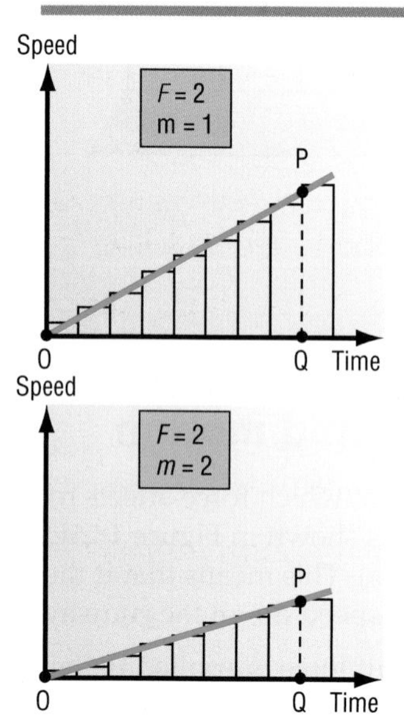

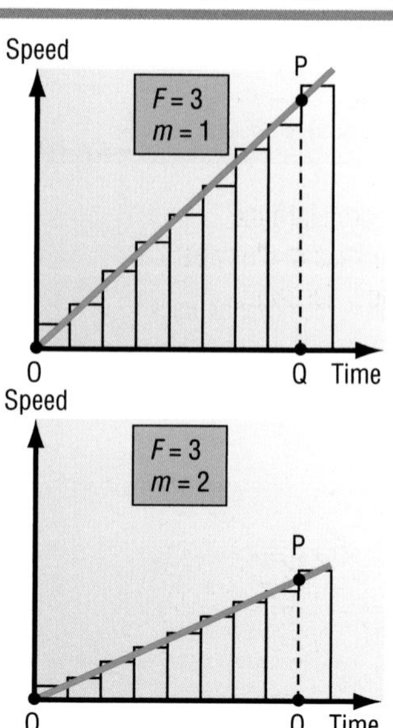

Figure 18.5E ▲ Tapecharts for $F = ma$

What do these results show? What would be the acceleration of a triple deck trolley pulled by two elastic bands? For this amount of force, Mass = Acceleration should be about 24. So the acceleration should be 8 for three trolleys.

The results show that

Force is proportional to Mass × Acceleration.

This link was another of Newton's discoveries, known as **Newton's Second Law**. It is used to define the unit of force, the newton.

The **newton, N,** is defined as **the force that will give a 1 kg mass an acceleration of 1 m/s².**

Force (N)	=	Mass (kg)	×	Acceleration (m/s^2)

Covering a symbol shows how it is related to the other two symbols

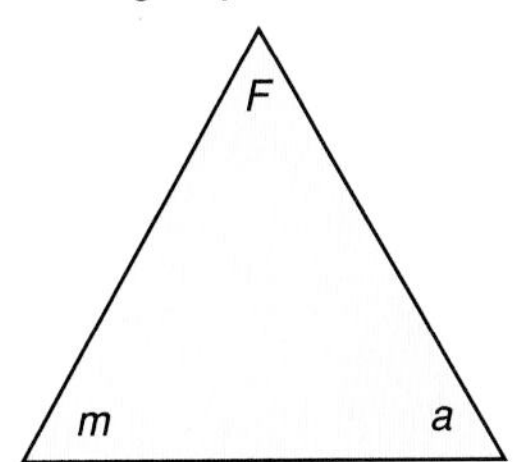

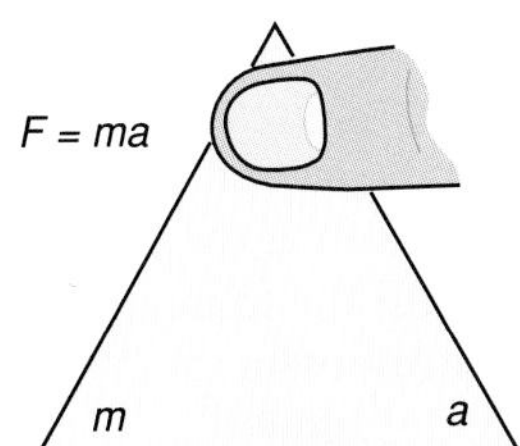

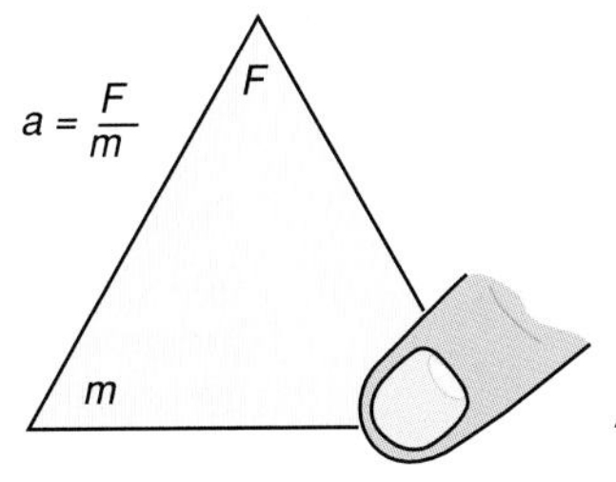

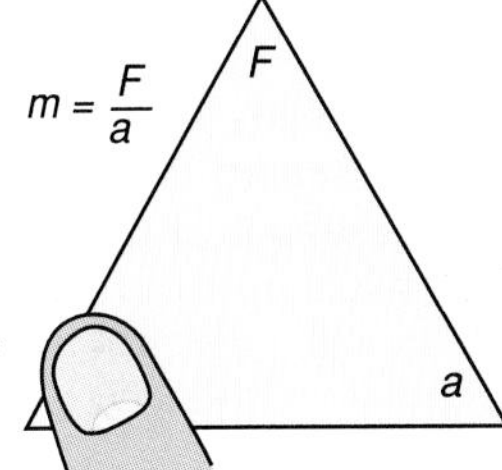

Figure 18.5F ▲ Using Newton's Second Law

The force-mass-acceleration triangle in Figure 18.5F shows how these quantities are related. Cover one symbol to see how it is related to the other two.

Road grip

Motorists need to take extra care on wet roads. Tyres grip road surfaces best in dry conditions where there is maximum friction between the tyres and the road. When the road surface is wet, friction is reduced and cars may slip or skid, especially if the tyre treads are worn. The worn tyre in the photograph is illegal, because the treads are not deep enough. Research into the design of tyres has made tyres safer and longer-lasting.

Worked example 1 A car of mass 800 kg accelerates from rest to a speed of 8.0 m/s in 20 s. Work out (a) the acceleration of the car, (b) the force needed to produce this acceleration.

Solution

(a) $\text{Acceleration} = \frac{\text{Change of speed}}{\text{Time taken}} = \frac{8.0 - 0.0}{20} = 0.40\ m/s^2$

(b) Force = Mass × Acceleration = 800 kg × 0.40 m/s^2 = 320 N.

Worked example 2 A certain type of car has a braking force of 6000 N. Its total mass with four occupants is 1200 kg. How long does it take to stop from a speed of 30 m/s when the brakes are applied?

Solution To work out the deceleration, use $F = ma$ with $F = 6000$ N and $m = 1200$ kg.

Thus $6000 = 1200 \times a$ which gives $a = \frac{6000}{1200} = 5.0\ m/s^2$.

To work out the time taken, t, use $a = (v - u)/t$ with $v = 0$, $u = 30$ m/s and $a = -5.0\ m/s^2$. (−ve sign denotes deceleration.)

Then $-5.0 = \frac{(0 - 30)}{t}$

which gives $-5.0t = -30$

and hence $t = \frac{-30}{-5.0} = 6.0$ s.

Weight is the force of gravity on an object. Objects in free fall accelerate due to gravity. Newton's Second Law tells us that Force = Mass × Acceleration. Hence the force of gravity on an object in free fall is its Mass x g, where g is the acceleration due to gravity. Thus an object's weight can be worked out from its mass using the formula:

$$\text{Weight} = \text{Mass} \times g$$

An object that falls without any support is sometimes said to be weightless during its fall. For example, a person in free-fall from a plane has no support and could be described as weightless. However, this is misleading since gravity continues to pull the person downwards. It would be more accurate to describe the person as *unsupported*.

See www.keyscience.co.uk for more about forces and motion.

Drag forces and terminal speed

For an object in a fluid acted on by a constant force F, the drag force increases until it equals F. The speed is then constant.

A falling object has an acceleration equal to g, provided air resistance is negligible. If air resistance is significant, the force due to air resistance drags on the object. This drag force increases as the object speeds up, until the force becomes equal and opposite to its weight. The acceleration becomes zero because the resultant force on the object becomes zero. The speed therefore becomes constant; this value is referred to as the **terminal speed**. Figure 18.5G shows how the speed of such an object increases.

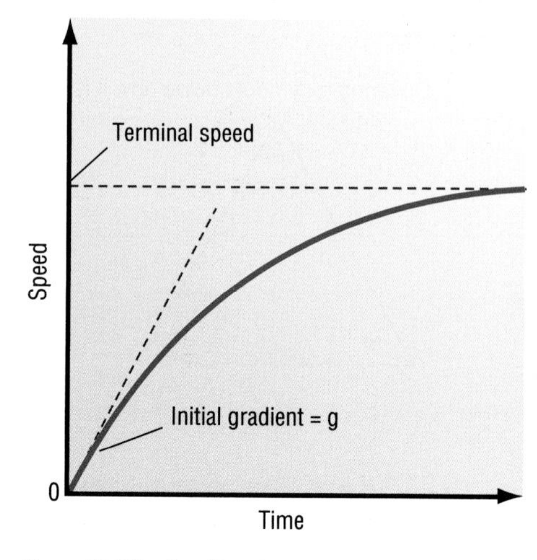

Figure 18.5G ▲ Terminal speed

Vehicle bodies are streamlined to reduce drag. For a constant engine force, the resultant force becomes less as the speed increases. This is because the engine force is opposed by drag which increases with speed. When the drag force becomes equal to the engine force, the vehicle has reached its top speed and it cannot move any faster. Streamlining the body of a vehicle reduces the drag force and therefore enables the vehicle to reach a higher top speed.

Figure 18.5H ▲ Streamline tests

The hull of a boat is streamlined to reduced drag when the boat is moving. Boats are unable to attain the same high speeds as land vehicles or aircraft because the drag forces are much greater in water than in air.

Power is wasted due to drag forces. The wastage increases considerably with increased speed. This is why the rate of fuel consumption of a vehicle increases with speed.

1 The energy lost by the drag force each second = the drag force × its speed, since its speed is the distance it moves each second.
2 The work done by the engine force each second = engine force × its speed.
3 For constant speed, the drag force is equal and opposite to the engine force. Therefore, the work done by the engine is wasted completely due to the drag force.

SUMMARY

The fastest speed of an object is called its terminal speed. For a falling object moving at its terminal speed, the drag force on it is equal and opposite to its weight. For a powered object moving at its terminal speed, the drag force is equal and opposite to the engine force.

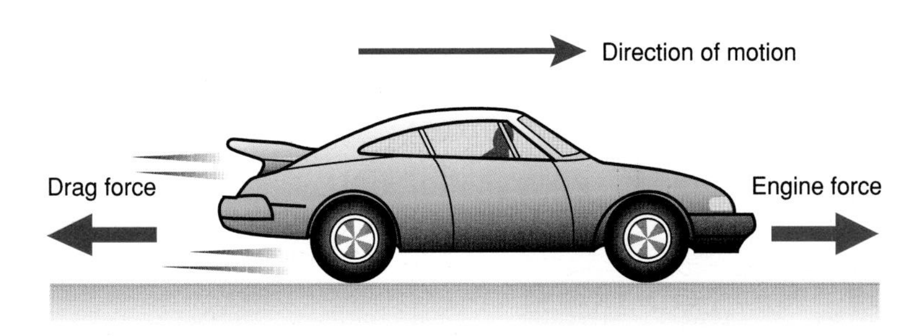

Figure 18.5l Power wasted

CHECKPOINT

➧ 1 (a) Why is it difficult to walk on an icy pavement?
(b) Why does oiling the wheel bearings of a bicycle make it easier to use?
(c) Why does streamlining the shape of a car reduce its fuel consumption?

➧ 2 Work out each of the following using Newton's Second Law.
(a) The force needed to give a 5.00 kg mass an acceleration of 0.30 m/s^2.
(b) The acceleration of a 0.20 kg mass when a force of 5.00 N is applied.
(c) The mass of an object that accelerates at 3.50 m/s^2 when a 14.0 N force is applied to it.
(d) The mass that must be added to a 0.80 kg trolley to give it an acceleration of 0.40 m/s^2 when a 0.50 N force is applied to it.

➧ 3 (a) A car of mass 800 kg is capable of reaching a speed of 20 m/s from rest in 36 s. Work out the force needed to produce this acceleration.
(b) What is the weight of the car?
(c) Work out the ratio of the car's accelerating force to its weight.

➧ 4 The engines of a ship have broken down. The ship is being towed into port by a tugboat at a steady speed of 4.0 m/s. The towing cable pulls the ship with a force of 22 500 N.
(a) The ship experiences a 'drag' force due to the water. This force opposes the motion of the ship. What is the size of the drag force?
(b) The total mass of the ship is 8.0×10^6 kg. When the cable is released, estimate how long the ship takes to stop, assuming the drag force does not change.

➧ 5 A parachutist of mass 70 kg supported by a of mass 20 kg descends vertically at a constant speed of 4.0 m/s.
(a) Explain why the parachutist falls at constant speed.
(b) Calculate (i) the total weight of the parachutist and her parachute, (ii) the force of air resistance on the parachute, (iii) the loss of potential energy of the parachutist each second, (iv) the energy transferred each second to the surrounding air by the force of air resistance.
(c) Explain why the answers to (b)(iii) and (iv) differ.

18.6 Kinetic energy

FIRST THOUGHTS

In this section you will find out how to calculate the kinetic energy of an object from its mass and speed. The formula is valid provided speeds do not approach the speed of light.

Imagine you are cycling along a level road and there is a steep hill ahead. The hill will be easier to climb if you speed up as much as possible before you come to it. Speeding up increases your kinetic energy. As you climb the hill, you lose speed. Your kinetic energy (KE) is converted to potential energy (PE). If you start the hill climb at a slow speed, you need to use your muscles much more to increase your potential energy.

Figure 18.6A ▲ Cycling uphill

Investigating kinetic energy

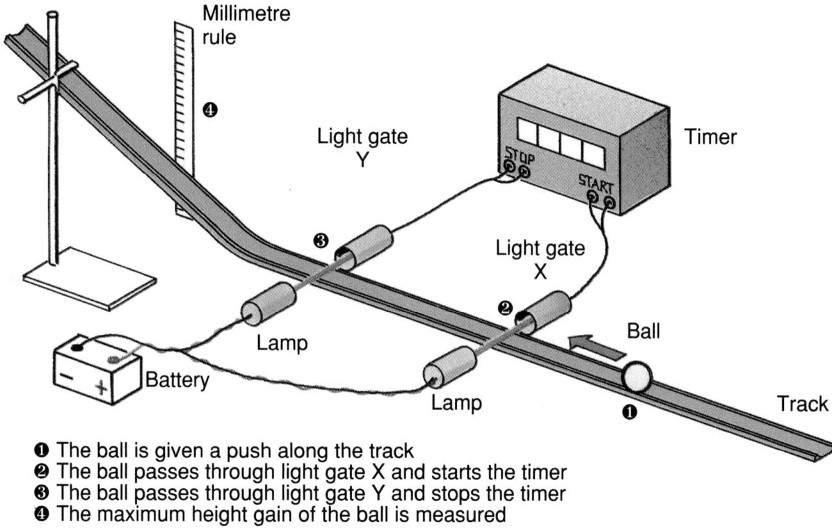

Figure 18.6B ▲ Investigating kinetic energy

When you cycle up a hill, how does your height gain depend on your initial speed? Figure 36.6B shows one way to investigate this. The ball is timed, using the light gates, over a measured distance XY before it reaches the 'hill'. The height gained is measured. Some measurements with XY = 1.0 m for different speeds are shown below.

Table 18.3 ▼ Investigating kinetic energy

Height gained (m)	0.05	0.10	0.16	0.20
Time over XY (s)	0.98	0.72	0.58	0.50
Speed (m/s)	1.02	?	?	?

Work out the speed in each case. The first value has been worked out for you. *Can you see a link between speed and height gain?* Double the speed and the height gain increases by four times. The height gain is proportional to the $(\text{speed})^2$. Check the other measurements to see if they fit this rule.

Now let us see if we can explain this link. The potential energy change is given by the equation Weight × Height change. Thus, for a mass m, the change of PE is given by the equation

> Change in PE = mgh
> where mg = weight and h = change in height

All the initial KE is converted into PE, so the height gain must be proportional to the initial KE. Since the experiment shows that the height gain is proportional to the $(\text{speed})^2$, then the KE must be proportional to the $(\text{speed})^2$.

To see the exact link between KE and speed, consider an object of mass m, initially at rest, acted on by a constant force F. Figure 18.6C shows the idea.

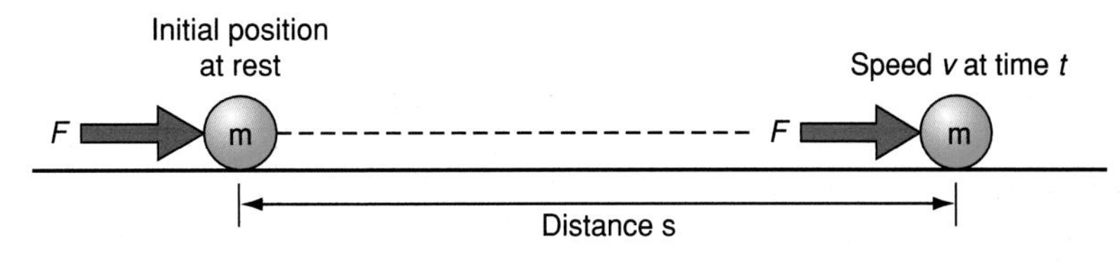

Figure 18.6C ▲ Gaining kinetic energy

Learn the formulas for potential energy and kinetic energy.

In time t, the speed of the object increases from zero to v. The distance travelled, $s = \frac{1}{2}(u + v)t$

$$s = \frac{1}{2}(0 + v)t = \frac{1}{2}vt$$

$$\text{Acceleration, } a = \frac{(v - u)}{t} = \frac{(v - 0)}{t} = \frac{v}{t}$$

EXTENSION FILE ASSIGNMENT

Using Newton's Second Law, Force = Mass × Acceleration,

$$F = ma = \frac{mv}{t}$$

Now Work done = Force × Distance

$$= \frac{mv}{t} \times \frac{1}{2}vt$$

$$= \frac{1}{2}mv^2$$

Since the gain of kinetic energy is due to the work done, then

$$\text{Kinetic energy} = \frac{1}{2} \times \text{Mass} \times (\text{Speed})^2$$

$$KE = \frac{1}{2}mv^2$$

SUMMARY

Kinetic energy $= \frac{1}{2} \times \text{Mass} \times (\text{Speed})^2$

Change of potential energy = Mass × g × Height change

CHECKPOINT

1. (a) Consider the experiment shown in Figure 18.6B. Use the results given to plot a graph to check the link between height and speed.
   (b) Why is the cyclist going uphill unlikely to convert all the KE into PE?
2. A car uses more fuel per kilometre when it keeps having to stop and start than when it travels at steady speed on a motorway. Why?
3. Work out the kinetic energy of each of the following:
   (a) a 0.5 kg ball moving at a speed of 5.0 m/s,
   (b) a 1000 kg car moving at a speed of 20 m/s,
   (c) a 50 000 kg aeroplane moving at a speed of 200 m/s.
4. A hot air balloon of mass 250 kg at a height of 150 m descends slowly into a field.
   (a) Work out the PE of the balloon before the descent.
   (b) What happens to this PE as a result of the descent?
   (c) If the balloon burst at 150 m height, what would be its speed of impact at the ground?
5. A trolley of mass 45 kg is released at a height of 4.2 m at the top of a ramp.
   (a) Work out its PE at the top of the ramp.
   (b) What happens to its initial PE when it rolls down the ramp?
   (c) Work out its speed at the bottom of the ramp.
6. A cyclist moving at a speed of 8.0 m/s along a level road reaches a steep hill. The mass of the cyclist and cycle is 55 kg.
   (a) Work out the initial KE of the cyclist.
   (b) What is the maximum height the cyclist can gain without pedalling? Assume $g = 10\ \text{m/s}^2$.

18.7 Collisions and explosions

FIRST THOUGHTS

A collision is when objects meet; an explosion is when they move apart. The same principle explains both collisions and explosions.

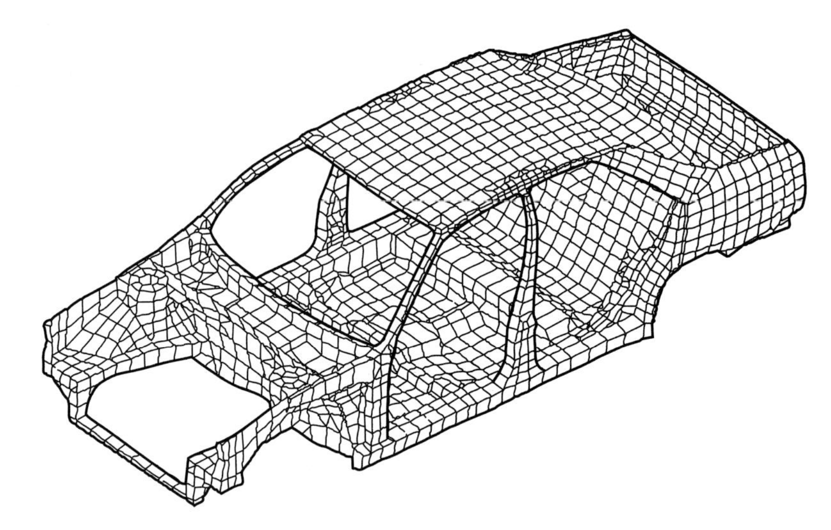

Figure 18.7A Car safety cage design

Safety is an important feature of car design. The structure of a well-designed car should protect the occupants in the event of a crash. The 'safety cage', where the occupants sit, is strengthened to protect those inside. The 'crumple zones' of a car are meant to lessen the force of an impact. Car manufacturers test the safety of their cars by driving them by remote control into brick walls.

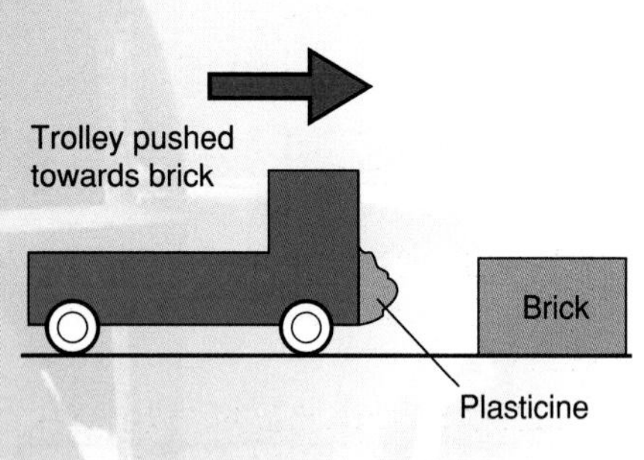

Figure 18.7B Investigating impacts

Why are impacts lessened by using crumple zones? Consider the collision between a trolley and a brick as shown in Figure 18.7B. The Plasticine® flattens on impact, making the impact time longer. *Why does this lessen the impact?* Newton's Second Law has the answer.

Let the initial speed of the trolley be u. Assume its final speed v is zero. Suppose the time for the impact is t.

$$\text{Deceleration due to the impact} = \frac{(v - u)}{t} = \frac{-u}{t}$$

Using Newton's Second Law, Force = Mass × Acceleration

$$F = -\frac{mu}{t}$$

The minus sign tells us that the impact force is in the opposite direction to the initial velocity. The equation shows that making the impact time longer (increasing the value of t), makes the impact force smaller. Crumple zones in cars are designed to make impact times longer so impact forces are reduced.

The **momentum of a moving object** is defined as its Mass × Velocity. The unit of momentum is kg m/s.

The initial momentum of the above trolley is $m \times u$. Its final momentum is zero. The impact force is given by

$$\text{Impact force} = \frac{\text{Change of momentum}}{\text{Time taken}}$$

Momentum = mass × velocity

Force = change of momentum per second

Worked example A car bumper is designed not to bend in impacts at less than 4 m/s. It was fitted to a car of mass 900 kg and tested by driving the car into a wall. The time of impact was measured and found to be 1.8s. Work out the impact force.

Solution

Initial momentum of car = Mass × Initial speed = 900 × 4 kg m/s,
Final momentum = 0 (assuming it is stopped by the impact).
Change of momentum = 900 × 4 = 3600 kg m/s.

$$\text{Force of impact} = \frac{\text{Change of momentum}}{\text{Time taken}} = \frac{3600}{1.8} = 2000\ \text{N}$$

Investigating collisions between moving objects

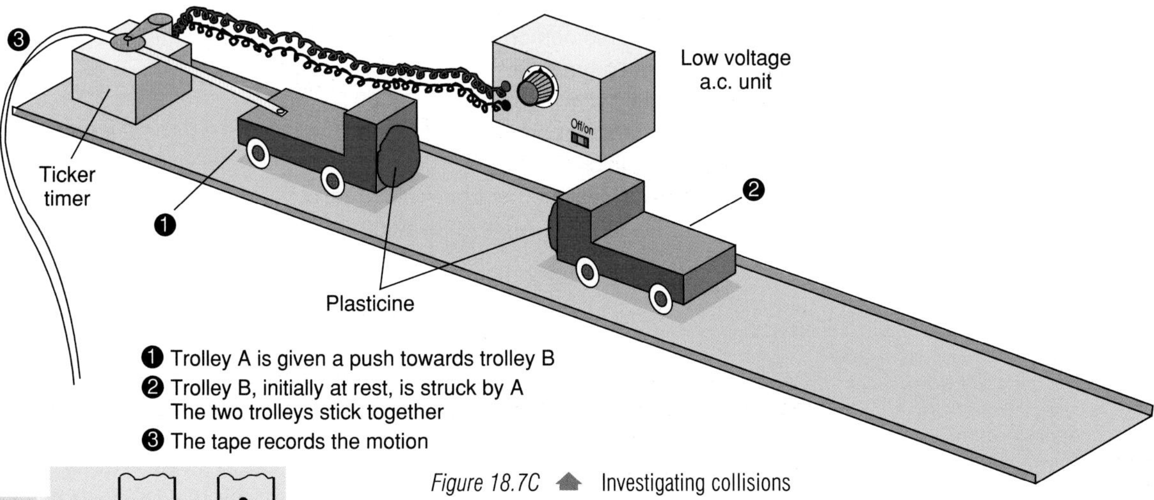

Figure 18.7C ▲ Investigating collisions

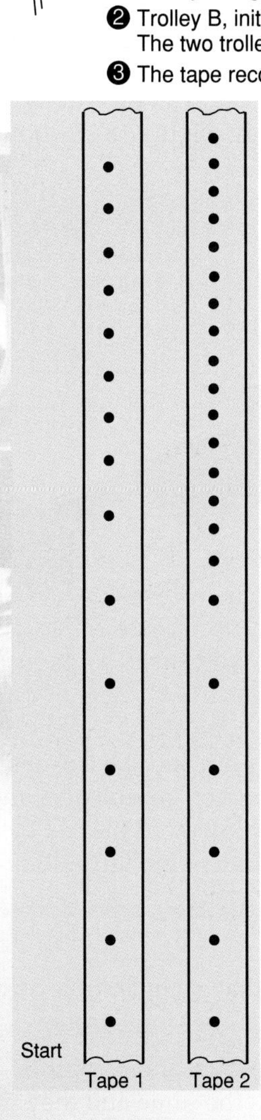

Figure 18.7D ▲ Collision tapes

In Figure 18.7C one trolley is given a push so it collides with another. The two trolleys stick together after they collide. The tape records their motion, as shown on Tape 1 in Figure 18.7D.

Can you see on the tape where the trolleys collided? The spacing between the dots is less after the impact. The dots are twice as far apart before the impact as afterwards, which means the speed has been halved.

Let the trolleys A and B each have the same mass, m. The velocity of A at impact is v. Therefore the momentum of A before the impact is mv.

After the impact the mass is doubled and the speed is halved. The momentum of A and B after this impact is

$$(m + m)\,(\tfrac{1}{2}v)$$
$$= 2\,m \times \tfrac{1}{2}$$
$$= mv$$

Thus the momentum before the impact is the same as the momentum afterwards.

What would the final speed be if a single trolley were pushed into a double trolley so they stuck together? In this case the combined mass would be three times the initial mass. Tape 2 in Figure 18.7D shows the result. The final speed is one third of the initial speed. Once again, the momentum is unchanged. Momentum is said to be **conserved**.

Prove for yourself that the momentum lost by A is equal to the momentum gained by B. Newton's second law tells us that force is equal to the change of momentum/the contact time. The change of momentum of either trolley is therefore equal to the impact force F x the contact time t. In other words, trolley A loses momentum equal to Ft and trolley B gains momentum equal to Ft. The quantity Ft is defined as the **impulse** of the force.

Vehicle damage in collisions would be reduced if all vehicles were fitted with elasticated 'bumpers' like those fairground 'dodgem' cars have. However, in a low speed impact, the occupants would probably suffer more though because the vehicles would rebound instead of losing kinetic energy in the impact. Would they be worse off in elastic vehicles in a high speed impact?

Momentum is conserved when objects interact, provided no external forces act on them. This principle applies to any type of collision. It also applies to explosions where objects fly apart.

If the total kinetic energy immediately after the collision is the same as immediately before it, the collision is said to be **elastic**. If the total kinetic energy is not the same immediately before and after the collision, the collision is said to be **inelastic**.

Worked example A rail wagon A of mass 3000 kg moving at a speed of 2.0 m/s collides with a stationary wagon B. The two wagons couple together and move at a speed of 1.2 m/s after the collision. What is the mass of the second wagon?

Solution Let the mass of the second wagon = M

Initial momentum of A = $3000 \times 2.0 = 6000$ kg m/s

Initial momentum of B = 0

Therefore Total initial momentum = 6000 kg m/s

Total final momentum of A and B = $(3000 + M) \times 1.2$ kg m/s

Since momentum is conserved

Total final momentum = Total initial momentum

$$(3000 + M) \times 1.2 = 6000$$

$$(3000 + M) = \frac{6000}{1.2} = 5000$$

Hence $M = 2000$ kg

Momentum is conserved when two or more bodies interact.

Investigating explosions

When a bomb explodes, fragments of metal fly off in all directions. Their kinetic energy is produced from chemical energy. Figure 18.7E shows a rather more controlled explosion using trolleys.

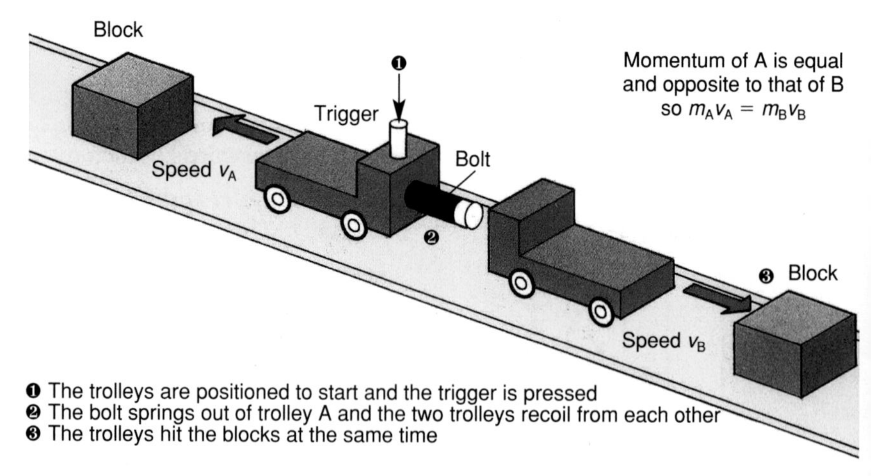
Block
Momentum of A is equal and opposite to that of B so $m_A v_A = m_B v_B$
❶
Trigger
Bolt
Speed v_A
❷
❸ Block
Speed v_B
❶ The trolleys are positioned to start and the trigger is pressed
❷ The bolt springs out of trolley A and the two trolleys recoil from each other
❸ The trolleys hit the blocks at the same time

Figure 18.7E ▲ Investigating explosions

See www.keyscience.co.uk for more about objects flying apart.

When the trigger rod is tapped, the bolt springs out and the trolleys recoil from each other. In order to compare the recoil speeds of A and B, blocks are positioned by trial and error on the runway so the trolleys reach them at the same time. Some results are shown in Figure 18.7F.

Two single trolleys travel the same distance in the same time. This shows that they recoil at equal speeds.

A double trolley only travels half the distance that a single trolley travels in the same time interval. Its speed is half that of the single trolley.

These results show that the trolleys recoil with equal and opposite momentum. The mass × speed of each trolley is the same and they recoil in opposite directions.

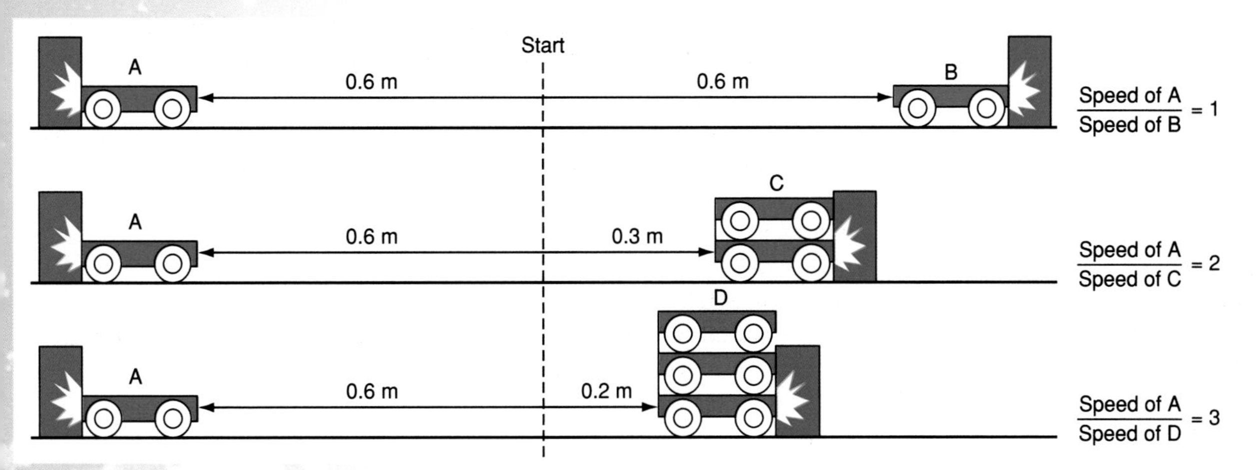

Figure 18.7F ▲ Using different masses

Momentum is conserved in an explosion. In the trolley examples, the initial momentum is zero. The total final momentum is also zero because the recoiling trolleys carry away equal amounts of momentum in opposite directions, which cancel each other out.

Worked example A bullet of mass 0.005 kg is fired from a rifle of mass 2.5 kg. The rifle recoils at a speed of 0.40 m/s. Work out the speed of the bullet when it leaves the rifle.

Solution Momentum of bullet = $0.005 \times V$ where V is the bullet speed. Momentum of rifle = Mass × Recoil velocity = $1.5 \times 0.40 = 0.60$ kg m/s. Since the momentum of the bullet is equal and opposite to the momentum of the rifle, then

$$0.005 \times V = 0.60$$

$$V = \frac{0.60}{0.005} = 120 \text{ m/s}$$

SUMMARY

Momentum is defined as Mass × Velocity. In situations where no external force acts, momentum is conserved.

CHECKPOINT

1. Explain each of the following:
   (a) An asphalt surface for a playground is safer than a concrete surface.
   (b) Catching a cricket ball is easier if you move your hand back as you catch the ball.
   (c) The soles of sports shoes are softer than normal shoe soles.
2. (a) Work out the initial momentum of an 800 kg car travelling at 7.5 m/s.
   (b) What force is required to stop the car in (i) 12 s, (ii) 1.2 s?
3. Simon is about to leap from a rowing boat on to the shore. His friends shout 'We are too far out at the moment!' Simon replies that he can easily jump the gap. He didn't. Why did he misjudge the jump?
4. A railway wagon of mass 800 kg moving at a steady speed of 2.5 m/s collides with another wagon of mass 1000 kg. The two wagons couple together after the collision. Work out the final speed and the loss of kinetic energy if the second wagon was:
   (a) stationary,
   (b) moving at a steady speed of 2.0 m/s in (i) the same direction as the first wagon, (ii) the opposite direction to the first wagon.
5. (a) Kim's skateboard recoils when she jumps off it. Explain why this happens.
   (b) The mass of the skateboard is 1.5 kg and Kim's mass is 40 kg. Work out the recoil speed of the skateboard if she jumps off at a speed of 1.2 m/s.

18.8 Rockets and satellites

FIRST THOUGHTS

The laws of motion are universal; they apply everywhere. They can be used to work out the flight paths of rockets and satellites.

Rockets

'Ten, nine, eight, seven, ..., two, one, ignition ... we have lift off!' A rocket launch is a spectacular event, the result of work by many scientists and engineers.

At lift-off, a mixture of fuel and liquid oxygen fed to the rocket engines is ignited. The gases produced by combustion are expelled downwards from the rocket engine at high speed. They thrust the rocket upwards. While the fuel continues to burn, the rocket gains speed as it lifts off the launch pad. This principle is illustrated in Figure 18.8B.

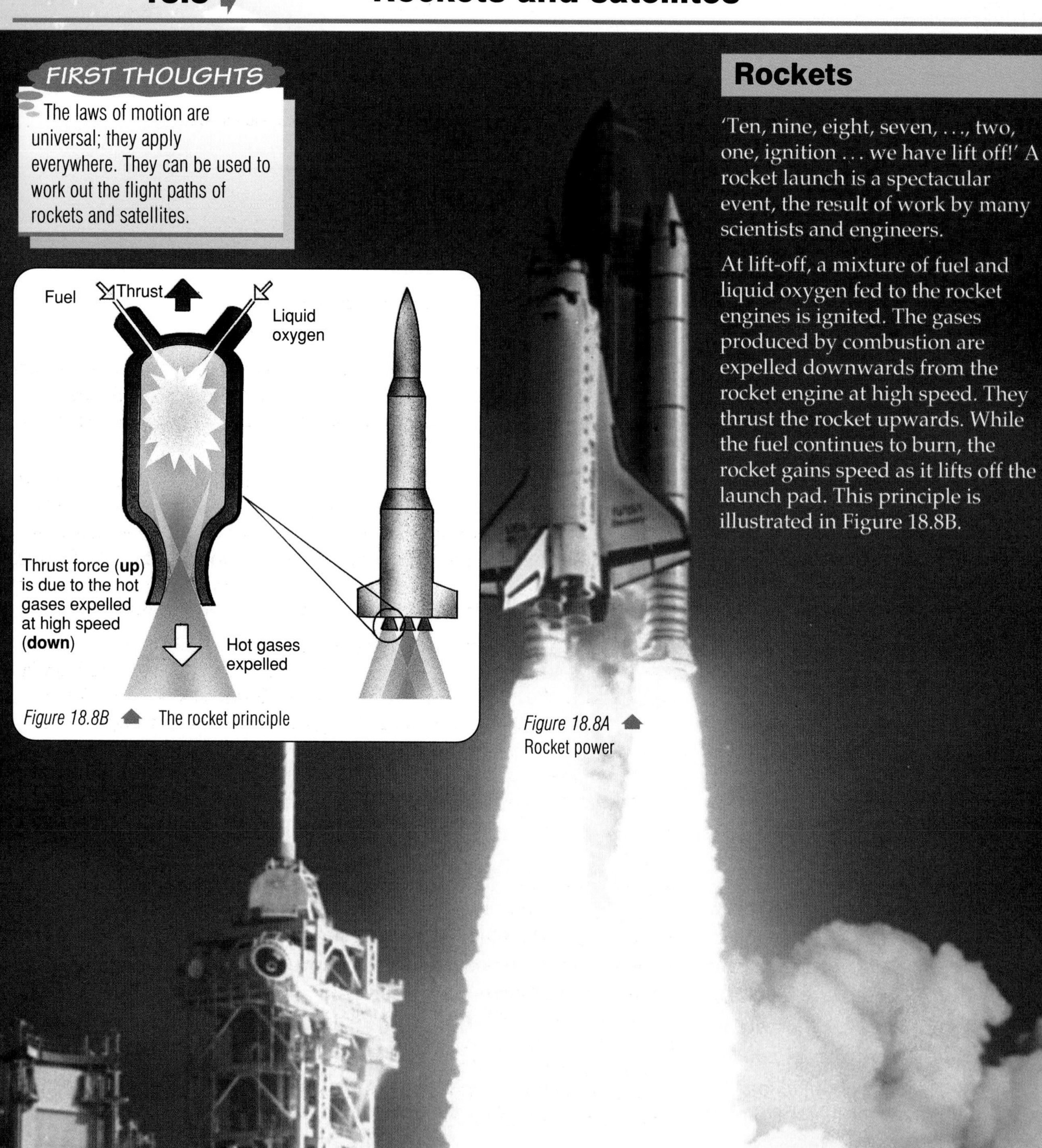

Figure 18.8B The rocket principle

Figure 18.8A Rocket power

Rockets were used as weapons in China many centuries before they were first used in Europe. The gunpowder used as fuel burned very quickly after ignition so these rockets could not attain great heights. In 1895, the principles of modern space flight were set out by Konstantin Tsiolovsky in Russia. He predicted multistage rockets using liquid fuel (see Figure 18.8C).

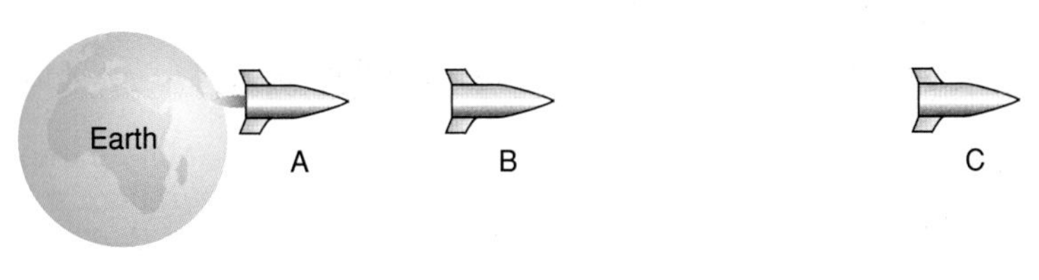

A → B Rocket uses fuel to attain high speed and to move away from Earth.

Chemical energy → KE + PE of rocket

B → C Rocket slows down as it moves away from Earth

KE → PE of rocket

Figure 18.8D ▲ Energy changes in a rocket leaving Earth

Tsiolovsky's ideas were first put into practice by Robert Goddard in 1926 in the US. His first rocket travelled just 55 m. In comparison, *Saturn V*, the most powerful rocket to date, is capable of carrying a 50 tonne cargo to the Moon.

Why do rockets for space travel need to be so powerful? Energy has to be used to move any object against the force of gravity. The force of the Earth's gravity stretches far into space, becoming weaker further from Earth. A rocket travelling into space has to use energy to move against the force of gravity. Chemical energy from the rocket fuel is converted into potential energy as the rocket rises above the Earth. Figure 18.8D explains this.

How powerful are the engines of a rocket? The mass of the Saturn V rocket that carried the first mission to the Moon was 3000 tonnes on the launch pad, almost all due to its fuel payload. It accelerated skywards off the launch pad because the thrust of its engines exceeded its weight. For a steady acceleration of $0.5\ \text{m/s}^2$, a 3000 tonne rocket would require a resultant force of 1.5×10^6 N ($= 3000\,000\ \text{kg} \times 0.5\ \text{m/s}^2$). The engine thrust would need to be at least 31.5×10^6 N (since the resultant force = the engine thrust − its weight).

A rocket engine or a jet engine achieves its thrust as a result of ejecting hot gas at high speed, as explained on p. 000. The rate of ejection of hot gas and its speed of ejection determines the thrust force. If the mass loss per second is M/t and the speed of ejection is v, the momentum loss per second is vM/t. Since force equals change of momentum per second, the thrust force is therefore equal to $v\,M/t$. In the example above, a speed of ejection of about 6000 m/s and a rate of loss of mass of about 5000 kg/s would be needed to generate a thrust of 30×10^6 N. Work out for yourself how long 3000 tonnes ($= 3 \times 10^6$ kg) of fuel would last at a rate of burn of 5000 kg/s. Then prove that a rocket would travel 90 km in this time at an acceleration of $0.5\ \text{m/s}^2$ from rest.

Present rocket technology is limited because chemical fuels supply little more than about 50 MJ per kg – not enough to escape from the Earth's gravity in a single stage. This is why multistage rockets are used as each stage provides a higher launch pad for the next stage. A rocket journey to Mars takes many months at present. A new type of rocket is necessary for faster space travel.

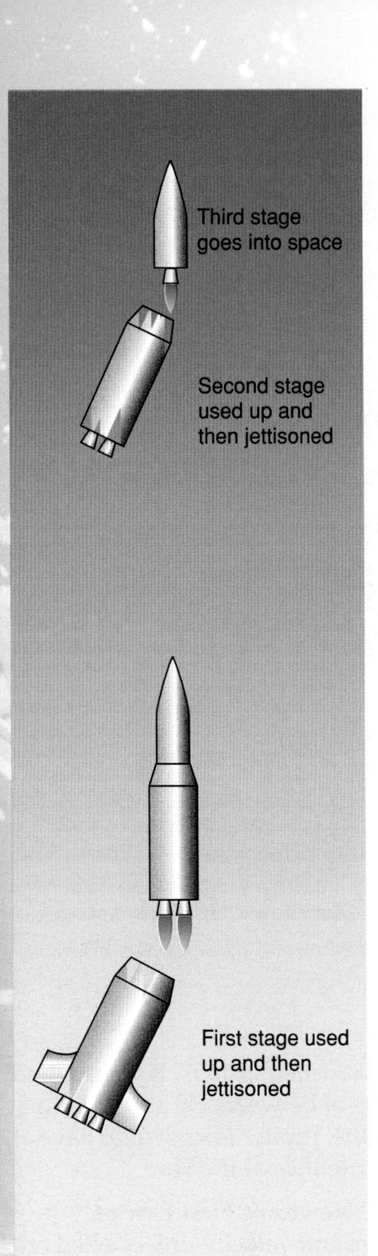

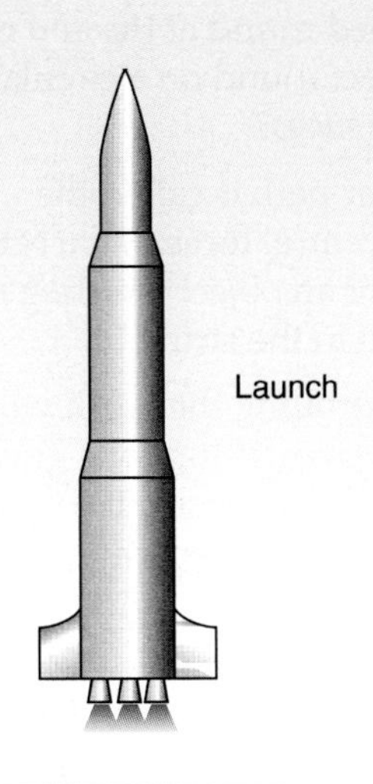

Figure 18.8C ▲ A multistage rocket

Satellites

Imagine launching a satellite into orbit round the Earth from the top of a very tall mountain, as in Figure 18.8E. If the satellite's initial speed is too low, it will fall to the ground. If its initial speed is too great, it will fly off into space. At the 'correct' speed, it orbits the Earth.

See www.keyscience.co.uk for more about satellites and space probes.

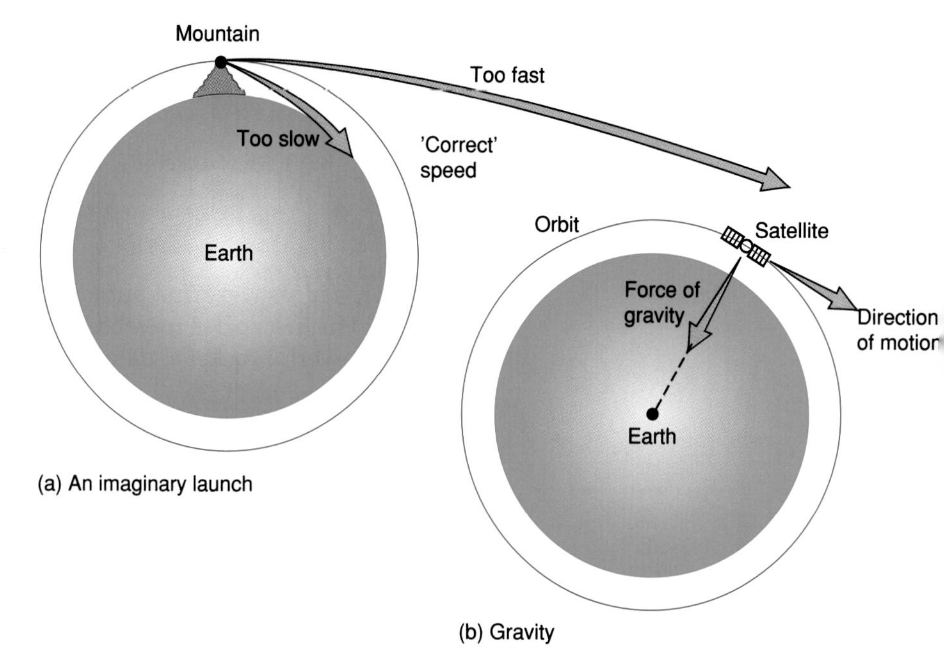

Figure 18.8E ▲ Satellites

The higher a satellite's orbit above the Earth, the longer it takes to go round.

The first artificial satellite, *Sputnik 1*, was launched in 1957. It went round the Earth once every 96 minutes at a height of between 230 and 950 km. The Moon is the Earth's only natural satellite. Jupiter is known to have at least 15 natural satellites. The planets are satellites of the Sun.

What keeps a satellite going round and round? **Newton's First Law of Motion** tells us that an object travels at constant velocity unless acted on by a force. Velocity is speed in a given direction. An object going round in a circle at a steady rate does not have a constant velocity as its direction keeps changing. Therefore there must be a force acting on the object to keep it in circular motion.

Consider the motion of an object being whirled round at the end of a string. The tension in the string pulls the object round on a circular path. *What happens to the object if the string suddenly snaps*?

The force pulling an object round on a circular path is called the **centripetal** force which means 'towards the centre' force. For a satellite, the centripetal force is the force of gravity. For an object whirling round on a string, the centripetal force is the tension in the string.

If gravity suddenly 'switched off', a satellite orbiting the Earth would fly off at a tangent. This is true of any satellite system; if Jupiter's gravity suddenly disappeared, its moons would fly off into space. If the Sun's gravity disappeared, all the planets would fly off into space.

Newton's theory of gravity

Sir Isaac Newton put forward the theory that there is a force of attraction between any two objects. For two point objects separated by distance r, Newton established that this force, which he referred to as the force of gravity, was proportional to:

1 the mass of each object

2 the inverse of the square of the separation, i.e. $\frac{1}{r^2}$

He wrote these statements as a single equation, known as Newton's Law of Gravitation.

For any two point masses m_1 and m_2 at separation r

$$\text{Force of attraction, } F = \frac{Gm_1m_2}{r^2},$$

where G is called the Universal constant of gravitation.

This type of equation is called an inverse square law. This is because the force is proportional to the inverse of the square of the distance. Using this equation, Newton was able to explain the motion of the planets, the Moon and the comets. He was also able to calculate the effect of the Moon's gravity on the Earth's oceans and thus explain the origin of the tides each day.

The planets are natural satellites of the Sun. The pull of gravity from the Sun provides the centripetal force needed to keep a planet in orbit. Figure 18.8F shows how the strength of the force of gravity between the Sun and a planet decreases with distance.

The speed of a satellite depends on the force of gravity between it and its 'parent body', (i.e. the object about which it orbits). The time per orbit of a satellite is greater, the further the satellite is from its parent body. This is because:

1 its orbit is bigger

2 it moves more slowly than it would if it was nearer the parent body.

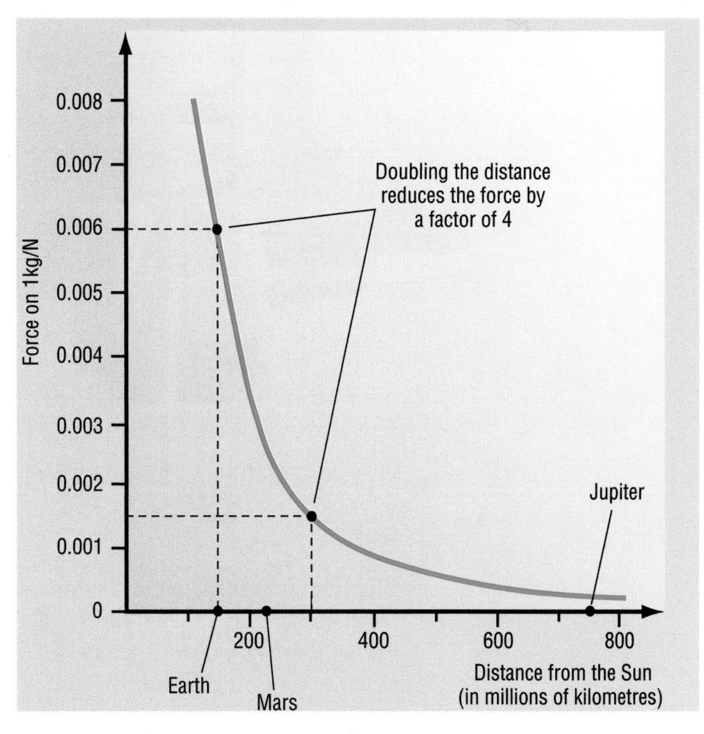

Figure 18.8F ▲ The inverse square law

Figure 18.8G ▲ (a) A Space Shuttle returns safely to Earth

Figure 18.8G ▲ (b) Space Shuttle in flight

The Space Shuttle is a reusable space vehicle, launched at high altitude from the top of a much larger aircraft. After launch, the shuttle flies under its own power into orbit round the Earth. In orbit, no fuel is used, since it stays at constant height. To return to the ground, the shuttle descends from orbit gradually and lands on a specially lengthened airstrip. The Shuttle is used to carry satellites into space.

Satellite equations

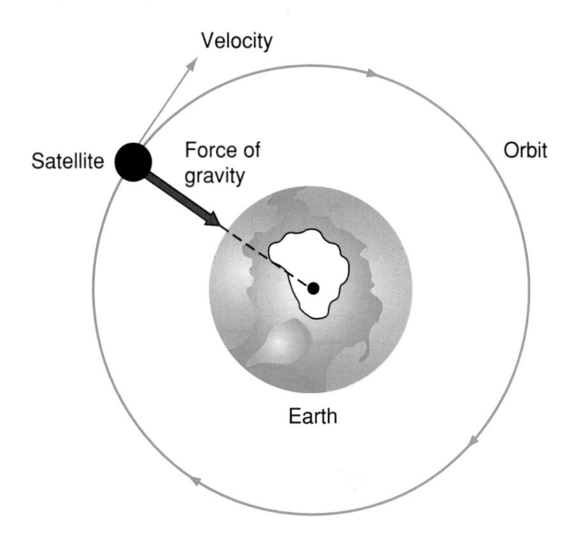

Figure 18.8H ▲ Satellite motion

- **The speed of a satellite** in a circular orbit round the Earth can be worked out if the radius of orbit, r, and the time for one complete orbit, T, are known. The circumference of a circle of radius r is $2\pi r$, therefore the speed v can be worked out from

$$\text{Speed} = \frac{\text{Distance travelled}}{\text{Time taken}}$$

hence

$$v = \frac{2\pi r}{T}$$

Worked example A satellite at a height of 200 km orbits the Earth every 86 minutes. Work out the speed of the satellite. The radius of the Earth is 6400 km.

Solution

$$\text{Radius of orbit, } r = 6400 + 200 \text{ km} = 6\,600\,000 \text{ m}$$

$$\text{Time of orbit, } T = 86 \text{ minutes} = 86 \times 60 \text{ s} = 5160 \text{ s}$$

$$\text{Speed } v = \frac{2\pi r}{T} = \frac{2 \times \pi \times 6\,600\,000 \text{ m}}{5160 \text{ s}} = 8036 \text{ m/s}$$

For an object in circular motion at constant speed, its velocity is tangential and its acceleration is towards the centre.

- **The acceleration of a satellite** moving at constant speed in a circular orbit is directed towards the centre of the Earth. This is because the force of gravity on a satellite is directed towards the Earth's centre and the acceleration is in the same direction as the force.

The acceleration, a, can be worked out from the speed, v, and the radius of orbit, r, using the equation

$$\text{Acceleration} = \frac{(\text{Speed})^2}{\text{Radius of orbit}}$$

$$a = \frac{v^2}{r}$$

In low orbits (i.e. radius of orbit not much more than the Earth's radius) the acceleration is equal to the acceleration of free fall. However, the strength of gravity decreases with distance from the Earth, as explained on p. 273. Therefore, the centripetal acceleration of a satellite is smaller the higher a satellite is. The speed is therefore less and the satellite takes longer to orbit the Earth than it would in a low orbit.

SUMMARY

A rocket's engines expel hot gases at high speed to give it thrust. The thrust enables the rocket to overcome gravity. A satellite in orbit is pulled round on its path by the force of gravity. The greater the radius of orbit of a satellite, the longer it takes for one orbit.

Communications satellites

These are satellites that take exactly 24 hours to orbit the Earth directly above the Equator. The radius of orbit is approximately 42 000 km. At this distance from the centre of the Earth, they go round exactly once every 24 hours. Since they orbit Earth far above the atmosphere, they are unaffected by air resistance, so they stay in orbit indefinitely. Satellites that orbit closer to Earth are affected by air resistance and so eventually come back to Earth.

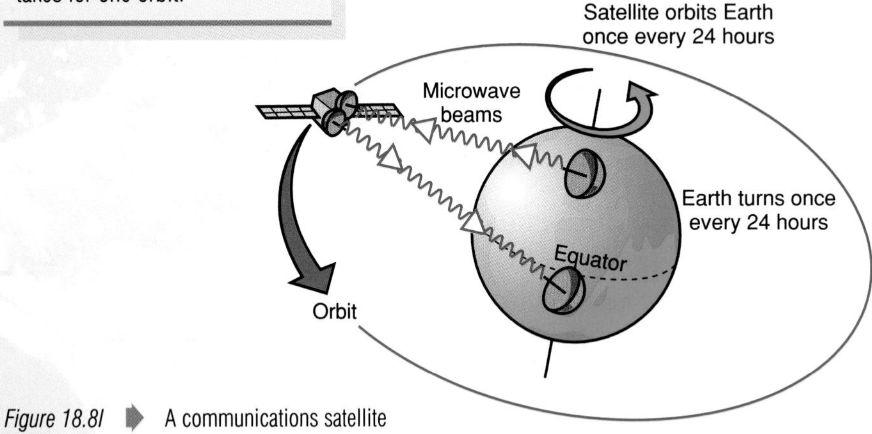

Figure 18.8I ➧ A communications satellite

Because these satellites orbit at the same rate as the Earth spins, they are always over the same point on the Equator. Hence, a microwave beam from a transmitter on the ground can be aimed permanently at the satellite. The satellite detects the beam and transmits a second beam to a receiver station on the ground (see Figure 18.8I). A satellite that transmits and receives signals is said to be an ***active*** satellite. A satellite that only receives signals is said to be *passive*. See p 202–205 for more about satellites used for communications.

CHECKPOINT

1 (a) A rocket is used to launch an astronaut in a space capsule into orbit round the Earth. Describe the forces on the astronaut after lift-off.

(b) What are the advantages and disadvantages of using a rocket in comparison with a space shuttle?

2 A rocket uses liquid hydrogen as its fuel. This is mixed and burned in the rocket engine with liquid oxygen.

(a) What are the likely products of combustion?

(b) Why doesn't a satellite carry fuel tanks?

3 *Apollo 11* was the first successful mission to land astronauts on the Moon. The lunar module carried the astronauts back to the command module to return to Earth. Why did the lunar module not need huge fuel tanks to lift off the Moon?

4 In 1618 Johannes Kepler established the laws of satellite motion by observing the motion of the planets round the Sun. He measured how long each planet took to go round the Sun and its distance from the Sun. His results are shown below.

	Mercury	*Venus*	*Earth*	*Mars*	*Jupiter*	*Saturn*
Time per orbit (years)	0.25	0.61	1.00	1.84	11.7	29.1
Distance (relative to Earth-Sun distance)	0.40	0.73	1.00	1.53	5.20	9.53

(a) Kepler worked out that $(\text{Time})^2/(\text{Distance})^3$ is the same for all the planets. Check this is so using the above values.

(b) The planet Uranus goes round the Sun once every 84 years. Use the link in (a) to work out its distance from the Sun.

5 Imagine you are planning a mission to Mars. The journey is likely to take six months from Earth to Mars. Use the information in question 4 to make a scale drawing of the orbits of the two planets and sketch a possible flight path.

Topic 19 Machines and engines

19.1 Making jobs easier

FIRST THOUGHTS

Read on to find out how machines makes jobs easier although they use more energy.

Kevin works for a builders' merchants and he has been told to load a truck with bags of cement. There are ten bags, each weighing 300 N. He struggles to lift the first one on to the truck. 'There must be an easier way than this' he thinks. He sees some planks of wood in the yard and decides to use them as a ramp (Figure 19.1A). He drags the next bag across the ground and up the ramp, which is slightly easier. Then he sees a wheelbarrow in a corner of the yard. Using this and the ramp, he completes the job much more easily than lifting each one directly.

Figure 19.1A Using a ramp

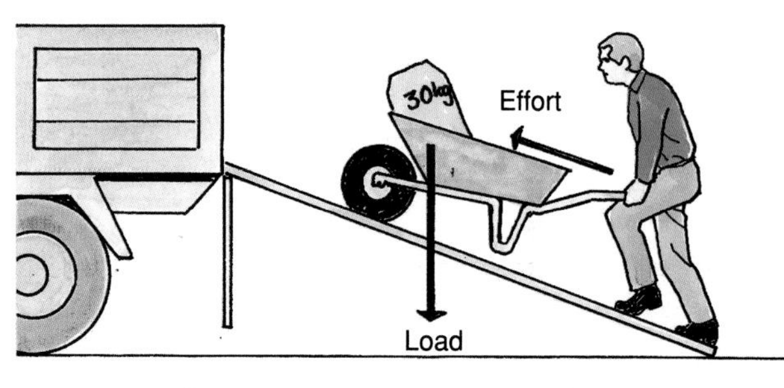

Figure 19.1B Making an effort

In the above example, the force that Kevin applies to the wheelbarrow is the **effort** (Figure 19.1B). The weight of the wheelbarrow and its contents is the **load**. The effort is much less than the load. That is why the job is easier using the ramp and wheelbarrow.

Pulleys can also be used to lift heavy objects. On the building site, Kevin's cement is mixed with sand and water to make mortar for bricks. The mortar has to be lifted to the top of some scaffolding. It could be carried by hand up the ladders. However, a winch consisting of two pulleys has been installed (Figure 19.1C). The bucket of mortar is the load and the pull on the rope is the effort. For a well-oiled pulley-system, the effort is much less than the load.

Some different pulley systems are shown in Figure 19.1D. The easiest one to use is the one with most sections of rope between the upper and lower pulleys. In effect, an equal fraction of the load is raised by each section.

Figure 19.1C Using pulleys

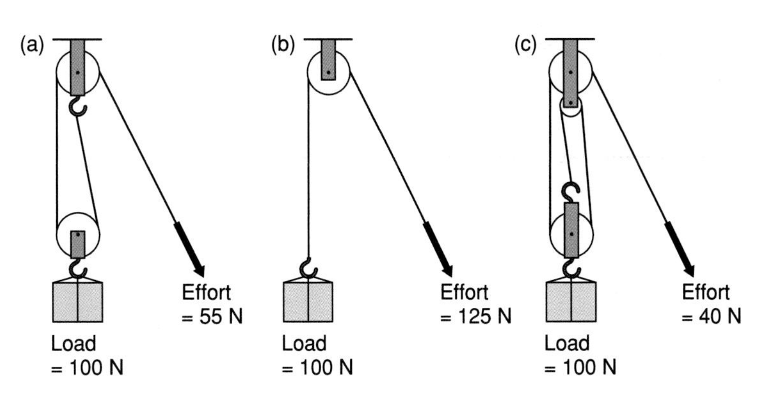

Figure 19.1D Different pulley systems

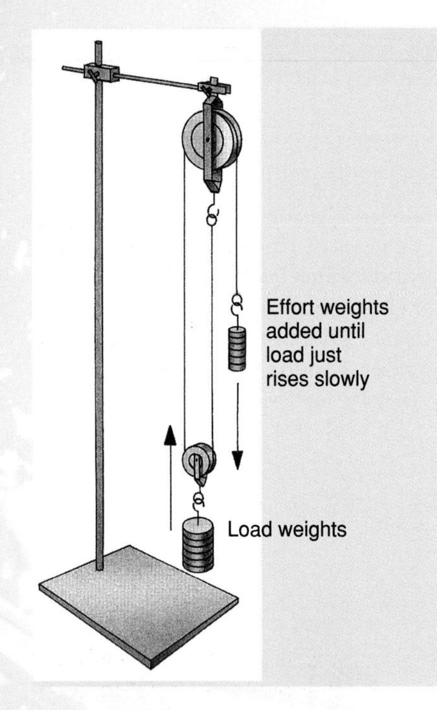

Figure 19.1E ▲ Investigating pulleys

Reducing friction

Energy costs money. Machines that waste energy waste money. Friction between moving parts is one cause of energy waste in machines. Lubricating moving parts with oil helps to reduce friction. Scientists in the motor industry have developed motor oils that do not 'thicken' in cold weather. These oils make cars more reliable in winter.

Investigating pulleys

Is a pulley system an energy saver? Pulleys make lifting easier because the effort is less than the load. *But is less energy used with the pulley system than without?* Figure 19.1E shows how this can be investigated.

The effort is measured by adding weights to the effort weight until the load just rises slowly. The height gain *H* of the load is measured. The corresponding distance moved down by the effort weights is also measured. Results for the pulley system shown in Figure 19.1E are as follows

Load = 9.0 N   Height gain, H, of load = 0.10 m
Effort = 6.0   Distance moved by effort, D = 0.20 m

These measurements show the effort has to move twice as far as the load, but the load is just 1.5 times the effort.

Work done by the effort = Effort × Distance = 6.0 × 0.20 = 1.2 J
PE gain of the load = Load × Height gain = 9.0 × 0.10 = 0.9 J

The calculations show that 1.2 J of work is done using the pulleys to lift the load. This is more than the PE gain of the load. So 0.3 J of energy is wasted using the pulleys. The waste energy is lost to the surroundings as heat.

The **efficiency** of a machine is defined as the percentage of the work done by the effort that is used to move the load.

$$\text{Efficiency} = \frac{\text{Useful energy supplied to load}}{\text{Work done by effort}} \times 100\%$$

No machine can ever be more than 100% efficient because the work done on the load can never be more than the work done by the effort. The useful energy from the machine is its **energy output**. The work done by the effort is the **energy input**. The energy output can never be more than the energy input.

In the example above, 0.9 J of useful energy is supplied by the pulley system to raise the load. The work done by the effort is 1.2 J, so the efficiency is $0.9/1.2 \times 100 = 75\%$. The remaining 25% is wasted energy, lost to the surroundings as heat.

Worked example To test the efficiency of a ramp, the force needed to pull a loaded trolley up the ramp was measured. The height gain and length of the ramp were measured. The measurements were as follows.

Weight of loaded trolley = 350 N   Height gain = 1.2 m
Effort (i.e.pull force) = 140 N   Length of ramp = 4.5 m

(a) Work out the efficiency of the ramp.
(b) How much energy was wasted in the test?

Solution

(a) Useful energy supplied to load = PE gain = Load × Height gain
= 350 × 1.2 = 420 J

Work done by effort = Effort × Length of ramp
= 140 × 4.5 = 630 J

$$\text{Efficiency} = \frac{\text{PE gain}}{\text{Work done by effort}} \times 100$$

$$= \frac{420}{630} \times 100 = 67\%$$

(b) Wasted energy = 630 − 420 = 210 J

Figure 19.1F ▲ Bicycle gears

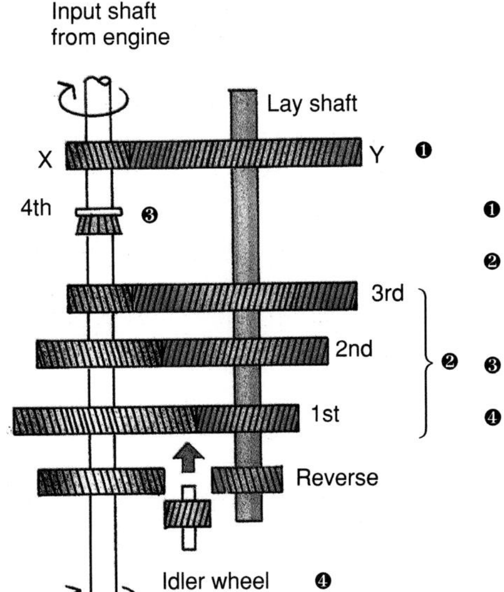

❶ The lay shaft is driven from the input shaft via gearwheels X and Y
❷ When 1st, 2nd or 3rd gear is selected, the appropriate gearwheel on the lay shaft is 'locked' on to the shaft. The other gearwheels turn freely.
❸ When 4th gear is chosen, the output shaft is directly coupled to the input shaft
❹ When reverse gear is chosen, the idler wheel is pushed forwards as shown to make the lay shaft turn the output shaft

Figure 19.1G ▲ How car gears work

SUMMARY

Machines make jobs easier by reducing the effort needed to move a given load. The efficiency of a machine is the percentage of the work done by the effort that is used to move the load. No machine can ever be more than 100% efficient.

Gears are another mechanism for making jobs easier. A cyclist on a hill climb should use the lowest possible gear. This allows the cyclist to keep moving with the least possible effort. However, the work done by the effort is always more than the PE gain of the cyclist. Cycling is easier using gears but some energy is wasted. The cyclist's muscles do not need to pull as hard in low gear, although they have to work longer.

How the gears in a car work is explained in Figure 19.1G.

CHECKPOINT

▶ **1** 'Cliff railways' are a tourist attraction in several seaside towns. Most were designed with twin tracks so that as one car climbs the track the other descends. An electric winch hauls the ascending car and its passengers up the track (see the diagram).
   (a) Why were they designed in this way?
   (b) The first cliff railway did not use an electric winch. Instead, each car included a 4500 litre water tank. The tank was filled with water when the car was at the top and then emptied when it reached the bottom. A steam-driven pump was used to pump water up to the top of the track. Explain why the system worked.

Winch
Ascending car
Descending car

▶ **2** Figure 19.1G shows the construction of part of a car gear box. There are four forward gears and reverse gear. The driver engages each gear using the gear lever. The gear system allows the input shaft to turn the output shaft easily.
   (a) In 4th gear, the output shaft is driven at the same speed as the input shaft. Does the output shaft turn faster or slower in 1st gear than in 4th gear?
   (b) Why is the ider gear essential to make the output shaft turn in reverse?

▶ **3**  A conveyor belt is used to transport boxes from the ground floor to the first floor of a factory. Each box weighs 150 N and is lifted through a height of 5.0 m. The belt is driven by a 2000 W electric motor and carries 15 boxes per minute up to the first floor.
   (a) How much energy is supplied by the 2000 W motor in one minute?
   (b) What is the gain of PE of each box?
   (c) Work out the efficiency of the system.

4 (a) The mechanical advantage of a machine is the ratio load/effort. Work this out for each of the pulley systems in Figure 19.1D. Which of the pulleys is easiest to use (has the greatest mechanical advantage)?
(b) The distance moved by the effort equals the height gain by the load x the number of sections of rope between the two pulleys. Why?
(c) Which of these pulleys systems is the most efficient? Work out its efficiency.

5 John Black Ltd have just bought an 'easilift' pulley system, guaranteed to be 90% efficient if it is maintained correctly. It is fitted to a steel jib at the top of a building, as shown. The manager decides to test it by lifting a crate of weight 1000 N through a height of 1.0 m. The effort he put in was 220 N.
(a) The distance moved by the effort was 5.0 m. How much work was done by the effort? What was the PE gain of the load?
(b) Work out the efficiency and compare it with the guaranteed value.
(c) What was the total force on the jib, ignoring the weight of the pulley, when it was used to lift the crate?

19.2 Engines

FIRST THOUGHTS

An engine uses fuel to work. The steam engine and the petrol engine revolutionised transport. This section tells you how they work.

Figure 19.2A Mill steam engine

Steam engines revolutionised the world in the nineteenth century. No longer did people have to rely on muscle power, windmills or waterwheels to make things. In 1769 James Watt invented the steam engine which was later to power factories and mills throughout the world (Figure 19.2A). The fuel used was coal, which became a vital resource and has been ever since. Transport was also revolutionised by the invention of the steam locomotive and the steamship.

Figure 19.2B ▲ Turbines in a power station

Different types of engines include:

- steam engines
- turbines
- internal combustion engines
- jet engines
- rocket engines.

What has become of the steam engine? You may find steam locomotives still used as museum exhibits but most trains are now pulled by electric locomotives. Factories use electric motors supplied from the National Grid instead of steam engines. Electricity is produced in power stations by means of huge electricity generators.

What keeps these generators turning? **Steam turbines** are used in most power stations (Figure 19.2B). The means of producing steam depends on the type of power station. Coal-fired and oil-fired power stations burn their fuel to heat water to make steam. **Nuclear power stations** are designed to raise steam too, although the heat necessary is released as a result of nuclear fission. How a turbine works is illustrated in Figure 19.2C.

The **petrol engine** is used in most cars. Its fuel is a mixture of petrol and air. It is called an **internal combustion engine** because the fuel is burned inside the engine. It works on a four-stroke cycle, as shown in Figure 19.2D. Its efficiency is about 25%, which means it wastes about 75% of the energy from the fuel. Petrol engines also produce many pollutants in the exhaust gases, including carbon monoxide, sulphur dioxide and lead compounds. New cars are fitted with engines designed to use 'unleaded' petrol because of public concern about the harmful effects of lead compounds in the air we breathe.

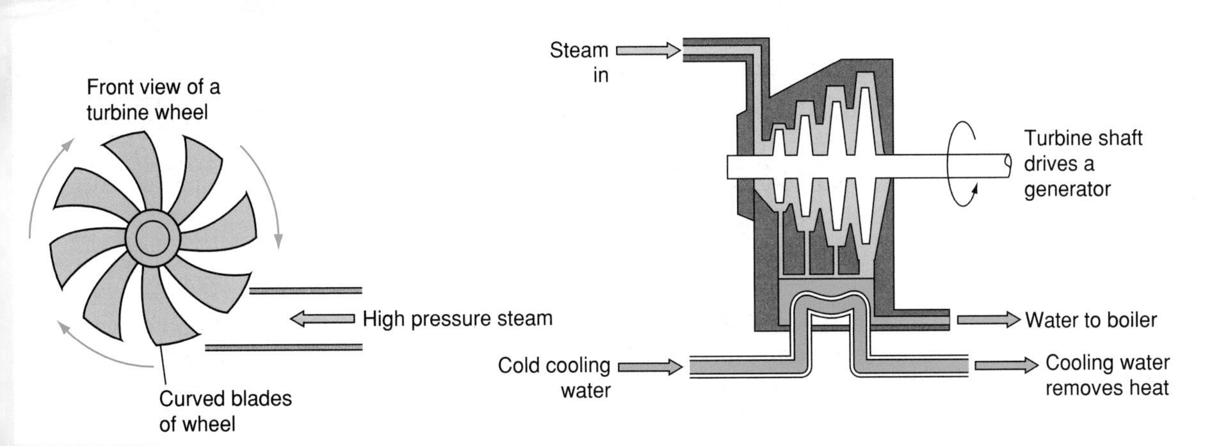

Figure 19.2C ▲ How a turbine works

SUMMARY

Engines produce useful energy from fuels. The petrol engine is used in most cars. The steam turbine is used in generators in power stations to produce electricity.

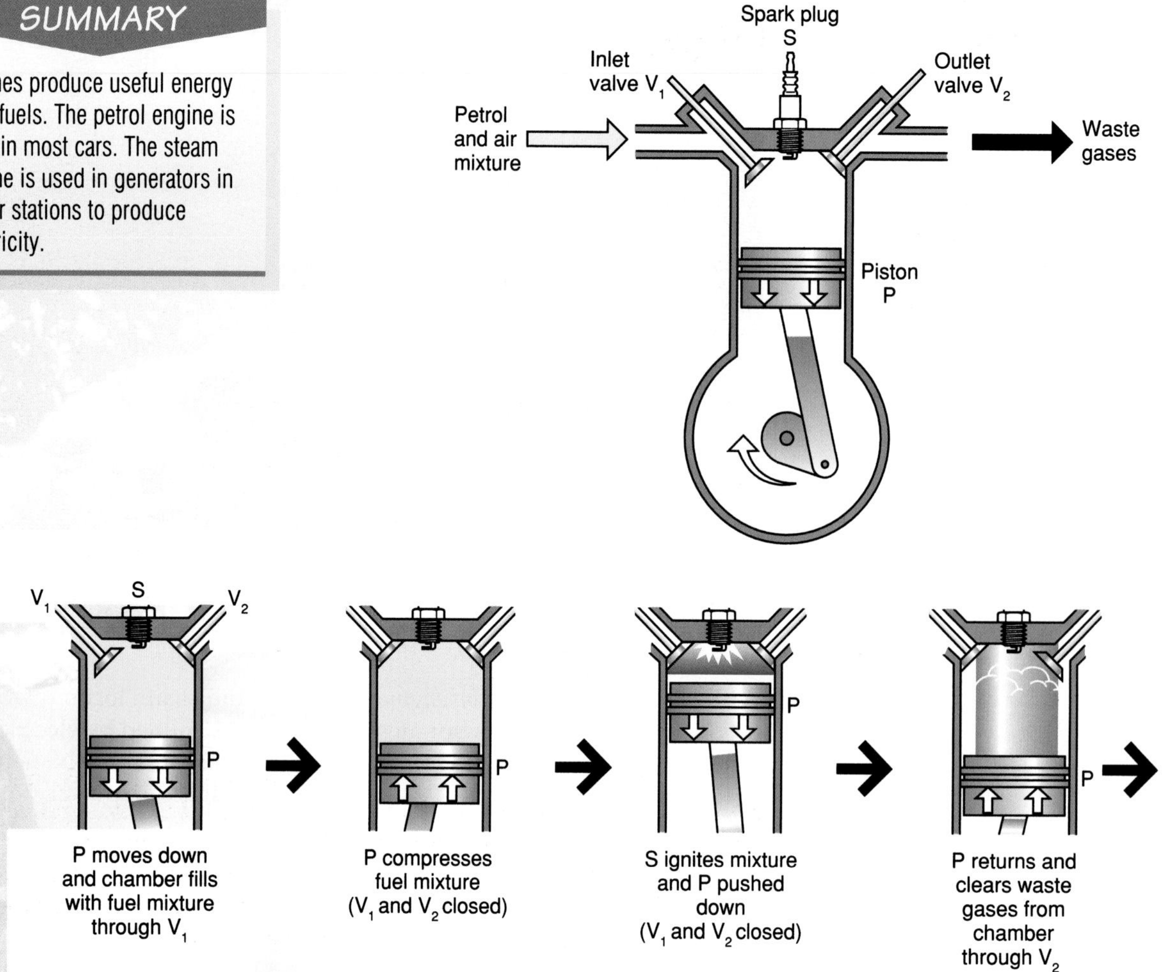

Figure 19.2D ▲ The four-stroke engine

CHECKPOINT

1. Steam engines use coal and produce a lot of smoke and dust. Electric trains are more economical and produce no waste products. Explain why.
2. Steam turbines in power stations must be supplied with cooling water to take away waste heat.
   (a) If the flow of cooling water became blocked, what would happen to the pressure in the turbine?
   (b) Why are power stations often sited on the coast or by a river or lake?
3. How would your life be changed if petrol supplies suddenly ceased everywhere?
4. The use of low density materials in cars saves fuel.

   (a) A car travels 12 000 km in one year and it uses 1 litre of petrol every 12 km travelled on average. How much fuel does it use in one year?
   (b) How much fuel would have been saved in one year if the car had travelled 14 km per litre on average?
5. Air pollution due to cars is a major problem in many countries. Controls on the emission of waste gases from car exhausts have been introduced in some of the worst-affected areas. Suppose you live in an area where air pollution is a problem, despite these controls. What other measures could be taken to deal with the problem effectively?

19.3 ▸ Flight

FIRST THOUGHTS

If you have ever flown in an aeroplane, you may know about rolling and diving. In this section you will find out how the motion of an aeroplane is controlled.

Figure 19.3A ▲ Taking off from London Heathrow Airport

In summer, millions of holidaymakers travel from northern Europe to countries bordering the Mediterranean Sea. At the busiest airports planes take off and land every few minutes. The skies above busy airports are crowded with planes. Air traffic controllers use powerful computers to keep track of them. The Wright brothers made the first powered flight in 1903. Now millions of people every year make routine journeys by aeroplane.

Airships were developed long before 1903. The balloon of an airship contains a gas such as helium, which is less dense than air, so the balloon is pushed upwards by the atmosphere. The force of gravity on the balloon is opposed by the upward force of the atmosphere on it. This force is called the **lift**. To make an airship float upwards, the lift must be greater than the total weight of the balloon and its load.

For an airship at constant height, the lift force is equal and opposite to its weight.

Hot air balloons get lift by trapping hot air. Hot air is less dense than cold air, so it rises. To make the balloon ascend, a burner is used to heat the air beneath the balloon.

Figure 19.3B ▲ (a) Airship

(b) Hot air balloons

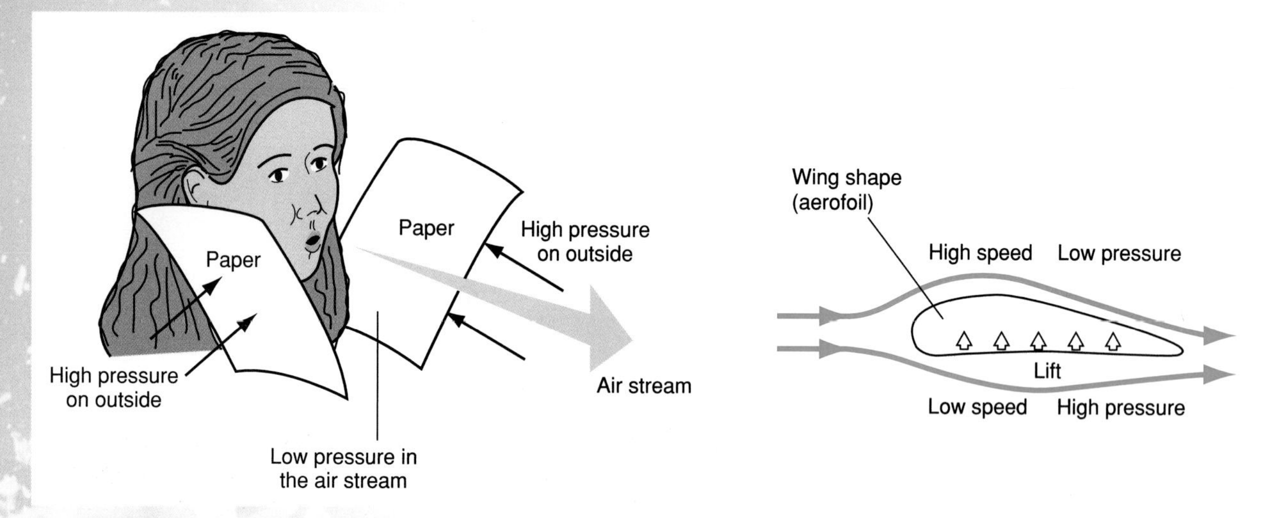

Figure 19.3C ▲ Making low pressure

Figure 19.3D ▲ How an aerofoil creates lift

Aeronautics (flight) is possible because the pressure of a stream of air depends on the air speed. This can be demonstrated by blowing between two pieces of paper, as in Figure 19.3C. The pressure in the fast-moving airstream is less than the atmospheric pressure outside, so the two pieces of paper are forced together. This is an example of the Bernouilli principle at work (see p. 243).

The shape of an aeroplane wing is called an **aerofoil**. As the wing moves forwards through the air, the airstream is faster above than beneath it (Figure 19.3D). The air pressure above the wing is less than the pressure beneath it and the wing lifts. The lift on both wings opposes the weight of the aeroplane.

The amount of lift increases with speed, as the air moves faster over the wing, increasing the pressure difference. This is why an aeroplane needs to achieve a high speed to take off. The aeroplane's engines supply the thrust to enable it to reach the take-off speed. Figure 19.3E shows how a **jet engine** works.

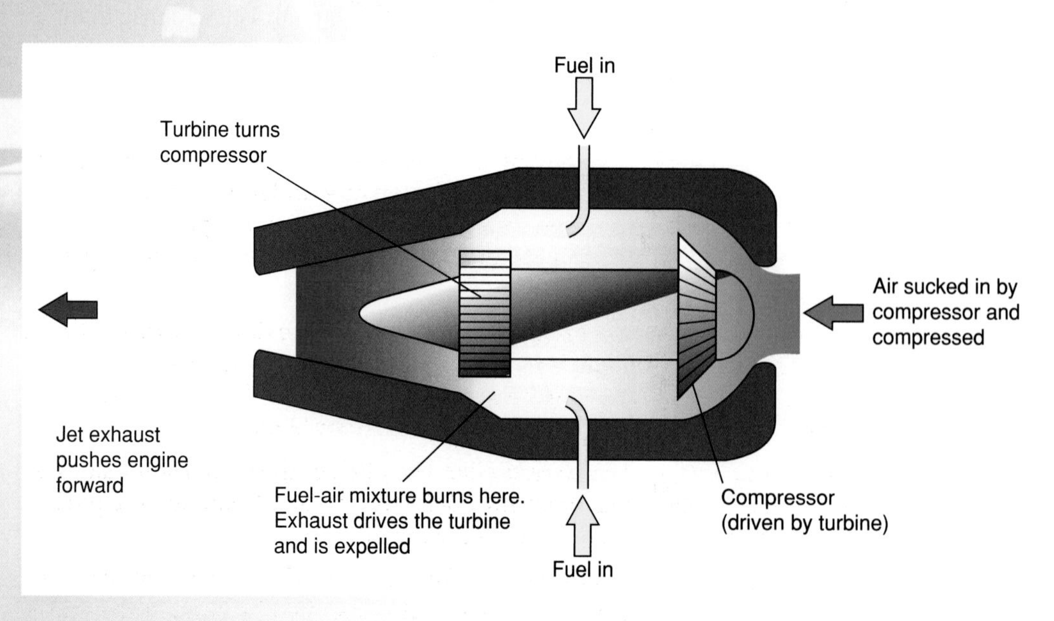

Figure 19.3E ▲ The jet engine

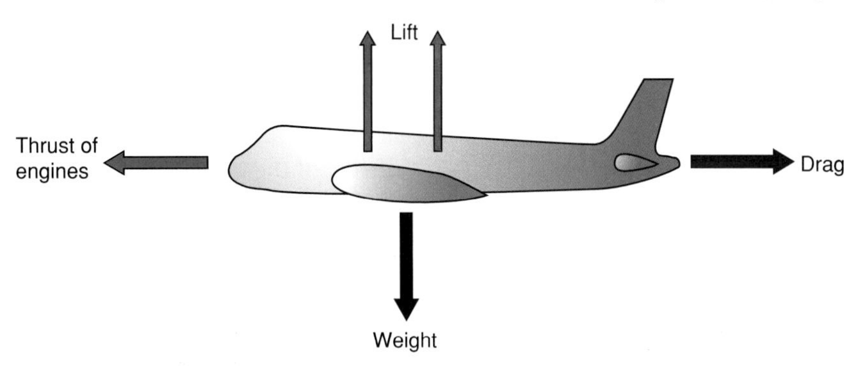

Figure 19.3F ▲ The forces on an aeroplane

The thrust of the engines is opposed by **drag** on the aeroplane due to air resistance. In level flight at steady speed, the drag is equal and opposite to the thrust and the lift is equal to the weight (see Figure 19.3F). The overall force on the aeroplane is zero so it continues to move at constant velocity.

How does an aeroplane gain height? The pilot pulls back on the control column to tilt the tailplane elevators as shown in Figure 19.3G. This increases the lift, so the aeroplane climbs. However, if the elevators are tilted too much, the aeroplane stalls. Altering the position of the rudder and ailerons makes the aeroplane turn and roll.

It's a fact!

Bird flight differs from powered flight because the thrust is supplied by the wings beating. A bird's wings are shaped to produce lift and the tail is for steering. Birds like the golden eagle, which can fly slowly, have slots in their wings. These slots make stalling less likely.

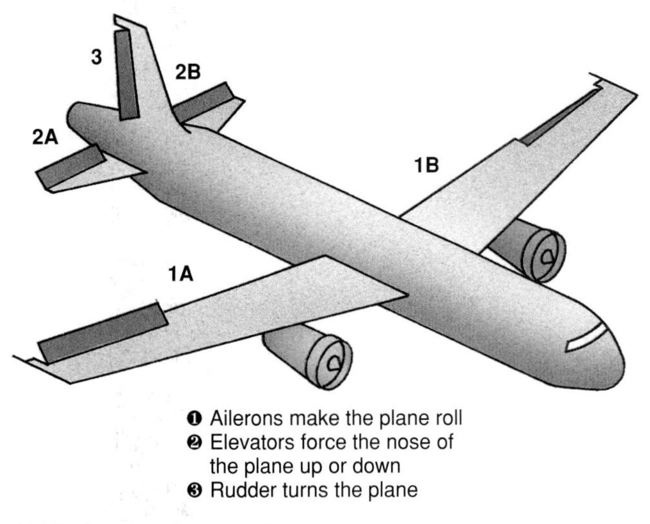

❶ Ailerons make the plane roll
❷ Elevators force the nose of the plane up or down
❸ Rudder turns the plane

Figure 19.3G ▲ Aeroplane controls

Helicopters gain lift by forcing air downward. The main rotor blades are rotating wings, shaped to create high pressure under the blades as the blades rotate. This pressure creates the lift force that supports the helicopter in flight. The tail rotor is necessary to prevent the helicopter body from rotating.

High pressure created under the blades thrusts air downward, exerting a downward force on anything or anyone under the helicopter. A helicopter hovering above the sea flattens the waves because of the downward force of the airstream from its rotor blades.

How does a helicopter move vertically? The pitch of the main rotor blades is controlled by a system of levers, operated by hand. Changing the pitch changes the lift just as with a fixed wing aircraft.

How does a helicopter move horizontally? The axis of the main rotor can be tilted using another system of levers, also operated by hand. If the axis is tilted forwards, the helicopter moves forwards because the lift force is

Make sure you can explain how lift is created in:

- an airship
- an aircraft
- a helicopter

tilted forwards. By adjusting the tilt of the axis to the required direction, the helicopter is made to move in that direction.

How does the helicopter swing round? Another system of levers operated by foot pedals alters the pitch of the tail rotor blades. This changes the thrust of the tail rotor, allowing the helicopter to turn one way if the thrust is increased or the opposite way if the thrust is decreased.

See www.keyscience.co.uk for a world in motion..

Figure 19.3H ▲ A hovering helicopter

SUMMARY

Lift is essential for flight. Wings are shaped to produce lift. If the wings are at too great an angle to the airstream, the aeroplane or bird will stall.

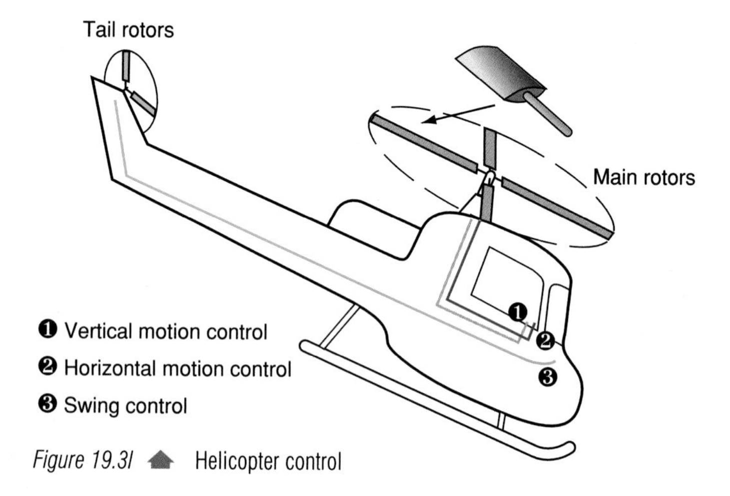

Figure 19.3I ▲ Helicopter control

CHECKPOINT

1. (a) Explain why an airship can gain height by jettisoning unnecessary weight.
   (b) How can an airship lose height?
2. Imagine you are at the controls of an aeroplane. How would you use the controls to (a) gain height, (b) turn in a horizontal circle?
3. Make a paper aeroplane. Bend the wings so it flies up.
   (a) Why does it stall when it goes too high?
   (b) How can you make it turn in a circle? Make a sketch showing how you did this.
4. How does a bird in flight manage to land safely? Observe a bird landing and describe how it uses its wings to do this.
5. (a) Sketch a helicopter hovering above the ground and show the direction of each of the following forces: (i) the lift force, (ii) its weight, (iii) the tail rotor thrust.
   (b) Repeat part (a) for a helicopter moving forwards.

Theme Questions

TOPIC 16

1 The figure below shows a special type of spanner called a mole wrench. It is being used to unfasten a nut from a rusted bolt. The average force applied to the mole wrench is 20 newtons (N) and this force moves 30 centimetres (cm) when one quarter turn is made.

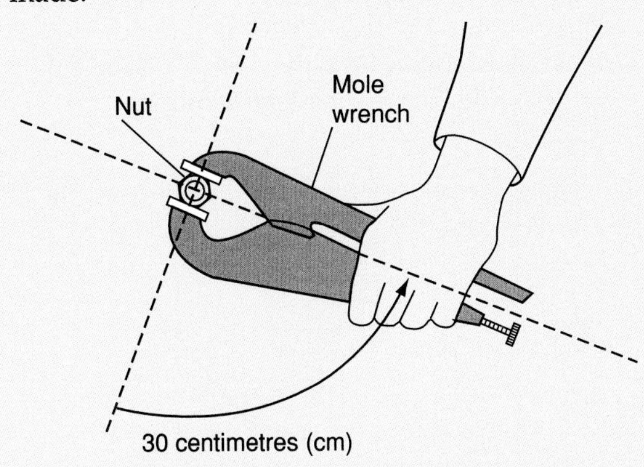

(a) (i) What is the pattern which links the energy transferred, distance moved and force applied?
(ii) What is the distance moved in metres (m) by the force when **one full turn** is made?
(iii) Calculate the energy transferred by the mole wrench when it makes **one full turn**. You should state the unit of energy in your answer.

(b) Suggest the name of the force which is **opposing** the rotation of the mole wrench. (SEG)

TOPIC 17

2 The diagram shows part of the disc brake system on a car.

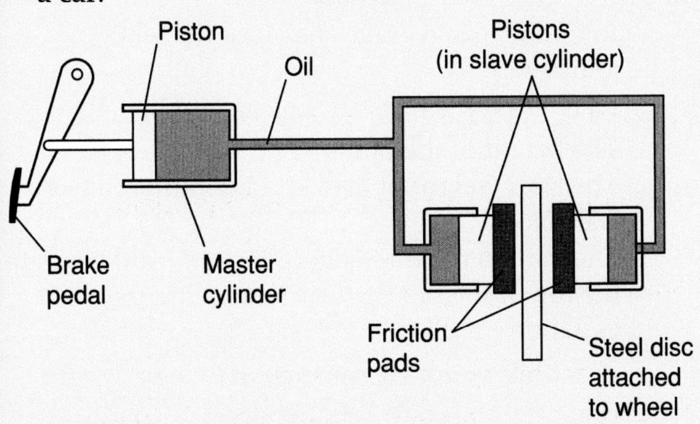

A force is applied to the brake pedal. This causes a force to act on the piston in the master cylinder.

(a) Explain how the friction pads are pushed against the rotating disc.

(b) Why do the brakes not work well if there is air in the oil? (MEG)

TOPIC 18

3 A ticker timer is a device that makes dots on a strip of paper at a steady rate of 50 dots every second. In an experiment one end of a long strip of paper was pinned to a baby's clothes. As the baby walked forward, she pulled the paper strip through a ticker timer. At the end of the experiment the marked paper was cut into 10-space pieces.

(a) There were twelve 10-space pieces of paper. They were stuck on to a chart in order. The figure below shows the chart.

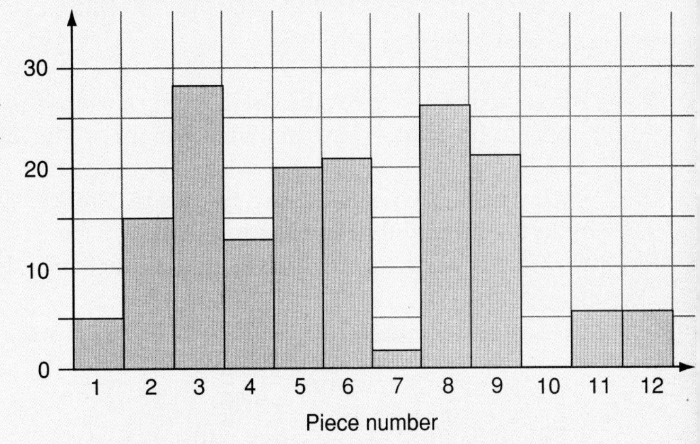

(i) Write down the number of the piece which shows the greatest speed.
(ii) Which **group of three** pieces shows when the baby was slowing down?

(b) The figure below shows the arrangement of dots on piece number 8.

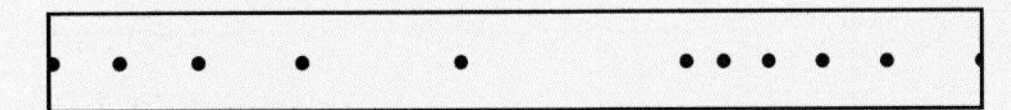

(i) What length of **time** is shown by piece number 8?
(ii) Use the pattern

$$\text{Average speed} = \frac{\text{Distance}}{\text{Time}}$$

to calculate the average speed at which the baby moved to produce piece number 8.
(iii) Does the value you calculated in (ii) represent the speed at which the baby was actually moving during that time? Explain how the evidence in piece number 8 supports your conclusion. (SEG)

4 A man jumps from a balloon at a height of 400 m and his parachute opens immediately. Air resistance causes an upward force of 300 N to be exerted on him. The man has a mass of 80 kg. His parachute is very light and you can ignore its mass in this question.

(a) The gravitational field strength is 10 N/kg. What is the man's weight?
(b) Make a copy of the diagram and draw labelled arrows to show the forces acting on the man.
(c) What is the size of the resultant force acting on the man?
(d) Calculate his acceleration.
(e) Suppose that the air resistance force remains constant at 300 N until the man reaches the ground.
   (i) Calculate the work done against air resistance as he falls 400 m.
   (ii) Calculate also his loss of potential energy.
   (iii) Hence find his kinetic energy as he reaches the ground.
(f) In practice the man will reach the ground with a far lower kinetic energy than you have just calculated. Explain this. (MEG)

5 Mustafa usually cycles the short distance from his home to his school. The combined weight of Mustafa and his bicycle is 600 N. It is divided equally over each wheel. Each of Mustafa's tyres is in contact with the ground through a strip 12 cm by 2 cm.
(a) Calculate the area of each tyre in contact with the ground.
(b) Use the equation

$$\text{Pressure} = \frac{\text{Force}}{\text{Area}}$$

to calculate the pressure which Mustafa and his bicycle exert on the ground.
The graph below shows how Mustafa's velocity changes through the journey.

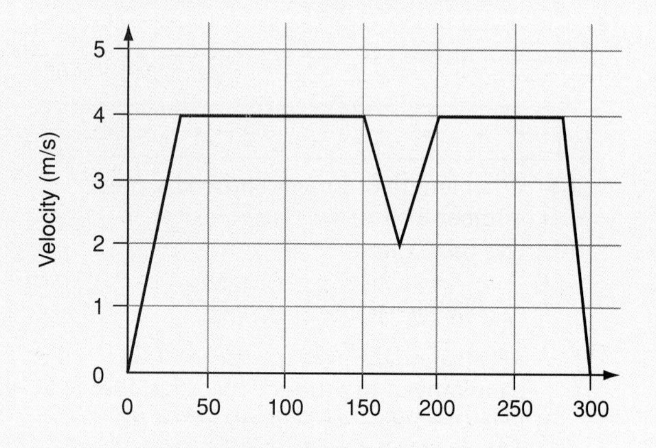

(c) Use the graph to find:
   (i) the greatest velocity at which Mustafa travels,
   (ii) the time taken for the whole journey.
(d) There is a pedestrian crossing in between Mustafa's home and the school. Sometimes he has to slow down or stop at it.
(i) How long after the start of his journey did Mustafa start to slow down for the pedestrian crossing?
(ii) Did Mustafa stop at the crossing? Explain how you worked out your answer.
(e) Use the equation

$$\text{Acceleration} = \frac{\text{Change of velocity}}{\text{Time}}$$

to calculate Mustafa's acceleration during the first thirty seconds of his journey.
(f) Calculate the distance between Mustafa's house and his school.

(LEAG)

6 Jean has a motor bike which she uses to get from her home to her work.

On the way to work Jean has to drive up a steep hill and sometimes she has to stop at a roundabout. The graph below shows how the distance travelled changes with time during Jean's journey.

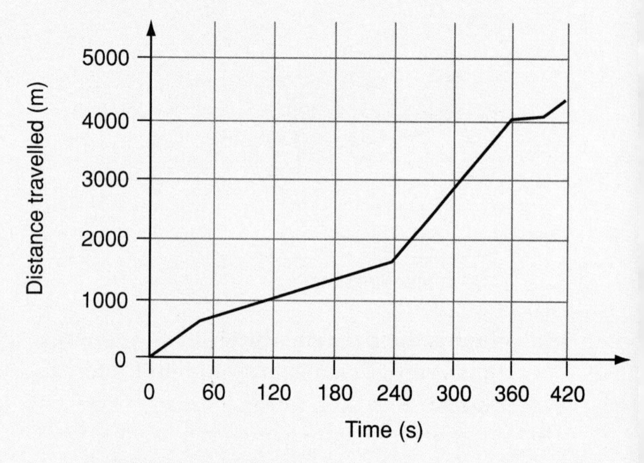

(a) Make a copy of the graph on graph paper. On your copy mark the letter:
   (i) S where Jean is at the start of the hill,
   (ii) F where Jean is at the finish of the hill,
   (iii) R where Jean is stopped at the roundabout.
(b) Use the graph to find:
   (i) the distance from Jean's home to her work.
   (ii) Use the equation

$$\text{Speed} = \frac{\text{Distance}}{\text{Time}}$$

to find Jean's average speed for the whole journey.
(c) Here are some facts about Jean and her journey.
The height of the hill is 35 m.
The total weight of Jean and her motorbike is 2000 N.
The gain in potential energy when Jean drives to the top of the hill can be found using the equation

$$\text{Gain in potential energy} = \text{Force} \times \text{Height}$$

When 1 g of petrol is burnt in the motorbike it produces 45 000 J of energy. The motorbike uses 4 g of petrol climbing the hill.

$$\text{Efficiency} = \frac{\text{Gain in potential energy}}{\text{Energy obtained from petrol}}$$

Calculate the efficiency of the motorbike.

(LEAG)

7 The stopping distance for a vehicle is made up of the driver's thinking distance and the braking distance.

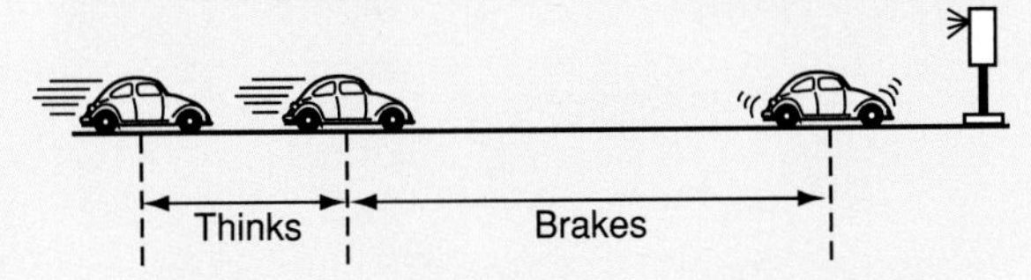

(a) (i) What is meant by the term 'thinking distance'?
(ii) For a vehicle driven along a straight, dry road, the thinking distance in metres is equal to $0.6 \times$ the speed of the vehicle in m/s. Calculate the thinking distance for a speed of 30 m/s.
(iii) If the driver is not in an alert state of mind, why is the thinking distance likely to be greater?
(b) For a speed of 30 m/s on a level dry road, the braking distance of the vehicle in (a) is 160 m.
(i) Assuming uniform deceleration, calculate the time the vehicle would take to stop.
(ii) Hence show that the decleration of the vehicle would be $2.8\,m/s^2$.
(iii) Explain why the braking distance needs to be greater if the road is wet.
(c) For the vehicle in (a) travelling at a speed of 20 m/s, calculate
(i) the thinking distance,
(ii) the braking distance, assuming the same deceleration on a dry level road,
(iii) the stopping distasnce
(d) The mass of the vehicle in (a) is 900 kg.
(i) Calculate its kinetic energy at 30 m/s.
(ii) Calculat the loss of kinetic energy per metre when it stops in a distance of 160 m from a speed of 30 m/s.
(iii) Calculate the force needed to stop the vehicle in this distance from this speed.

TOPIC 19

8 An electric winch is used to raise bricks and other materials at a building site. The winch consists of a 500 W electric motor and a pully system as shown in the diagram.

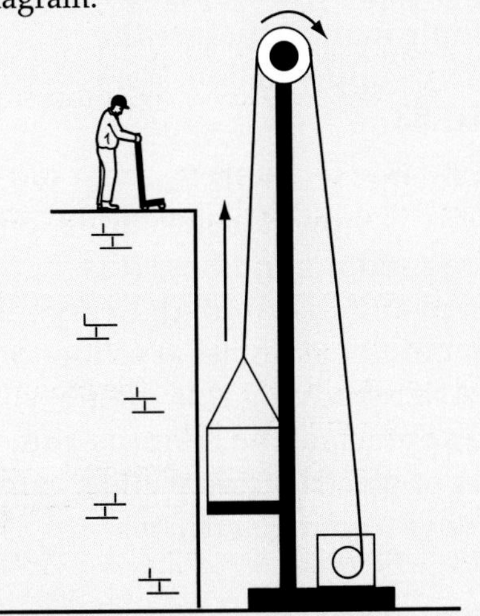

(a) The winch takes 25 s to raise a load of 100 N by a height of 6.5 m. Calculate the gain of potential energy of the load.
(b) Calculate the electrical energy supplied to the electric motor in 25 s.
(c) Give a reason why the gain of potential energy is less than the electrical energy supplied to the motor.
(d) Calculate the efficiency of the winch.

9 In a rocket engine, the rocket fuel and oxygen carried by the rocket react together producing very hot exhaust gases.

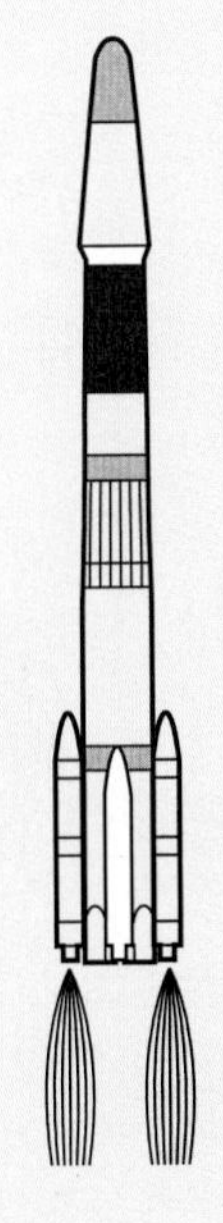

(a) Explain why a force is exerted on the rocket engine as a result of the hot exhaust gases leaving the engine.
(b) The engine expels exhaust gases at a rate of 20 kg/s at a speed of 1200 m/s relative to the rocket.
(i) Calculate the momentum loss from the rocket each second due to the exhaust gas.
(ii) Hence calculate the force on the rocket due to this engine.
(iii) The total mass of the rocket is 6000 kg and it has 4 identical engines. Calculate its acceleration when it is moved vertically upwards with all 4 engines operating.

10 The table shows information about five cars.

engine capacity in litres	1.8	2.0	2.5	3.5	4.0
maximum speed in km/h	168	204	224	248	248
efficiency	0.35	0.33	0.32	0.28	0.27

(a) (i) Describe how efficiency changes as engine capacity increases.
(ii) Suggest what happens to most of the energy wasted by the engines.
(b) (i) A car being tested on a track accelerates at $-3\,m/s^2$.
Explain precisely what is meant by 'accelerates at $-3\,m/s^2$'.
(ii) A car accelerates from 5 m/s to 15 m/s over a distance of 50 m. Calculate the acceleration of the car. Use the equation below.
You *must* show how you work out your answer:
$v^2 = u^2 + 2as$

(OCR)

Electricity

In winter, a power cut in the electricity supply to our homes makes life very difficult. *If you knew that a power cut was to be made in your area tonight, what plans would you make?* Get some candles for lighting, use overcoats to keep warm, check the batteries in your radio, fill a flask with a hot drink before the power cut. Electricity keeps factories, shops, hospitals and offices operating. How did people manage before the electricity supply network was established?

In this theme, you will find out what electricity is, how it is measured, what it can be used for and how it is generated and distributed. Everyone uses electricity at home, at school or at work. After studying this theme, you should appreciate the benefits and dangers of electricity as well as gaining knowledge about electricity.

Topic 20 Electric charges

20.1 Charging things up

FIRST THOUGHTS

Tiny sparks created by static electricity can be very dangerous in the wrong place. Read on to find out why.

Comb your hair with a plastic comb and then see if the comb can 'pick up' small bits of paper. Running a comb through your hair charges the comb up so it attracts paper. If you rub a plastic rule on a dry cloth, the rule and the cloth become charged. If you sit in a plastic chair, especially if your clothes are made from nylon materials, your clothes and the chair become charged when you get up from the chair. Electric charge on an object is often called **static electricity**.

Objects charge up because they gain or lose electrons. The electron is the smallest particle of the **atom**. The electron carries a fixed negative charge. Every atom has a positively charged nucleus that is surrounded by electrons (Figure 20.1A). An atom is uncharged because it contains equal amounts of positive and negative charge.

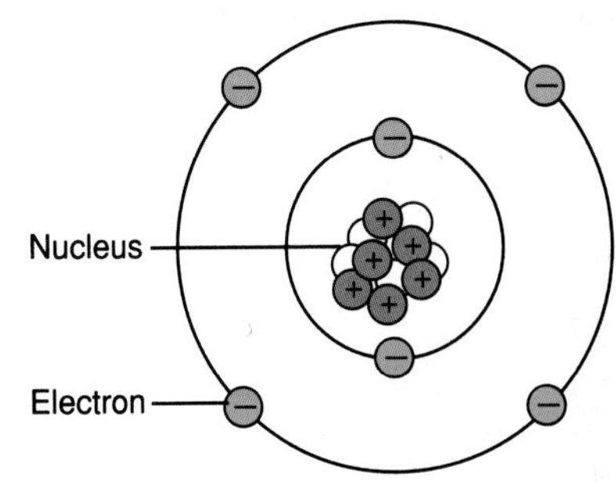

Figure 20.1A ▲ An uncharged atom – the positive charges balance the negative charges

An uncharged object that gains electrons becomes negatively charged. Objects that lose electrons become positively charged. Certain **insulators** like polythene become negatively charged when rubbed with a dry cloth; electrons are transferred from the cloth to the polythene by rubbing. Other insulators (e.g. perspex) become positively charged when rubbed with a dry cloth because electrons are transferred from the perspex to the cloth during rubbing (Figure 20.1B).

A positively charged object has lost electrons. A negatively charged object has gained electrons.

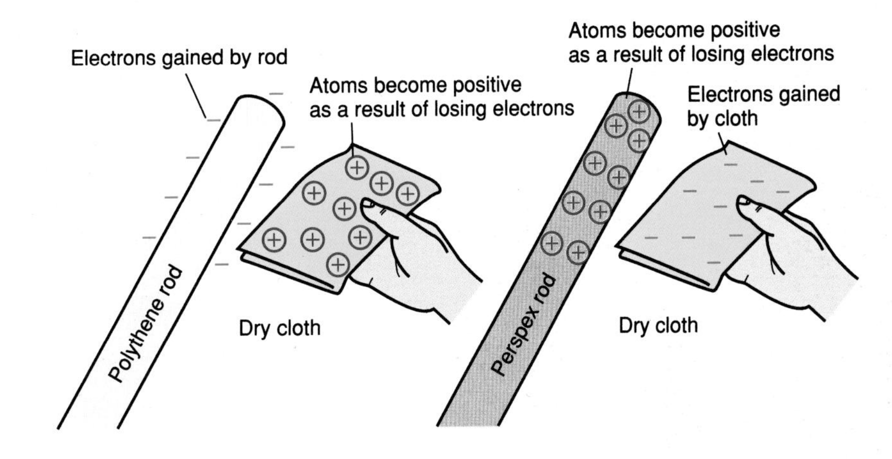

Figure 20.1B ▲ Charging by friction

Conductors such as metal objects can be charged if they are insulated from the ground. A conductor in electrical contact with the ground is said to be electrically 'earthed'. Electrons can move freely through

It's a fact!

A silo containing chocolate powder exploded as it was being filled. The explosion destroyed the silo and severely damaged the silo building. Three people were injured in the explosion and were taken to hospital. The silo had recently been modified by replacing the metal inlet pipe with a plastic pipe. Investigators think that pinpricks found on this pipe were caused by sparks, which made the powder explode.

conductors and pass to or from the ground if the conductor is earthed. Charge cannot therefore build up on a conductor which is earthed. However, if a conductor is insulated from the ground, electrons cannot pass from it to earth. A conductor which is insulated from the ground is therefore capable of holding electric charge. An effective method of earthing a metal object is to connect a wire from the object to a metal pipe running into the ground.

Investigating static electricity

Charged objects exert forces on each other due to their charge. Objects with the same charge repel each other; objects with opposite charge attract each other. Figure 20.1C shows how to test this.

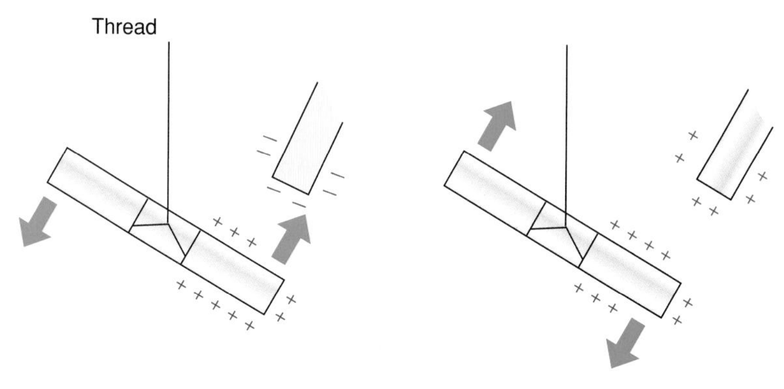

Figure 20.1C The law of force for charges
(a) Opposite charges attract (b) Like charges repel

SUMMARY

There are two types of charge: positive and negative. Objects carrying the same type of charge repel one another. Objects carrying opposite types of charge attract one another. Insulated conductors may be charged by direct contact or by induction from a charged insulator.

An insulated conductor can be charged either by **direct contact** or by **induction** from a charged insulator. Direct contact between the charged insulator and the conductor transfers charge to the conductor from the charged insulator. It gives the conductor the same type of charge as the insulator. Figure 20.1D shows charging by induction. Charging by induction gives the conductor an opposite charge to the insulator and no charge is gained or lost by the insulator.

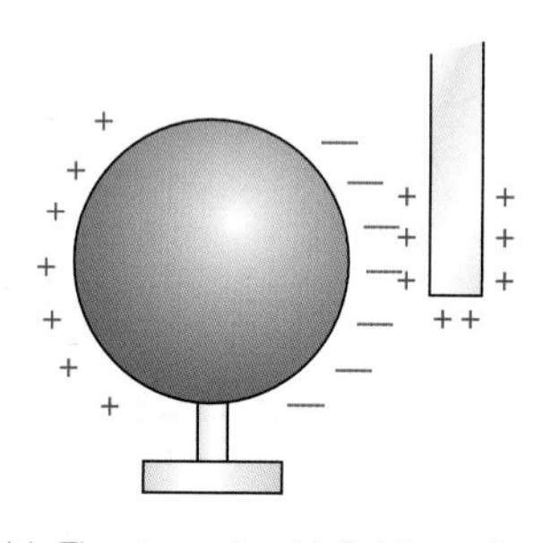

(a) The charged rod is held near the sphere

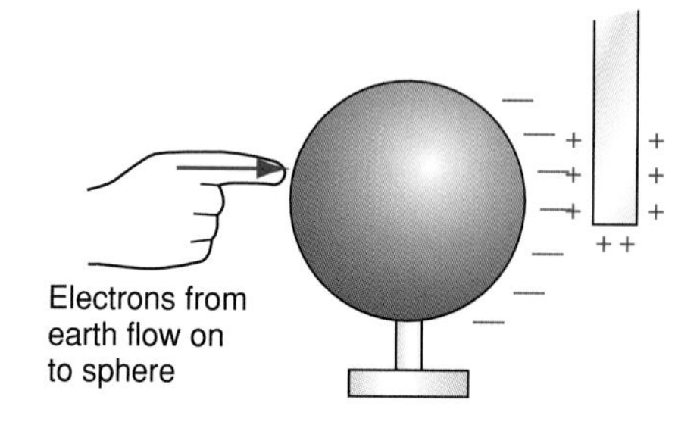

(b) The sphere is earthed briefly

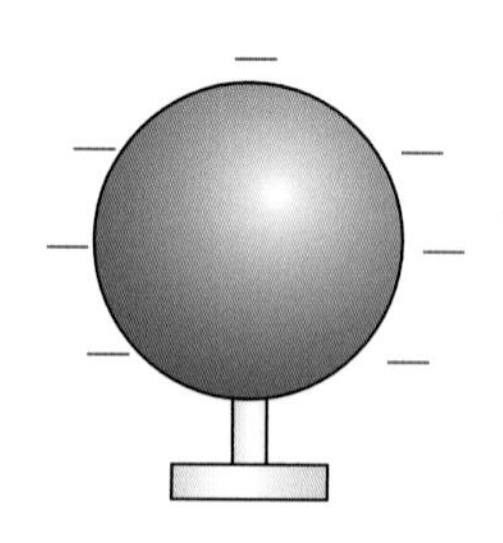

(c) The rod is removed. The sphere is left with an opposite charge to the rod

Figure 20.1D Charging by induction

The gold leaf electroscope can be used to find out if an object is charged and what type of charge it carries. The electroscope itself may be charged by direct contact or by induction. Figure 20.1E shows how the electroscope works.

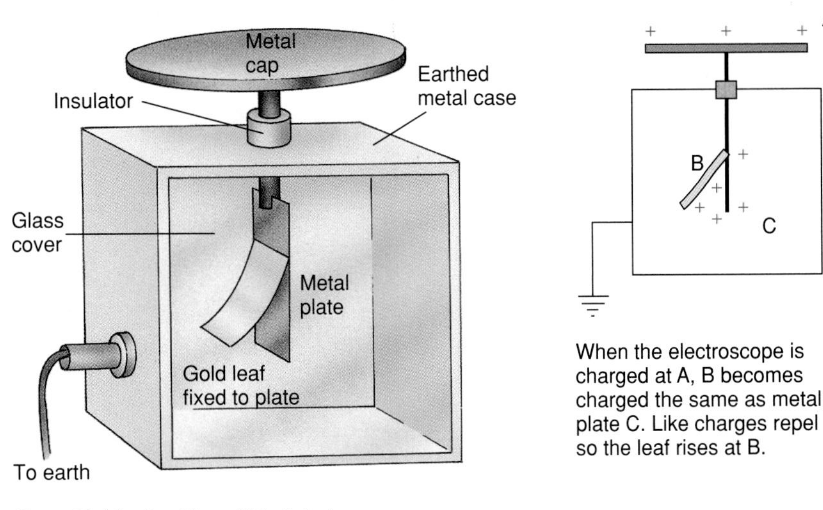

Figure 20.1E ▲ The gold leaf electroscope

See www.keyscience.co.uk for more about electricity.

Electrostatic hazards

Flow through pipes

Some liquids and powders pumped through pipes create static electricity. In Figure 20.1F, the electroscope becomes charged as powder is poured into the can through the pipe. The grains of powder rub against the pipe and are charged by friction. A plastic or insulated metal pipe would also be charged in the process. A charged pipe is dangerous not only because anyone who touched it would receive a shock; a tiny spark from the pipe could ignite the powder and cause an explosion. To avoid this happening, earthed metal pipes and earthed metal tanks should be used.

A tiny spark in the presence of a fine powder or a vapour of a flammable liquid can cause an explosion.

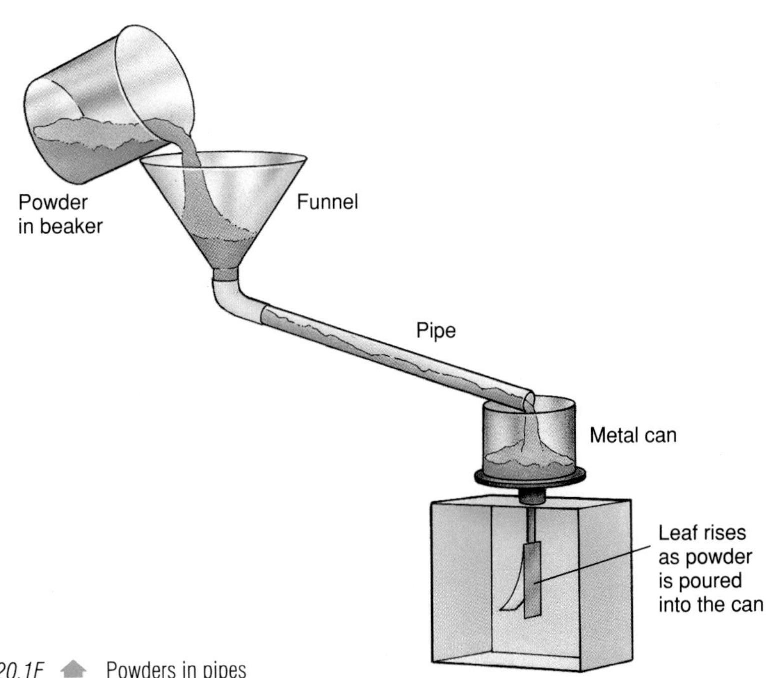

Figure 20.1F ▲ Powders in pipes

Storage tanks

Steel tanks used to contain flammable liquids or powders need to be earthed, otherwise a tiny spark might set off a massive explosion. When bulk carriers used to ship oil or ore are cleansed using flammable liquids, great care must be taken to ensure no sparks are produced, otherwise the vapour of the flammable liquid will ignite and explode.

Figure 20.1G ▲ Disaster at sea

Hospital theatres

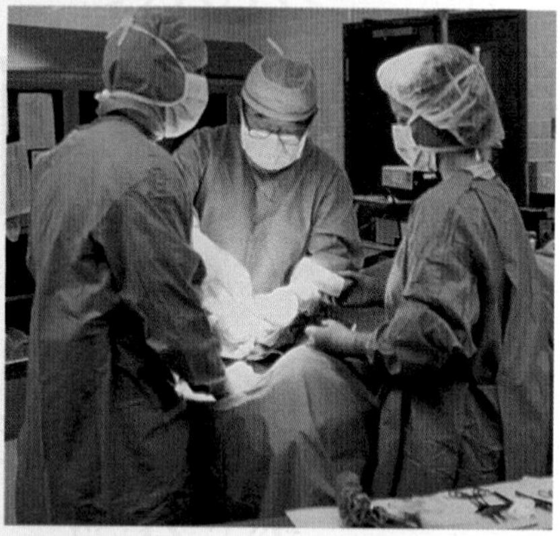

Operating theatres in hospitals are fitted with antistatic floors. Theatre staff and equipment might otherwise become charged and produce sparks. Charging occurs when clothing fabrics rub together or on insulated metal objects. Sparks in an operating theatre would be highly dangerous because anaesthetic vapour can be explosive. An antistatic floor covering conducts charge sufficiently to prevent the staff or equipment becoming charged.

Figure 20.1H ▲ In an operating theatre

Microchips

The metal pins of a microchip must not become charged, otherwise the chip will be ruined. The voltage on the pin due to static electricity would be sufficient to damage the circuits in the chip beyond repair. Contact between a pin and a charged object will therefore obviously damage the chip. Touching a pin by hand in the presence of a nearby charged object is sufficient to damage a chip, even though the charged object does not actually come into contact with the pin. This is because the pin would become charged by induction, as shown in Figure 20.1I.

People handling microchips wear antistatic clothing and they work in rooms fitted with antistatic floors. Microchips are stored in antistatic packets and handled only with special tools.

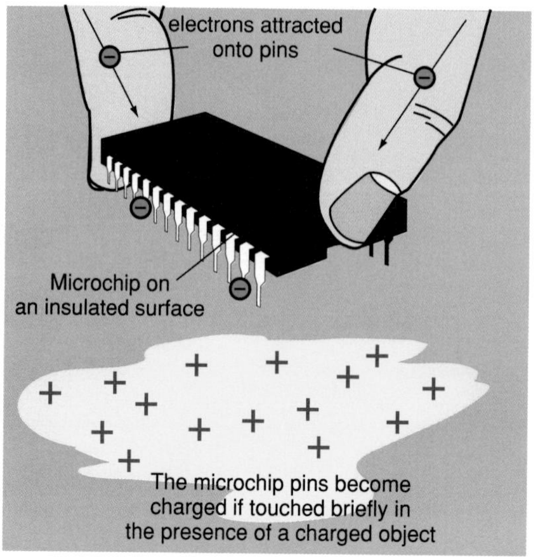

electrons attracted onto pins
Microchip on an insulated surface
The microchip pins become charged if touched briefly in the presence of a charged object

Figure 20.1I ▲ Microchip damage

SUMMARY

A spark due to static electricity can cause an explosion of a flammable vapour or a powder. Metal tanks used to store flammable liquids and powders should be earthed. Antistatic clothing and flooring is used in operating theatres and where microchips are handled.

CHECKPOINT

▶ **1** Explain each of the following:
  (a) If you walk on a nylon carpet and then touch a metal radiator you may get a shock.
  (b) Getting out of a car seat can give you a shock.
  (c) Taking off a pullover can produce lots of crackles and tiny sparks.

▶ **2** Graham is testing the charge gained by powdered milk when it passes through a plastic pipe. He uses a gold leaf electroscope as in Figure 42.1F, and he charges it by pouring the powder into the can through the pipe.
  (a) When a negatively charged rod is brought near the electroscope, the leaf falls. What type of charge is on the electroscope?
  (b) What type of charge is gained by (i) the powder, (ii) the pipe?
  (c) A spark from the pipe during pumping could ignite the powder and make it explode. An explosion is more likely to happen with fine powders than with coarse powders. Why?

▶ **3** In a factory, workers near a conveyor belt complain of electric shocks. Tests show that this is because the belt becomes charged as it passes round a roller.
  (a) Explain why anyone standing near the charged belt can become charged by touching an earthed bench near the belt.
  (b) What measures would you recommend to overcome this problem?

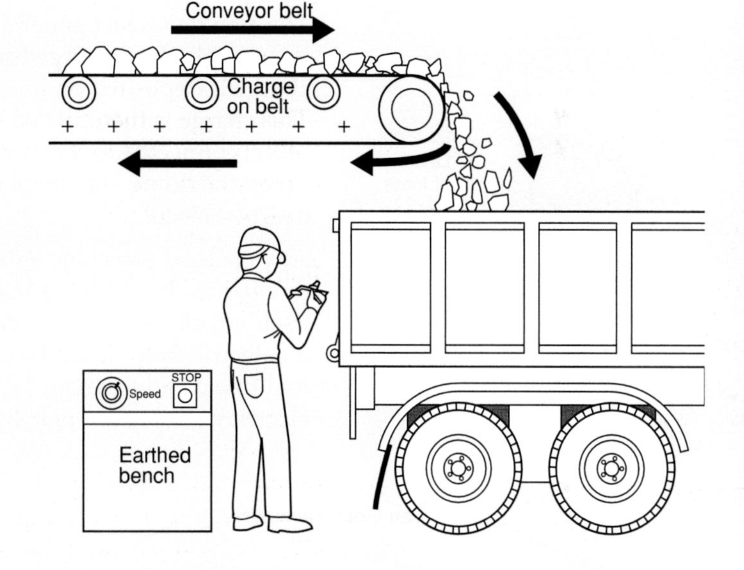

▶ **4** (a) Oil flowing from a pipe can produce static electricity due to friction between the pipe and the oil. Why must the pipe be earthed to ensure that static electricity does not build up on it?
  (b) Things that are charged up soon become dusty. Why?

▶ **5** In hot weather, a car on the move becomes charged unless it is fitted with a conducting strip trailing from its chassis.

  (a) Why does the conducting strip need to touch the ground?
  (b) (i) Without a conducting strip fitted, anyone touching the car would receive a shock. Why?
  (ii) Why is a car without a conducting strip unlikely to become charged when it is raining?

20.2 Electric fields

FIRST THOUGHTS

A stroke of lightning is a massive discharge of electricity to the ground from a thundercloud. In this section, you will find out how thunderclouds are discharged safely.

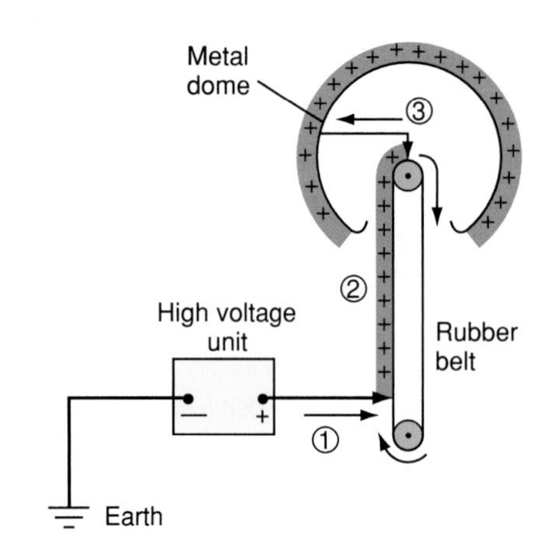

Figure 20.2A The Van de Graaff generator

The **Van de Graaff generator** is a machine for charging things up. When it is switched on, charge builds up on its dome. This happens because charge is deposited on the bottom of the belt, as shown in Figure 20.2A. This charge is then carried up to the dome by the belt. Any insulated object connected to the dome is charged too. If too much charge builds up on the dome, the dome discharges itself by letting sparks fly to any nearby object.

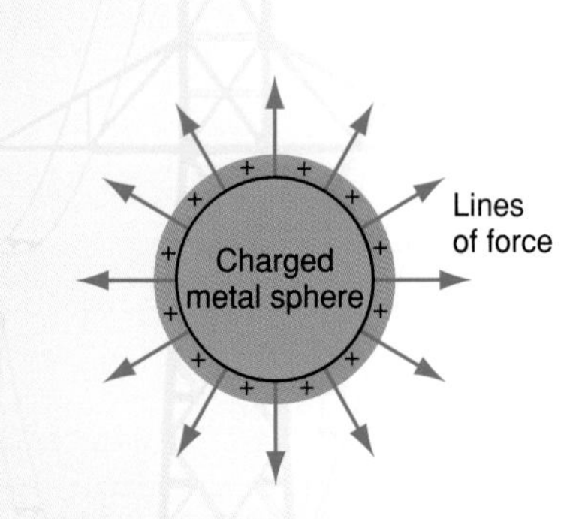

Figure 20.2B The electric field near a charged sphere

Any charged object brought near a charged dome experiences a force due to the dome. This is because the dome creates a force field round itself which acts on any other charged object near it. This field is called an **electric field**. It can be pictured as **lines of force**. These are the paths that free positive charges would take in the field. Figure 20.2B shows the electric field pattern near a charged dome.

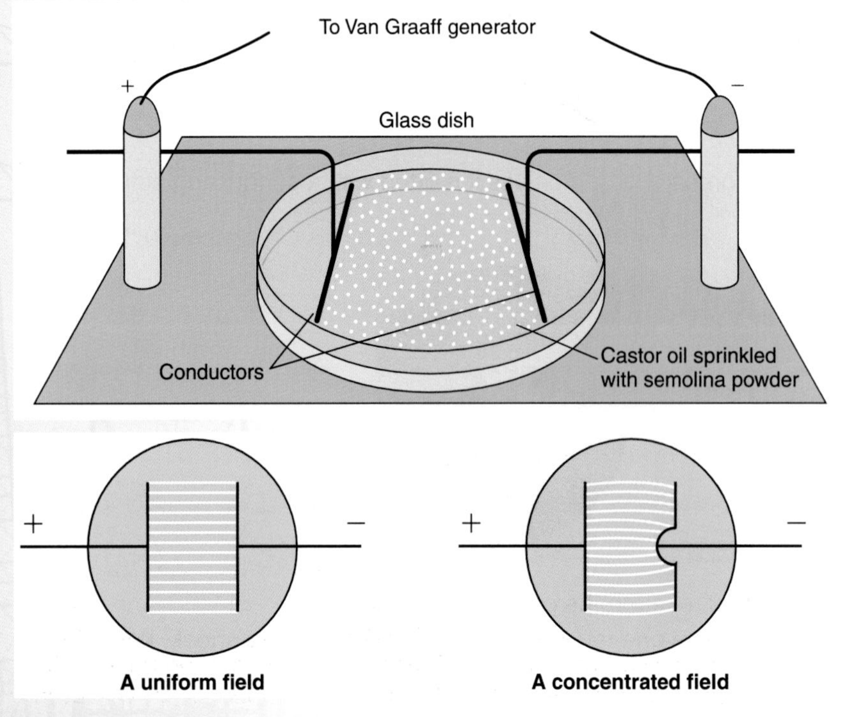

Figure 20.2C Field patterns (a) A uniform field (b) A concentrated field

Electric field patterns

You can see the patterns of an electric field using the apparatus in Figure 20.2C. Two conductors, connected to a Van de Graaff generator are submerged in castor oil sprinkled with semolina powder. The semolina powder forms patterns because the powder grains line up along the lines of force of the field. The field between two oppositely charged parallel plates is said to be uniform because the lines are parallel to one another. If a conductor is curved the lines of force are concentrated where the curve is greatest, because this is where the charge is most concentrated.

Lightning conductors under thunderclouds create very strong electric fields in the surrounding air. Air molecules near the tip of the conductor become **ionised** due to electrons being pulled off. The ions then discharge the thundercloud so no lightning flash is produced (Figure 20.2D).

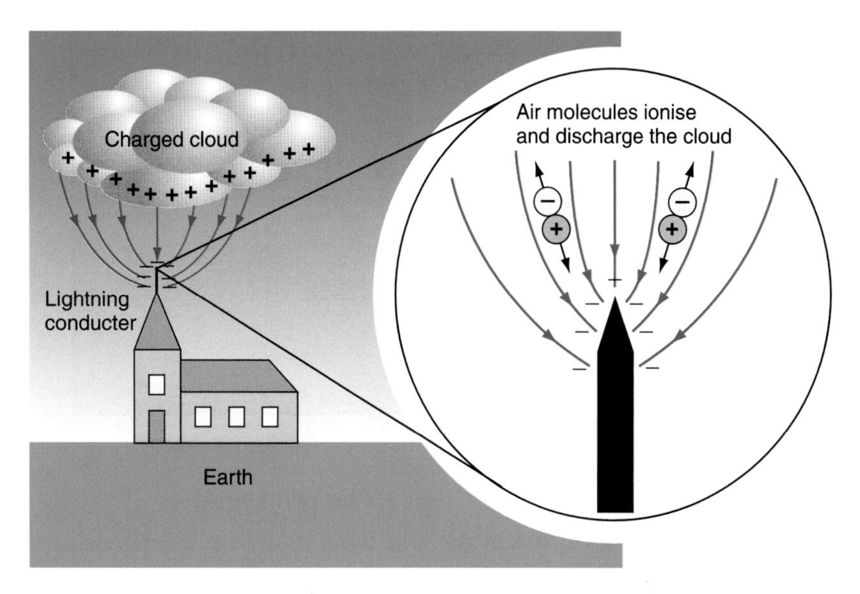

Figure 20.2D ▲ The lightning conductor

SUMMARY

The charge on a conductor is most concentrated where the conductor surface is most curved. If the electric field near a charged object is strong enough, air molecules in the field become ionised and can discharge the object.

The action of a pointed conductor such as a lightning conductor can be shown using a Van de Graaff machine. A drawing pin is fixed to the uncharged dome, as shown in Figure 20.2E. When the machine is switched on, the dome does not charge up as it would without the drawing pin. Instead the charge on the dome concentrates at the tip of the pin, making the electric field near the tip very strong. Air molecules near the tip become ionised and discharge the dome.

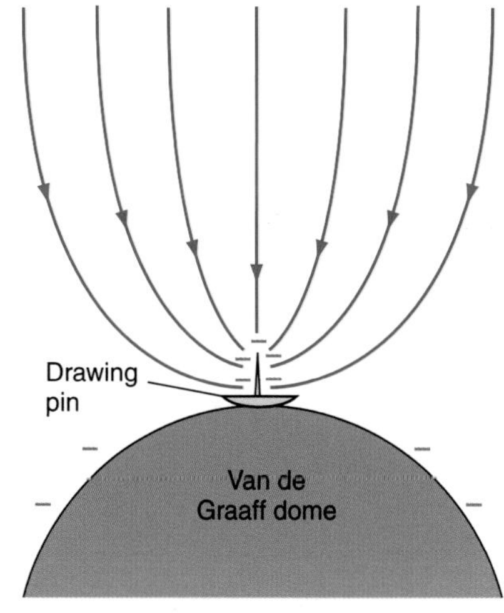

Figure 20.2E ▲ Discharging due to ions

Electric fields at work

The electrostatic paint spray

Paint sprays like the one in Figure 20.2F are used to coat car body panels on assembly lines. The spray gun is designed to produce tiny charged droplets of paint. The paint spray nozzle is connected to one terminal of an electrostatic generator. The other terminal is connected to the metal panel which is earthed. As a result, the charged droplets are attracted to the car body panel. This gives a uniform coating of paint on the panel. Also, the droplets can travel along the lines of force of the field to reach hidden parts of the panel.

Figure 20.2F ▲ An electrostatic paint spray

See 'The Ink Jet Printer' on p 305

The electrostatic precipitator

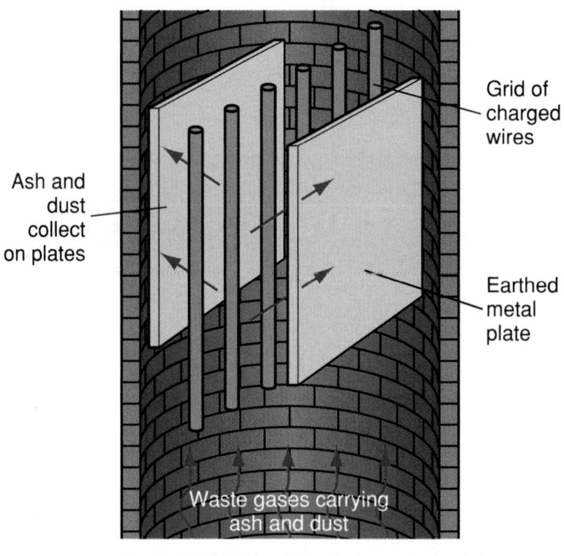

Figure 20.2G The electrostatic precipitator

Coal-fired power stations produce vast quantities of ash and dust. Electrostatic precipitators are used to remove them from the flue gases so they are not released into the atmosphere. The particles of ash and dust pass through a grid of wires in the precipitator, as shown in Figure 20.2G. The grid is at a high voltage so it charges the particles. They are then attracted on to the metal plates instead of passing through the precipitator with the waste gases.

EXTENSION FILE ASSIGNMENT

The photocopier

To use a photocopier, all you need to do is to put the sheet to be copied in the machine and press a button. Within seconds, out comes a perfect copy. Although a photocopier is simple to use, the process of making a photocopy is quite complicated, as shown in Figure 20.2H.

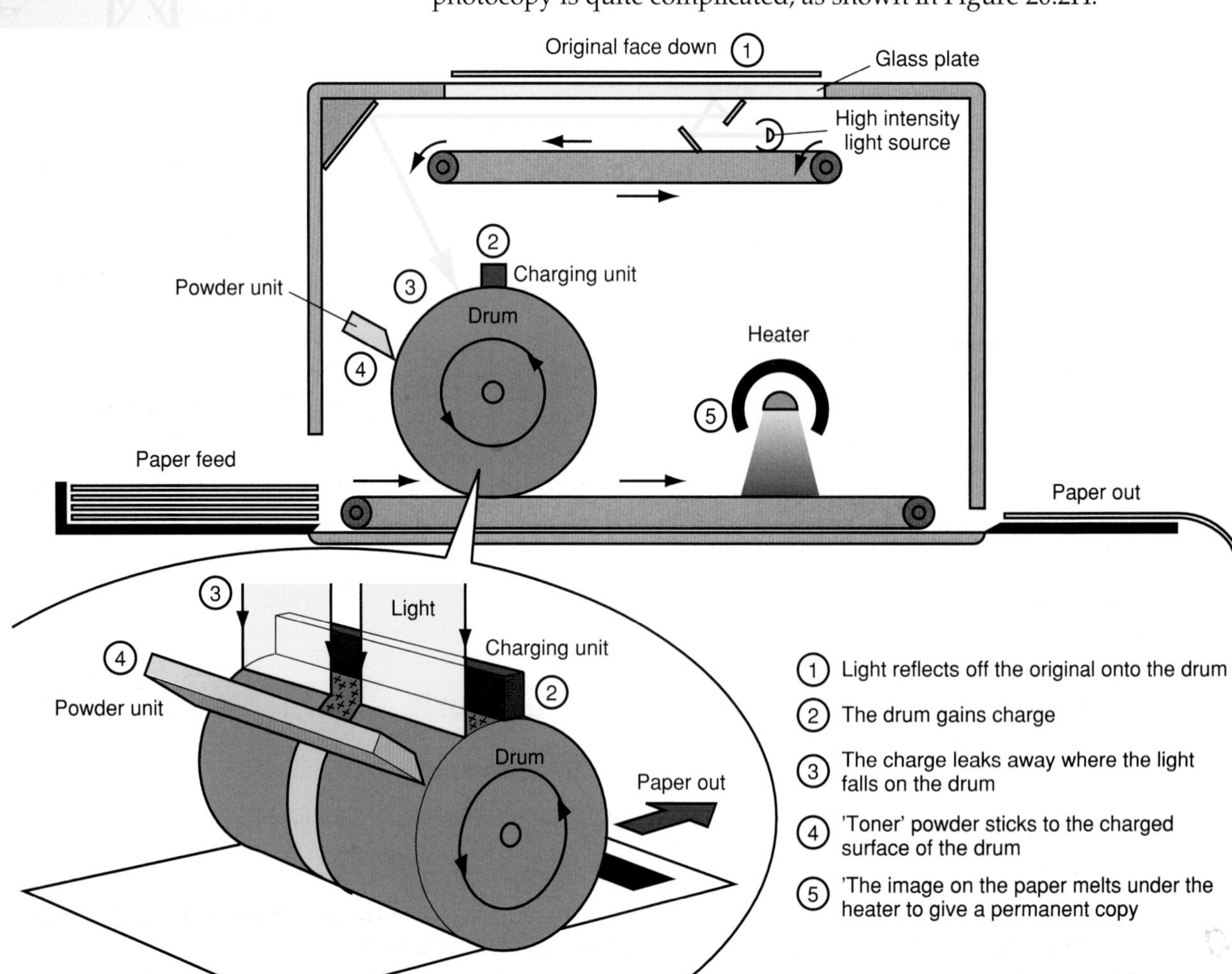

Figure 20.2H The photocopier

SUMMARY

Electric fields are used in various devices, including paint sprayers, electrosotatic precipitators, photocopiers and video cameras.

The video camera

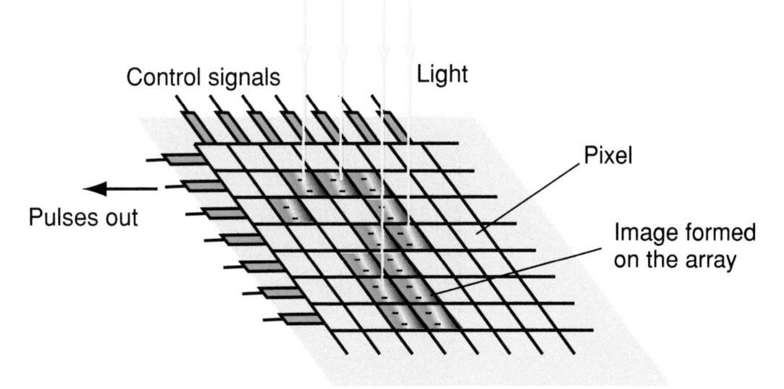

Figure 20.2H ▲ Inside a video camera

The light-sensitive part of a video camera is called a **charge-coupled device** (CCD). This device consists of an array of cells, referred to as 'pixels'. The camera lens forms a real image on the array. Each pixel on which light falls generates a tiny amount of charge. Charge from adjacent pixels is transferred automatically as a stream of electrical pulses which can then be recorded and used to reconstruct the image on a TV monitor. Low-cost video cameras are used for security purposes as well as home entertainment; amazing images of distant galaxies have been obtained by astronomers using expensive CCD arrays containing over 50 000 pixels per square millimetre fitted to telescopes.

Figure 20.2I ▲ Dust and gas clouds in space seen though the Hubble telescope

CHECKPOINT

1. (a) Why is it dangerous to stand under a tree during a thunderstorm?
   (b) Explain why a lightning conductor fitted to the top of a building reduces the risk of a lightning strike damaging the building.
2. Susan is investigating the gold leaf electroscope. She can make the leaf rise by charging the electroscope from her plastic comb. She discovers that the leaf falls slowly if there is an upturned drawing pin on the cap of the electroscope. She also discovers that holding a burning match near the electroscope makes the leaf fall.
   (a) Explain why the leaf rises or falls in each case.
   (b) Susan's teacher then demonstrates to the class that radioactivity near the electroscope makes the leaf fall. What do you think radioactivity does to the air near the electroscope?
3. Dust extractors in power station chimneys are very effective at stopping ash being carried by flue gases into the atmosphere. The flue gases are passed through a negatively charged wire grid. The grid is fixed inside an earthed metal tube. Particles of ash passing through the grid become negatively charged from the grid. The diagram shows the idea.

   Plate
   \+ + + + + + + + + + + + + + + + +
   Negative wire
   Dust particles in flue gas
   \+ + + + + + + + + + + + + + + + +
   Plate
   Dust particles are charged negative by the wire. The negative charged particles are then attracted to the plates

   (a) Why are the particles attracted to the metal tube?
   (b) Copy the diagram and sketch the lines of force between the wire and the plates.
4. A student on work experience is using a photocopier.
   (a) He notices that the powder refill indicator is blinking. If the powder cassette is not refilled, how will the appearance of the photocopies being produced change?
   (b) He notices that the copies are sometimes charged. Why do you think this happens?

20.3 Charge and current

FIRST THOUGHTS

An electric current in a wire is a flow of charge due to electrons moving along the wire. Read on to find out how charge is measured and how much charge is carried by a single electron.

The ping-pong ball experiment in Figure 20.3A shows that an electric current is a flow of charge. The ball has a coat of conducting paint. When the electrostatic generator is switched on, the metal plates become oppositely charged. The ball moves to touch one plate and becomes charged the same as the plate. Since like charged objects repel, the ball is then forced to the opposite plate. It therefore continues to move to and fro between the plates.

An electric current is a flow of charge.

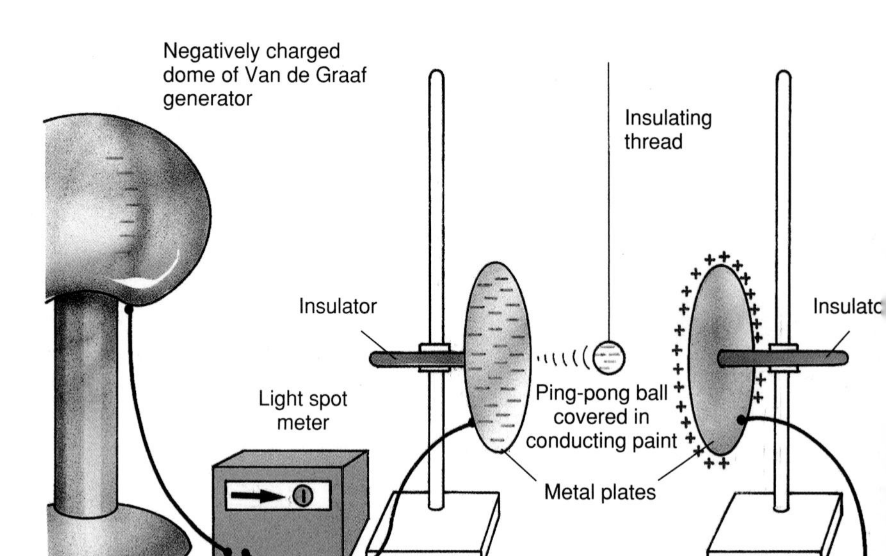

Figure 20.3A The ping-pong ball experiment

The ball transfers charge from one plate to the other each time it moves across the gap. Figure 20.3A shows how this happens. The meter registers a tiny current because electrons pass through it from the generator to replace the electrons carried from the negative plate by the ball.

How do you think the meter reading changes if the plates are moved closer together? The ball bounces to and fro more rapidly. This makes the reading increase because the ball is 'ferrying' charge across the gap at a greater rate than before. The current through the meter is due to the flow of charge from the generator to the plates.

Electric current is measured in **amperes,** A. The ampere is defined in terms of the magnetic effect of an electric current. All other electrical units are derived from the ampere.

Electric charge is measured in **coulombs,** C. One coulomb is defined as the amount of charge passing a point in a circuit each second when the current is one ampere. In other words, a current of one ampere is equal to a rate of flow of charge of one coulomb per second.

For a steady current in a circuit

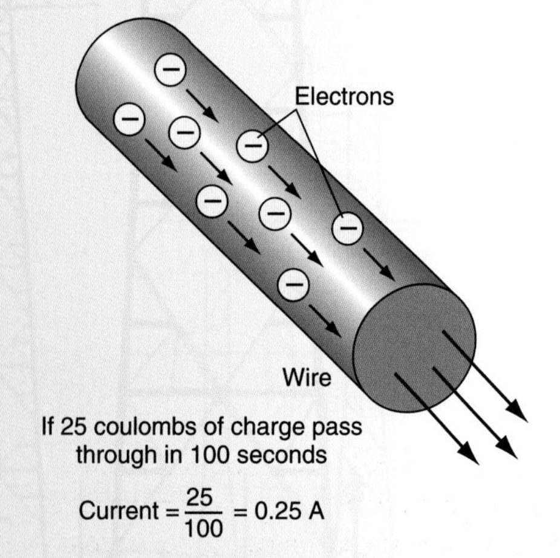

Figure 20.3B Current equals charge flow per second

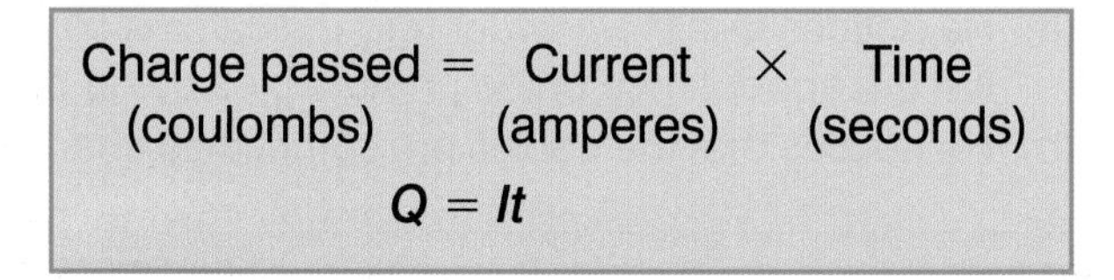

Charge passed (coulombs) = Current (amperes) × Time (seconds)
$Q = It$

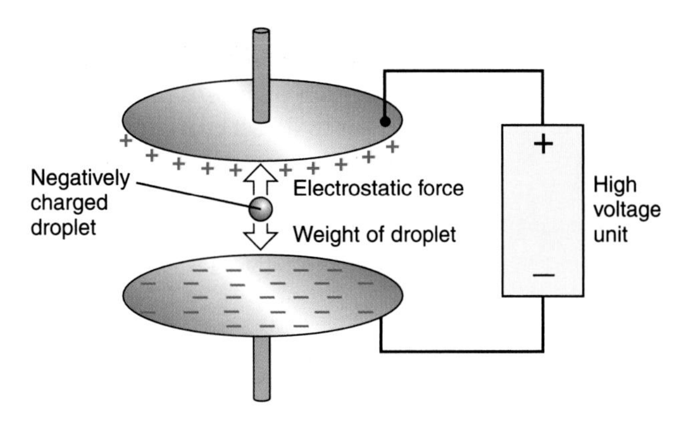

Figure 20.3C ▲ An oil droplet in balance

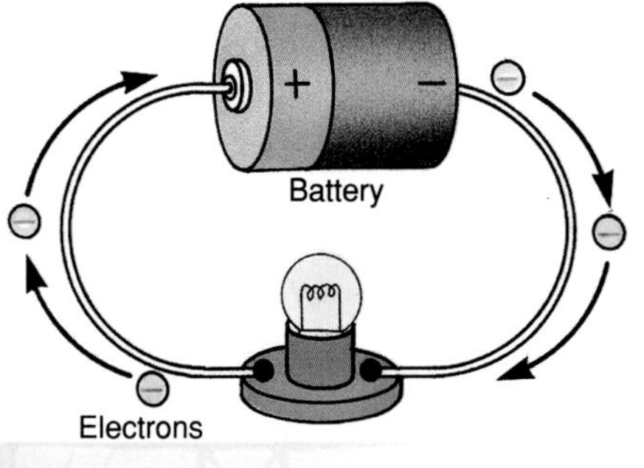

Figure 20.3D ▲ Charge flow

The charge carried by an electron was first measured exactly by Robert Millikan in 1915. He studied the motion of charged oil droplets between oppositely charged parallel plates, as shown in Figure 20.3C. From his measurements, he worked out that the electron carried a charge of 1.6×10^{19} C.

Work out for yourself how many electrons must pass each second along a wire carrying a current of one ampere. *In other words, how many electrons are needed to make a total charge of one coulomb?* You should find 6.2×10^{18} electrons are needed.

Storing charge

Capacitors are used to store charge. They are used in radio circuits, in power supply circuits and in electronic timing circuits. A simple capacitor consists of two insulated parallel metal plates. When a battery is connected to the two plates, as in Figure 20.3E(a), electrons from the negative terminal go onto one plate, making it negative. Electrons leave the other plate and flow into the battery at its positive terminal. This plate becomes positive. The amount of charge on the plates depends on the voltage of the battery.

A capacitor is a device designed to store charge.

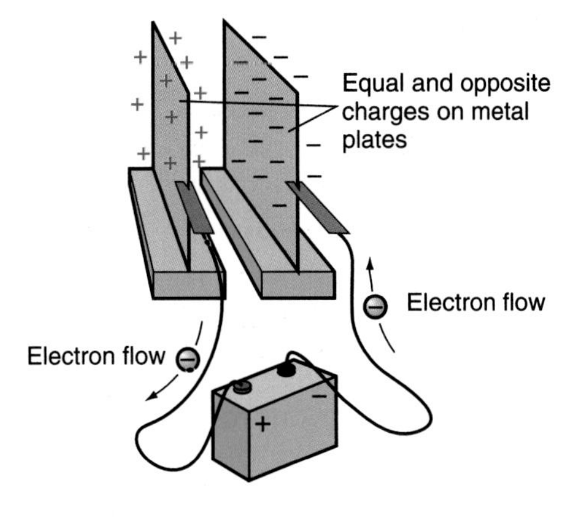

Figure 20.3E ▲ Capacitors (a) A simple capacitor (b) symbol

Practical capacitors are made from strips of metal foil, insulated from each other and rolled up, as shown in Figure 20.3F. Large capacitors are used in computer power supply units. Without a large capacitor across the output terminals of the power supply unit, unexpected voltage interruptions would cause havoc in the circuits of a computer. With a large capacitor across the output terminals, the capacitor takes over from

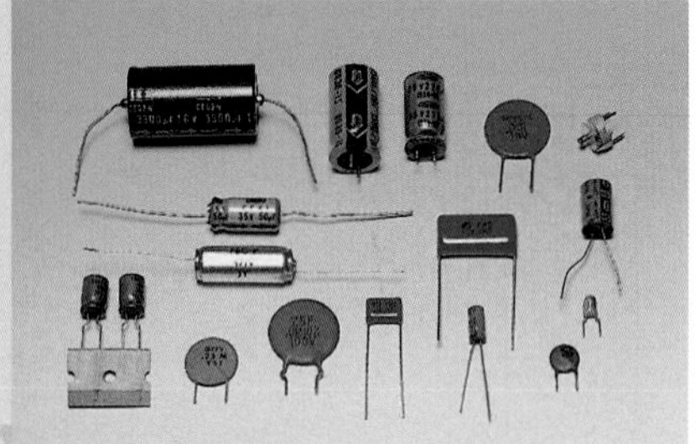

(a) Different types

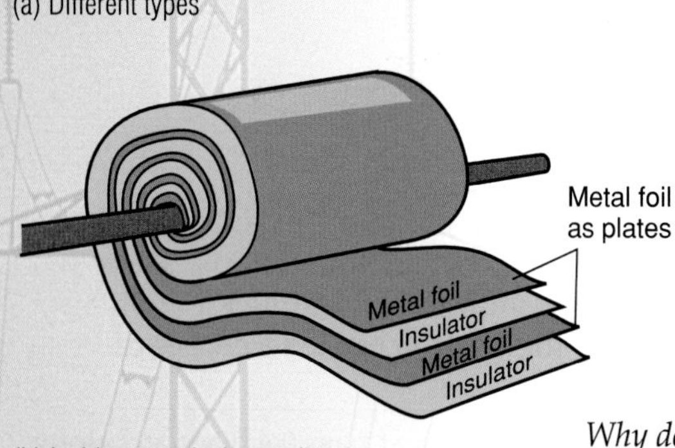

(b) Inside a practical capacitor

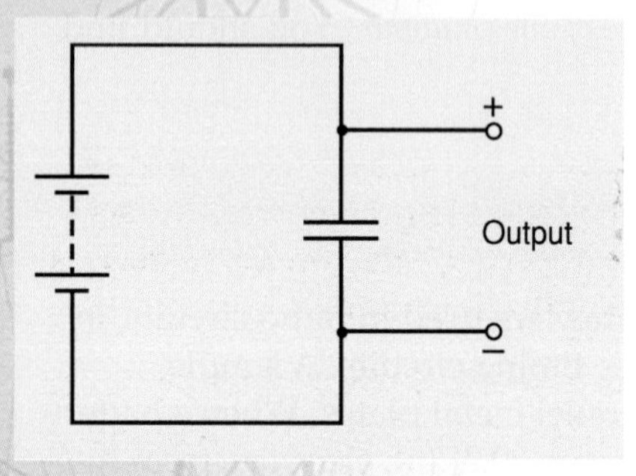

(c) A capacitor in use

Figure 20.3F Practical capacitors

the power supply if the supply is interrupted. In other words, the capacitor smooths supply variations out. Capacitors are also used for smoothing purposes in rectifier circuits which convert alternating current to direct current, (see p. 351). They are also used in electronic timing circuits, as explained on p. 368.

Electrolysis

Electricity can be transferred between France and Britain along undersea cables. These cables are waterproof because sea water conducts electricity. Conduction in a liquid can be tested using the circuit shown in Figure 20.3G. The terminals which pass electricity into the liquid are called **electrodes**. The ammeter reading shows whether or not the liquid conducts electricity. Tests like this show that certain liquids such as oil, paraffin and pure (i.e. distilled) water do not conduct electricity. Other liquids such as acids, salt solutions, molten ionic compounds, tap water and sea water do conduct electricity.

Why do certain liquids conduct electricity while others do not? One clue is to see what happens at the electrodes. Bubbles of gas may be seen at the electrodes, depending on which liquid is in the beaker. When this happens, elements leave the liquid. The liquid is said to **decompose**. Decomposing a liquid by passing current through it is called **electrolysis**. The beaker and the two electrodes make up an **electroytic cell**. The liquid is called an **electrolyte**.

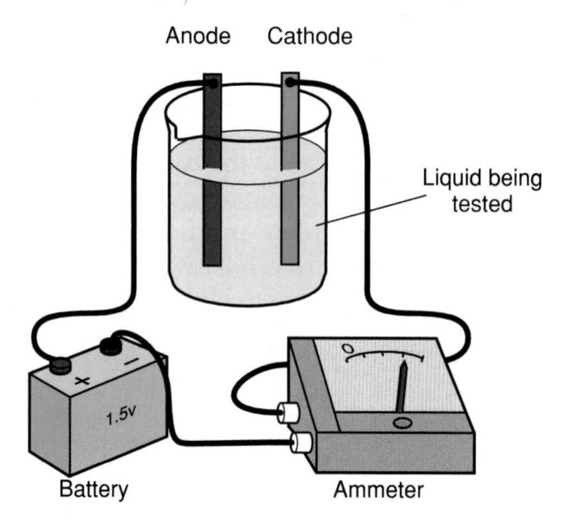

Figure 20.3G Conduction in liquids

The ions in the electrolyte are attracted to the oppositely charged electrode.

What happens in an electrolyte when it conducts? Electrolytes contains positive and negative ions.

1. The positive ions are attracted to the negative electrode, the **cathode**. Positive ions reaching the cathode accept electrons from the cathode. They are 'discharged' and become neutral atoms. If an electrolyte contains more than one type of positive ion, only one type is discharged. This preferential discharge occurs because each type of ion requires a different amount of energy to be discharged. The type discharged requires least energy.
2. The negative ions are attracted to the positive electrode, the **anode**. Negative ions reaching the anode give up electrons to the anode. They are 'discharged' and become neutral atoms. If an electrolyte contains more than one type of negative ion, preferential discharge of one type only occurs.

Figure 20.3H shows what occurs in a beaker of dilute hydrochloric acid when an electric current is passed through it. Hydrogen ions are attracted to the cathode because they are positively charged. Chlorine ions are attracted to the anode because they are negatively charged.

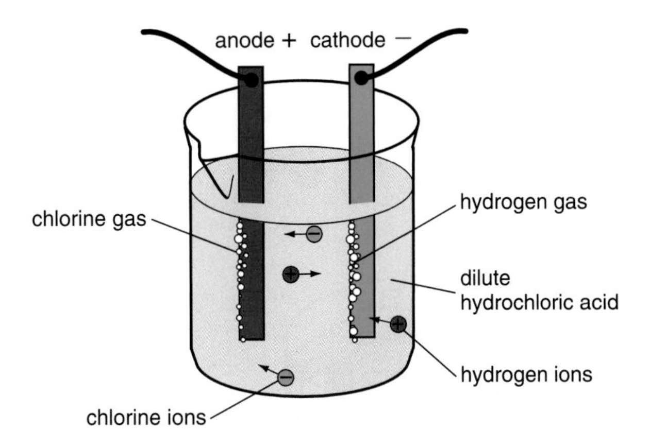

Figure 20.3H ▲ In an electrolyte

Metal plating

Electrolysis can be used to plate a conducting object with a coating of metal. For example, chromium plating is used to protect steel surfaces from wear and rust. A layer of copper is electroplated on the surface as an undercoat before it is electroplated with chromium.

In copper plating, each copper ion carries a double positive charge (+2e) and is attracted to the cathode where the ion is neutralised as a result of gaining 2 electrons.

An object to be metal-plated is used as the cathode in an electrolytic cell containing ions of the metal to be plated on the object. Metal ions are always positive so the metal ions are attracted to the object. They are discharged at the cathode and stick to it, to form a layer of metal on the object. Figure 20.3I shows how a metal object can be copper-plated by using the object as the cathode in an electrolytic cell containing copper sulphate solution. Copper ions from the solution are attracted to the cathode where they are discharged and form a layer of copper on the cathode. If the anode is a copper plate, copper ions from the anode go into solution to keep the concentration of copper ions in the solution constant.

What factors control the thickness of the layer of copper deposited on the cathode? In a factory where metal ornaments are copper-plated, too little copper deposited on the ornaments would make them unattractive and unsaleable; too much copper would make the ornaments expensive and uneconomic.

The thickness of the copper layer depends on the number of copper ions attracted to the cathode and on the surface area of the cathode. Each copper ion arriving at the cathode is discharged by gaining two electrons from the battery. Thus a steady flow of electrons round the circuit means that the mass of copper on the cathode increases at a steady rate.

The electric current caused by the flow of electrons round the circuit can be measured using an **ammeter** as shown in Figure 20.3I. Table 20.1 gives some measurements made by several groups in a class investigating how the mass of copper deposited depends on the current and time taken. The measurements are plotted in Figure 20.3J.

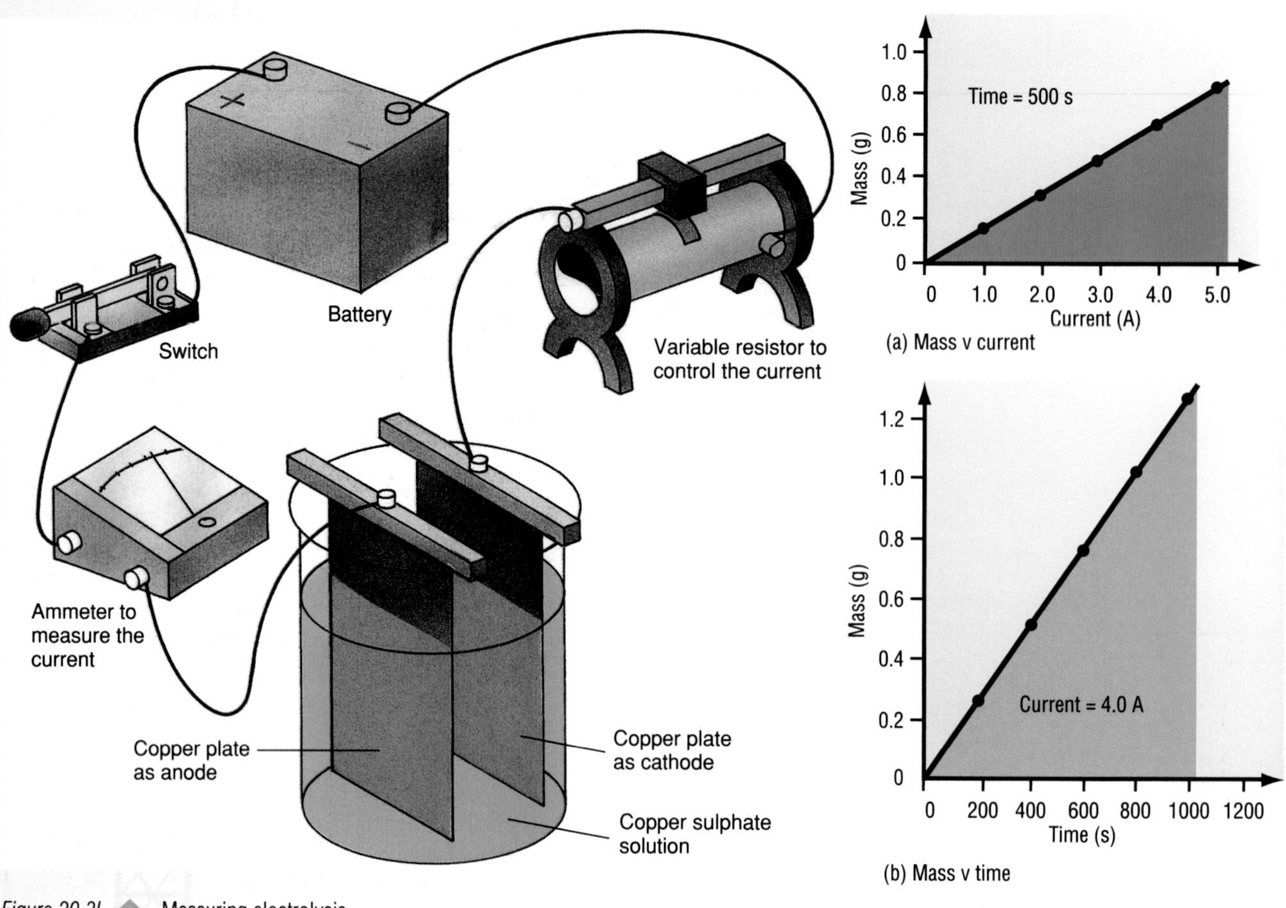

Figure 20.3I Measuring electrolysis

Figure 20.3J Typical results

Table 20.1 Measuring electrolysis

Mass deposited in 500 s (g)	0	0.16	0.33	0.48	0.65	0.82
Current (A)	0	1.0	2.0	3.0	4.0	5.0
Mass deposited at 4.0 A (g)	0	0.26	0.53	0.78	1.05	1.31
Time taken (s)	0	200	400	600	800	1000

What conclusions can you draw from these results? The graphs show that the mass deposited is proportional to the current and the time taken.

This can be written as an equation

$$\text{Mass deposited} = \text{Constant} \times \text{Current} \times \text{Time}$$

However, current × time gives the charge passed, so the mass deposited is proportional to the charge passed

$$\text{Mass deposited} = \text{Constant} \times \text{Charge passed}$$

Use the graphs in Figure 20.3J to work out the mass deposited per unit of charge. This is the constant in the equation above.

To understand what the equation means, remember each copper ion discharged at the cathode increases the mass of copper on the cathode and takes two electrons from the battery. Thus the mass deposited is proportional to the number of copper ions discharged. This is proportional to the number of electrons and hence the charge from the battery. Hence the mass deposited is proportional to the charge from the battery.

Electron beams

Television tubes and computer VDUs (visual display units) use electron beams to build up pictures on the screen. Figure 20.3K shows how an electron beam is produced by an **electron gun** in an electron deflection tube, which is a special tube designed to show some of the properties of electron beams. The filament is heated by passing an electric current through it. Electrons in the hot filament gain sufficient energy to leave the filament. This is called **thermionic emission**. These electrons are attracted towards the anode A, which is a metal plate with a hole in it. Some of the electrons pass through the hole and emerge as a narrow beam. There must be a vacuum in the tube, otherwise gas atoms would stop the electrons from the filament reaching the anode.

The beam current $I = ne$, where n is the number of electrons passing a point along the beam each second, and e is the charge of an electron which is equal to 1.6×10^{-19}C.

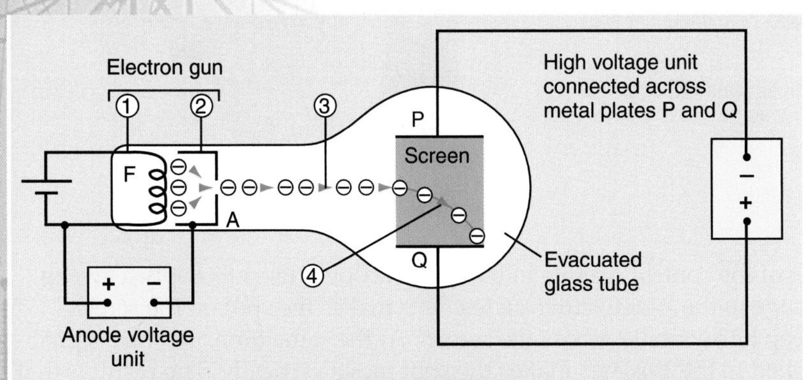

Figure 20.3K ▲ An electron deflection tube

A high-voltage unit is connected across the deflecting plates P and Q in Figure 20.3K. The electron beam is attracted towards the positive plate. *What does this tell you about the type of charge carried by an electron?* Another way to deflect an electron beam is to use a magnetic field. Television tubes and VDUs contain magnetic deflecting coils.

The path of the beam of electrons between the deflecting plates is the same shape as the path of a projectile acted on by gravity. As explained in Topic 18.4, the force of gravity on a projectile causes constant acceleration downwards. If an object is launched horizontally, its path is a curve known as a **parabola**. An electron directed into a uniform electric field follows a parabolic path because the force on the electron due to the electric field is constant in magnitude and in direction.

The Ink Jet Printer contains an 'ink gun' which directs a jet of ink particles at a sheet of printer paper. The particles of ink are charged by the gun and they pass between a pair of deflecting plates before reaching the paper. When a potential difference is applied between the plates, the ink particles are deflected as they pass between the plates because they are charged. The potential difference is controlled by signals from the computer to which the printer is linked. Thus, the inkjet produces characters and graphics on paper at high speed.

The speed of an electron in the beam depends on the anode voltage, V_A. Each electron is attracted from the filament to the anode, gaining kinetic energy as a result. The work done on each electron accelerated from the filament to the anode is equal to (electron charge e × anode voltage A). This is because voltage is equal to work done per unit charge. It follows that the kinetic energy of each electron from the electron gun is given by the equation

$$\tfrac{1}{2}mv^2 = eV_A$$

The oscilloscope

This is an instrument used to display electrical signals on a screen. Figure 20.3L shows the internal construction of the oscilloscope tube. An electron beam passes between two pairs of deflecting plates and hits the screen, producing a spot of light on the screen. The electrical signal is supplied to the input terminals which are connected to the Y-deflecting plates. Different signals deflect the electron beam by different amounts.

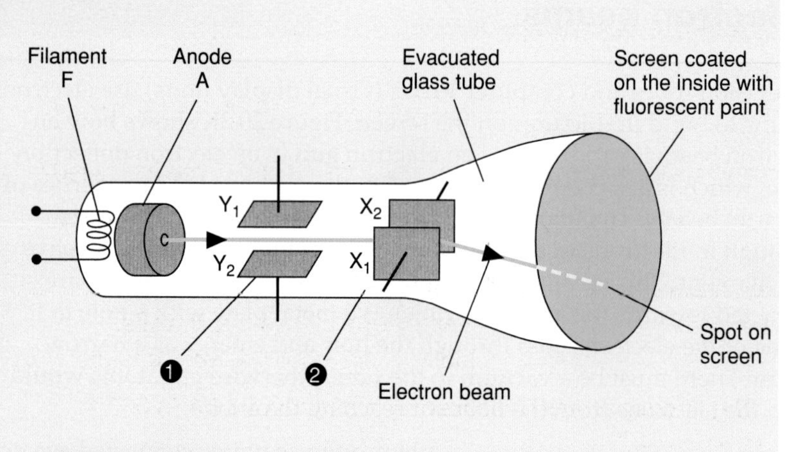

Figure 20.3L (a) The construction of an oscilloscope tube (b) An oscilloscope in use

SUMMARY

An electric current is a flow of charge. Current is measured in amperes. The unit of charge is a coulomb. Metals conduct electricity due to the passage of electrons. Television tubes, VDUs and oscilloscopes use electron beams to produce their displays.

One of the control circuits in the oscilloscope is used to apply a **sweep** voltage to the X-deflecting plates. This makes the spot on the screen sweep horizontally across the screen. At the same time, the 'test' signal applied to the Y-plates makes the spot move vertically. The result is that the spot produces a trace on the screen which shows how the voltage varies with time (Figure 20.3M).

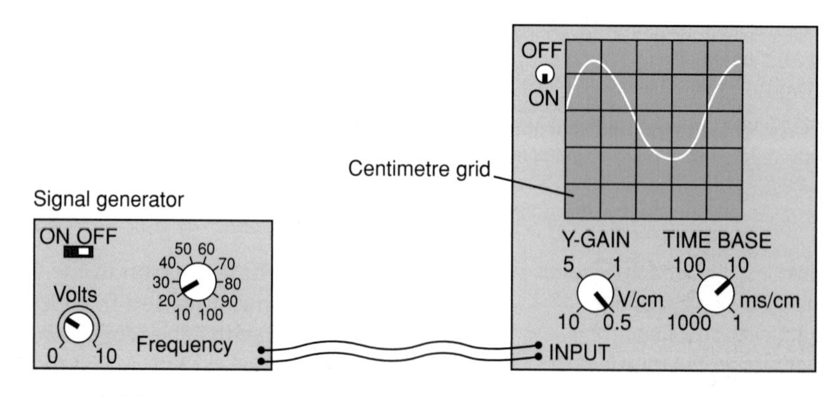

Figure 20.3M Using an oscilloscope

CHECKPOINT

1 Look at the ping-pong ball experiment in Figure 20.3A.
(a) What would be the effect on the meter reading of moving the plates further apart?
(b) Explain in terms of electrons how the ball transfers charge from one plate to the other as it swings back and forth between the plates.

2 (a) A wire carries a steady current of 3.5 A. How much charge passes along the wire in (i) 1 s (ii) 1 minute (iii) 10 minutes?
(b) A charge of 200 C is to be passed along a wire in 40 seconds. What will be the average current?
(c) What would the current be if the same charge were passed in 10 seconds?
(d) How long would it take to pass a charge of 500 C along a wire if a steady current of (i) 0.5 A (ii) 1.0 mA were passed along the wire?

3 Consider the electron deflection tube shown in Figure 20.3K.
   (a) Why does the beam bend towards plate Q?
   (b) Increasing the voltage at the anode makes the electrons in the beam travel faster. How would this affect the trace on the screen?
   (c) How would the trace on the screen be affected if the voltage across plates P and Q was increased?

4 Figure 20.3M shows an oscilloscope used to display an alternating signal of constant frequency and amplitude.
   (a) What is the amplitude (i.e. height above the centre) of the trace in millimetres?
   (b) The Y-gain control knob is set at 0.5 V/cm. Work out the amplitude of the trace in volts.
   (c) How would the trace appear if the frequency was doubled with the amplitude unchanged?

5 In a copper-plating experiment, 0.40 g of copper was deposited by a current of 2.0 A in 600 s.
   (a) How many coulombs of charge were used?
   (b) What mass of copper would have been deposited by (i) 4.0 A in 600 s (ii) 2.0 A in 300 s (iii) 4.0 A in 300 s?
   (c) What mass is deposited by 1 C of charge?
   (d) The mass of a copper ion is 1.1×10^{-22} g. Use the information above to work out its charge. How many electrons is the copper ion short of?

20.4 Batteries

FIRST THOUGHTS

In this section, you will find out how a battery works and what a rechargeable battery is. You will even find out how to save money when you buy batteries.

Figure 20.4A Different batteries

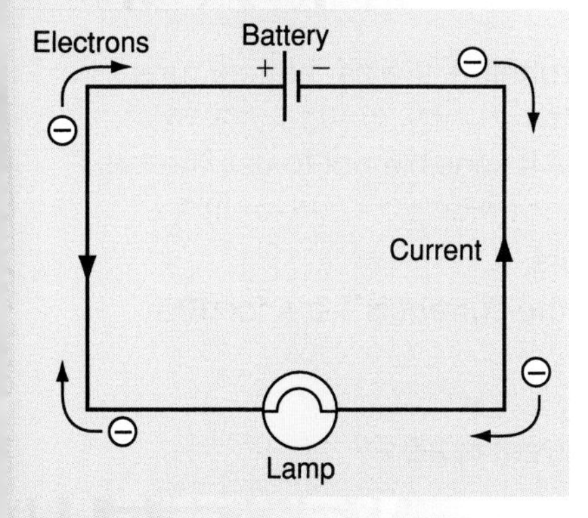

Figure 20.4B The current convention

How many electrical items do you have that use batteries? Calculators, digital watches, cameras, radios and cassette recorders all use batteries. They are used in cars, hearing aids, torches, toys and many other items. Batteries vary in size from the tiny batteries used in digital watches to heavy duty car batteries.

A **cell** produces electricity. Lots of cells joined together form a **battery** of cells. A battery pushes electrons round a circuit from the negative terminal to the positive terminal. Early scientists did not know about electrons and they imagined the charge from a battery flowed round the circuit from the positive pole to the negative pole. That is why current directions in circuits are always marked from 'positive to negative'.

A battery is necessary in a circuit to push electrons round the circuit. In Figure 20.4B, electrons deliver energy from the battery to the lamp.

The **voltage** of the battery is a measure of the energy delivered by the electrons to the lamp. Battery voltage is sometimes called electromotive force or e.m.f.

The unit of voltage is the **volt**. A battery with a voltage of one volt is able to deliver one joule of energy for each coulomb of charge that passes through it. In other words, the voltage of a battery is the number of joules per coulomb that the battery can deliver when it is connected in a circuit.

Primary cells do not produce electricity when the chemicals in the cell are used up. They must be replaced by fresh cells. A simple primary cell can be made using a copper plate and a zinc plate, as shown in Figure 20.4C. Zinc is more reactive than copper in dilute sulphuric acid. Zinc atoms leave the plate and go into the solution as ions. The zinc plate becomes negatively charged and the copper plate becomes positively charged. Hydrogen ions from the sulphuric acid (which are positive) are attracted to the copper plate where they gain electrons from the copper plate and are discharged as gas. The bubbles of hydrogen gas stop further hydrogen ions reaching the copper plate. As a result, the cell voltage drops.

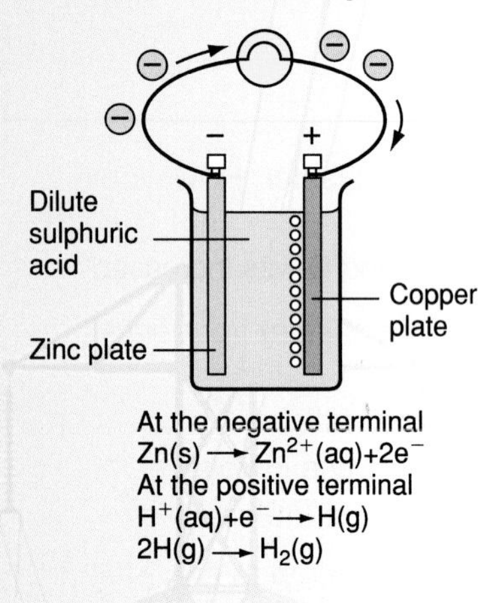

Figure 20.4C ▲ A primary cell

Some more primary cells are shown in Figure 20.4D. The Leclanché cell contains manganese(IV) oxide to stop hydrogen gas forming. The dry cell is based on the Leclanché cell but the electrolyte is in the form of a paste.

Secondary cells are rechargeable. This means that after use, a secondary cell can be recharged to be used again. When it is recharged, electrical energy is converted back to chemical energy inside the cell. The lead-acid accumulator is one of the most common secondary cells. Each cell has a voltage of 2 V. A 12 V car battery can be made using six of these cells. Nickel-cadmium cells are also rechargeable. These are much lighter than lead acid cells and they last longer before needing to be recharged.

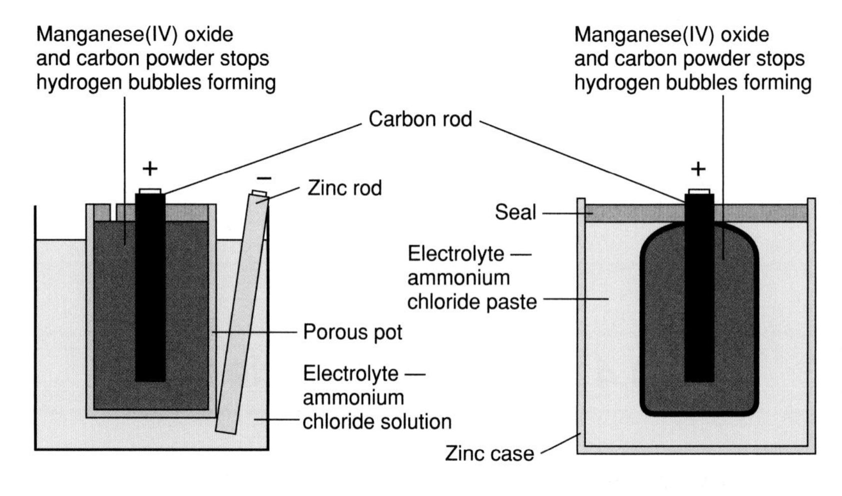

Figure 20.4D ▲ More primary cells (a) The Leclanché cell (b) A dry cell

SUMMARY

Batteries and cells convert chemical energy into electrical energy. Primary cells are not rechargeable, whereas secondary cells are. The voltage of a cell is the number of joules per coulomb that the cell can deliver in a circuit.

CHECKPOINT

1. Make a list of electrical devices you have that are battery-operated. Write down if the batteries are rechargeable and what type they are.
2. Marie needs to replace the two 1.5 V batteries in her radio cassette. She is thinking of buying a battery charger at £6.50 and two 1.5 V rechargeable batteries at £1.50 each. The cheapest non-rechargeable 1.5 V batteries cost 50p each but need to be replaced every four weeks.
   (a) How much will she save over a year if she buys rechargeable cells and the battery charger?
   (b) Long-life non-rechargeable cells are on sale in the shop when she buys the rechargeable cells. Each long-life cell lasts six times longer than an ordinary cell although it costs £3.50. Do you think she should have bought these instead?
3. Milk floats are electric vehicles that run off batteries. List some of the advantages and disadvantages of this type of electric vehicle.
4. (a) If the headlamps of a car are switched on without the engine running, the car battery runs down. Why is it difficult to start the car then?
   (b) Cars are often difficult to start on cold, damp mornings. Why is it sensible not to use the car heater until the car engine has been started?
5. A 12 V battery is capable of delivering 1.0 A of current for 100 hours.
   (a) How many coulombs of charge pass round a circuit in which the current is 1.0 A for 100 hours?
   (b) How much energy is available from the battery?
   (c) How long will it take to recharge the battery if the charging current is 2.0 A?

Topic 21 Circuits

21.1 Batteries and bulbs

FIRST THOUGHTS

The simple circuits in this section must be mastered before attempting the more complicated circuits in later sections.

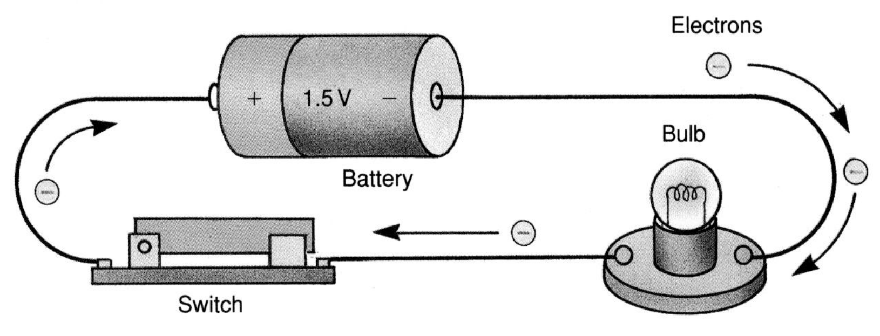

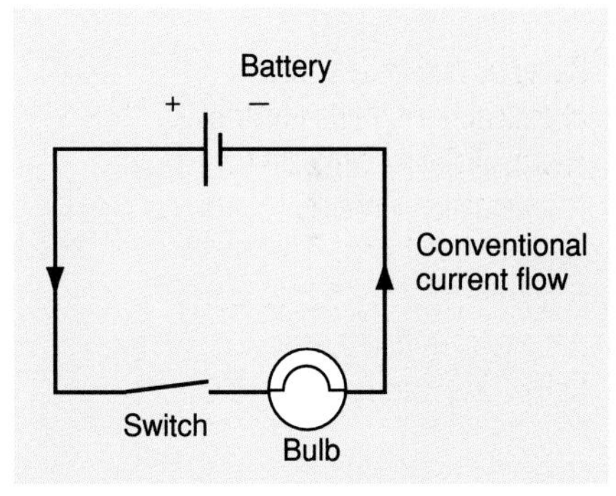

Figure 21.1A Lighting a torch bulb (a) Making connections (b) Circuit diagrams

Figure 21.1A shows how to make a circuit to light a lamp bulb using a 1.5 V dry cell. When the switch is closed, electrons are forced round the circuit by the battery. The electrons deliver energy from the battery to the lamp bulb. They pass from the negative terminal round the circuit to the positive terminal, where they re-enter the battery. However, the convention is for electric current to be considered as flowing from positive to negative. The convention was decided before the discovery of electrons.

Note that 1 volt equals 1 joule per coulomb.

The **voltage** across the bulb is the amount of energy delivered to the bulb by each coulomb of charge passing through it. This is sometimes called **potential difference**. One volt of potential difference is equal to one joule of energy per coulomb of charge. The term 'potential difference' (or p.d.) is used for the voltage between two points in a circuit.

Investigating simple circuits

Set up each of the four circuits in Figure 21.1B using bulbs and dry cells identical to the ones used in Figure 21.1A. *How does the brightness of each bulb compare with the bulb in the circuit in Figure 21.1A?*

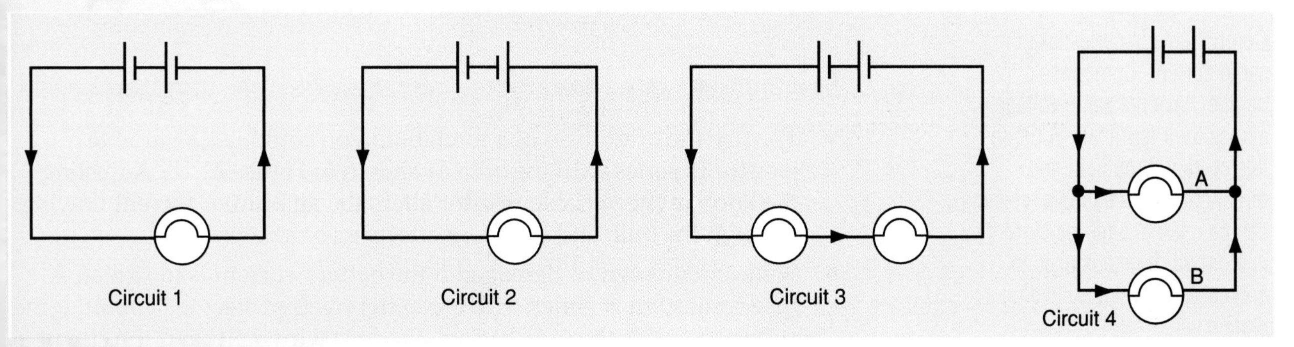

Figure 21.1B Comparing brightnesses

It's a fact!

The nerves in your body send their messages to your brain electrically. When a nerve sends a signal, the voltage of each section is changed by the previous section. In this way, a 'voltage pulse' travels along the nerve fibre.

- **Circuit 1** *Why is the bulb much brighter than the bulb in the Figure 21.1A circuit? How much voltage is across the bulb?*
- **Circuit 2** This has two batteries and a lamp bulb like Circuit 1, yet the bulb does not light up. *Can you explain why?*
- **Circuit 3** The same current passes through each bulb. Unscrew one of the bulbs from its holder. The other bulb goes out too. The two bulbs are connected **in series**. Components in series always take the same current. The bulbs light normally even though two batteries are used. The battery voltage is shared or divided between the two bulbs.
- **Circuit 4** The current through bulb A is not the same as through bulb B. Unscrew one bulb and the other bulb remains lit. The two bulbs are connected **in parallel**. Each bulb is connected across the cell terminals. The p.d. across each bulb is therefore the same. Circuit components in parallel always have the same p.d.

Circuit training

More circuit components are shown in Figure 21.1C. *Can you think where they may be used?*

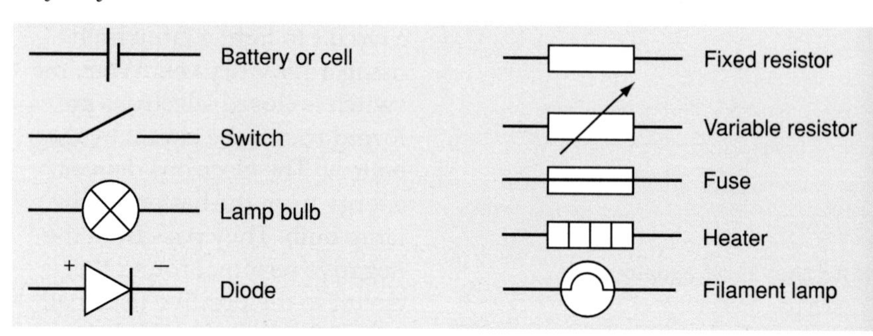

Figure 21.1C Circuit components

- You would damage a portable radio if you put the batteries in the wrong way round (so current flowed in the wrong direction). For a simple torch bulb circuit it does not matter which way the current flows, but radios contain more complicated components. A diode could be used as shown in Figure 21.1D, then if the battery were put into the circuit incorrectly the diode would not let any current pass.

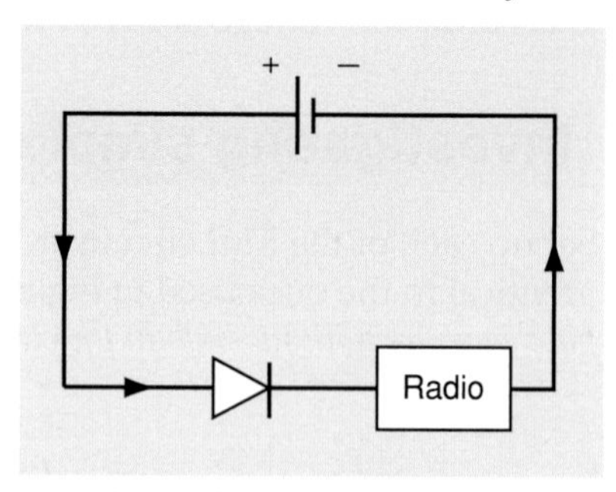

Figure 21.1D Using a diode

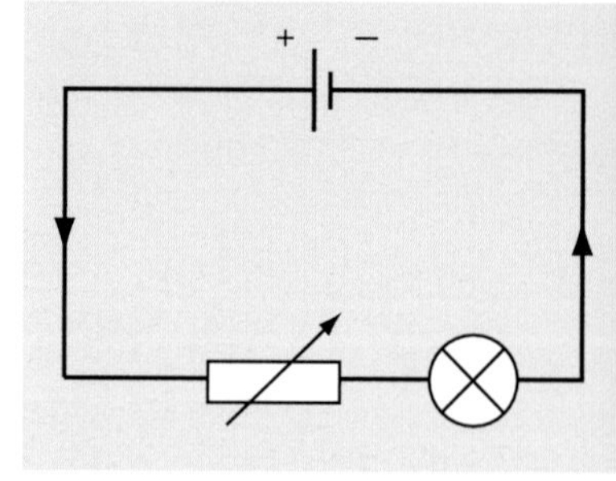

Figure 21.1E Using a variable resistor

- To vary the brightness of a torch bulb you could use a variable resistor in series with the bulb as shown in Figure 21.1E. Adjusting the knob of the variable resistor alters the amount of current flowing through the bulb and therefore affects its brightness.
- Some circuits can be damaged if the battery current is too great. A fuse connected in series with the battery will protect the circuit. If the current exceeds the fuse 'rating', the fuse wire melts and the circuit is broken so no current flows.

SUMMARY

Components in series take the same current. Components in parallel have the same voltage across their terminals. A diode lets current through in one direction only. A variable resistor can be used to alter the current in a circuit. A fuse is used to prevent the current exceeding a certain value.

CHECKPOINT

1. (a) Set up each of the circuits shown in Figure 21.1B.
   (b) How could you check that the bulbs are identical and that each cell has the same voltage?
   (c) Which circuit would use up chemical energy most quickly? Explain your answer.
2. In each of the circuits shown below the switch is used to 'short-circuit' one of the components. This means the current passes through the short-circuit instead of through the component. What happens to the brightness of each torch bulb in each circuit when the switch is closed?

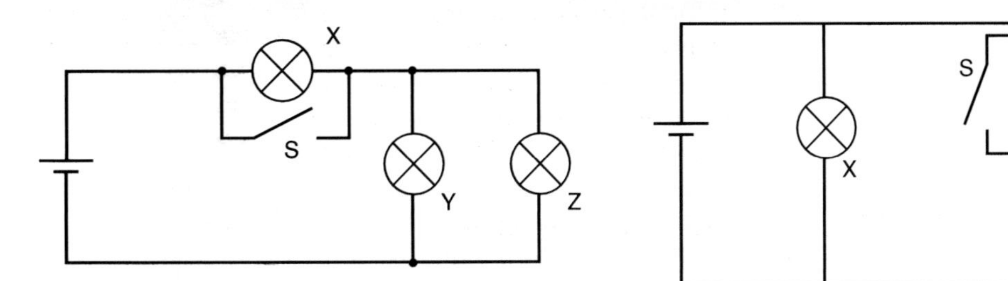

3. Design and test a circuit that will:
   (a) light up two bulbs so that one bulb is brighter than the other,
   (b) light up two bulbs, one with constant brightness and the other with variable brightness.
4. Simon tests a diode by connecting it in series with a bulb, a battery and a variable resistor. The bulb does not light at first so he adjusts the variable resistor. It still does not light so he reverses the connections to the diode. The bulb still does not light so he adjusts the variable resistor once more. Now it lights up.
   (a) Sketch the circuit diagram for the successful circuit.
   (b) Why did the bulb not light up when he first adjusted the variable resistor?
   (c) Why did the bulb not light up when he reversed the diode before adjusting the variable resistor again?
   (d) Sketch the circuit he started with.

21.2 Electricity in the home

FIRST THOUGHTS

Mains electricity is dangerous. Study this section carefully to appreciate the safety features of the electric circuits in the home.

Figure 21.2A Household machines

Make a list of all the electrical appliances in your home. *Which ones are not battery-powered?* These appliances need to be plugged in to the 'mains' to work. In the UK, the voltage of the mains is 240 V. Voltages above about 50 V can be lethal. **Never fiddle with mains appliances and fittings.**

Heating appliances include electric kettles, immersion heaters, electric irons, hair driers and electric fires. All heating appliances contain a heating element that converts electrical energy into heat energy.

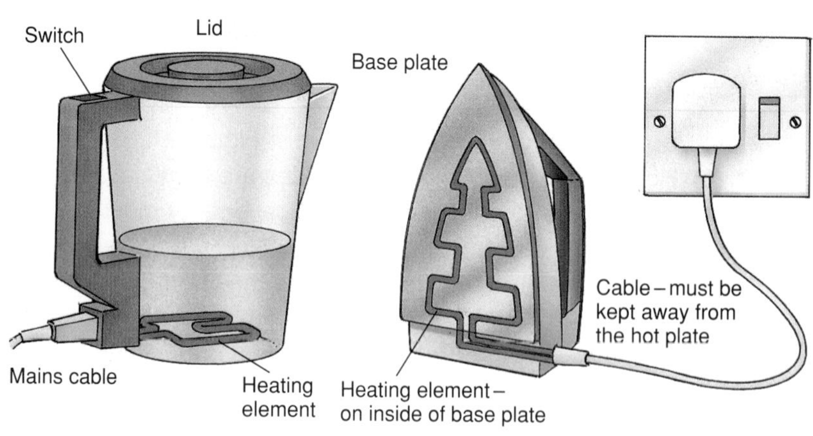

Figure 21.2B Heating elements

These appliances must be plugged into a mains socket. Electric cookers are always connected permanently to the mains using special wiring because they take much more current than other appliances.

Washing machines, fridges, tumble driers and electric drills each contain a **motor** that converts electrical energy into kinetic energy. Washing machines and tumble driers also contain heating elements. These appliances also have to be plugged into the mains.

Note that 1 volt is also equal to 1 watt per ampere.

Light sockets and switches are on a separate circuit to the wall plug sockets. The currents in the lighting circuit are much smaller than the currents in the wall socket circuit.

Power

How much energy per second does each mains appliance use? Most appliances are labelled to tell you. For example, an electric kettle may be labelled '240 V, 3 kW'. This tells you that the kettle operates from 240 V mains and uses 3000 watts (i.e. 3 kW) of electrical power. This means it uses 3000 joules of electrical energy per second.

The current passed by a mains appliance can be worked out if the power and voltage are known, using the following equation

$$\underset{\text{(watts)}}{\text{Electrical power}} = \underset{\text{(amperes)}}{\text{Current}} \times \underset{\text{(volts)}}{\text{Voltage}}$$

The voltage is the number of joules per coulomb delivered to the appliance by charge passing through it.

The current is the number of coulombs per second passing through the appliance.

So $\text{Current} \times \text{Voltage} = \dfrac{\text{Joules}}{\text{Coulombs}} \times \dfrac{\text{Coulombs}}{\text{Seconds}} =$ Power in watts (J/s)

Fuses

Every mains appliance has a fuse in its plug. The fuse is designed to cut off the current if the appliance passes more current than it is supposed to. To work out the **fuse rating** of an appliance, first calculate the current it passes. For example, a 3 kW, 240 V electric kettle normally passes 3000 W/240 V = 12.5 A of current. The mains plug of the kettle should therefore be fitted with a 13 A fuse, which will melt if a current greater than 13 A passes through it.

EXTENSION FILE ACTIVITY

Household circuits

Any mains appliance is supplied with electrical energy from the electricity mains via two wires, the **live** wire and the **neutral** wire. The live wire is at 240 V and the neutral wire is earthed at the electricity supply station.

The ring main supplies current to an appliance through each part of the ring main.

When you switch an appliance on or off at home, no other appliance is affected. For example, switching your electric kettle off does not affect your television. This is because household sockets are wired in parallel with each other. The wall sockets are part of a circuit called the **ring main**. Figure 21.2C shows part of the ring main for a typical house. Light sockets are part of the lighting circuit, which is separate from the ring main. Separate circuits supply electricity to an electric cooker or an electric shower.

- A **domestic electricity meter** is fitted in every home to record the electrical energy supplied to all the appliances and lights in the home.
- The **main switch** disconnects all the circuits in the home from the electricity supply. This switch must be turned off whenever any of the mains circuits or sockets is being checked, repaired or extended.
- The **distribution box** is where the electricity supply is connected to all the circuits in the home. Each circuit is protected by a **fuse** in the live wire at the distribution box.
- The **ring main** consists of the live wire, the neutral wire and a third wire, the earth wire which is earthed at the home. Electricity passes through an appliance via the live wire and the neutral wire. The voltage of the live wire alternates between +325 V and –325 V at a frequency of 50 Hz. In terms of power, this is equivalent to a direct voltage of 230 V.

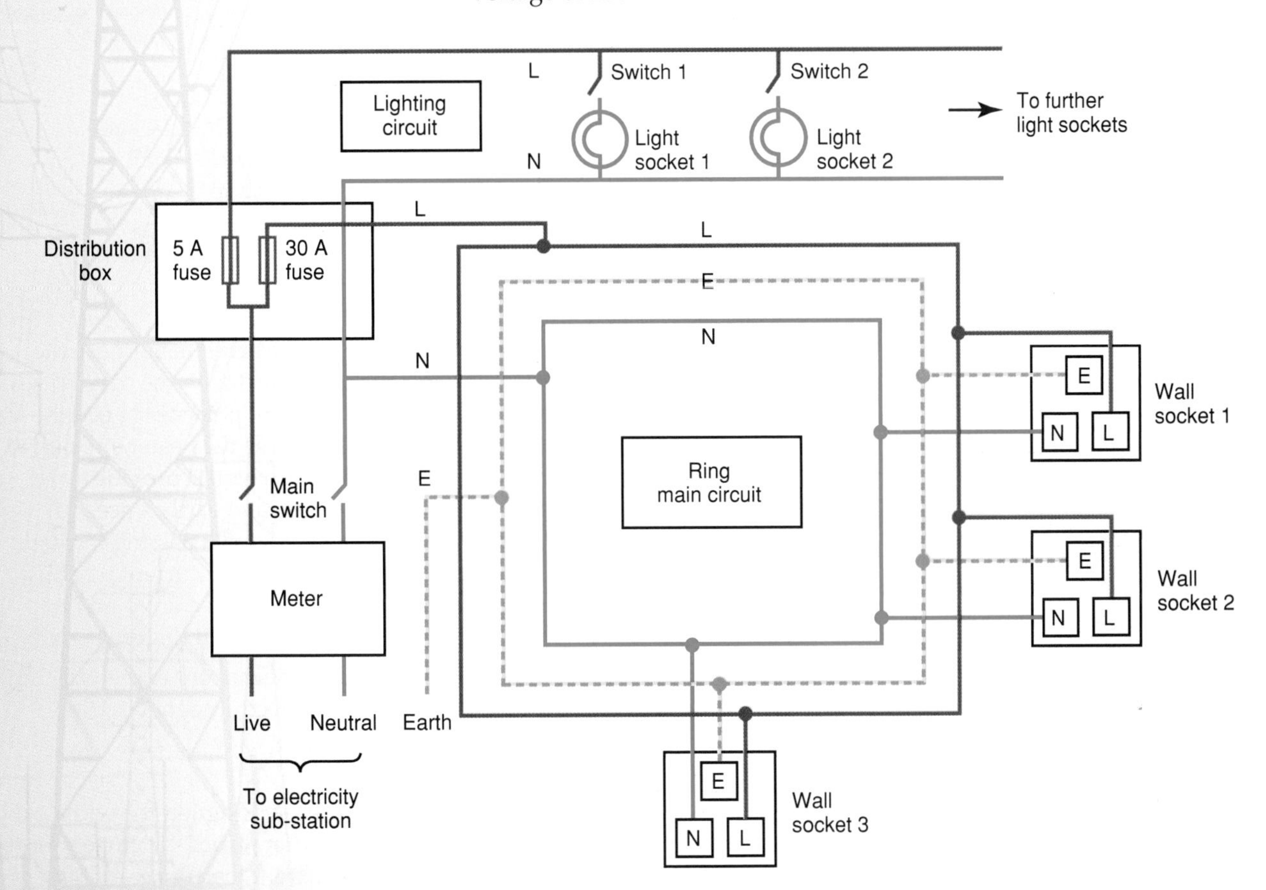

Figure 21.2C ▲ Home circuits

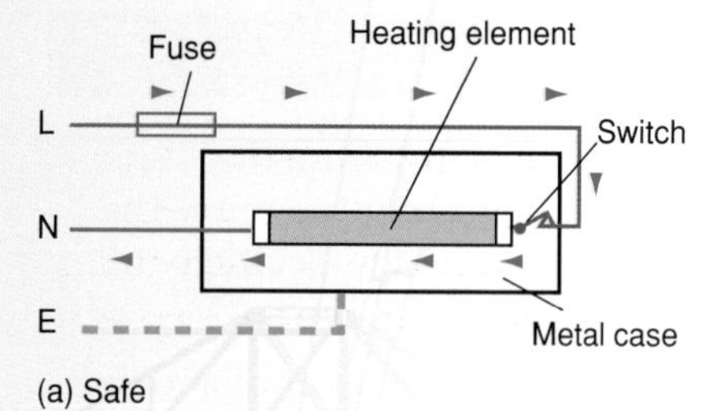

(a) Safe

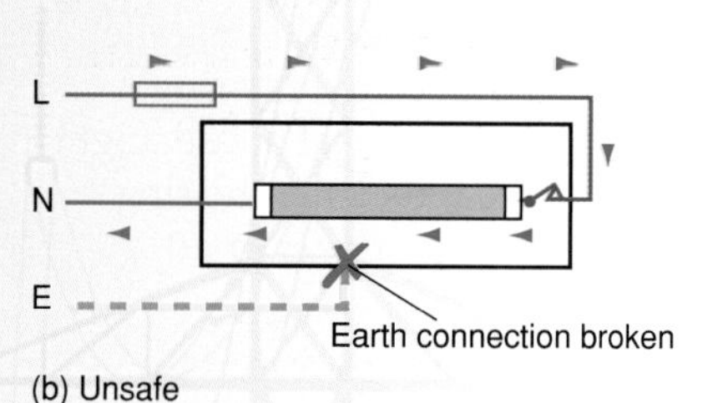

(b) Unsafe

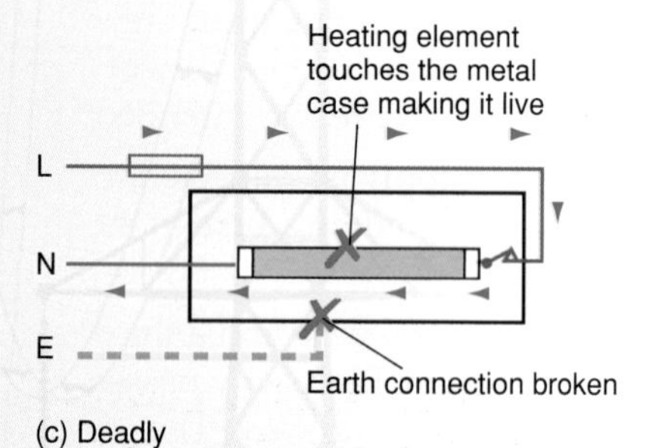

(c) Deadly

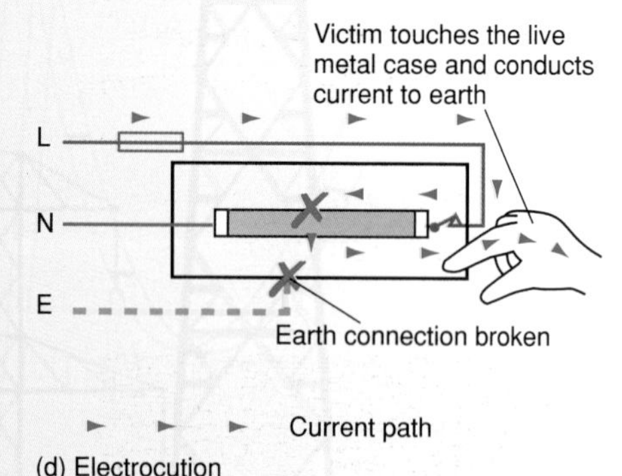

(d) Electrocution

Figure 21.2D The importance of earthing

Figure 21.2F Double insulation

Safety matters

The **earth wire** is intended to ensure any mains appliance with a metal case is safe to touch. Without the earth wire, if the live wire in an appliance broke and touched the metal case, anyone touching the case would be electrocuted because the electric current would pass through the body to earth. To prevent the case from becoming 'live', it is connected to the earth wire of the ring main.

Double insulation is an important safety feature of any mains appliance which does not have a metal case. The casing consists of a double layer of a tough, stiff, non-conducting material and carries the double insulation symbol shown in Figure 21.2F.

Fuses are designed to protect the circuit wires from passing too much current which can make the wires over-heat. For example, suppose the live wire to a light socket becomes disconnected from the socket and touches the neutral wire. This is an example of a **short-circuit**. The current passing along the wires increases and the fuse blows, cutting the socket off from the live wire. If the fuse did not blow, the increase in current would make the wires very hot, which could cause a fire.

The **3-pin plug** is used in the UK to connect an appliance to the ring main. The fuse in it is designed to protect the appliance and is always connected in the live wire. If it melts, it cuts the appliance off from the live wire of the ring main.

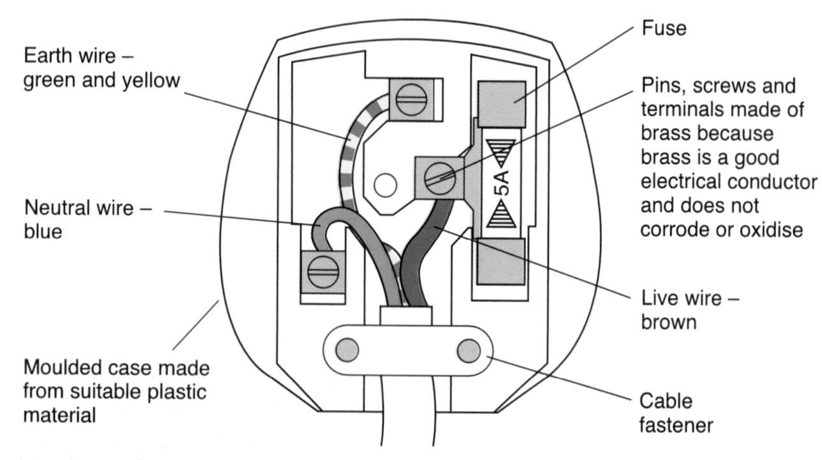

Figure 21.2E Inside a three-pin plug

Fuses are **not** designed to protect users from electrocution. Suppose the heating element of an electric radiator makes contact with the metal case of the radiator, which is not earthed correctly. Current passes between the live and the neutral wire through the heating element so the heater appears to work safely. However, someone touching the case would provide a pathway for current to pass between the live case and earth; since electrocution can result from currents as small as 50 mA passing through the body, the fuse would not prevent such a small extra current passing along the live wire. If the case had been earthed correctly, it would not become live as a result of the heating element touching it (Figure 21.2D).

Circuit breakers are designed to switch off if the current exceeds a specified value. A circuit breaker contains an electromagnetic 'trip switch' which opens automatically if the current exceeds a certain value. The switch must be reset manually.

The Residual Current Circuit Breaker (**RCCB**) is designed to switch off automatically if the current in the live wire differs from the current in the neutral wire. This might happen if a small current leaks to earth, (see p. 336).

Costing electricity

A domestic electricity meter measures electricity used in **kilowatt-hours.** This is the amount of energy used by a one kilowatt appliance in one hour and is equal to 3.6 million joules. Electricity bills always tell you the 'unit' price of electricity. This is the cost of one kilowatt-hour of electrical energy.

The energy used by an appliance can be worked out if its power rating is known. For example, a three kilowatt heater switched on for two hours would use six kilowatt-hours. This would make the meter reading increase by six 'units'. Table 21.1 gives some more examples.

See www.keyscience.co.uk for more about domestic electricity.

Table 21.1 Using electricity

Appliance	Power (kW)	Time appliance is used for (hours)	Number of units used
Heater	3.0	2.0	6.0
Lamp	0.1	24.0	2.4
Hair drier	0.2	0.5	0.1
Cooker oven	5.0	2.5	12.5

To check the amount of electricity used by a particular appliance, switch off all the other appliances in the house and read the meter. Then use the appliance for a given time and read the meter again after that time. The increase in the meter reading should equal the number of kilowatts multiplied by the number of hours (i.e. the number of kilowatt-hours).

Worked example Figure 21.2G shows two photographs of a domestic electricity meter taken one week apart. How much electricity did this household use in one week?

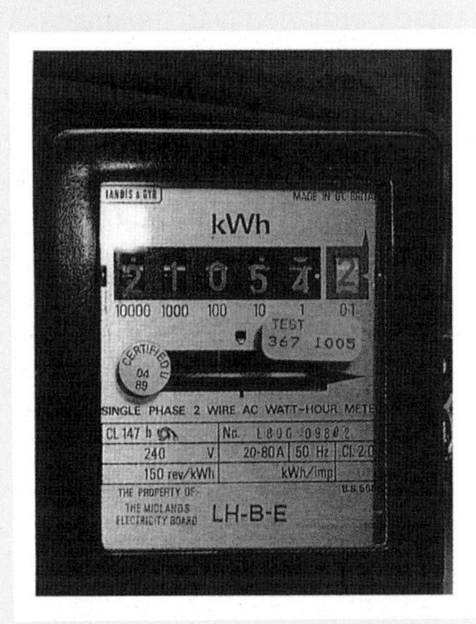

Figure 21.2G (a) Initial meter reading (b) Meter reading after one week

Solution

Initial meter reading = 21 054.2 units

Reading after one week = 21 778.0 units

Number of units used in one week = 723.8 units

SUMMARY

The current passed by an appliance can be worked out from the equation

Power = Current × Voltage

The fuse rating is determined by the current. The energy used can be calculated using the equation

Energy = Power × Time

The unit of electrical energy measured by domestic electricity meters is the kilowatt-hour, which is equal to 3.6 MJ

CHECKPOINT

1 Explain each of the following:
   (a) A mains plug should always be fitted with the 'correct' fuse.
   (b) An appliance with a frayed mains cable should not be used until the cable has been replaced.
   (c) The cable of an electric iron must never be allowed to touch the iron when the iron is hot.

2 Prakesh's father has asked him to fit a plug and a fuse to a new electric iron. The appliance is rated at 240 V, 1000 W.
   (a) What current does the appliance take when it uses 1000 W of power?
   (b) Prakesh has a 3 A fuse, a 5 A fuse and a 13 A fuse. Which one should he use in the plug?
   (c) The iron has a control dial for different fabrics, as shown in the diagram. Explain how it works.

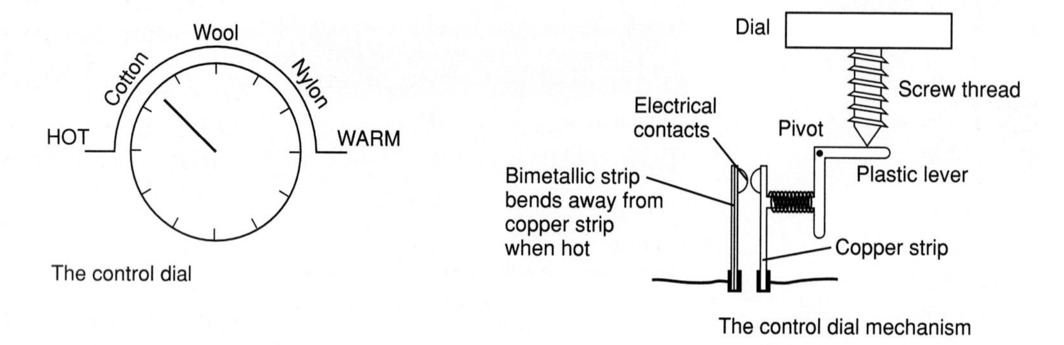

The control dial

The control dial mechanism

   (d) What is likely to happen if the iron is used on full power on a nylon shirt?

3 (a) Work out the number of 'units' of electricity used when
   (i) a 3 kW electric heater is used for 4 hours,
   (ii) a 100 W lamp is used for 10 hours,
   (iii) a 300 W hair drier is used for 10 minutes,
   (iv) a 2.5 kW electric kettle is used for 5 minutes.
   (b) If the unit price of electricity is 5.5p, work out the cost of using the heater for two hours, the lamp for five hours, and the kettle four times for five minutes each time.

4 Electric cookers are connected permanently to the mains. The fuse is in the fuse box. Usually this is a 30 A fuse. How much power does a cooker use when it takes 30 A at 230 V? How many units of electricity would it use in three hours and what would be the cost at 5.5p per unit?

5 To check the power rating of an electric heater, Colin has switched every electrical appliance in the house off. He notes the meter reading at 56712.3 and then switches the heater on for exactly two hours. After this time, he switches the heater off and notes the meter reading again. This time it reads 56717.5.
   (a) How many kilowatt hours did the heater use in this time?
   (b) Work out the power supplied to the heater in kW.
   (c) What would the meter reading become if he then used an 800 W microwave oven for exactly 15 minutes without switching any other appliances on?

21.3 Electrical measurements

FIRST THOUGHTS

In this section, you will find out how to measure currents and voltages in low voltage circuits.

Measuring current

Debbie has brought a battery-operated vacuum cleaner to school to ask her science teacher to check it. The cleaner is rated at 12 V, 85 W and is for use in her parents' caravan. It does not work when connected to the 12 V socket in the caravan.

When it is connected to a 12 V battery, the cleaner should take 7.1 A. You should be able to work this out from the voltage and power ratings. An ammeter can be used to measure the current passing through the cleaner. Debbie's teacher connects the cleaner in series with an ammeter and a 12 V battery, as shown in Figure 21.3A. The ammeter reads 7.0 A when the cleaner is switched on. The teacher suggests that the wiring in the caravan might be at fault. *Can you think of any other reasons why the cleaner will not work in the caravan?*

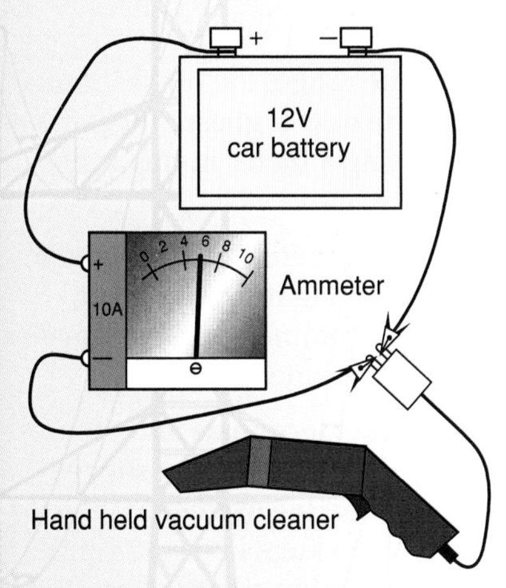

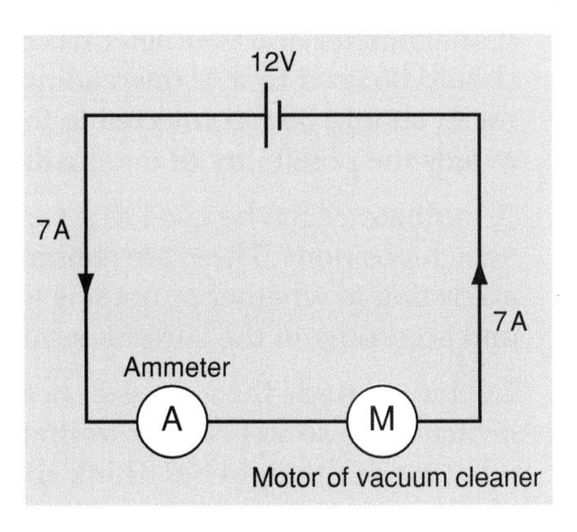

Figure 21.3A Using an ammeter
(a) Checking a low voltage vacuum cleaner
(b) Circuit diagram

Does it matter which side of the cleaner plug the ammeter is connected to? All the components are in series so they pass the same current. The ammeter position does not matter. Figure 21.3B shows how two ammeters could be used to show this.

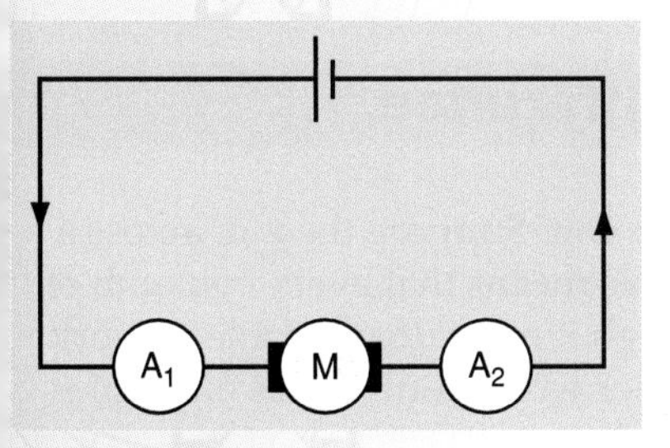

Figure 21.3B Components in series pass the same current

The circuit diagram for the caravan is shown in Figure 21.3C. In addition to the cleaner, the circuit includes two 12 V lamps. Debbie borrows the ammeter to test the lamps. She connects the ammeter in series with the battery, as shown in Figure 21.3C, and tests each lamp in turn. Her results are as follows

Lamp bulb X only ON	Ammeter reading = 2.0 A
Lamp bulb Y only ON	Ammeter reading = 1.0 A
Both lamp bulbs ON	Ammeter reading = 3.0 A

With both lamp bulbs on, the ammeter registers the current taken by X and the current taken by Y. The current entering junction Z from the ammeter is 3.0 A; at the junction, this current splits into 2.0 A that passes through X and 1.0 A that passes through Y.

The total current entering a junction is always equal to the total current leaving the junction

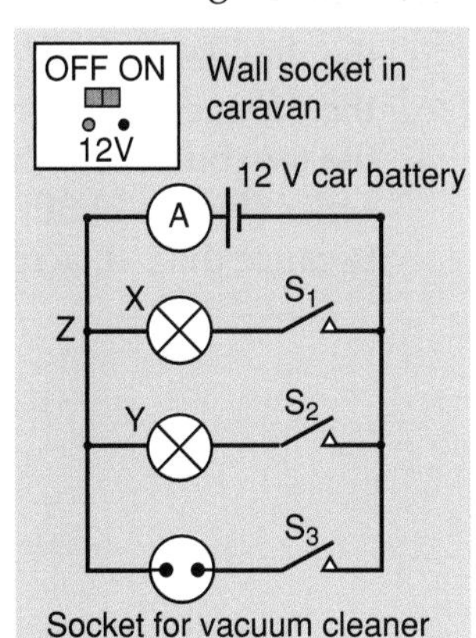

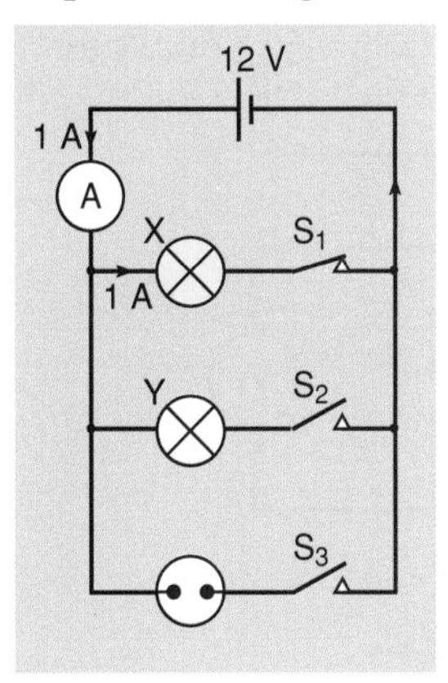

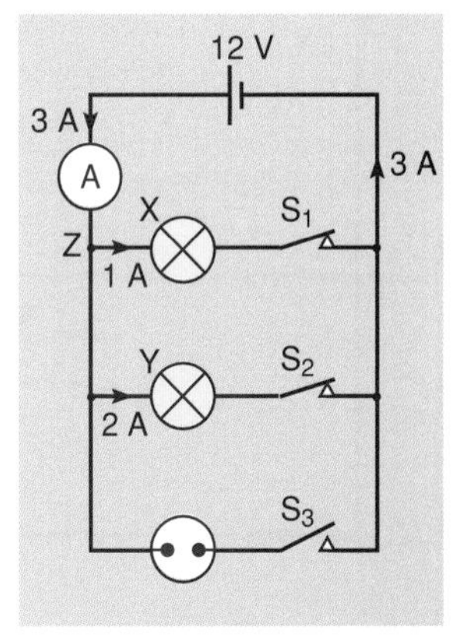

Figure 21.3C The caravan circuit diagram

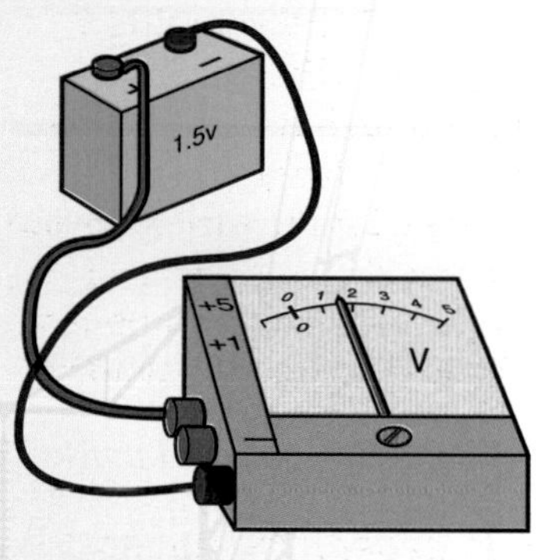

(a) Using a voltmeter

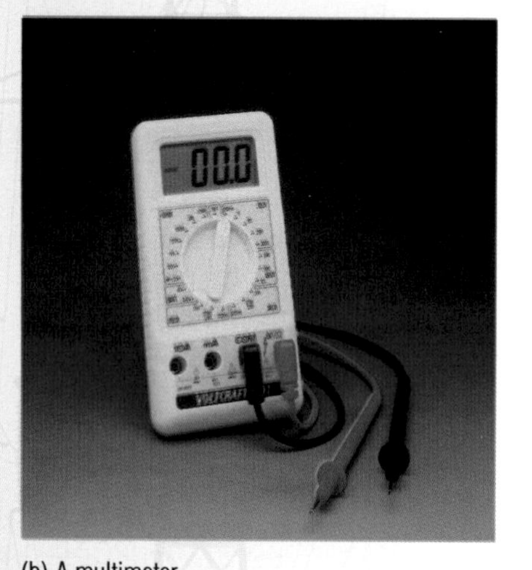

(b) A multimeter

Figure 21.3D Electrical measurements

Measuring potential difference

In the caravan, the lamp bulbs work so the battery must be working. The cleaner worked in school but does not work in the caravan. Perhaps the voltage of the 12 V battery is not reaching the socket of the cleaner. Debbie borrows a voltmeter from school to find out.

A voltmeter measures the voltage (i.e. potential difference) between two points in a circuit. This is a measure of the work done to push charge round the circuit. It is similar to the pressure in the water pipes of a central heating system. A p.d. across the socket terminals is necessary to push current through the vacuum cleaner when it is plugged in.

If an ammeter or a voltmeter has more than one range, the highest range should be used first. If the reading is within the next highest range, the meter should be reconnected to that range. Using a meter in this way avoids the possibility of overloading it.

A multimeter can be used as an ammeter or a voltmeter, according to its switch positions. These are chosen before connecting it to the circuit, according to whether or not it is to be used as an ammeter or a voltmeter and according to the current or voltage range to be measured.

Debbie connects the voltmeter between the socket terminals and then switches the socket on. The voltmeter still reads zero. Can you suggest what the fault might be? Think about a central heating system in which no pressure is getting through to a radiator. A blockage in a water pipe could cause this. What could prevent the voltage from a battery reaching a socket?

Investigating potential difference

The unit of p.d. is the volt. Suppose the p.d. across a resistor is one volt; this means that every coulomb of charge passing through it must do one joule of work to get through. In other words, one volt is one joule per coulomb.

Let's consider a practical problem to understand more about the meaning of p.d. Figure 21.3E shows two bulbs and a dimmer (i.e. variable resistor) connected in series with each other and a 6 V battery. Claire sets the circuit up and adjusts the dimmer so that the bulbs are at normal brightness. Graham then uses the voltmeter and measures the p.d. across each bulb and the dimmer as 2.5 V across each bulb and 1.0 V across the dimmer. The battery voltage is shared between the two bulbs and the dimmer because they are in series (Figure 21.3E). Each bulb lights up normally because the p.d. across each is 2.5 V.

Figure 21.3E Sharing voltage

Why do the p.d.s across the bulbs and dimmer add up to the battery voltage? The p.d. across a component is the number of joules delivered to the component by each coulomb of charge passing through it. Since the two bulbs and the dimmer are in series, each coulomb of charge from the battery passes through each component, delivering 2.5 J to each bulb and 1.0 J to the dimmer. Hence each coulomb of charge delivers a total of $2.5 + 2.5 + 1.0 = 6$ J of energy from the battery to the circuit components. The battery therefore supplies 6 J of electrical energy to each coulomb of charge passing through it, which is 6 V.

For components in series connected to a battery, the battery voltage is equal to the sum of the potential differences across each component.

High speed trains

Railway electrification of the main east coast line from London to Edinburgh has decreased journey times considerably. The electric locomotives used to pull the trains are powered by 25 kW electric motors.

Measuring Power

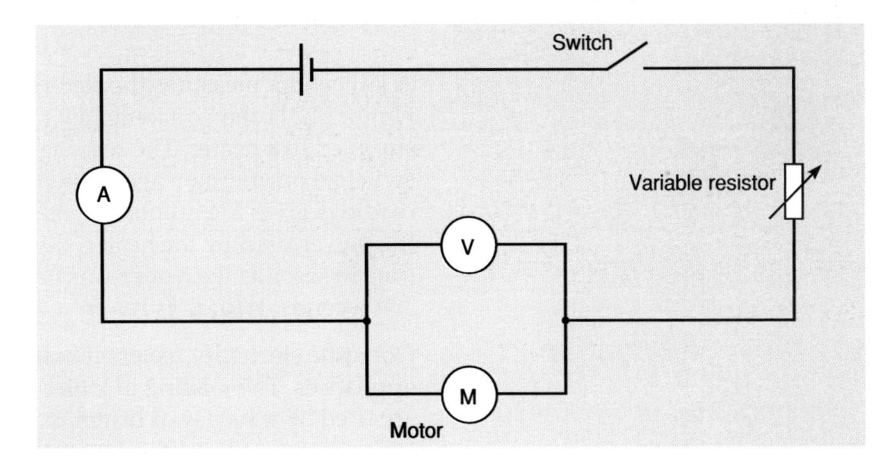

Figure 21.3F ▲ Measuring electrical power

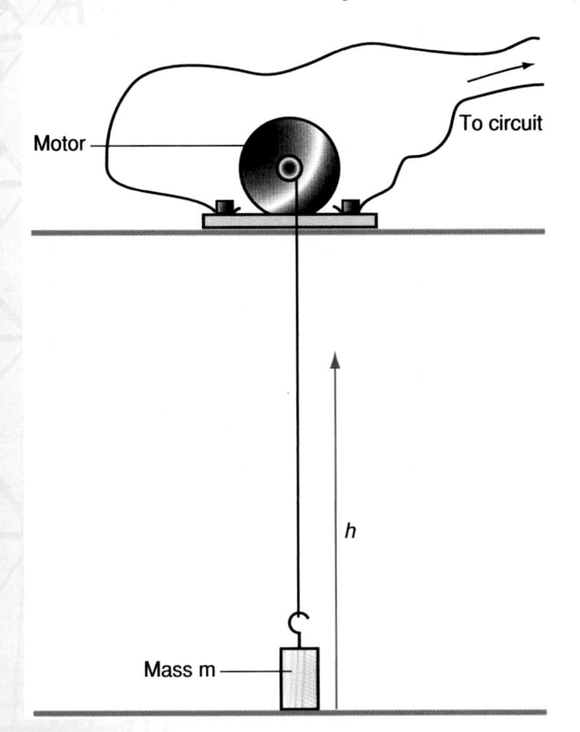

Figure 21.3G ▲ Measuring the work done by an electric motor

- The **power** used by an electrical appliance can be determined if the current through the appliance and the voltage applied to it are known. Figure 21.3F shows how the power used by a 6 V model electric motor may be measured. The variable resistor is adjusted until the voltmeter reads exactly 6.0 V. The power of the motor is given by

  Power = Current × Potential difference

- The **efficiency** of an electric motor can be measured by timing it to raise a known weight, as shown in Figure 21.3G. At the same time, its current and voltage are measured so that the electrical power supplied to it can be calculated.

1. For an object of mass m raised through height h, its gain of potential energy $= mgh$. This is the work done by the motor.
2. The electrical energy supplied to the motor = electrical power × time $= IVt$, where I is the current, V is the voltage and t is the time taken.

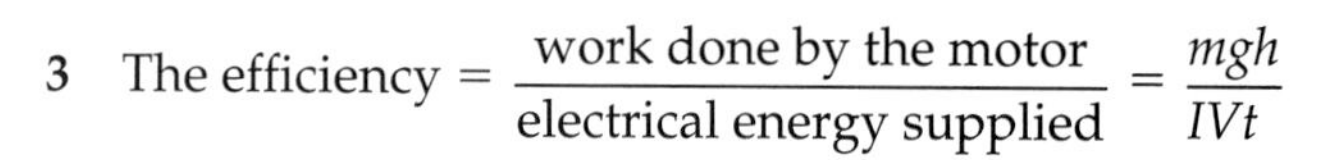

3 The efficiency $= \dfrac{\text{work done by the motor}}{\text{electrical energy supplied}} = \dfrac{mgh}{IVt}$

4 No machine can be 100% efficient. Friction at the bearings and resistance heating are the main causes of inefficiency in an electric motor.

Worked example A 12 V electric winch is used to raise a 50 N weight through a height of 1.5 m to test its efficiency. The motor current was 4.0 A and it took 5.0 seconds to raise the weight. Calculate (a) the electrical energy supplied to the motor, (b) the gain in potential energy of the weight and (c) the efficiency of the motor.

Solution

(a) Electrical energy supplied = IVt = 4.0 A × 12.0 V × 5.0 s = 240 J

(b) The gain in potential energy of the weight = mgh = 50 N × 1.5 m = 75 J

(c) The efficiency = $\dfrac{75}{240}$ = 0.31 = 31%

A joulemeter measures the electrical energy used directly in joules. Figure 21.3H shows a joulemeter being used to measure the energy supplied to a heater. The reading must be taken before the heater is switched on and then again after it is switched off. The difference in the readings gives the number of joules supplied to the heater. To calculate the power used by the heater, divide the number of joules used by the time in seconds the heater was on. Suppose the heater used 7210 joules in 300 seconds. *What is its power?*

Domestic electricity meters measure the electrical energy used by mains appliances. The reading in joules would be enormous. How many joules are used by a 1000 watt heater in one hour? The heater uses 1000 joules each second, so in 1 hour it would use 3.6 million joules. This amount of energy is defined as one kilowatt-hour. Domestic meters are marked in 'units' of kilowatt hours.

SUMMARY

Current is measured using an ammeter. An ammeter must always be connected in series with the appliance taking the current to be measured.

Potential difference is measured using a voltmeter. Voltmeters are always connected in parallel with the appliance whose potential difference is being measured.

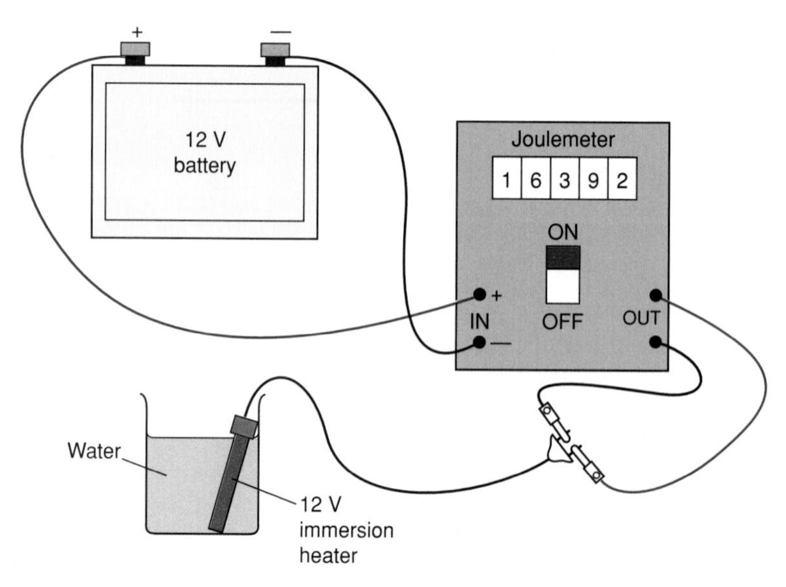

Figure 21.3H ▲ Using a joulemeter

CHECKPOINT

1 (a) Work out the current through each lamp bulb in circuits A-C below.
(b) What is the p.d. across each lamp bulb in circuits D-F below?

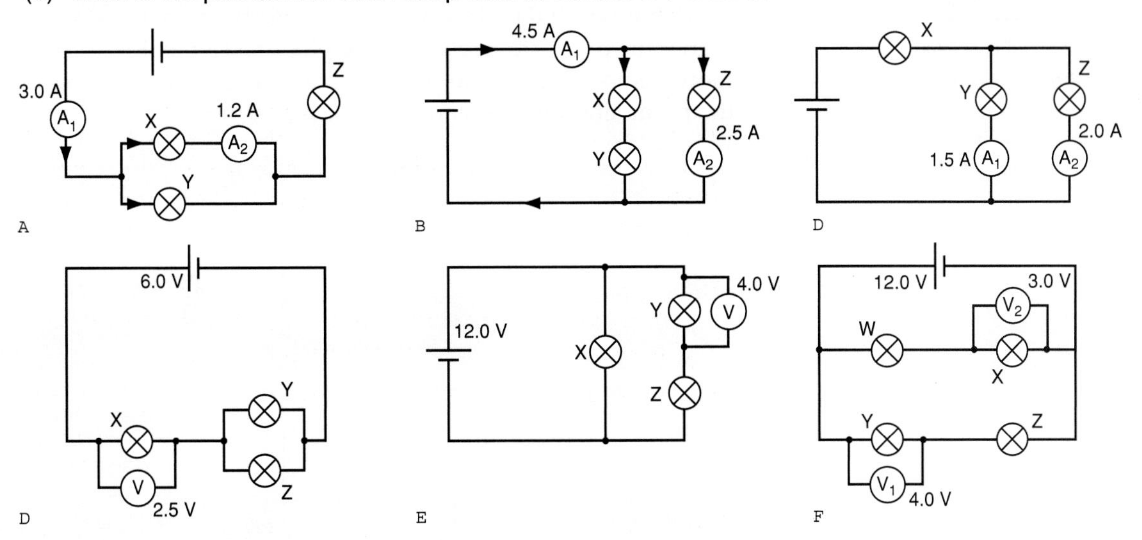

2 Claire has repaired the faulty wire in the caravan circuit and she decides to see if the lamp bulbs both work when the vacuum cleaner is switched on. The circuit is shown in the diagram below.

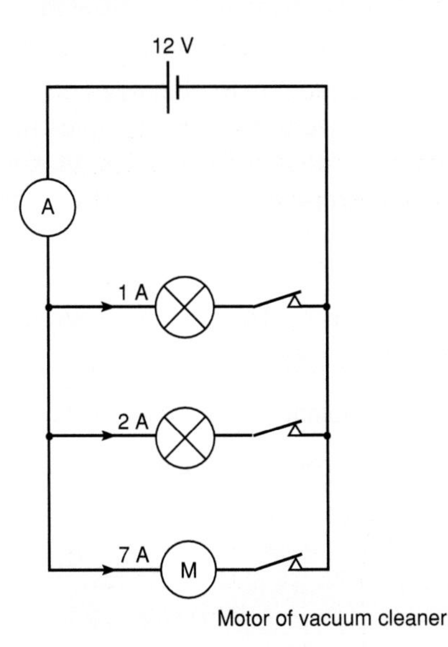

(a) What do you think the ammeter will read when all three appliances are switched on?
(b) How much energy does each appliance use per second when switched on?
(c) How much energy does the battery supply each second when all three appliances are switched on?

3 The circuit below is to test car headlamp bulbs.
(a) What is the reading on the ammeter when (i) S1 is closed, (ii) S2 is closed, (iii) both switches are closed?
(b) How much charge passes through each bulb each second when both switches are closed?
(c) The battery has a voltage of 12 V. How much energy is delivered to each bulb each second when both switches are closed?

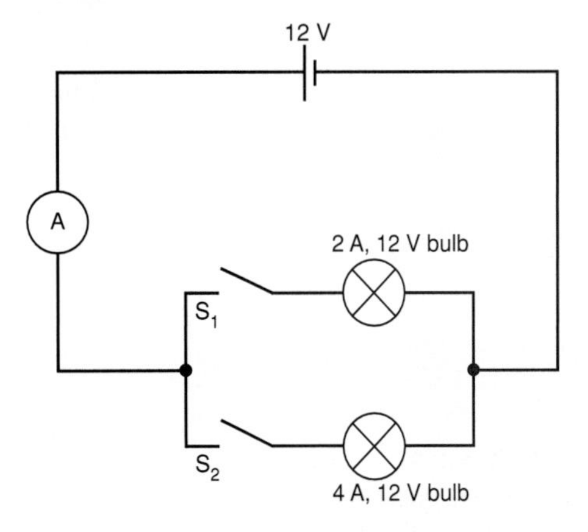

4 Julie has a low voltage immersion heater designed to run off a car battery to warm water in a cup. The heater is rated at 12 V, 5 A.
(a) How much energy should it use in (i) 1 second, (ii) 5 minutes?
(b) How could Julie check its energy usage using a joulemeter and a 12 V battery?

21.4 Resistance

FIRST THOUGHTS

After studying this section, you should know how to work out currents, voltages and resistances in circuits.

In a central heating system, the pump forces water through the radiators and the pipes. If the pipes are narrow, the pump has to do more work than if the pipes are wide. Narrow pipes resist the flow of water much more than wide pipes do. The flow rate depends on the pressure from the pump and the 'resistance' of the pipes.

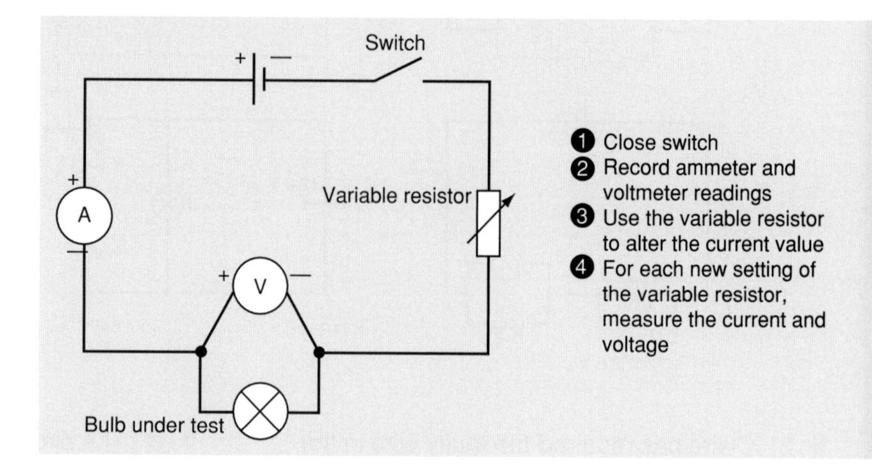

Figure 21.4A Investigating components

Note that 1 ohm equals 1 volt per ampere.

The blood system of the body is like a 'water circuit'. The heart forces blood through the arteries, capillaries and veins. The flow rate depends on the pressure from the heart and the 'resistance' of the blood vessels.

In an electric circuit, the current depends on the battery voltage and the components in the circuit. Figure 21.4A shows how to investigate the link between current and voltage for different components. The measurements can be plotted on a graph of current (on the vertical axis) against voltage. Figure 21.4B shows some typical results for a lamp bulb, diode and wire.

Resistance is defined as voltage/current. The unit of resistance is the *ohm* (Ω, the Greek letter 'omega').

$$R = \frac{V}{I}$$ where V = potential difference in volts

I = current in amperes

R = resistance in ohms

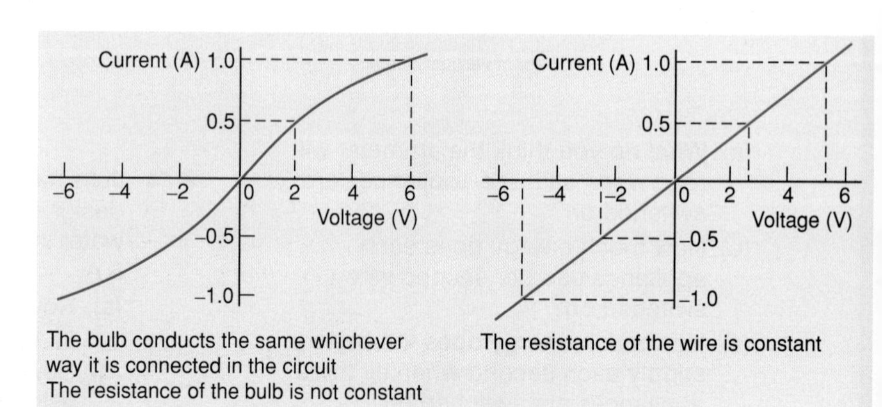

Figure 21.4B (a) Testing a lamp bulb (b) Testing a wire

It's a fact!

The resistance of the human body can be as low as 1000 Ω. If the hands are wet, electrical contact is more effective than when they are dry, since water conducts electricity, and the resistance can be even lower than 1000 Ω. Shocks can be lethal if the current passed through the body is greater than about 50 mA. This means that voltages over about 50 V (= 0.050 A × 1000 W) can be dangerous.

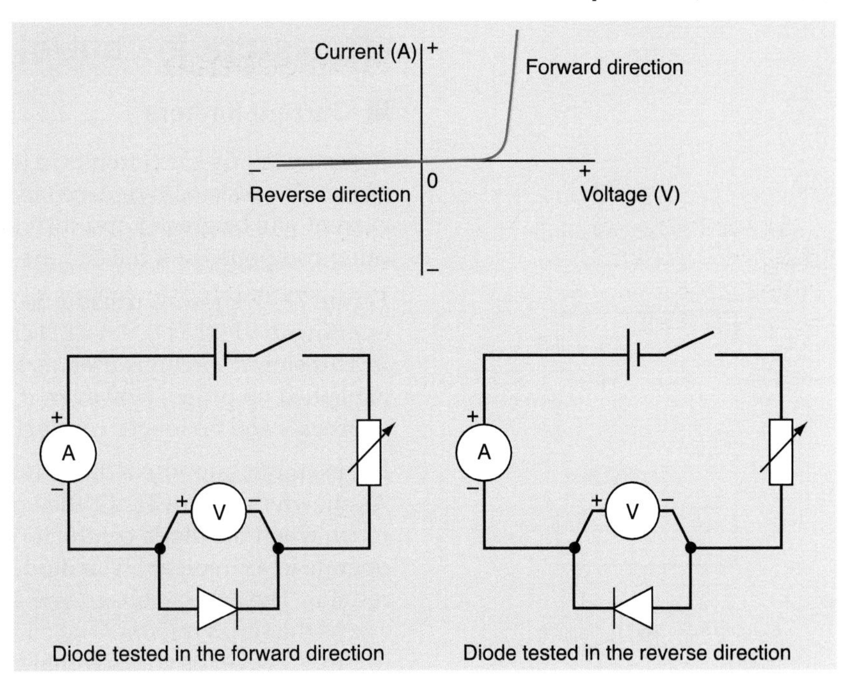

Figure 21.4C ▲ Testing a diode

The equation can be used to work out R or V or I if the other two quantities are known. For example, suppose a heater has a resistance of 8.0 Ω and it is to be connected to a 12 V battery. The current taken is given by p.d./resistance = 12 V/8 Ω = 1.5 A.

Consider the graphs in Figure 21.4B and C. *What can you tell from them about the resistance of each component?*

- The diode has a much greater resistance when connected in its 'reverse' direction compared with its 'forward' direction. In reverse, very little current passes through it.
- The lamp bulb has more resistance as the current increases. From Figure 21.4B, I = 0.5 A when V = 2.0 V and I = 1.0 A when V= 6.0 V. The resistance at 0.5 A is 4.0 Ω (= 2.0 V/0.5 A); at 1.0 A, the resistance is 6.0 Ω (= 6.0 V/1.0 A).
- The wire has constant resistance. This does not depend on the current passed. From Figure 21.4B, I = 0.5 A when V = 2.5 V and I = 1.0 A when V = 5.0 V. Both these readings from the graph give 5.0 Ω for the resistance of the wire.

Ohm's Law states that the current through a metallic conductor at constant temperature is proportional to the p.d. thus p.d./current is constant.

In other words, the resistance of a metallic conductor is constant, provided the temperature is constant. The resistance of a metal increases as its temperature increases. The filament of a lamp bulb is a metal wire. As the current increases it gets hotter and so its resistance increases.

A **resistor** is a component designed to have a specific resistance. Resistors can be made from metal wires or from carbon. Figure 21.4D shows the resistor colour code. A resistor with 4 coloured bands which are *red red orange gold* in order would have a value of 22000 ohms with a tolerance of 5%.

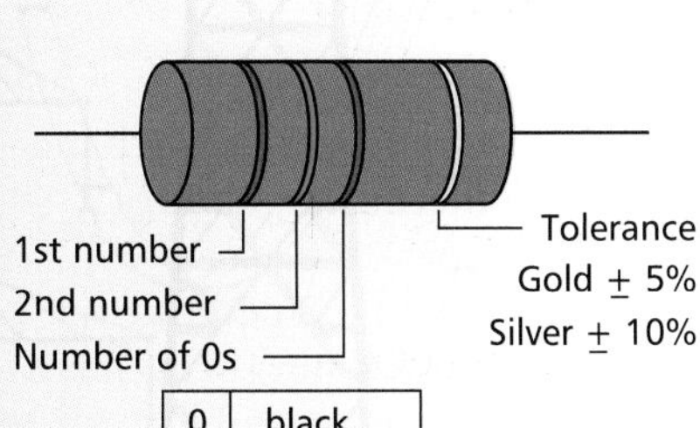

0	black
1	brown
2	red
3	orange
4	yellow
5	green
6	blue
7	violet
8	grey
9	white

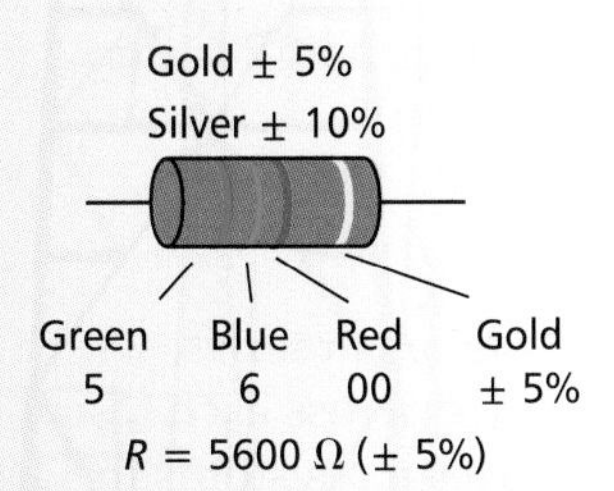

Figure 21.4D ▲ The resistor colour code

Resistors in control

Current limiters

Resistors are used in circuits to limit the current in each branch of a circuit. If a resistor is replaced by one of lower resistance in a circuit, the current will be greater and this may cause overheating and damage to other components in the circuit.

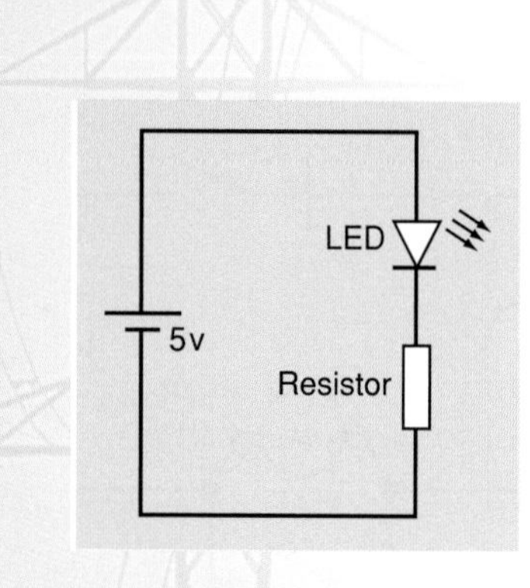

Figure 21.4E ▲ A limiting resistor

Figure 21.4E shows a resistor used to limit the current through a light-emitting diode (LED). An LED emits light when current passes through it. This simple circuit is used as an indicator in an electronic circuit, as explained on p. 360. However, if too much current passes through it, it overheats and no longer conducts or emits light.

For example, suppose a diode is designed to operate at a current of 5 mA. As shown in Figure 21.4C, the voltage across a diode does not change much when the diode conducts. Most diodes operate at about 0.6 V. To operate in a circuit at 5 V, a diode must therefore be in series with a resistor. The excess voltage of 4.4 V (= 5.0 – 0.6 V) would be dropped across the series resistor. The current through it would need to be limited to 5 mA. Prove for yourself that the resistance would therefore need to be about 900 ohms (=4.4 V/0.005 A).

Variable resistors

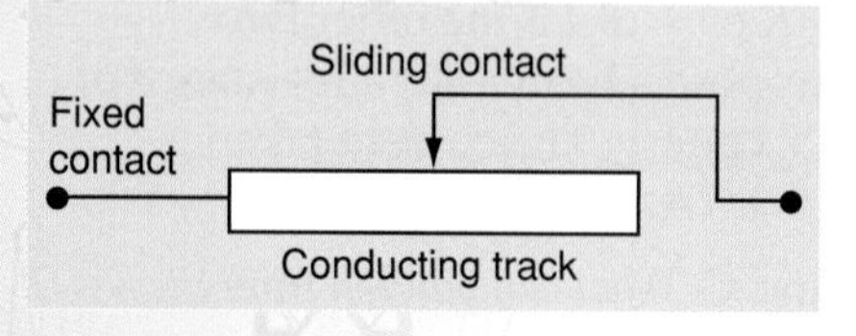

Figure 21.4F ▲ A variable resistor

A variable resistor is used in a circuit to change the current. For example, a variable resistor in series with a motor could be used to control the speed of the motor. A variable resistor consists of a conducting track of resistance material with a fixed contact at one end and a sliding contact on the track. Moving the contact along the track changes the length of material and hence the resistance between the contacts.

Resistors in sensors

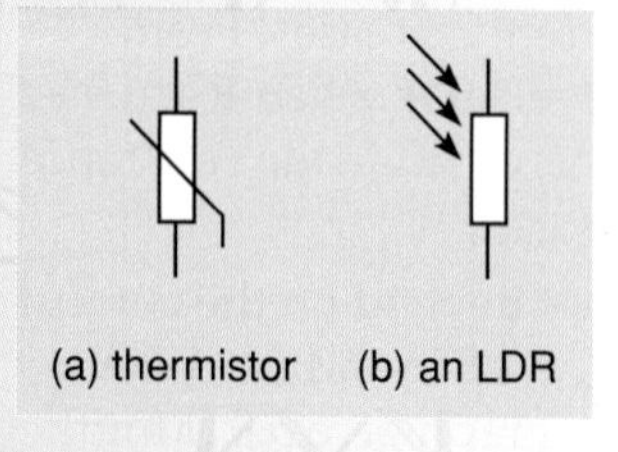

Figure 21.4G ▲ More symbols

The **thermistor** is a temperature-dependent resistor. Its resistance changes if its temperature is changed. The resistance of a thermistor made of a semiconducting material falls with increasing temperature, (see p. 326). A thermistor can be used in a potential divider to make a temperature sensor, as explained on p. 364. The **light dependent resistor** (LDR) has a resistance which decreases with increasing light intensity. An LDR can be used in a potential divider to make a light sensor, as explained on p. 364.

Electrical heating

The heating element in an electrical heater is essentially a length of resistance wire. Figure 21.2B shows the heating element of an electric kettle and an electric iron. The electrical energy supplied to the heating element is almost all converted to heat. Hence the heat per second produced by a heating element can be calculated using the following equation:

$$\text{Heat/second} = IV = I^2R$$

where I is the heater current, V is the heater voltage and R is the resistance of the heating element.

The potential divider

A potential divider consists of two or more resistors in series, with a fixed voltage across the combination. The voltage across each resistor is a fraction of the total voltage. The voltage across each resistor is in proportion to the resistance of the resistor since the current is the same.

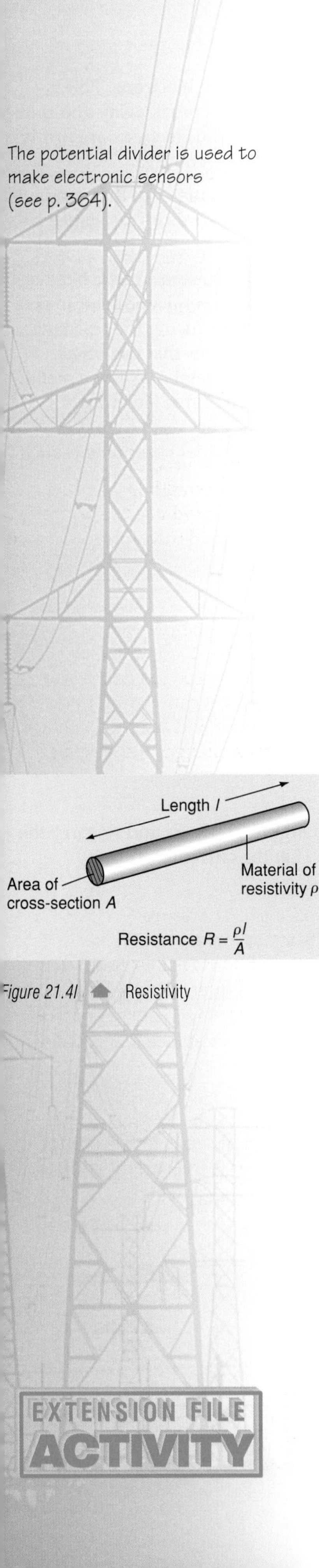

The potential divider is used to make electronic sensors (see p. 364).

Figure 21.4I ▲ Resistivity

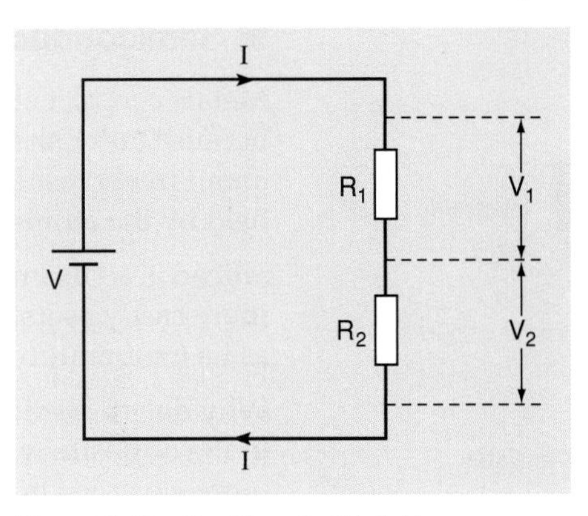

Figure 21.4H ▲ The potential divider

For two resistors R_1 and R_2 connected to a fixed voltage V, as in Figure 21.4H, the current I through each resistor is given by the equation $I = V/(R_1 + R_2)$

Hence the voltage across resistor R_1, $V_1 = VR_1/(R_1 + R_2)$

and the voltage across resistor R_2, $V_2 = VR_2/(R_1 + R_2)$

Therefore $V_1/V_2 = R_1/R_2$

A potential divider can be formed from a conducting track of resistance material with a fixed contact at **each** end and a sliding contact on the track. With a fixed voltage across the end contacts, the voltage between either end and the sliding contact is changed by moving the sliding contact. This is how a variable voltage can be obtained from a fixed voltage.

Resistivity

The resistance of a uniform conductor depends on the length, the area of cross-section and the type of material.

- The longer the length of a conductor, the greater its resistance.
- The narrower a conductor is, the greater its resistance.
- Metals conduct much better than non-metals. Copper is the best conductor.

Tests show that the resistance of a uniform conductor is

1 proportional to its length,
2 inversely proportional to its area of cross-section.

In other words, its resistance $= \dfrac{\text{constant} \times \text{length}}{\text{area of cross-section}}$

The constant depends on the type of material and is referred to as the **resistivity** of the material. The unit of resistivity is the ohm metre (Ω m). The symbol for resistivity is the Greek letter ρ, pronounced 'rho'. If the resistivity of a material is known, the resistance of a given length of material of known area of cross-section can be calculated.

Note: for a wire of diameter d, *the area of cross-section* $= \pi d^2/4$

Worked example Calculate the resistance of a uniform copper wire of length 10 m and of diameter 1.5 mm². The resistivity of copper is 1.7×10^{-8} Ω m.

Solution

Area of cross-section $= \pi (1.5 \times 10^{-3})^2/4 = 1.8 \times 10^{-6}$ m²

Using the equation for resistivity gives

$$\text{resistance} = \frac{\text{resistivity} \times \text{length}}{\text{area of cross-section}} = \frac{1.7 \times 10^{-8} \times 10}{1.8 \times 10^{-6}} = 0.1\ \Omega$$

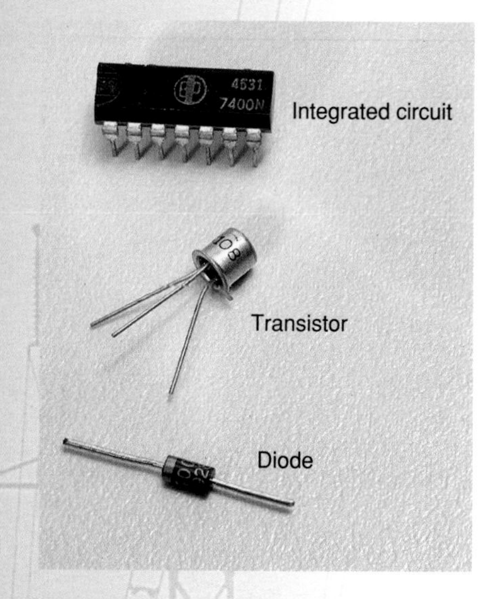

Figure 21.4J Semiconductor components

Semiconductors

Metals conduct electricity because they contain free electrons. This is because outer shell electrons from the metal atoms break away and move about freely inside the metal. In an insulator, all the electrons are firmly held by the atoms so insulators cannot conduct electricity.

Silicon is an example of a **semiconductor**, that is a material that conducts more easily as its temperature is raised. In other words its resistance falls as its temperature is raised.

Why does a semiconductor behave in this way whereas a metal behaves in the opposite way? Raising the temperature of a semiconductor makes more electrons break away from the atoms thus reducing the resistance of the material. However, in a metal, all the electrons that can break away have done so; raising the temperature of a metal makes the metal atoms vibrate more and this hinders the passage of free electrons. Semiconductors are very important because they are used to make integrated circuit chips for computers as well as other components such as diodes and transistors.

Resistor combination rules

Resistors in series

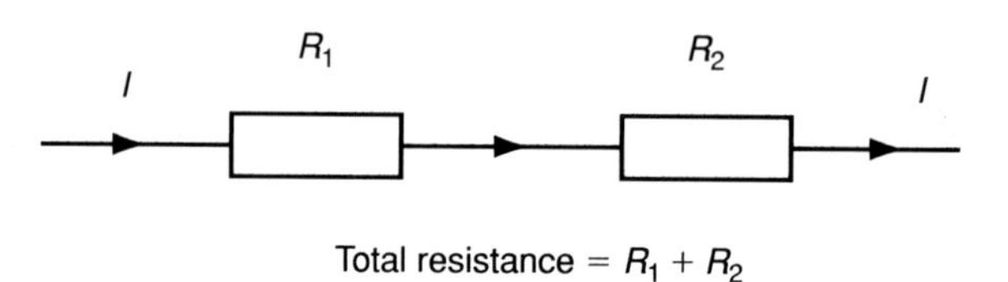

Total resistance $= R_1 + R_2$

Figure 21.4K Resistors in series

Consider the two resistors R_1 and R_2 in Figure 21.4K. R_1 and R_2 carry the same current because they are in series. Let the current $= I$

Hence p.d. across R_1, $V_1 = I R_1$
and p.d. across R_2, $V_2 = I R_2$

Since Total p.d. across R_1 and $R_2 = V_1 + V_2$
then Total p.d. $= I R_1 + I R_2$

$$\text{Total resistance, } R = \frac{\text{Total p.d.}}{\text{Current}} = \frac{I R_1 + I R_2}{I} = R_1 + R_2$$

For two resistors R_1 and R_2 in series, the total resistance R is equal to the sum of the individual resistance.

Series resistor rule:
$R = R_1 + R_2$

Worked example A 6V battery is connected in series to resistors of 1Ω, 2Ω and 3Ω as shown in Figure 21.4L. What is the current and p.d. for each resistor?

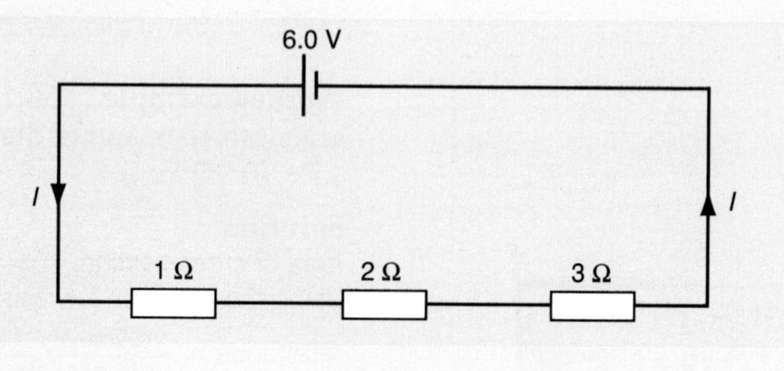

Figure 21.4L

Solution The total resistance = 1 + 2 + 3 = 6 Ω (using the above rule). Hence

$$\text{Battery current, } I = \frac{\text{Battery voltage}}{\text{Total resistance}} = \frac{6\text{V}}{6\Omega} = 1\text{ A}$$

Each resistor passes this current. The p.d. across each resistor can be worked out from current = resistance. Hence

p.d. across 1 Ω = 1 × 1 = 1 V
p.d. across 2 Ω = 1 × 2 = 2 V
p.d. across 3 Ω = 1 × 3 = 3 V

Note that the battery voltage is equal to the sum of the resistor voltages because they are in series.

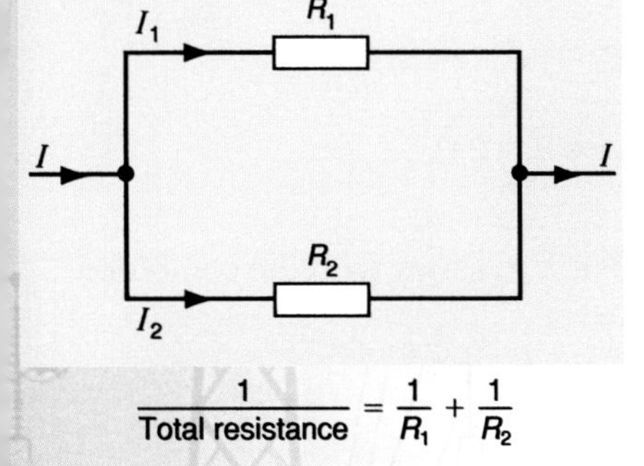

$$\frac{1}{\text{Total resistance}} = \frac{1}{R_1} + \frac{1}{R_2}$$

Figure 21.4M ▲ Resistors in parallel

Resistors in parallel

Consider two resistors R_1 and R_2 connected in parallel as shown in Figure 21.4M. The voltage is the same across the two resistors. Let this voltage = V

Hence Current through R_1, $I_1 = \frac{V}{R_1}$

and Current through R_2, $I_2 = \frac{V}{R_2}$

Therefore Total current $I = I_1 + I_2 = \frac{V}{R_1} + \frac{V}{R_2}$

Since Total resistance, $R = \frac{\text{Voltage}}{\text{Total current}}$

then $= \frac{1}{R} = \frac{I}{V} = \frac{V/R_1 + V/R_2}{V} = \frac{1}{R_1} + \frac{1}{R_2}$

Parallel resistors rule:

$$\frac{1}{R} = \frac{1}{R_1} + \frac{1}{R_2}$$

For two resistors R_1 and R_2 in parallel, the total resistance R is given by

$$\frac{1}{R} = \frac{1}{R_1} + \frac{1}{R_2}$$

Worked example 1 A 4 Ω resistor and a 6 Ω resistor are joined in parallel and then connected to a 12 V battery, as shown in Figure 21.4N. Work out the total resistance and the battery current.

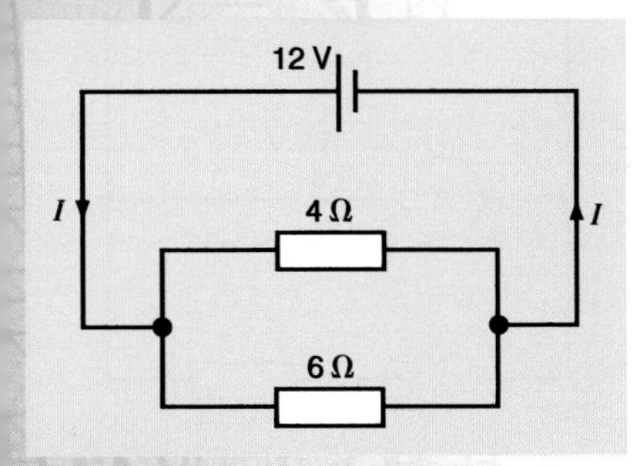

Figure 21.4N ▲

Solution The total resistance R is given by

$$\frac{1}{R} = \frac{1}{4} + \frac{1}{6} = \frac{6+4}{(4 \times 6)} = \frac{10}{24}$$

Therefore $R = \frac{24}{10} = 2.4\,\Omega$

Battery current $I = \frac{\text{Battery voltage}}{\text{Total resistance}}$

$= \frac{12}{2.4} = 5.0\text{ A}$

Worked example 2 For the circuit shown in Figure 21.40, work out (a) the total resistance and (b) the battery current.

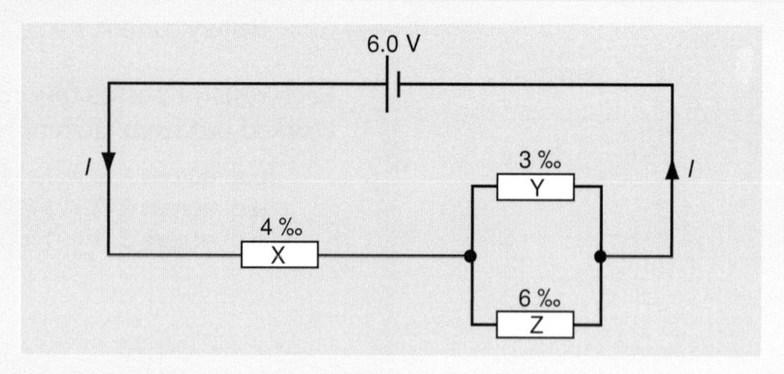

Figure 21.40

Solution (a) Work out the resistance of Y and Z in parallel first, using the parallel combination rule.

$$\frac{1}{R} = \frac{1}{3} + \frac{1}{6} = \frac{1}{2}$$

So Combined resistance R = 2 Ω

The total resistance is the resistance of X in series with the combined resistance of Y and Z.

Hence Total resistance = 4 + 2 = 6 Ω

(b) Battery current, $I = \frac{\text{Battery voltage}}{\text{Total resistance}} = \frac{6}{6} = 1.0\text{ A}$

SUMMARY

Resistance is defined as voltage/current. The unit of resistance is the ohm. The resistance of a metal increases with increasing temperature. The resistance of a semiconductor decreases with increasing temperature. The resistance of any combination of resistors can be calculated using the resistor combination rules.

CHECKPOINT

1 Consider each of the circuits (a)-(c) in the diagram below. What is the ammeter reading when:
   (i) switch S1 alone is closed,
   (ii) switch S2 alone is closed,
   (iii) both switches are closed?
   (iv) What is the total resistance of the circuit when both switches are closed?

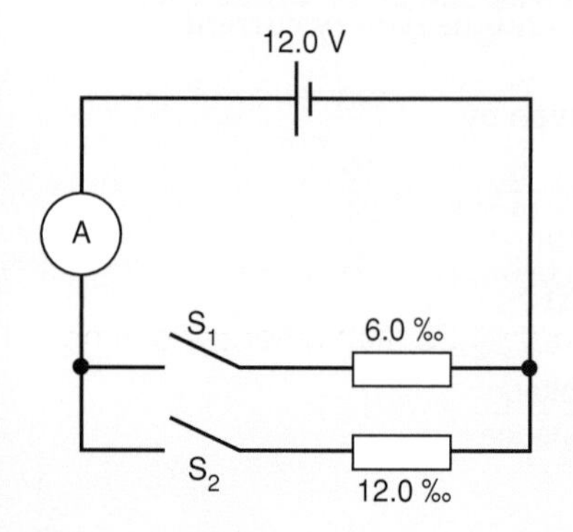

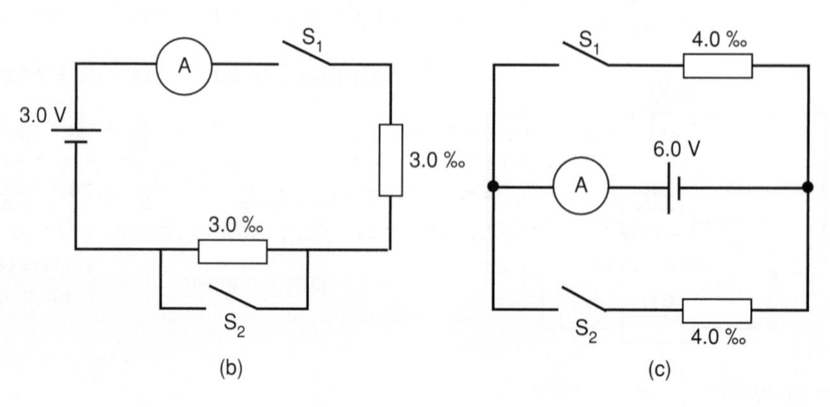

2 (a) What is the total resistance of a 3 Ω and a 6 Ω resistor when they are joined (i) in series, (ii) in parallel?

(b) What is the total resistance of a 3 Ω, a 5 Ω and a 6 Ω resistor when they are joined (i) in series, (ii) in parallel?

(c) The diagram below shows another way to join the three resistors together. Show that the total resistance is 7 Ω.

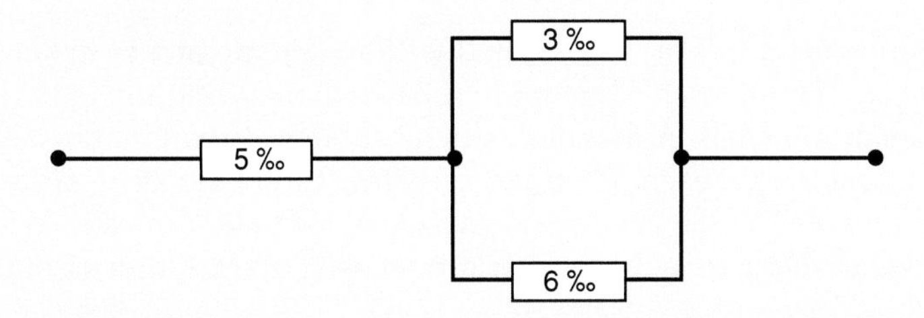

(d) What would the total resistance be if a 6 Ω resistor was used in place of (i) the 5 Ω resistor, (ii) the 3 Ω resistor?

3 The table shows results taken in an experiment to determine the resistance per metre of a wire.

Length of wire (m)	0	0.20	0.41	0.60	0.79	1.00
Resistance (Ω)	0	1.2	2.4	3.5	4.8	6.1

(a) Plot a graph of resistance (on the *y*-axis) against length.

(b) What is the resistance per metre of the wire?

(c) What length of wire would pass a current of 0.5 A when joined in series with a 10 Ω resistor and a 6 V battery?

4 Rachel wants to fix a broken set of fairy lights. There are 20 lights in all, joined in series. They are all supposed to light up when the set is plugged in to 240 V mains.

(a) Before she plugs the set in to the mains, she checks the condition of the wires. Why is this important?

(b) While checking the wires, she notices that bulbs are missing from two of the light sockets. Her father suggests the sockets of the missing bulbs could be 'short-circuited'. Rachel tells him that it is dangerous to use the set with fewer than the correct number of bulbs. Why?

(c) Rachel buys some new bulbs, which she fits to the missing sockets. When she plugs the set in, they still do not work. With the set unplugged, she removes each bulb in turn from its socket to test it using a battery. What should the voltage of the battery be?

(d) Each bulb is designed to take a current of 0.06 A. Work out its resistance and the total resistance of the whole set of 20 bulbs.

5 (a) Sketch a circuit diagram showing a light-emitting diode in series with a resistor and a 1.5 V cell, with the diode in the forward direction.

(b) The LED operates normally at 5 mA and 0.6 V. Calculate the resistance of the resistor for normal operation of the diode in this circuit.

6 (a) Sketch a circuit diagram showing a thermistor in series with a 1000 Ω resistor and a 4.5 V battery.

(b) The resistance of the thermistor is 2000 Ω at 20 °C.

(i) Calculate the total resistance of the circuit at this temperature.

(ii) Hence calculate the voltage across the 1000 Ω resistor at 20 °C.

(c) Describe and explain how this voltage would change if the temperature of the thermistor was increased.

Topic 22 Magnetism

22.1 Permanent magnetism

FIRST THOUGHTS

Have you ever used a magnetic compass to find your bearings? Read on to find out the difference between Magnetic North and True North.

Cupboard doors are often fitted with magnetic catches to keep them closed. These catches are more reliable than spring-loaded catches which sometimes snap. A magnetic catch contains a small magnet. Magnets can attract or repel each other and can attract iron and steel. *How would you design a pair of magnetic catches to keep a cupboard door shut?* You would need to make sure the catches attract each other. Otherwise, the door would never stay closed.

Investigating bar magnets

When you dip a bar magnet into iron filings you find that the filings cling to the ends of the magnet. The ends are called **poles** because that is where the magnetism seems to be concentrated.

Figure 22.1A shows a bar magnet suspended from a string. The magnet lines up with one end pointing north and the other pointing south. This is because of the Earth's magnetism.

The north pole of a bar magnet is the end that points north when the bar magnet is free to turn horizontally.

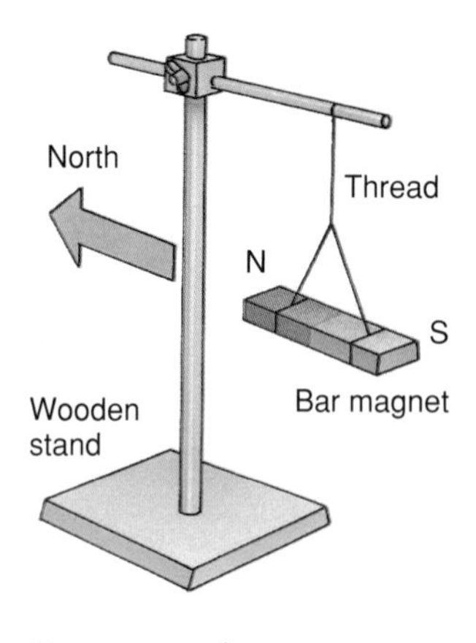

Figure 22.1A ▲ Direction finding

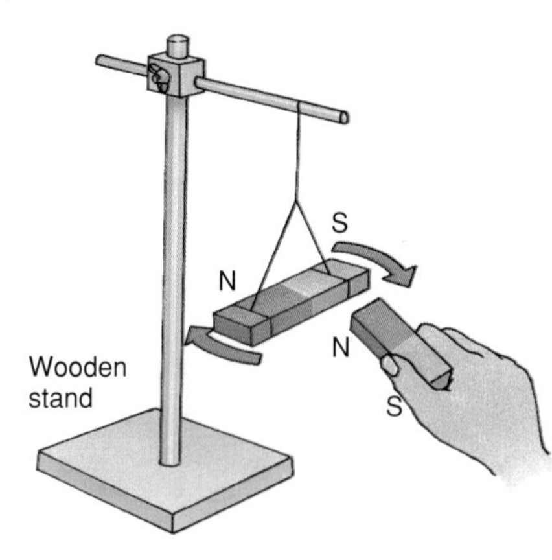

Figure 22.1B ▲ Like poles repel; unlike poles attract

When another magnet is held near one end of the suspended magnet, as in Figure 22.1B, the ends either attract or repel each other. The magnets attract each other when a north pole is held near a south pole, and repel each other if similar poles are held near each other.

Like poles repel; unlike poles attract

When a nail is held near either end of the suspended bar magnet the magnet attracts the nail. Any iron or steel object is attracted by a magnet. See how many paper clips you can hang in a chain from the pole of a bar magnet. *How would you use this to compare the strength of two magnets?*

Figure 22.1C shows how to magnetise an iron nail using a bar magnet. *How can you find the polarity of each end of the nail?*

Cobalt and nickel can also be magnetised. Oxides of these metals in powder form are used to make ceramic magnets and to coat magnetic tapes.

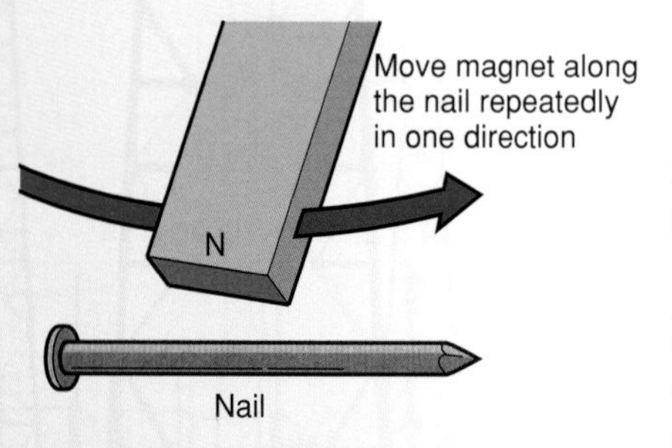

Figure 22.1C ▲ Making a magnet

Magnetic fields

Iron is much easier to magnetise or demagnetise than steel. Iron is therefore used in electromagnets (see p.333) and steel is used for permanent magnets.

Put a sheet of paper on a bar magnet and sprinkle some iron filings on the paper. The filings form a pattern which is shown in Figure 22.1D. The space round a magnet is called a **magnetic field**. Any other magnet in this space experiences a force due to the first magnet. In Figure 22.1D, the filings form lines that end on or near the poles of the magnet. These are called **lines of force**. By convention, the direction of these lines is always from the north pole to the south pole of the magnet.

Some further magnetic field patterns are shown in Figure 22.1E. A horseshoe magnet has a very strong field between its poles.

Lines of force point from the north to the south pole of a magnet.

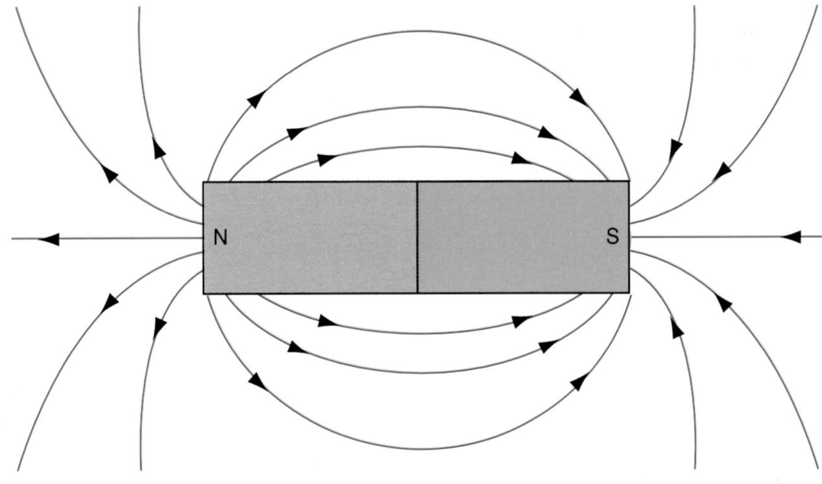

Figure 22.1D The magnetic field near a bar magnet

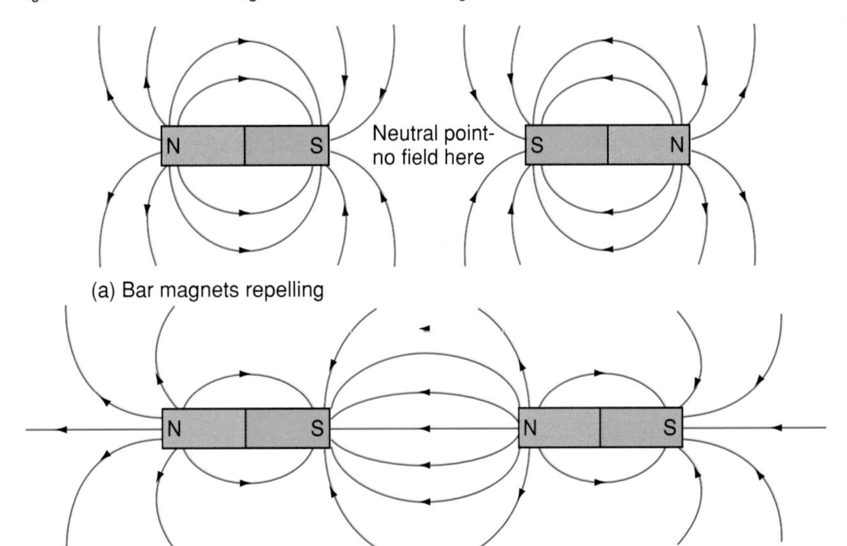

Lines of force may be plotted using a plotting compass because a plotting compass always points along a line of force.

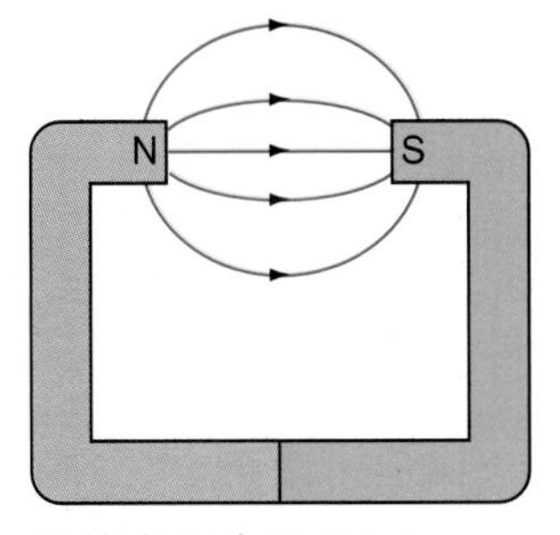

Figure 22.1E Magnetic field patterns

It's a fact!

Homing pigeons have amazing navigational skills. Scientists have found that a pigeon carrying a small magnet is not as good at flying home. It is thought that part of the brain of a pigeon is magnetic. Pigeons are affected by the Earth's magnetism. This helps them to get home.

The Earth's magnetism

Have you ever used a magnetic compass on your travels? A compass always points north. It is used as a direction finder. Compasses have been used for thousands of years by travellers.

Compasses do not actually point to the Earth's North Pole (True North). They point to Magnetic North. This is where the Earth's magnetism is concentrated in the Northern hemisphere. Map readers using a compass need to know the **angle of variation** for their locality. This is the angle between True North and Magnetic North.

Scientists have plotted the Earth's magnetic field accurately. The pattern is like that of a bar magnet. However, since it is thought the Earth is partly molten inside, the idea of a giant bar magnet cannot explain the Earth's magnetism.

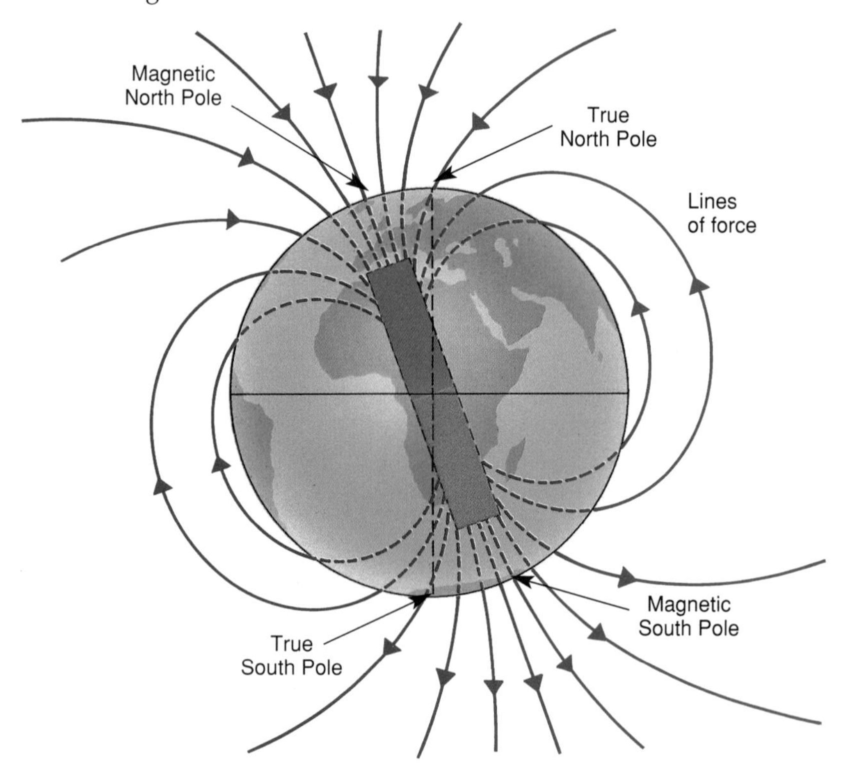

Figure 22.1F The Earth's magnetic field

SUMMARY

Magnets are made from certain metals and alloys like iron or steel. When two magnets are held near each other, like poles repel and unlike poles attract. A compass is a magnet on a pivot. It is used to find the direction of magnetic north.

CHECKPOINT

1. (a) How would you find out which pole of a bar magnet is its north pole?
   (b) How would you find out if a steel ruler is magnetised?
   (c) Given two bar magnets, how would you find out which is strongest?
2. (a) How would you magnetise a steel needle?
   (b) How would you test that it is magnetised?
   (c) Is it possible to make a magnet with a single pole? Cut a magnetised steel needle in half. You will find each half is a bar magnet. What will happen if you cut one of the halves in two? Does this produce a magnet with a single pole?
3. How could you compare the strengths of two bar magnets using a plotting compass and a metre rule? With the aid of one or more diagrams, outline your method to find out which magnet is the strongest.

22.2 Electromagnetism

FIRST THOUGHTS

In this section you will find out how electricity is used to create magnetism. This is called electromagnetism.

When insulated wire is wrapped round an iron nail and the ends of the wire are connected to a battery the nail becomes capable of picking up iron filings and paper clips. This is a simple **electromagnet**. The nail has been magnetised by the current in the wire. Disconnect the wire from the battery and the paper clips fall off. The nail loses most of its magnetism when the current is switched off.

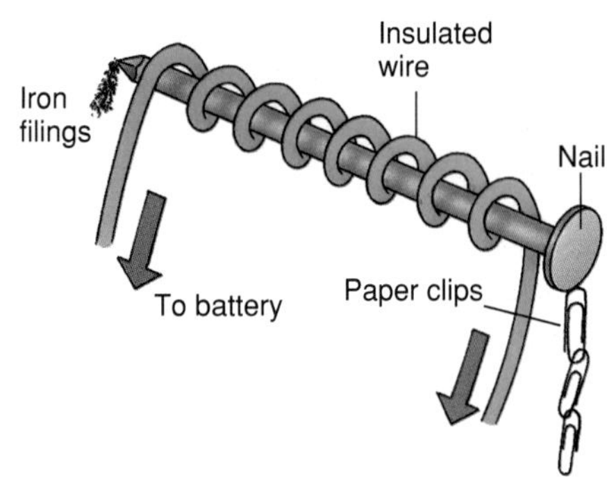

Figure 22.2A ▲ Investigating electromagnetism

The passage of an electric current along a wire creates a magnetic field around the wire. Figure 22.2B shows how the pattern of the magnetic field around a long, straight wire can be seen by using iron filings and a plotting compass. The lines of force due to a straight current-carrying wire are circles, centred on the wire. The field is strongest near the wire. The direction of the field is reversed if the direction of the current is reversed.

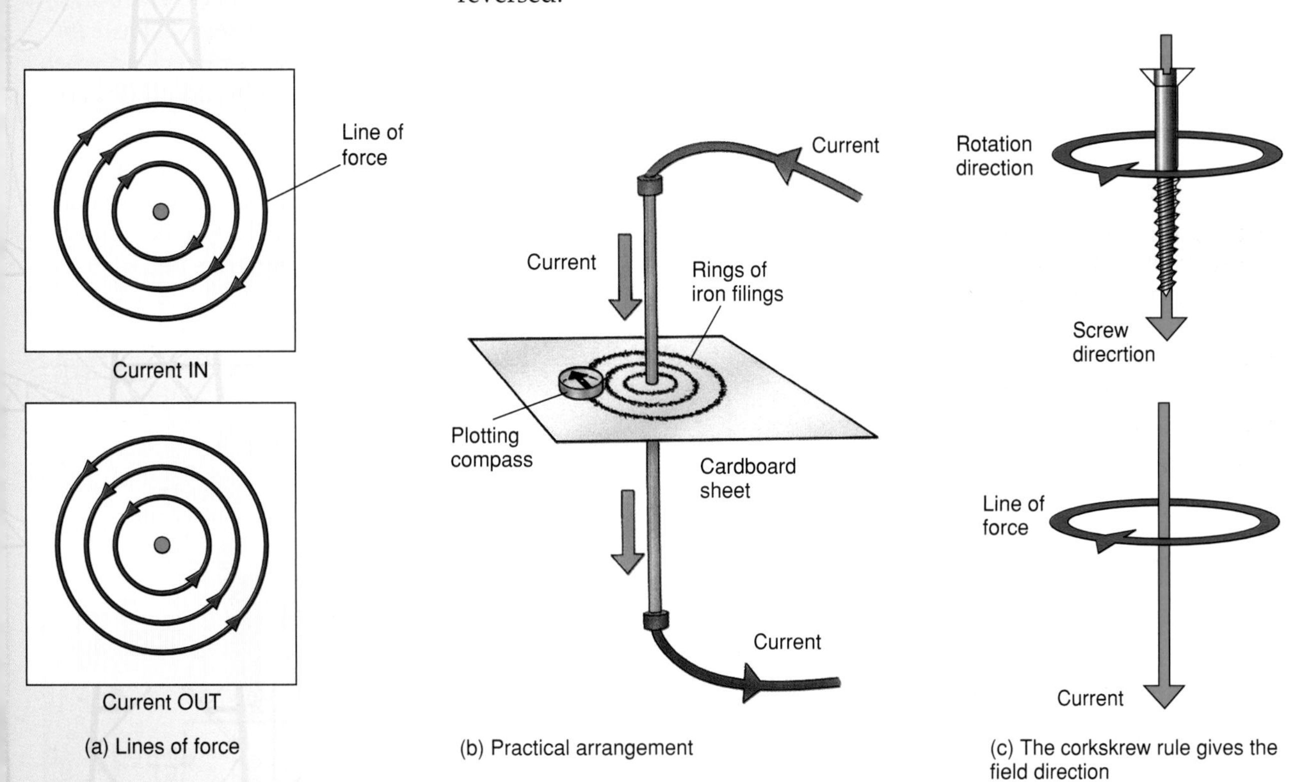

Figure 22.2B ▲ The magnetic field near a long straight wire

A **solenoid** is a long coil of wire. The magnetic field pattern created by a current-carrying solenoid is like that of a bar magnet. However, unlike a bar magnet the field lines pass through the solenoid along its axis (Figure 22.2C). *How can the field created by the solenoid be made stronger?* Using more windings or a bigger current are two ways. Inserting an iron bar into the centre of the solenoid is another way.

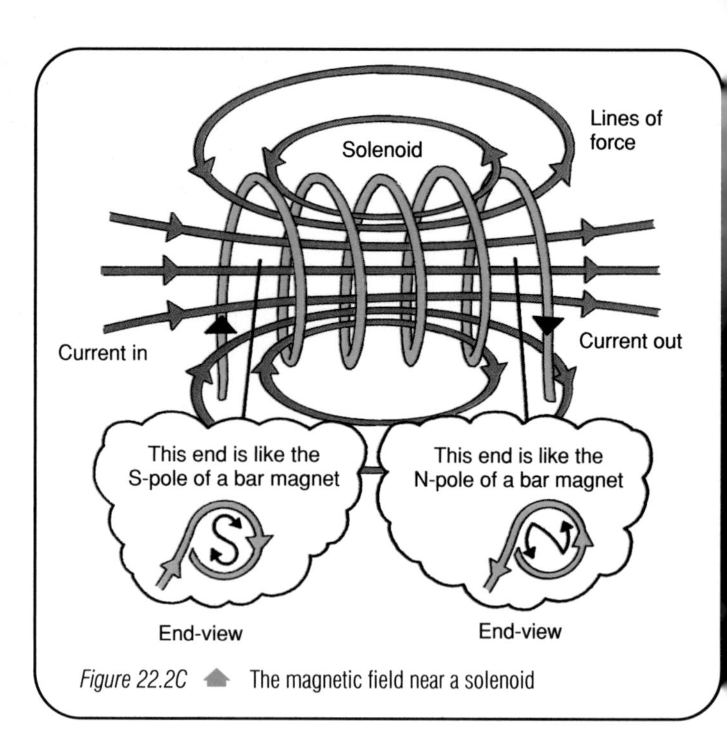

Figure 22.2C The magnetic field near a solenoid

Figure 22.2D Using an electromagnet

Materials that can be magnetised are described as 'Ferromagnetic'. Placing a Ferromagnetic material in a direct current solenoid will magnetise it. Demagnetisation can be achieved by withdrawing the material gradually from an alternating current solenoid.

Electromagnets are used in scrap yards to lift car bodies. The iron core of the electromagnet must lose its magnetism to release its load when the current is switched off (Figure 22.2D).

Bells and buzzers contain an electromagnet that operates a **make-and-break** switch. A make-and- break switch that operates a bell is shown in Figure 22.2E.

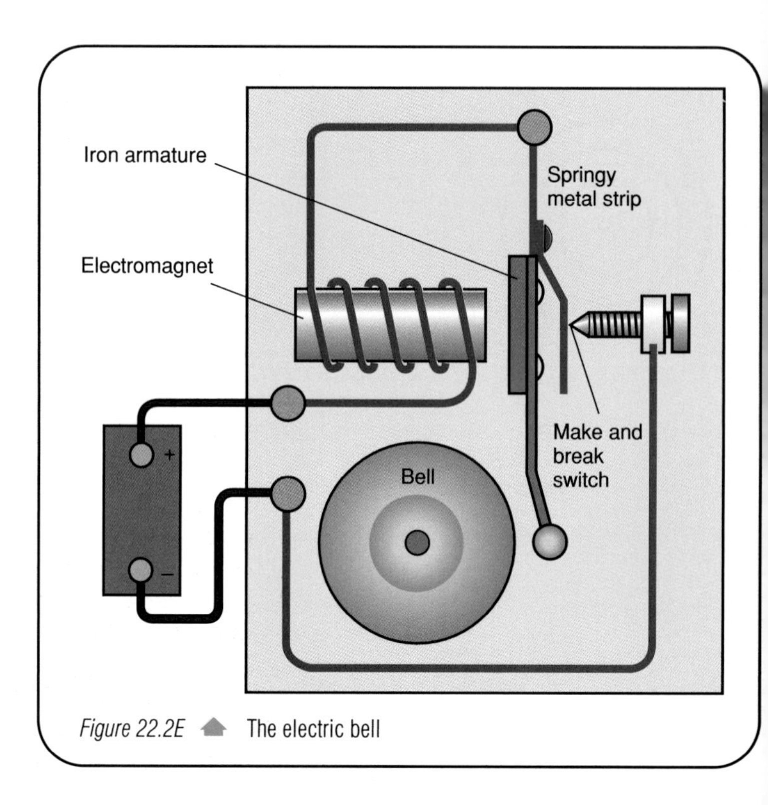

Figure 22.2E The electric bell

When the bell is connected to a battery, the iron armature is pulled on to the electromagnet. This opens the make-and-break switch and the electromagnet switches off. When the electromagnet switches off, the armature springs back and the make-and-break switch closes again so the whole cycle repeats itself.

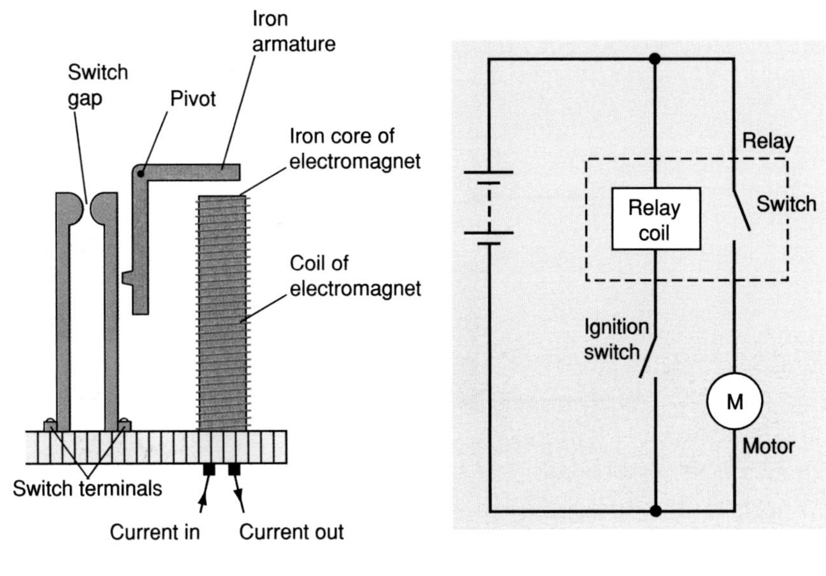

Figure 22.2F ▲ A relay (a) Construction (b) Circuit

- A relay is used to switch current on and off in a different circuit.
- A circuit breaker is used to switch excessive current off in the same circuit.

Relays are used in control circuits to switch machines on or off. Figure 22.2F shows the construction of a relay. When current passes through the electromagnet the iron armature is pulled on to the electromagnet. The armature turns about the pivot and closes the switch gap. In this way, a small current can be used to switch on a much greater current. For example, when the ignition switch of a car is operated, a current of a few amperes passes through a relay coil and the relay switch closes. This lets a much greater current pass through the starter motor.

Circuit breakers are used to prevent excessive currents. A standard circuit breaker can be reset after it has been 'tripped' by excessive current, once the cause of the excessive current has been removed. In comparison, a fuse needs to be replaced once it has been blown by an excessive current.

See www.keyscience.co.uk for more about electromagnetism.

1 **The standard circuit breaker** shown in Figure 22.2G (a) consists of an electromagnetically-operated switch in series with the electromagnet. The switch is held closed by a spring. If the current exceeds a certain value, the electromagnet pulls on the switch sufficiently to make the switch spring open. The circuit is now broken so there is no current. The switch is designed to stay open even though there is no longer any force from the electromagnet itself. It must be reset manually. A thermal circuit breaker has a bimetallic switch that springs open as a result of the heating effect of an excessive current.

2 **The residual current circuit breaker (RCCB)** is designed to protect the user of a mains appliance. An RCCB will trip if the current in the live wire differs from the current in the neutral wire. This could happen if the live wire on a mains cable becomes exposed and someone touches the cable at that point. An RCCB contains an iron core with two oppositely wound coils. One coil is in series with the live wire and the other is in series with the neutral wire. In normal operation, the magnetic effects of the two coils cancel out so the electromagnet is unaffected. However, if the live current differs from

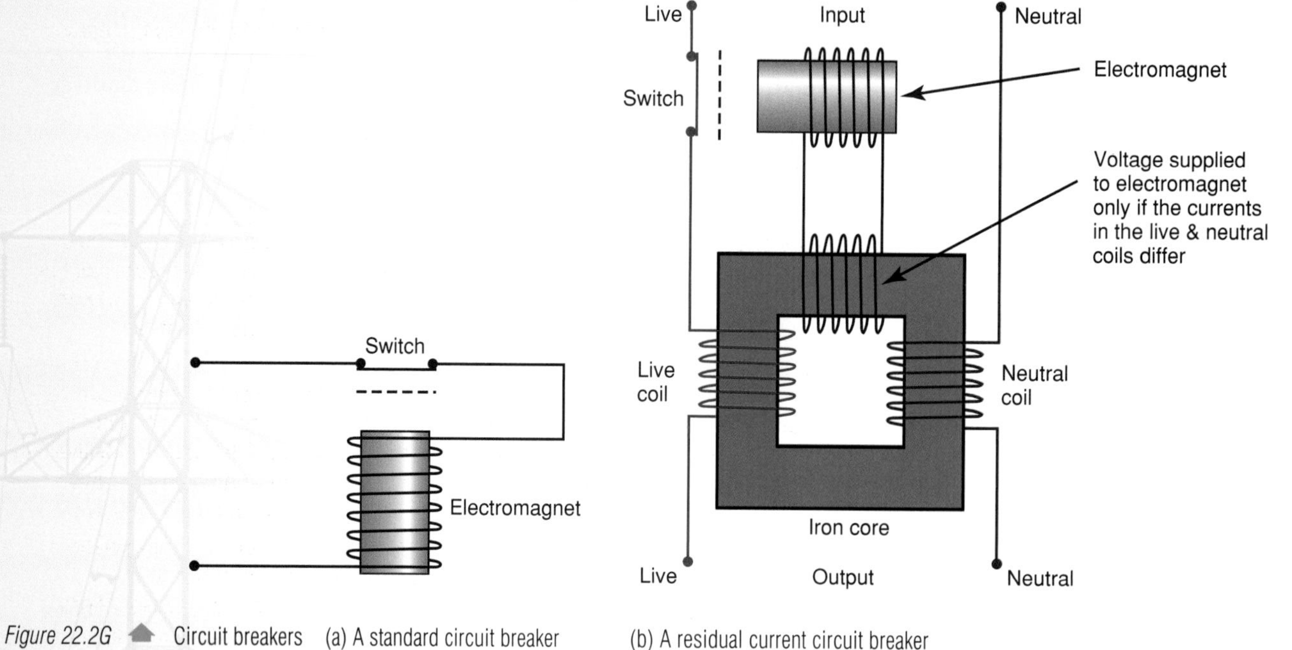

Figure 22.2G Circuit breakers (a) A standard circuit breaker (b) A residual current circuit breaker

the neutral current, the magnetic effects of the two coils do not cancel out. The result is that the electromagnet is magnetised and therefore opens the switch in series with the live wire.

A current of more than 30 mA through the body could be lethal yet it is not enough to blow a fuse. An RCCB is designed to trip if the live and the neutral current differ by more than 30 mA.

Magnetic tape and discs are used to record computer programs and sound. When you record sound, a microphone in the recorder converts the sound into an electrical signal. This signal is supplied to an electromagnet in the recording head. The tape passes against the recording head and is magnetised by the electromagnet according to the signal. To play the signal back, the tape is passed against the replay head. This produces an electrical signal that 'drives' a loudspeaker (see p. 372).

SUMMARY

A current carrying wire creates a magnetic field around itself. An electromagnet is a coil of wire wound round an iron core. Electromagnets are used in bells, buzzers, relays, circuit breakers and tape recorders.

KEY SCIENTIST

Current matters

The direction of current on a circuit diagram is always shown from 'positive to negative'. This rule was put forward by André Ampère in the early nineteenth century. Ampère showed how to measure the direction and size of an electric current.

How did he decide on 'positive to negative'? Electricity is a difficult topic and you can perhaps appreciate why this is so when we look at how famous scientists like Ampère struggled to understand electricity.

The attractive power of amber when rubbed was known to the philosophers of Ancient Greece. William Gilbert was a sixteenth century physician who discovered that other materials such as glass, ebonite and resin are also capable of attracting bits of straw and paper when they are rubbed. Gilbert introduced the word 'electricity' from the Greek word for 'amber' to describe the attractive power of these materials. The idea that rubbing any of these materials 'charges' it (i.e. fills it) with electricity dates from the mid-seventeenth century. The term 'electric charge' was introduced at this time.

What happens when charged materials are brought near to each other?

Further experiments were carried out by Charles Dufay in France in the early eighteenth century. He found that:

- glass repels glass,
- ebonite repels ebonite,
- ebonite attracts glass,
- resin repels resin,
- resin repels ebonite.

What do you think happens with resin and glass?

Dufay showed that all electrified materials could be placed in one of two lists which he called 'resinous' and 'vitreous'; any material in either list repels all the materials in the same list and attracts all the materials in the other list. In other words, Dufay showed that like charges repel and unlike charges attract.

Benjamin Franklin, an eighteenth century American scientist, investigated these effects in more detail. He deduced that the two types of electric charge can be described as positive and negative because they cancelled each other out in equal amounts. The charge on glass rubbed with silk was defined as positive. He thought of electricity as a positive fluid and ordinary matter as being negative. Too much electricity in an object made it positive, too little made it negative.

Machines for creating electric charge and producing artificial lightning were invented during the eighteenth century. In 1798, a different principle for producing electricity was discovered by an Italian scientist called Alessandro Volta. He invented the first battery and used it to make electricity pass through a conductor. Figure 22.2G (b) shows the construction of Volta's battery.

Volta used the electricity from the silver end to charge an electroscope. He discovered that the electroscope became charged when it was touched with a wire joined to the silver end of the battery with the other end of the battery earthed. When a negatively charged rod was brought near the electroscope, the electroscope leaf rose showing that the charge on the electroscope was also negative. Volta deduced that the silver end of the battery was therefore negative and the other end positive. In honour of Volta's contributions to science, the unit of potential difference, the volt, is named after him.

The unit of electric current is named after André Ampère.

The invention of the battery led to further investigations into the properties of current electricity. Hans Oersted, a Danish scientist, discovered the magnetic effect of an electric current in 1820. This led to the invention of the electromagnet by William Sturgeon in England in 1826 and the invention of further electromagnetic devices such as the relay and the electric bell.

André Ampère was a French scientist who discovered that an electric current has direction and size. Before Ampère, some scientists thought that positive electricity went round an electric circuit from the positive pole of the battery and negative electricity went the other way round from the negative pole at the same time. Ampère observed that a magnetic compass near a wire was deflected when the current was switched on.

Leather disc soaked in sodium chloride solution

1 Wire touched briefly on the electroscope cap

Silver disc

2 Negative rod brought near the cap makes the leaf rise further

Wire to earth

Zinc disc

Figure 22.2H (a) Alessandro Volta (b) Volta's battery

More importantly, he noticed that the deflection was in the opposite direction when the battery was reversed in the circuit. This cannot be explained if there is a flow of positive electricity and a flow of negative electricity at the same time. Ampère realised that one type of electricity only flows round a circuit.

Which type of electricity is responsible for current electricity through wires?

Ampère had no evidence to decide this so he suggested that an electric current is a flow of positive charge. This is why the direction of electric current in a circuit is always considered to be from positive to negative. He went on to show how the magnetic effect of an electric current can be used to measure the size of the current. The importance of his work was recognised by naming the unit of electric current after him.

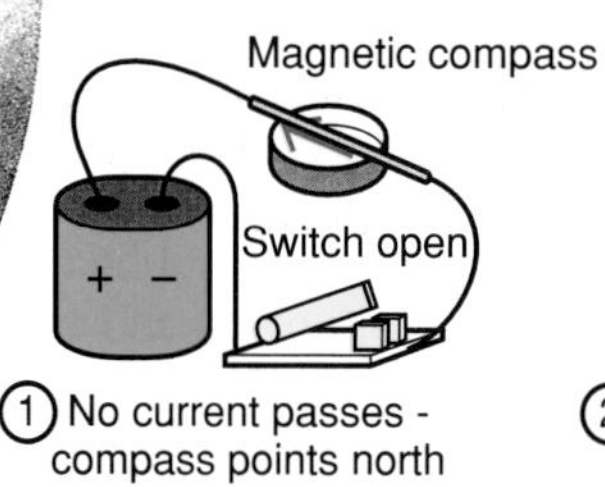
Magnetic compass
Switch open
+ −
① No current passes - compass points north

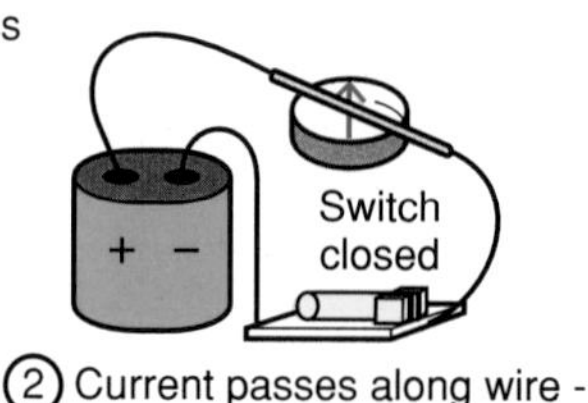
Switch closed
+ −
② Current passes along wire - compass points north-east

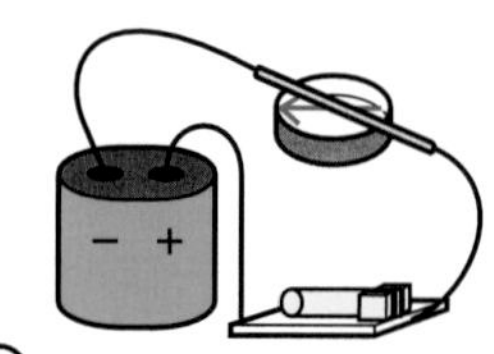
− +
③ Current passes along wire in opposite direction - compass points north-west

Figure 22.21 (a) André Ampère (b) Current directions

CHECKPOINT

1 Sketch the pattern of magnetic lines of force for each of the following:
  (a) a vertical wire carrying current upwards,
  (b) an air-filled solenoid,
  (c) a solenoid with an iron core.

2 (a) Current is passed through an air-filled solenoid. An unmagnetised iron bar is then inserted into the solenoid. How does this affect a plotting compass near the end of the solenoid?
  (b) A bar magnet is held near the end of the solenoid to repel the solenoid. What happens:
  (i) if the magnet is turned round,
  (ii) if the current is switched off,
  (iii) if the current is switched off then reversed in direction?

3 Results from an experiment to find out how the weight supported by an electromagnet varies with current are given below.

Current (A)	0	0.5	1.0	1.5	2.0	2.5	3.0
Weight (N)	0	0.5	1.5	2.4	3.3	3.8	4.0

  (a) Use these results to plot a graph of weight (on the vertical axis) against current.
  (b) Do you think there is a limit to weight that can be supported?

4 Karen and Paul have made a 'shaky hand' tester (shown opposite) to use at the school fair. The idea is that the bell rings if contact is made between the metal ring and the copper wire. Unfortunately the bell is not working.
  (a) Karen tests the electromagnet coil of the bell with a battery and light bulb. What did she do to make this test?
  (b) The electromagnet does work. Paul notices that the make and break contacts are dirty. He cleans them with a small file. Why does this make a difference?
  (c) Now the bell works. Explain how it works.

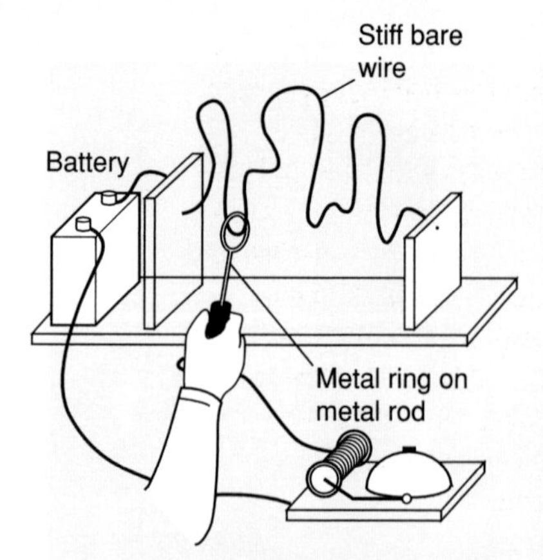
Stiff bare wire
Battery
Metal ring on metal rod
The bell rings if the metal ring touches the bare wire

22.3 Motors and meters

FIRST THOUGHTS

A current-carrying wire creates magnetism. How is this magnetism affected by permanent magnets? Read on to find out.

How many appliances can you think of that contain an electric motor? At home, electric motors are used in appliances such as vacuum cleaners, washing machines, food mixers and electric shavers. *Can you think of any other appliances in your home which contain electric motors?*

The motor effect

A current-carrying wire creates a magnetic field. *What happens if a magnet is held near the wire?* Figure 22.3A(a) shows a horseshoe magnet held near a current carrying wire. The wire is pushed up by the magnet. This is called the **motor effect**.

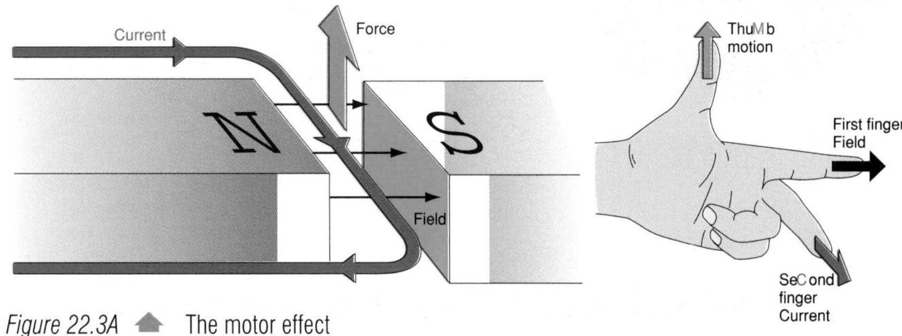

Figure 22.3A The motor effect
(a) The force on a current-carrying conductor (b) The left-hand rule

If the magnet is turned round so that its north and south poles swap positions, the wire is pushed down by the magnet. Reversing the direction of the magnetic field reverses the direction in which the wire moves. Reversing the current also has this effect. The **left-hand rule** tells you the direction of the force which moves the wire when you know the direction of the current and the direction of the magnetic field. Figure 22.3A(b) shows you how to remember the left-hand rule.

Loudspeakers, meters and motors are designed to make use of the motor effect. All these devices work because current-carrying wires are being moved by magnets.

The loudspeaker

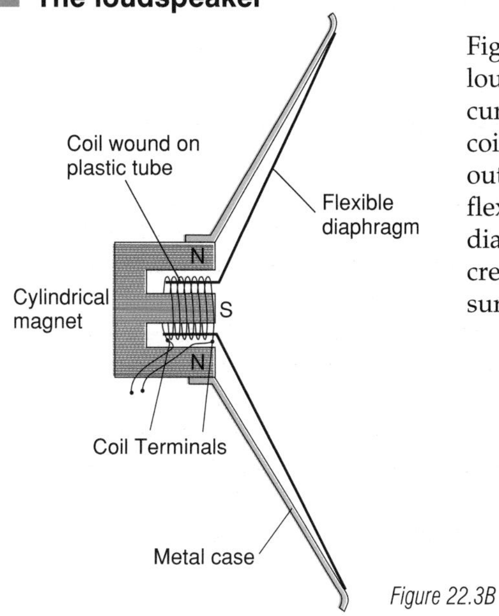

Figure 22.3B shows a loudspeaker. When varying current is passed through the coil, the magnet is forced in and out of the coil. This makes a flexible cone called the diaphragm move in and out, creating sound waves in the surrounding air.

Figure 22.3B A loudspeaker

Small speakers

The headphones of a personal stereo work on the same principle as the loudspeaker. Each headphone consists of a small moving coil attached to a flexible diaphragm. A tiny magnet makes the coil vibrate when the current in the coil changes. The coil is very light so it responds easily at high frequencies.

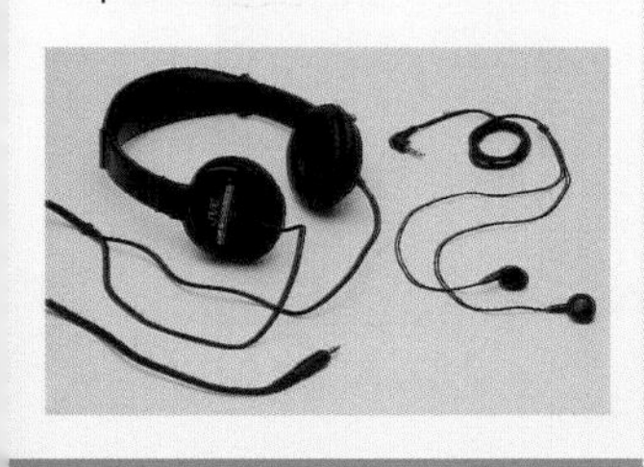

A solar challenge

Lightweight electric motors are being developed by scientists in several countries. Any car has several electric motors, including its starter motor, its windscreen wiper motors and its heater fan motor. Using lighter motors will reduce the overall weight of the car and save fuel. Lighter magnets are being developed to make electric motors lighter.

Sunraycer, the eventual winner of the first international solar-powered car race was equipped with a lightweight electric motor.

The moving coil meter

A model moving coil meter is shown in Figure 22.3C. When current is passed through the coil, the coil turns. Each side carries a current across the lines of the magnetic field so each side experiences a force.

Because the current is in opposite directions on each side, the force on one side is in the opposite direction to the force on the other side. If the current is increased, the coil turns more. This is how the moving coil meter works.

When the current is switched off, the coil returns to its starting position. When the current is reversed, the coil turns in the opposite direction.

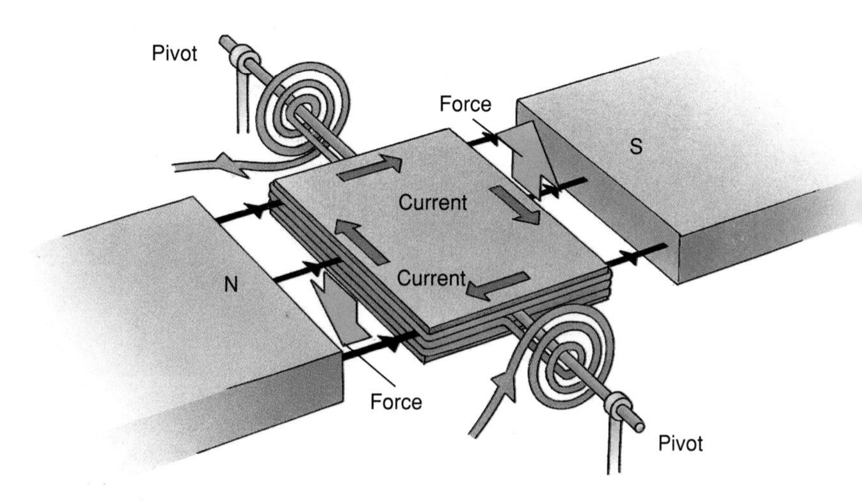

Figure 22.3C ▲ Making a meter

Figure 22.3D shows the construction of an accurate moving coil meter. Most ammeters and voltmeters with a pointer are constructed like this. Current enters the coil via the upper hairspring and leaves via the lower hairspring. The magnetic force on each side of the coil makes it turn, winding up the hairsprings. When the current is switched off, the hairsprings unwind and the coil returns to its 'zero' position. The iron drum produces a radial magnetic field which makes the coil deflection proportional to the current. This gives a linear scale.

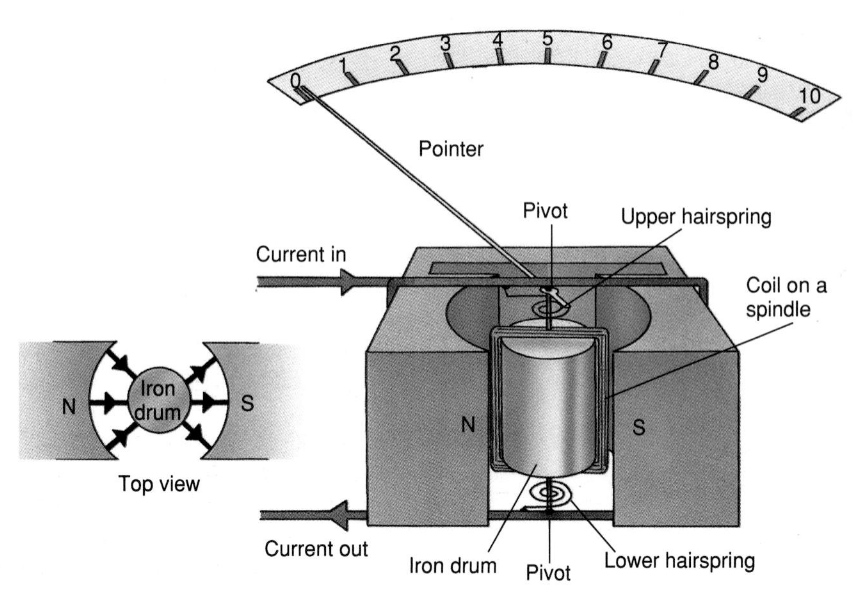

Figure 22.3D ▲ The moving coil meter

The moving coil motor

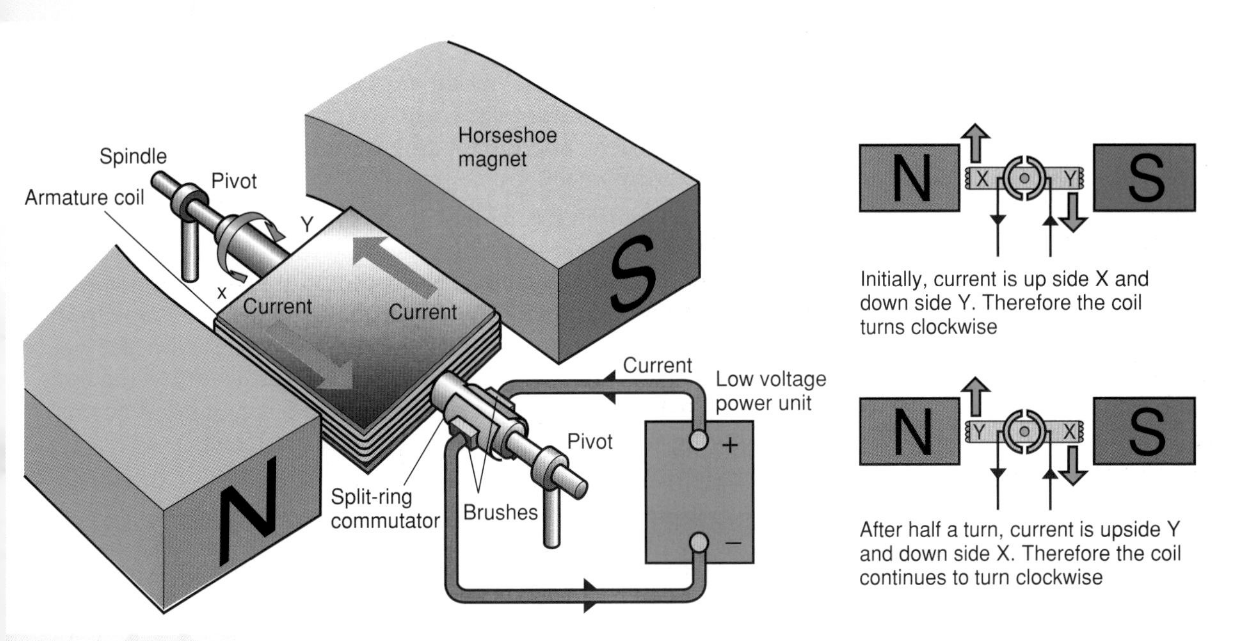

Figure 22.3E ▲ The simple electric motor

A simple electric motor is shown in Figure 22.3E. When current is passed through the **armature coil**, side X of the coil is forced up and side Y down. This is because each side carries current in opposite directions in the magnetic field.

The **commutator** ensures that the coil turns continuously in the same direction. After each half-turn, the commutator reverses the direction of current round the coil. Since the sides of the coil move through 180° each half-turn, the overall effect is to give the coil a push in the same direction every half-turn.

Figure 22.3F ▲ A practical motor

The motor can be made to turn in the opposite direction by reversing the battery connections or by turning the magnet round. The motor will not turn at all if it is connected to an alternating current power supply. This is because the coil will be forced one way then the other way each time the alternating current reverses its direction.

The speed of the motor can be increased by increasing the current or by using a stronger magnet or by using a coil with more turns. If the load on an electric motor is increased, the motor current increases to enable the motor to move the extra load. For example, when an electric drill is used to bore a hole in a wall, friction between the tip of the drill and the wall increases. This causes an extra load on the motor of the drill. The motor current then increases so the drill overcomes the extra friction on it due to the wall. However, if the motor current increases too much, the armature coil may burn out through overheating.

Practical motors

1. A mains electric motor contains an **electromagnet** in place of a permanent magnet. The electromagnet is usually in parallel with the armature coil. The motor turns continuously in one direction when it is connected to an alternating current power supply. This is because the magnetic field of the electromagnet reverses at the same time as the armature current reverses each time the power supply changes its

polarity. As the field and the armature current both reverse their directions, the direction of the turning effect on the armature is unchanged.

2 The **brushes** in a practical motor are made of graphite. They are held against the commutator by springs. Graphite is used because it conducts electricity and the rotating commutator slides over it easily without losing contact.

3 Commercial motors are made with several coils on the same armature. The coils are equally spaced on each armature. Each coil is connected to its own section of the commutator. This design ensures that the motor runs smoothly. The armature is usually made from iron which makes the magnetic field much stronger. Also, the armature is usually cylindrical in shape. This ensures that the force on each coil side is always at right angles to the cylinder diameter, thus giving the maximum turning effect on each coil.

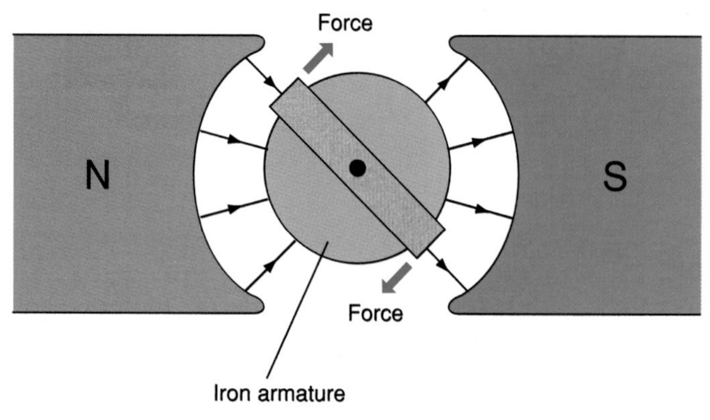

Figure 22.3G The armature

The definition of the ampere

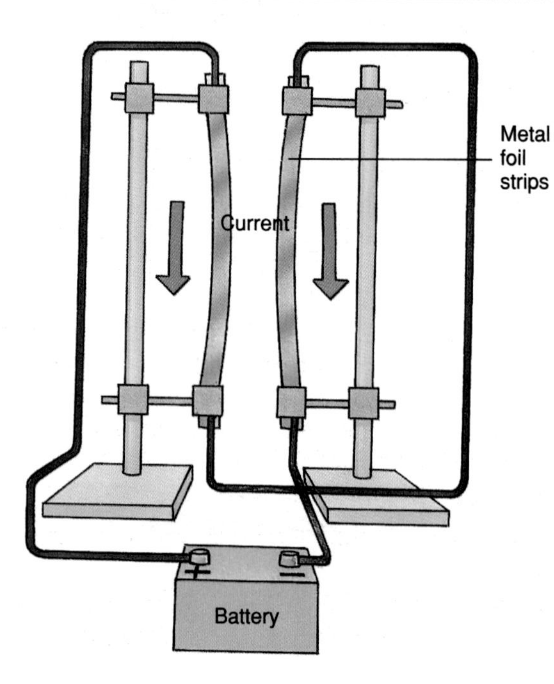

Figure 22.3H The force between two current-carrying wires

Do current-carrying wires exert forces on each other? Figure 22.3H shows how to test this using strips of metal foil as lightweight wires. The strips attract each other when the current is the same direction in both. They repel when the current is in opposite directions.

The ampere, the unit of current, is defined from this effect. One ampere is that current in two infinitely-long parallel wires 1.0 m apart that causes a force of 2×10^{-7} N on each metre of wire.

SUMMARY

Any wire carrying current across the lines of force of a magnetic field experiences a force due to the field. This is called the motor effect. Loudspeakers, meters and motors all make use of the motor effect.

CHECKPOINT

1 (a) A loudspeaker converts energy from one form to another. What form does it use and what form is this turned into?

(b) Outline the energy changes that take place when an electric motor is connected to a battery.

(c) A battery-operated toy car is driven by a simple electric motor. What would be the effect on the motion of the car if (i) a lighter battery of the same voltage was used, (ii) a heavier battery of the same voltage was used, (iii) a motor with a stronger magnet was used?

2 (a) How many electric motors are there in your home? Make a list of all the appliances in your home that contain electric motors.

(b) The power supplied to an electric motor is equal to the product of its current and its voltage. An electric drill operating at 230 V is used to drill into a wall. It takes several minutes to do this and the motor becomes quite warm. Why does it become warm when used in this way?

3 An electric mixer has a three-speed control switch and an on/off switch. This is achieved using two identical resistors, as shown in the figure below.

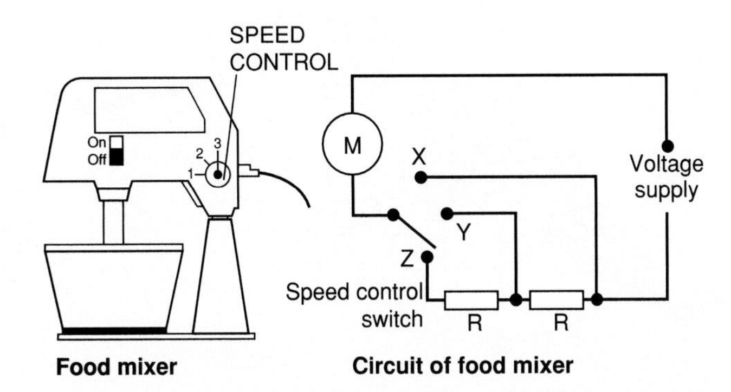

(a) The speed control switch can be set at X or at Y or at Z. Which position gives the lowest speed 'slow', which position gives 'medium' speed and which position gives 'fast'?

(b) The figure does not show the on/off switch. Sketch the circuit and show where the switch should be.

4 An electric motor contains a permanent magnet. The motor is to be connected to a battery. Describe and explain how the motion of the motor would be affected if:

(a) the magnet is reversed,

(b) a stronger magnet is used,

(c) a resistor is connected in series with the motor and the battery.

5 (a (i) A small electric motor in a battery-operated toy car has a permanent magnet. How would the motion of the car be affected if the battery was reversed in its holder?

(ii) How would the toy car be affected if the magnet gradually become weaker and weaker? Explain your answer.

(b) (i) A mains electric motor in an electric drill has an electromagnet instead of a permanent magnet. What is the advantage of using an electromagnet instead of a permanent magnet?

(ii) Would it be better for the armature of an electric motor to be made from iron instead of from a plastic material? Explain your answer.

6 (a) Make a labelled sketch of a moving coil meter and explain how it works.

(b) In a moving coil meter, describe the function of (i) the iron drum, (ii) the hairsprings, (iii) the zero adjuster.

(c) A moving coil meter is made so that its pointer deflects exactly to the end of the scale when the current is 1.0 A. However, its magnet gradually becomes weaker. How does this affect its accuracy?

22.4 The television tube

FIRST THOUGHTS

Thousands of technicians and engineers work in television. Perhaps your future might lie in making television programmes.

- **The demonstration electron tube** in Figure 22.4A shows a beam of electrons passing over a special screen inside the tube. The electrons cause a visible trace along the screen. When a magnet is brought near the tube, as shown, the electrons are pushed downwards. Placing a current-carrying wire in a carrying coil near the tube would have the same effect. Electrons moving across the lines of force of a magnetic field are pushed by the field. The same happens to electrons moving along a current-carrying wire in a magnetic field. The force on the wire is because the moving electrons are pushed by the field.

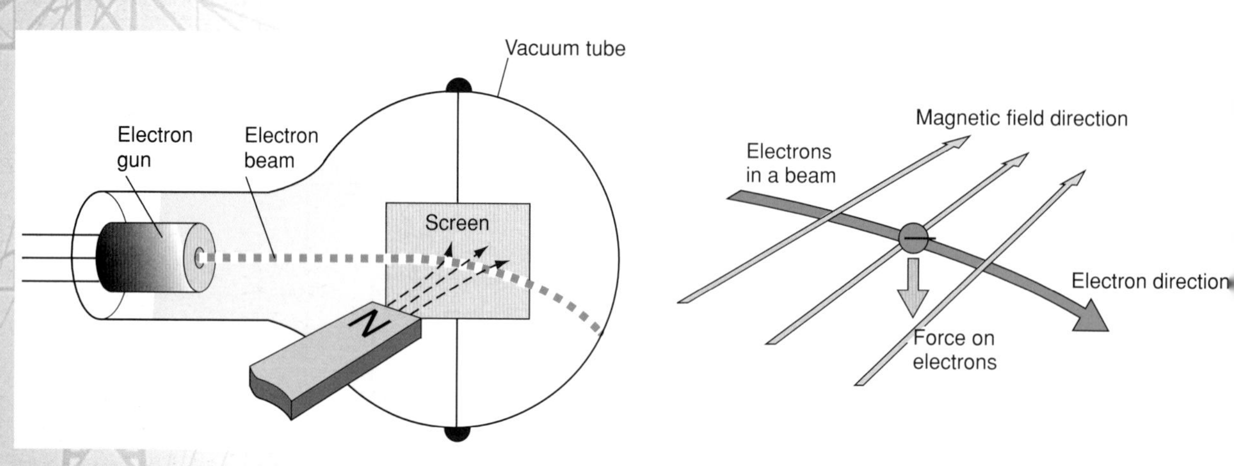

Figure 22.4A Electron beams in a magnetic field (a) The electron tube (b) Magnetic force on moving electrons

A beam of electrons in a uniform magnetic field follows a circular path. This is because the magnetic force is at right angles to the direction of the beam. Satellites move round the Earth in circular orbits because the force of gravity is always at right angles to the direction of motion.

You can use the left-hand rule as in Figure 22.3A to work out the direction of the force on a beam of electrons in a magnetic field — but remember that the convential direction of current is in the opposite direction to the electron direction.

- **When you watch television**, you are watching pictures built up on the television screen by beams of electrons. The screen is the flat end of a special electron tube called the **television tube**. The screen is formed by coating the inside of the tube at the flat end with a special chemical. A spot of light is emitted from the screen at the point where an electron beam strikes it.

Each beam is produced by an **electron gun** at the narrow end of the tube. A black and white television tube contains a single electron gun. A colour tube contains three electron guns.

The electron beam is deflected by magnetic coils at the narrow end of the tube so that the spot traces lines across the screen. In a black and white television tube, the television signal is used to vary the brightness of the spot as it moves across the screen. Figure 22.4B shows how a picture is built up on the screen in this way.

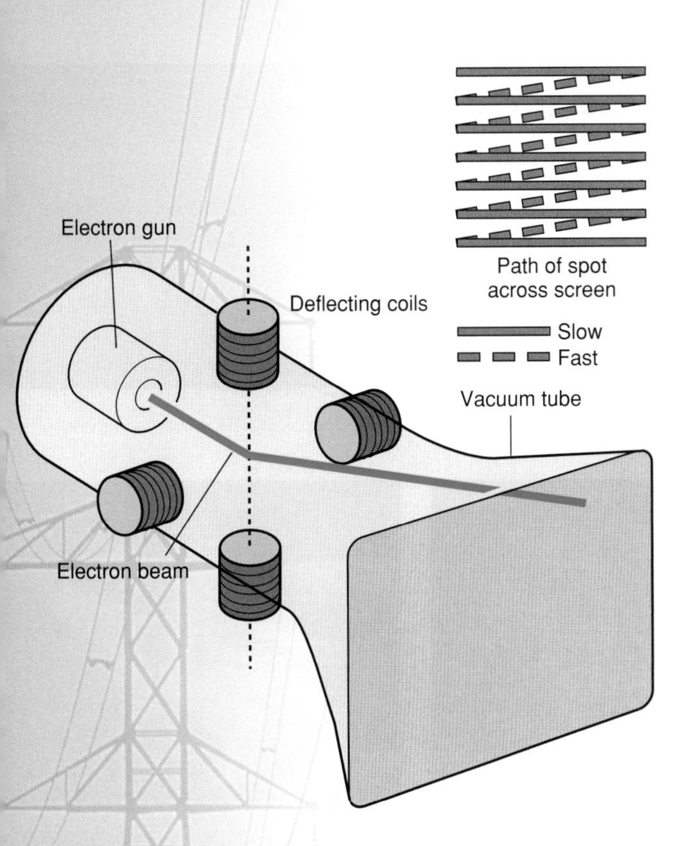

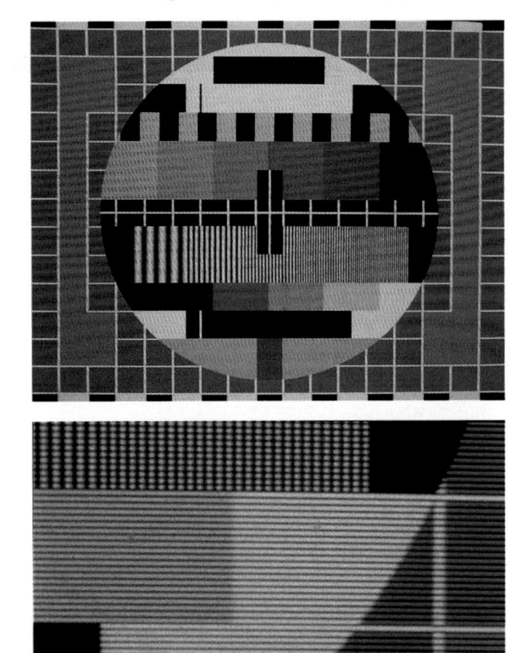

Figure 22.4B ▲ The television tube

Colour television tubes use three electron guns. The screen is designed so that one gun makes it emit red light, one gun makes it emit green light and the remaining gun makes it emit blue light. Red, blue and green are the primary colours of light. Different mixtures of these colours give any other colour. This is how the picture on a colour television is formed.

Figure 22.4C shows how a colour television works. The screen of a colour television is a matrix of dots of special chemicals. There are three types of dots. Each type of dot produces a primary colour of light when electrons in a beam hit it. A shadow mask is used to ensure that each type of dot can only be hit by electrons from one particular gun. In this way, one gun produces a red picture, one gun produces a green picture and one gun gives a blue picture. The pictures overlap so the viewer sees the correct colours on the screen (Figure 22.4D).

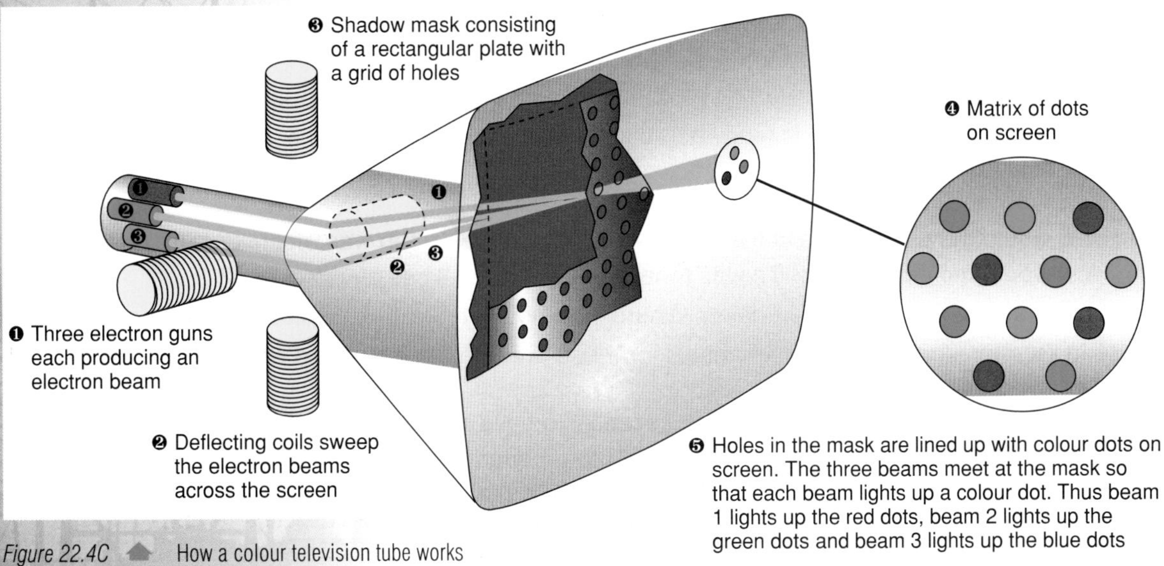

Figure 22.4C ▲ How a colour television tube works

It's a fact!

On screen

Closed circuit television (CCTV) is now in common use for security purposes to observe, monitor and record people. The number of recorded crimes has fallen in towns and cities where CCTV cameras are in use. Few objections have been raised to the limited use of CCTV in public places in Britain because most people recognise that the benefits of security outweigh the possible loss of privacy.

(a) Red

(b) blue

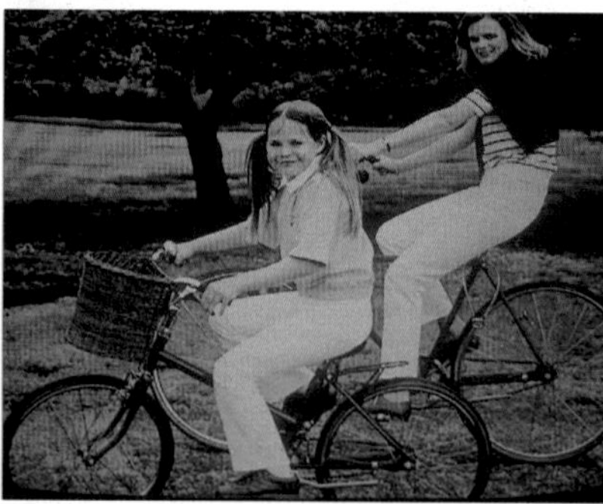

(c) Green

(d) Full colour

Figure 22.4D ▲ Colour pictures

SUMMARY

Television pictures are built up by beams of electrons. A colour television tube contains three electron guns, producing pictures in red, green and blue light. These pictures overlap to produce the actual colour picture on the screen.

CHECKPOINT

1 (a) When a television is switched off, a spot that fades away is sometimes seen on the screen. What causes the spot and why does it fade away?

(b) Why does static electricity build up on television screens?

2 (a) A colour television shows a picture of the Union Jack. How many of its three electron guns are used to produce (i) the red part, (ii) the blue part, (iii) the white part of the picture?

(b) Make a sketch of what you would see on the screen if the red gun stopped working.

3 The picture on a colour screen is renewed 25 times each second. Each time, the electron beam traces out 625 lines on the screen.

(a) How long does the beam take to trace out a single line?

(b) How would the picture differ if the number of lines on the screen was much less than 625?

(c) How would the picture differ if it was renewed at just fives times each second?

4 Most homes have at least one television. Where else would you expect to find televisions in use?

5 Satellite TV offers a wide range of channels for those prepared to invest in a receiver dish and a decoder.

(a) To receive satellite television, why is it essential to have a suitable dish?

(b) The television signal from the satellite must be decoded. This is because the broadcasting station put the signal into its own code. Why is the signal coded by the television station?

22.5 Electromagnetic induction

FIRST THOUGHTS

In this section, you will find out how the electricity you use at home is produced.

Electricity provides us with instant energy in our homes at the flick of a switch. Distant power stations generate electricity which is then transmitted to our homes through a network of cables called the **National Grid**. How did people manage before electricity reached the home? If you have ever suffered a power cut at home, you should know the problems of coping without electricity.

The principle of generating electricity was discovered by Michael Faraday in 1831. He showed that when a magnet is pushed in and out of a coil, a voltage is 'induced' in the coil. If the coil is part of a complete circuit, the induced voltage drives a current round the circuit. This effect is known as **electromagnetic induction** (Figure 22.5A).

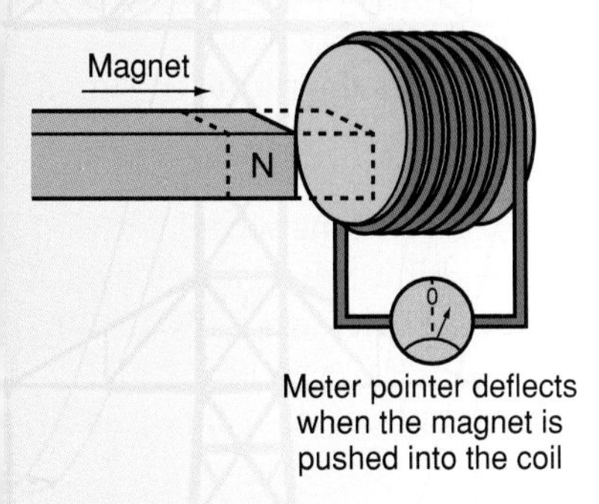

Figure 22.5A (a) Electromagnetic induction (b) Michael Faraday

Investigating electromagnetic induction

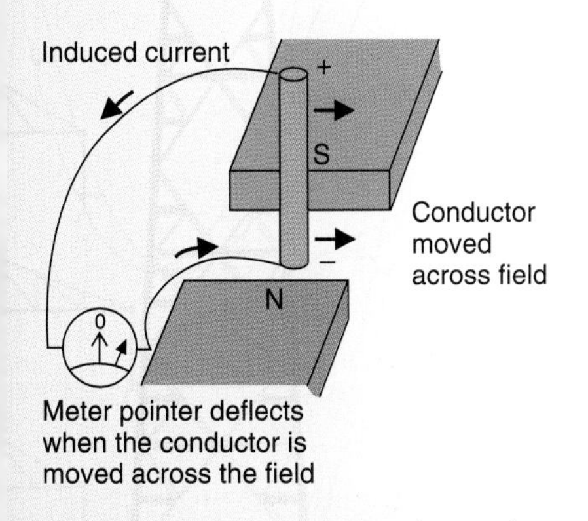

Figure 22.5B Induced voltage in a moving conductor

Figure 22.5B shows another way to induce a voltage in a wire. When the wire is moved between the poles of the horseshoe magnet, a voltage is induced across its ends. The induced voltage pushes current round the circuit through the meter.

Why should a voltage be induced when a wire moves across a magnetic field? A wire contains many free electrons because it is made of metal. When the wire is moved across the field as in Figure 22.5C, the electrons are moved with it. But the field pushes the electrons along the wire. So a voltage is created across the ends of the wire.

What happens if the wire stops moving? No voltage is induced in a wire that is stationary relative to the magnet. Move the magnet but keep the wire fixed and a voltage is induced. This happens whenever the wire moves relative to the magnet.

How can the induced voltage be made larger? You could use a stronger magnet or use more wire or move the wire faster.

What happens if the wire is moved along the lines of force of the field rather than across them? No voltage is produced now. The direction of motion of the wire must cut across the lines of force.

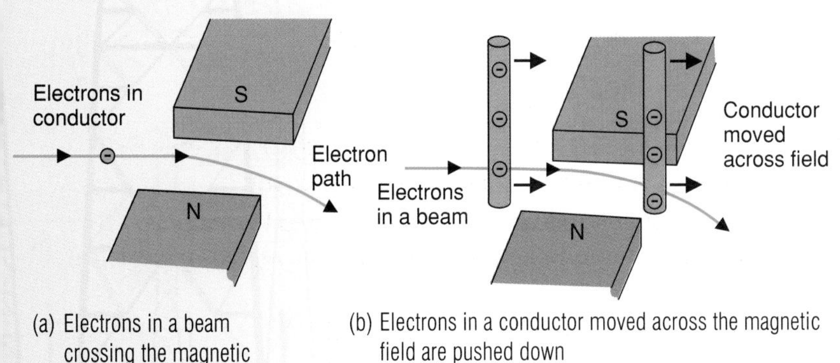

(a) Electrons in a beam crossing the magnetic field pushed down (b) Electrons in a conductor moved across the magnetic field are pushed down

Figure 22.5C Explaining induced voltage

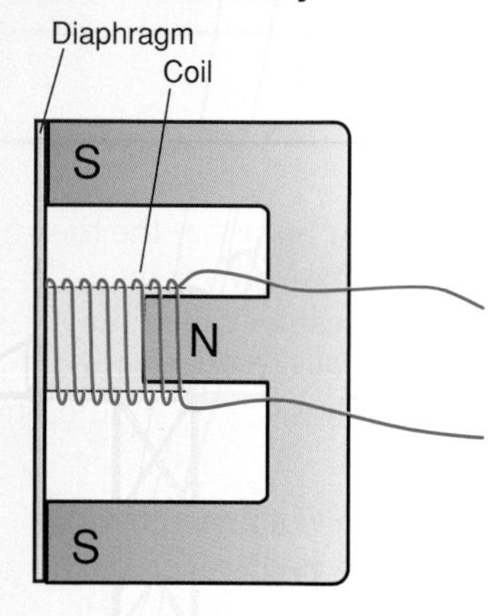

Figure 22.5D ▲ The microphone

Electromagnetic induction at work

1 **The microphone** uses the principle of electromagnetic induction to convert sound waves into electrical waves. The sound waves make a diaphragm in the microphone vibrate. A small coil attached to the diaphragm is in the field of a magnet. The varying motion of the coil relative to the magnet induces an alternating voltage in the coil. Figure 12.1A shows the voltage from a microphone being displayed on an oscilloscope.

2 **Magnetic recording** is achieved by passing a magnetic tape or disc against a recording head which consists of an electromagnet supplied with the electrical signal to be recorded. Once recorded, the signal is retrieved by passing the tape or disc against an electromagnet referred to as the replay head. Changes of magnetism on the tape as it passes the replay head induce a voltage in the replay head electromagnet. This induced voltage regenerates the recorded signal (see p. 372).

3 **The electromagnetic flow meter** consists of a turbine wheel with small magnets in its rim. Fluid flows through the wheel and makes it spin. The small magnets in the rim spin past a 'pick-up' coil outside the pipe. Each magnet that passes the coil generates an electrical voltage in the coil. The number of electrical pulses per second can be counted, giving a measure of the flow speed.

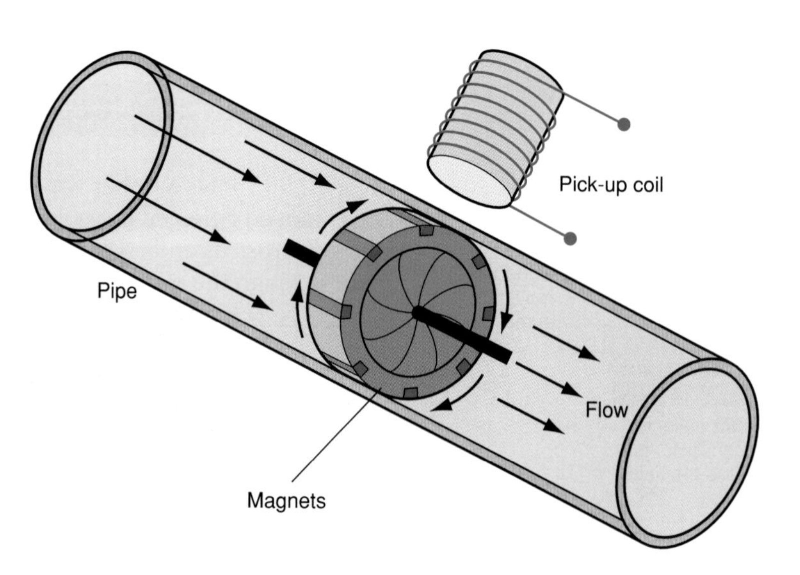

Figure 22.5E ▲ The electromagnetic flow meter

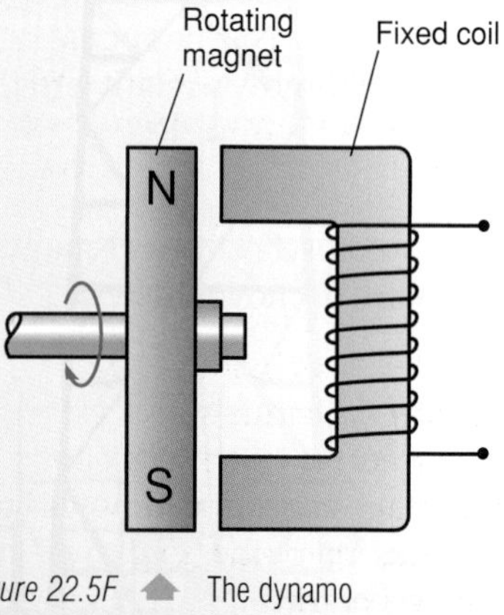

4 **The dynamo** consists of a magnet which is forced to rotate near one end of a coil. The repeated motion of each pole of the magnet past the end of the coil induces an alternating voltage in the coil. The faster the magnet spins, the bigger the induced voltage is. Work done on the dynamo to keep it turning is converted to electrical energy by the dynamo.

Figure 22.5F ▲ The dynamo

The alternating current generator

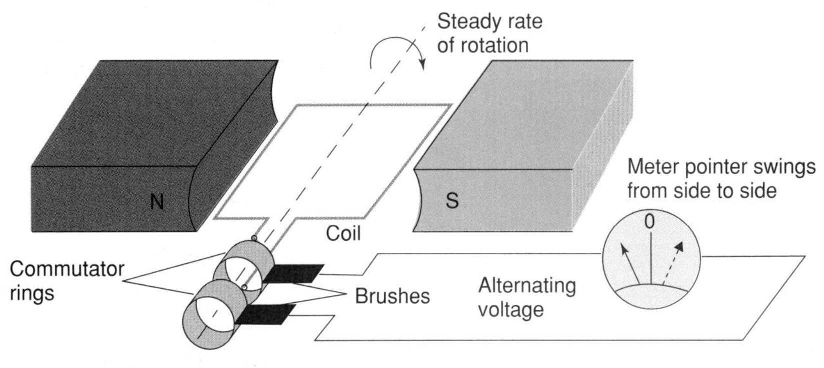

(a) Construction

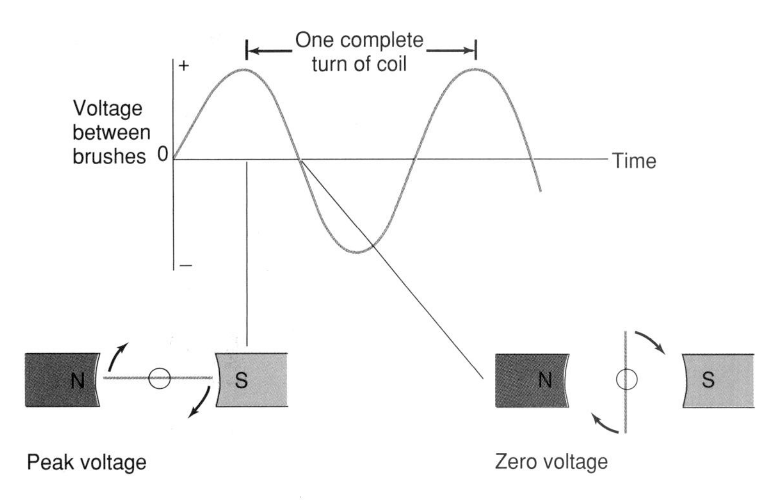

(b) Alternating voltage

Figure 22.5G The a.c. generator

See www.keyscience.co.uk for more about electricity generation.

The peak voltage is when the plane of the coil is parallel to the field.

Figure 22.5G(a) shows a coil in a magnetic field. The field is connected to a sensitive centre-reading meter. When the coil is turned at a steady rate, the meter pointer swings repeatedly from one side of the zero mark to the other side. The current through the meter repeatedly changes from one direction to the other and back. This is known as **alternating current** as opposed to direct current which passes in one direction only.

The effect happens because the voltage induced in the coil alternates, as in Figure 22.5G(b). This type of generator is sometimes called an **alternator**. If the coil is turned too fast, the meter pointer cannot keep up with the changes. But a lamp bulb would remain lit.

The voltage is biggest when the coil sides cut through the lines of force of the field. As the coil turns from this position, the voltage drops to zero and then reverses polarity as the coil sides cut across the field lines once more. Graphite brushes press against the commutator rings to make continuous contact with the coil.

- A **direct current generator** is the same as the alternating current generator except it has a split-ring commutator, as in Figure 22.5 H, instead of two separate slip rings. The split-ring commutator reconnects the coil in the opposite direction when it turns through 180°. As a result, the output voltage does not reverse direction as in the a.c. generator. The induced voltage varies from zero to a maximum twice each cycle, never changing polarity.

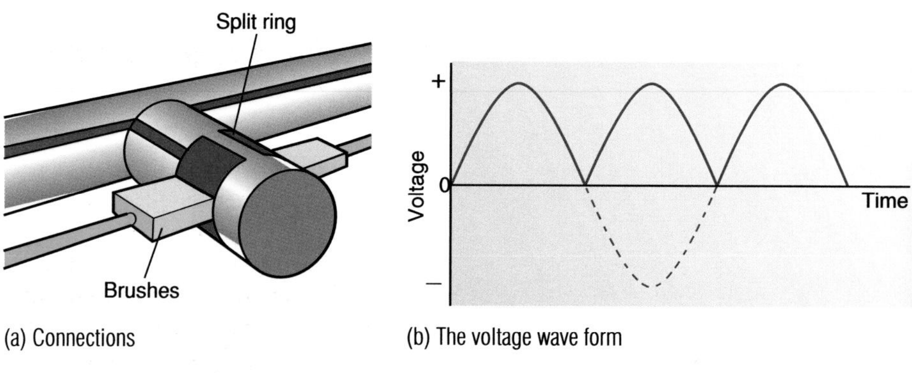

(a) Connections (b) The voltage wave form

Figure 22.5H The d.c. generator

- A **power station alternator** consists of three pairs of stationary coils and a rotating electromagnet, as in Figure 22.5I. The electromagnet is supplied with direct current from a separate d.c. generator. An alternating voltage is induced in each pair of stationary coils.

When a wire cuts across the lines of force of a magnetic field, a voltage is induced across the ends of the wire. If the wire is part of a complete circuit, the induced voltage causes current to pass round the circuit.

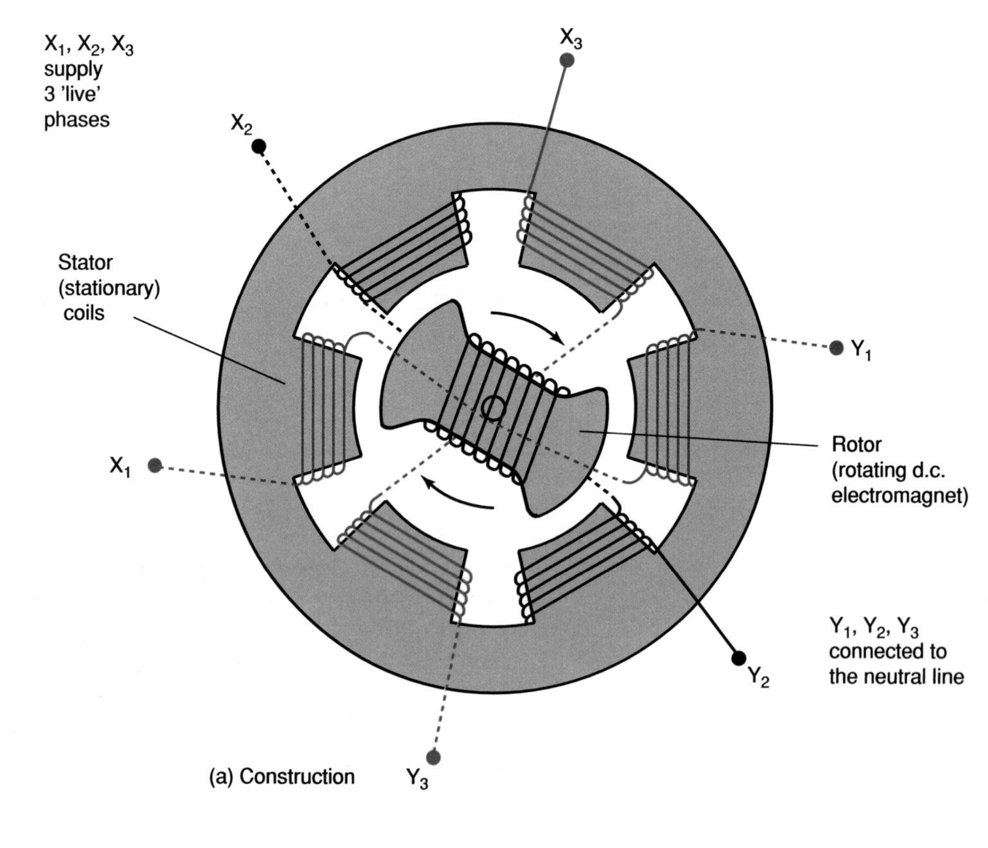

(a) Construction

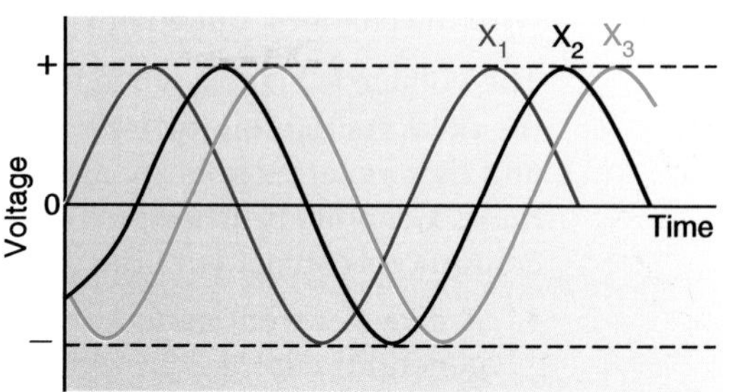

(b) Voltage waveform for a 3-phase supply. Each consumer is connected to one phase only.

Figure 22.5I A power station alternator

Converting alternating current to direct current

1 *Half-wave rectification*

A diode passes current in one direction only. Figure 22.5J shows the variation in current with time for a circuit consisting of a single diode, a resistor and an a.c. supply. When the supply polarity causes the diode to be reverse-biased, no current passes until the supply polarity changes back. This waveform is referred to as half-wave a.c. and it can be displayed on an oscilloscope as shown in Figure 22.5J.

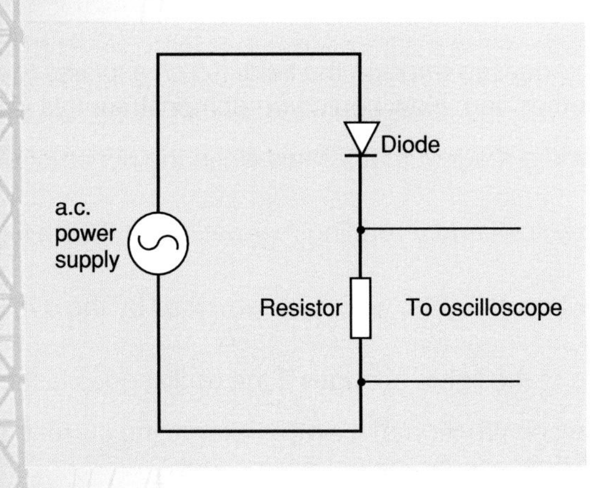

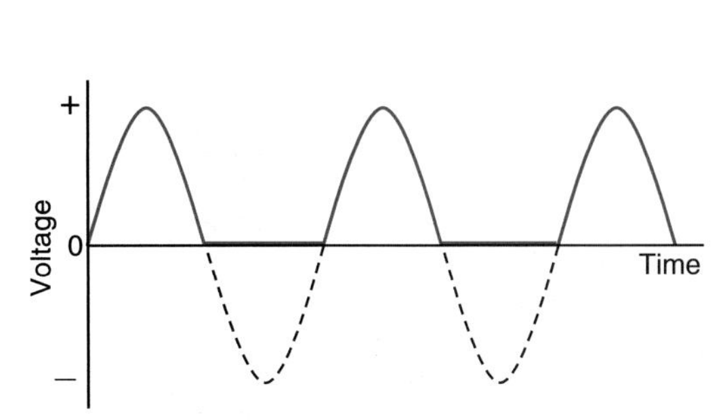

Figure 22.5J Half-wave rectification (a) Circuit diagram (b) Displaying the waveform

2 *Full-wave rectification*

This can be achieved using four diodes in a circuit referred to as a **bridge rectifier**, as shown in Figure 22.5K. Opposite diodes in the bridge conduct on alternate half-cycles. The result is that the current through the resistor is always in the same direction. The output voltage varies from zero to a maximum value each half-cycle. A capacitor across the output terminals can be used to smooth the output voltage to produce a steady direct voltage.

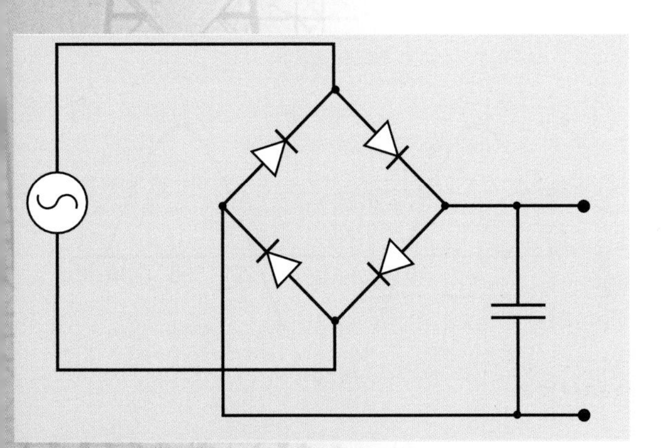

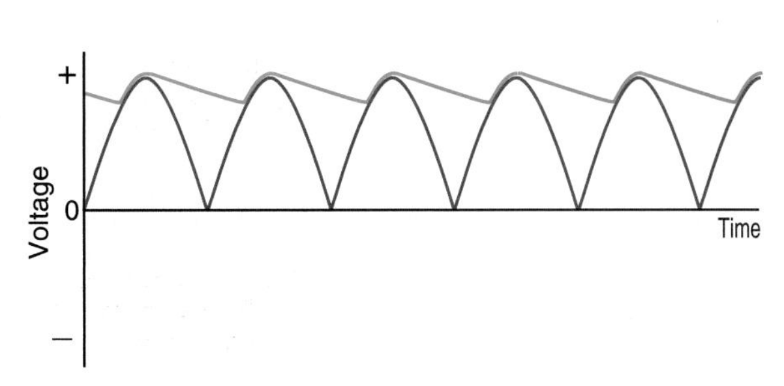

Figure 22.5K Full-wave rectification (a) A bridge rectifier (b) Full-wave a.c.

CHECKPOINT

1 The north pole of a bar magnet is pushed into a coil, as in Figure 22.5A. The meter pointer deflects to the right as shown. Which way would it deflect if:
   (a) the north pole is then withdrawn from the coil,
   (b) a south pole is pushed into the coil instead,
   (c) a south pole is withdrawn from the coil?

2 Karen is showing Sarah how to generate a voltage by passing a wire between the poles of a horseshoe magnet. The wire is connected to a centre reading meter. The meter pointer kicks to the left of centre when the wire is moved sharply through the field.
   (a) What happens when the wire is moved (i) sharply in the reverse direction, (ii) slowly in the original direction?
   (b) The wire is doubled back on itself and passed through the field. Karen explains to Sarah that one part of the wire cancels out the other part. How could you demonstrate this is so?

3 A cycle dynamo consists of a fixed coil and a magnet that rotates when the dynamo is driven by the motion of the wheel.
   (a) What is the advantage of having a fixed coil and a rotating magnet rather than a fixed magnet and a rotating coil?
   (b) Why is pedalling the bicycle harder when its lights, which are powered by the dynamo, are switched on?
   (c) State and explain how the brightness of the lights changes if the cyclist goes faster.

4 (a) Draw a labelled diagram showing the construction of a simple alternating current generator which consists of a coil that can spin between the poles of a magnet.
   (b) Sketch a graph to show how the output voltage of an a.c. generator varies with time over at least two cycles.
   (c) Explain why the output voltage is (i) at a maximum when the coil is parallel to the field lines, (ii) zero when the coil is at 90° to the field lines.

5 The diagram shows how the output voltage of an electromagnetic flow meter varies with time when a fluid passes through the meter at a steady rate. Each pulse is produced when a magnet in the turbine wheel passes the coil.
   (a) (i) How many pulses are produced each second?
       (ii) The turbine wheel has eight equally spaced magnets. How many complete turns does the wheel make each second?
   (b) Sketch how you would expect the voltage to vary with time if the wheel is made to rotate twice as fast.

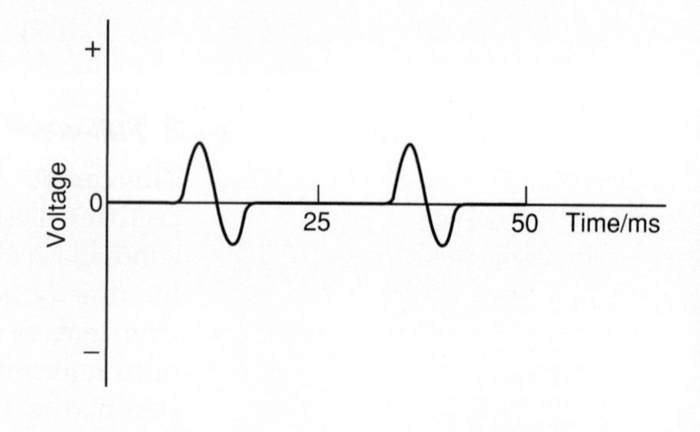

6 The diagram opposite shows how the output voltage of an alternating current generator varies with time.
   (a) Estimate from the graph (i) the peak-to-peak voltage, (ii) the number of cycles per second.
   (b) Copy the diagram and sketch how you would expect the voltage to vary if the speed of rotation of the coil is halved.

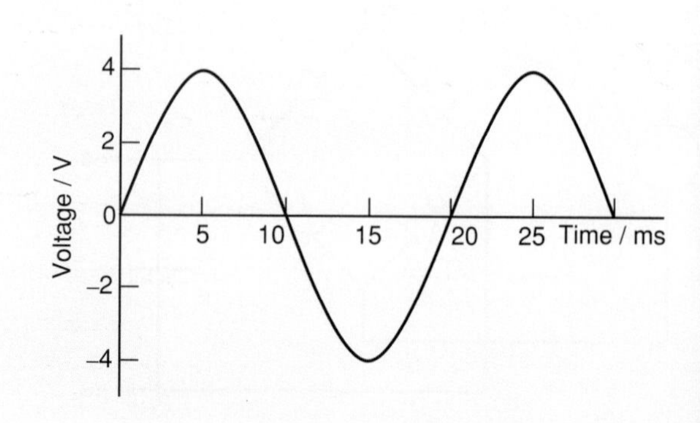

22.6 Transformers

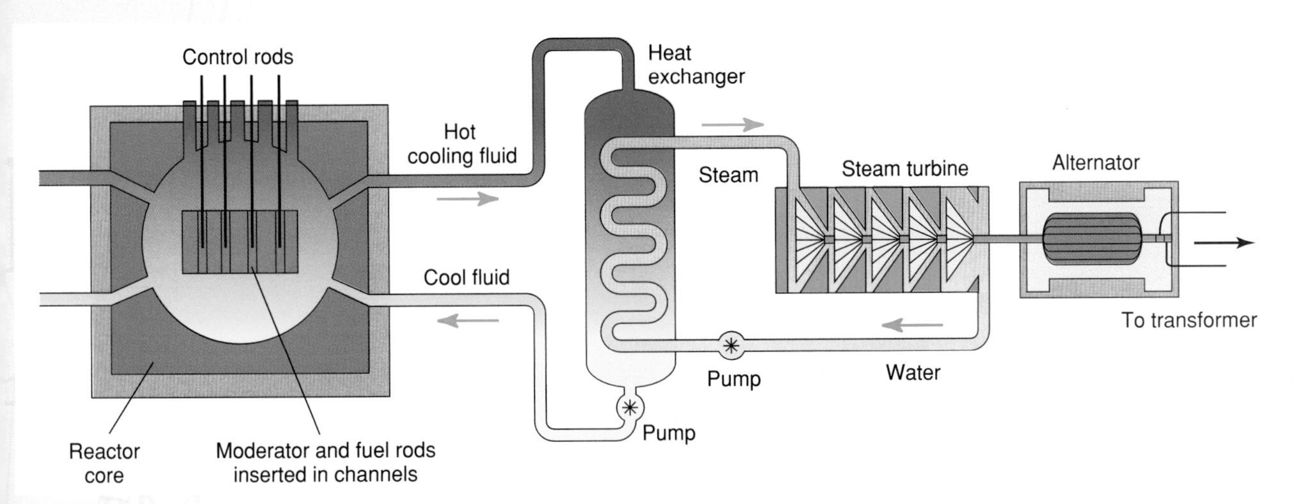

Figure 22.6A A nuclear power station

FIRST THOUGHTS

How does electricity from a power station reach your home? Read on to find out.

It's a fact!

Before the National Grid was set up, each town had its own power station. Voltage levels and frequencies differed from one town to the next. Some towns had constant voltage levels instead of alternating voltage. The overall system was very inefficient and was rationalised in 1926 when the National Grid was set up.

Power station alternators are designed to deliver alternating current to the National Grid. The frequency of the current must be exactly 50 Hz. This is achieved by making the alternator coil rotate steadily at a precise rate of 50 turns per second. In most power stations, **turbine engines** drive the alternators round at steady speed, making the alternators deliver electrical energy to the National Grid system.

Most power stations use fuel to heat water to create steam. The high-pressure steam is then used to turn the turbine wheels. The rotating turbine shaft is used to make the alternator turn steadily. Thus chemical energy from the fuel is converted to electrical energy at the power station and the electrical energy is delivered through the National Grid to consumers far away from the power station.

The advantage of the National Grid system for supplying energy is that the fuel is burned at power stations. Before the National Grid system, factories obtained power from coal-fired steam engines at each factory. They made the atmosphere in industrial areas very polluted.

However, by burning fossil fuels, power stations increase carbon dioxide levels in the atmosphere, thereby contributing to the greenhouse effect. Power stations emissions contain dust and gases that dissolve in water in the atmosphere to produce **acid rain**.

Nuclear power stations use enriched uranium to heat water to create steam. The uranium nuclei split apart, releasing fast-moving neutrons which then split more uranium nuclei, creating a chain-reaction. The fast-moving neutrons and fragmented nuclei heat the reactor core and a cooling fluid is pumped through the core to transfer heat from the core to a heat exchanger where steam is produced (Figure 22.6A). See p. 142.

Nuclear reactors do not produce carbon dioxide or acid rain. However, the spent nuclear fuel is highly radioactive and must be stored in safe conditions for hundreds of years until harmless. *Can you think of any further drawbacks to nuclear power stations?*

Hydroelectric power stations use water pressure to drive the turbines. A hydroelectric power stations is usually sited near the base of a mountain with an upland lake so the water from the lake can be piped down to the station. *Can you think of any environmental problems associated with hydroelectric power stations?*

The National Grid

The National Grid takes alternating current from power stations and delivers it to factories and homes. The output voltage from a power station is usually about 25 000 V. The voltage in a house is 230 V. **Transformers** are used to change the voltage levels.

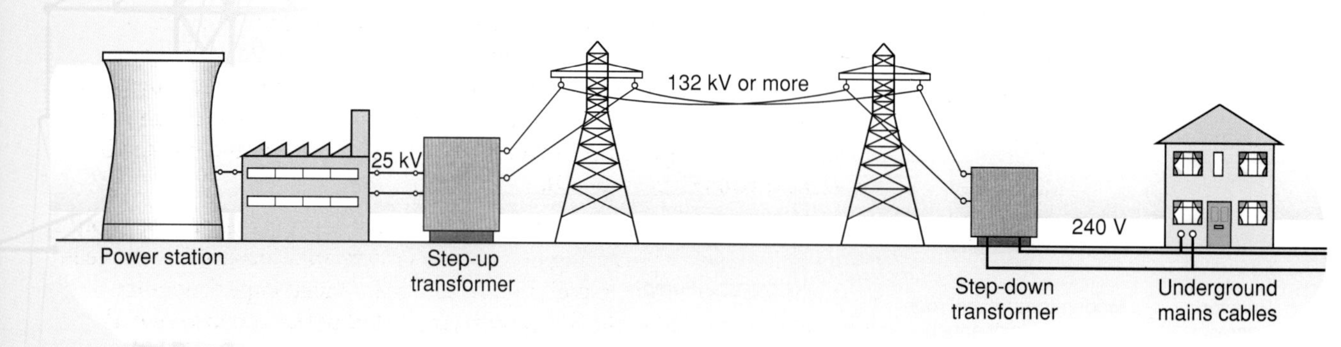

Figure 22.6B The National Grid

The output voltage of a power station must be 'stepped up' by transformers before being supplied to the National Grid. High voltages are better than low voltages for transmitting electrical power. At the consumer end of the National Grid, the voltage level must be stepped down before being supplied to homes and factories. The reason why high voltage transmission is more efficient than low voltage transmission is because a small current at high voltage delivers the same power as a large current at low voltage. For example, 100 W of electrical power can be delivered at 10 V by a current of 10 A or at 100 V by a current of 1 A. However, the heating effect of a large current due to the cable resistance is much greater than with a small current. As a result a larger fraction of the power supplied is wasted if the power is transmitted at low voltage and high current. In the above example, if the cable resistance is 0.1 Ω, prove for yourself that the power wasted is 10 W for a current of 10 A and 0.1 W for a current of 0.1A (see p. 325 if necessary).

Figure 22.6C shows a model of a power line at high voltage and a model of a power line at low voltage. The high voltage line is much more efficient than the same power line at low voltage.

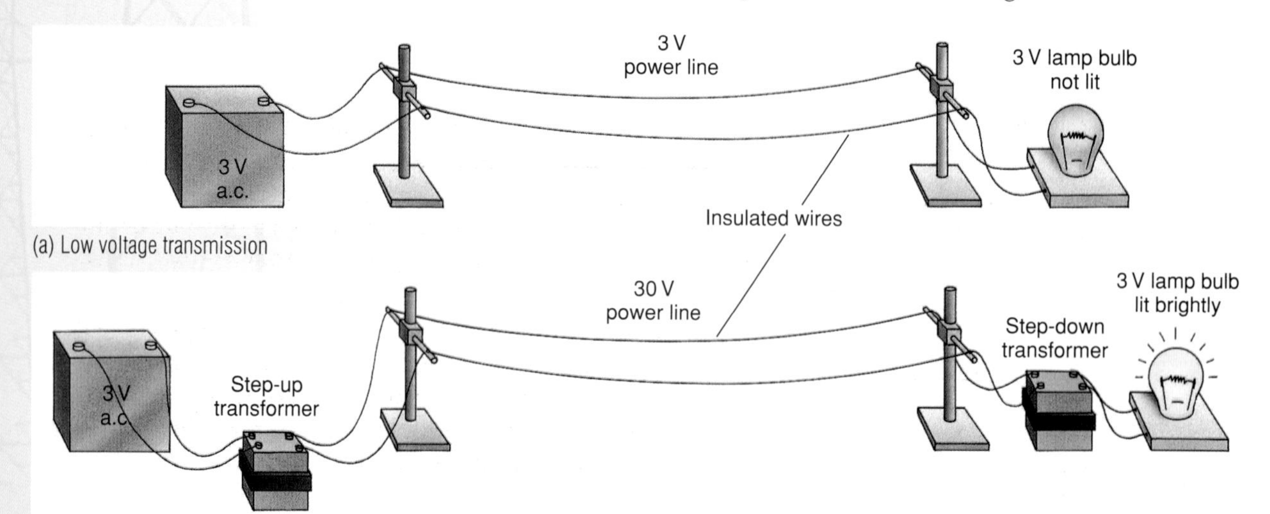

(a) Low voltage transmission

(b) High voltage transmission

Figure 22.6C Model power lines

Investigating transformers

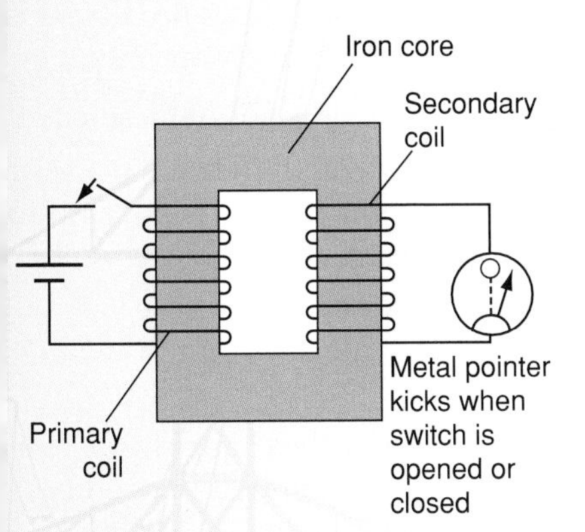

Figure 22.6D ▲ Investigating transformers

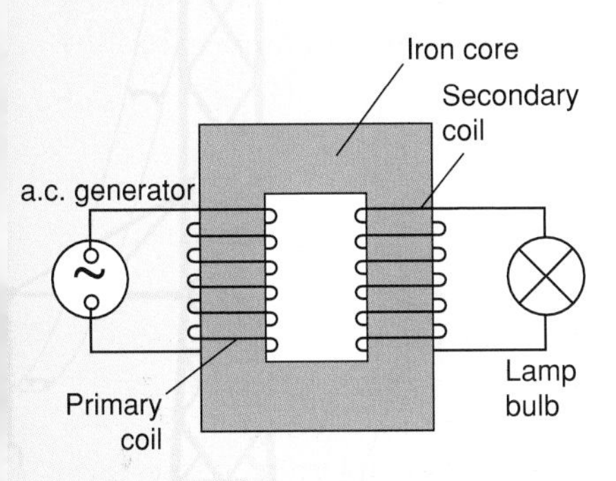

Figure 22.6E ▲ Using a transformer

Wind two coils of wire on an iron core. Connect one coil in series with a battery and switch; connect the other coil to a centre-reading meter. Open and close the switch repeatedly. You will find the meter pointer kicks one way then the other way each time the switch is opened and closed.

Why does this happen? Each time the switch is closed, the iron core is magnetised. When the switch is opened, the core loses its magnetism. The magnetism passes through the second coil. Each time the magnetism changes, a voltage is induced in the second coil. This makes the meter pointer kick. Changing the current in the first coil induces a voltage in the second coil (Figure 22.6D).

The coils are called the **primary** and **secondary** coils. If the primary coil is connected to an alternating current generator, an alternating voltage is induced in the secondary coil. Figure 22.6E shows an experiment to demonstrate the idea. The lamp bulb connected to the secondary coil lights up when the primary coil is connected to the power unit. If the two coils are separated, the lamp bulb goes dim because the magnetism from the primary coil no longer passes through the secondary coil.

How can the lamp bulb in Figure 22.6E be made brighter?

- The primary current could be increased. This makes the magnetism of the coil stronger giving a bigger induced voltage.
- Increasing the frequency is another way. Remember Faraday's discovery that the faster the magnet is moved, the greater the induced voltage. A higher frequency makes the magnetism in the core change faster, giving a bigger induced voltage in the secondary coil.
- Using more turns on the secondary coil increases the lamp brightness. Each turn 'picks up' a certain voltage. The more turns that are used, the greater the induced voltage. You could use voltmeters to show that the ratio of the secondary voltage to the primary voltage is the same as the turns ratio. This is called the transformer rule.

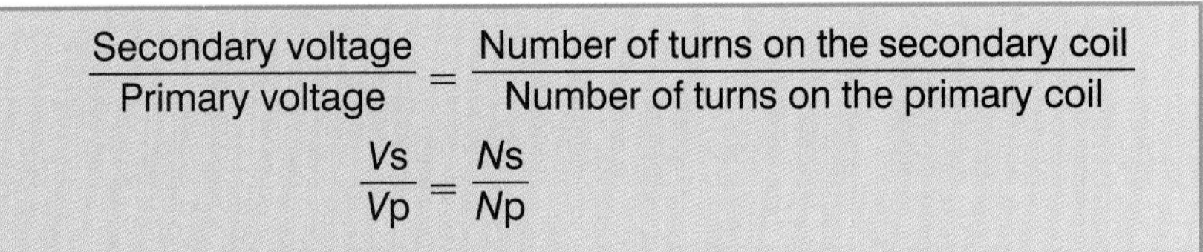

$$\frac{\text{Secondary voltage}}{\text{Primary voltage}} = \frac{\text{Number of turns on the secondary coil}}{\text{Number of turns on the primary coil}}$$

$$\frac{Vs}{Vp} = \frac{Ns}{Np}$$

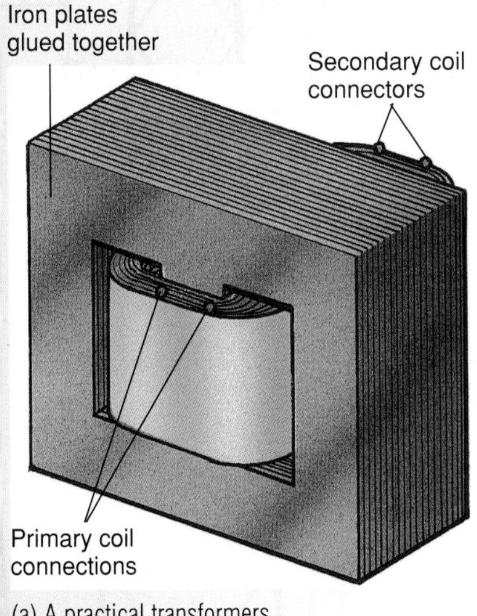

(a) A practical transformers

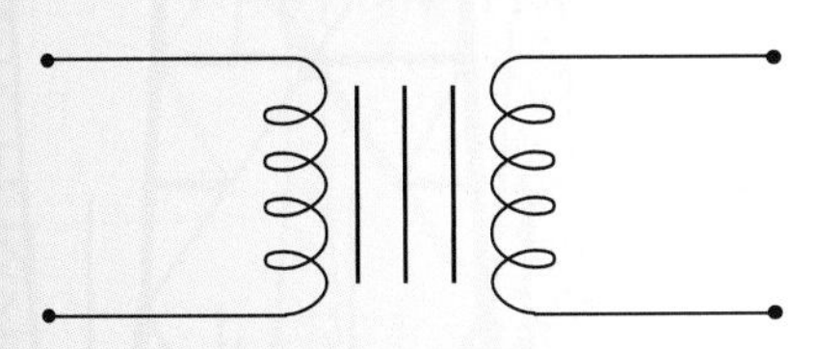

(b) Circuit symbol

Figure 22.6F ▲ Transformers in circuits

Practical transformers are designed to step alternating voltages up or down. Figure 22.6F shows the construction and circuit symbol for a practical transformer. A low voltage power unit contains such a transformer to step the mains voltage of 230 V down to the output level of the power unit. A television or radio operating from the mains also contains a step-down transformer.

The efficiency of most practical transformers is very high. For an ideal transformer delivering power, the power it can supply from its secondary coil is equal to the power supplied to its primary coil. Since electrical power is given by current multiplied by voltage, the following equation holds for an 'ideal' transformer.

Primary current × Primary voltage = Secondary current × Secondary voltage

$$Ip\ Vp = Is\ Vs$$

The equation shows that if the voltage is stepped down, the current is stepped up. Also, if the voltage is stepped up, the current is stepped down. This is why voltages are stepped up on power lines. The current is stepped down so there is much less electrical heating of power lines.

There are several reasons why a practical transformer is not 100% efficient.

1 **Resistance heating** occurs in both coils even though they are made of low-resistance copper wires. The resistance of both coils is not zero so some heat will be produced when current passes through each coil.
2 **Magnetisation** of the iron core wastes energy as the iron core is repeatedly magnetised and demagnetised by the changing primary current.
3 **Induced currents** are created in the iron core by the changing currents in the coils. These induced currents are referred to as **eddy currents** as they waste energy just as eddies produced in flowing water do. The iron core is constructed from alternate layers of iron and insulating material to reduce eddy current losses. This layered structure is usually referred to as a **laminated** structure.

The core is made from iron because iron makes the magnetic field linking the coils much stronger.

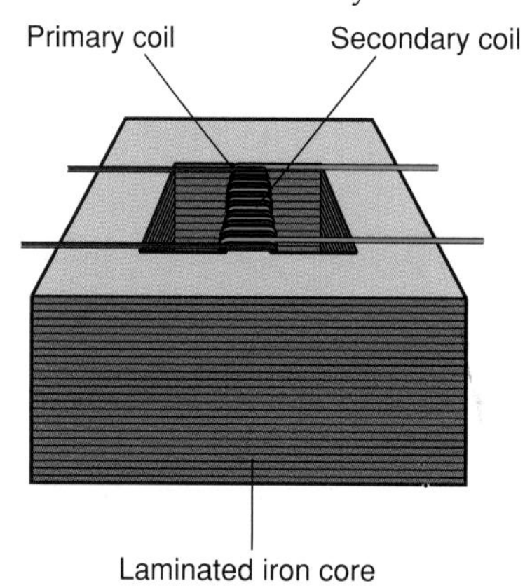

Figure 22.6G ◀ A laminated core

Worked example A transformer is to be used to 'step down' an alternating voltage from 240 V to 12 V. The primary coil has 1200 turns.
(a) Work out the number of turns on the secondary coil.
(b) The secondary coil is used to light a 12 V, 24 W bulb. What will be the current in the primary coil, assuming the transformer is 100% efficient?

Solution

(a) Primary voltage = 240 V, Secondary voltage = 12 V, Primary turns = 1200

Using the transformer rule $\frac{V_S}{V_P} = \frac{N_S}{N_P}$

$$\frac{12}{240} = \frac{N_S}{1200}$$

Therefore Number of secondary turns, $N_s = \frac{12 \times 1200}{2400} = 60$

(b) The power of the lamp is 24 W. Since power = current = voltage, the lamp current is therefore 2 A (= 24W/12V). This is the secondary current. To work out the primary current, use the equation

$$I_P V_P = I_S V_S$$

From the equation above $I_P \times 240 = 12 \times 2$

So Primary current, $I_P = \frac{12 \times 2}{240} = 0.1$ A

SUMMARY

A transformer is used to step alternating voltages up or down. The voltage ratio is equal to the turns ratio. For an ideal transformer, the primary current x the primary voltage equals the secondary current x the secondary voltage.

CHECKPOINT

1 The circuit diagram shows a 12 V a.c. power supply connected to the primary coil of a transformer and a 240 V, 100 W light bulb connected to the secondary coil.

12 V a.c. power supply
240 V 100 W light bulb

(a) The primary coil has 60 turns. How many turns must the secondary coil have if the bulb is to light normally?

(b) (i) Calculate the current through the secondary coil of the transformer when the bulb lights normally.

(ii) Assuming the transformer is 100% efficient, calculate the primary current when the bulb lights normally.

2 A 6 V a.c. power supply for an electronic synthesiser consists of a transformer that steps the mains voltage down from 240 V to 6 V.

(a) (i) Sketch the circuit diagram for this arrangement.

(ii) The primary coil has 1200 turns. How many turns are on the secondary coil?

(b) The transformer is designed to deliver a current of 1 A at 6 V.

(i) How much power is the secondary coil designed to deliver?

(ii) Calculate the primary current needed to deliver this amount of power, assuming the transformer is 100% efficient.

3 A power station alternator produces an alternating voltage at 25 kV. This is stepped up to 150 kV by a transformer.

(a) Calculate the turns ratio of the transformer.

(b) The alternator is designed to deliver 1000 kW of electrical power. Calculate the current in the primary winding of the transformer when the alternator delivers this amount of power.

(c) The transformer is 90% efficient. Calculate (i) the output power of the transformer, (ii) the current in the secondary coil when it delivers this amount of power.

4 (a) The figure opposite shows the voltage output of an alternator as a graph of voltage against time. Sketch similar graphs for the same alternator turned (i) more quickly, (ii) more slowly.

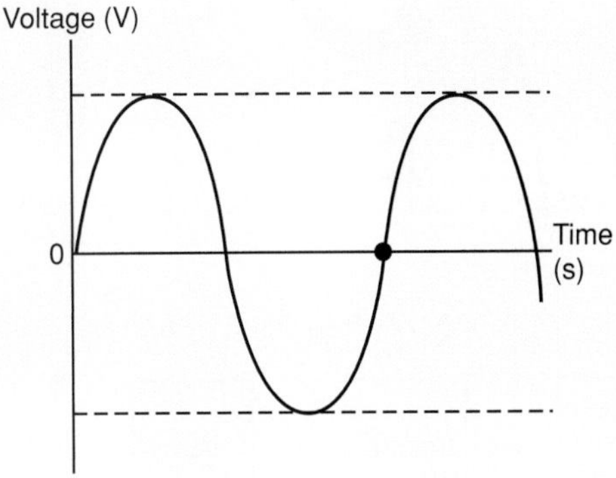

(b) The output voltage from a power station is stepped up to supply power to the National Grid. This means that the currents passing through the grid cables are smaller than they would be if the voltage had not been stepped up. Why is less power wasted by stepping the voltage up?

(c) A transformer with 1000 turns on its primary coil and 50 turns on its secondary coil is used to step down an alternating voltage of 240 V. The primary coil is in a circuit with a 1 A fuse. Work out:

(i) the secondary voltage,

(ii) the maximum current in the secondary coil when it is in a circuit,

(iii) the maximum power the transformer can deliver.

5 A section of the grid system consists of cables of total resistance 20 Ω. This section is used to deliver 10 MW of electrical power to a sub-station.

(a) If this power is delivered to the sub-station at a voltage of 100 kV, calculate (i) the current through the cables, (ii) the power wasted due to resistance heating in the cables.

(b) Hence show that 2% of the power supplied is wasted due to resistance heating in the cables.

(c) If the same power was delivered at 20 kV, show that one third of the power supplied would be wasted due to resistance heating of the cables.

Topic 23 Electronics

23.1 Electronic systems

FIRST THOUGHTS

Electronics has revolutionised many activities. In this section, you will find out the difference between digital and analogue electronics.

Make a list of all the electronic gadgets that you have used. Perhaps you have a video recorder or microcomputer to put on your list. Do not forget the phone or the television remote control unit. You may have been shopping at a supermarket with laser bar-code readers at the tills or perhaps have seen a cash dispensing machine used outside a bank.

These are all examples of electronic systems that handle information. The development of **microelectronics** has made all these systems possible. Twenty or more years ago, electronic circuits used **valves** to move information round. Valves are bulky glass vacuum tubes that need high voltages to work and consequently get hot. Valves have been replaced by **silicon chips**. A single silicon chip no bigger than a thumbnail contains the equivalent of hundreds of valves. Chips do not get hot and they use very little power so they can work off small batteries.

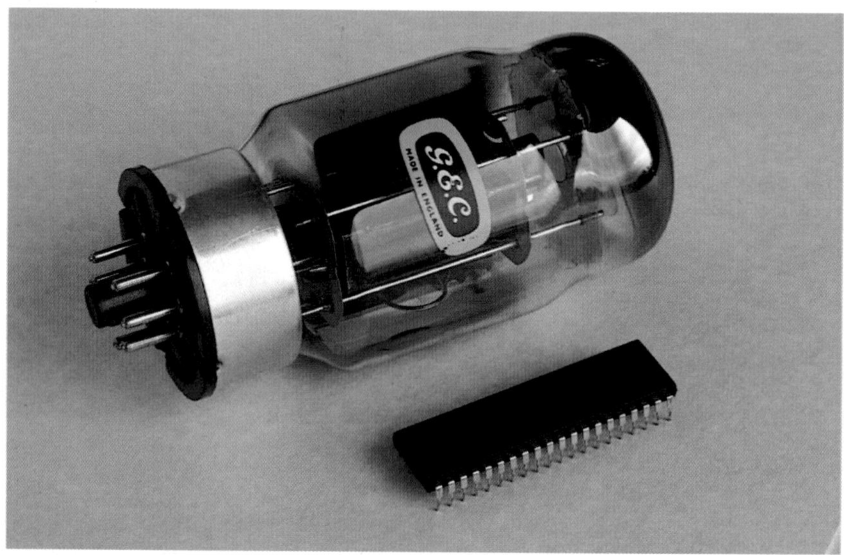

Figure 23.1A A valve and a silicon chip

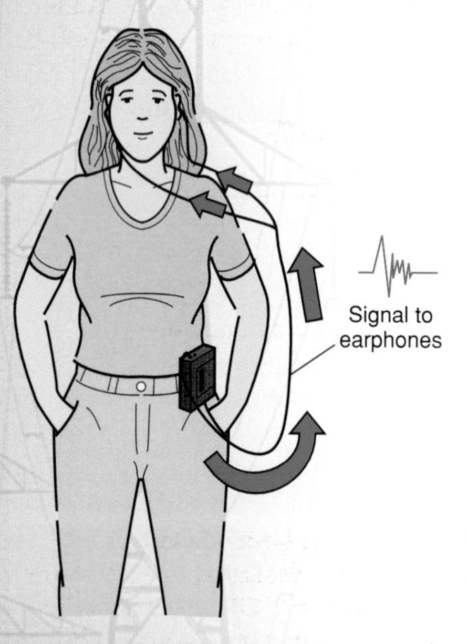

(a) An analogue system

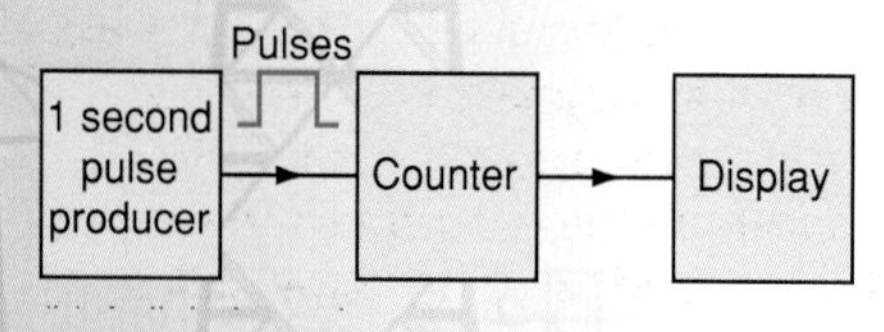

(b) A digital watch system

Figure 23.1B

Which two electronic gadgets do you and your friends use most often? Digital watches and personal stereos probably top the list. A personal stereo radio uses one or more chips to boost the tiny radio signal picked up by the aerial inside the stereo. The circuit in the stereo which is used to make the signal stronger is called an amplifier. The circuit is an example of an **analogue** system. Analogue means that voltages in the system can be at any value between the limits set by the power supply.

A digital watch contains a complicated electronic system designed to count and display voltage pulses produced at a rate of one per second. A small battery is all that is needed to supply power to the system.

The electronic circuit in the watch is an example of a **digital** system. Digital means that the voltage at any point in the system is either zero (i.e. low) or at a fixed positive value (i.e. high). No other values of voltage are possible in a digital system. Digital circuits are useful for making decisions, storing information, counting and many other things.

SUMMARY

Digital systems have just two possible voltage levels, referred to as high and low. In analogue systems, the voltages can be at any value between the limits set by the power supply.

23.2 Electronic logic

FIRST THOUGHTS

What was the last decision you made? Did you make a logical decision? Read on to find out how electronic circuits make logical decisions.

Decisions are made by each one of us all the time. Each decision involves making a choice of some sort. Using **logic** allows us to make decisions. Suppose you are about to go out and you are thinking about whether or not you should wear a coat. Your decision would probably depend on your answer to two questions

Question A: Is it raining?
Question B: Is it cold outside?
Decision to be made: Shall I wear a coat?

If you answer YES to A or B, then you would take your coat. This situation where the 'output' depends on the 'inputs' can be shown by a **truth table**. Figure 23.2A shows the truth table for this situation. In the truth table, YES is shown by 1 and NO by 0. So the output is 1 if A OR B is 1.

INPUTS *Is it raining?*	*Is it cold?*	OUTPUT *Shall I wear a coat?*
0	0	0
0	1	1
1	0	1
1	1	1

1 = YES 0 = NO

Figure 23.2A ▲ A truth table

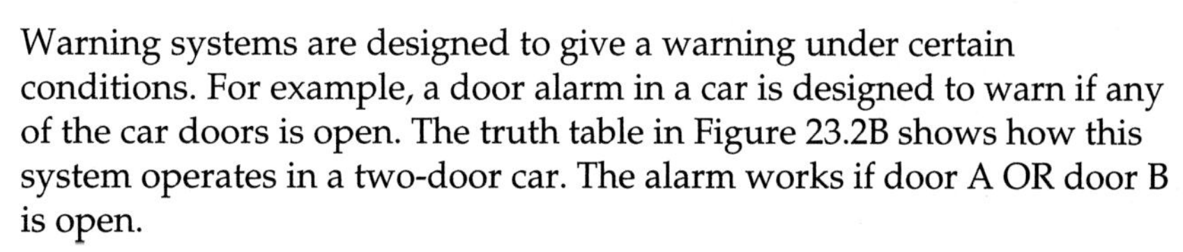

Warning systems are designed to give a warning under certain conditions. For example, a door alarm in a car is designed to warn if any of the car doors is open. The truth table in Figure 23.2B shows how this system operates in a two-door car. The alarm works if door A OR door B is open.

The alarm in Figure 23.2B is ON or OFF depending on the settings of the doors. A switch in each door provides an input signal to an electronic circuit that controls an alarm buzzer. The output signal from the circuit switches the buzzer on or off. This depends on the condition of the input signals. In other words, the output state is decided by the input states.

INPUTS *Door A*	*Door B*	OUTPUT *Buzzer*
0	0	0
1	0	1
0	1	1
1	1	1

0 = Door closed 0 = Buzzer off
1 = Door opened 1 = Buzzer on

Figure 23.2B ▲ A warning system

Gate	Symbol	Function (High voltage = 1, Low voltage = 0)	Truth Table INPUTS A	B	OUTPUT
OR	A, B → OUTPUT	OUTPUT = 1 if A OR B = 1	0 0 1 1	0 1 0 1	0 1 1 1
AND	A, B → OUTPUT	OUTPUT = 1 if A AND B = 1	0 0 1 1	0 1 0 1	0 0 0 1
NOR	A, B → OUTPUT	OUTPUT = 0 if A OR B = 1	0 0 1 1	0 1 0 1	1 0 0 0
NAND	A, B → OUTPUT	OUTPUT = 0 if A AND B = 1	0 0 1 1	0 1 0 1	1 1 1 0
NOT	INPUT → OUTPUT	OUTPUT = 1 if INPUT = 0 OUTPUT = 0 if INPUT = 1	0 1		1 0

Figure 23.2C Logic gates

Logic gates

The circuit controlling the car door alarm in Figure 23.2B is called a logic gate because it is designed to make a decision. Any logic gate has one or more input terminals and an output terminal. The voltage at the output (i.e. the output state) depends on the voltages at the input terminals (i.e. the input states). This may be described by a truth table. We will only consider what a logic gate does: not how it does it. The circuit symbols and truth tables of some common logic gates are shown in Figure 23.2C.

Figure 23.2D shows how the car door alarm control circuit is made using an OR gate. Check for yourself that its truth table is the same as in Figure 23.2B. *What would happen if an AND gate was used instead?*

Logic gates are made in chip form. Figure 23.2E shows an AND chip. It contains four AND gates, each with two input terminals and an output terminal. The chip has twelve pins for these terminals and two pins for the supply voltage.

An indicator is used to display the output state of a logic gate. Figure 23.2F shows how an indicator may be made using a light-emitting diode (LED) and a resistor. The LED lights up when a '1' is applied to the input terminal of the indicator.

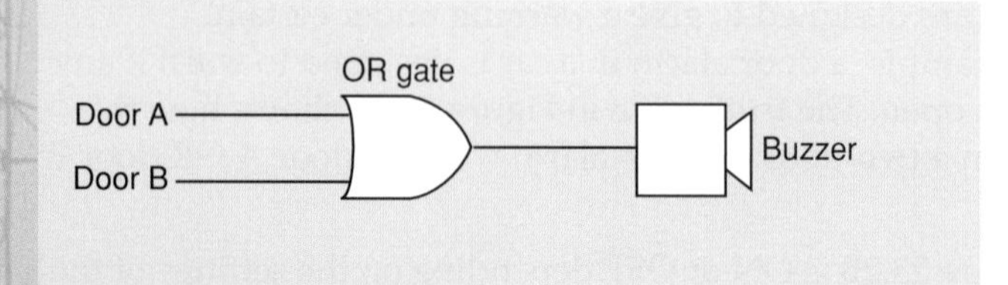

Figure 23.2D Door alarm

To supply a '1' to a logic gate input terminal, the terminal can be connected to the positive terminal of the gate's voltage supply unit. Connecting the input to the negative terminal (which is usually earthed) applies a '0' signal at the input. An indicator is connected to the output terminal to observe the output state.

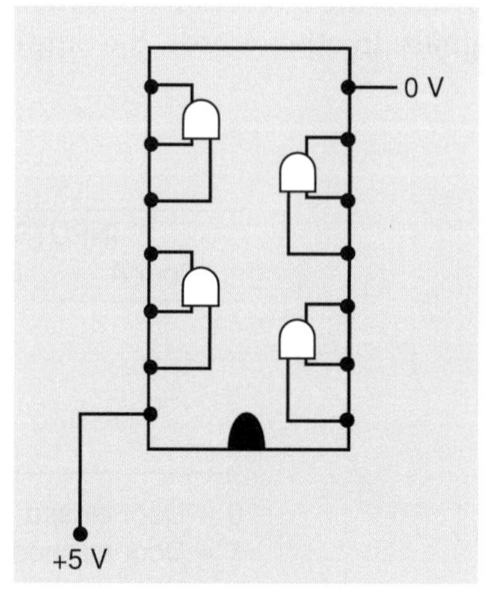

Figure 23.2E An AND chip

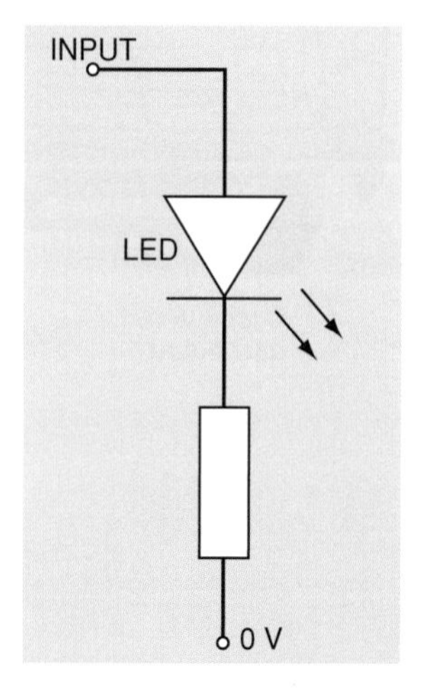

Figure 23.2F A logic indicator

SUMMARY

Only two voltage levels are possible in a digital circuit. The output voltage of a logic gate depends on the voltage level at each input terminal of the gate.

CHECKPOINT

1 Copy and complete the truth table opposite for each of the logic systems (a) to (j) shown here.

A	B	OUTPUT
0	0	
0	1	
1	0	
1	1	

(a)

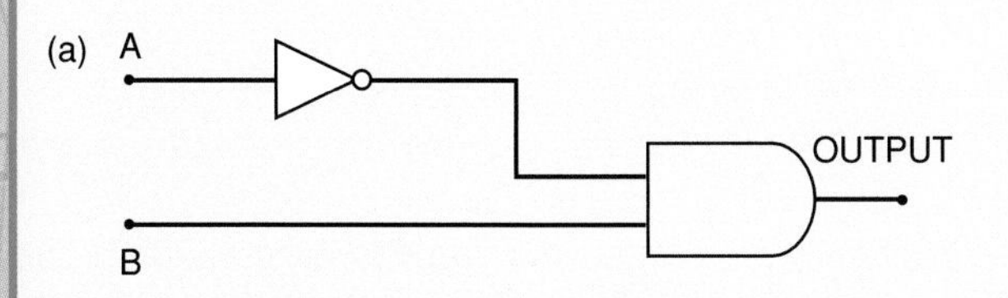

(b)

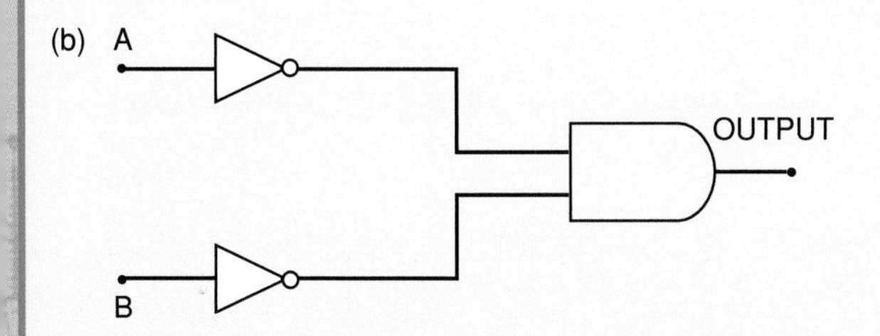

(c)

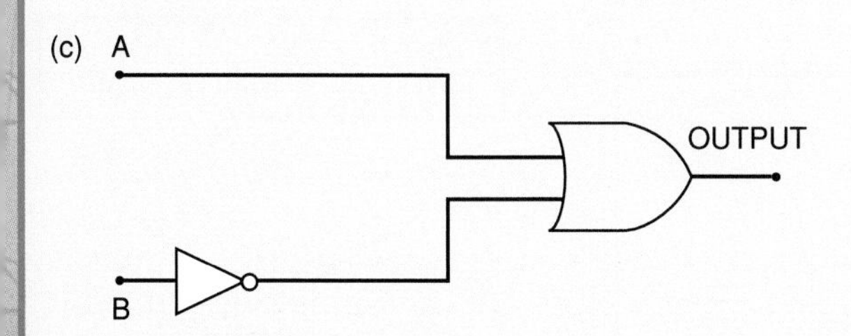

(d)

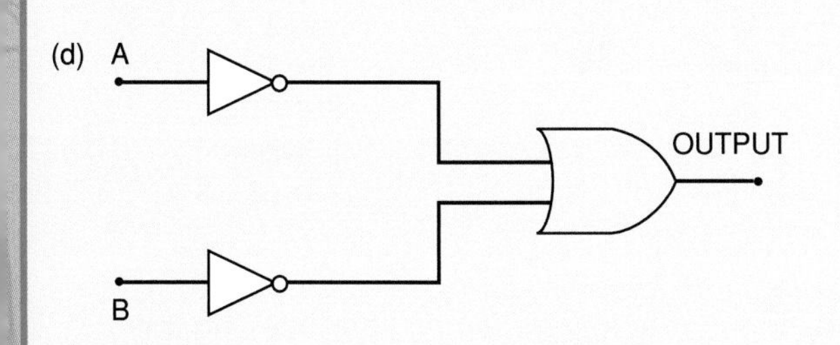

(e)

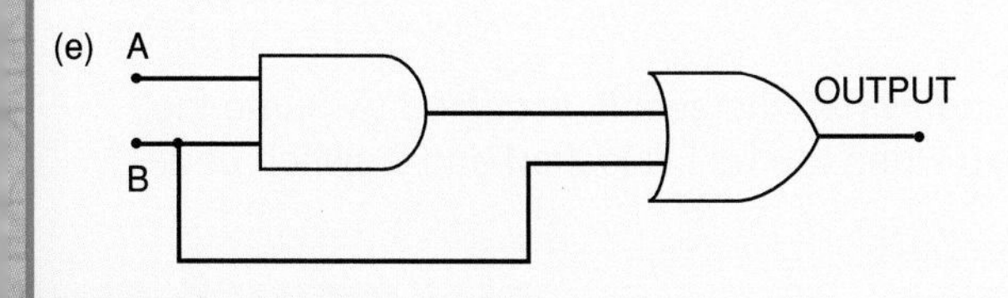

(f)

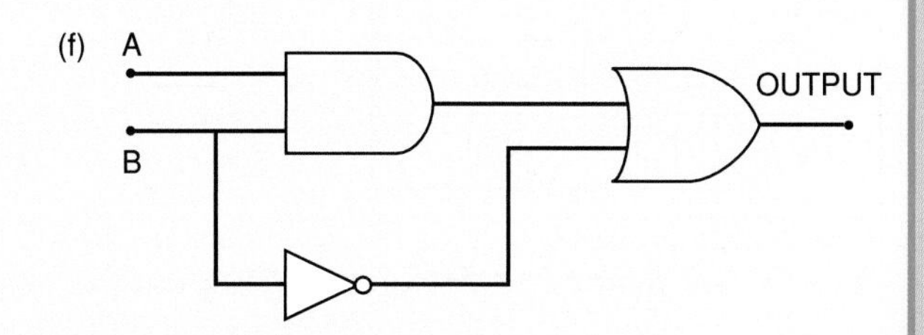

(g)

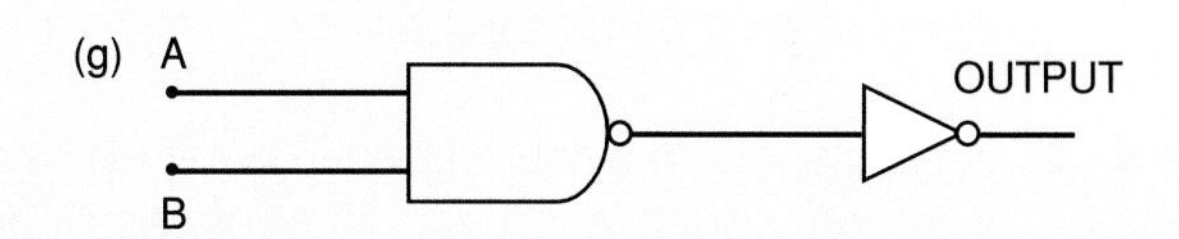

(h)

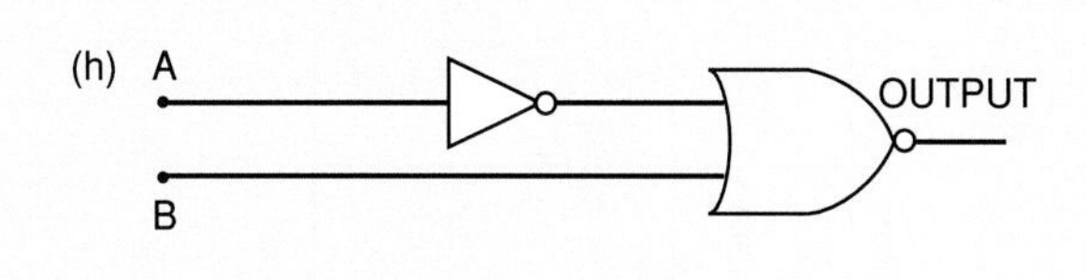

(i)

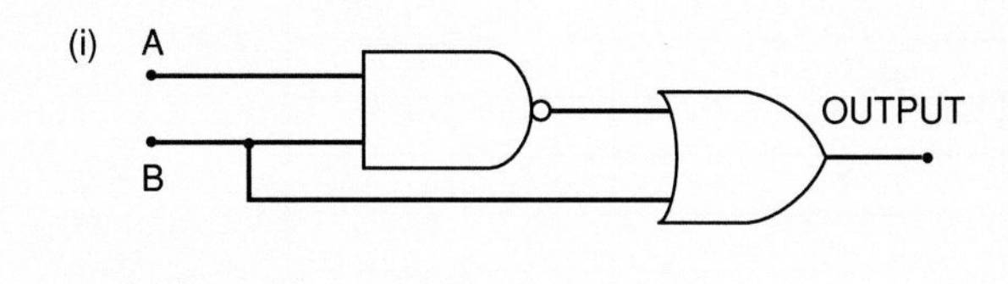

(j)

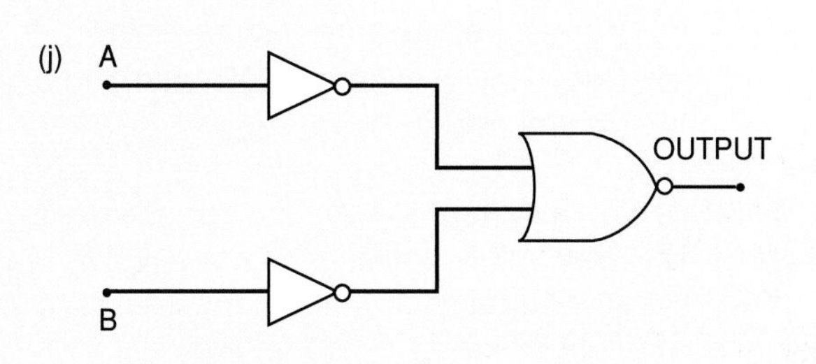

2 A comparator is a logic circuit that gives a '1' at its output terminal only when the logic state of its input terminals are the same. The diagram shows its truth table.

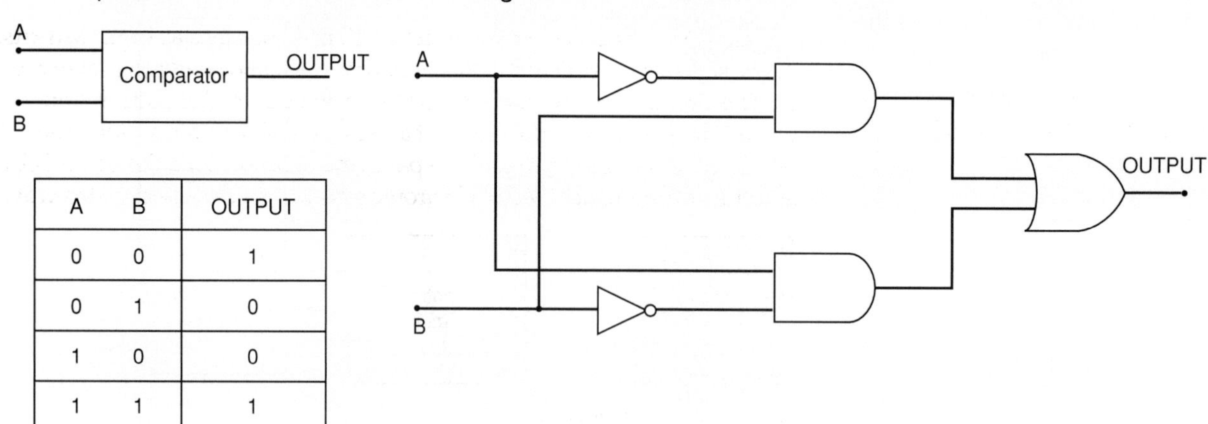

A	B	OUTPUT
0	0	1
0	1	0
1	0	0
1	1	1

(a) Simon has designed a comparator using some logic gates. His design is shown above. Unfortunately, there is a fault in his design. Make a truth table for the system he has designed.

(b) Redesign Simon's system so it works correctly.

3 Design a logic system, using two-input gates, for a four-door car to warn if any of the doors are open when the handbrake is off.

4 A burglar alarm system fitted in an apartment is designed so that the alarm is activated if a key-operated switch is 'on' and the entrance door is open or a pressure pad under the carpet behind the door is activated. The figure below shows the system and part of its truth table.

INPUTS			OUTPUT
Door switch	Key switch	Pressure pad	Alarm
OPEN = 1 CLOSED = 0	ON = 1 OFF = 0	ON= 1 OFF = 0	ON = 1 OFF = 0
0	0	0	0
1	0	0	
0	0	1	0
1	0	1	
0	1	0	0
1	1	0	
0	1	1	1
1	1	1	

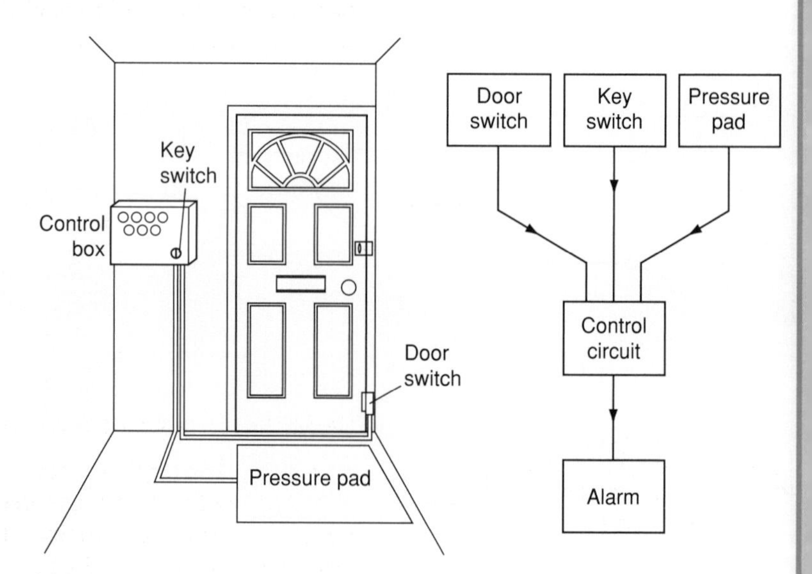

(a) Copy and complete the truth table.

(b) Anyone entering the apartment has 20 seconds to turn the key switch off after opening the door or else the alarm is activated. Why is this delay necessary?

(c) Why is the pressure pad essential?

(d) Show how a two-input OR gate and a two-input AND gate may be connected together as the control circuit.

5 Design a system to warn if a car seat belt is unfastened when the seat is occupied. Assume the seat belt unit supplies a '1' to a suitable logic system when the belt is locked and a switch under the seat supplies a '1' when the seat is occupied.

23.3 Electronics at work

FIRST THOUGHTS

In this section, you will find out how electronic sensors work and what they can be used for.

The car door alarm system in Figure 23.2B uses switches as **input sensors**. These signals sense (detect) when a door is open and send a signal to the control circuit. The circuit operates the alarm buzzer. The buzzer is an **output device**, designed to convert an electrical signal into sound. The whole system is designed in three parts, the input sensors, the control circuit and the output device. Many electronic systems are designed in this way.

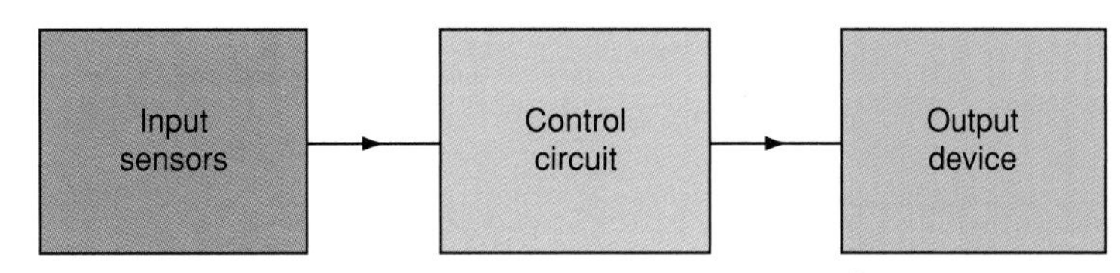

Figure 23.3A ▲ A control system

Input sensors

A sensor is a device that produces an electrical signal in response to a change in a physical variable such as temperature or light intensity, pressure, angle of inclination (i.e. tilt) or the presence of a magnetic field or of moisture.

A simple pressure sensor is shown in Figure 23.3B. With sufficient pressure on the switch, the switch closes and connects the output terminal of the sensor to the positive terminal of the voltage supply. The LED lights up when this happens. The same circuit with the pressure switch replaced by a 'tilt' switch which closes when it is tilted too much will operate as a tilt sensor.

Remember an LED is a light-emitting diode.

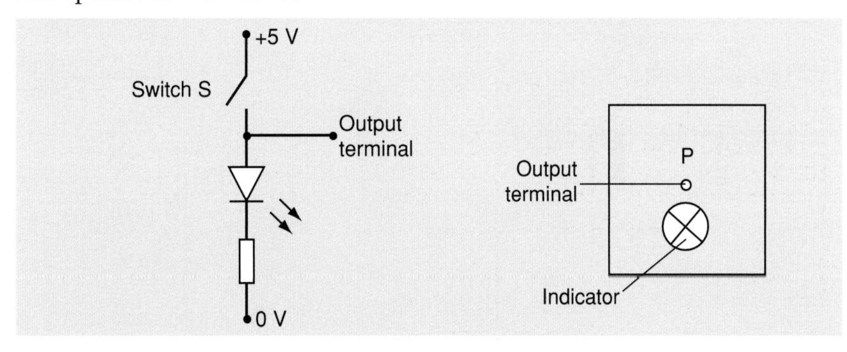

Figure 23.3B ▲ A pressure sensor (a) Circuit (b) Symbol

See www.keyscience.co.uk for more about electricity and electronics.

A moisture sensor has a gap in its circuit, as shown in Figure 23.3C. The gap is formed by two wires fixed on an insulator. When the gap becomes moist, it conducts electricity. The result is that the voltage at the output terminal of the sensor rises.

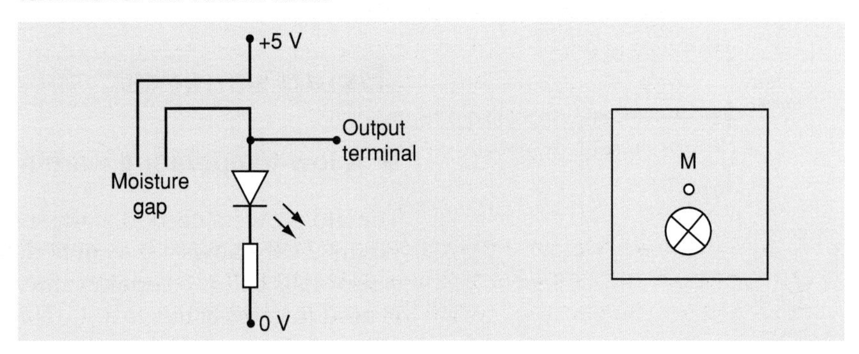

Figure 23.3C ▲ A moisture sensor (a) Circuit (b) Symbol

The light sensor in Figure 23.3D uses a **light dependent resistor** (LDR) and a resistor to form a potential divider. The LDR's resistance drops when the intensity of light on it increases. This makes the output voltage of the sensor rise. In darkness, the voltage drops.

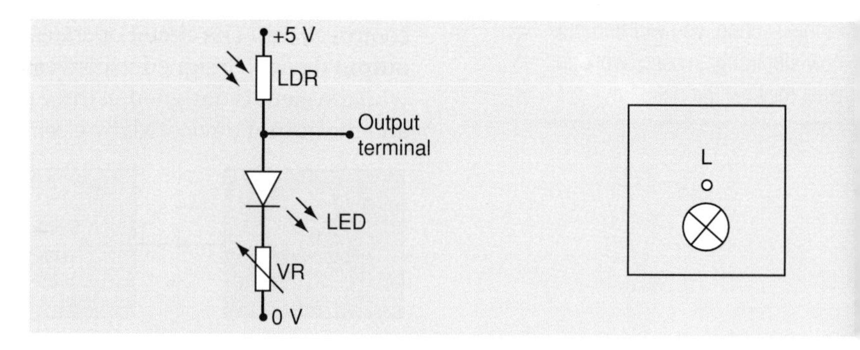

Figure 23.3D A light sensor *(a)* Circuit *(b)* Symbol

EXTENSION FILE ACTIVITY

It's a fact!

Science at work
Domestic alarm systems often include magnetic switches fitted to opening windows and doors. When the window or door is opened, a small magnet in the window panel or door is moved away from a magnetic switch fitted to the frame. As a result, the switch opens and breaks the low-voltage circuit which it is in, activating the alarm if the alarm system is on.

The LDR is made from silicon. When light falls on silicon, electrons break free from the silicon atoms. This is why the resistance drops. Extra electrons become available to carry current.

Figure 23.3E shows a temperature sensor. Temperature sensors use a thermistor in series with a resistor. The resistance of a thermistor drops as its temperature rises. This makes the output voltage of the sensor rise. Falling temperature makes the voltage fall.

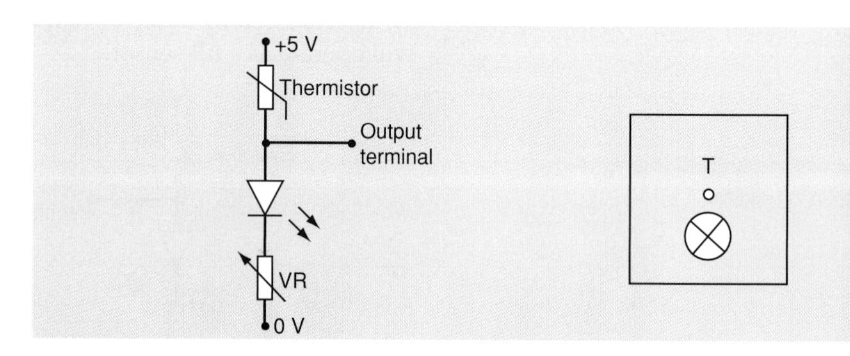

Figure 23.3E A temperature sensor *(a)* Circuit *(b)* Symbol

A thermistor, like an LDR, is made from silicon. Raising the temperature of silicon enables electrons to break free from silicon atoms. The thermistor must be in a light-proof cover. *Why?*

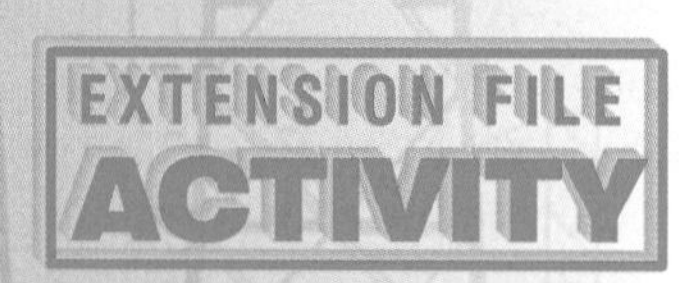

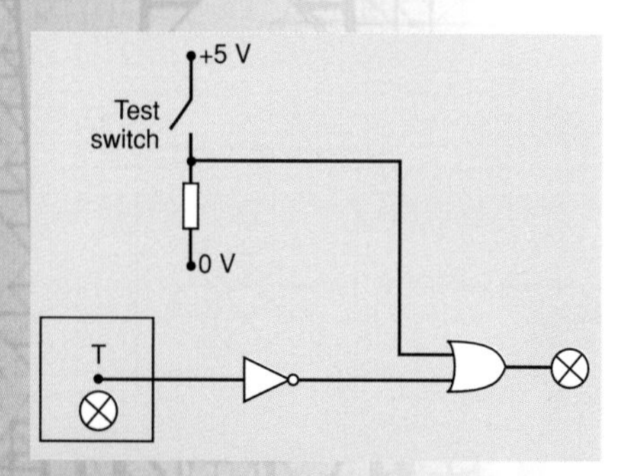

Figure 23.3F A low temperature indicator

Using sensors

A low-temperature warning indicator

Tomato growers need to make sure that plants are protected from frost. Figure 23.3F shows a system to do this. When the temperature of the sensor falls below a certain value, the sensor output is zero. A NOT gate is used to convert this to a '1'. This is supplied to one of the inputs of an OR gate. The other input is supplied by a 'test' switch. This is used to check the indicator works.

SUMMARY

Electronic systems that respond to changes in physical conditions like temperature and pressure may be designed using sensors and logic gates.

A night-time rain alarm

Your parents might find this useful when wet clothes are left outside on the washing line at night. In darkness, the light sensor supplies a '0' to the NOT gate which therefore supplies a '1' to one of the inputs of the AND gate. The moisture sensor supplies the other AND input. The buzzer sounds the alarm if rain falls on the moisture sensor at night. Figure 23.3G shows the night-time rain alarm system.

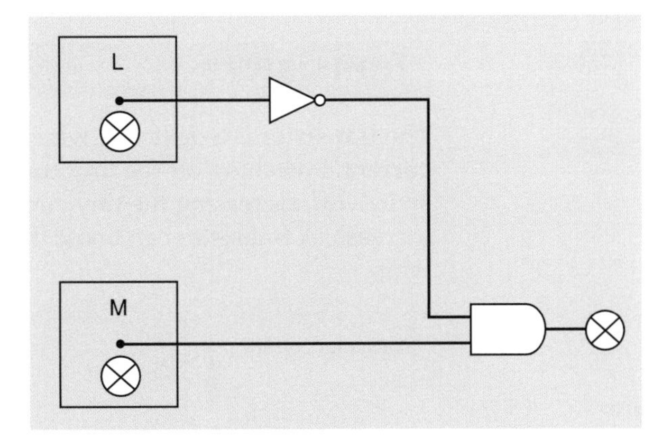

Figure 23.3G ▲ A night-time rain alarm

CHECKPOINT

1. In Figure 23.3E, the variable resistor VR in the temperature sensor can be adjusted to alter the temperature at which the indicator switches on.
   (a) Suppose the resistance of VR is increased. Does this raise or lower the temperature at which the indicator switches on when the thermistor is cooled?
   (b) If the NOT gate in Figure 23.3F is removed what happens now when the thermistor is cooled?
2. (a) Suppose you are responsible for a rare plant that can only survive if the soil temperature stays between 10 °C and 30 °C. How would you use two temperature sensors, a NOT gate and an AND gate to warn if the soil temperature goes outside this range?
   (b) How would you check that your system does operate exactly as required?
3. Design a high temperature alarm to operate a buzzer when the temperature goes above a certain value. Include a 'test' switch to check if the buzzer works when the switch is closed.
4. A certain camera is fitted with a light sensor to warn if there is not enough light in the room. This causes a LED to light up in the viewfinder when a button switch on the camera is pressed.
   (a) Unfortunately the system shown in the figure opposite is designed incorrectly. What makes the LED in the system shown light up?
   (b) Redesign the system shown so it operates as intended.

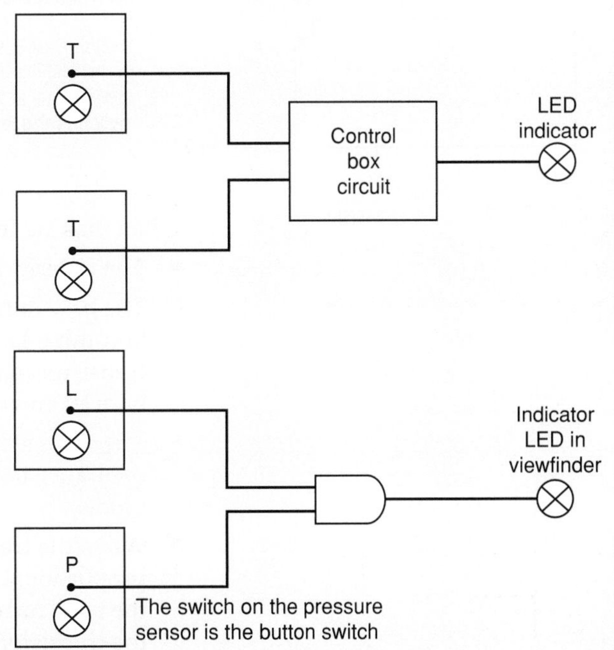

23.4 Transistors and relays

FIRST THOUGHTS

A garage that opens its doors automatically when lit up by car headlights is fitted with light sensors. The door is opened by electric motors. The sensors are connected to transistors and relays that switch the motors on.

An LED indicator or a low-power buzzer may be used to display the output state of a logic gate. When the voltage at the output terminal of the gate goes 'high', the buzzer or the LED is activated. However, a lamp or a motor cannot be activated directly by a logic gate. The gate is unable to supply sufficient current.

Transistors

The **transistor** is a device in which a tiny current controls a much larger current. Switching off the tiny current makes the much larger current switch off; increasing the tiny current makes the much larger current increase. A transistor can boost the current from a logic gate to operate a lamp.

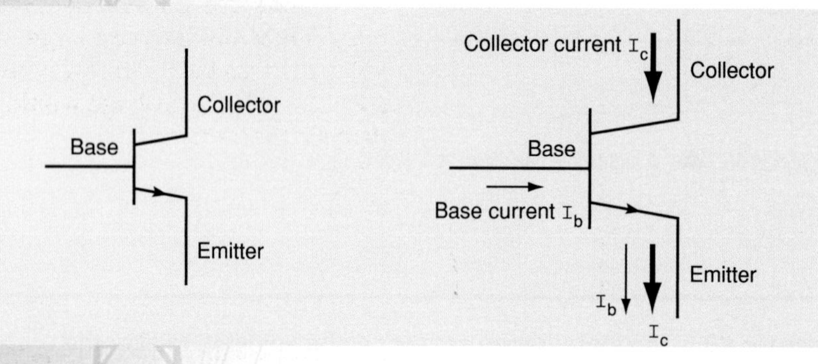

Figure 23.4A The transistor (a) Symbol (b) Current flow

Transistors are made from silicon. A transistor has three terminals, called the **collector, base** and **emitter**. When it is in use, current goes in at its collector and out from its emitter. The important feature of a transistor is that the collector current is controlled by a much smaller current that goes in at the base (and out from the emitter). The transistor is said to be switched ON when current enters its collector (Figure 23.4A).

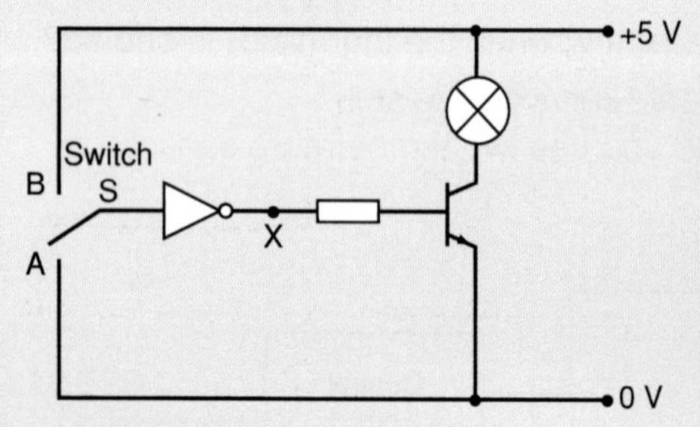

Figure 23.4B Using a transistor

The circuit in Figure 23.4B shows a transistor connected to a logic gate to operate a lamp. When the output voltage from the logic gate is 'high', a tiny current enters the base of the transistor to switch the transistor on. Hence current passes through the lamp into the collector of the transistor (and out from the emitter).

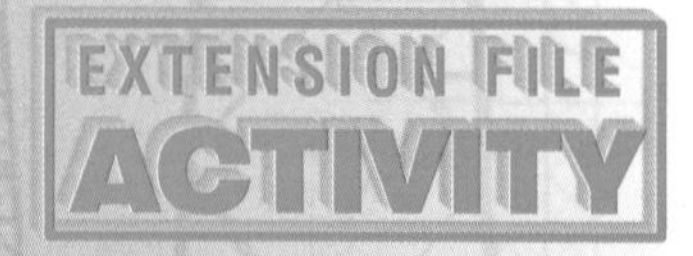

When making transistor circuits, two key rules must be observed:

- The polarity of the voltage supply must be correct.
- The base current should be limited to prevent the collector current becoming large enough to overheat the transistor. If this happens, the transistor may stop working altogether. A resistor in series with the base terminal is used to limit the base current.
- The voltage between the base and emitter stays at about 0.7 V when the transistor is on. This is because the base–emitter junction is like a diode.
- When the transistor is switched on, the collector current I_c is proportional to the base current I_b. The ratio of the collector current to the base current is constant, and is referred to as the current gain of the transistor.

$$\text{Current gain} = \frac{\text{Collector current } I_c}{\text{Base current } I_b}$$

SUMMARY

A transistor is a three-terminal device in which a large current (the collector current) is controlled by a small current (the base current). The base current may be supplied by a logic gate. The transistor can be used to switch a lamp or a relay on or off.

Relays

A light-operated garage door would be very useful for lazy motorists. The car headlamps could activate a light sensor. This would then send a signal to a control circuit to operate an electric motor which would then open the door. However, the motor could not be switched on directly by a transistor; the motor needs much more current than a transistor can pass. A **relay** is used to switch the motor on. The relay coil is connected to the transistor, as shown in Figure 23.4C.

When the transistor in Figure 23.4C is switched on (by the output of the logic gate going 'high'), current passes through the relay coil into the collector of the transistor. The relay coil is therefore energised and thus closes the relay switch. This turns the motor on.

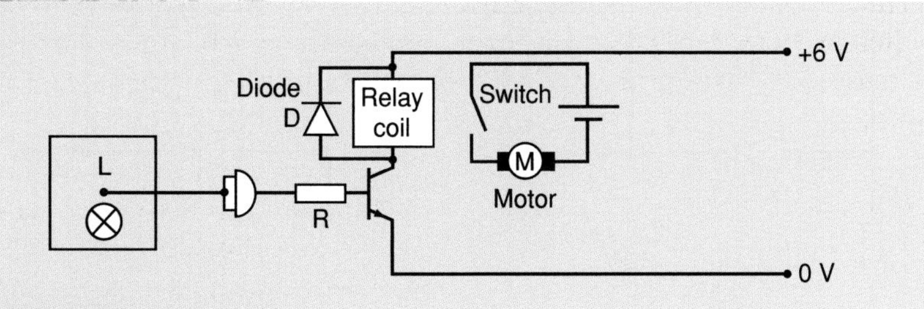

Figure 23.4C ▲ Using a relay

A diode must be connected in parallel with the relay coil as shown. This is to protect the transistor when it is switched off. If the diode was not included, the relay coil would induce a large voltage across the transistor when switched off. This voltage would damage the transistor.

CHECKPOINT

1 The figure below shows a circuit with two lamps labelled X and Y.

(a) When the switch is closed, Y lights up but X does not. Explain why this happens.

(b) What would you expect to see if X and Y were interchanged?

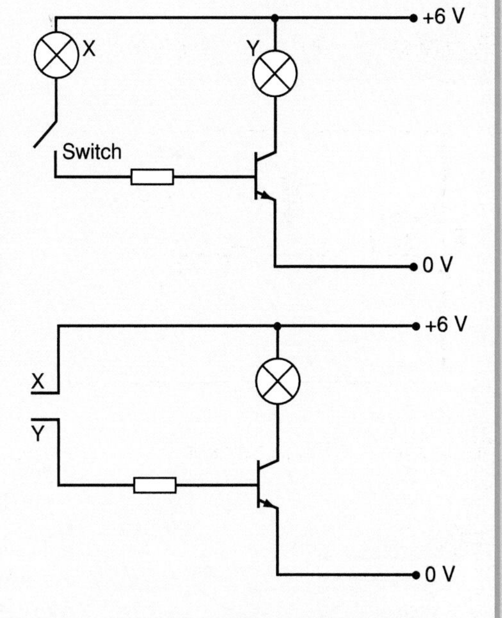

2 Marvin the Magician is showing an audience a trick using the circuit shown opposite. He has connected an LDR between X and Y and the room lights have been dimmed so the lamp is off.

(a) He strikes a match and holds it near the LDR. The lamp lights up. Why?

(b) The match goes out but the lamp stays on. Can you explain this?

(c) He covers the lamp briefly with his hands and the lamp goes out. Why?

3 Transistors can fail to work correctly. Brendan has made the circuit shown in question 2 but it doesn't seem to be working correctly.

(a) The lamp does not come on when the gap XY is short-circuited. Brendan thinks the lamp may have failed but it could be the transistor at fault. How could he test the lamp without taking it out of the circuit?

(b) Brendan discovers that there is a piece of solder joining the base and emitter of the transistor. Why would this stop the circuit working?

4 A motorised washing line that brings the washing in if it rains or if it becomes too cold would be very useful. Design a circuit that will switch a motor on either if it rains or if the temperature falls below a certain value.

23.5 Electronic memory

FIRST THOUGHTS

Computer memories in tiny chips enable instant access to information. How do these memories work? Read on to find out.

Delayed action circuits

Do you ever fall asleep after your bedside alarm has woken you up? If so, you need an alarm that sounds twice – the second time a few minutes after the first time. This type of alarm has a **time-delay circuit**.

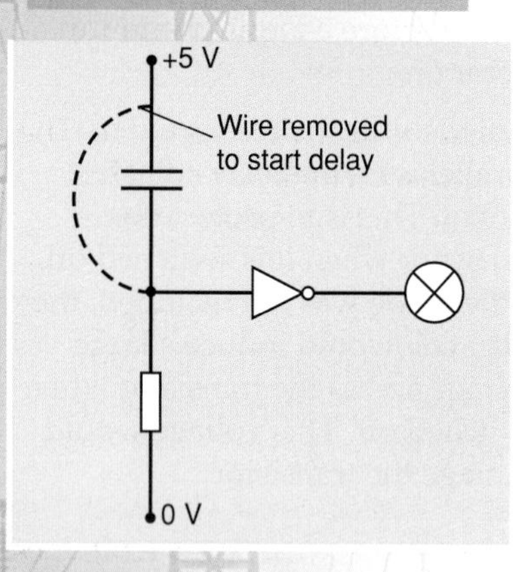

Figure 23.5A A time delay circuit

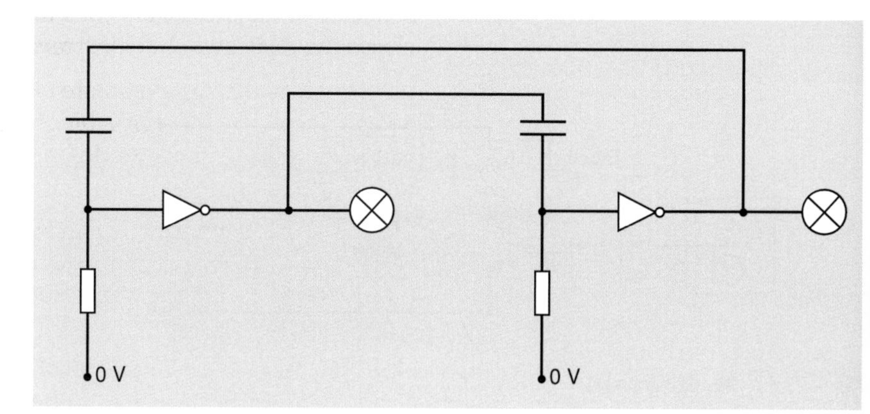

Figure 23.5B An astable circuit

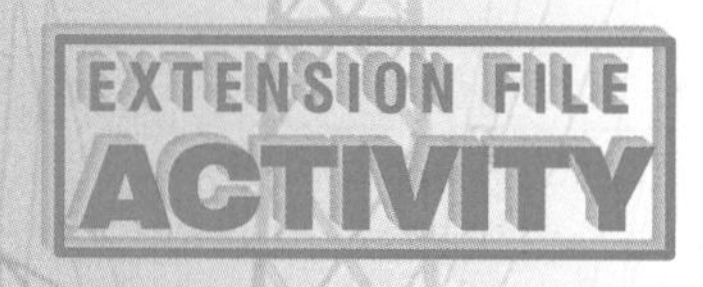

A time-delay circuit works by charging a **capacitor** up. A capacitor is a device that stores charge (see p.296). The capacitor is in series with a resistor. As the capacitor charges up, the voltage across the capacitor rises and the voltage across the resistor falls. The voltage changes can be used to change the output state of a logic gate after a delay. The gate switches an indicator or a buzzer on. Figure 23.5A shows a time-delay circuit.

To start the circuit working, the capacitor is discharged by connecting a wire across its terminals. The indicator goes off because the NOT gate has a '1' applied to its output. As soon as the wire across the capacitor is removed, the capacitor starts charging up and the voltage at the NOT gate input gradually falls. After a certain time, this voltage becomes low enough to switch the indicator back on.

Increasing the resistance increases the time delay before the indicator switches back on.

The circuit 'remembers' to switch on after a delay. It is sometimes called a **monostable** circuit. This is because there is only one stable state for its output.

An indicator is easier to notice if it switches on and off repeatedly. This can be achieved using two time-delay circuits linked to each other, as in Figure 23.5B. This type of circuit cannot remember which indicator to keep on. The circuit is sometimes called an **astable** circuit. Each indicator repeatedly switches from one state to the other automatically.

When one indicator comes on, it switches the other one off for a time; when this one comes back on, it turns the first one off for a time and so on. The output voltage at each gate repeatedly switches on then off. Each on-off cycle is a single pulse. The circuit is a pulse-producing circuit.

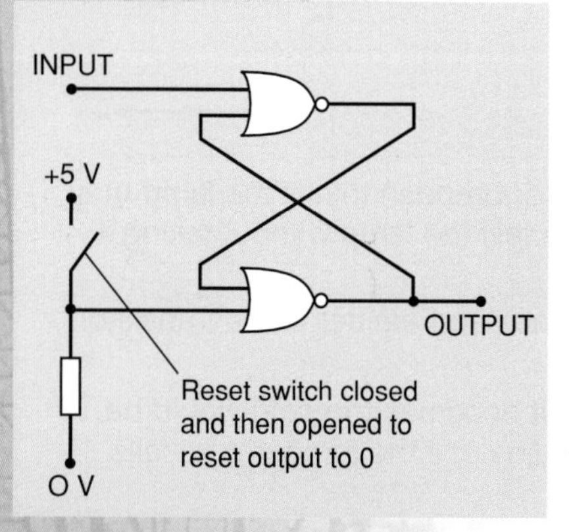

Figure 23.5C A simple latch

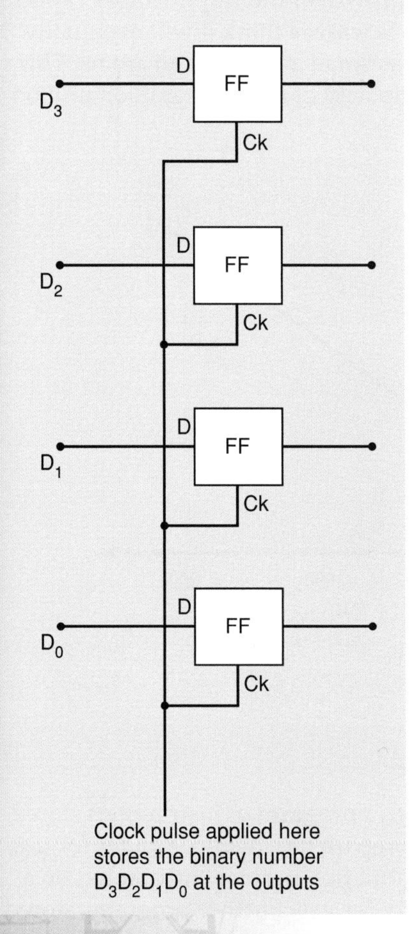

Figure 23.5D Using a flip-flop

Storing and counting circuits

Karen and Rubinda are designing an alarm system activated by opening a door or stepping on a pressure pad. They have made and tested the circuit satisfactorily; however, they have just realised that an intruder just needs to step off the pressure pad and the alarm stops. They want a circuit that will 'remember' if the pressure pad is stood on, even after the pressure is removed.

Figure 23.5C shows how this can be achieved using two NOR gates. When a '1' is applied to the input, the output becomes '1' and stays at '1', even after the input drops to '0'. The circuit is sometimes called a **latch**. In the alarm circuit, the latch continues to supply a '1' to the warning buzzer, even after the pressure on the pad is removed. The '1' is stored by the latch.

Calculators and computers store bits of data in the form of '0's and '1's. The type of circuit used to store a bit of data is called a **bistable** circuit. It has two stable states, one for storing a '0' and one for a '1'. It is sometimes called a **flip-flop**.

The flip-flop shown in Figure 23.5D stores a bit of data when a pulse is applied to its 'clock' input. To store a four-bit binary number (e.g. 1100), each bit is applied to the data input of a separate flip-flop. When a clock pulse is applied, the four bits are 'latched' on to the outputs.

Flip-flops are made using logic gates manufactured on chips. The exact design of the logic circuits is not important to the user; what the circuit does is what matters.

Memory chips

A microcomputer stores data in its memory chips. Each memory chip contains many flip-flops for storing bits of data. The chip is designed like a stack of boxes, with each box able to hold eight bits. An eight-bit binary number is called a **byte**. Each box has an address which must be used to store a byte of data in that box. Figure 23.5E shows the organisation of a simple memory chip.

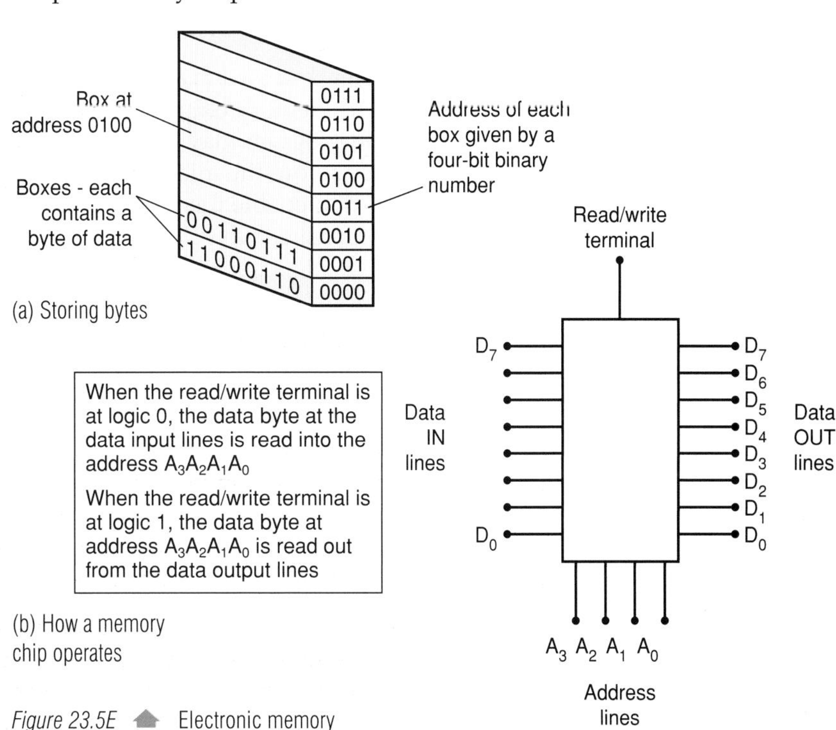

Figure 23.5E Electronic memory

(c) A memory chip

The capacity of a microcomputer's memory is usually expressed in **kilobytes** (= 1024 bytes) or **megabytes** (= 1024×1024 bytes). Before the invention of memory chips, data was stored on tapes; to retrieve data from a tape, the location of data on the tape must be found and this takes time. However, data can be retrieved from memory chips almost instantly because all the locations or boxes are equally accessible.

Your memory is like a memory chip because you can remember things fast and easily. If you couldn't, you would need to keep looking things up in reference books. That is how computers had to work before memory chips were invented. Memory chips enable instant access to data and that is why microcomputers and wordprocessors work so fast.

The graph in Figure 23.5F shows the growth of the chip memory capacity since the memory chip was invented. Scientists think it will eventually be possible to make a single flip-flop as small as a hundred atoms. This would give a single chip a memory capacity of a million million million bytes or more.

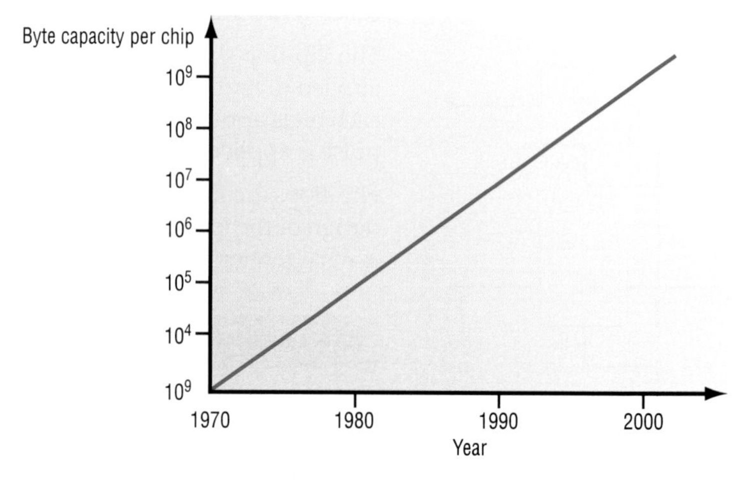

Figure 23.5F The growth of electronic memory

Discs and tapes

When you switch a microcomputer on, a program of instructions stored in its memory chips is carried out to prepare the microcomputer for use. The program may need to call on information permanently stored on a disc before the microcomputer is ready for use. Further programs stored on disc enable the microcomputer to be used as a data base, a spreadsheet, a wordprocessor or a scientific data recorder.

Figure 23.5G A microcomputer in use

SUMMARY

- The output of an astable circuit repeatedly changes from 0 to 1 and back automatically.
- When the output of a monostable circuit is changed from 0 to 1, it reverts back to 0 after a delay.
- A flip-flop is a bistable circuit which has two stable states and may be used to store a bit of data.
- A memory chip contains many flip-flops.

Magnetic disc storage

A computer disc consists of a plastic disc coated with a thin magnetic film. The disc is turned on a spindle through very small steps by an electric motor and a recording head is moved radially within the magnetic range of the disc. Before any data can be stored on the disc, the disc must be 'formatted' into a fixed number of concentric tracks, each track to be used as a storage location. Data is stored on each track as a result of applying voltage pulses to the magnetising coil of the recording head as the disc turns under the head. Each voltage pulse stores a '1' on the film by energising the magnetising coil which then magnetises the film near the read/write head.

To read data from a computer disc, the head detects changes of magnetism as the disc turns under it. Each time the head moves from a '1' to a '0' or vice versa, the change in magnetism induces a voltage pulse in the coil of the head. In this way, data on the disc is converted to a sequence of voltage pulses.

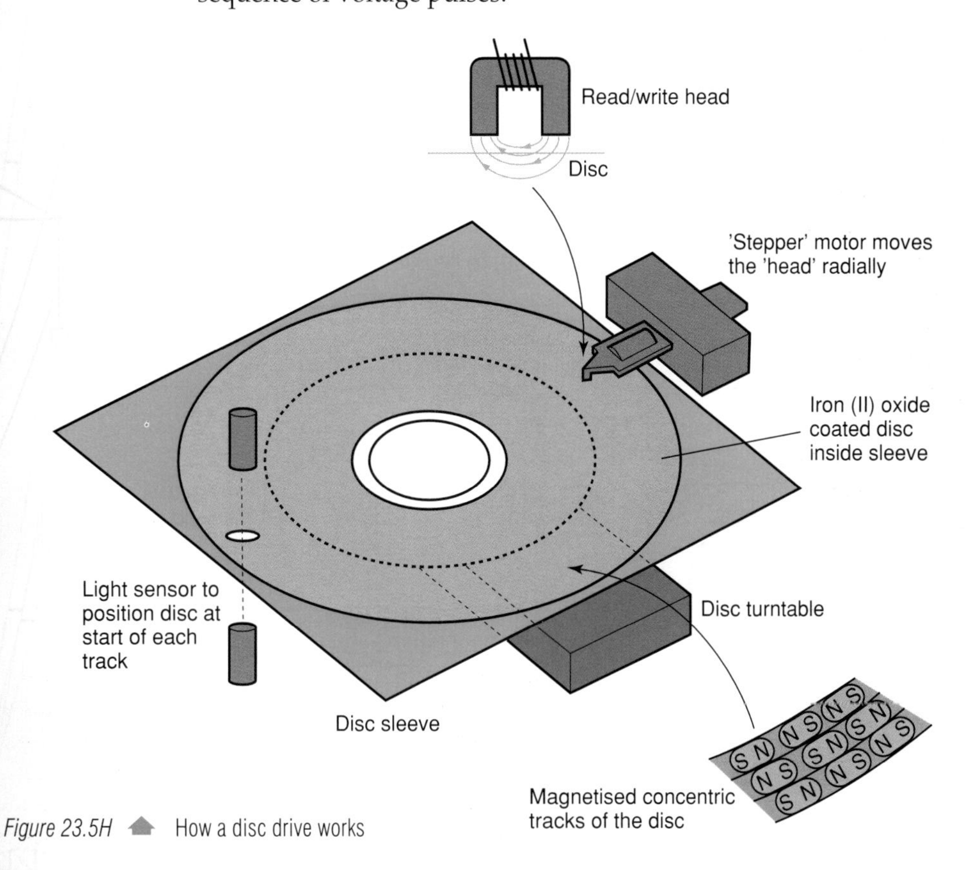

Figure 23.5H ▲ How a disc drive works

1 gigabyte = 1000 megabytes
1 megabyte = 1000 kilobytes
1 kilobyte = 1000 bytes

The **capacity** of a computer disc is expressed in terms of how many bytes can be stored on the disc. For example, a microcomputer with a '1.2Gb hard disc' contains an internal disc drive fitted permanently with a disc that can store 1200 megabytes of data. The computer will also be fitted with an external disc drive into which a disc may be inserted to load programs into the computer or to store data from the computer. The capacity of such a disc is much smaller than the capacity of the hard disc.

Magnetic tape storage

Magnetic tape was used to store computer data before high-capacity magnetic discs became available. The principle is the same as that of disc storage. However, magnetic tape storage is slow in comparison with disc storage, even with high tape speeds. Winding and rewinding a tape to find a data location is time consuming!

Video recorders use magnetic tape because enormous amounts of information must be stored to recreate a sequence of TV pictures. The longer the tape, the greater the amount of information that can be stored. Video tapes are much broader than audio tapes because the recording head scans up and down the tape as the tape moves across the head.

To estimate the storage capacity of a video recorder, remember that a colour TV picture is renewed 25 times each second. Since a colour TV has about half-a-million colour dots used to create the picture, the signal needs to supply about 12 million pulses every second to generate the picture. The storage capacity of a 1 hour video tape is therefore about 40 000 million 'bits' (approximately 4000 megabytes).

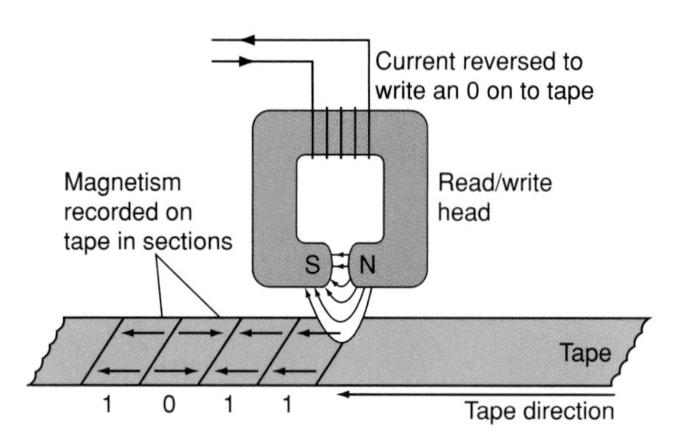

(a) Writing on the tape

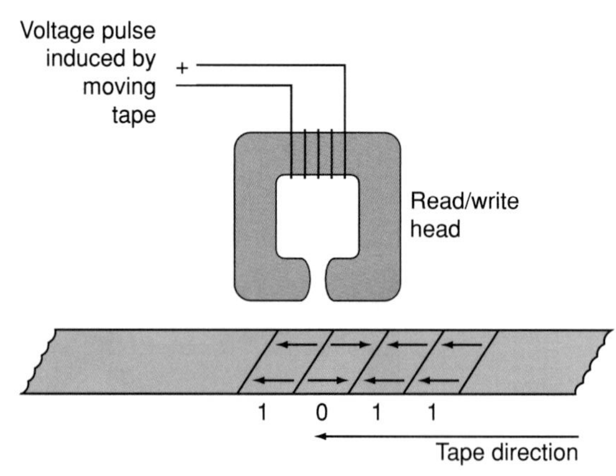

(b) Reading the tape

(c) Audio and video tapes

Figure 23.5l Magnetic tapes

Compact discs

If you have a musical 'ear' you will be able to compare the quality of recorded music from a compact disc player with other forms of recorded music. However, you cannot record your own music on a compact disc as you can with a cassette tape; you can only 'read' a compact disc.

The surface of a compact disc is covered with a transparent film of plastic to protect the disc track which spirals out from the centre to the rim. Pits along the track represent bits of data and the disc rotates slowly under a reading head that moves gradually towards the centre. The reading head detects the pits optically and a continuous sequence of electrical pulses representing '1's and '0's is produced. These pulses are then used to recreate the audio signal that made the pits when the disc was manufactured.

It's a fact!

The spiral track of a compact disc is about 5 km in length. The size of a pit is about 5 mm. That gives about 1000 million pits along the track. A compact disc designed to last about 1 hour would therefore involve detecting about 250 000 pits per second.

(a) A compact disc on its loading platform

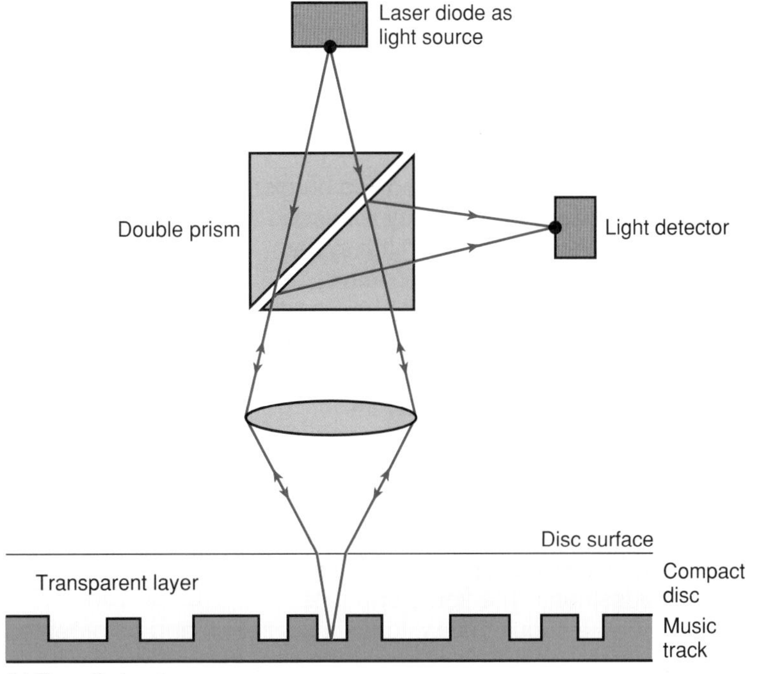

(b) The optical system

(c) Microscopic view of CD

Figure 23.5J ▲ CDs

The reading head is an array of photodiodes which detect light from a laser diode reflected by the compact disc surface. No light is reflected when the recording head is over a pit; as the track moves under the head, a continuous sequence of light pulses is incident on the recording head.

A CD ROM system is a compact disc system designed for a microcomputer. The term 'ROM' means 'Read Only Memory'. A colour TV picture consists of about half-a-million dots so a large compact disc with about 10 000 million pits could provide about 20 000 TV pictures – useful as a resource base but too short for TV programmes! However, by using light of shorter wavelength, discs could be made with storage capacity equivalent to or better than video tapes. Manufacturers would benefit from such a development since compact discs, unlike tapes, cannot be recorded over!

SUMMARY

- A flip-flop has two stable states and may be used to store a bit of data. A memory chip contains many flip-flops and is used to store data that can be retrieved rapidly.
- Discs store much more data than memory chips at present but take longer to retrieve data from.
- Tapes store much more data than discs or chips but take too long to retrieve data from. TV programmes are recorded on tape rather than on disc at present because a disc has insufficient storage capacity.

Figure 23.5K CD ROM

CHECKPOINT

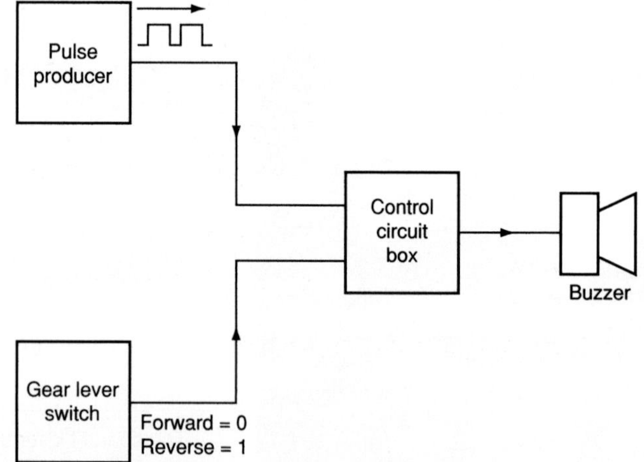

1 A time-delay circuit is needed in a burglar alarm system. This time delay is needed to allow the user time to switch the system on inside the house and then leave.
(a) What would happen if the time delay was too short?
(b) When the system is installed, the time delay is adjusted by the installation engineer to suit the user. How can the time delay for the circuit in Figure 23.5A be made longer?

2 (a) When a bus or lorry reverses, an alarm sounds very loud to warn pedestrians. The figure opposite shows a circuit that switches a buzzer on and off repeatedly when the gear lever is put into reverse. What type of logic gate should be in the control box?
(b) At a pedestrian crossing, a bleeper sounds when the 'cross now' sign lights up. Design a system to light an LED and sound a bleeper for a fixed time when a button is pressed.

3 (a) Eight flip-flops are needed to store a single byte. How many flip-flops are contained in a 16 kilobyte chip?
(b) Write down the binary number for (i) 31, (ii) 12.
(c) Give your date of birth and then convert it into binary code as two bytes.

4 (a) Use Figure 23.5F to estimate when a chip with a memory capacity of some million million bytes is likely to be developed.
(b) Present memory chips can retrieve a data byte within microseconds. How long would it take to search through all the bytes in a one megabyte chip?
(c) Scientists are trying to find ways of making chips work faster. Why is this important?

5 To create a TV picture, about half-a-million bits of data are needed. A TV programme displays 25 pictures per second.
(a) Work out how much data storage capacity is needed for a 1 hour TV programme.
(b) How many 200 page books could be stored by a device with the same capacity as in (a)?

23.6 Electronic control

FIRST THOUGHTS

Computer controlled devices are programmed to follow sets of instructions. You need to have studied the previous sections to grasp the topics in this section.

Programmable control

Lights that flash on and off help to enliven the atmosphere at a disco. The lights could be switched on and off **manually**, but the operator would soon get fed up and would probably disappear into the crowd. Using electronics, the lights can be **programmed** to switch on and off automatically in any chosen sequence.

Figure 23.6A shows how this can be done using a memory chip. The program is stored in the first eight addresses of the chip's memory. Each address contains eight bits of data. When each address is read, the four bits, D_3, D_2, D_1 and D_0, are used to switch the four indicators on or off. The addresses are read repeatedly in sequence to make the indicators switch on and off repeatedly.

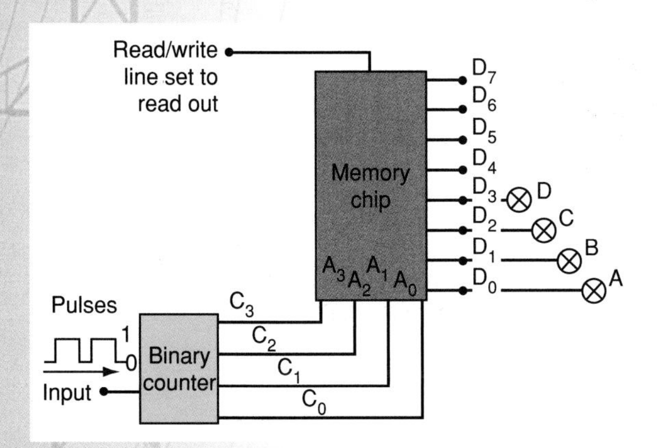

Figure 23.6A ▲ Programmable memory

Sample program

Number of pulses counted	Address A_3 A_2 A_1 A_0	Data byte read out D_7 D_6 D_5 D_4 D_3 D_2 D_1 D_0
0	0 0 0 0	0 0 0 0 0 0 0 1
1	0 0 0 1	0 0 0 0 0 0 1 1
2	0 0 1 0	0 0 0 0 0 1 1 1
3	0 0 1 1	0 0 0 0 1 1 1 1
4	0 1 0 0	0 0 0 0 0 0 0 1
5	0 1 0 1	0 0 0 0 0 0 1 0
6	0 1 1 0	0 0 0 0 0 1 0 0
7	0 1 1 1	0 0 0 0 1 0 0 0

Pulses 0–3 switch indicators A, B, C, D on one by one
Pulses 4–7 switch the indicators on one at a time

Traffic lights are programmed to operate in a particular sequence. Figure 23.6B shows how this is done using a memory chip. The chip is programmed by 'writing' bits of data into each address. These bits are then 'read' out to switch the indicators on or off as programmed.

Microcomputers can be programmed to switch devices on and off. Data can be supplied to or from the computer at terminals called **ports**.

Number of pulses counted	Address A_3 A_2 A_1 A_0	Data byte read out D_7 D_6 D_5 D_4 D_3 D_2 D_1 D_0	State of indicators D_2(red)	D_1 (amber)	D_0 (green)
0	0 0 0 0	0 0 0 0 0 1 0 0	ON	OFF	OFF
1	0 0 0 1	0 0 0 0 0 1 1 0	ON	ON	OFF
2	0 0 1 0	0 0 0 0 0 0 0 1	OFF	OFF	ON
3	0 0 1 1	0 0 0 0 0 0 1 0	OFF	ON	OFF
4	0 1 0 0	0 0 0 0 0 1 0 0	ON	OFF	OFF

Figure 23.6B ▲ Traffic lights program

A port has a number of **lines**. The logic state of each line can be set at 0 or 1 by means of a suitable program. Each device is connected to an interface circuit which is then connected to a port. The interface is to protect the computer's circuits. The microcomputer can be programmed to supply bits of data to the port in any desired sequence to switch devices on or off. Figures 23.6C and D show how a microcomputer may be used to control a buggy driven by two electric motors.

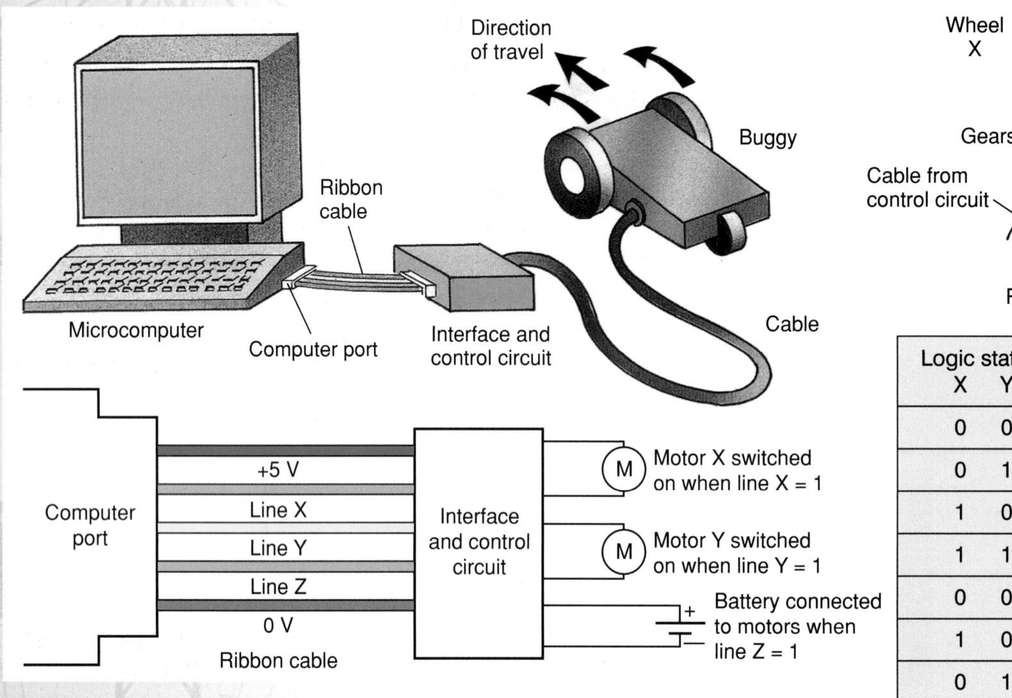

Logic state of line X Y Z	Direction of travel of buggy
0 0 0	Stationary
0 1 0	Stationary
1 0 0	Stationary
1 1 0	Stationary
0 0 1	Stationary
1 0 1	Turn left
0 1 1	Turn right
1 1 1	Forward

Figure 23.6C A computerised buggy

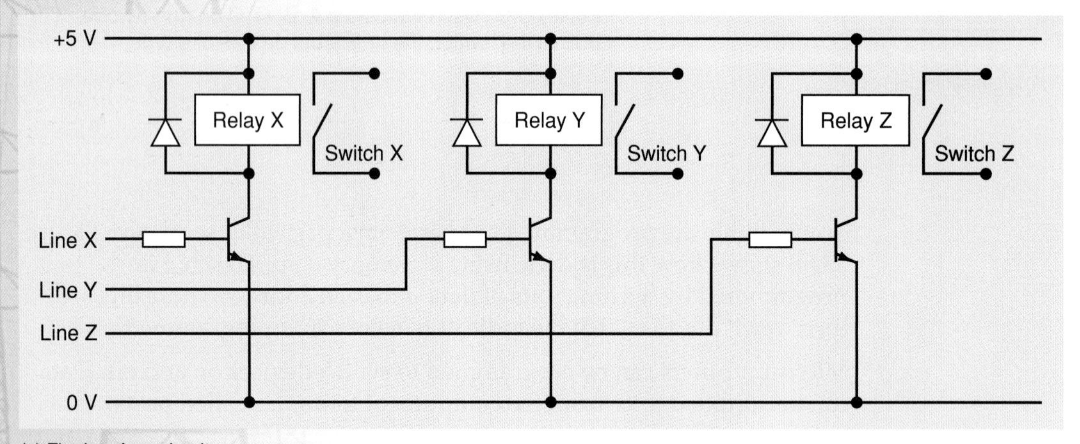

(a) The interface circuit

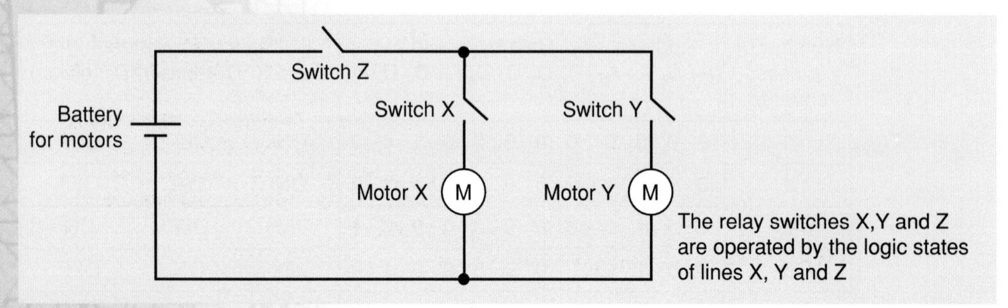

(b) The control circuit

Figure 23.6D Inside the interface and control circuit

FIRST THOUGHTS

The human body has a remarkable feedback mechanism that maintains body temperature at 36.4 °C, winter or summer. Temperature 'sensors' in the skin send signals to the brain. This sends signals back to the blood system to control heat loss from the body.

Feedback

In factories, lots of jobs are repetitive and are done by machines. These machines are often called **robots**. They are programmed to perform routine jobs repeatedly without stopping. For example, a lifting robot may be used to fix the windscreen on to a car. *How does the robot know when the windscreen is correctly in place?* Pressure sensors feed a signal back to the control unit of the robot when the windscreen is in place.

Radio-controlled model cars also use feedback. If the car moves in the wrong direction, the controller uses a joystick on the control unit to make the car move back on to the correct path. The controller feeds a signal back to the car. This is an example of **manual feedback** because someone must watch the car's path and correct it if necessary. The lifting robot uses **automatic feedback** using sensors; no one is needed to watch the windscreen being placed.

Automatic feedback can be used to make a buggy follow a line drawn on the floor. Two LDRs are used as 'eyes' looking at the floor. If the buggy strays from its path, one of the LDR's moves off the line and sends a signal to one of the motors. This moves the buggy back on course. (See Figure 23.6E.)

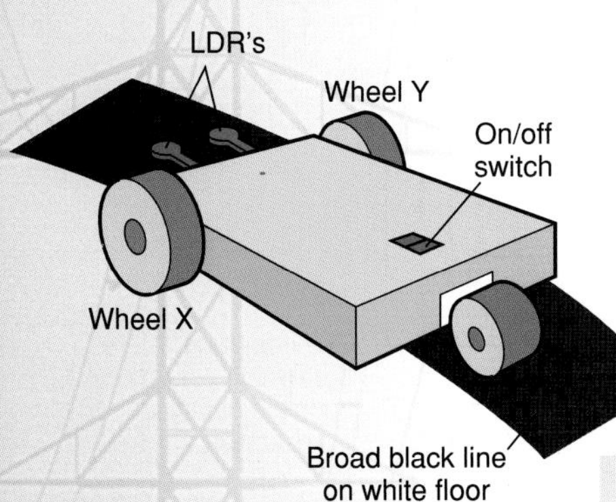

LDR X switches motor Y on or off
LDR Y switches motor X on or off

(a) Using LDRs

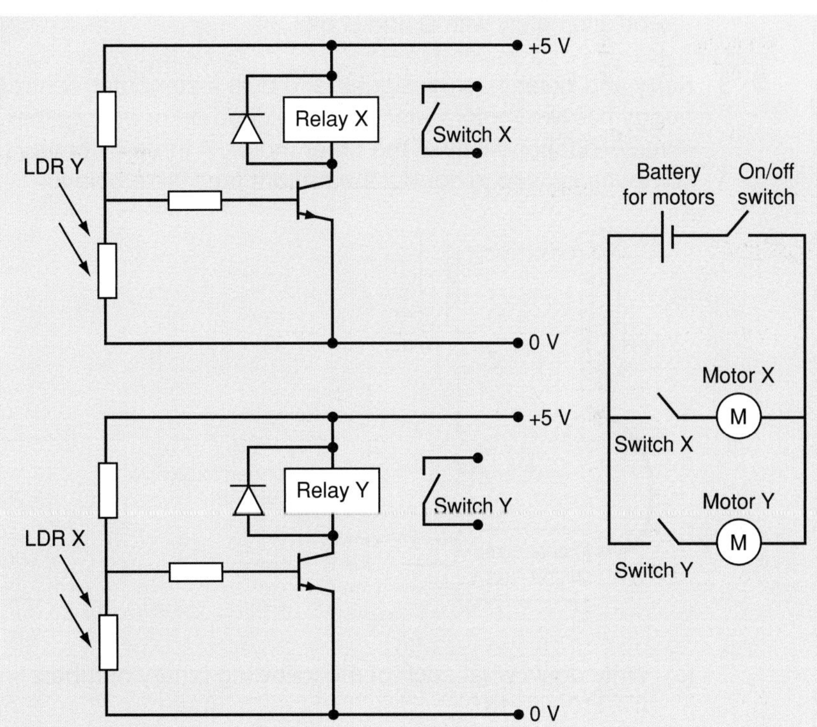

(b) The control circuit

LDR X	LDR Y	Direction of travel of buggy
Black	Black	Straight ahead (Both motors ON)
Black	White	Turn left (Motor X OFF, motor Y ON)
White	Black	Turn right (Motor X ON, motor Y OFF)

When each LDR is over the black strip, its resistance is high so the relay in its circuit is energised.
If one of the LDRs moves off the black strip, its resistance falls because it receives more light from the white background. This turns the relay in its circuit off.

(c) LDR control

Figure 23.6E ▲ An automated buggy

SUMMARY

Computers can be used to control machines automatically. Feedback is necessary to ensure machines do not go out of control.

CHECKPOINT

1 (a) In Figure 23.6A, state which indicators are switched on by each of the following four-bit addresses (i) 0 1 0 1, (ii) 0 0 0 0, (iii) 0 1 1 0.

(b) Write down the sequence of four-bit addresses that would make A flash on and off repeatedly, keep B off throughout, make C off every time A flashes on, and keep D on throughout.

2 A computer is used to switch motorway warning lights on. Each set of lights consists of four indicators at the corners of a square sign, as shown in the figure opposite.

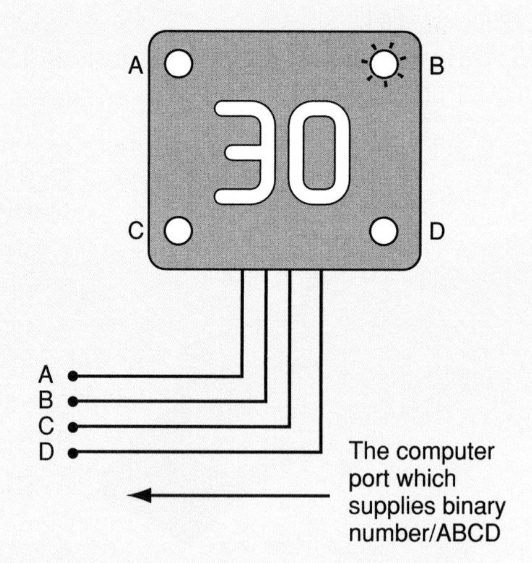

(a) What binary number must be supplied to the four lines of the output port to switch all the indicators (i) on, (ii) off?

(b) What would be seen if these two numbers were supplied alternately and repeatedly?

(c) What binary numbers should be supplied to switch A and B on and C and D off?

(d) How could A and B be flashed on and off alternately with C and D off?

3 Kerry and Helen have designed and built a computer-controlled as buggy as shown below. The buggy has two motors which are switched on or off together. Each motor can go either forward or reverse independently of the other motor. A three-bit binary number supplied to the port by the computer is used to control the motors (see table below).

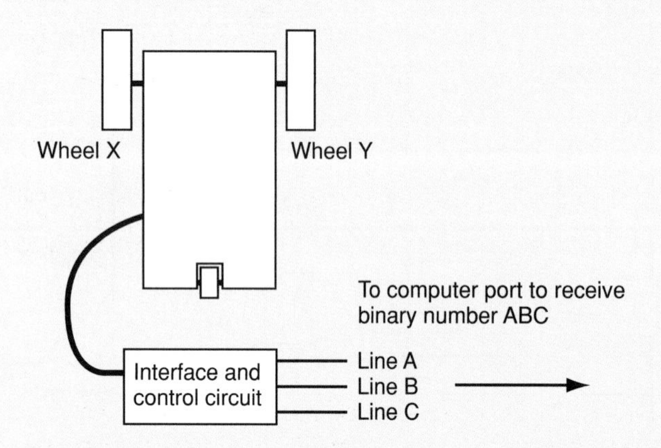

	A	B	C
Logic 0	Motors off	Wheel X forwards	Wheel Y forwards
Logic 1	Motors on	Wheel X reverse	Wheel Y reverse

(a) Write down what each of the following binary numbers would do to the buggy (i) 111, (ii) 000, (iii) 110, (iv) 101.

(b) Write down the sequence of numbers that would make the buggy travel clockwise round the sides of a square.

4 A factory manager is considering ordering a computer controlled lathe to make machine parts. The machine costs £55, 000. It can make 20 parts per hour compared with the five parts per hour a manual operator can make on a lathe costing £10,000. The manual operator costs £10 per hour.

(a) How many manual lathes are needed to produce 20 parts per hour?

(b) How much extra does the computer controlled lathe cost compared with the manual lathes for the same production?

(c) Why is the real cost of the manual lathes much higher?

23.7 Using microcomputers in science

FIRST THOUGHTS

In this section, you will find out some of the immediate benefits of microcomputers.

With so much work to do for GCSEs, you might perhaps find it helpful if a microcomputer could do some of it for you. No doubt you use a calculator when doing maths; perhaps you have used a word processor when writing up an essay or report. In the science laboratory, it is possible to use a microcomputer to make measurements automatically and to work out calculations. Meanwhile, you could be catching up on some overdue homework.

Suppose you have to investigate the acceleration of a trolley down a runway to find out how the slope of the runway affects the motion. You could use a **ticker timer**. A much easier method would be to use a microcomputer to time the trolley as it passes through each of two light gates, as in Figure 23.7A. The computer can also be programmed to time how long the trolley takes to go from one gate to the other and to work out the acceleration for you. All you need to do (after setting the experiment up and programming the computer) is to release the trolley and then write down its acceleration from the microcomputer's visual display unit (VDU) afterwards.

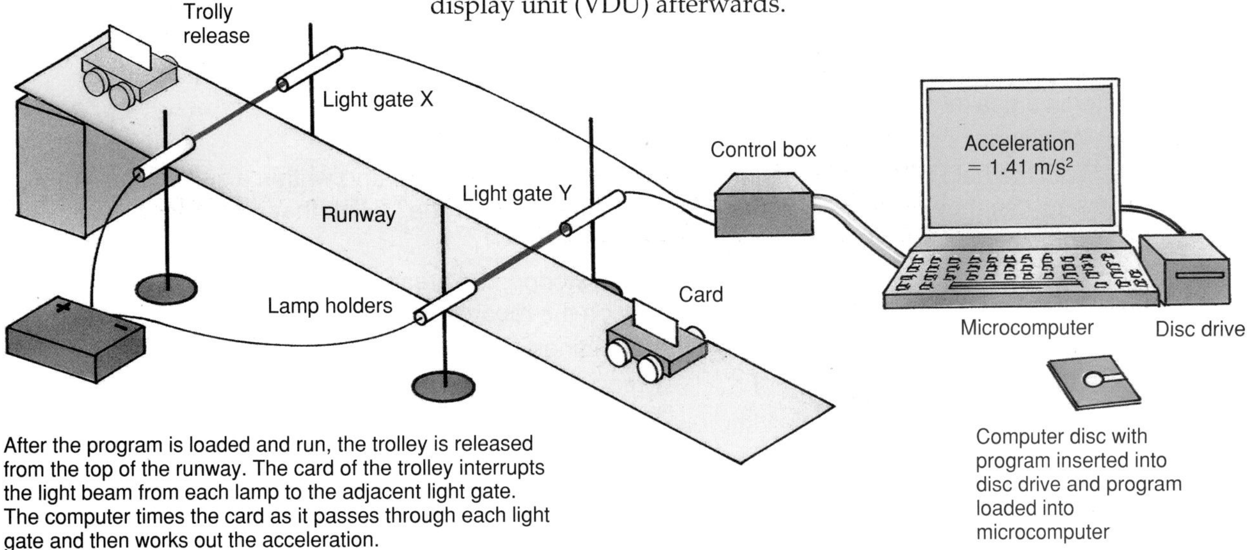

After the program is loaded and run, the trolley is released from the top of the runway. The card of the trolley interrupts the light beam from each lamp to the adjacent light gate. The computer times the card as it passes through each light gate and then works out the acceleration.

Figure 23.7A Using a microcomputer to determine acceleration

Microcomputers can be programmed to receive data from suitably designed sensors. For example, Figure 23.7B shows a microcomputer being used to measure and display temperature. The sensor contains a **thermistor** connected into an electronic circuit. The circuit supplies a voltage to the microcomputer that varies with temperature. The microcomputer can be programmed by the user to measure the temperature at regular intervals and to plot a graph of temperature against time on its VDU.

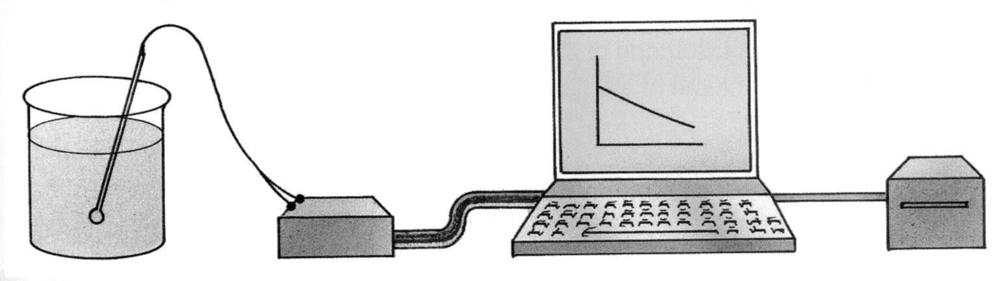

Figure 23.7B Measuring temperature

SUMMARY

Microcomputers and data recorders can be connected to specially designed sensors to measure and record physical quantities such as temperature, light intensity, pH values, etc. A sensor must be connected to a microcomputer via an interface circuit.

Data recorders are pre-programmed to record data. This means that the user does not have to supply a program because the program is already stored in one of the recorder's memory chips. The user needs only to select the required program using the keypad.

Many physical quantities such as light intensity, pH value, speed, pressure can be measured using a microcomputer or a data recorder. For use with a microcomputer, the sensor has to be connected to one of the microcomputer's ports via an interface circuit. This is a circuit designed to protect the microcomputer.

Figure 23.7C ▲ pH sensor connected to a microcomputer

CHECKPOINT

1 Does adding weight to a trolley make it accelerate faster down an incline? How could you investigate this, using the apparatus shown in Figure 23.7A and some additional weights?

2 Simon and Linda want to use a microcomputer to investigate whether the temperature of the water in a beaker changes when common salt is dissolved in the water.
   (a) What type of sensor is needed?
   (b) With the help of their teacher, they have set the apparatus up and written a suitable program. How can they check the accuracy of the temperature probe? (Hint: they must measure something of known temperature.)

3 A certain data recorder is capable of taking and storing 1024 readings at any one of several different rates. These rates, selected by a dial on the recorder, are listed below.

A	1000 readings per second	E	10 readings per minute
B	100 readings per second	F	1 reading per minute
C	10 readings per second	G	10 readings per hour
D	1 reading per second	H	1 reading per hour

   Select the most appropriate range for each of the following tasks:
   (a) Recording the changing light intensity in a greenhouse over 12 hours from sunrise to sunset.
   (b) Investigating the current growth in a light bulb when the bulb is switched on.
   (c) Recording the growth of a plant using a movement sensor.
   (d) Recording the temperature of a certain liquid near its freezing point when it suddenly solidifies due to a 'seed' crystal being dropped into it.

4 The figure opposite shows the results from an experiment in which an alkaline solution was added drop-by-drop to some dilute sulphuric acid in a beaker. A microcomputer was used to measure and display the pH value of the contents of the beaker against the volume of alkali added.
   (a) Explain the shape of the curve.
   (b) What volume of alkali neutralised the acid exactly?
   (c) Sketch the curve you would expect for pH-value against volume added if the acid had been added to the alkali.

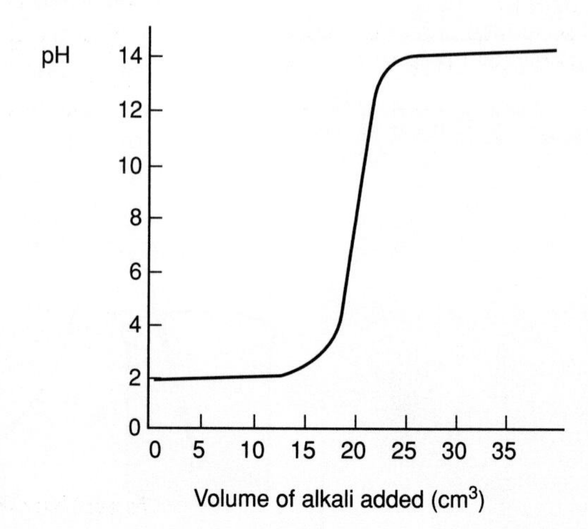

23.8 Electronics and communications

FIRST THOUGHTS

Whatever career you choose, you will undoubtedly use electronics to communicate with other people.

It's a fact!

The Internet is a communications network which developed from a communications system devised by the US Government in the Cold War to withstand a nuclear attack. Instead of linking the computers in the system to a single control centre, a network of powerful computers are linked to each other permanently to provide many routes between any two computers. The Internet consists of a permanent network of internet service providers (ISPs), each connected to the telephone system. A message from one computer to another is chopped into a sequence of packets, each sent independently to the same destination by any route. In contrast, a phone call or a fax is sent along a route which is the same throughout the transmission. E-mail is a transmission control system (referred to as a 'protocol') recognised by ISPs for addressing and transmitting information. The World Wide Web (WWW) was invented in 1992 by physicists at CERN, the European Centre for Nuclear Research, to enable physicists around the world to gain access via the Internet to scientific papers filed electronically. Within a few years, WWW became the indispensable centrepiece of commerce, industry and entertainment and has revolutionised almost every aspect of our lives.

Figure 23.8A ▲ Using a fax machine

Stephanie is an architect working in her office at home on the design of a sports centre to be built in a nearby town. The builder has just telephoned her to find out how her work is progressing. Stephanie uses her fax machine to send a drawing of the roof design to the builder.

Using the telephone, you can speak to relatives and friends locally and internationally. A fax machine uses a telephone line to send a written message or a drawing to any other fax machine in just a few minutes. For many people, electronic mail (e-mail) is now the preferred method of sending information to other people, using personal computers linked to the Internet via 'modems'.

We send and receive information to and from each other all the time. We get most of our information from what we see and hear. The 'mass media' (radio, television, newspapers, etc.) supply information to millions of people. Books, computer data bases and websites on the World Wide Web store information which can be referred to when required. Digital television and websites inform, educate and entertain subscribers on demand.

Information is stored by computers as bytes of data, each byte consisting of a sequence of 0's and 1's. Figure 23.8B shows a simple computer program to store and retrieve the phone numbers of your friends.

```
10  INPUT "NAME TO BE FOUND"; N
20  READ L $, L%
30  IF N$ = L$ THEN 50
40  GOTO 20
50  PRINT "NUMBER IS"; L%
60  DATA CLAIRE, 52713, JOHN, 25413, RUBINDA, 36214, SARAH, 42138
70  DATA DATA MICHELLE, 76344, SEPHINDA, 44213
```

Note: Use QBASIC, see p. 162. Write your information for lines 60, 70, etc.

Figure 23.8B ▲ A personal telephone directory

(a) The Highway Code: what does this mean?

(b) A sorting machine reading postcode dots

(c) Reading bar codes
Figure 23.8C

To retrieve a number, type in the appropriate name on the keyboard and the number is displayed on the VDU. As each letter of the name is typed in, an eight-bit byte is sent from the keyboard to the microcomputer's memory. All the letters and symbols on the keyboard are coded so that each gives a different eight-bit byte. The code is known as ASCII (American Standard Code for Information Interchange).

Many different codes exist for sending information. The reason for using a code may be to give rapid transfer of information (as with ASCII) or perhaps to protect information (as with secret codes). Some more codes are described below.

- **The Highway Code** is a list of regulations that must be obeyed by road users. Traffic signs supply information at a glance to motorists. The motorist must remember what each sign means. (Figure 23.8C(a).)
- **Postal codes** allow letters and parcels to be sorted and delivered rapidly. When you send a letter, be sure to write on the post code of the destination. After your letter has been collected from the post box, it is taken to the nearest sorting office where the post code is read and then coded as a sequence of dots printed on the envelope. These dots glow in ultraviolet light and can be read 'automatically' by sorting machines which channel your letter to its destination. Your letter may pass through several sorting machines in different offices before it is delivered to its destination. (Figure 23.8C(b).)
- **Bar codes** are used in many supermarkets at the check-outs. Each item on the shelves has a bar code printed on it. The assistant at the cash-till passes each item's bar code across the bar-code reader; the description and price of the item is displayed and printed out on the till receipt. The supermarket manager can keep track of stocks of items and the consumer gets a print-out of the description and price of each item bought. The information about each item is held in the memory of a computer which is linked to every till in the supermarket. (Figure 23.8C(c).)

Communication links

Computers in different towns can be linked using the telephone system. To send information from one computer to another, each byte is converted into a stream of pulses of light or electricity. The pulses are transmitted down telephone lines or optical fibre lines or using microwave beams (see p. 202).

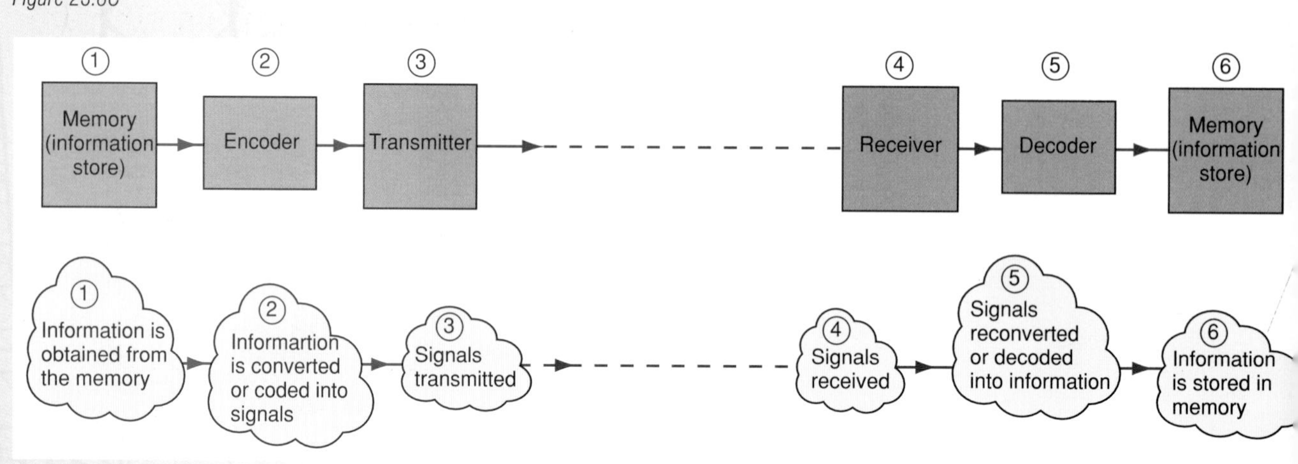

Figure 23.8D Sending information

It's a fact!

The Data Protection Act gives each of us the right of access to check personal information held in computer records.

Safety on the line

The high speed rail line between London and Edinburgh requires high speed signals. Signal boxes in the major cities along the line will be linked by microwave towers. Each piece of signalling equipment along the line will receive identical commands from the nearest two signal boxes. These commands will be transmitted from each box along optical fibre cables. If one of the cables supplying a piece of equipment fails, the signals will still arrive via the other cable.

Computer records are data bases holding vast amounts of information. The Inland Revenue computers store information about every working adult in the country. Hospital computers hold personal medical records. Banks and building societies use computers to store financial information about customers. Ministry of Defence computers hold vital information about national defence.

How secure is the information held in computer data bases? If computers send information using the phone network, what safeguards prevent unauthorised access to such information? Security codes are used to stop illegal entry to data bases. To gain access, the correct security code must be keyed in.

How secure is information when it is being transmitted? Optical fibres and microwave links are gradually replacing copper wires in the phone network. Figure 23.8E shows how a phone conversation is carried by a pair of copper wires. If several pairs of wires are twisted together, 'crossed lines' may occur and phone callers may hear other calls in the background. In fact, such phone lines can easily be 'tapped' to find out what is being said. *Are optical fibres and microwave links more secure from tapping than copper wires?*

Figure 23.8E ▲ How a phone call is carried

A single pair of wires can be made to carry hundreds of phone calls at the same time. This uses a technique known as pulse code modulation (PCM). Each audio signal is sampled (i.e. measured) 8000 times per second. Each sample is converted into a binary number and transmitted as a sequence of very brief pulses. In the time between one pulse and the next, pulses from other calls are transmitted (Figure 23.8F). A single pair of wires is capable of transmitting about two million pulses each second.

A microwave beam is capable of carrying thousands of phone calls at the same time. Microwaves are electromagnetic waves with frequencies of approximately 10^{10} Hz. As explained on p. 204, a microwave beam can be switched on and off extremely fast, enabling it to carry more than 1000 million pulses per second. Using PCM, many phone calls can be carried at the same time. A single microwave beam can even carry all the signals for a television channel.

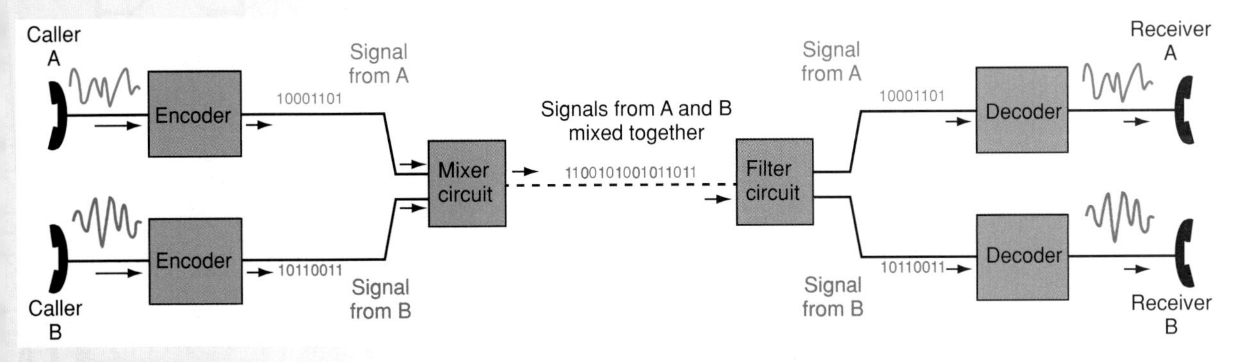

Figure 23.8F ▲ Pulse code modulation

Microwave towers in cities and towns are used to transmit and receive microwave beams. The microwaves travel in a straight line between the towers and are unaffected by weather conditions. Satellites in geostationary orbits receive and transmit microwave beams from ground stations to carry international phone calls.

Optical fibres are used to guide light over long distances. The frequency of light used is approximately 10^{14} Hz which is 10 000 times higher than typical microwave frequencies. Thus, a light beam can be used to carry many more pulses each second than a microwave beam. A single optical fibre, using PCM, can carry more than 30 000 phone calls or 30 television channels at the same time.

The glass in an optical fibre is extremely clear. A semiconductor laser is used to send light pulses along an optical fibre. Where the fibre bends, the light rays are **totally internal reflected** where they touch the fibre boundary.

Optical fibres can carry much more information than copper wires and glass is much cheaper than copper. New developments such as home shopping, local television stations and video phones will become possible as optical fibres replace copper wires where demand is likely to be high such as between telephone exchanges. More efficient communications circuits known as 'broadband' will connect to homes using existing copper wires thus cutting installation costs for new services.

Figure 23.8G Microwave traffic
(a) The BT tower, 189 m high, is used to send and receive microwaves to and from other towers throughout the UK

(b) The UK microwave network

(c) A microwave beam carrying data

SUMMARY

Computers store and transfer information as bytes of data, each byte consists of a sequence of 1's and 0's. The phone network is used to link computers in different locations. Copper wires in the network are being replaced by optical fibres and microwave beams which can carry much more information.

CHECKPOINT

▶ **1** Morse code was invented to send information in the form of electrical pulses along telegraph wires. The code is shown below. Each dot represents a short pulse and each dash a long pulse.

(a) Write your name in Morse code.

(b) Write a short sentence and convert it to Morse code.

(c) Send your sentence in Morse code to a friend who then has to decode it and send a reply back in Morse code.

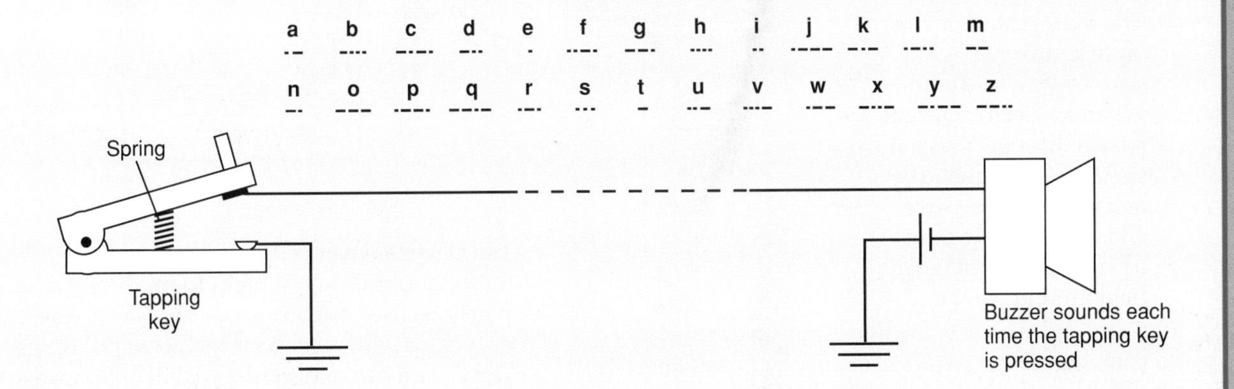

▶ **2** Make a list of ten points of information you have gained over the past 24 hours. If possible, include in your list at least one point from the radio, one from the TV, one from a friend and one using neither sight nor sound. Write down how you gained each point on your list.

▶ **3** To obtain money using a bank card, the holder puts the card into the dispensing machine and then keys in his or her personal identification number (PIN) before keying in the amount of money required.

(a) Under no circumstances should the PIN be written on the bank card. Why?

(b) The card holder is usually allowed three attempts to key in his or her PIN number. After that, the machine keeps the card. Why does the bank not allow unlimited attempts?

(c) If the holder loses his or her card, why is it important to inform the bank as soon as possible?

▶ **4** Braille enables blind people to read. It consists of raised dots on paper, as shown below.

(a) Use a pin to write a short message on a sheet of paper in Braille.

(b) Ask a friend to decode the message and write a reply.

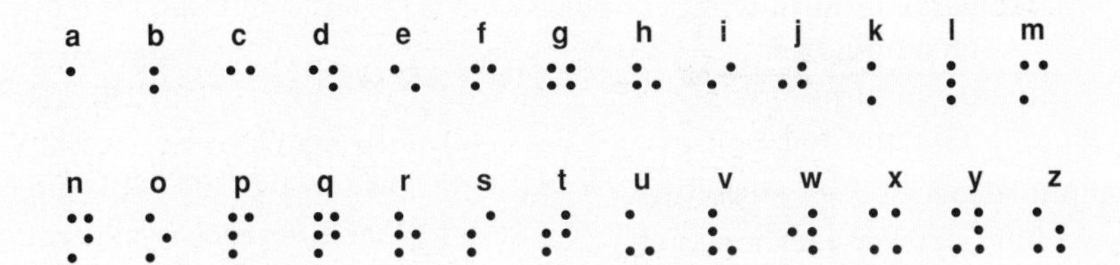

▶ **5** (a) How is it possible for light to be guided along an optical fibre that is bent?

(b) Why do optical fibres need to be made from extremely clear glass?

(c) Why is an optical fibre link more secure from being 'tapped' than (i) a link using copper wires, (ii) a microwave link?

Theme Questions

1 (a) When a plastic ruler is rubbed on piece of wool it becomes charged. Give one simple test you could use to show the ruler was charged.

(b) The hoses used at petrol filling stations to transfer petrol from the pumps to cars are made of a special type of rubber which conducts electricity. Explain why.

(c) The figure shows a direct electric current passing down a length of wire. When an electric current passes through a wire it produces a magnetic field around the wire. Copy and complete the figure by drawing the magnetic field around the wire.

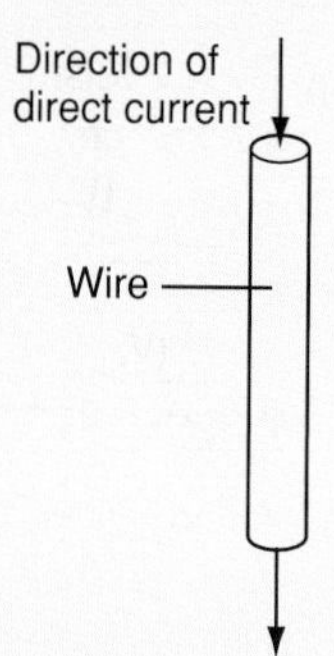

(d) Copy and complete the two graphs below by showing how the voltage of an a.c. current and a d.c. current vary with time.

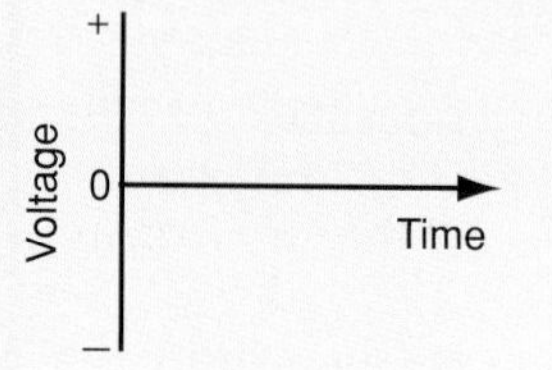

Alternating current

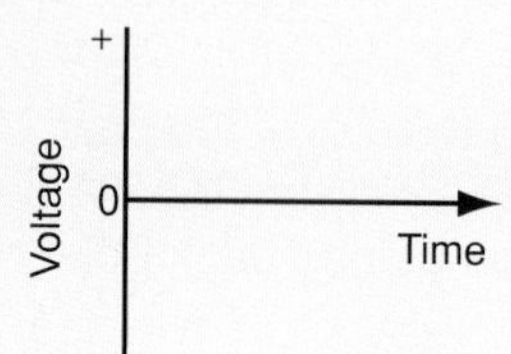

Direct current

(e) Billy uses a 'power pack' to provide electrical energy for his train set. It contains a transformer like the one shown.

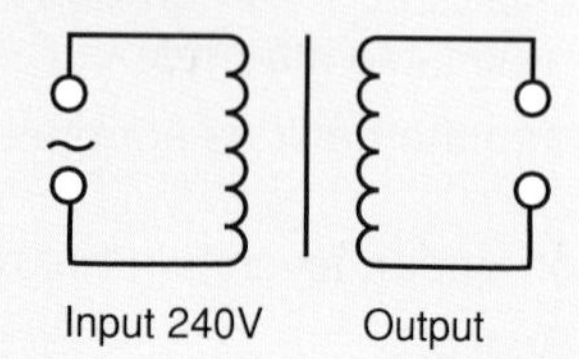

Input 240V Output

1000 turns of wire 50 turns of wire

Use the formula

$$\frac{\text{Output voltage}}{\text{Input voltage}} = \frac{\text{Number of turns of wire on output coil}}{\text{Number of turns of wire on input coil}}$$

to calculate the output voltage of this transformer.

(f) A circuit diagram of Billy's power pack is shown below.

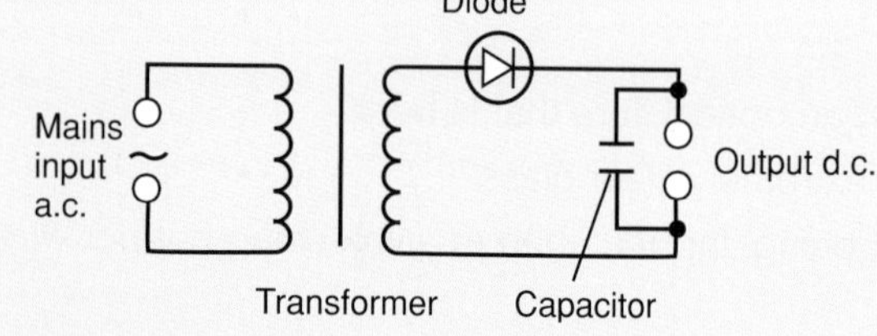

Explain how this circuit is able to convert an a.c. into a smoothed d.c. current. (LEAG)

2 The information below is taken from the instruction book for an electric cooking pot. Use it to help you answer the questions which follow it.

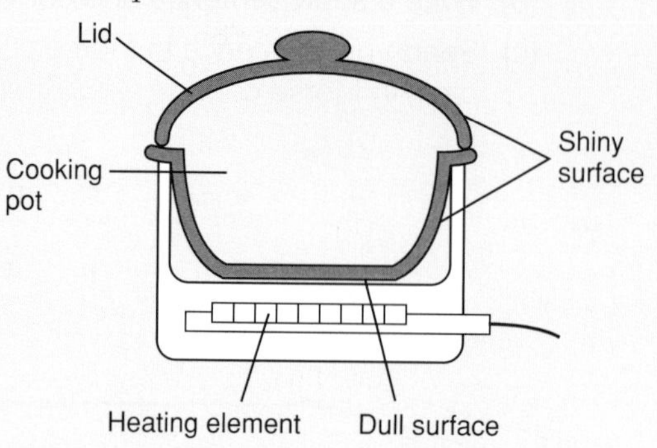

The Devon cooker is designed to cook food slowly throughout the day or night to produce tender food. It may be used to cook soups and stews directly in the pot, or as a water bath for steam puddings.
The Devon has two settings: HIGH: 115 watts (W)
LOW: 95 watts (W)
The Devon must be fitted with a three-pin plug that has the normal 13 amp fuse replaced by a 3 amp fuse. The Devon works on the normal 240 volts mains supply.

(a) One recipe needs the Devon cooker set on HIGH (115 W) for two hours.
(i) How much electrical energy will the Devon use in one second?
(ii) Calculate the amount of electrical energy the Devon will use while cooking the food.
(iii) Explain how the amount of energy gained by the food in the cooked recipe compares with your answer to (ii).

(b) Suggest an explanation for the instruction: *The Devon must be fitted with a three-pin plug that has the normal 13 amp fuse replaced by a 3 amp fuse.* (Any calculation should be shown clearly.) (SEG)

3 An electric kettle is marked

3 pints/1.7 litres
2400 watts
230/240 volts

(a) If the mains voltage is 240 V what current will flow through the kettle after it is switched on?
(b) What difference would you notice if the kettle was used in a country where the mains voltage is 120 V? (MEG)

4 The purpose of an electricity sub-station is to reduce the voltage at which electrical energy is transmitted down to the 240 volts which we use in out homes.
(a) What name do we give to the device which achieves this in a sub-station? Draw a simple labelled diagram to show how such a device works.

(b) If one hundred homes, all connected to one such device, arrange to draw a current of 20 amperes each, what is the minimum current which must flow into the sub-station from the grid? (Assume that the transmission voltage is 10 000 volts.)

(c) When a table lamp is switched on, it gives what seems to be a steady light. But the voltage applied to the bulb is changing many times each second from 240 V to 0 V. Explain in as much detail as you can how you would find out whether the brightness of the bulb changes over a short period of time. What results would you expect? (MEG)

5 Mr Dewar controls the temperature of the room where his wines ferment using an electric convector heater. The heater is controlled by the following circuit.

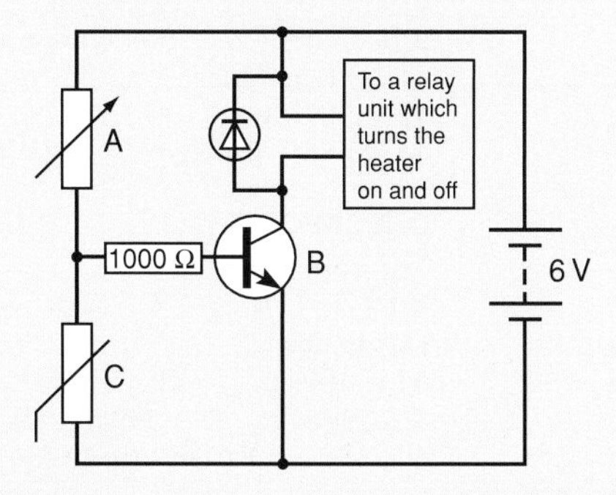

(a) Explain how heat is transferred around the fermenting room.

(b) Some electric heaters have a fan built into them. What advantage is this?

(c) Name the electronic components labelled A, B and C in the figure above.

(d) How does the resistance of component C vary with temperature?

(e) Explain how the circuit in the figure operates. You may assume that the room is colder than Mr Dewar wants it when he installs the system. (LEAG)

6 A student designed a circuit for an alarm. It is shown in the diagram below.

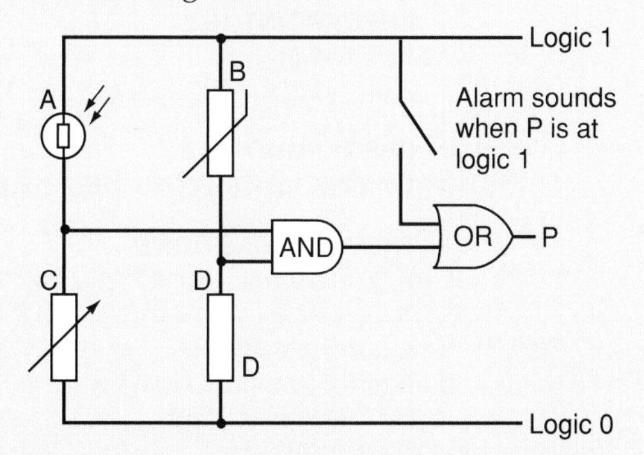

(a) Name the components labelled A–D in the circuit. Select names from the following list: switch resistor variable resistor cell thermistor lamp diode light dependent resistor light-emitting diode

(b) Which logic gate in the circuit is responsible for the fact that:

(i) the alarm sounds when a settee is burning vigorously, but not when a light is turned on in the room,

(ii) the switch can be used to test the alarm in a cold, dark room.

(c) Explain why this alarm might fail to sound when there is a fire. Assume that there is no electrical fault. (MEG)

7 The table shows corresponding values of potential difference across a torch bulb and the current passing through it.

Potential difference (V)	0	0.02	0.1	0.5	1.0	1.65	2.3	3.1	4.0
Current (A)	0	0.04	0.08	0.12	0.16	0.20	0.24	0.28	0.32

(a) Draw a circuit diagram of a circuit which could have been used to obtain these data.

(b) On graph paper, plot a graph of current on the y-axis against potential difference on the x-axis.

(c) (i) Use the graph to find the potential difference across the bulb when the current through it was 0.25 A.

(ii) Calculate the resistance of the bulb filament when the current through it was 0.25 A.

(d) (i) State how the resistance of the bulb filament changes when the current through it is increased.

(ii) Why does the resistance change in this way?

(e) In the circuit shown below, assume that the battery and ammeters have negligible resistance and that the voltmeter draws a negligible current. Calculate;

(i) the resistance of the two resistors in parallel,

(ii) the total resistance of the circuit,

(iii) the reading on A_1,

(iv) the reading on V,

(v) the reading on A_2,

(vi) the power dissipated in the 2 ohm resistor.

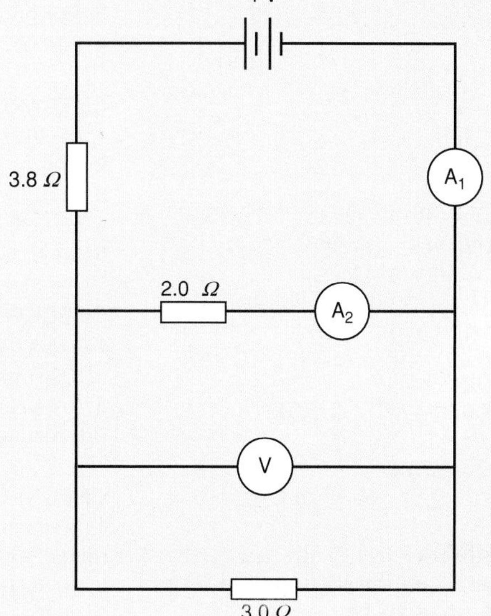

(NEAB)

Numerical Answers

CHECKPOINT 1.2
3 (a) 12° (b) 92° approx

CHECKPOINT 1.3
3 (a) 0.52, 2.52 (b) 0.26, 1.72
4 (a) 4.2 = Distance from Earth to the Sun

CHECKPOINT 1.4
2 2062

CHECKPOINT 1.5
3 (a) 1° (b) 7°

CHECKPOINT 1.6
1 (a) 1957 (b) 1969

CHECKPOINT 2.2
3 (a) 5 cm/s (b) 50 cm/s

THEME A QUESTIONS
3 (a) 10 hours ago (b) 1.4 s ago
7 (a) greater (c) (iii) 50 km/s
8 (c) 21 000, 40 000, 62 000
(d) (i) 17.5, 17, 17
(e) 18 000 million light years

CHECKPOINT 3.2
1 2.7 g/cm^3
2 7500 g (7.5 kg)
3 2720 g (2.72 kg)
4 A 0.86 g/cm^3: floats B 4.33 g/cm^3: sinks
5 (a) (i) 0.024 m^3 (ii) 24 000 cm^3
(b) 2500 kg/m^3
6 (a) 0.034 m^2, 4.0×10^{-4} m^3 (b) 7.2 kg

CHECKPOINT 3.5
8 (a) (i) 4.10×10^{-5} m^3 (ii) 0.057 kg
(b) 2.50 kg, 2.82 kg, 2.56 kg
9 (b) 6.67 N (c) 1400 N

CHECKPOINT 4.3
1 (a) 3600 J (b) 1.08 MJ (c) 900 kJ
(d) 216 kJ
2 (b) 400 kJ (c) 1.6 kW
3 (a) 75 J (b) 25 W

CHECKPOINT 4.4
3 (a) 14 976 J (b) 50 W

CHECKPOINT 5.1
2 (b) 3.75 mV
3 (b) 33 °C (c) –3°C

CHECKPOINT 5.2
3 (a) 150 mm (b) 10 mm
4 (a) 2.5×10^{-5} m^3 (b) 30 cm^3
5 (a) 80 cm^3 (b) 133 kPa
6 (a) 4.0 m^3 (b) 300 K (c) 267 kPa
(d) 267 K (e) 3.0 m^3

CHECKPOINT 5.3
2 (a) 54 kJ (b) 8.4 MJ (c) 510 kJ
(d) 506 kJ
3 (a) 1260 J (b) 0.42
4 (a) 900 kJ, 672 kJ (b) 33.6 kJ, 56 J/s

CHECKPOINT 5.4
2 (b) 0.044 kg (c) 15 000 J (d) 341 kJ/kg
5 (a) 90 kJ (b) 0.039 kg

CHECKPOINT 5.6
3 (a) 69.1 MJ (b) £1.04 (c) 77 days

CHECKPOINT 5.8
4 (a) 25.6 J/s, 42 J/s (b) 1352 J/s (c) 1.65 p/hr

CHECKPOINT 6.2
3 (a) 15

CHECKPOINT 6.3
1 (b) 3500 MJ/m^2/year (c) 57.1 m^2
5 (a) 350 watts (b) 2.9 m^2
6 (a) 40 MW (b) 300

THEME B QUESTIONS
3 (c) aluminium
4 (a) 300 N (b) 360 J
5 (c) 40%
7 (c) (i) 3 °C (ii) 31.5 kJ (d) 7 kg min^{-1}
(e) (i) 60 m^3 (ii) 10 min 43 s

CHECKPOINT 7.2
3 (a) E (b) C (c) D (d) A, B

CHECKPOINT 8.2
2 9p, 9e, 10n
3 (a) 17, 35 (b) 27, 59 (c) 50, 119

CHECKPOINT 8.5
3 N 7, 7, 7; Na 11, 11, 12; K 19, 20, 19;
U 92, 143, 92

CHECKPOINT 9.1
2 (a) 29.5 c.p.m. (b) 355.7 c.p.m.
(c) β-radiation
3 (a) 92p, 146n (b) 91p, 143n
(c) 89p, 138n
4 (a) $^{234}_{90}$Th; $^{238}_{92}\text{U} \rightarrow {}^{4}_{2}\text{He} + {}^{234}_{90}\text{Th}$
(b) $^{234}_{92}$U, $^{234}_{91}\text{Pa} \rightarrow {}^{0}_{-1}\text{e} + {}^{234}_{92}\text{U}$

CHECKPOINT 9.2
1 (a) 4.0 mg (b) 2.0 mg (c) 0.031 mg
2 (a) 5 years (b) 25 years
3 19.5 hours
4 (c) 1.55 hours

CHECKPOINT 9.3
1 (b) 4000 years
2 β, strontium
3 Bromine-82
4 Use cobalt-60
6 (a) (i) 8 (ii) 18

CHECKPOINT 9.9
1 (a) 1860 units (b) 521 units (c) medical
(d) 2381 units (i) 0.125% (ii) 0.420%
5 (b) (i) air (ii) nuclear power
(iii) cosmic, ground, food, drink, air

CHECKPOINT 10.1
1 (a) 6p, 6n (b) (i) 6p, 8n (ii) β^- radiation
(iii) $^{14}_{6}\text{C} \rightarrow {}^{0}_{-1}\beta + {}^{14}_{7}\text{N}$
2 (a) 8p, 6n (b) 8p, 8n (c) 8p, 11n
3 (a) $^{16}_{8}$O (b) $^{14}_{8}$O is a β^+ emitter,
$^{19}_{8}$O is a β^- emitter

CHECKPOINT 10.2
1 (a) 27p, 33n (b) 28p, 32n
2 (a) 2u + 1d (b) 6u + 6d
3 (a) 2 (b) 4
4 (a) 1p + 1n (b) 1p + 2n
5 (a) +1e (b) 0

THEME C QUESTIONS
4 11p, 12n
5 82p, 124n, No
6 (a) 1p, 1p + 1n, 1p + 2n (b) 1 (c) 3
7 (a) Z = 82, A = 208 (b) 17
8 (a) 375 c.p.m. (b) 6 c.p.m.
9 Sodium-24, γ-emitter with suitable half life
11 (c) 12 minutes
12 (b) (i) 0.45 s^{-1} (iii) 2.50 s^{-1}
14 (a) 143 (b) 48p + 74n (c) β^- particles

CHECKPOINT 11.3
1 (b) 2040 m
2 (a) 200 kHz (b) 97.7 MHz (c) 92.4 MHz
(d) 648 kHz
3 (a) 0.50 Hz (b) 20 m, 10 m/s
4 (a) 0.664 m (b) 0.113 m (c) 13.6 mm
5 (a) 1.0 cm, 6.0 cm (b) 12 cm/s

CHECKPOINT 12.2
1 (a) 330 m/s
5 270 m

CHECKPOINT 13.2
4 2 mm

CHECKPOINT 13.3
1 (a) 4 m (b) 1.5–1.9 m (c) $a > 2b$

CHECKPOINT 14.1
3 (b) 3.0 m (c) The top half

CHECKPOINT 14.3
1 (b) 1.5

CHECKPOINT 14.4
7 (a) 30 cm, 60 cm, 30 cm

CHECKPOINT 15.2
1 (c) 0.12 m
2 (a) 50 kW h (b) 12.5 kW h (c) 37.5 kW h

CHECKPOINT 15.3
1 (a) 16.7 ms (b) 2.53 s (c) 4.2 years

THEME D QUESTIONS
1 (a) (i) x (ii) w (b) 4 (d) 2 Hz (e) 120
2 (b) 9000 m (c) 1500 m/s (d) 3000 m
3 (a) (i) 126 dB
7 (c) (i) 6 minutes (ii) 0.1 h
(iii) 7.2 kW h (iv) 36p
8 (a) (i) 200 kHz

CHECKPOINT 16.1
2 1.0 N, 2.5 kg, 600 N, 0.02kg
4 1.8 N
5 (b) (i) 2.5 N (ii) 0.025 N (c) 5.75 N
(d) 0.575 kg (e) 0.66 J

CHECKPOINT 16.3
2 (a) 10 N to the left (b) 50 N up (c) 500 N up the ramp
4 (a) 2.40 N (b) 4.80 N (c) 7.20 N
5 3900 N

CHECKPOINT 16.4
2 (a) 1.5 N (b) 5.67 N (c) 1.5 N (d) 5.25 N
3 (b) 340 N (c) 1.9 m from the fulcrum

CHECKPOINT 17.1
3 (a) 1750 Pa (b) 35 kPa

CHECKPOINT 17.2
2 (a) 30 N (c) 300 N
4 18 kN
5 (a) 800 N

CHECKPOINT 17.3
3 (a) 7×10^6 Pa
4 (a) 900 Pa (b) 0.54 N

CHECKPOINT 17.4
1 (a) Greater (b) Same (c) 2500 Pa
2 20 m
3 16.3 kPa

CHECKPOINT 17.5
2 10 m
4 10.8 kN

CHECKPOINT 17.6
3 (a) 5.0 N (b) 5.0 N (c) 0.5 kg

CHECKPOINT 18.1
1 (a) 4.67 km/h (b) 1.30 m/s
2 (a) 1.67 m/s
4 (b) 2.67 m/s
5 (a) 20 km (b) 6 hours 40 minutes

CHECKPOINT 18.2
1 (b) 1 s (a) (i) 27 cm/s (ii) 0.27 m/s
2 (a) 0.30 m/s^2 (b) 540 m
4 (b) 2 m/s^2 (c) (i) 400 m (ii) 400 m
5 3150 m, 7 m/s^2
6 (a) 18 m (b) 5.0 s (d)75 m

CHECKPOINT 18.3
1 6.67 m/s^2, 7.50 m
2 (a) 10 cm (b) 10 cm/s (c) 20 cm/s^2
3 (b) (i) 0.125 m/s^2, 400 m (ii) –0.25 m/s^2, 200 m (c) 5 m/s
4 (a) (i) 30 m/s (ii) 0 m/s (b) 15 m/s (c) 5 s, –6 m/s^2
5 (a) 109 m/s (b) 219 m/s (c) 6.83 m/s^2

CHECKPOINT 18.4
2 (b) (ii) 0.21 s
3 (a) 15 m/s (b) 11.25 m
4 (a) 2.1 s (b) 0 (c) 21 m/s
5 (b) Simon's (c) 45 m (d) 20 m/s

CHECKPOINT 18.5
2 (a) 1.5 N (b) 25 m/s^2 (c) 4.0 kg (d) 0.45 kg
3 (a) 444 N (b) 8 kN (c) 0.056
4 (a) 22 500 N (b) 23.7 minutes
5 (b) (i) 900 N (ii) 900 N (iii) 2800 J/s (iv) 3600 J/s (c) The rate of loss of p.e. of the parachute = 800 J/s

CHECKPOINT 18.6
3 (a) 6.25 J (b) 200 kJ (c) 1000 MJ
4 (a) 375 kJ (c) 55 m/s
5 (a) 1890 J (c) 9.2 m/s
6 (a) 1760 J (b) 3.2 m

CHECKPOINT 18.7
2 (a) 6000 kg m/s (b) (i) 500 N (ii) 5000 N
4 (a) 1.11 m/s, 1390 J (b) (i) 2.22 m/s, 56 J (ii) 0, 4500 J
5 (b) 32 m/s

CHECKPOINT 18.8
4 (b) 19.1

CHECKPOINT 19.1
3 (a) 120 kJ (b) 750 J (c) 9.4%
4 (a) 1.82, 0.8, 2.5 (c) A, 91%
5 (a) 1100 J, 1000 J (b) 91% (c) 1220 N

CHECKPOINT 19.2
4 (a) 1000 l (b)143 l

THEME E QUESTIONS
1 (a) (ii) 1.20 m (iii) 24.0 J
3 (a) (i) 3 (ii) 8,9 and 10 (b) (i) 0.2 s (ii) 1.3 m/s
4 (a) 800 N (c) 500 N (d) 6.25 m/s^2 (e) (i) 120 kJ (ii) 320 kJ (iii) 200 kJ
5 (a) 0.0024 m^2 (b) 125 kPa (c) (i) 4 m/s (ii) 300 s (d) (i) 150 s (e) 0.13 m/s^2 (f) 1050 m
6 (d) (i) 4200 m (ii) 10 m/s (e) 39%
7 (a) (ii) 18 m (b) (i) 10.7 s (c) (i) 12 m (ii) 71 m (iii) 83 m (d) (i) 405 kJ (ii) 2530 J/m (iii) 2530 N
8 (a) 6500 J (b) 12500 J (d) 0.52
9 (b) (i) 24 000 kg m/s (ii) 24 000 N (iii) 6.2 m/s^2
10 (b) (ii) 2.0 m/s^2

CHECKPOINT 20.1
2 (a) + (b) (i) + (ii) –

CHECKPOINT 20.3
2 (a) (i) 3.5C (ii) 210 C (iii) 2100 C (b) 5.0 A (c) 20.0 A (d) (i) 1000 s (ii) 500 000 s
3 (b) Smaller deflection (c) Larger deflection
4 (a) 15 mm (b) 0.75 V
5 (a) 1200 C (b) (i) 0.80 g (ii) 0.20 g (iii) 0.40 g (c) 3.33×10^{-4} g (d) 3.3×10^{-19} C, 2

CHECKPOINT 20.4
2 (a) £3.50
5 (a) 360 000 (b) 4.32 MJ (c) 50 h

CHECKPOINT 21.2
2 (a) 4.2 A (b) 5.0 A
3 (a) (i) 12 (ii) 1 (iii) 0.05 (iv) 0.21 (b) 40.3p
4 6.9 kW, 20.7, £1.14
5 (a) 5.2 kWh (b) 2.6 kW (c) 567 17.7

CHECKPOINT 21.3
1 (a) (i) X = 1.2 A, Y = 1.8 A, Z = 3.0 A (ii) X = Y = 2.0 A, Z = 2.5 A (iii) X = 3.5 A, Y = 1.5 A, Z = 2.0 A (b) (i) X = 2.5 V, Y = Z = 3.5 V (ii) X = 12.0 V, Y = 4.0 V, Z = 8.0 V (iii) W = 9.0 V, X = 3.0 V, Y = 4.0 V, Z = 8.0 V
2 (a) 10.0 A (b)12 J, 24 J, 84 J (c) 120 J
3 (a) (i) 2.0 A (ii) 4.0 A (iii) 6.0 A (b) 2 C, 4 C (c) 24 J, 48 J
4 (a) (i) 60 J (ii)18 kJ

CHECKPOINT 21.4
1 (a) (i) 2.0A (ii) 1.0 A (iii) 3.0 A (iv) 4.0 Ω (b) (i) 0.5 A (ii) 0 A (iii) 1.0 A (iv) 6.0 Ω (c) (i) 1.5 A (ii) 1.5 A (iii) 3.0 A (iv) 2.0 Ω
2 (a) (i) 9 Ω (ii) 2 Ω (b) (i) 14 Ω (ii) 1.43 Ω (d) (i) 8.0 Ω (ii) 8.0 Ω
3 (b) 6.0 Ω/m (c) 0.33 m
4 (c) 12 V (d) 200 Ω, 4000 Ω
5 (b) 180 Ω
6 (b) (i) 3000 V (ii) 1.5 V

CHECKPOINT 22.4
2 (a) (i) 1 (ii) 1 (iii) 3
3 (a) 64 μs

CHECKPOINT 22.5
5 (a) (i) 40 (ii) 5
6 (a) (i) 8.0 V (ii) 50 Hz

CHECKPOINT 22.6
1 (a) 1200 (b) (i) 0.42 A (ii) 8.3 A
2 (a) (ii) 30 (b) (i) 6 W (ii) 25 mA
3 (a) 1; 6 (b) 40 A (c) (i) 900 kW (ii) 6 A
4 (c) (i) 12.0 V (ii) 20 A (iii) 240 W
5 (a) (i) 100 A (ii) 200 kW

CHECKPOINT 23.5
3 (a) 128 k (b) (i) 11111 (ii) 1100
4 (a) 2000–2010 (b) approximately 1 s
5 (a) approx. 5000 Mbytes (b) 450

CHECKPOINT 23.6
4 (a) 4 (b) £15,000

CHECKPOINT 23.7
4 (b) 20 cm^3

THEME F QUESTIONS
1 (e) 12 V
2 (a) (i) 115 J (ii) 828 kJ
3 (a) 10 A
4 (b) 48 A
7 (c) (i) 2.5 V (ii) 10 Ω (e) (i) 1.2 Ω (ii) 5.0 Ω (iii) 0.8 A (iv) 0.96 V (v) 0.48 A (vi) 0.46 W

Index

Page numbers in bold indicate the main entry for that topic.

A

B

C

D

E

F

M

N

O

T

U

V

W

X

Y